Y0-BXH-226

Acoustical Imaging

Volume 8

Ultrasonic Visualization and Characterization

Acoustical Imaging

Volume 1 Proceedings of the First International Symposium, December 1967, edited by A. F. Metherell, H. M. A. El-Sum, and Lewis Larmore

Volume 2 Proceedings of the Second International Symposium, March 1969, edited by A. F. Metherell and Lewis Larmore

Volume 3 Proceedings of the Third International Symposium, July 1970, edited by A. F. Metherell

Volume 4 Proceedings of the Fourth International Symposium, April 1972, edited by Glen Wade

Volume 5 Proceedings of the Fifth International Symposium, July 1973, edited by Philip S. Green

Volume 6 Proceedings of the Sixth International Symposium, February 1976, edited by Newell Booth

Volume 7 Proceedings of the Seventh International Symposium, August 1976, edited by Lawrence W. Kessler

Volume 8 Proceedings of the Eighth International Symposium, May 29–June 2, 1978, edited by A. F. Metherell

Acoustical Imaging

Volume 8

Ultrasonic Visualization and Characterization

Edited by
A. F. Metherell
University of California at Irvine
Orange, California

PLENUM PRESS · NEW YORK AND LONDON

The Library of Congress cataloged the first volume of this title as follows:

International Symposium on Acoustical Holography.

Acoustical holography; proceedings. v. 1-
New York, Plenum Press, 1967-

v. illus. (part col.), ports. 24 cm.

Editors: 1967- A. F. Metherell and L. Larmore (1967 with H. M. A. el-Sum)
Symposiums for 1967- held at the Douglas Advanced Research Laboratories, Huntington Beach, Calif.

1. Acoustic holography–Congresses–Collected works. I. Metherell. Alexander A., ed. II. Larmore, Lewis, ed. III. el-Sum, Hussein Mohammed Amin, ed. IV. Douglas Advanced Research Laboratories. v. Title.
QC244.5.I 5 69-12533

Library of Congress Catalog Card Number 69-12533
ISBN 0-306-40171-1

Proceedings of the Eighth International Symposium on Acoustical Holography and Imaging, held in Key Biscayne, Florida, May 29–June 2, 1978

A Division of Plenum Publishing Corporation
227 West 17th Street, New York, N.Y. 10011

Printed in the United States of America

PREFACE

This volume contains the Proceedings of the Eighth International Symposium on Acoustical Imaging, held in Key Biscayne, Miami, Florida May 29th to June 2nd, 1978. The title of the Symposium was changed again this year by dropping the word "Holography" to reflect the further emphasis on the general imaging aspects of the Symposium and the de-emphasis on the Holographic aspects. Because of this continued changing nature of the Symposium Series this volume has undergone the title change from ACOUSTICAL HOLOGRAPHY to ACOUSTICAL IMAGING.

The 47 papers presented here illustrate the continued growth in this dynamic field. There has been a large emphasis on Array Technology as well as Underwater Applications, Seismic Applications, Transducers, New Methods, Acoustic Microscopy, Non-destructive Testing, Computer Tomography Techniques, Medical Applications as well as Tissue Characterization.

The meeting was a great success and a stimulating experience for all concerned due principally to the enthusiasm and contributions of all of the authors represented here. The editor wishes to extend his appreciation and thanks to each and every one of them. The editor also wishes to thank the members of the Program Committee who helped in selecting the papers and giving their able advice on the details of the meeting. The Program Committee consisted of Pierre Alais, University of Paris, Byron B. Brendon, Holosonics, Inc., C. B. Burckhardt, Hoffman-LaRoche and Co., Basle, Switzerland, Philip S. Green, Stanford Research Institute, Menlo Park, California, B. P. Hildebrand, Battelle Northwest, Richland, Washington, Joie Pierce Jones, University of California, Irvine, Lawrence W. Kessler, Sonoscan, Inc., Bensenville, Illinois, A. K. Nigam, Horizons Research Laboratory, Fort Lauderdale, Florida, Jerry L. Sutton, Naval Ocean Systems Center, San Diego, California, Frederick L. Thurstone, Duke University, Durham, North Carolina, Olaf Von Ramm, Duke University, Durham, North Carolina, Robert C. Waag, University of Rochester Medical Center, Rochester, New York, Glen Wade, University of California, Santa Barbara, and Keith Wang, University of Houston, Texas. The editor also wishes to thank the

Session Chairmen, Bill O'Brien, Pierre Alais, Keith Wang, Lewis Larmore, Jerry L. Sutton, Ken Erickson, B. Percy Hildebrand, Fred Kremkau, Frederick L. Thurstone, Robert C. Waag, H. Dale Collins, Mahfuz Ahmed.

The editor also wishes to especially thank the Office of Naval Research for sponsoring this meeting and particularly Dr. Lewis Larmore of the Office of Naval Research who has personally supported this Symposium from the very beginning. The followers of this Symposium Series will recognize Dr. Lewis Larmore as one of the Co-Editors of ACOUSTICAL HOLOGRAPHY, VOLUME I and it was through his personal support that the very first Symposium took place. When the first Symposium was held in December 1967 we never realized that it would develop into a continuing series and be the success it has become in over 10 years. The Symposium was also held in cooperation with the American Institute of Ultrasound in Medicine as well as the IEEE group on Sonics and Ultrasonics and the IEEE Group on Biomedical Engineering.

The help and assistance of Eric N. C. Milne, M. B., Ch.B., F.R.C.R., Professor and Chairman, Department of Radiological Sciences University of California, Irvine is also very gratefully acknowledged. Through him the secretarial support of Ms. Jennifer Scott and Marie Burrell was obtained and their help and assistance with all of the secretarial work and arrangements is much appreciated.

The Ninth International Symposium on Acoustical Imaging is scheduled to be held in Houston, Texas in December 1979 under the Chairmanship of Dr. Keith Wang. The Tenth International Symposium on Acoustical Imaging will be held in Cannes, France in October 1980 under the Chairmanship of Dr. Pierre Alais and Dr. A. F. Metherell. The Eleventh International Symposium on Acoustical Imaging will be held in Monterey, California in 1981 under the Chairmanship of Dr. John P. Powers.

Alexander Metherell

CONTENTS

TRANSDUCERS

METHODS

TISSUE CHARACTERIZATION

COMPUTER TOMOGRAPHY

ACOUSTIC MICROSCOPY/NDT

NONDESTRUCTIVE TESTING

MEDICAL APPLICATIONS

DIGITAL SCAN CONVERSION AND SMOOTHING FOR A REAL TIME LINEAR ARRAY IMAGING SYSTEM

James T. Walker

Stanford Electronics Laboratories

AEL 211, Stanford University, Stanford, CA 94305

INTRODUCTION

Background

One form of real time ultrasonic imaging system uses a linear array of transducer elements operated in their near field [1]. Sequential transmission and reception from the elements provides a limited number of distinct ultrasonic reflection patterns from which a B-scan cross-sectional image is assembled. Prior systems utilized an XY oscilloscope as the display device, where the X deflection position is a repetitive sawtooth function plotting echoes linearly as a function of time after transmission, and the Y deflection corresponds to the element location and therefore the ultrasonic beam. Intensity modulation with the received echo signal amplitude then produces a B-scan image. Since the ultrasound beams are generally parallel and equally spaced, proper adjustment of the deflection factors preserves the geometric image properties and permits dimensional measurements.

Statement of Problem

One important performance limitation of the linear array system as described is the resultant image quality. Two particular problem areas are lack of gray scale and raster line coarseness. General-purpose XY display units are often optimized in design for criteria such as deflection speed, positional accuracy, and sensitivity. Gray scale performance, and spot defocusing with intensity changes, take on a secondary importance. Use of a display technique allowing display units optimized for gray scale performance would alleviate this problem.

Raster line coarseness results from writing only one line of data on the display for each ultrasonic beam location. If the signal reception delay and number of beams permits it, the apparent raster line spacing may be reduced by impulsing the same transducer several times in succession and displaying the result as adjacent raster lines on the display. However, in the case of long signal delays, the number of possible raster lines may be insufficient to permit complete elimination of the raster structure. A striation effect due to the line duplications will additionally be visible. Use of an interpolation process would reduce the image striations.

Proposed Solution

A standard television monitor provides an excellent display unit with many advantages. If the ultrasonic data is put into the form of an RS170 video signal, then a single coaxial cable carries all necessary formation to the display unit. Television format signals may be easily recorded for later moving image analysis, and display monitors are commonly available.

A scan conversion process can be performed on the ultrasonic data to transform the low rate ultrasonic video to the higher rate need by a television display. During the conversion, data lines can be duplicated as needed to utilize all of the television raster lines and remove visible raster structure. Interpolation performed during the line duplication procedure then removes striations in the image.

The proposed system therefore consists of a high speed analog-to-digital converter (ADC), followed by a digital data memory. Output from the data memory goes through an interpolator and is reconverted to an analog signal in a television format.

LINEAR ARRAY ULTRASONIC SYSTEM

Data Collection Scan Format and Timing

Figure 1 shows a block diagram of the linear array ultrasonic system under consideration [2]. This system was developed for B-scan cross-sectional imaging of the human eye. A linear array containing 35 elements scanned in groups of four provides 32 ultra-

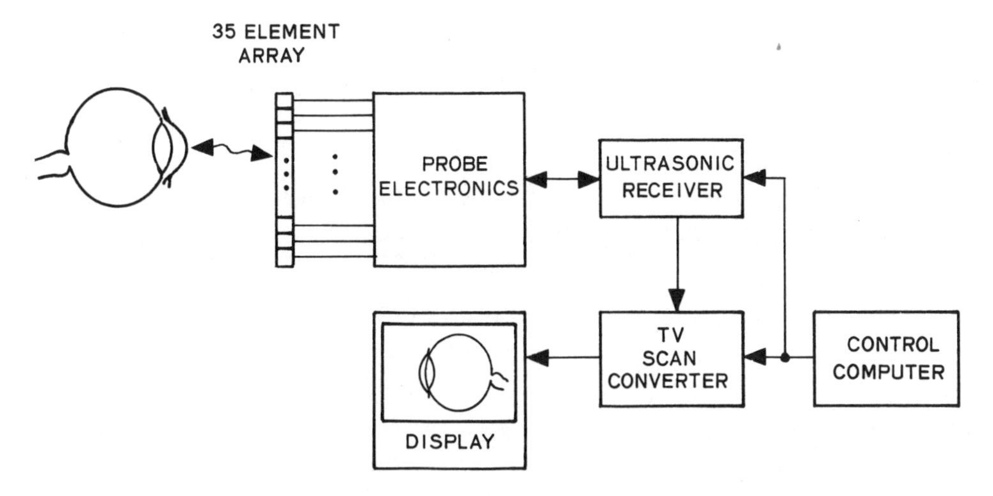

FIGURE 1. Linear array system simplified block diagram.

sonic beams for near field imaging. Electronic multiplexing with custom high voltage DMOS integrated circuits developed at Stanford sequentially connects a single transmitter and receiver section to the various transducer groups. The scanned tissue cross section is 3.2 cm wide by 6.0 cm deep, requiring 78 microseconds for ultrasonic echoes to return from the farthest depths. Some additional time between transmissions in excess of 78 μsec should be allowed for ultrasonic reverberations to dissipate. The received ultrasonic data has a video bandwidth of 2 MHz maximum.

Display Scan Format

A standard television video format meeting RS170 specifications is desired for the display monitor. This requires a vertical field frequency of 60 Hz and a horizontal scanning frequency of 15,634 Hz. Therefore, approximately 240 raster lines are available in each field, each with an unblanked length of 54 μsec. Use of a B-scan image area containing 128 lines vertically in each field, with an active length of 40 μsec each, gives a properly proportioned image on the display. The geometric proportions correspond well to the actual 3.2 by 6.0 cm target area. The remaining screen area is used for distance reticles, image boundaries, A trace data, range markers, and alphanumeric annotation. Thus, a photograph or video recording of the display provides a comprehensive record of the examination conditions and results. Use of a standard television format provides ease of signal distribution to multiple displays and of moving image recording and playback.

Digital Scan Conversion Process

When the ultrasonic B-scan display orientation is chosen so that the depth direction extends horizontally, a digital scan conversion process requires only partial image memory. Two tasks must be performed: (1) time compression of the ultrasonic data by a 2:1 ratio, and (2) vertical raster line interpolation. The 2:1 time compression allows 78 μsec of received data to fit within the desired 40 μsec image area width. Vertical raster line interpolation expands the available data in a smooth manner so that 32 ultrasonic data lines give a filled-in 128 line image.

The scan conversion process follows the procedure shown in Fig. 2. Incoming ultrasonic video goes first to a high speed ADC, where it is digitized at a 7.19 MHz word rate. The resultant 6 bit words then go to a line memory unit where the 560 pixels of each active line length are temporarily stored. After two adjacent lines of data are collected, they are repetitively read at a 14.38 MHz rate into the digital interpolator. The 8 bit interpolator output words then go to a DAC and are converted into video to intensity modulate the display.

A total of 4 shift register sections, each 560 words long, are used in the line memory for ultrasonic data storage. Thus, while one register is acquiring data at 7.19 MHz, two others are feeding the interpolator at 14.38 MHz and the last is temporarily storing the A trace data for later display. Shift registers were used because of the high data rates required, with three way interleaving to reduce the effective bit rate. Since four television raster lines occur for each ultrasonic data line, the transmitter operates at intervals of 254 μsec, and only when new data is needed. During the portion of the display not containing B-scan data, the ultrasonic transmitter is inactive. Acquisition and conversion of data only as needed for display permits a substantial reduction in hardware requirements [3].

Digital Image Interpolation

For each received line of ultrasonic data, four raster lines must be generated on the television display. Use of an interpolator circuits permits smooth transitions between adjacent transducer

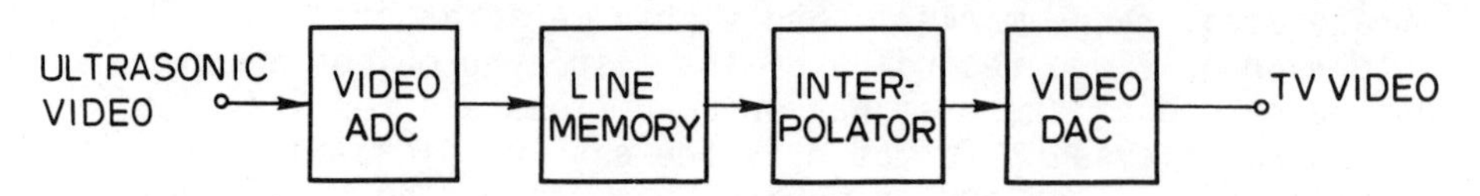

FIGURE 2. Scan conversion process block diagram.

responses to improve the image quality. Figure 3 shows a block diagram of the interpolator section. Three line memories, marked A, B, and C, provide the temporary data storage. Thus, one is acquiring data while the remaining two send data to the interpolator. The three memories exchange functions in a cyclic or rotating manner as the display progresses vertically. Three multiplexers at the line memory outputs connect the interpolator inputs to the proper sources for each display line.

The interpolation process uses a digital adder with three inputs for weighting the ultrasonic data. As the display line count proceeds between the lines coinciding with received ultrasonic data, the line memory outputs are routed to the adder inputs as needed. Varying the weighting sequence controls the nature of the interpolation function. Curves of the intensity response in the vertical direction resulting from three different interpolation modes appear in Fig. 4. These curves are for no interpolation (replication only), linear interpolation, and cosine interpolation. Dots mark the intensity levels of the discrete raster line locations.

SCAN CONVERSION SYSTEM PERFORMANCE EVALUATION

Test Format and Interpolation Modes

Performance evaluation of the digital interpolation is done by utilizing a special circuit to replace the interpolation controller at the bottom center of Fig. 3. A counter sequences through four states as the display is generated from left to right, with its output selecting various control ROM signals for the adder input multiplexers. Additionally, for linear interpolation, the bottom adder section is bypassed with a multiplexer to cause weights of 1/4, 1/4, and 1/2.

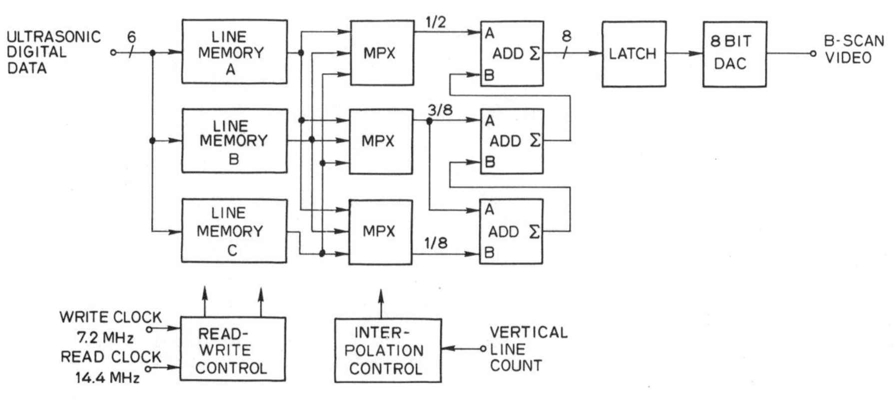

FIGURE 3. B-Scan memory and interpolation block diagram.

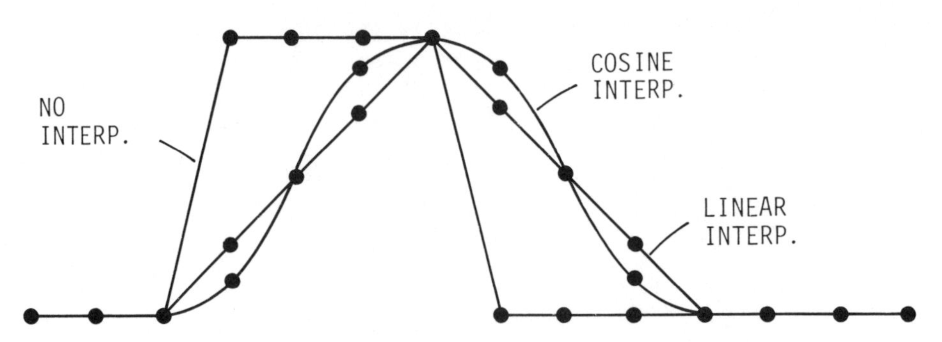

FIGURE 4. Ultrasonic data line interpolation curves.

The counter states divide the B-scan area into four regions in a horizontal direction. The leftmost region has only one intensified raster line per ultrasonic data line, as would occur with the simplest XY monitor. In the next region, the intensified raster lines merely repeat the ultrasonic data in a vertical direction, giving rise to horizontal stripes of equal intensity. In the third region, linearly weighted interpolation is used vertically to provide a smoothly varying intensity between adjacent ultrasonic data lines. In the last region to the right, cosine weighted interpolation is used.

An artificial test signal replaces the normal ultrasonic video data for testing the interpolator. A combination of several signal generators provides a train of four pyramidal voltage pulses horizontally, with their maximum amplitude varying vertically as a cosine square pulse envelope. Alternatively, the vertical function may be a short pulse, intensifying only one ultrasonic data line, or a square wave, intensifying alternate ultrasonic data lines.

Interpolation Performance

The interpolation results with a vertical cosine square function appears in Fig. 5. The top part shows the intensity modulated B-scan area, while the bottom contains the A trace cut horizontally through the center of the B-scan area. The leftmost pattern has 32 equally spaced lines with blanked raster between them, and the next pattern has the ultrasonic data lines merely replicated vertically. The third pattern from the left has linear vertical interpolation, and the last at the right has cosine weighted interpolation. It is seen that the separated raster line image structure obscures the overall intensity pattern. Vertical replication of the ultrasonic data alleviates most of this problem, but horizontal striations occur due to the intensity steps at the

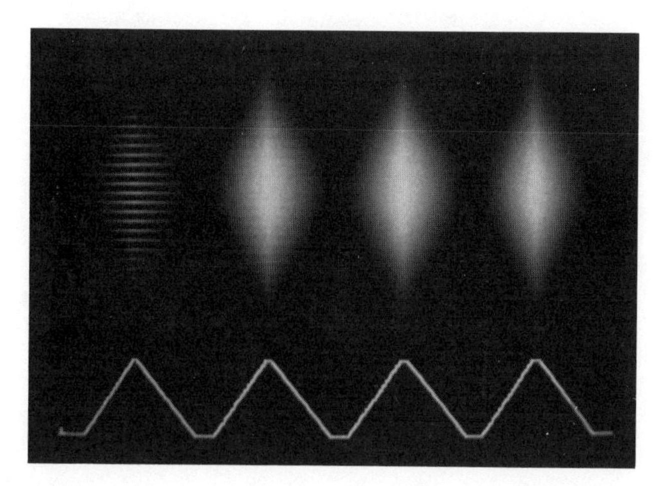

FIGURE 5. Video response with four vertical interpolation modes and simulated ultrasonic data. Left: original data only with blanked lines. Second from left: original data replication only. Second from right: linear weighting. Right: cosine weighting.

adjacent edges of raster line groups. Linear vertical interpolation produces a smooth intensity function on both axes with no apparent discontinuities or other disadvantages. Cosine weighting produces an image with quality lying between the vertical replication and linear interpolation cases. Some horizontal striation is still visible, but with less severity than in the vertical replication case.

Another method of evaluating the interpolation performance is to make a vertical cross section cut through each of the four B-scan patterns in Fig. 5. A cross section of the leftmost pattern appears in Fig. 6. Each vertical line corresponds to one ultrasonic data line of the original B-scan, and its height is proportional to the B-scan intensity. The horizontal axis here corresponds to the vertical axis of Fig. 5. The cross section in Fig. 7 corresponds to the B-scan area where raster lines are duplicated vertically. This gives greater average intensity, with steps in amplitude corresponding to the horizontal striations. Figure 8 shows a cross section of the region using linear interpolation. Here, the waveform envelope shows a smooth replica of the original cosine square vertical function before the quantizing effects of sampling. The envelope closely follows the original function. Finally, Fig. 9 shows a vertical cross section of the rightmost intensity pattern in Fig. 5. Roughness on the envelope edges corresponds to horizontal striations shown previously and results from the cosine weighted interpolation between adjacent ultrasonic data lines.

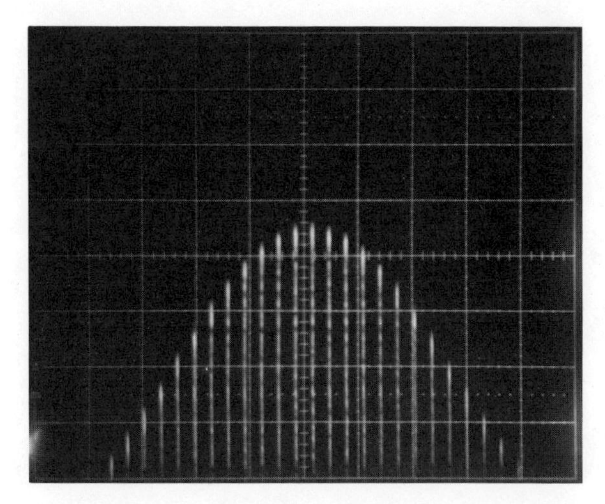

FIGURE 6. Vertical cross section through original data of Fig. 5.

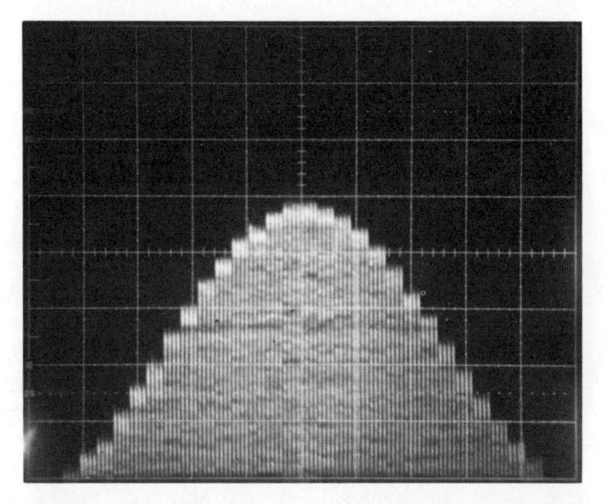

FIGURE 7. Vertical cross section through duplicated raster lines of Fig. 5.

Thus, it appears from the above data that, of the two point interpolation schemes presented, linear weighting gives the best performance on smooth objects. A question of significance then is whether any cases exist for which cosine weighted interpolation appears best. The most extreme opposite of the smooth cosine square test signal used previously is an impulse, simulated by pulsing ON only the center ultrasonic data line. Figure 10 shows the resulting A- and B-scan displays. In this figure, the right-most response corresponding to cosine weighting appears narrower

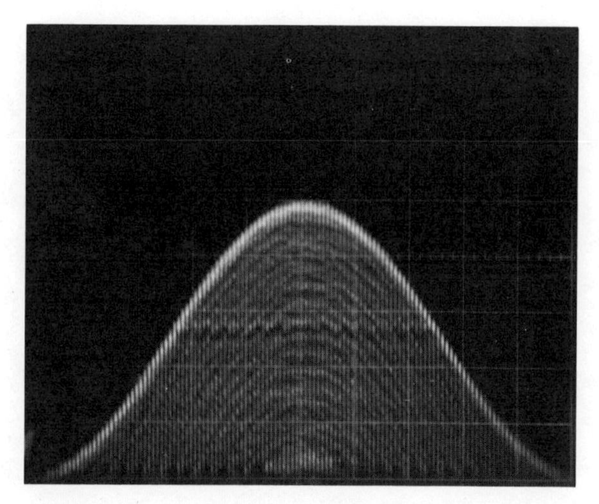

FIGURE 8. Vertical cross section through linearly weighted video of Fig. 5.

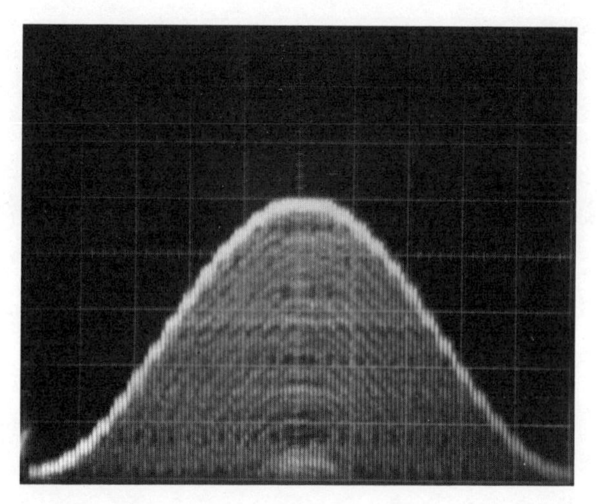

FIGURE 9. Vertical cross section through cosine weighted video of Fig. 5.

vertically than the linear weighted response. In actuality, both responses have an equal number of nonzero intensity raster lines, as demonstrated in Fig. 11. For this figure, a digital threshold circuit in the system which precedes the output digital-to-analog converter caused all image pixels identically equal to zero to be blanked. Now, both responses are seen to have the same vertical width. The greater steepness of the cosine weighting function causes a more narrow apparent width.

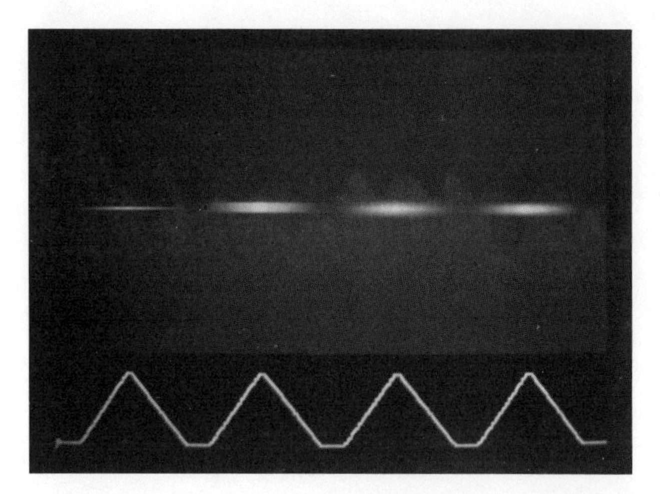

FIGURE 10. Interpolation mode comparison for single line of ultrasonic data. Modes are same as Fig. 5.

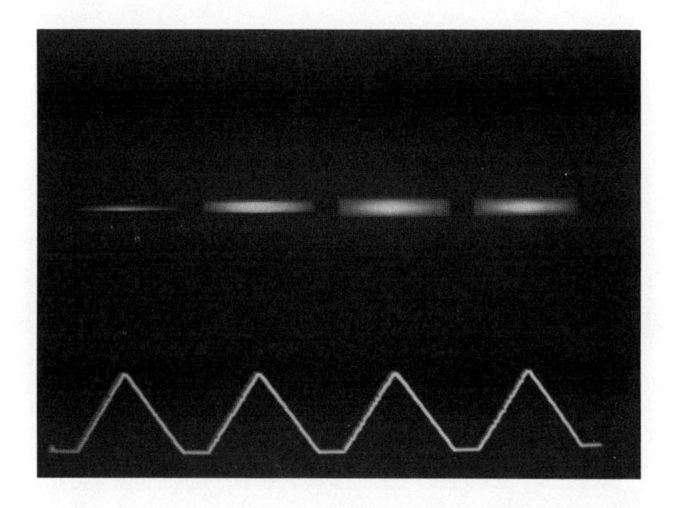

FIGURE 11. Interpolation mode comparison of Fig. 10 with video threshold used to accentuate non-zero pixels.

Another demonstration of this effect is obtained by using a square wave for the vertical function so ultrasonic data lines are alternately intensified and blanked. Figure 12 shows the result, with cosine weighting giving greater apparent sharpness than linear weighting. In fact, though, interleaving two sets of lines as in Fig. 12 gives uniform illumination for both cases. So, it appears that cosine weighting may give greater subjective sharpness for targets containing predominately high spatial frequencies.

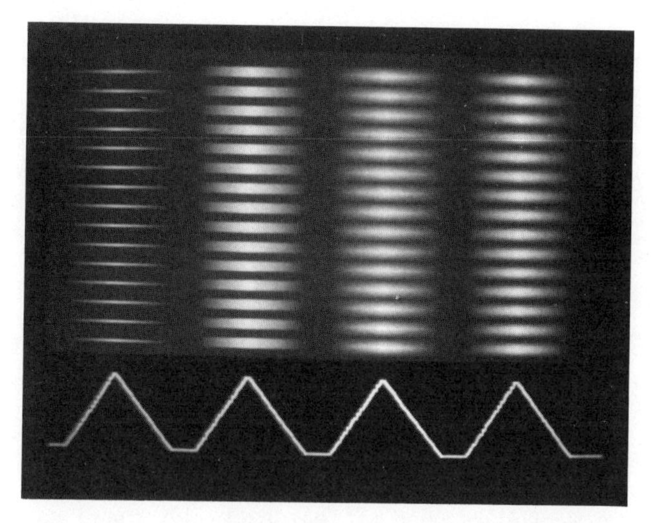

FIGURE 12. Relative interpolation sharpness comparison with alternate ultrasonic data lines pulsed ON.

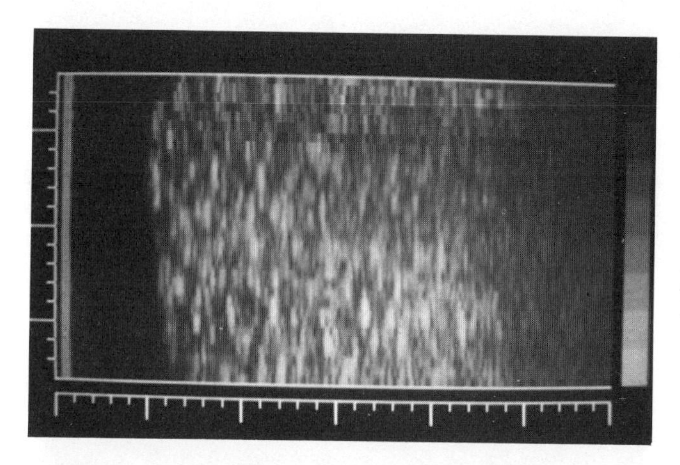

FIGURE 13. Ultrasonic B-scan of a reticulated foam target with three interpolation modes. Top: line replication only. Center: linear interpolation. Bottom: cosine interpolation.

Experimental Performance

In order to test the various interpolation modes further, the complete ultrasonic system was used to image a section of reticulated foam sponge, with different parts of the B-scan receiving the various interpolation treatments. Figure 13 shows the B-scan region divided vertically into three horizontal strips. The first, corresponding to the top two squares of the right-hand gray scale, has replicated scan lines only without any interpolation. The second, corresponding to the next three gray scale squares, uses linear interpolation, while the last, corresponding to the bottom three squares, uses cosine interpolation. Objectionable striations occur in the top part of the B-scan, while the central region appears slightly more fuzzy vertically than the bottom one. So, cosine weighting appears better for this type of target. In clinical use of the ultrasound system, however, linear interpolation gives generally more pleasing images.

Figure 14 shows an ultrasonic echo image of a phantom eye target. Linear interpolation is used, and the resulting real time image moves smoothly as the transducer probe is scanned. Large tick marks on the distance scales for each axis denote centimeters of actual object size. Another example of linear interpolation performance appears in Fig. 15. This is the usual type of photographic record made during a patient exam and includes alphanumeric annotation to record the examination conditions. The ultrasonic eye cross section appears in the B-scan area, with the cornea at the left and the iris to its right. A small specular reflection occurs to the right of the iris from the back lens surface. At the right side of the B-scan is seen the echo pattern due to fat behind the eye. This patient has a detached retina, which appears at the right side of the clear area due to the vitreous as two lines of echoes extending into the vitreous. Figure 15 demonstrates the image display improvements that interpolation can give when working with a limited ultrasonic data base.

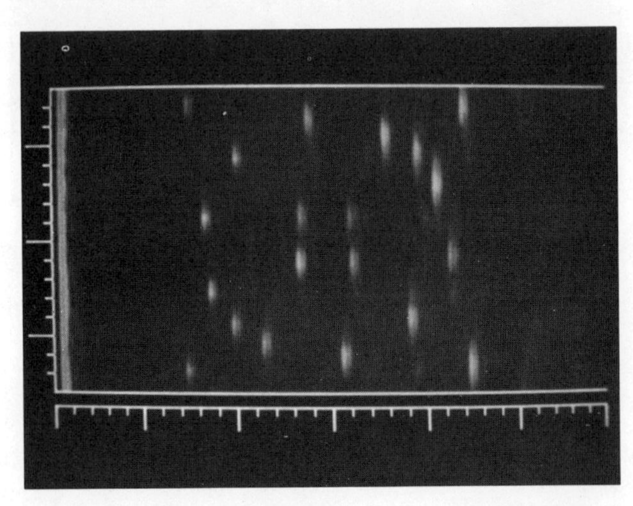

FIGURE 14. B-scan image of a wire eye phantom target using linear interpolation.

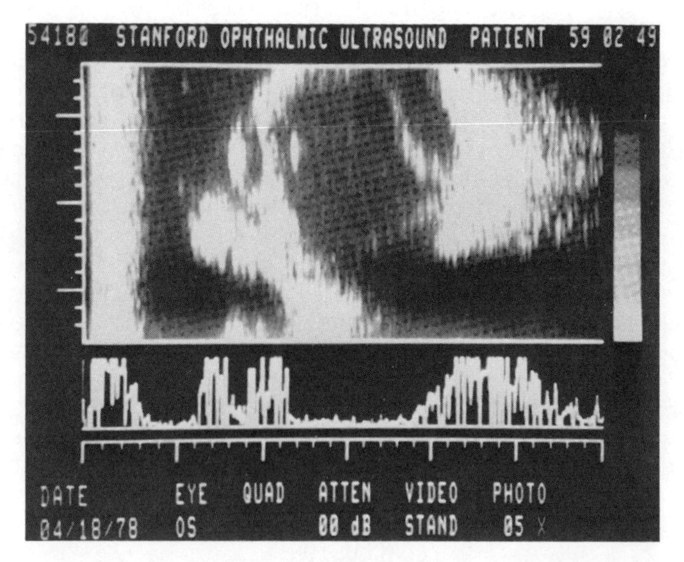

FIGURE 15. Eye cross-sectional image from a patient examination.

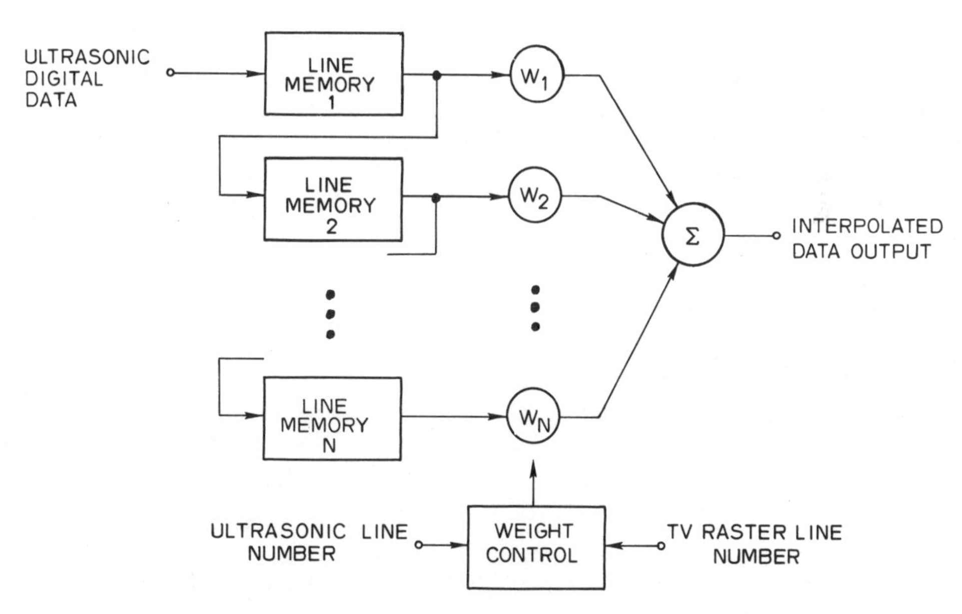

FIGURE 16. Generalized higher order interpolation method block diagram.

Interpolation Improvement

A possible improvement in the interpolation performance would be use of more than two ultrasonic data lines at a time [4]. Figure 16 shows a generalized interpolation block diagram. Here, a number of line memories store the required ultrasonic data, and weighting controls W_i determine what fraction of each ultrasonic line adds into the final result. The weight control W_i vary as a function of the display raster line. They could conceivably also vary as a function of range to implement an aperture correction algorithm for beamwidth reduction. Care must be taken to preserve the zero phase nature of this transversal filtering process by proper choice of the weighting coefficients. The next step up in performance from linear interpolation is a four point Lagrangian interpolation function. Implementation of this function should give the apparent sharpness of the cosine interpolation shown previously without its drawback of excessive image striation.

CONCLUSION

It has been demonstrated with both synthetic signals and actual ultrasonic data that digital scan conversion and interpolation can produce high quality images with a limited data base. Linear interpolation produces the most generally pleasing images, while cosine interpolation provides better apparent resolution with some targets.

REFERENCES

[1] N. Bom, C.T. Lancee, J. Honkoop, and P.G. Hugenholtz, "Ultrasonic Viewer for Cross-Sectional Analysis of Moving Cardiac Structures," Biomedical Engineering, Nov. 1971, p. 500.

[2] J.T. Walker, A.L. Susal, and J.D. Meindl, "High Resolution Dynamic Ultrasonic Imaging," Proc. of the SPIE, 152, Aug. 1978 (in press).

[3] J.A. Vogel, C. Ligtvoet, N. Bom, G. van Zwieten, and P.G. Hugenholtz, "Processing Equipment for Two-Dimensional Echocardiographic Data," Ultrasound in Medicine and Biology, 2(3), 1976, pp. 171-179.

[4] R.W. Schafer and L.R. Rabiner, "A Digital Signal Processing Approach to Interpolation," Proc. of IEEE, 61(6), June 1973, pp. 692-702.

This work was supported under NIH grant P01 GM 17940.

DOPPLER WAVEFRONT DISTORTION EFFECTS DUE TO CONTINUOUS DYNAMIC ARRAY FOCUSING

James T. Walker

Center for Integrated Electronics in Medicine

Stanford University, Stanford, CA 94305

INTRODUCTION

Use of continuous dynamic focusing in an ultrasonic linear phased array system can cause doppler distortion of the received signals. This distortion results from variation of the effective beamformer delay during signal reception, and can affect both the received signal frequency and synthesized beamwidth.

Consider the linear array system in Fig. 1. A line of transducer elements first transmits a plane wavefront at an angle θ from the array axis so as to illuminate the target T, at a range R_0 from the array center. Circularly curved echo wavefronts return from the target to the array, producing element responses at various times. A set of analog delay lines then inserts appropriate true time delays into the transducer paths so as to produce a maximum echo output from the RF summation. The echo processor is therefore electronically synthesizing an ultrasonic beam response pattern focused on object T_1. Variation of the delay line values during the time after transmission permits the focal point to follow the target range variation with time [1]. This dynamic focus concept causes all targets in the object space to be effectively in focus simultaneously on the ultrasonic image.

One implementation of a dynamically focused linear array appears in Fig. 2 [2,3]. This system uses an SISO CCD for the analog delay in each transducer element channel. Signal time delay in the CCD varies inversely with the CCD clock rate, with a multiplicative approximation used for the control algorithm. A voltage controlled oscillator (VCO) provides the basic clock frequency, and is swept in frequency during the signal reception time

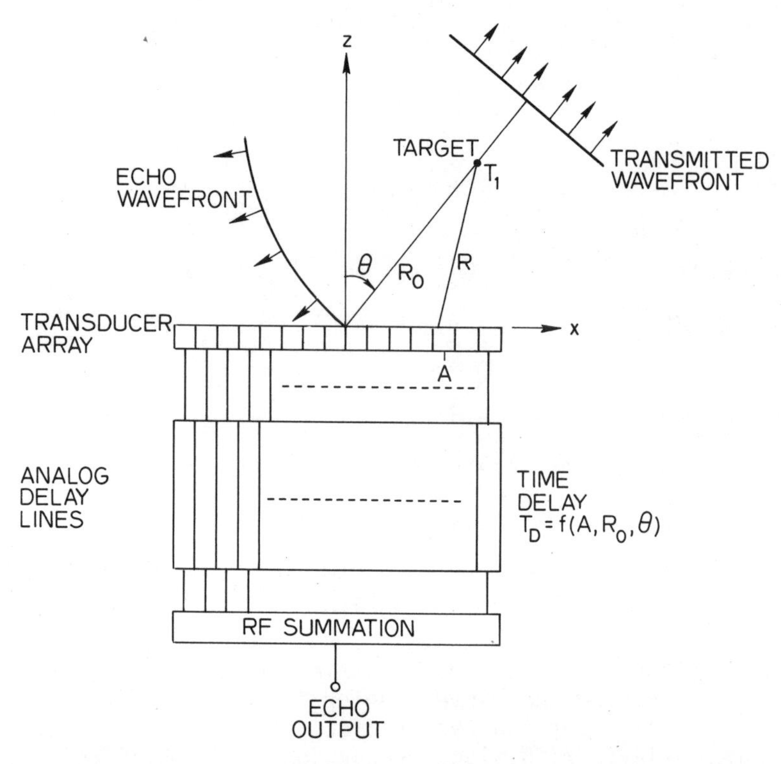

FIGURE 1. General linear array system with analog delay.

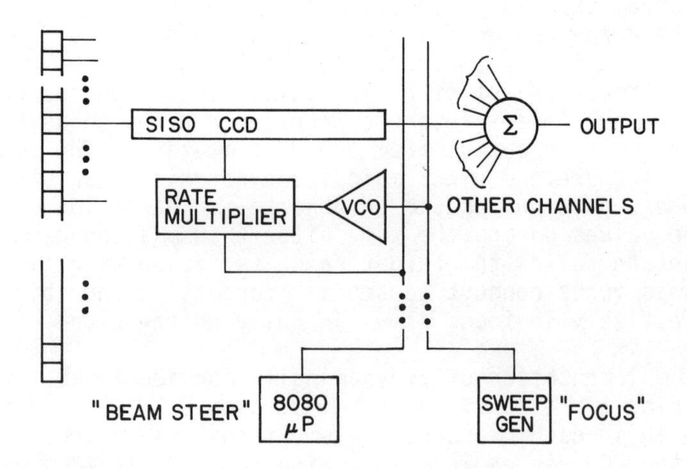

FIGURE 2. Phased array beamformer with serial input-serial output (SISO) CCD delay lines.

to produce the dynamic focusing. The VCO output pulse rate is multiplied by a coefficient determined by the beam pointing angle θ. Due to the approximation used, only the beamsteering coefficients fed into the rate multipliers vary during scanning, and the VCO frequency sweeps remain unaltered.

Experimental observation of the doppler distortion effect first occurred during preliminary alignment of this system. During this procedure a target containing small diameter wires is placed on the array axis. With the beam deflection angle at $\theta = 0$, the various VCO frequency sweep curves are adjusted for time coincidence of the CCD outputs at the RF summation unit. Figure 3 shows an oscilloscope photograph made at that time. Four groups of two traces appear; the bottom trace in each group is from the CCD output ($A = 0$) of the array center, and the top trace is from the combined output of CCDs at $A = \pm 11$ mm. The wire range for each echo pair is noted at the right. The objective of the alignment is to cause the positive-going echo cycles of each pair (located 0.6 division to the right of the vertical centerline) to coincide in time. Although this requirement is met in the 2.5 cm traces, a definite phase shift occurs as time progresses to the right. A similar effect of lesser magnitude can be observed in the trace pairs of the other three echo ranges. The phase shift appears as a doppler that lowers the RF pulse carrier frequency with a corresponding envelope lengthening. Reducing the magnitude of the doppler shift by increasing the range corresponds to less CCD focus variation at far range. The doppler shift was observed

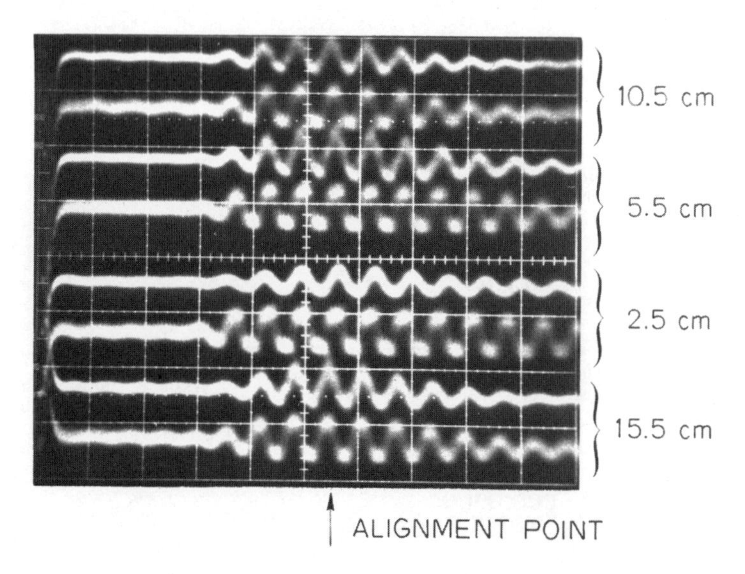

FIGURE 3. Experimentally observed doppler shift.

to be greatest at short ranges and the array ends, becoming unmeasurable at long ranges.

This doppler shift has two significant effects on the beamformer performance. First, since the CCD outputs are always lower in frequency, the average frequency of the summed RF pulse is reduced and the pulse length correspondingly increased. Second, lower frequency at the array edges implies increased time displacement rates as the true target position moves across the focal point. Therefore the apparent -3 dB synthesized beamwidth is slightly reduced.

Doppler Effect Derivation

Consider the linear array and target geometry shown in Fig. 4. We are interested in the condition where the array is statically or dynamically focused on its axis, and an echo target is present at some angle θ off of the array axis. For focusing, it is necessary that the CCD delay times are controlled so that the ideal delay

$$T_D(A) = \frac{1}{C}\left[R_0 - \left(R_0^2 - 2R_0 A \sin\theta + A^2\right)^{1/2}\right] \tag{1}$$

is satisfied everywhere in the target space. In the case of dynamic focusing, the substitution

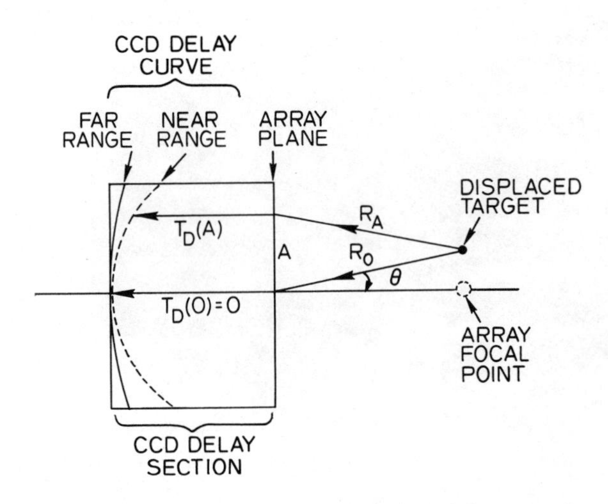

FIGURE 4. Linear array and target geometry.

$$R_0 = \frac{ct}{2} \tag{2}$$

is made (where t is the time after transmission and c is the velocity of sound in the medium), and the derivative of Eq. (1) is

$$\frac{dT_D(A)}{dt} = \frac{1}{2}\left\{1 - \frac{(ct/2) - A \sin\theta}{\left[(c^2t^2/4) - Act \sin\theta + A^2\right]^{1/2}}\right\} \tag{3}$$

Recalling that $T_D(A)$ is the added time delay experienced by an echo returning to the array center (A = 0) at time t, $dT_D(A)/dt$ is interpreted as the percentage change in instantaneous period encountered by echoes returning to the array center at t. Because the entire array uses the differential delay in Eq. (1) to cause time coincidence of the various element signals from the desired target, Eq. (3) applies to all element signals from that target at (R_0,θ).

Examination of Eq. (3) reveals that, when A = 0, the time expansion factor $dT_D(A)/dt = 0$ because, in Eq. (1), $T_D(A) = 0$ for all values of focus range R_0. Signals from the center array element, therefore, are not distorted in time.

A problem of greater significance is the time distortion of ultrasonic wavefronts at the CCD output [Eq. (3)]. To calculate representative wavefront positions, the echo path lengths are defined in Fig. 4 and the array is assumed to be dynamically focused on an axis normal to its length.

If α represents the time delay after transmission when echoes impinge on the element at location A, then

$$\alpha = \frac{R_A}{c} + \frac{R_0}{c} + \gamma \tag{4}$$

where γ is the relative time displacement along the ultrasonic echo pulse from the nominal pulse center. Geometrically,

$$R_A = \left(R_0^2 - 2AR_0 \sin\theta + A^2\right)^{1/2} \tag{5}$$

so that, if the substitution $R_0 = ct/2$ (true for dynamic focusing) is made,

$$\alpha = \frac{t}{2} + \left(\frac{t^2}{4} - \frac{At}{c}\sin\theta + \frac{A^2}{c^2}\right)^{1/2} + \gamma \tag{6}$$

Inversion of this expression to solve for t as a function of α on axis ($\theta = 0$) with $\gamma = 0$ yields

$$t = \alpha - \frac{A^2}{\alpha c^2} \tag{7}$$

Substitution of Eqs. (2) and (7) into Eq. (1) then defines $T_D(A)$ as a function of the echo reception time α.

$$T_D(A) = - \frac{A^2}{\alpha c^2} \tag{8}$$

This value of $T_D(A)$ is the time delay in the CCD when signals arrive at element A at time α after transmission, and the array is assumed to be focused along the $\theta = 0$ axis.

To derive the actual ultrasonic signal phase fronts emerging from the CCD when a coherent echo is of nonzero length, the value of α must be determined. Calculation of echoes from targets displaced an angle θ from the array focal point will define the dynamic focusing effects. If T_δ denotes the time of a wavefront emergence from the CCD at element location A relative to the same wavefront emergence at the array center, then

$$T_\delta = T_D(A) + \frac{R_A}{c} - T_D(0) - \frac{R_0}{c} \tag{9}$$

The task is to determine the terms in Eq. (9). First, $T_D(0) = 0$ because of the arbitrary choice made when deriving Eq. (1). Equation (5) defines R_A, and R_0 and θ designate the displaced target location desired. Echo reception time α is defined by Eq. (6). Substituting the α value in Eq. (6) into Eq. (8) obtains the value of $T_D(A)$ in Eq. (9).

The above manipulations are a mathematical simulation of dynamically focusing the array on its $\theta = 0$ axis, followed by calculation of the phase-front time displacements for an off-axis target relative to the array center-element response. Variation of γ determines the effects at various parts of an extended pulse length. The parameter R_0 sets the effective target range, and the negative calculated values for T_δ and $T_D(A)$ indicate the necessity of adding an equal fixed positive time delay to all CCD channels.

Computed CCD Beamformer Wavefronts

The following parameters, typical of the system described here, were selected for producing plots of T_δ.

array width	31 mm
pulse length	5 μsec
RF frequency	1.54 MHz
target range	20 to 200 mm

These parameters were used in Eq. (9) to plot the equiphase front at the CCD output for various positions in the received echo pulse.

Figure 5 is a graph of the resultant phase fronts for a simulated target range of 20 mm, with the array dynamically focused at $\theta = 0$. The actual target angle corresponds to the -3 dB response level on one side of the focused point response. Time τ during the received pulse extends from left to right, and the vertical axis represents the position across the array width.

The parameter τ designates relative time measured along the pulse at the CCD output (in contrast to γ at the CCD input), with $\tau = 0$ at the peak signal envelope. It can be defined, therefore, as

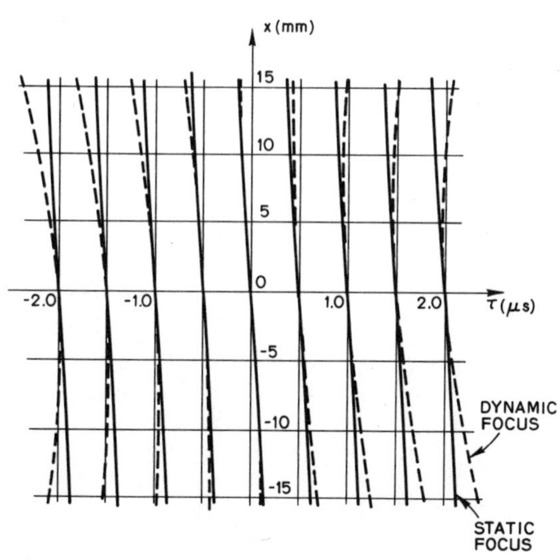

FIGURE 5. Calculated wavefronts at beamformer output with target at 20 mm range.

$$\tau = t - \frac{2R_0}{c} \tag{10}$$

A distortion between τ and γ is a result of the time-varying CCD delay. The phase fronts plotted in Fig. 5 are for uniformly spaced values of γ that are 0.5 μsec apart.

The vertical solid lines denote time intervals of 0.5 μsec, and the peak echo response ($\tau = 0$) is at the center. Curved dot-dash lines indicate the wavefronts for dynamic focusing as predicted in Eq. (9), and the solid lines represent the CCD array with static focusing at the target range. The difference between these curves reveals the effect of dynamic focusing; these differences disappear when $A = 0$ [where $T_D(0) = 0$ for all τ] and when $\tau = 0$ (where, by definition of dynamic focusing, there can be no error).

Of significance is that the dynamically focused wavefront has a slightly greater tilt for the curves closest to $\tau = 0$, which indicates that dynamic focusing may improve angular resolution. In addition, these same wavefronts uniformly curve away from $\tau = 0$. Summation of the CCD outputs, therefore, should reveal a slight pulse lengthening, with a corresponding doppler-frequency drop.

For comparison, curves similar to Fig. 5 appear in Fig. 6 at a 30 mm range. As the target range increases, dynamic focusing

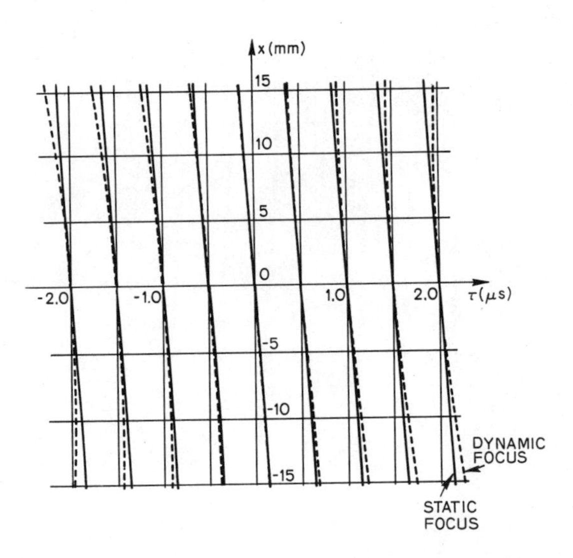

FIGURE 6. Calculated wavefronts at beamformer output with target at 30 mm range.

decreases rapidly, as could have been predicted by Eq. (3). The angle and range effects lessen in magnitude, and the visible wavefront curvature is reduced.

In Fig. 7, the assumed target displacement from the focal axis of $\theta = 0.1$ rad at a range of 20 mm causes highly sloped phase fronts; however, the dynamic-focusing errors correspond well with those in Fig. 5 for similar values of A and τ because they are time-delay errors that add linearly. Note that the dynamically and statically focused wavefronts coincide wherever they cross the $\tau = 0$ and $A = 0$ axes. As a result, much of the central echo-pulse region has small delay errors.

Best Fit Wavefront Derivation

The next step is to vary the parameters of a statically focused echo response to produce the best fit to the dynamically focused response. For this purpose, the echo-pulse envelope in Fig. 8 was assumed. The criterion to obtain a good fit is to minimize the weighted mean-square error Y calculated over the pulse length and array width. The formula for Y is

$$Y = \int_{-W/2}^{W/2} \int_{-L/2}^{L/2} \left(\frac{L}{2} - |\gamma|\right) (T_\delta - T_\varepsilon)^2 \, d\gamma dA \tag{11}$$

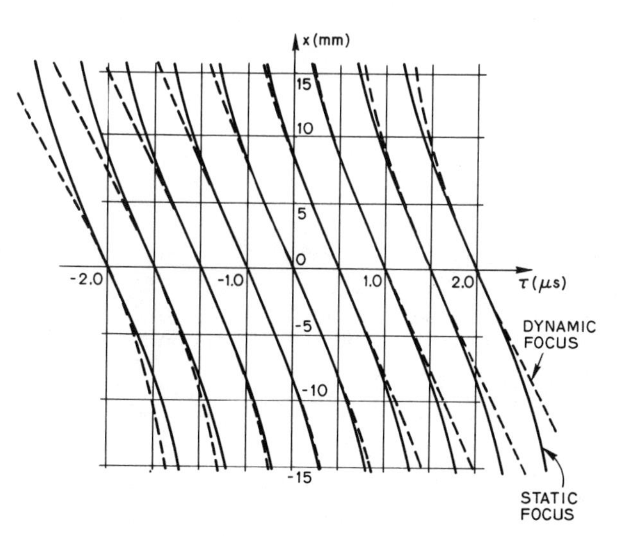

FIGURE 7. Calculated wavefronts at beamformer output with target at 20 mm range displaced 0.1 rad from array axis.

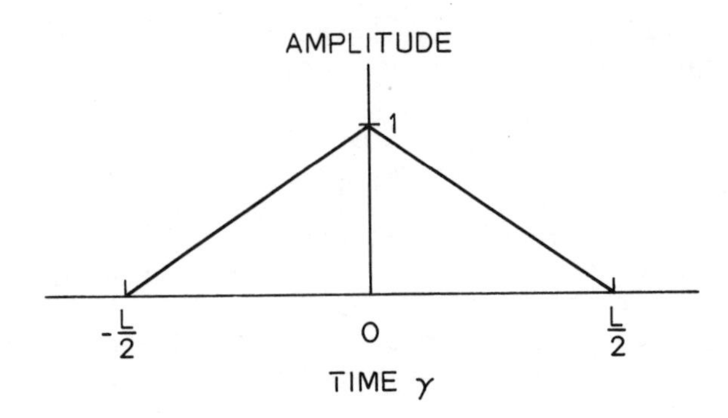

FIGURE 8. Assumed echo pulse envelope.

where T_δ is defined in Eq. (9), and T_ε denotes the wavefront delays for statically focused operation. Two additional parameters (M and ϕ) are inserted into the definition of T_ε to obtain the best fit on both the γ- and A-axes. The factor 1 + M represents the incremental increase in the time scale to simulate the doppler effects. The new beam angle Φ determines the static-response point-angle equivalent to the dynamically focused response. The definition of T_ε, therefore, is

$$T_\varepsilon = T_F + \frac{R_F}{c} - T_{DO} - \frac{R_0}{c} - M\tau \tag{12}$$

Here, T_F is $T_D(A)$ [from Eq. (1)] evaluated with a fixed R_0 and $\theta = 0$, and R_F is R_A [from Eq. (5)] with a fixed R_0 and the new angle Φ substituted for θ. The added term $M\tau$ stretches the statically focused wavefront response along the τ-axis with the expansion factor M. Because T_F is not a function of γ, there is a linear relationship between γ and τ. As a result, the time axis of T_ε is proportional to $\gamma + M\tau = (1 + M)\gamma$, and doppler stretching of the received echo pulse can now be simulated.

Variation of M and Φ in T_ε minimizes the weighted mean-square error Y in several sample cases. In the cases chosen, the CCD array is dynamically focused along the $\theta = 0$ axis, with a simulated target at θ corresponding to the -3 dB response point, and the target ranges are 20, 30, 50, 100, and 200 mm. After determination of the optimal M and Φ values, the equiphase wavefronts were plotted.

Figure 9 is a plot of the statically focused wavefronts that obtain the best fit to the dynamically focused wavefront, with a target at 20 mm. Note that the chosen value of Φ has caused the wavefronts closest to $\tau = 0$ to coincide closely, and M has

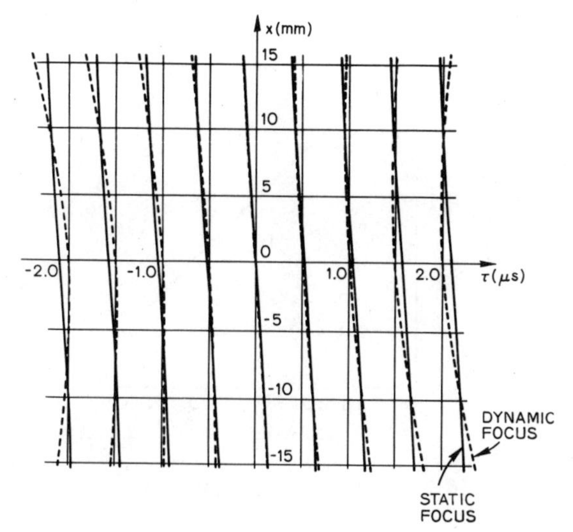

FIGURE 9. Static focused wavefronts with best fit to dynamic focused beamformer output for 20 mm range.

stretched them along the τ-axis so as to simulate the average displacement caused by the dynamically focused wavefront curvature. This criterion appears to work well for all values of τ simultaneously.

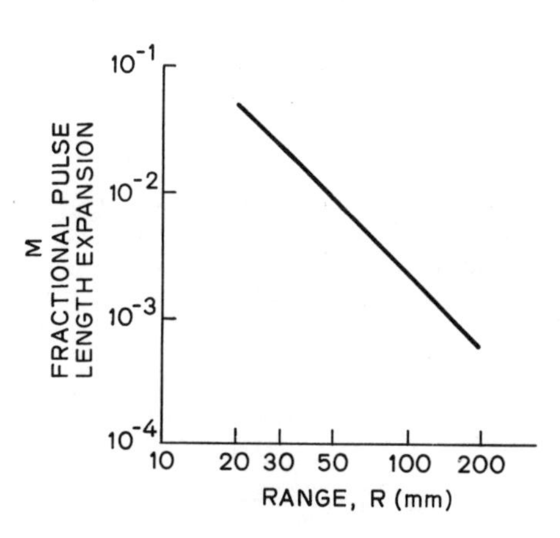

FIGURE 10. Fractional pulse length expansion.

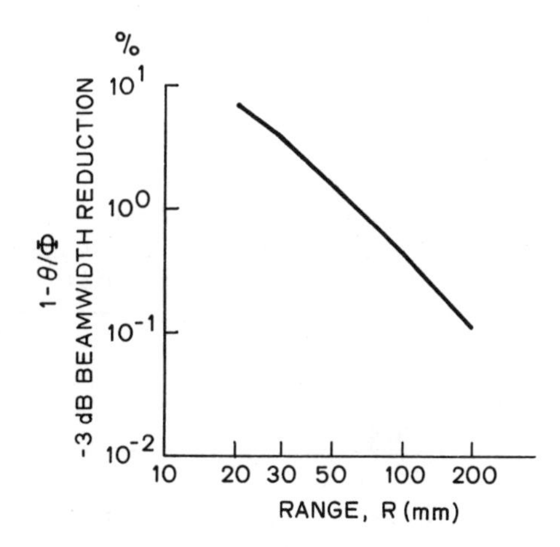

FIGURE 11. Beamwidth reduction.

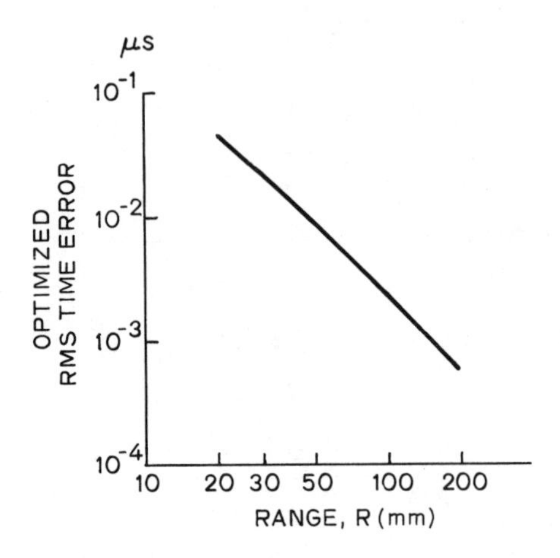

FIGURE 12. RMS time error.

Parametric Effects

The above optimization was performed at the five representative target ranges, and the results were plotted on a log-log set of axes to produce the curves in Figs. 10, 11, and 12. In each curve, the independent parameter drops rapidly in value in proportion to the square of the target range. As a result, the effects caused by dynamic focusing are only significant at short ranges (on the order of 20 mm); even there, changes in azimuth and resolution amount to less than 10 percent and will not seriously affect system performance.

CONCLUSION

It has been shown that use of continuous dynamic focusing of a linear phased ultrasonic transducer array causes reduction of the average received pulse frequency due to time stretching or elongation, in conjunction with a narrowing of the effective beamwidth. These effects occur predominately at short ranges and rapidly diminish as the target range increases.

ACKNOWLEDGMENT

This work was supported under NIH grant P01 GM 17940.

REFERENCES

[1] O.T. von Ramm, "A Real-Time Digitally Controlled Ultrasound Imaging System," Duke University, 1973.

[2] J.T. Walker and J.D. Meindl, "A Digitally Controlled CCD Dynamically Focused Phased Array," Proc. IEEE Ultrasonic Symp., 1975, pp. 80-83.

[3] J.T. Walker, "A CCD Phased Array Ultrasonic Imaging System," Stanford University, 1977.

A 1-MHz LINEAR PHASED ARRAY ULTRASOUND SYSTEM FOR INTRACRANIAL IMAGING

Ralph W. Barnes and Ward A. Riley

Bowman Gray School of Medicine

Winston-Salem, North Carolina 27103

In the Cerebrovascular Research Center at the Bowman Gray School of Medicine, there has been a long term interest in the noninvasive evaluation of intracranial cerebrovascular disease. One approach to this evaluation is to image the major intracranial arteries in a real time two dimensional image format using a 1MHz linear phased array ultrasound system.

SYSTEM DESCRIPTION

The 1MHz transducer array shown in Figure 1 consists of 20 lead zirconate titanate elements each 1.54mm (one wavelength) wide and 15mm (10 wavelengths) long. Interelement spacing is 0.19mm or about one-eighth wavelength. This small interelement spacing prevents the formation of significant grating lobes in the acoustic beam when adjacent elements are excited during the transmit operation.

The ultrasound beam is formed by exciting ten consecutive elements to form an effective aperture of about 17mm x 15mm. Appropriate excitation phasing is used to nominally focus the beam at a 7cm range. An image is formed by sequentially exciting groups of ten elements. Line 1 is formed by elements 1 through 10, line 2 by elements 2 through 11, etc. The resulting image is about 2cm wide by 15cm in range.

Each element is excited by a square wave pulse 500ns

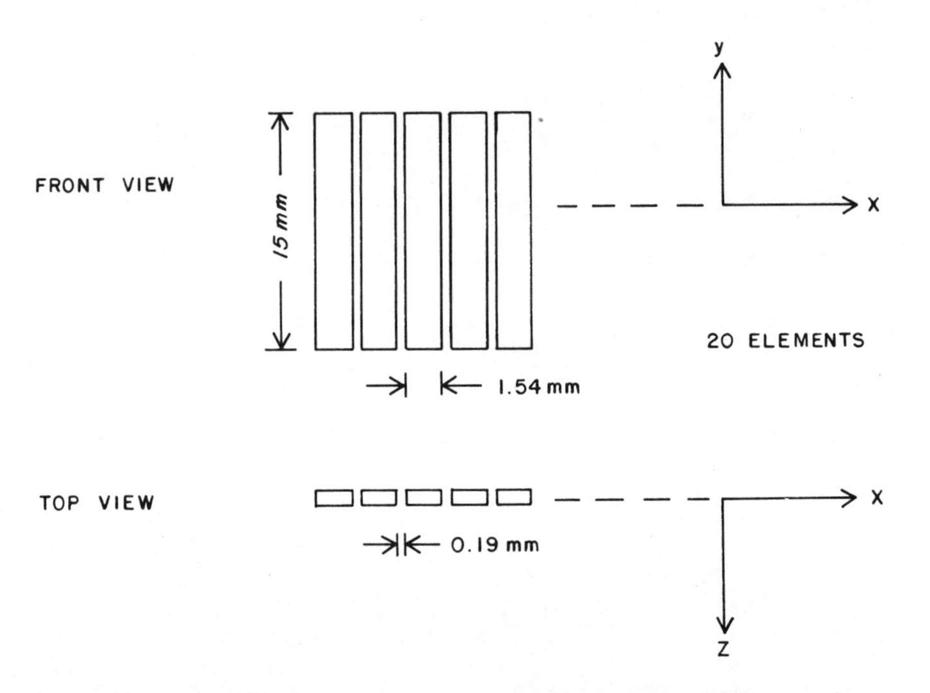

Fig. 1 Geometry of the 1MHz transducer array.

in time duration. Pulse amplitude for intracranial imaging is about 180 volts. The acoustic pressure waveform 6mm in front of a single element as measured by an acousto-optical method when the element is excited by a 100 volt pulse is shown in Figure 2. The time waveform consists essentially of two cycles of 1MHz followed by small amplitude reflections from within the casing.

The -6dB beamwidth for a single element as a function of range was measured and compared to the theoretical -6dB beamwidth for an ideal element. The acoustic field produced by a single element was found to be about one-half as wide as the theoretical beamwidth at the -6dB point.

The time waveform changed essentially only in amplitude when a 6mm thick section of human skull was placed 5mm in front of the transducer element. The time waveform measured 15mm in front of the element with and

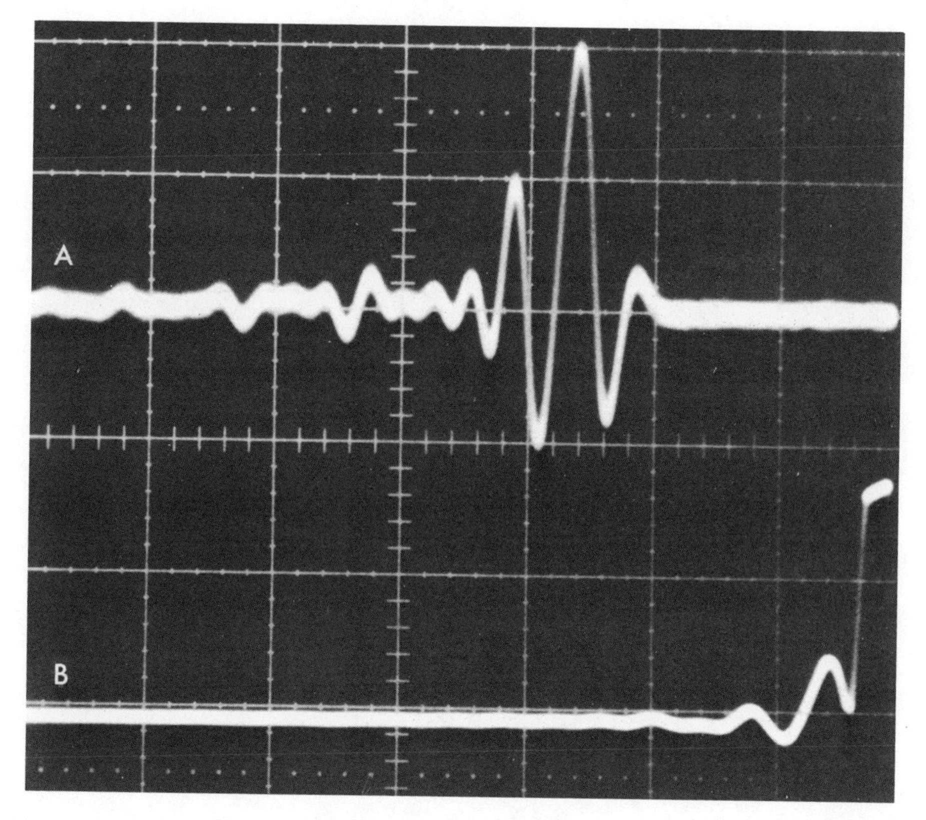

Fig. 2 A. Top: 100 volt excitation pulse to a single element. Approximate duration is 500ns.
B. Bottom: Transmitted acoustic pressure time waveform observed 6mm from the face of one element using an acousto-optical method. Time scale is 2μs per division.

without the skull present is shown in Figure 3. The frequency spectrum for each waveform is shown in Figure 4. The spectra are very similar except for the overall attenuation introduced by the skull. The pressure profile of the beam, measured with a piezoelectric microprobe, with and without the skull present is shown in Figure 5. With the skull present there is only a slight increase in the -6dB beamwidth from 5.8 to 6.0mm.

Figure 6 shows the transmit beam profile at a range of 6cm with and without the skull present when ten consecutive elements were excited and appropriately phased

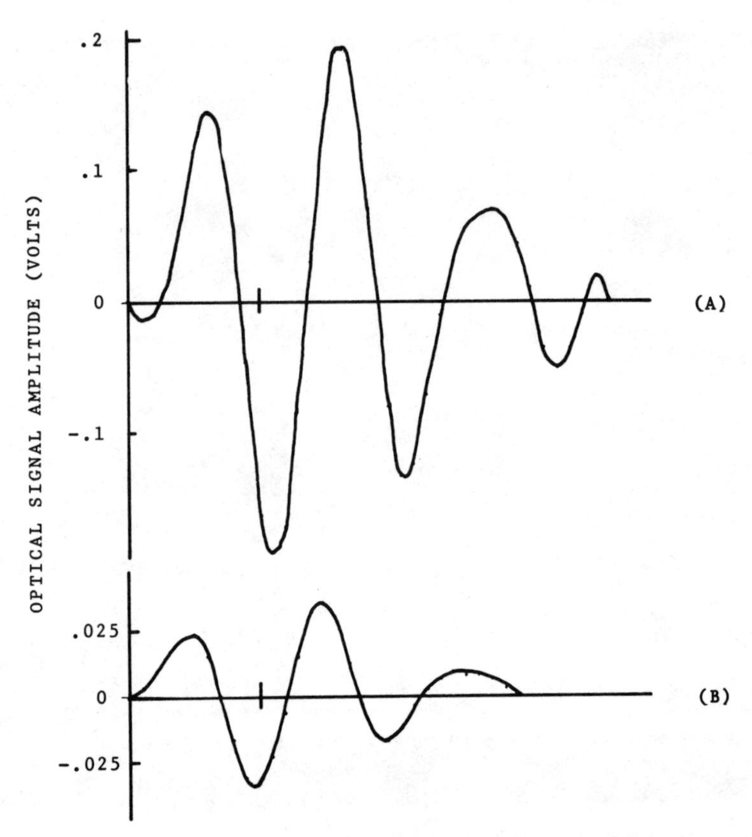

Fig. 3 Time variation of the acoustic pressure transmitted by one element of the array at a depth of 15mm. (A) In distilled water with no skull present. (B) With a 6mm thick skull sample inserted into the beam. Time scale is 500ns per division.

for a nominal focus at 7cm. At the-6dB points, the transmit beamwidth was 9mm with no skull present and 12mm with the skull present.

Figure 7 shows the block diagram of the entire system as constructed. The transmit excitation pulses are derived from the transmit focus generator and distribution which actuate the pulses driving the array elements. In the receive mode, the array is connected directly to line drivers, a 10dB amplifier and an emitter follower to isolate the element from coaxial cable loading. The line drivers feed preamplifiers which have a maximum voltage gain of about 60dB. Preamplifier gain is controlled by a programmable time gain control which can provide up to 60dB of gain control for each 1cm range.

Dynamic focussing in the receive mode is accomplished by selecting time delays from wideband delay lines. Time delay values are selected in 50ns steps up to a maximum delay of 500ns. The 50ns time delay steps provides receive focus timing accuracy of ±25ns or about 9° of phase.

The receive focus outputs are fed to a linear summing amplifier, a linear 50 ohm driver, and a detector and video amplifier. These linear outputs are used for interfacing with other signal processing units.

The receive focus outputs are also fed to logarithmic amplifiers to form the receive focus beam. The logarithmically compressed signals are summed, fed to a 50 ohm driver, and to a square law detector and video amplifier. Video signals are also logarithmically compressed.

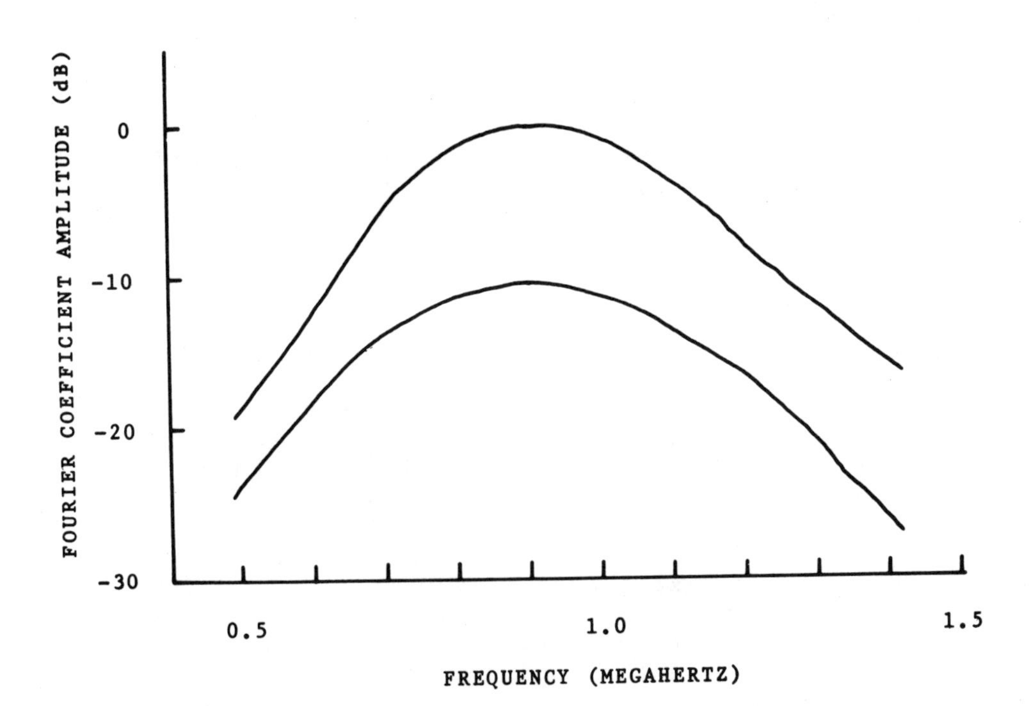

Fig. 4: Fourier spectra of pulses shown in Fig. 3. TOP (A) BOTTOM (B)

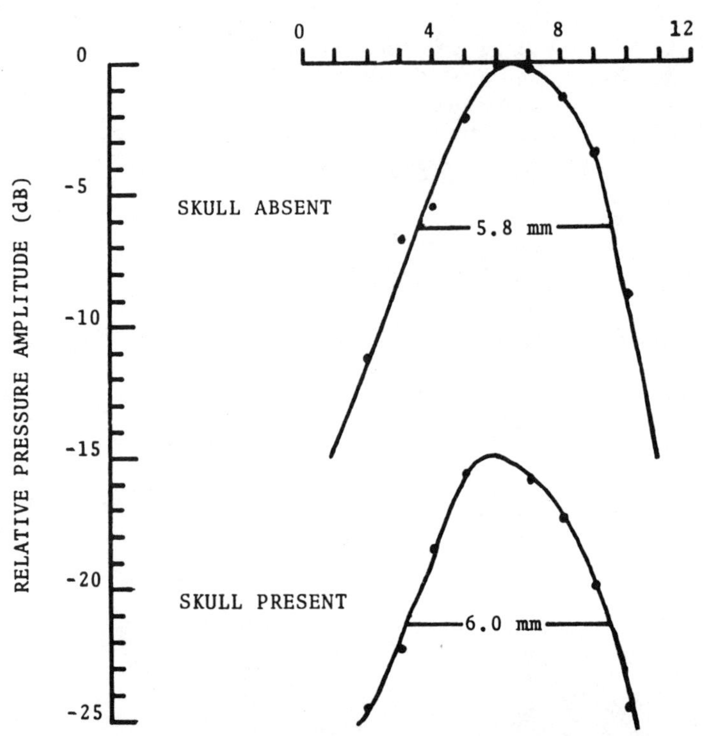

Fig. 5: Measured pressure profile of beam transmitted by one array element into distilled water at a depth of 15mm with and without a skull present.

The overall system response for exciting ten consecutive elements nominally focussed at a range of 7cm on transmit, dynamically focussed on receive, and the output taken from the logarithmic video output at the–6dB points varied from 3 to 6mm over the range of 4 to 12cm with no skull present. With the skull sample placed about 5mm in front of the array face, the system response was 4 to 7mm over the same range of 4 to 12cm.

It was our experience that with this 1MHz linear phased array system operated with the skull sample about 5mm in front of the array face, for the on-axis transmit and receive modes, the dominant effect of the skull was attenuation, with only minor changes in the pulse temporal waveshape, frequency spectrum and beam pressure profile.

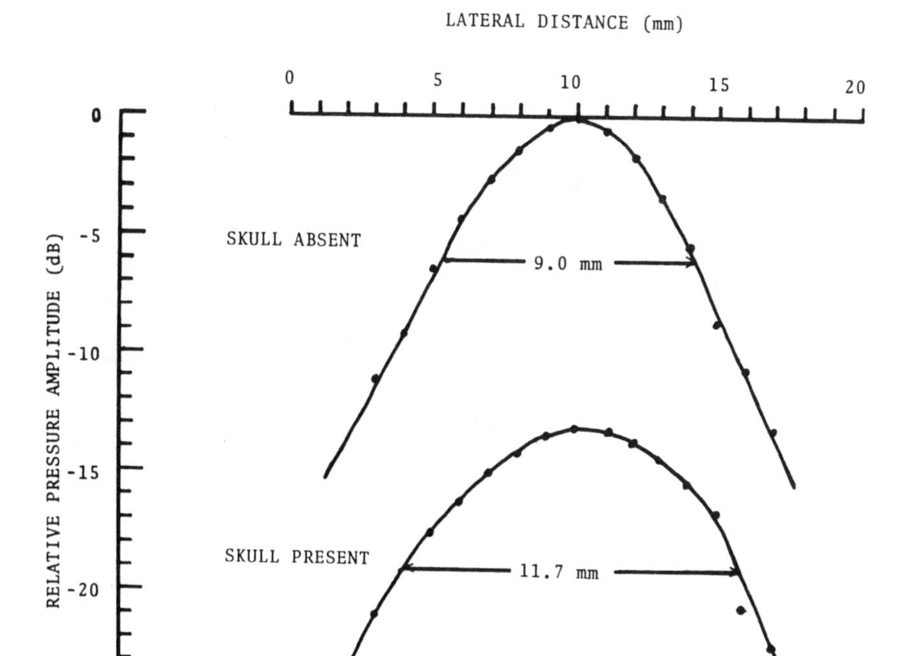

Fig. 6: Measured pressure profile of beam formed by 10 consecutive elements of the array phased for a nominal focus at a depth of 7cm. Measurements were made at a depth of 6cm with and without a skull sample present.

PRELIMINARY RESULTS

The major intracranial arteries of interest included the internal carotid arteries in their intradual course, the middle cerebral arteries as they travel laterally from the internal carotid arteries, the basilar artery as it travels along the ventral aspect of the midbrain parallel to the clivus, and the posterior cerebral arteries as they travel along the lateral aspect of the midbrain.

An example of imaging the intracranial arteries is the internal carotid artery. The transducer array is placed on the superior surface of the cranium about 1cm anterior and 1cm lateral to the intersection of the vertical line joining the two external auditory meatus

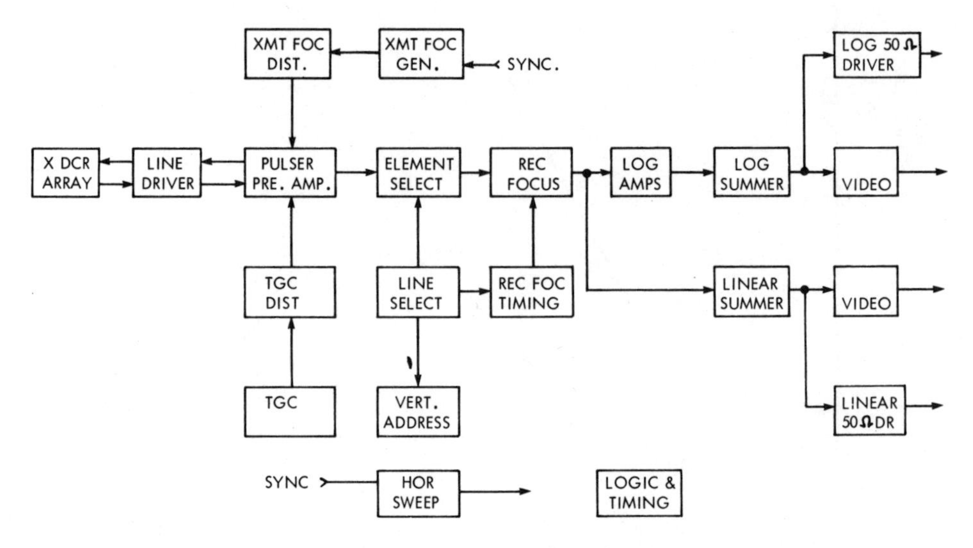

Fig. 7: Block diagram of the 1MHz system.

and the vertical line connecting the nasion and inion. The transducer array is directed inferiorly and slightly medially.

Echoes from the internal carotid artery in its intradual course are detected about 8 to 10cm from the transducer, as shown in Figure 8. (Both arterial walls are detected, and in the real time two dimensional display were observed to move in opposition to each other during the systole time of the blood pressure pulse. The diameter of the artery is about 4mm.) The arterial echoes are identified by their pulsatile activity synchronous with the cardiac cycle, while other nearby bone echoes showed no pulsatile activity.

Similar results were obtained from the other major intracranial arteries.

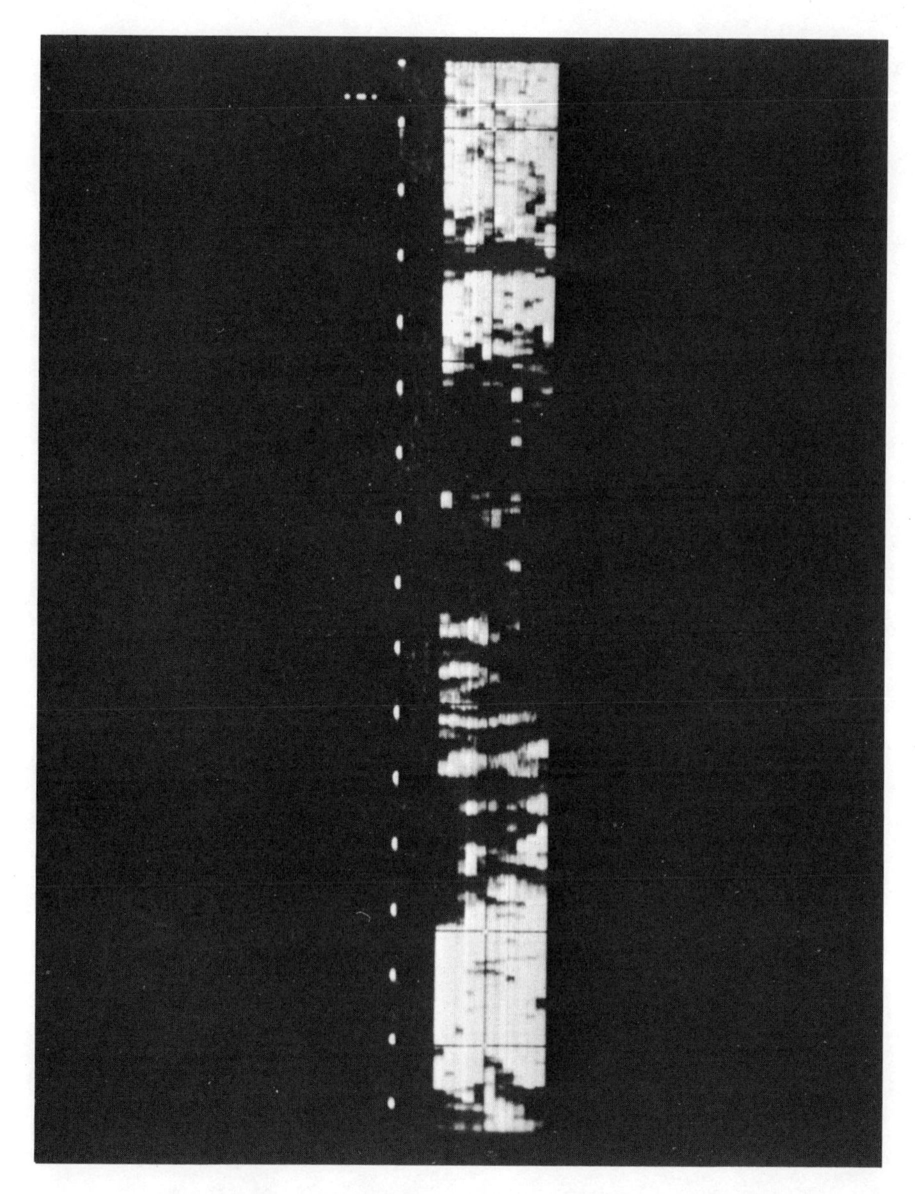

Fig. 8: Example of results obtained for imaging of the internal carotid artery. Vertical range marks are spaced at 1cm intervals. The arterial lumen is centered at a depth of about 9cm.

CONCLUSION

A 1MHz linear phased array ultrasound system has been designed, constructed, and used in preclinical trials to successfully image the major intracranial arteries. Noninvasive evaluation of intracranial vascular disease now appears to be a real possibility.

Acknowledgement

This work was supported by Cerebrovascular Research Center Grant (NS-06655).

A DIGITAL SYNTHETIC FOCUS ACOUSTIC IMAGING SYSTEM

P.D. Corl, G.S. Kino, C.S. DeSilets, and P.M. Grant

Stanford University

Stanford, California 94305

ABSTRACT

A new real time synthetic focus digital acoustic imaging system has been developed. It operates by exciting, with an impulse, one element of a transducer array, digitizing the return echo and storing it in a Random Access Memory. This process is repeated for all the array elements and using the focus information which has been loaded from the mini computer, the system generates a series of swept-focus lines perpendicular to the array face. In comparison to earlier computer based systems, our processor handles data at rates sufficient to generate real time images.

As only one transducer at a time is excited, it has been necessary to develop a high efficiency broadband transducer array with quarter wavelength matching layers. The array we have developed has an 11 dB return loss, a 2.7 — 4.3 MHz frequency range with an impulse response approximately 2-1/2 cycles long. The digital processor operates at a 10 - 16 MHz sample rate with 8-bit quantization. Theoretical and experimental results are presented for a system with a 96 line display employing 8 and 32 active transducer elements.

1. INTRODUCTION

1.1 Synthetic Focus Imaging

This paper describes the design and initial operation of a new real-time synthetic focus or synthetic aperture digital acoustic imaging system. The system is functionally equivalent to a tomographic imaging system with filtered back-projection, operating in real-time. The basic principles of a very closely related system have already been demonstrated by Johnson, et al,[1] using relatively slow computer reconstruction techniques. We have obtained high speed operation by performing the synthetic focus processing in dedicated digital hardware which is capable of operating at up to 16 MHz data rates.

In this system, we transmit from one element at a time and receive the return signal on the same element. The received signal passes through an analog multiplexer, an amplifier and an Analog-to-Digital (A to D) converter before storage in the signal memory, Fig. 1. This operation is repeated for successive array elements, with the analog multiplexer selecting the desired element. To implement this synthetic aperture imaging system we must be able to store a complete set of signals, one from each transducer element. In order to do this, we use a video A-to-D converter and semiconductor Random Access Memories (RAM). To provide adequate sampling of amplitude and phase we must operate the A-to-D converter at a sampling rate greater than twice the upper cutoff frequency of the transducer elements. Thus if the upper cutoff frequency of the transducer elements is 4.3 MHz, the system clock rate must be greater than 8.6 MHz. Once we have stored the signals from a sufficient number of elements, we can reconstruct an entire two-dimensional image by adding the information from the appropriate locations in the signal memories. Equal time delays to the point

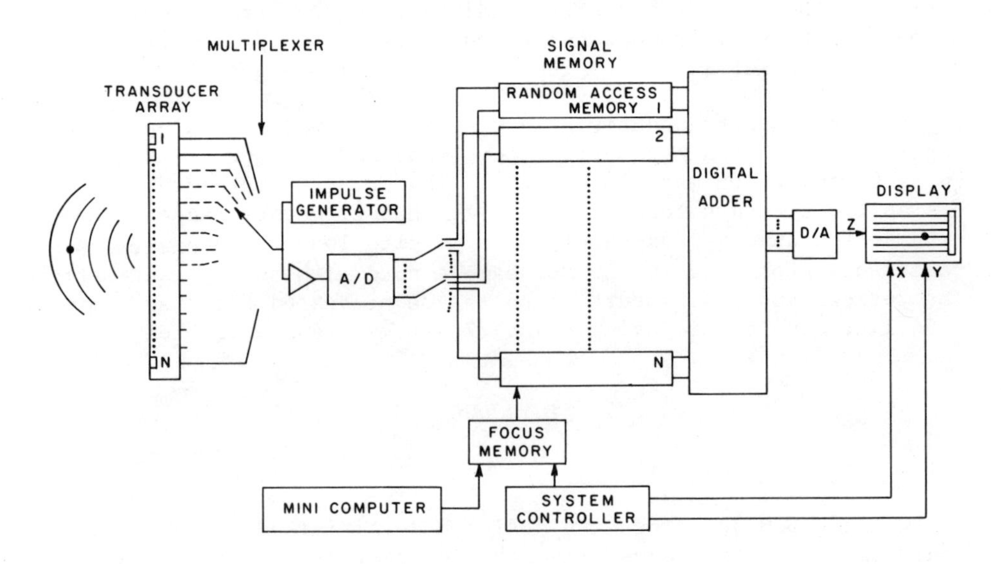

FIGURE 1

of interest, are inserted during the display process, Fig. 2, to reconstruct points in the image plane.

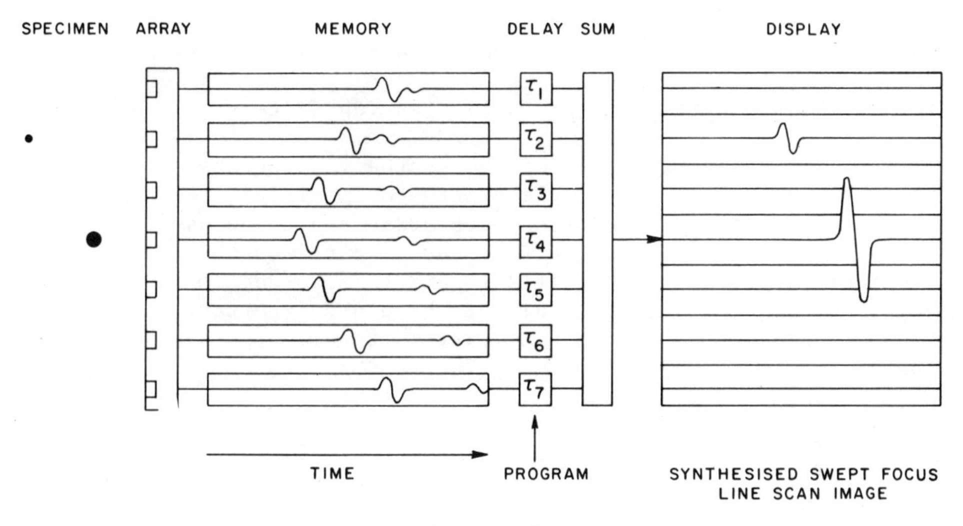

FIGURE 2

The reconstruction of a two-dimensional image from the set of signals stored in the digital memories is a formidable computational task and to accomplish this we have implemented the back-projection method which requires a set of geometric calculations to control the addressing of the signal memories. However, once the required addressing information has been computed, it can be stored in table form in a high speed memory which we refer to as the focus memory, Fig. 1. In a typical 32-element system, the focus memory will be about 1/4 the size of the signal memory. Alternatively, the focus memory can be implemented using a Programmable Read Only Memory (PROM) which is much denser than the RAM we are using for signal memory. However, the initial advantage of using a RAM rather than a PROM is that the scan format can be programmed at will from a computer or microprocessor. With this technique we can generate scan lines perpendicular to the array, perform a radial sector scan, or synthesize any other desired scan format. We can also vary the spacing between scan lines either by reprogramming the focus memory from a microprocessor, or by interpolation techniques. We have chosen to display our image as a raster scan with lines perpendicular to the array face.

One reason for this choice of scan format is the possibility of carrying out integration of the image on a scan converter, much as in a conventional mechanical B scan system. Suppose, for instance, 128 elements in total are employed and the image from 32 elements is read into a scan converter. Then by passing to the next 32 elements a new image is formed, and read into the scan converter. With the scan format employed, integration will take place on scan lines common to both images. Repeating this process with an array whose elements possess a wide angle of acceptance, produces an image which is equivalent to a very wide aperture system.

1.2 Comparison with Other Acoustic Imaging Systems

An important advantage of the synthetic aperture approach is that it requires only a single front-end amplifier, regardless of how many elements of the transducer array are to be used. This means that a great deal of effort can be put into the design of the front-end amplifier with little regard for its complexity, number of adjustments, expense, etc., all of which are important considerations in a system where an amplifier is required for each element of the array.

Another important advantage of the synthetic aperture approach over other acoustic imaging systems is that the transverse resolution is twice as good as that of an equivalent system in which a parallel beam is transmitted and the system is focused on receive. This is due to the fact that the time and phase difference to a point, Z , from a transducer element is doubled, because the signal travels to the image point and back. Thus our system has a transverse resolution which is equivalent to that of a conventional imaging system operating at twice the frequency. The range resolution is essentially determined by the pulse length (bandwidth) as with other imaging techniques. The system therefore provides the same improvement in transverse resolution capability which has already been demonstrated in scanned holographic imaging.[2] But, in addition, as we are using time delay rather than phase delay techniques to reconstruct the image, we should also obtain excellent range resolution.

One problem with most electronically scanned acoustic imaging systems is that the display line and frame times are not usually compatible with a TV monitor. This is a disadvantage as the grey scale image quality of magnetically deflected cathode ray TV tubes, Figure 5(b), is superior to that of the electrostatic deflection tubes used in oscilloscopes, Figure 5(a). Our present design can easily be made compatible with the line time of a TV display by controlling the speed at which the focused lines are read out. However, the number of lines currently employed, 96, is far less than in a TV display (525). So the image will not look continuous. One technique to eliminate this difficulty is to use only part of the screen. Another is to use interpolation routines for filling in.

We have carried out an analysis to predict the sidelobe levels of the system. We assume that the system is excited by a pulse of the form $F(t) \exp j\omega t$, and is focused on the point x_o, z_o. Then after suitable time delays have been introduced, the sum of the delayed signals returning to the transducers at $x_n, 0$ is of the form

$$G(t) = \sum_{x_n} F(t - 2R_n/v) \exp j\omega(t - 2R_n/v) \tag{1}$$

where

$$R_n = \sqrt{z^2 + (x_n - x)^2} - \sqrt{z_o^2 + (x_n - x_o)^2} \tag{2}$$

and V is the acoustic velocity in the medium.

By making the paraxial approximation that $(x_n - x)^2 \ll z^2$, taking $x_o^2 \ll z_o^2$, for simplicity, we can write

$$R_n \approx \Delta z - x_n \Delta x / z_o \tag{3}$$

where $\Delta z = z - z_o$, $\Delta x = x - x_o$. It follows that the range resolution, i.e., the result with $\Delta x = 0$ is determined by the function $F(t - 2\Delta z/v)$ for all transducers. So the range resolution is determined by the pulse length.

On the other hand, the transverse definition and sidelobe levels $(\Delta z = 0)$ is determined by the sum

$$G(t) = \sum_{x_n} F(t + 2x_n \Delta x / z_o v) \exp j\omega(t + 2x_n \Delta x / z_o v) \quad . \tag{4}$$

If $F(t)$ is a long pulse the response $G(t)$ will vary as

$$\left| \frac{\sin (\pi x / d_s)}{\sin (\pi x / d_g)} \right| , \tag{5}$$

where d_s, the 4 dB definition of the system, is

$$d_s = \lambda z_o / 2D \tag{6}$$

and

$$d_g = \lambda z_o / 2\ell \tag{7}$$

is the grating lobe spacing, with $D = N\ell$ the width of N elements of the array and ℓ is the element spacing.

On the other hand, if the pulse extends from a time $-T/2$ to $T/2$, the maximum value of x_0 for which all transducers contribute to the response is

$$|x_n(\max)| = z_o vT/4\Delta x \quad . \tag{8}$$

For a signal 1 rf cycle long $(T = 2\pi/\omega)$ with $\Delta x = d_s$, it follows that $|x_n \max| = D/2$, while for $\Delta x = d_g$, $|x_n \max| = \ell/2$.

It therefore follows that beyond the first zero of the main lobe, not all the elements contribute, but the behavior of the focusing system near the main lobe is like that of a conventional lens operating with signals of wavelength $\lambda/2$. For Δx large, however, only one element contributes at a time, and so the response falls off by a factor $1/N$ where N is the number of elements. Thus by using only a short rf pulse grating lobes should be eliminated. We would therefore expect that with a 32 element system, the far out sidelobe level would be approximately -30 dB, and that because of the absence of grating lobes, relatively sparse wide aperture arrays can be used to give improved resolution.

2. IMAGING SYSTEM DESIGN

2.1 Transducer Array

Because only one element is excited at a time, the transducer array must have extremely high efficiency with as broad a bandwidth as possible, i.e., short duration impulse response, and broad angular beamwidth. Thus we designed the array with quarter-wave acoustic matching techniques[3] in conjunction with tall, narrow, piezoelectric ceramic elements.[4] Proper application of quarter-wave acoustic matching allows highly efficient transduction of acoustic energy into the low impedance load medium, typically water $(Z = 1.5 \times 10^6\ \mathrm{kg/m^2\text{-}sec})$, from the high acoustic impedance ceramic $(Z = 29.7 \times 10^6)$ over octave frequency bandwidths. With these characteristics in mind, a 180-element quarter-wave matched array was designed and built to operate with fully-slotted elements at a 3.8 MHz center frequency. The measured minimum round-trip insertion loss of the array was 11 dB at 3.85 MHz, Fig. 3, and the 3 dB fractional bandwidth was 45%, when an additional 2.2 dB was subtracted from the experimental data to account for the reflected signal which was incident upon the gaps between the elements. The impulse responses of 32 transformer impedance matched elements were measured by reflecting a signal off a thin 0.18 mm diameter wire target. The excitation was a 0.17 μsec wide square pulse. A 3.5 MHz 5 half-cycle (2.5 full cycles) impulse response, consistent with the measured 45% bandwidth is observed in Fig. 4 for each of the connected 32 elements. Excellent uniformity from element to element should be noted.

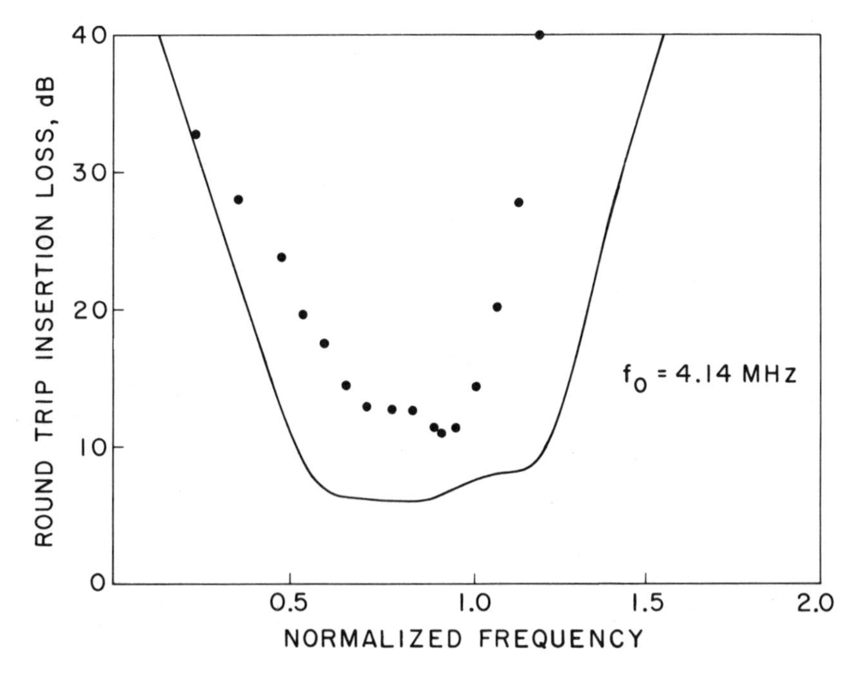

FIGURE 3. Comparison of theoretical and experimental two-way array insertion loss.

2.2 Control Electronics

In our initial experiments, we have constructed an 8-element system, to verify the basic principles of operation, and to gain experience with the hardware design. With the transducer array described above, the system has a field of view in water which is 5 cm wide and 4-8 cm deep dependent on the system operating frequency. The initial range can be varied under software control. Using shear waves in metal, the field of view will be 5 cm wide and 8-14 cm deep.

The system hardware comprises 32 printed circuit boards, the associated power supplies and card racks. The signal memory associated with each transducer element consists of a RAM board incorporating eight Intel 2125-AL 1K × 1 static RAM's arranged to give 1024 8-bit bytes of serial storage. When used in conjunction with the Tektronix ADC-820T A-to-D converter, the system is capable of digitizing and storing signals at up to a 16 MHz rate. The eight channel system we have constructed uses eight equally spaced transducer elements from the center 32 elements of the array. The system is designed to display a 96 line raster scanned image with a line-to-line spacing which is 1/4 of the element-to-element spacing.

Typically, we use every fourth element of the array which gives an interelement spacing of 2.03 mm, and a line-to-line spacing of 0.508 mm. For reasons of flexibility and simplicity, we chose to use the same RAM boards for the focus memory. For the 96 line display we have described, we require 128 lines of focus information. Fortunately, due to the symmetry about the center of the array, it is necessary to store only 64 lines of focus information. Since each line of focus information is 1024 × 1 bit, we can store the required amount of focus information on eight of the 1024 byte RAM boards. In operation, the system is designed to acquire the 8 signals in approximately 2.5 msec and display the complete image in approximately 7.5 msec which will allow a frame rate of approximately 100 Hz. For a larger system, using more transducer elements and displaying more lines, the frame rate would be correspondingly reduced; however, a 32-element system, displaying 400 lines at a 30 Hz frame rate is feasible using this technique. It should be noted that at present, our analog multiplexer is still being built. At the moment it is a mechanical switch, and hence the signal acquisition is not currently being performed in real time.

Figure 4 shows the signals as received on all 32 transducer elements from a single isolated wire target in a watertank. The difference in timing of the stored signals for the range $Z = 70$ mm is clearly seen. When focused on the scan line through the target all these delay differences are compensated by the focus information, giving a focused image of amplitude N times the individual signal amplitudes (N equals the number of active transducers). On other scan lines the information is not correctly timed and hence does not add coherently to form an image. Measurement of a fine wire target in water with the focused system at $Z = 62$ mm gave a 4 dB resolution of ~ 0.9 mm. For a 1.3 mm diameter wire we obtained a focused image on 2 or 3 adjacent scan lines which agrees closely with this theory.

Figure 5(a) shows the focused image, obtained with our 8 element experimental system, on a single 1.3 mm diameter wire target at $Z = 62$ mm range when sampling at 14 MHz. The sidelobe structure, comprising 8 distinct arcs, are clearly visible in Figure 5(b) which was obtained by loading the sampled data into our minicomputer via a Biomation recorder, and focusing in the computer prior to display on the scan converter. The far out sidelobe level is -14 dB, compared to a theoretical -18 dB (for 8 active transducers). Two grating lobes introduced by the 2.5 cycle impulse response, occur 13 scan lines each side of the focused image with relative amplitude -8 dB, compared to the theoretical value 2.5/N (-10 dB). Figure 6 shows, for comparison, the image which would be obtained if all 32 transducers were used. This image, which was again obtained by computer processing, has a measured sidelobe level of -20 dB compared to a theoretical -30 dB. It is most encouraging to achieve these low sidelobe levels at this early stage of system

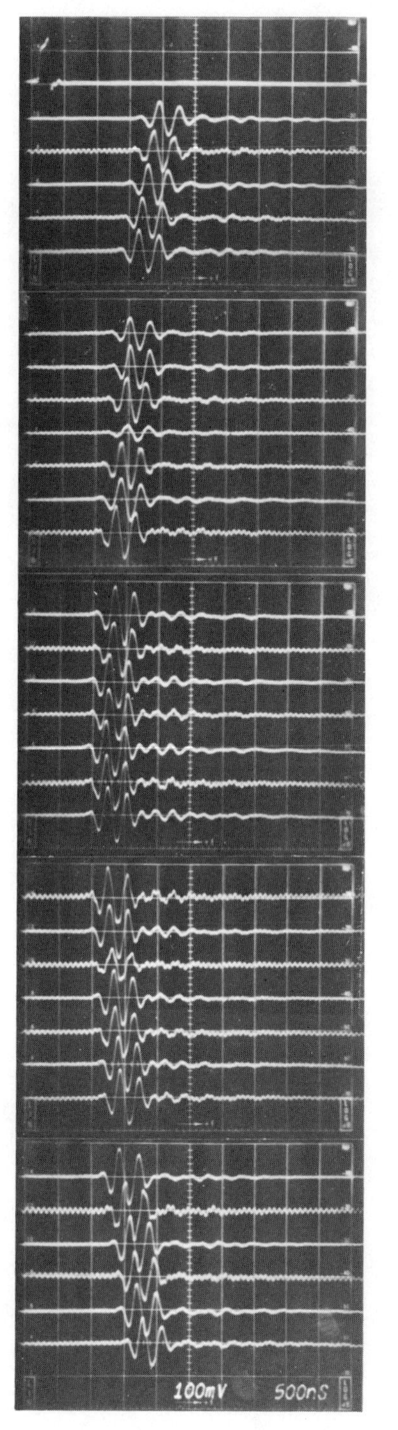

TRANSDUCER RESPONSES TO SINGLE TARGET

TRACE 1
INPUT IMPULSE (VERTICAL SCALE 50 V/DIV)

TRACES 2 to 33
SIGNAL RETURNS ON INDIVIDUAL TRANSDUCER ELEMENTS FROM A SINGLE 0.18 mm DIAMETER WIRE TARGET LOCATED NORMAL TO TRANSDUCER CENTER AT RANGE Z = 70 mm. RECEIVED SIGNALS AMPLIFIED BY 20 dB AND DELAYED BY 92 μs PRIOR TO DISPLAY.

DIGITAL IMAGING SYSTEM

FIGURE 4

development, as our previous system[5] has taken considerable time and effort accurately matching the gain of the transmit and receive electronics between channels to reach this sidelobe level.

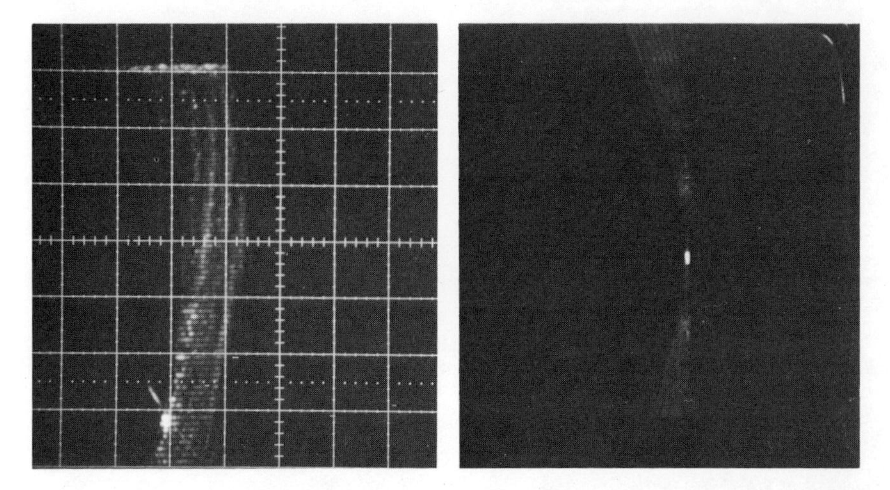

FIGURE 5. (a) Image obtained with our experimental eight transducer system on a single 1.3 mm diameter wire target in a water bath at $Z = 62$ mm range.
(b) Target moved to center of field and image processed in computer and displayed on TV monitor via scan converter.

FIGURE 6. Image obtained by computer processing to simulate, from real transducer data, the effect of incorporating 32 active transducers.

3. FURTHER DEVELOPMENTS

3.1 Sidelobe Response

One problem with all imaging systems is that associated with sidelobes, as the sidelobe level controls the dynamic range. One approach[6] to overcome this problem is to compress the dynamic range of the signals at the system inputs and expand again at the output after summing the signals prior to display. Because the synthetic aperture system requires only a single front-end amplifier, it is ideally suited to the implementation of such a compression technique, since we need not worry about matching the characteristics of a large number of compression circuits. We have the choice of implementing the compression with either analog circuits or a digital table look up.

To show the improvement obtainable using a compression technique, consider the case where we take the square root of the amplitude of the signal, but leave the sign or phase of the signal unchanged. Suppose the ratio of a main lobe to sidelobe is M. Then if one point on the object of amplitude a excites the main lobe and another point of amplitude b excites the sidelobe the ratio of their outputs will be Ma/b. Now suppose we take the square root of the output of each transducer before summing. The main lobe to sidelobe ratio will still be M but the ratio of the two outputs will be $M\sqrt{a/b}$. If we now square the summed output the ratio of the two signals will be M^2a/b and the original linear relation of the signals is restored, but the effective sidelobe levels have changed from M to M^2. This gives a consequent decrease in sidelobe level by a factor of 2 in dB, which is of major importance for detecting a small target in close proximity to a large reflector. However, this improvement is only obtained at the expense of halving the SNR. Thus we intend to investigate, with a digital table look up approach, the operation of our prototype system with square root and other weaker nonlinearities.

Figure 7 shows the results of incorporating square root compression on our 8 element imaging system. As the hardware is not currently complete we have again taken real image data from our transducer and loaded it, via the biomation recorder, into a mini computer. The input square root, focusing and output squaring have all been calculated off line and the resulting image read into an analog scan converter for display on a conventional TV monitor. In comparison with Fig. 5, the input compression is seen to reduce the displayed sidelobe levels from -14 dB to -22 dB at the edge of the field of view. We now intend to study the theoretical and practical performance trade offs which can be achieved with various different levels of input signal compression.

FIGURE 7. As Figure 5(b) with nonlinear processing to reduce the sidelobe levels.

3.2 Resolution Improvement

We have been performing additional analysis of the system to optimize its performance. We have concluded that the conflicting requirements of a short transmitted pulse, for good range resolution, with ideally a CW transmission for optimum transverse resolution can best be satisfied with a transducer impulse response comprising a single cycle sinusoid at its resonant frequency. We have already shown in Fig. 4, that our array currently falls short of this requirement, resulting in the introduction of the grating lobe.

One promising method of achieving the required waveform appears to be an inverse filter[7] implemented with a weighted tapped delay line. Computer simulations, Figure 8, using real transducer data gathered from wire target reflections have shown that after inverse filtering we can achieve the desired single cycle sinusoid response with -12 to -18 dB spurious levels. The tap weights were calculated with a Least Mean Squares (LMS)[8] adaptive filter algorithm, which gave convergence after ~100 iterations.

We intend to implement the inverse filter at the output of our processor as this will require the tap weights to be set only once to optimize to the composite impulse response of the transducer array. This impulse response will have to be loaded into the computer and the calculated tap weights read out and stored within the filter. We anticipate that the inverse filter can be realized with a hybrid digital analog approach, Fig. 9. It will comprise a 16 stage × 8 bit digital shift register with 16 multiplying D to A

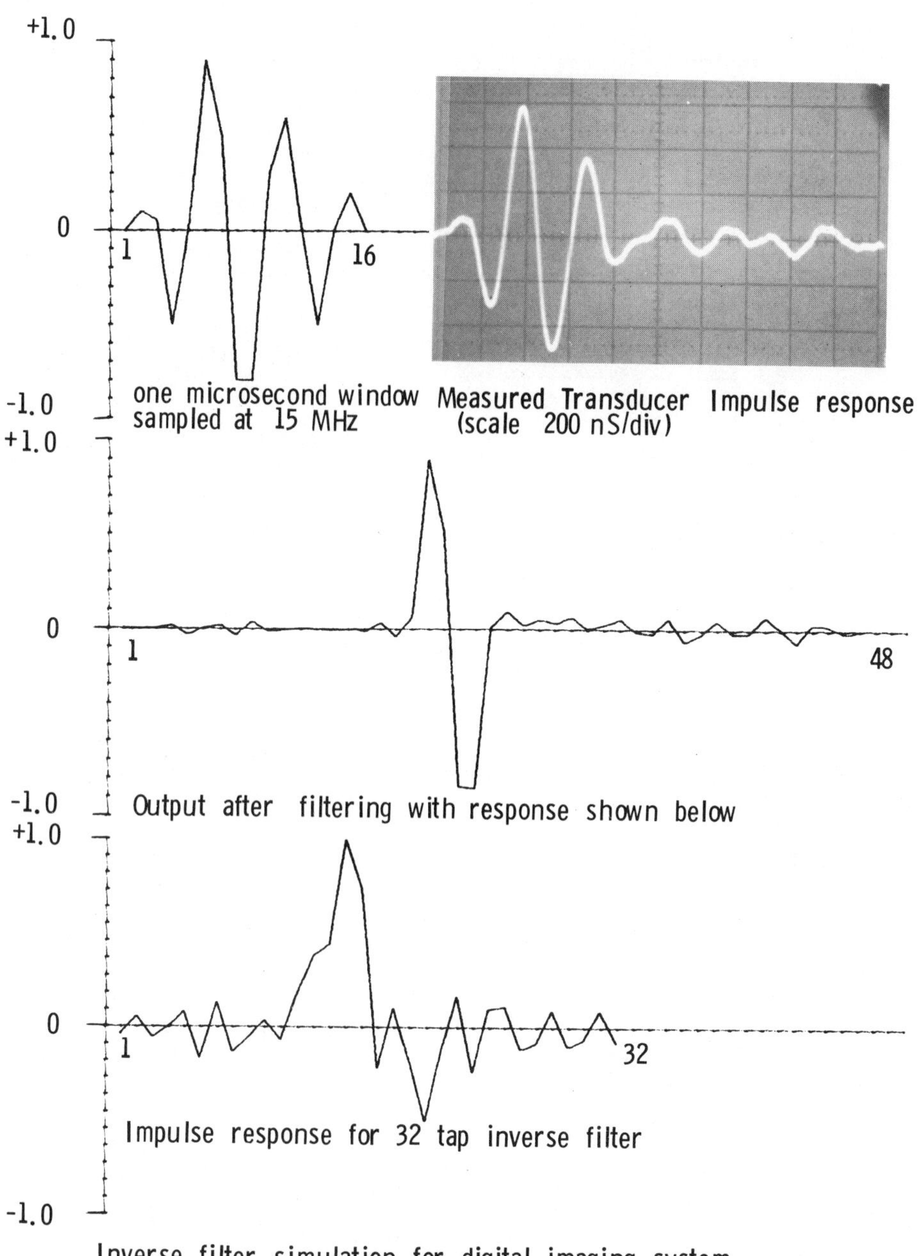

Inverse filter simulation for digital imaging system

FIGURE 8

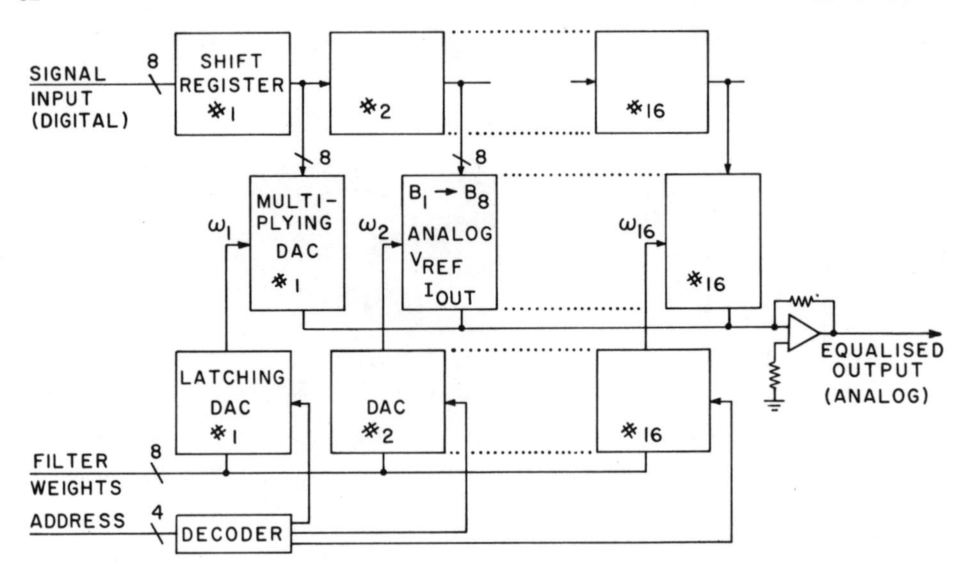

FIGURE 9 Digital implementation of inverse filter.

converters and a summing amplifier, and will occupy an additional four printed circuit cards.

Conclusions

A new design approach for acoustic imaging has been described and initial practical results on a small prototype system have been presented. Incorporation of nonlinear (gain compressed) processing and inverse filtering is predicted to improve significantly the early results presented here. The reduction in sidelobe levels by using gain compression is of major importance, for if compression is as effective as we predict, then acceptable sidelobe levels (-30 dB) may be achievable with a system comprising only 8 or 16 active elements.

One advantage of the synthetic aperture approach to imaging is that it becomes practical to consider a two-dimensional sparse array of 32 elements to obtain good definition in all three dimensions. As the sidelobe level in this case should be on the order of $1/N$, three-dimensional imaging with good definition in all 3 directions does appear to be possible using this approach.

Another advantage of the time delay focusing approach is that we can use a very wide transducer aperture for improved transverse resolution with a scan converter for obtaining the electronic equivalent of a mechanically scanned B-scan system, but with focusing. The additional electronics required for this approach involves only a larger multiplexer and the use of a scan converter.

A final possibility is to attempt post processing on the complete image. Schemes such as two-dimensional inverse filtering might be attempted to improve the resolution and reduce further the sidelobe level. Thus we confidently predict that digital synthetic focus processing will become increasingly attractive as systems are developed further.

Acknowledgements

This work was sponsored by the Center for Advanced NDE operated by the Science Center, Rockwell International for the Advanced Research Projects Agency and the Air Force Materials Laboratory under contract RI 74-20773.

REFERENCES

1. S. A. Johnson, J. F. Greenleaf, F. A. Duck, A. Chu, W. R. Samayou, and B. K. Gilbert, "Digital Computer Simulation Study of a Real-Time Collection, Post Processing Synthetic Focusing Ultrasound Cardiac Camera," Acoustical Holography, Vol. 6, Plenum Press (1975), p. 193.

2. H. Dale Collins, "Acoustical Interferometry Using Electronically Simulated Variable Reference and Multiple Path Techniques," Acoustical Holography, Vol. 6, Plenum Press (1975), p. 597.

3. C. S. DeSilets, J. D. Fraser, and G. S. Kino, "Design of Efficient, Broadband Transducers," IEEE Trans. Sonics and Ultrasonics, SU-25, No. 3, (May 1978), pp. 115-125.

4. C. S. DeSilets, J. Fraser, and G. S. Kino, "Transducer Arrays Suitable for Acoustic Imaging," 1975 IEEE Ultrasonics Symposium Proceedings, pp. 148-152.

5. T. M. Waugh and G. S. Kino, "Real Time Imaging with Shear Waves and Surface Waves," Acoustical Holography, Vol. 7, Plenum Press (1976).

6. F. L. Thurstone and O. T. Von Ramm, "A New Imaging Technique Employing Two-Dimensional Beam Steering," Acoustical Holography, Vol. 5, Plenum Press (1973).

7. G. L. Kerber, R. M. White, and G. R. Wright, "SAW Inverse Filter for NDT," Proc. IEEE Ultrasonics Symposium (1976), pp. 577-581.

8. B. Widrow, et al., "Stationary and Nonstationary Learning Characteristics of the LMS Adaptive Filter," Proc. IEEE, Vol. 64, No. 8 (August 1976), pp. 1151-1162.

A FLEXIBLE, REAL-TIME SYSTEM FOR EXPERIMENTATION IN PHASED-ARRAY ULTRASOUND IMAGING

M.D. Eaton, R.D. Melen and J.D. Meindl

Center for Integrated Electronics in Medicine

Stanford University, Stanford, California 94305

ABSTRACT

An experimental phased-array ultrasound imaging system using C3D electronic lenses is being developed in the CIEM at Stanford. This real-time system is designed to extend the imaging capability of the phased array systems used in modern clinical practice. The general purpose machine uses a modular architecture with a very high degree of computer control of important system variables. The flexibility of this system allows the experimenter great latitude in testing imaging schemes. The machine is comprised of five major computer controlled modules: (1) a 32-element linear and 32-element annular transducer array; (2) a bank of individually gain controlled preamps; (3) a transmitter bank with individual control of power, firing time and transmit waveform; (4) a matrix of C3D electronic lenses; and (5) a display frame memory for raster line interpolation and sidelobe deconvolution. System variables are easily accessible by the researcher through a CRT terminal and disk drive. As a first application, the theoretical predictions that aperture apodization should reduce sidelobe response has been experimentally verified.

INTRODUCTION

Sector B-scans from phased-array ultrasound imaging systems have now proven to be of great value in diagnostic medicine [1]. This imaging modality has been particularly effective in cardiac imaging where real-time display is important for clear visualization. Indeed, a recent Department of Health, Education and Welfare study found that "cardiovascular disease remains the number one killer in the nation" [2].

The major problem with present day real-time scanners is a limited information acquisition ability. Consequently, physicians describe the resultant images as only "adequate." It is difficult to precisely determine which system parameters specifically limit the image quality. Generally, clinicians use the words "noisy," "muddled" and "streaky" to describe images (Figure 1) from phased-array scanners. Additionally they refer to the images as "low resolution" compared with other imaging technologies such as x-ray. These resolution limitations are well understood. Our goal is to significantly improve the information gathering/processing ability of phased-array systems by developing a machine with a wide variety of easily changed system parameters.

There are five major technological approaches to improve system information gathering that may be explored using this machine: (1) the Theta- array, a new transducer configuration for high volumetric resolution imaging; (2) the C3D, a charge-coupled device recently invented at Stanford; (3) "front-end" aperture tailoring information gathering techniques; (4) a precision, variable-gain custom pre-amplifier IC; and (5) two-dimensional digital filtering of the detected video for raster line interpolation and side lobe deconvolution. Each of these topic areas are the subject of in-depth research in the CIEM. The remainder of this paper covers the hardware capability of the system in general and introduces some initial results of the aperture tailoring research.

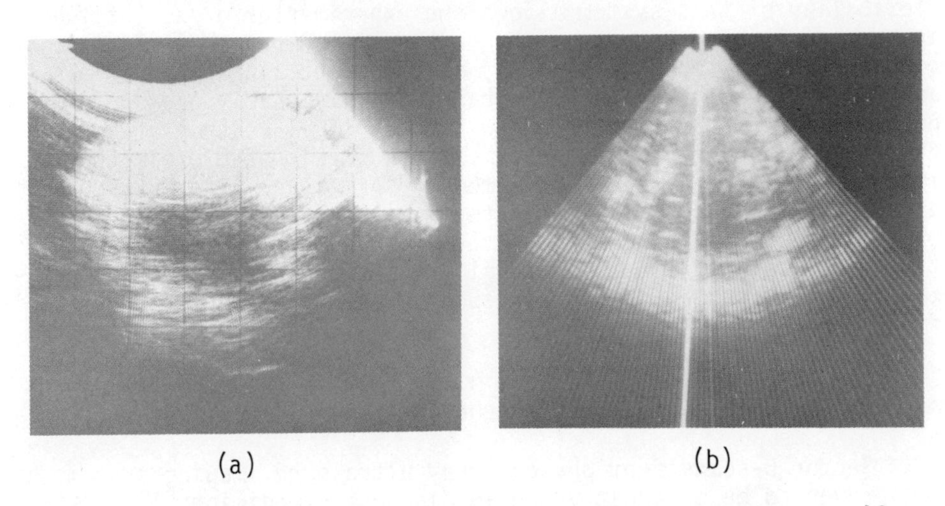

(a) (b)

FIGURE 1. Typical B-scans having a "streaky" quality, as if smeared circumferentially. 10 MHz compound scan (a) of excised dog heart in water tank; 2.3 MHz real-time phased-array, _in vivo_ image (b) of adult human heart.

SYSTEM DESCRIPTION

We will not delve into the basic operation of phased-array imaging which is already well described in the literature [3,1]. Instead we proceed immediately to the difference between previous systems [4,5,6] and ours. Figure 2 is a simplified diagram of our system. The Theta-array [7] consists of an apodized, annular, transmitting array of 32 elements and an apodized, linear, receiving array (also containing 32 elements). "Apodization," also known as "shading," means that the array elements are summed with weighted gains. The transducer connects to the system through "apodizing circuits" on both transmit and receive. A digital transmit controller drives the apodizing circuits in the transmitter portion while the receiver apodizing circuits feed a Cascade Charge-Coupled Device, or C3D [8,9], for receiver steering and focusing. The C3D is a multiple input, single output device that is the equivalent of a bank of CCD delay lines -- one for each receiver array element. The C3D, which feeds a conventional CRT display, also provides RF summing of the individual channel signals. A commercial microcomputer controls all five of the blocks shown in Figure 2.

A more detailed diagram of the system appears in Figure 3. Because of its great flexibility and ease of changing system parameters we call the imaging system a Medical Hybrid Computer (MHC). The system may be configured in the matrix-C3D form shown, however our current plans are limited to a single C3D device.

The machine offers the user a number of easily changed system parameters, all under computer control:

(1) Receiving Portion

- individual gain on each channel
- wide range of steering angles

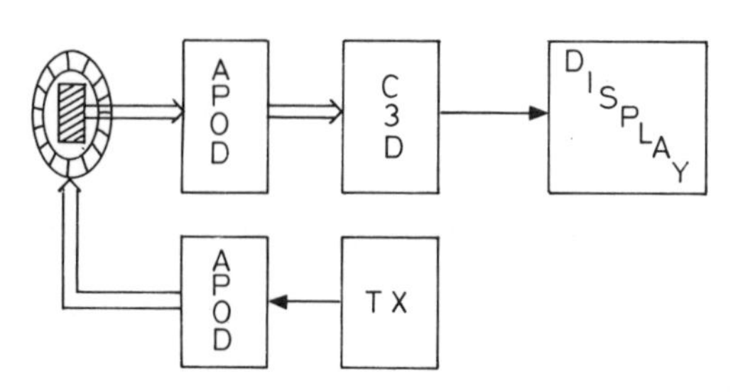

FIGURE 2. Simplified system diagram displaying "front-end" signal processing blocks.

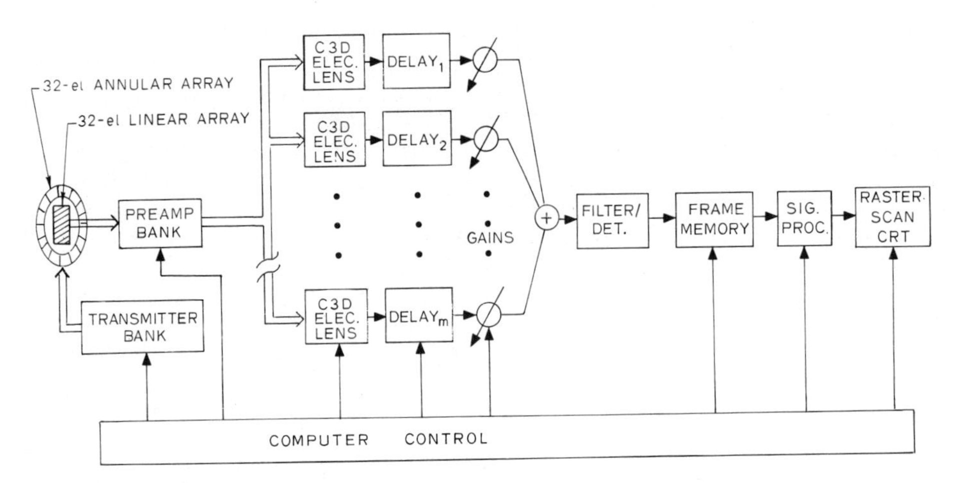

FIGURE 3. System architecture using very high degree of computer control. (Medical Hybrid Computer)

· precision, indexed focal lengths

· delay and weight on each C3D output

· arbitrary 2-D filtering of detected video

(2) Transmitting Portion

Each channel has individual control of:

· amplitude level

· firing time

· transmitted waveform

The design of the machine makes use of three central key design concepts:

(1) Modularity

· ability to "bring up" minimum configuration -- increase complexity later

- timely replacement of single-purpose modules as advanced technology (newest custom ICs) are developed

- ease of troubleshooting

(2) General Purpose Architecture

- great latitude for experimentation

- increased imaging research throughput

(3) High Degree of Computer Control

- software control programs

- quick, easy implementation of imaging experiments

- minimum interaction/high repeatability of hardware

Figure 4 is a detail of the components which constitute a single receiver channel. A transducer array element intercepts the incident sound wave. Apodization is accomplished by a computer gain-controlled preamp which amplifies the signal. A hardware time delay tweaker compensates for transducer/preamp phase mismatch. The channel signal is then passed to a computer controlled delay line which is one of 32 integrated into the C3D lens. Our present C3D has only 20 inputs but we are now laying out the masks for the 32-input device.

The current experimental configuration is depicted in Figure 5. In addition to the hardware previously described (MHC), we also have a disk drive for loading programs and data and a CRT terminal for setting apodization gains and C3D clock frequencies. The hand-held grey case of the receiving array transducer probe houses tiny transformers which match the transducers to the line impedance and preserves the full transducer bandwidth.

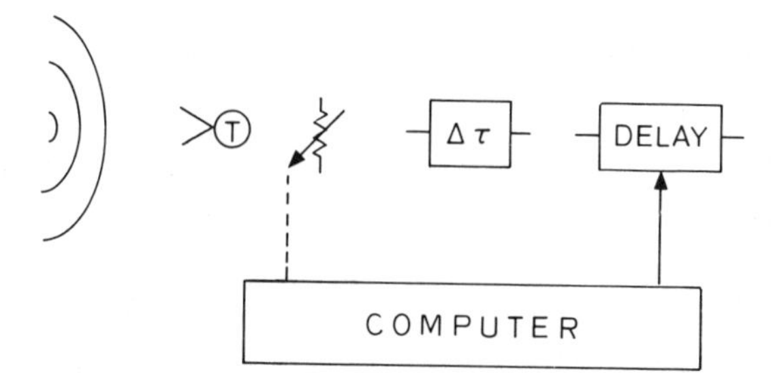

FIGURE 4. Channel receiving electronics for each transducer array element. Compucer adjusts gain and delay of each channel independently.

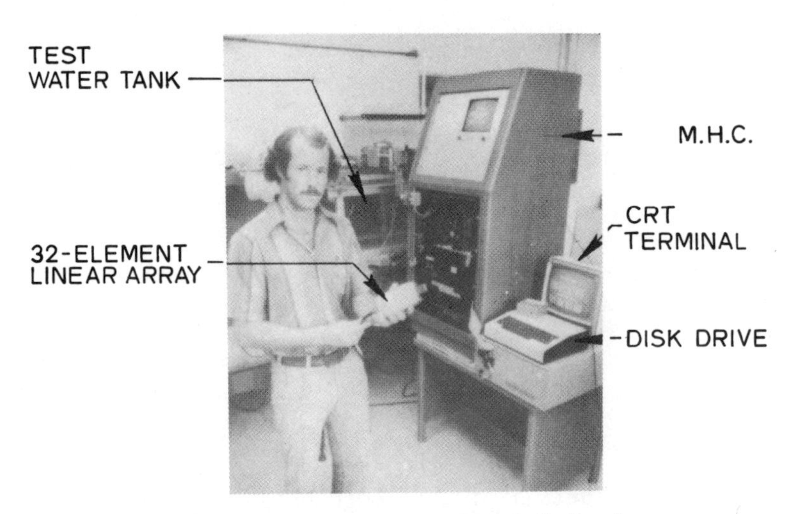

FIGURE 5. Current experimental hardware.

Figure 6 shows two of the 5" x 10" circuit cards used in the system. The board at top is the C3D electronic lens board with the 40-pin C3D IC in the center. This particular C3D, a four-clock section device with 20 inputs, uses two sections for steering and two sections for focusing. The circuit board has been designed to be upward compatible to accept the 32-input, six-clock section C3D we are developing.

The bottom circuit card is an eight-channel transmit/receive apodizer with each preamp channel implemented as a two-stage cascade of commercial, AGC amplifier ICs. Each transmitter channel consists of a push-pull driver powered from a voltage-controlled regulator IC. The driver is capable of delivering 400 Vp-p to the transducer. A bank of sample/hold circuits store the apodization gains.

Our system is also used as a test bed for evaluating advanced, custom integrated circuits in a medical electronics application. We first build the best circuits we can using available ICs. We then evaluate these circuits to establish a set of performance specifications for a custom IC design. Finally, we design and fabricate the IC and incorporate it into the system (e.g., the C3D electronic lens). We are currently developing a very low power version of the amplifier IC used in the preamp board of Figure 6 which also boasts dramatically improved gain control. This chip is a stepping stone to a much more complicated, multiple preamp IC with direct digital gain control. This next generation chip will replace the entire circuit card.

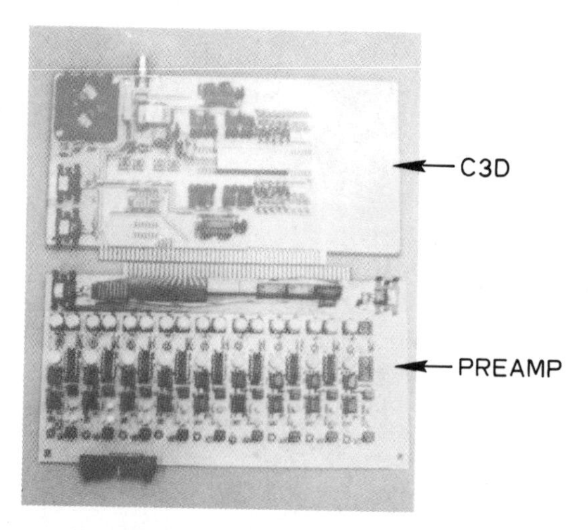

FIGURE 6. C3D electronic lens circuit board and eight-channel preamplifier/transmitter board. Amplified channel signals travel to C3D over back plane bus.

EXPERIMENTAL RESULTS ON APODIZED, WIDEBAND BEAM PATTERNS

As a first application of our system, we combined two of the preamp cards shown in Figure 6 and the linear transducer array with a linear peak detector. We investigated the effect of various aperture shading functions on the receive-only beam pattern. The theoretical foundation for this is, of course, Fourier Transform theory. A point source transducer at 25 cm range was used as the transmitter. In the experimental configuration of Figure 7 the angle θ is changed through linear translation of the point source. Note that the range to the array center varies in this geometry.

For our first beam pattern (Figure 8a) we adjusted the channel gains to give equal RF output amplitude from the preamps, which compensates for any mismatches of transducer amplitudes. We also used many cycles in the transmit burst, effectively producing a monochromatic acoustic wave. Under these conditions the expected beam pattern from simple theory is an absolute sinc function. Therefore, this plot served as a calibration to ensure our detector was truly linear. It's important to note that the first side lobes are both equal to 22% of the main lobe, exactly that predicted by theory. This agreement gives us high confidence that our measured wideband beam patterns are accurate. Other noteworthy aspects are

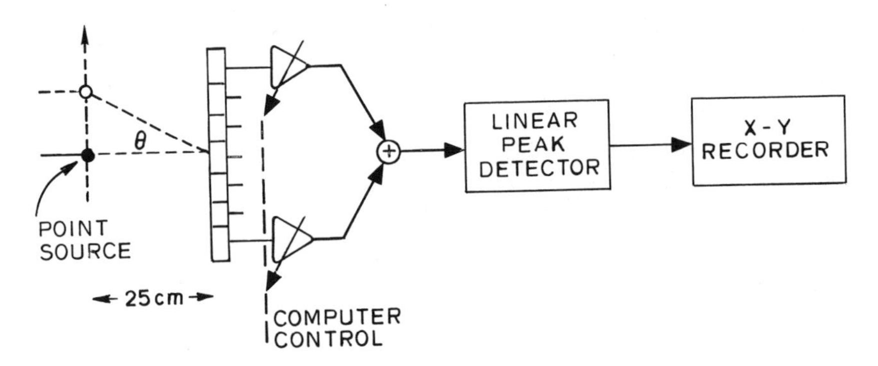

FIGURE 7. Experimental configuration for measurement of apodized, wideband, receive-only beam patterns using 16 elements of the linear array.

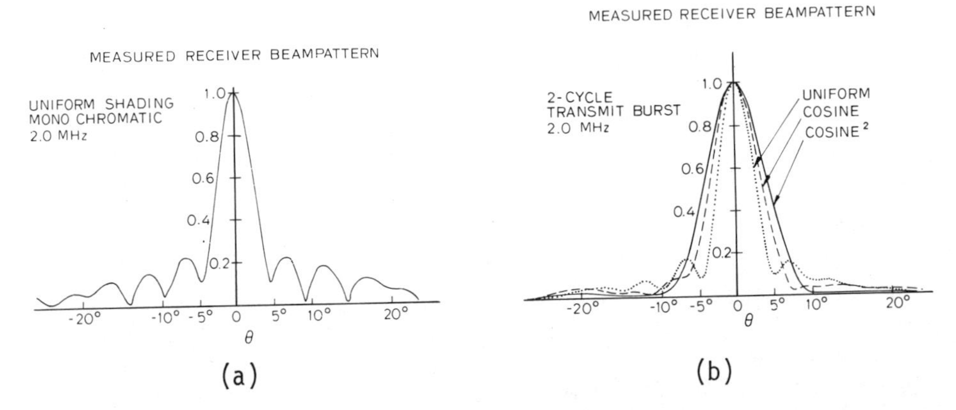

FIGURE 8. Beam patterns obtained from configuration of Figure 7 using computer controlled aperture shading. (a) Calibration of detector linearity using many cycles. (b) Wideband patterns with shading function as parameter.

that the pattern is very symmetrical and the width of the main lobe agrees exactly with theory. It is questionable whether the depths of the first nulls are due more to transducer phase errors or to the fact that the array is unfocused in these experiments. For reference the relevant array parameters are: 16 elements, 560 micron center-center spacing, 406 micron element width, 1.2 MHz bandwidth and 14 dB insertion loss. The dynamic range of these plots is 40 dB, limited by system noise.

In Figure 8b we have reduced the transmitted burst to only two cycles of RF to obtain the wideband beam patterns actually used in practice. We measured beam patterns with three different shading functions: uniform, cosine and cosine-squared. In the uniform aperture case (Figure 8b, dotted) the low order side lobes are largely unaffected by the short transmit burst, while the higher order side lobes are smeared into a uniform background level. This is predicted by wideband theory [10]. However, when the array is apodized (Figure 8b, dashed), both side lobe and background levels are reduced.

Dramatic improvement in the wideband beam pattern is obtained using the cosine-square apodization (Figure 8b, solid). In this case the sidelobes have dropped into the system noise indicating they are at least 40 dB down. If this receive-only pattern were combined with a similar transmit pattern, extremely low round-trip side lobe levels could be achieved.

DISCUSSION

Upon viewing the images from phased-array imaging systems we have frequently noticed a "streaking" effect and/or "muddled" picture quality (Figure 1). Typically this is manifest as one or more bright but fuzzy circular arcs in a sector scan display. The streaks appear intermittently and are highly dependent upon the transducer probe orientation. We believe the streaking is due to one or more unusually bright reflectors in the body which may possibly be calcified regions or specular reflectors. Whatever their origin, it seems probable that, at a given depth the dynamic range of the reflectance field may be very large. Range swept gain cannot compensate for this. To adequately cope with the problem, a beam pattern with very weak side lobes is necessary and front-end circuitry with broad dynamic range is required.

The medical imaging problem is very different from the radar or even sonar echo ranging problem. In the latter two cases the scene under study is more-or-less composed of discrete reflectors in a nonreflecting background. This is in marked contrast to medical imaging where it's desired to obtain an intensity map of a continuous, three-dimensional reflectance field. In this case the reflectance dynamic range may easily be 60 dB at a given depth. Thus, a sensing beam with more than 60 dB dynamic range is needed to obtain a good image from a phased-array sector scanner. These ideas have been reinforced by the typical adjustments made while using clinical scanners. Physicians at the Stanford University Hospital performing clinical patient evaluations generally preferred [11] these panel control settings: (1) linear display (no logarithmic compression); (2) display threshold set above the noise/side lobe floor; and (3) adjustment to let the strongest echoes saturate the display. These adjustments result in best signal detectivity [12] of the low level echoes which are so important for accurate diagnosis.

We view the discussion above as strong evidence that the apparent picture quality observed on the CRT screen is probably due more to the side lobe levels of the round trip beam pattern than to the classical 6 dB main lobe beamwidth. For example, "high resolution" is meaningless in a system having a 1° angular beamwidth if it also has side lobes only 35 dB down and is used in a clinical environment having a 50 dB reflectance dynamic range. Of course, such a system may be "doctored" in a test tank using a wire target to get impressive resolution. However, a binary reflector (wire target) is very different from the normal clinical environment with large reflectance dynamic range. Especially troublesome are relatively large side lobes that are substantially removed from the vicinity of the main lobe. These can cause both ghosts and nonlinear artifacts in the image and may very well be the origin of the "muddled" appearance of many sector scans.

Our preliminary experiments on using receiver aperture shading to reduce side lobes are done as a hopeful cure for the streaking problem. While some widening of the main lobe is sacrified, we greatly reduce the likelihood of receiving unwanted energy from angles far removed from the beam pointing angle. We view this as the principal benefit of using aperture apodization.

Simulation work done by Wang [13,14] indicates that amplitude-only (including polarity) array weighting gives performance within 3 dB of that obtained using optimal, complex array weighting. This is very fortunate since complex weighting requires impractical large amounts of front-end hardware.

Nonetheless the matrix-C3D architecture used in the MHC system makes possible the closest approximation to the general real-time, wideband "array processor" that is feasible with present technology.

Further reductions in side lobe levels are expected by using the Theta-array transducer configuration. A transmit beam with very small side lobes relieves the receiver front-end electronics from much of the burden of the requirement for wide dynamic range. This is of crucial importance in order to obtain round trip beam patterns having 60 to 70 dB of dynamic range.

FUTURE EXPERIMENTAL RESEARCH

We already have plans for a variety of other experiments making use of the MHC imaging system. First priority among these is thorough experimental inquiry into the performance limits imposed by the C3D lens device. At Varian Associates' Systems and Techniques Laboratory initial systems evaluation tests [15] on an early generation C3D met with mixed results; promising performance was obvious, however technical hardware difficulties masked any true measurements of fundamental device limitations. A full investigation is needed.

Initial characterization of single elements of the annular transmitting array has been completed. Pending completion of the digital transmit controller careful transmitter beam pattern measurements will be made, including: steering and focusing, array apodization, and effects of signal design (waveform).

Further in the future we plan to research the use of two-dimensional digital filtering of the detected video signal. The key component for this work is a digital frame memory which stores a complete CRT frame of video information. This memory is a fairly sophisticated subsystem due to the sector format of the input information, in contrast to the rectangular format needed for signal processing. Alternative engineering designs are now being evaluated for the frame memory. The benefits obtained from side lobe deconvolution [16] and raster line interpolation [17] will be examined through the use of the digital frame memory.

Incorporation of our 32-input, six clock section C3D lens promises to enhance the information gathering/processing capability of our system. Finally, new low power custom amplifier ICs will allow more precise gain control for effective receiver aperture apodization and a close tracking range swept gain function.

SUMMARY AND CONCLUSIONS

An experimental phased-array ultrasound imaging system using C3D electronic lenses is being developed. The purpose is to explore novel imaging techniques made possible through application of advanced, custom integrated circuits. Great versatility has been achieved through system design using modularity, a general purpose architecture and a high degree of computer control. Lastly, we have found that apodization greatly reduces wideband side lobes. We believe that very low side lobe levels may well be the key to significantly improved diagnostic B-scans.

ACKNOWLEDGMENTS

The authors wish to thank Ernest J. Wood for system software development and maintenance and Paul Jerabek for precision transducer fabrication. This work was supported through an IBM Corporation fellowship and by the Department of Health, Education and Welfare under PHS Research Grant P01 GM17940.

REFERENCES

[1] W.A. Anderson, J.T. Arnold, L.D. Clark, W.T. Davids, W.J. Hillard, W.J. Lehr and L.T. Zitelli, "A New Real-Time Phased-Array Sector Scanner for Imaging the Entire Adult Human Heart," Ultrasound in Medicine, 3B, 1977, pp. 1547-1558.

[2] San Francisco Chronicle, 13 March 1978, San Francisco, p. 1.

[3] J.C. Somer, "Electronic Sector Scanning for Ultrasonic Diagnosis," Ultrasonics, 6, July 1968, pp. 153-159.

[4] M.G. Maginness, J.D. Plummer and J.D. Meindl, "An Acoustic Image Sensor Using a Transmit-Receive Array," Acoustical Holography, 5, July 1973, pp. 619-631.

[5] F.L. Thurstone and O.T. von Ramm, "A New Ultrasound Imaging Technique Employing Two-Dimensional Electronic Beam Steering," Acoustical Holography, 5, July 1973, pp. 249-259.

[6] J.F. Havlice, G.S. Kino, J.S. Kofol and C.F. Quate, "An Electronically Focused Acoustic Imaging Device," Acoustical Holography, 5, July 1973, pp. 317-333.

[7] A. Macovski and S.J. Norton, "High Resolution B-Scan Systems Using a Circular Array," Acoustical Holography, 6, 1975, pp. 121-143.

[8] R.D. Melen, J.D. Shott, J.T. Walker and J.D. Meindl, "CCD Dynamically Focused Lenses for Ultrasonic Imaging Systems," Proc. 1975 Naval Electronics Laboratory Center Intl. Conf. on Application of Charge Coupled Devices, October 1975, pp. 165-171.

[9] J.D. Shott, "Charge-Coupled Devices for Use in Electronically Focused Ultrasonic Imaging Systems," Ph.D. dissertation, TR No. 4957-1, Stanford Electronics Laboratories, Stanford University, May 1978.

[10] S.J. Norton, "Theory of Acoustic Imaging," Ph.D. dissertation, TR No. 4956-2, Stanford Electronics Laboratories, Stanford University, December 1976, pp. 57-75.

[11] Dr. Barbara A. Carroll, Stanford University Medical Center, private communication.

[12] J.D. Meindl and A. Macovski, "Recent Advances in the Development of New Imaging Techniques," Recent Advances in Ultrasound in Biomedicine, 1, ed. D. White, 1977, pp. 180-181.

[13] H.S.C. Wang, "Interference Rejection by Amplitude Shading of Sonar Transducer Arrays," J. Acoust. Soc. Am., 61, No. 5, May 1977, pp. 1251-1259.

[14] H.S.C. Wang, "Rejection of Sonar Interference of Finite Bandwidth by Amplitude Shading of Transducer Arrays," J. Acoust. Soc. Am., 63, No. 1, January 1978, pp. 111-114.

[15] W.T. Davids, J.T. Arnold and M.D. Eaton, "CCD/V-3000 Scanner Evaluation Summary," unpublished work.

[16] D.B. Boyle, "Real-Time Lateral Edge Enhancement of Ultrasound Images Using CCD Delay Lines," unpublished work.

[17] J.T. Walker, "Digital Scan Conversion and Smoothing for a Real-Time Linear Array Imaging System," Acoustical Holography, 8, (in press).

MONOLITHIC SILICON-PVF_2 PIEZOELECTRIC ARRAYS FOR ULTRASONIC IMAGING

R. G. Swartz and J. D. Plummer

Integrated Circuits Laboratory

Stanford University, Stanford, California 94305

ABSTRACT

The piezoelectric polymer Polyvinylidene Fluoride (PVF_2) has recently become of great interest in acoustic imaging applications because of its strong piezoelectricity, its low acoustic impedance, and its physical flexibility. This paper will describe the development of linear and two-dimensional ultrasonic transducer arrays employing PVF_2. These arrays are based on integrated circuit technology, using silicon wafers containing multiplexers and amplifiers in a matrix configuration of unit cells. Following batch fabrication of the integrated circuit devices within the silicon wafer, a PVF_2 layer is physically bonded on top of the silicon. This results in a simple, yet high performance array structure, one which is suitable for implementation of very large and very dense arrays for medical imaging or nondestructive testing applications over a very broad frequency spectrum. A description of the fabrication details and experimental results from present arrays will be presented, along with a discussion of future capabilities.

I. INTRODUCTION

Presently, the preferred piezoelectric material for acoustic imaging transducers is one of the lead zirconate titanate (PZT) compounds. They offer high sensitivity, low impedance, and relatively low cost. For nondestructive testing of acoustically similar metals or ceramics, they are perhaps optimal. However, in medical systems designed for ultrasonic imaging of body tissues, use of these materials is inherently nonoptimal because of the large acoustic impedance mismatch between the ceramic transducer and the mostly liquid body constituents. The consequences of this mismatch can be loss of sensitivity, bandwidth, and acoustic wave acceptance angle. Use of carefully tailored front matching layers and backing layers can largely restore performance in single transducers [1]. In the design of linear arrays of small element transducers, a trade-off between sensitivity and bandwidth yet remains. In two-dimensional arrays, the major problem is the one of physically making contact with each of the elements in a dense array. Various solutions exist to the problem of forming a two-dimensional array of transducers [2], though none are as yet completely satisfactory.

The recent discovery of the piezoelectric polymer Polyvinylidene Fluoride (PVF_2) [3] offers the hope of solution to many of the above mentioned difficulties. PVF_2, because it is a plastic, has the relatively low acoustic impedance of approximately $4 \cdot 10^6$ kg/m^2-sec. This compares with 20 to 35 $\cdot 10^6$ kg/m^2-sec for ceramic piezoelectric materials and $1.5 \cdot 10^6$ kg/m^2-sec for water. Thus, the impedance mismatch between a PVF_2 transducer and water is much less severe than that between water and a ceramic. The consequence of this is improved bandwidth and increased sensitivity and acceptance angle, without the necessity of front matching layers. Another characteristic of PVF_2 is its mechanical flexibility. This allows it to conform to nonplanar substrate surfaces. Furthermore, the tolerance of the resulting transducer structure to high stress will become a function only of the strength of the substrate on which the PVF_2 is mounted.

These advantages are balanced partly by the smaller electromechanical coupling coefficient of PVF_2 relative to PZT. An additional problem is the low dielectric constant ($\epsilon/\epsilon_0 < 15$) of PVF_2, which necessitates the use of high voltages when the material is used as an acoustic wave generator.

This paper describes a new type of receive-only piezoelectric transducer, making use of PVF_2, in combination with integrated circuit technology. It is similar in concept to the "PI-FET" described by K. W. Yeh, R. S. Muller, et al [4,5], though offering superior sensitivity and usability in medical imaging applications. This follows in part from the use of PVF_2 rather than ZnO as the piezoelectric. Further advantages accruing from this new design will be discussed.

II. THEORY OF OPERATION

A. Description of Device

In a medium frequency (1 to 10 MHz) medical imaging system, the chosen mode of operation of the piezoelectric transducer is generally the thin plate, longitudinal mode. A simple possible design of a transducer operating as a receiver in this mode is shown in side view in Figure 1A. Here, a thin sheet of PVF$_2$ is bonded to the surface of a stiff, high density backing. If this backing is selected so that its acoustic impedance is several times that of the PVF$_2$, then the transducer will be acoustically resonant at a frequency where its thickness is one-fourth wavelengths. For example, a 30 micron thick sheet of PVF$_2$ would be resonant at approximately 17 MHz. The implications of this will be discussed shortly. Thin electrodes on the top and bottom sides of the PVF$_2$ are used to make electrical contact to the transducer.

In a minor modification of this scheme, the bottom electrode of the PVF$_2$ sheet may be sputtered or evaporated directly on the substrate instead of on the PVF$_2$ itself. In this case, instead of a direct connection to the transducer, the lower electrode is capacitively coupled to the PVF$_2$ via the glue layer. Figure 1B illustrates this modification. Now, the substrate has been chosen to be a silicon wafer mounted on a tungsten-loaded epoxy backing of the same acoustic impedance. An oxide (SiO_2) layer on the surface of the wafer is used to electrically isolate the lower transducer electrode from the silicon. The silicon has an acoustic impedance of $20 \cdot 10^6$ kg/m^2-sec which is sufficient to ensure the resonance of the PVF$_2$ layer at the $\lambda/4$ frequency.

Figure 1C shows a fundamental and significant extension of the concept. Here, n-type dopants have been diffused into the p-type silicon wafer at two locations. These form the source and drain diffusions of an MOS transistor. The lower electrode of the PVF$_2$ layer, over the insulating SiO_2 layer, now constitutes the gate of the MOSFET. The resulting structure: piezoelectric-oxide-semiconductor, field effect transistor, has been named the "POSFET." As can be seen from the figure, the upper electrode of the PVF$_2$ layer is connected to electrical ground. Operation of the device is now as follows: An incident longitudinal acoustic wave modulates the thickness of the PVF$_2$ layer. The electric charge resulting from piezoelectric action in the transducer will appear directly on the gate of the MOS transistor. The MOSFET is an uncommitted circuit element. It may be used to provide amplification of this electrical signal or as a multiplexer to select for further processing one signal from many such transducer elements.

Figure 2 illustrates a more usable form of this structure. Now, the MOSFET has been confined to only a small portion of the cell area. The metal gate has been extended to occupy most of the cell and acts as a charge collector for the relatively small transistor. Though the POSFET is shown here only in cross section, it is clear that the extended gate electrode may be of arbitrary shape and size. This lower electrode, since it is defined on the surface of the silicon wafer, requires only standard integrated circuit

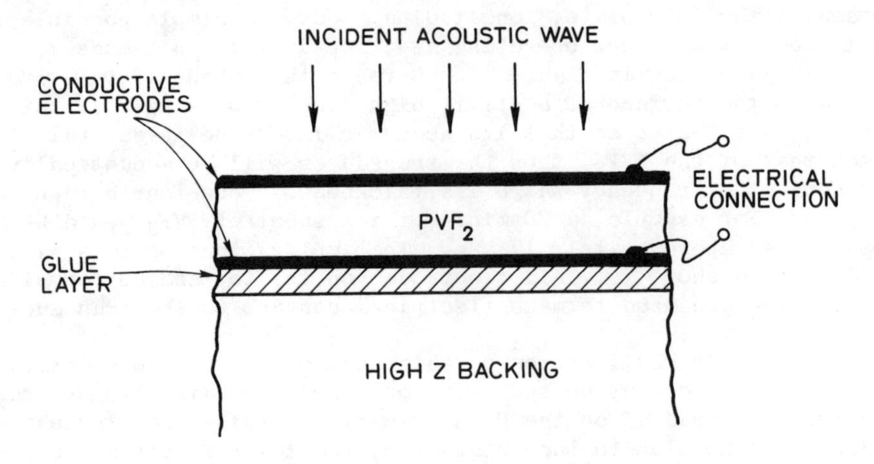

A. PVF_2 transducer configuration

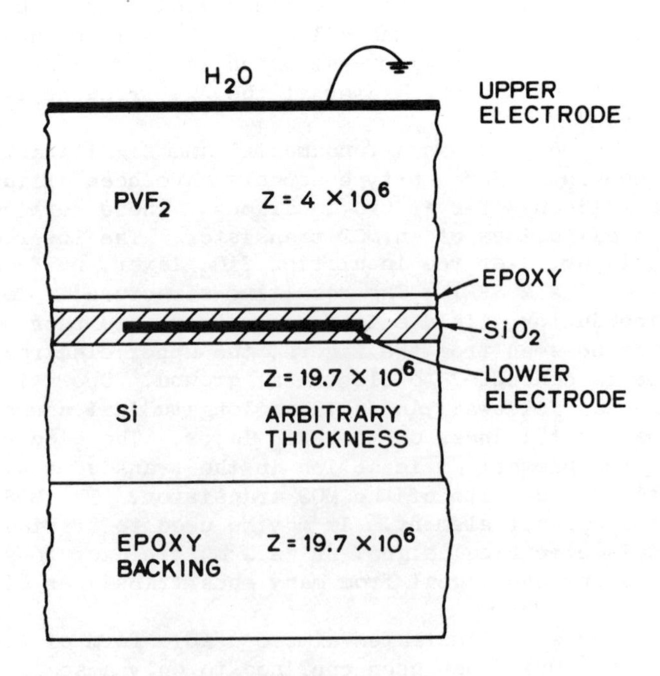

B. Silicon substrate PVF_2 transducer

Figure 1. EVOLUTION OF PVF_2 TRANSDUCER.

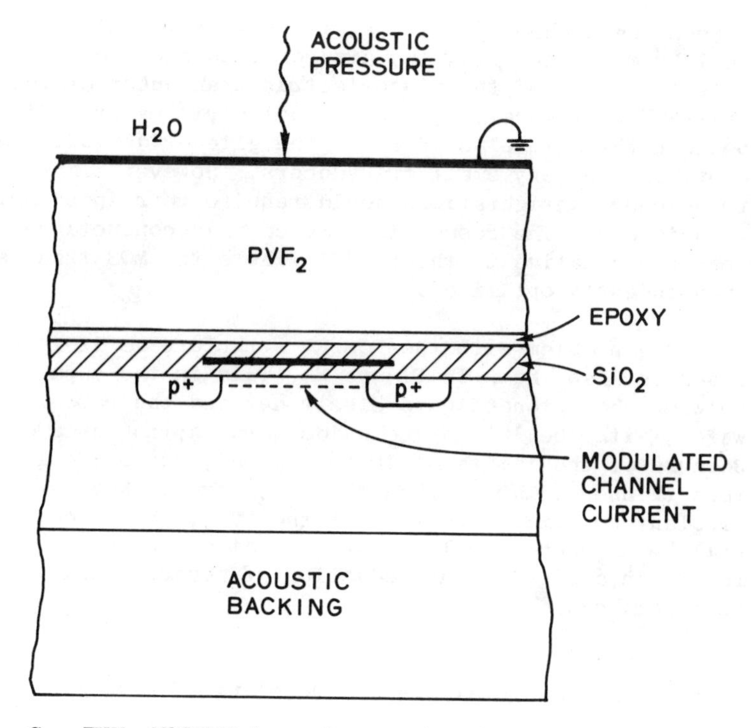

C. PVF_2-MOSFET transducer structure

Figure 1. CONTINUED.

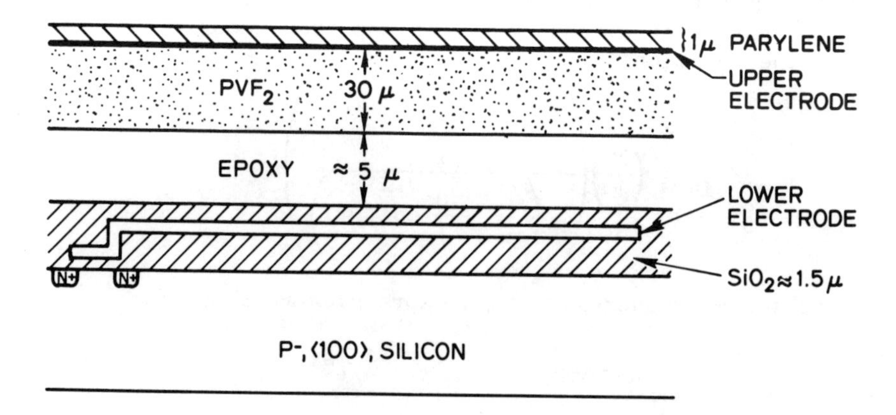

Figure 2. INITIAL POSFET EXPERIMENTAL STRUCTURE.

processing technology. Thus, its area may easily range from 10^{-2} to 10^{+2} mm^2. In turn, the area of this electrode determines the effective area of the piezoelectric transducer element. In the "PI-FET" configuration, the piezoelectric material is sandwiched between the gate electrode and the gate oxide [3]. This is acceptable for very small transducers. However, large geometry transducer configurations would require correspondingly large transistors. The result is reduced transconductance and frequency response relative to the POSFET, where the MOS transistor can be independently optimized.

One problem with the POSFET structure as it is presently defined is also shown in Figure 2. This is the capacitance existing between the extended gate electrode, and the conductive silicon wafer, with the insulating oxide layer acting as a dielectric. Because of the limits of IC technology, the oxide layer will be much thinner (~1.5 μ) than the PVF_2 layer. Having a dielectric constant comparable to that of the PVF_2, the oxide capacitance will be unacceptably larger than that of the PVF_2. This will result in a significant reduction of device sensitivity, as will be described.

B. Analysis and Modeling of Device

The electrical port of the PVF_2 transducer may be modeled in the ideal case as a Thevenin voltage source and series output impedance. The Thevenin voltage generator can be determined as a function of the input acoustic stress using the standard thin plate "thickness" mode Mason model [6]. The transducer, as described earlier, has a backing with acoustic impedance much greater than that of the PVF_2. Therefore, the model of Figure 3 applies. This lumped parameter circuit model of the transducer can be solved to give the relation:

$$V_{out}/T_{in} = \left(\frac{2h}{j\omega}\right)\left(\frac{1}{Z_w\left(1 - \frac{j\bar{Z}_0}{Z_w}\cot(\bar{\beta}\ell)\right)}\right) \tag{1}$$

At very low frequencies, this expression reduces to:

$$V_{out}/T_{in} = 2h\ell/v\bar{Z}_0\,, \qquad h = e/\epsilon^S \tag{2}$$

At resonance:

$$V_{out}/T_{in} = 4h\ell/jZ_w\pi v \tag{3}$$

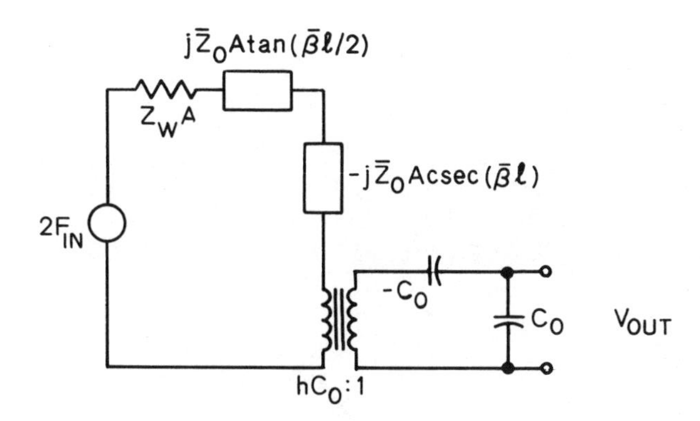

Z_w = acoustic impedance of water

A = surface area of transducer

F_{IN} = acoustic input force

ℓ = thickness of transducer

C_0 = clamped capacitance of transducer

$\bar{Z}_0$ = stiffened acoustic impedance of transducer

Figure 3. MASON MODEL EQUIVALENT CIRCUIT OF PVF_2 TRANSDUCER.

The ratio of these two quantities is given by:

$$\frac{(V_{out}/T_{in})_{\omega=0}}{(V_{out}/T_{in})_{\omega=\pi v/2\ell}} = \frac{j\pi Z_w}{2\bar{Z}_0} \cong j/2 \tag{4}$$

Thus, the decline in the magnitude of V_{out}/T_{in} is approximately 6 dB between DC and the PVF_2 resonant frequency. Note that resonance occurs where:

$$\bar{\beta}\ell = \pi/2 = 2\pi\ell f/v \quad \text{or} \quad \ell = v/4f = \lambda/4 \tag{5}$$

A ceramic transducer could be operated in the quarter wave resonant mode only with great difficulty because of the high acoustic impedance of the ceramic materials. As a result, they are generally operated near the $\lambda/2$ thickness resonance. This mode is inherently less sensitive to low frequency excitation because the transducer is generally stiffer than its backing. The reduced bandwidth can be regained partially by adding a matched acoustic backing to the ceramic transducer, at a cost of a substantial reduction in sensitivity. Figure 4 shows a normalized computer calculation of the voltage/stress transfer relation for a PVF_2 $\lambda/4$ transducer, resonant at 17 MHz, and for a typical 5 MHz ceramic transducer with

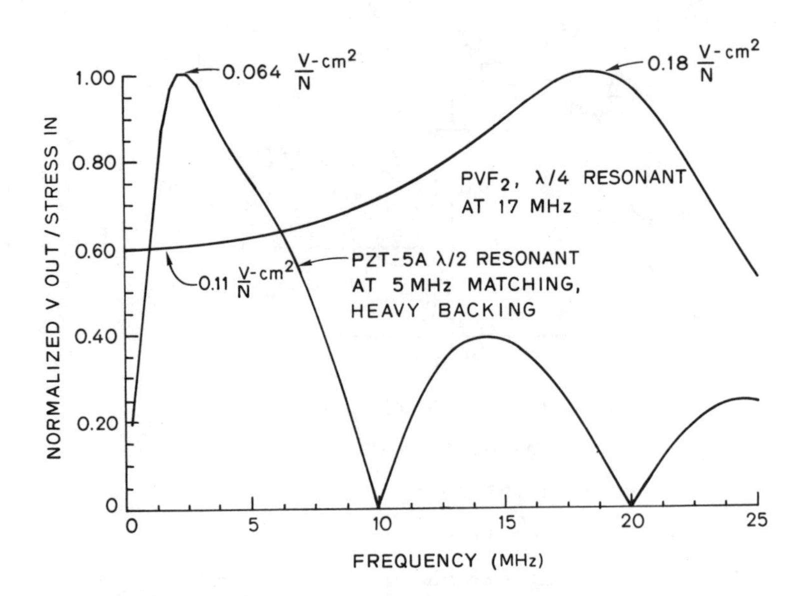

Figure 4. NORMALIZED RATIO OF OUTPUT VOLTAGE TO INPUT STRESS VS FREQUENCY FOR BROADBAND PZT TRANSDUCER AND POSFET STRUCTURE. (COMPUTER CALCULATION)

heavy backing for broadbanding, operating into a matched electrical load. Both curves apply for thin plate operation. Observe that, despite the lower k_T of the PVF_2 material relative to the ceramic (0.19 [7] vs 0.49 [8]), the PVF_2 transducer provides greater sensitivity. The ceramic transducer sensitivity is reduced not only by the heavy backing but also by the inherent, acute acoustic mismatch between the ceramic and water. PVF_2, with much lower acoustic impedance, comes closer to a match with any liquid environment. Note that a PVF_2 transducer will have a sensitivity and bandwidth superior to a ZnO transducer (with $k_T \leq 0.29$ and $Z = 36 \cdot 10^6$ kg/m^2-sec) for the same reasons.

The Thevenin series output impedance below resonance for the PVF_2 transducer can be calculated from Figure 3. This gives:

$$Z_{in} \cong \frac{\left(1 - k_T^2\right)}{j\omega C_0} \qquad \text{where} \quad k_T^2 = \frac{e^2}{c^E \epsilon^S + e^2} \tag{6}$$

Thus, the equivalent output impedance is simply a capacitor of value C_0 (to first order). Now, the importance of the MOSFET loading of the PVF_2 layer, inherent in the POSFET structure, becomes clear. Because of the negligible loading by the MOS transistor, the ideal POSFET structure, including glue bond, may be modeled as shown in Figure 5. We have:

$$V_G/V_{out} = \frac{C_0C_B}{C_0C_B + C_0C_{GSub} + C_BC_{GSub}} \quad (7A)$$

If the capacitance of the glue bond (C_B) is ignored, we have:

$$V_G/V_{out} = \frac{C_0}{C_0 + C_{GSub}} \quad (7B)$$

This relation is valid from DC to just below resonance. Furthermore, we have seen that, over this same frequency region, the Thevenin voltage generator V_{out} is nearly a constant and independent of frequency. Thus, with the commercially available 30 micron thick PVF$_2$, the POSFET structure offers the possibility of a frequency response flat within 6 dB from DC to 17 MHz and a sensitivity comparable to or greater than a broadband ceramic transducer used in the same mode. Observe from Eq. (7B), however, that this full sensitivity cannot be realized unless the extended gate electrode capacitance (C_{GSub}) is minimized.

Of fundamental importance is knowledge of the absolute device detectivity or minimum detectable signal level. This can be calculated for the PVF$_2$ transducer by modeling it electrically as a series real and imaginary impedance. The minimum detectable input acoustic power, then, is defined to be that which generates a Thevenin output voltage equal to the thermal noise voltage generated by the series real impedance. The principle source of thermal noise in a typical broadband ceramic transducer is due to coupling to the heavy backing used for broadbanding. This is not a problem

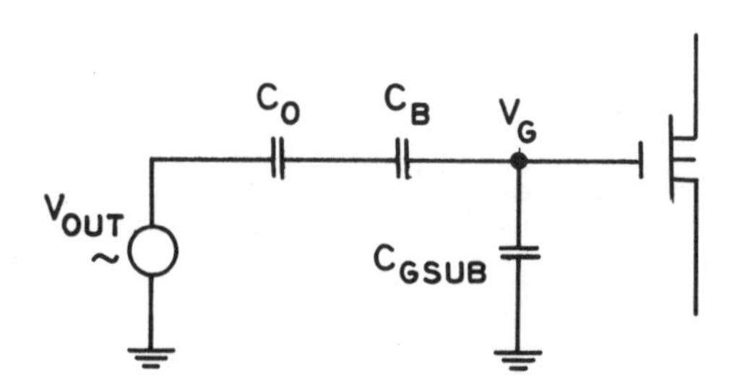

$$V_G = \left(\frac{C_0C_B}{C_0C_B + C_0C_{GSUB} + C_BC_{GSUB}}\right) V_{OUT}$$

Figure 5. POSFET ELECTRICAL EQUIVALENT OUTPUT CIRCUIT.

for a PVF_2 transducer. However, dielectric loss within the PVF_2 at frequencies above 1 MHz is responsible for considerable noise. The result is that typical minimum detectable acoustic intensities are on the order of $5 \cdot 10^{-13}$ W/cm^2 for PVF_2 with a 1 MHz bandwidth. This is a factor of approximately two worse than that for the broadband ceramic transducer. This result was obtained for 30 μ thick PVF_2. Better detectivities can be obtained by using thicker layers.

Additional reductions in detectivity arise, naturally, from the POSFET amplifier characteristics. Because of the integrated structure of the device, however (implying, for example, the absence of long electrical leads), the net transducer-amplifier detectivity for the POSFET may easily surpass that of a comparable ceramic transducer-amplifier system.

C. Low and Very Low Frequency Operation

Although the POSFET device is intended principally for operation above 1 MHz, the theoretical possibility of operation far below this range arises as a result of the $\lambda/4$ mode of operation described in the previous section. Examples of this might be low frequency sonar applications (1 to 100 kHz) or even very low frequency application as pressure transducers. Operation at low frequencies is complicated, however, by an additional factor. As described previously, broadband operation of the POSFET requires a high impedance backing. In the analysis given earlier, this backing was tacitly assumed to be many wavelengths thick. At audio frequencies and below, however, this can not be an automatic assumption. If we have, for example, a POSFET with an acoustic backing of silicon followed by tungsten loaded epoxy, with typical velocity of $2 \cdot 10^3$ m/sec, then, for a backing which is half-wave resonant at 10 kHz, we require a total backing thickness of 10 cm. The effect of the backing resonance can most easily be seen by supposing that an air medium ($Z = 415$ kg/m^2-sec) lies behind this backing. As the frequency of operation decreases, acoustic attenuation in the backing will become less pronounced, and the backing will begin to look like a short section of lossless, acoustic transmission line. At the $\lambda/2$ frequency of the backing, therefore, the PVF_2 transducer will see an <u>air</u> load (an acoustic short circuit) rather than a high impedance load. Transducer sensitivity will rapidly decrease as the $\lambda/2$ frequency and its integer multiples are approached, as shown in Figure 6. This effect will be significant at frequencies for which attenuation in the backing is minor. It will be less severe in a water medium but is still important.

In a very low frequency application as a pressure transducer, the requirement of a high backing impedance is satisfied by a rigid

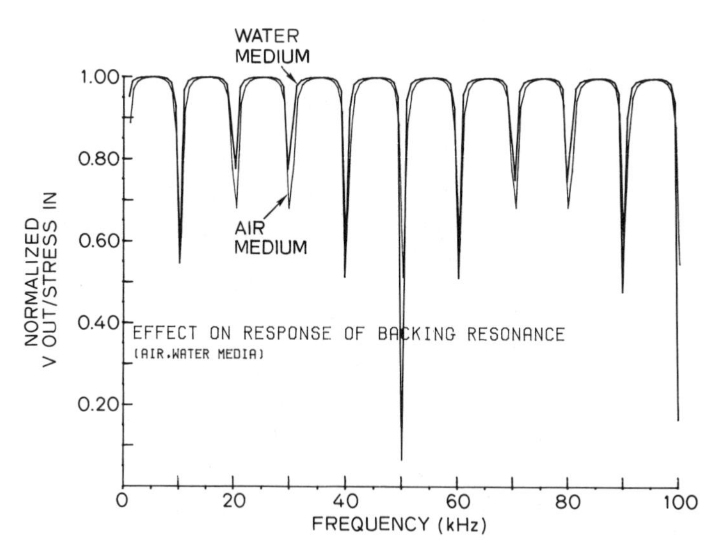

Figure 6. POSFET RESPONSE AT LOW FREQUENCY WITH SHALLOW LOSSLESS BACKING. (COMPUTER SIMULATION)

mechanical mount so that the transducer is compressed by incident stress.

D. One and Two Dimensional Receiving Transducer Arrays

Arrays of ultrasonic transducers are particularly important in real-time medical imaging systems. The POSFET structure is well-suited for the implementation of both linear and area arrays. The key to this characteristic is the fact that the lower electrode of the PVF_2 layer is defined on the silicon wafer and, in turn, completely determines the transducer element. Thus, construction of an array of POSFET transducers is accomplished merely by repeating the transistor geometry and accompanying metal pattern of a single cell across the silicon wafer. Multiplexing is optional in a linear array and is, in fact, unnecessary in a phased array system [9]. However, in an area array, some sort of multiplexing is essential [2]. Figure 7 illustrates schematically a typical two-dimensional array implementation of POSFET transducers. Here, the MOSFETs are used as multiplexers; and conductive polysilicon lines are routed between elements to carry control and signal data. After the electronics have been formed within the silicon wafer as shown, a single sheet of PVF_2 would be bonded to the surface of the wafer, creating thereby a receiving array with full three-dimensional imaging capability. The difficulty and expense would be comparable to that of processing an ordinary custom silicon integrated circuit.

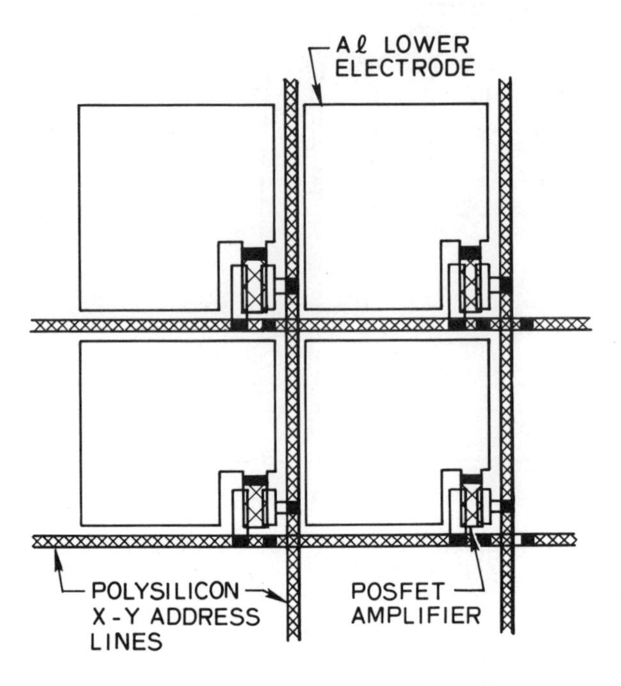

Figure 7. TWO-DIMENSIONAL POSFET ARRAY CONFIGURATION.

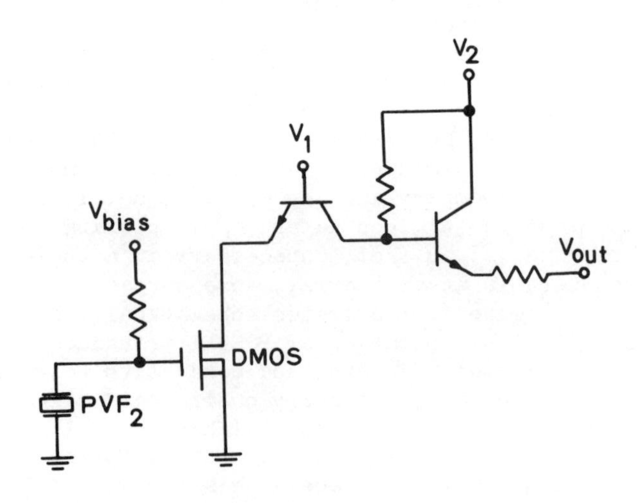

Figure 8. POSFET CASCODE AMPLIFIER.

In an electronically focused phased array system, an important characteristic of the transducer array is the lateral width of the element. It can be shown [10] that the requirement for the absence of grating lobes (subsidiary foci) in a phased array system is that the center-to-center element spacing be less than $\lambda/2$. In a water medium at 2 MHz, this condition becomes $\Delta x < 0.37$ mm. At 5 MHz, we must have $\Delta x < 0.15$ mm. Fabrication of conventional physically separated ceramic transducers satisfying this condition is difficult in a 2 MHz system and extraordinarily so at 5 MHz. Construction of POSFET elements of this size, however, should present little difficulty using standard integrated circuit technology.

E. On-Chip Signal Processing

The POSFET in its simplest form, as described previously, contains a single transistor per transducer cell. Obviously, silicon technology is not limited to this degree of complexity. Figure 8 shows a cascode amplifier designed for a voltage gain of 10 to 20, with a frequency response flat within 3 dB from 0.5 to 5 MHz, and a power dissipation of 15 mW. This circuit is designed to be processed within the individual POSFET cell so that the signal from each element in an array is amplified and buffered before being subjected to interference from external noise sources. In a linear array presently being fabricated, this electronic circuitry requires less than 15% of the cell area, leaving 85% for the lower extended gate electrode of the POSFET transducer.

Other signal processing circuitry may be added to the cell as well, depending on system requirements. This might include further amplification, complex multiplexing schemes, or even charge-coupled delay lines (CCDs) for use in electronic array focusing systems. The potential has yet to be fully explored.

F. Removal of Extended Gate Electrode Capacitance

As described earlier, the oxide capacitance of the extended gate electrode is a severe and principle source of degradation of sensitivity in the POSFET. As such, it is extremely desirable to minimize this capacitance. Many methods have been considered. The most promising ones at this time appear to be the replacement of the silicon under the lower electrode by high resistivity polysilicon or sapphire (Al_2O_3) substrates as illustrated in Figure 9. Small islands of single crystal silicon within the wafer will contain the active electronics. The remainder of the wafer will consist of insulating material. Using this technique, the effective transducer sensitivity should be increased by a factor of five or more.

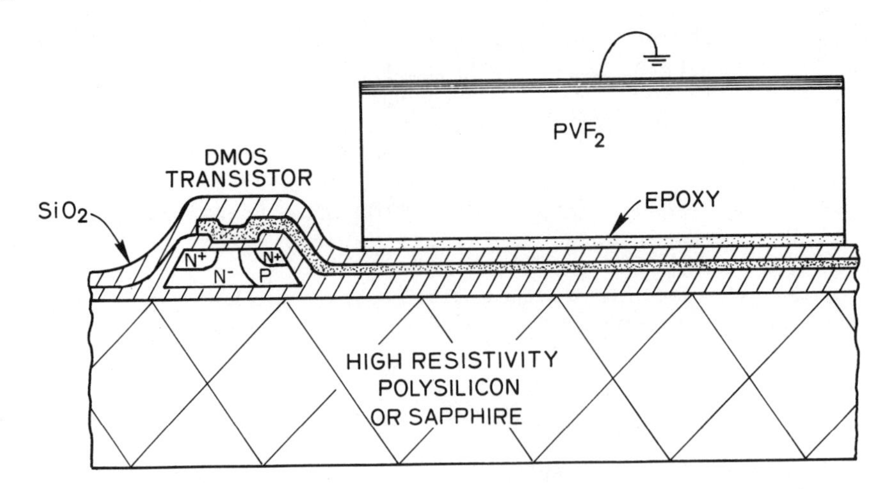

Figure 9. SCHEMATIC REPLACEMENT OF SILICON BY HIGH RESISTIVITY SUBSTRATE.

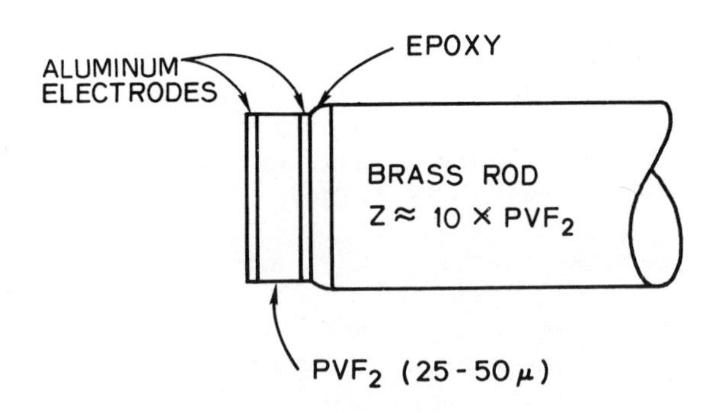

Figure 10. PVF_2 TEST STRUCTURE.

III. DEVELOPMENT

The development of the POSFET has proceeded in several well-defined steps. A quantity of the 30 μ thick PVF_2 film was obtained from Kreha Corporation of America. This film was, when received, already stretched, poled, and metallized with 600 to 800 Å thick aluminum electrodes. As a first step, a simple test structure was constructed after Bui, Shaw, et al [11] (Figure 10). A layer of the 30 μ PVF_2 was glued onto the end of a brass rod ($Z = 36 \cdot 10^6$ kg/m^2-sec) using Emerson & Cuming Stycast 1217 epoxy, with Catalyst 9. Because of the high acoustic impedance of the brass backing, this structure is analogous to that of Figure 1A. Tests were made with this device in a water medium, at frequencies between 2 and 10 MHz, using it both as a transmitter and a receiver. Qualitative measurements were made to establish the fact that the bandwidth of the device when used as a receiver was indeed exceptional. Furthermore, the sensitivity of the transducer when used as a receiver was similar to that of a commercial broadband transducer (Panametrics VIP series).

The next step was the fabrication of an actual POSFET. A photomicrograph of this device is seen in Figure 11A. The construction is similar to that of Figure 2. Most of the cell area is occupied by the rectangular extended gate electrode, with dimensions 2.25 × 3.0 mm. Figure 11B shows an expanded view of the MOSFET alone. After fabrication of these devices on a silicon wafer, the wafer was diced. Individual die were mounted in dual-in-line packages. A small rectangular piece of PVF_2 was glued to the surface of the device, again using Stycast 1217 epoxy. This epoxy was chosen because of its low viscosity, making possible very thin ($< 5\,\mu$) glue bonds. After the upper electrode of the PVF_2 layer was connected to electrical ground, using a wire bonded with silver-loaded conductive epoxy, the entire structure was coated with a 5 μ thick layer of parylene for protection against moisture and physical shock. The final device is shown mounted in Figure 12. Detailed quantitative tests were made using this device of such characteristics as bandwidth, sensitivity, and reverberation.

The previously described devices were used as single elements. Following characterization of these transducers, another batch of devices with the same geometry were processed. These were mounted as 3 × 1 linear arrays, again in DIP packages (Figure 13). Using this simple POSFET array, estimates could be made of the effect of inter-element cross-coupling. This was a matter of concern since the array elements are not physically separated from one another, as in a common ceramic array. The effect of any lateral cross-coupling in a transducer array would be reduced lateral resolution.

The next advance in the development of these devices is intended to be the construction of a 32 × 1 linear array of POSFET transducers. This will be used in a medical imaging system as the inner array of a "Theta" array transducer [12]. It is presently under development.

IV. EXPERIMENTAL RESULTS

Initial characterization of the POSFET involved the measurement of the ratio V_G/V_{out} (see Eq. (7)). For present devices, this was found to be 1/13.8. Subsequent sensitivity calculations, therefore, have been multiplied by the inverse of this factor to account for the sensitivity reduction originating from the input capacitive divider.

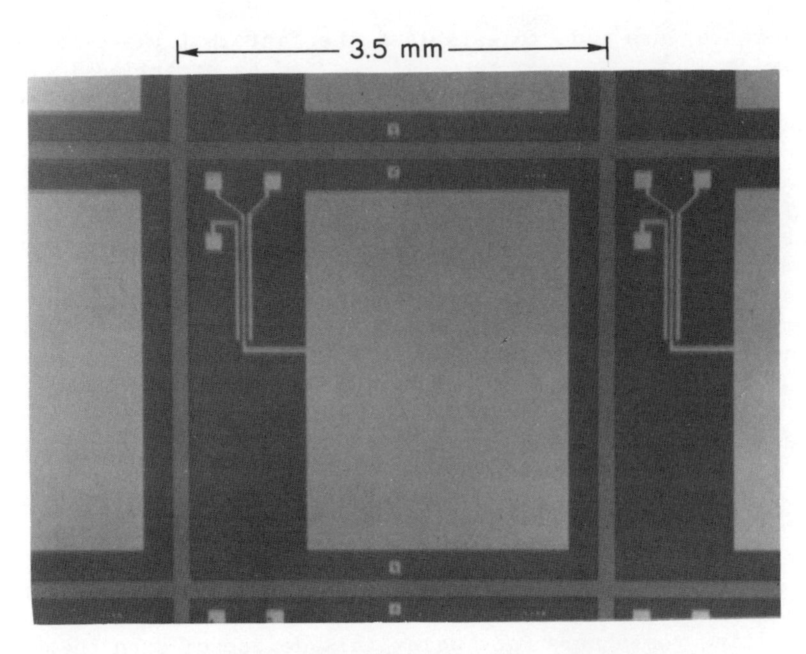

A. POSFET unit cell

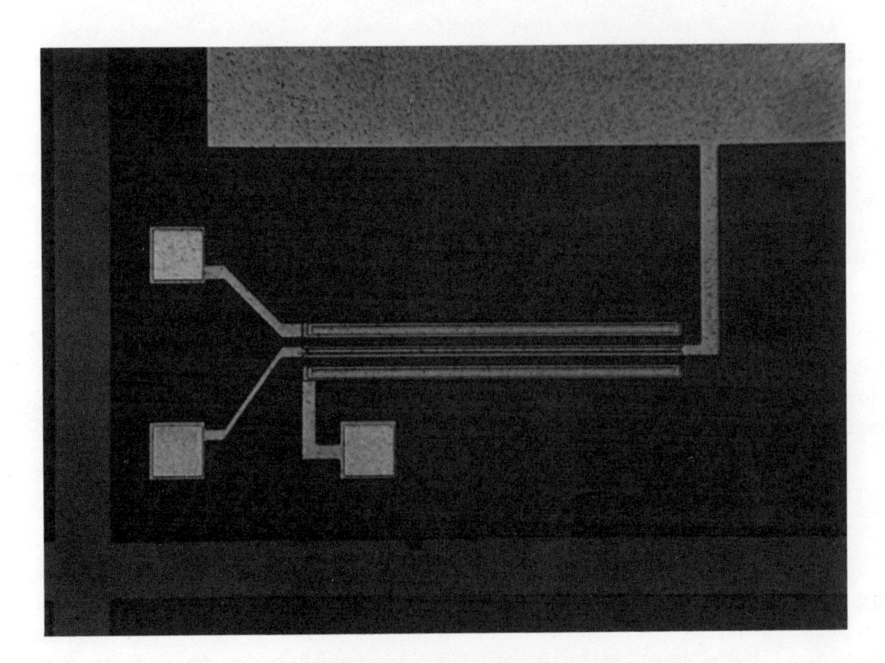

B. POSFET MOS transistor

Figure 11. SILICON WAFER VIEW OF POSFET.

Figure 12. POSFET MOUNTED IN DUAL-IN-LINE PACKAGE FOR EXPERIMENTAL EVALUATION.

Figure 13. THREE ELEMENT LINEAR ARRAY OF POSFET DEVICES MOUNTED IN DUAL-IN-LINE PACKAGE, PRIOR TO ATTACHMENT OF PVF_2 LAYER.

Since the POSFET is a receiving transducer element, it was necessary in its evaluation to have a known "reference" transmitting transducer. Selected for this purpose was the commercial Panametrics VIP-5-$\frac{1}{2}$I, a broadband ceramic transducer with center frequency at 5 MHz and a round trip insertion loss at that frequency of -38 dB. This transducer achieves its excellent frequency response (Figure 14) by means of a heavy backing layer. Lacking the front matching layer required for operation in water, its sensitivity is reduced. Thus, this transducer should correspond fairly well with the one whose response is modeled in Figure 4 in comparison with the POSFET. Sensitivity and frequency response measurements were then made with the POSFET relative to those characteristics of the Panametrics transducer.

A. Sensitivity and Frequency Response

The experimental setup is shown in Figure 15. The reference transducer, driven by a generator of known electrical output impedance, is used in a transmit/receive mode. A brass block is located in the near field of the transducer in a water medium (Figure 15A). The generator is electrically matched at a particular frequency to the transducer, which is then mechanically manipulated to give the maximum return signal. From the ratio of transmitted to received

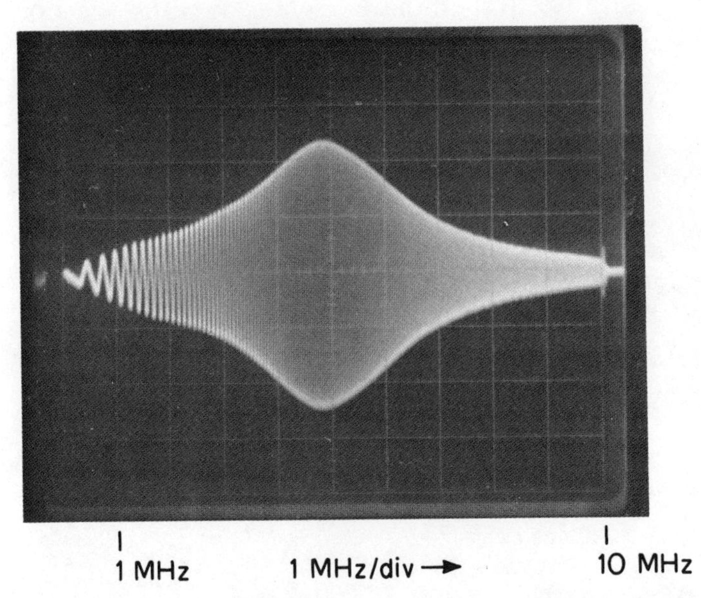

Figure 14. MEASURED PANAMETRICS VIP-5-$\frac{1}{2}$I TWO WAY FREQUENCY RESPONSE WITH 75 Ω ELECTRICAL LOAD.

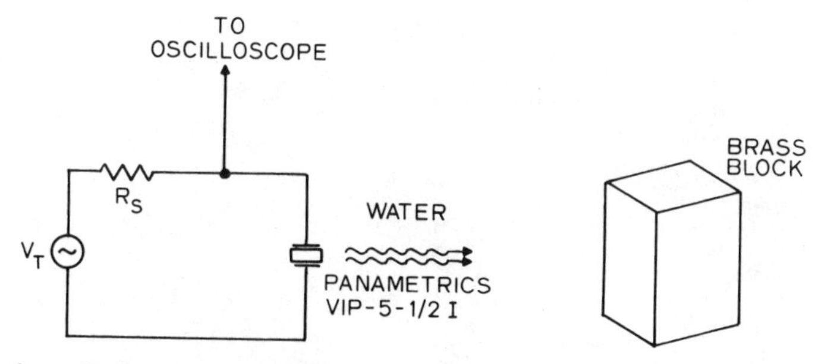

A. Reference transducer characterization system

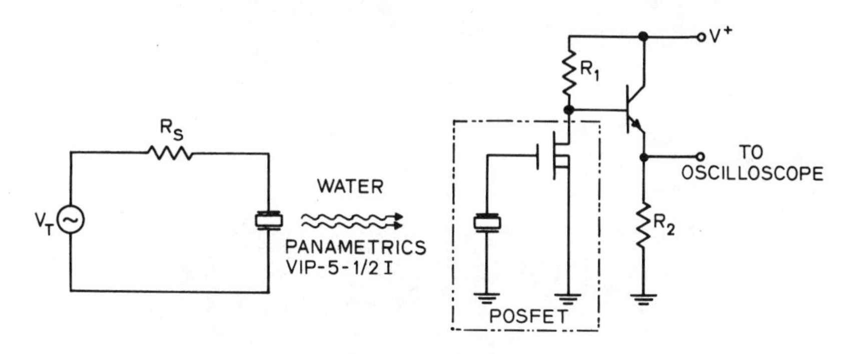

B. POSFET characterization circuit

Figure 15. EXPERIMENTAL SETUP FOR EVALUATION OF POSFET DEVICES.

energy, the insertion loss figure is calculated. From this result, the net acoustic power transmitted by the reference transducer is calculated as a function of frequency. The reference transducer is then used to transmit to the POSFET (Figure 15B). With a known acoustic power input to the POSFET (ignoring beam spreading), the projected electrical output voltage can be computed using Figure 4. This can then be compared with the measured output voltage vs frequency. Note that "insertion loss" as applied to the POSFET alone is not a meaningful quantity since no output power is delivered by the PVF$_2$ into the MOSFET load.

The device frequency response can be calculated following a similar procedure. Figures 16A and 16B show the frequency and impulse responses, respectively, of the reference transmitting transducer to POSFET receiver system. The POSFET transducer response alone can be extracted from this result when information about the two-way response of the reference transducer is obtained from the setup of Figure 15A.

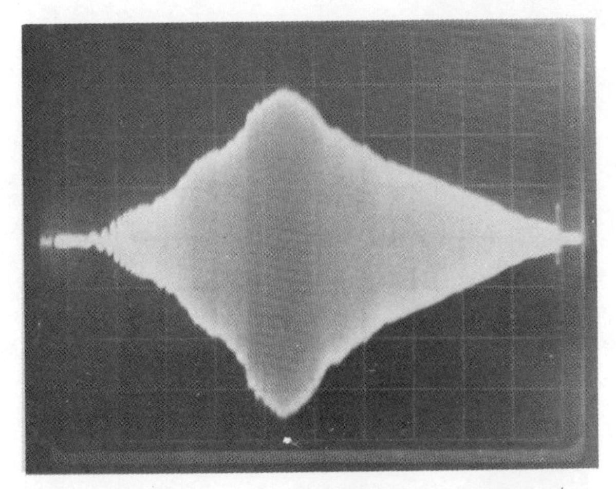

A. Frequency response

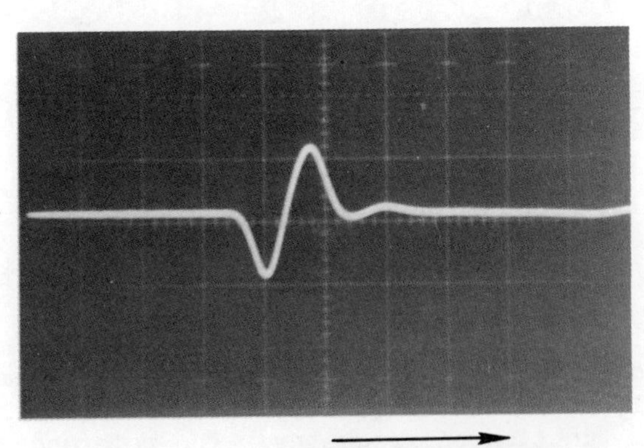

B. Impulse response

Figure 16. SYSTEM RESPONSE FOR PANAMETRICS VIP-5-½I BROADBAND TRANSDUCER TRANSMITTING TO PVF_2 ARRAY ELEMENT.

With knowledge of the sensitivity and spectral characteristics obtained from these experimental measurements, it is now possible to construct a plot of the actual V_{out}/T_{in} response of the POSFET. This is shown in Figure 17 in comparison with the theoretical result of Figure 4 over the frequency range of 1 to 10 MHz. The discrepancies originate from three sources:

(1) The theoretical response of Figure 4 for the POSFET was obtained assuming a uniform dielectric constant for PVF_2 of $5\epsilon_0$. Actually, over the frequency range 1 to 10 MHz, ϵ_{PVF_2} decreases approximately from $14\epsilon_0$ to $5\epsilon_0$. This is of consequence since the transfer function V_{out}/T_{in} is inversely proportional to the dielectric constant (refer to Eq. (2)).

(2) The roll-off at low frequencies in the measured response originates from the resonance of the backing. This will be discussed further.

(3) The discrepancy between the high frequency measured response and the theoretical values is most likely an experimental problem. The reference transducer response rolls off so quickly at high frequency that accurate measurement is difficult.

Nevertheless, the measured results, even on a linear scale, look very promising.

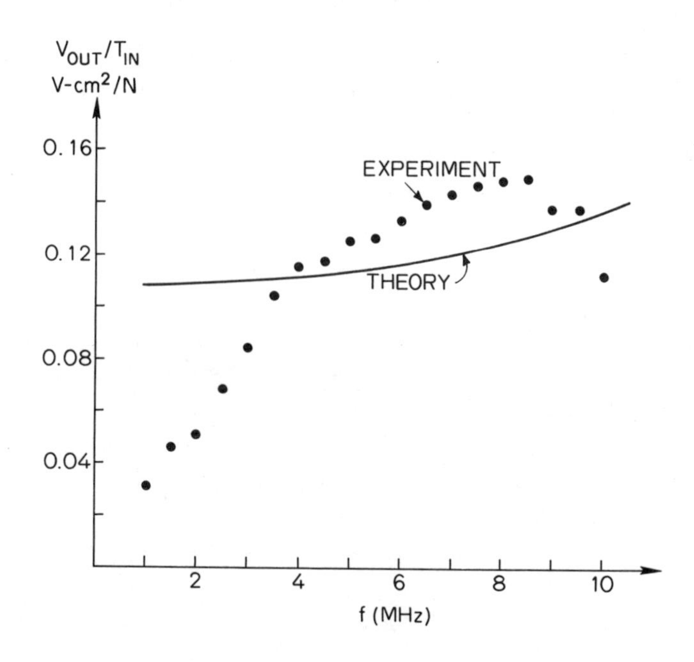

Figure 17. THEORETICAL AND EXPERIMENTALLY MEASURED VALUES FOR PVF_2 VOLTAGE/STRESS TRANSFER RELATION.

B. Reverberation

Reverberation in the POSFET transducer structure manifests itself in the notches and irregularities of the frequency response oscillograph of Figure 16A. These originate from the resonance of the transducer backing. The structure of present experimental POSFETs, mounted in integrated circuit dual-in-line packages, is diagrammed in Figure 18. Recall from the analysis of the THEORY OF OPERATION section that quarter-wave resonant operation of the PVF_2 transducer requires a backing of high impedance. However, considering the backing as an acoustic transmission line, we see that, at integral multiples of approximately 2 MHz, where the backing is one-half wavelengths thick, the PVF_2 acoustic load impedance goes to a very low value (air load). Correspondingly, the POSFET response goes to a minimum. In this sense, the POSFET acts like a high Q acoustic notch filter. A computer analysis of the POSFET response is shown in Figure 19, over the 1 to 10 MHz range, with the present nonideal packaging , in comparison with the theoretical result for an ideal high impedance backing. Proper acoustic backing in future designs (Figure 1) will substantially reduce this problem.

C. Acoustic Far-Field Radiation Pattern

As discussed in the DEVELOPMENT section, interelement cross-coupling in a transducer array will result in decreased lateral resolution in an imaging system. This is naturally of concern in the POSFET structure since individual elements are not discrete.

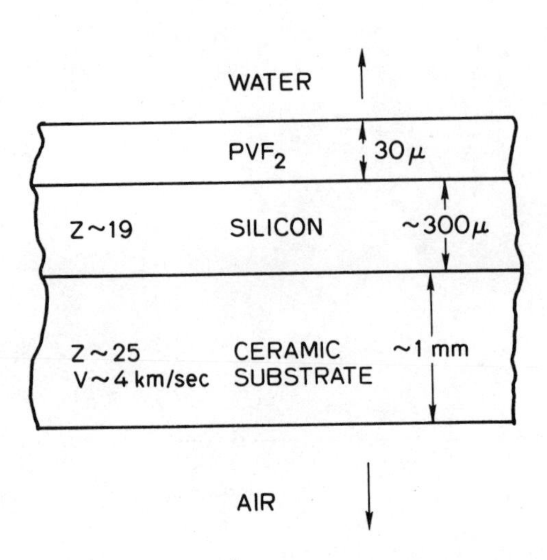

Figure 18. CROSS SECTION OF POSFET MOUNTED IN DUAL-IN-LINE PACKAGE.

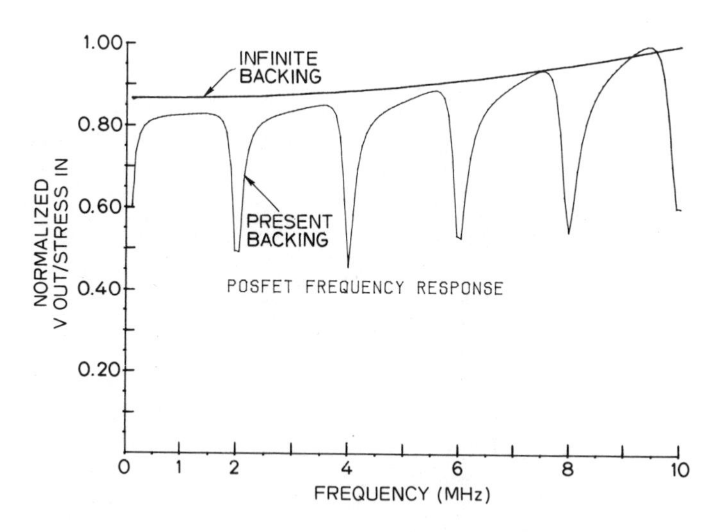

Figure 19. PREDICTED NORMALIZED VOLTAGE/STRESS TRANSFER RELATION FOR POSFET MOUNTED IN DUAL-IN-LINE PACKAGE.

Cross-coupling can be both acoustic and electrical and can be detected by an analysis of the far-field radiation pattern of the transducer. These patterns were obtained for each element of the POSFET 3×1 experimental array by sweeping a "point source" transmitting transducer (diameter = 0.5 mm, f = 2.0 MHz) laterally across the far-field of the POSFET. The resulting patterns are shown in Figure 20.

An ideal rectangular element will have a far-field radiation pattern varying in space as $\mathrm{sinc}(w/\lambda \sin\theta)$ [13], where w is the element width. The width of the radiating element can thus be obtained from the width of the ideal radiation pattern in the far-field at half amplitude, according to the formula:

$$w \cong 1.2\lambda z/\Delta x \qquad (8)$$

where Δx is the width of the pattern at the half-amplitude points and z is the perpendicular distance to the far-field measurement plane. While this formula is precise only for an ideal rectangular radiating element, it is nevertheless a useful guide to the acoustic response of arbitrary "real-world" elements. Cross-coupling will tend to make the element radiate from a larger area than in the ideal case. This will be manifest as a more narrow far-field radiation pattern and, correspondingly, a larger "equivalent width" as calculated from Eq. (8). Thus, the application of (8) to the patterns of Figure 20 gives equivalent widths of 3.55, 3.42, and 3.30 mm to elements 1, 2, and 3 of the experimental array. The actual element width as defined by the lower electrode of the

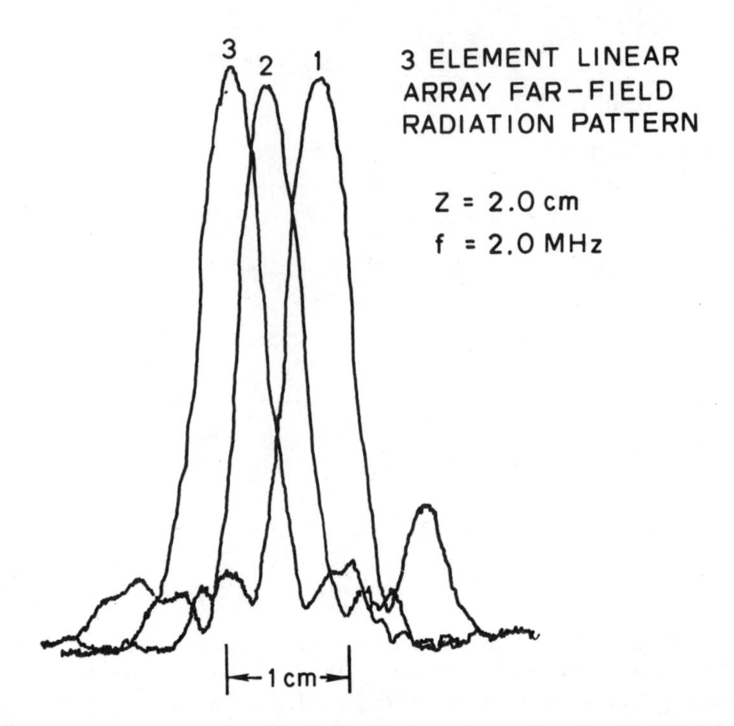

Figure 20. POSFET 3 × 1 ARRAY ELEMENT FAR-FIELD RADIATION PATTERN.

POSFET is 3.0 mm in each case. This differs little from the measured equivalent widths; so, it is reasonable to assume that acoustic cross-coupling, while present, is not severe.

Electrical cross-coupling appears in a somewhat different way. It arrives in the form of subsidiary peaks in the response of one transducer element at a position correspondingly to the main response of adjacent elements. This is observable in the plots of Figure 20, as can be seen by noting the abrupt widening of the pattern for a particular element at the same positions as the main peak or sidelobes of the adjacent element pattern. It is not a fundamental device limitation, however, and can be minimized by more careful isolation of device electrical leads. In a multiplexed array, however, electrical cross-coupling can become important [2].

V. CONCLUSION

In summary, the POSFET concept is described by the following features:

(1) A silicon wafer forms a rigid backing for a PVF_2 layer. This backing, with acoustic impedance $Z \cong 2 \cdot 10^7$ kg/m^2-sec, causes the transducer to be resonant at a frequency where the PVF_2 thickness is one-fourth wavelengths. An acoustic backing of tungsten-loaded epoxy can be placed behind the silicon to damp out backing reverberations and enhance low frequency response.

(2) Only the gate capacitance of an MOS transistor electrically loads the PVF_2. Therefore, usable output signal levels can be obtained despite the relatively lower k_T of PVF_2 compared to ceramics; i.e., a high impedance amplifier/multiplexer is well suited to the high electrical impedance PVF_2 transducer.

(3) The effective area and shape of an individual transducer are determined by the lower electrode of the PVF_2. An array of transistors can be fabricated in a silicon wafer using standard integrated circuit technology. Then, a single layer of PVF_2 is bonded to the surface of the wafer, providing a linear or two-dimensional array of receiving transducers. All of the critical geometry of the array structure is determined by the lithographic steps used in the fabrication of the silicon devices. There is no critical alignment required during the mounting of the PVF_2.

(4) Cell sizes from under 0.1 mm square to several cm square are attainable since cell geometry is determined only by the metalization pattern on the silicon wafer.

To date, discrete elements and small linear arrays of these transducers using the above technology have been fabricated using 30 micron thick PVF_2. Receiver sensitivity and bandwidth is superior to that of commercial broadband ceramic transducers. Far-field radiation patterns of the linear array show that array interelement cross-coupling, while present, will not substantially degrade array resolution.

The performance advantages of these transducers and the batch fabrication (integrated circuit) technology by which they are made point towards wide application of these devices in medical imaging and NDT in the future. It is currently planned to use the POSFET in the next generation of a real-time medical ultrasonic imaging system under development in the authors' laboratory.

VI. ACKNOWLEDGMENT

The authors would like to acknowledge the assistance of their colleagues in the Integrated Circuits Laboratory, in particular, P. Jerabek and M. Maginness.

VII. REFERENCES

1. C. DeSilets, J. Fraser, and G. S. Kino, "The Design of Efficient Broadband Piezoelectric Transducers," IEEE Trans. on Sonics and Ultrasonics, Vol. SU-25, No. 3, May 1978, pp. 115-125.

2. J. D. Plummer, R. G. Swartz, M. G. Maginness, J. R. Beaudouin, and J. D. Meindl, "Two Dimensional Transmit/Receive Ceramic Piezoelectric Arrays - Construction and Performance," to be published in IEEE Trans. on Sonics and Ultrasonics.

3. H. Kawai, "The Piezoelectricity of Polyvinylidene Fluoride," Japan. J. Appl. Phys., Vol. 8, 1969, p. 975.

4. K. W. Yeh, R. S. Muller, and S. H. Kwan, "Detection of Acoustic Waves with a PI-DMOS Transducer," Japan. J. Appl. Phys., Vol. 16, 1977, pp. 517-521.

5. S. H. Kwan, R. M. White, and R. S. Muller, "Integrated Ultrasonic Transducer," 1977 Ultrasonics Symp. Proc., pp. 843-847.

6. B. A. Auld, Acoustic Fields and Waves in Solids, Vol. I, John Wiley and Sons, New York, 1973.

7. H. Ohigashi, "Electromechanical Properties of Polarized Polyvinylidene Fluoride Films as Studied by the Piezoelectric Resonance Method," J. Appl. Phys., Vol. 47, No. 3, Mar 1976.

8. H. Jaffe and D. A. Berlincourt, "Piezoelectric Transducer Materials," Proc. IEEE, Vol. 53, No. 10, Oct 1965.

9. J. T. Walker and J. D. Meindl, "A Digitally Controlled CCD Dynamically Focussed Phased Array," 1975 IEEE Ultrasonics Symp., Los Angeles, Sep 1975, pp. 80-83.

10. B. D. Steinberg, Principles of Aperture and Array System Design, John Wiley and Sons, New York, 1976.

11. L. Bui, H. J. Shaw, and L. T. Zitelli, "Experimental Broadband Ultrasonic Transducers Using PVF_2 Piezoelectric Film," Electronics Letters, Vol. 12, 1976, p. 393.

12. A. Macovski and S. J. Norton, "High Resolution B-Scan Systems Using a Circular Array," Acoustic Holography, Vol. 6, edited by G. Wade, Plenum Press, New York, pp. 111-126.

13. A. Macovski, "Theory on Imaging with Arrays," Acoustic Imaging, edited by Glen Wade, Plenum Press, New York, 1976, pp. 114-118.

ACOUSTICAL HOLOGRAPHY MATRIX ARRAY IMAGING SYSTEM FOR THE UNDERWATER INSPECTION OF OFFSHORE OIL PLATFORM WELDMENTS

H. D. Collins, R. P. Gribble, T. E. Hall, W. M. Lechelt, J. T. Luebke, J. Spalek, E. M. Sheen and A. Stankoff†

Holosonics, Inc.
2400 Stevens Drive
Richland, Washington 99352

ABSTRACT

This paper describes a diver operated underwater optical-acoustical imaging system for the inspection of internal defects in offshore platform weldments. The two-dimensional acoustic array is electronically programmed with digital techniques to simulate focused and non-focused source-receiver scanning. The electronic simulated reference beam is programmable using erasable programmable read only memories (EPROMS). The imaging device consists of a diver hand held gun containing the acoustic array, miniature television camera and the visible light emitting diode (VLED) display array. The gun is connected via the diver pack and cable to the control unit, digital memory display and data recording units located in the submersible. The television camera provides an optical view of the external weld surface, identification marks, etc., which is integrated with the acoustical image on a standard television monitor. The acoustical array provides complete real-time inspection by electronically scanning and constructing multiple focused holograms through the entire weld volume. The defect images are presented in side, plan and pseudo three-dimensional views with the options of rotation, tilt and zoom magnification. The system has two permanent recording techniques. The focused holographic defect images and optical views are stored on videotape and the basic r-f data on digital tape. This system is now undergoing tests in the North Sea by International Submarine Services.

---

†A. Stankoff is with International Submarine Services, SA

INTRODUCTION

An operational underwater optical/acoustical imaging system employing focused holographic techniques is described. The system will be used in the inspection of offshore oil platform weldments in the North Sea. Preliminary experimental results are presented of defect images in simulated platform structure weldments, which demonstrate the unique capabilities of the system.

The real-time acoustical holography weld inspection system offers many advantages over conventional techniques, such as scanned pulse-echo system, magnetic particle, visual inspection, etc. The usual scanned acoustical system requires excessive diver time, and magnetic particle technique is limited to internal defects with surface protrusions for detection. The acoustical holography system with the flexible array circumvents the problem of excessive inspection time and detection of internal defects within the weld volume.

Figure 1 is the typical operational configuration employing a lock-out submersible attached to the platform with the diver at the weld inspection site. The external weld surface is cleaned and reference marks inserted to provide accurate identification of the inspection area. The entire weld volume is inspected and the data recorded in the submersible via digital and videotape systems.

The VLED optical array on the gun displays a C-Scan or plan view of the internal defect providing diver interaction with the imaging system. The diver now has the ability to view the defect and interact with the placement or position of the device.

The operator in the submersible directs the inspection and views the defect images on the television display. After returning to the surface, the defect data are analyzed by playback of the original acoustic information via the digital magnetic tape and image display systems.

The imaging device consists of a diver hand held gun containing the acoustic array, miniature television camera and the VLED display array as shown in Figure 2. The gun is connected via the diver pack and cable to the control unit, digital memory display and data recording units located in the submersible. The complete system including the gun, diver pack and three instrumentation units is shown in Figure 3. The television camera on the gun provides an optical view of the external weld surface, identification marks, etc. which is integrated with the acoustical images on a standard television monitor. The flexible acoustical array conforms to the complex cylindrical platform structure geometry and provides complete real-time inspection by digital electronic scanning techniques.

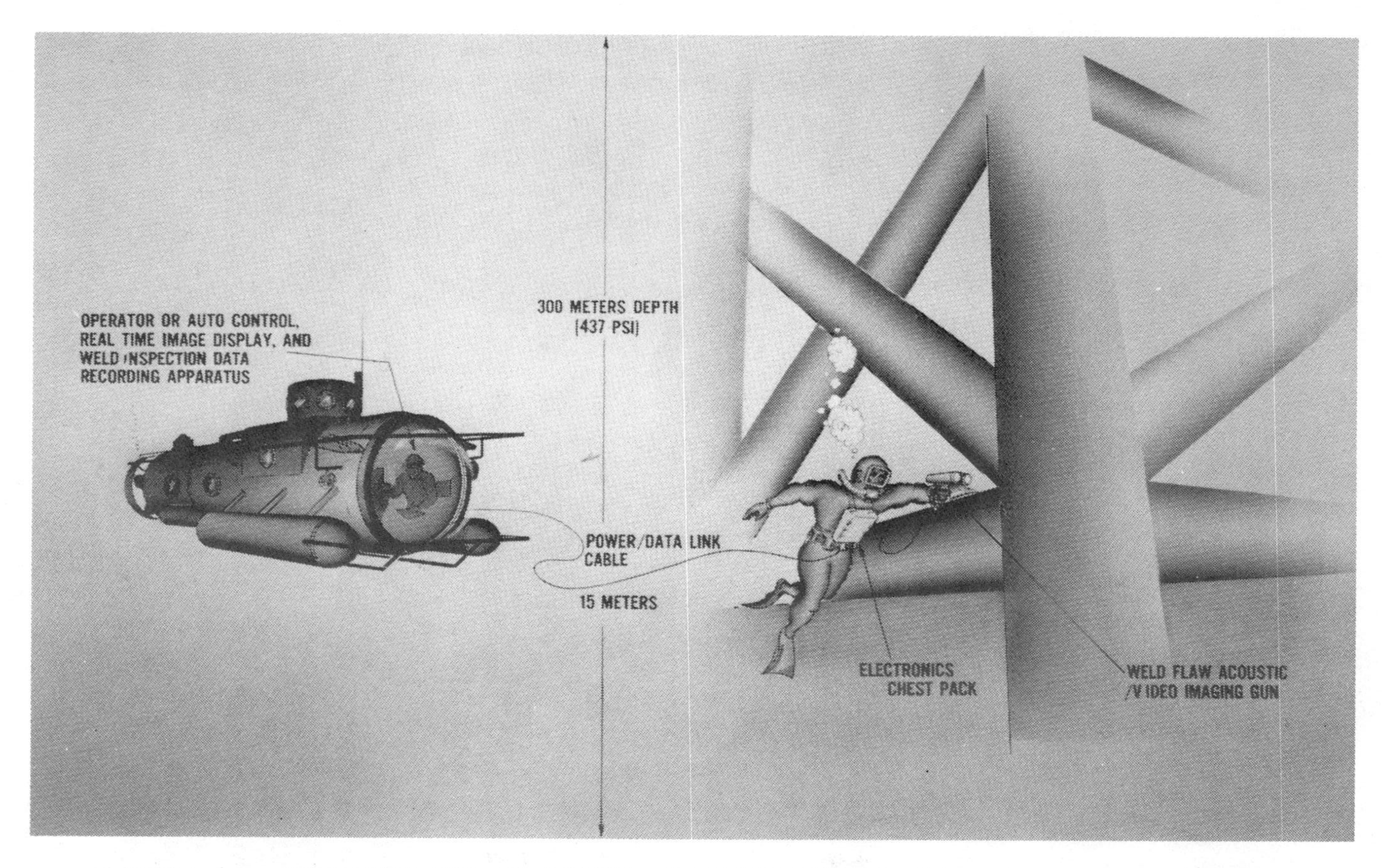

FIGURE 1. OFFSHORE OIL PLATFORM ACOUSTIC/VIDEO WELD INSPECTION SYSTEM

FIGURE 2. DIVER HAND HELD OPTICAL/ACOUSTICAL IMAGING DEVICE

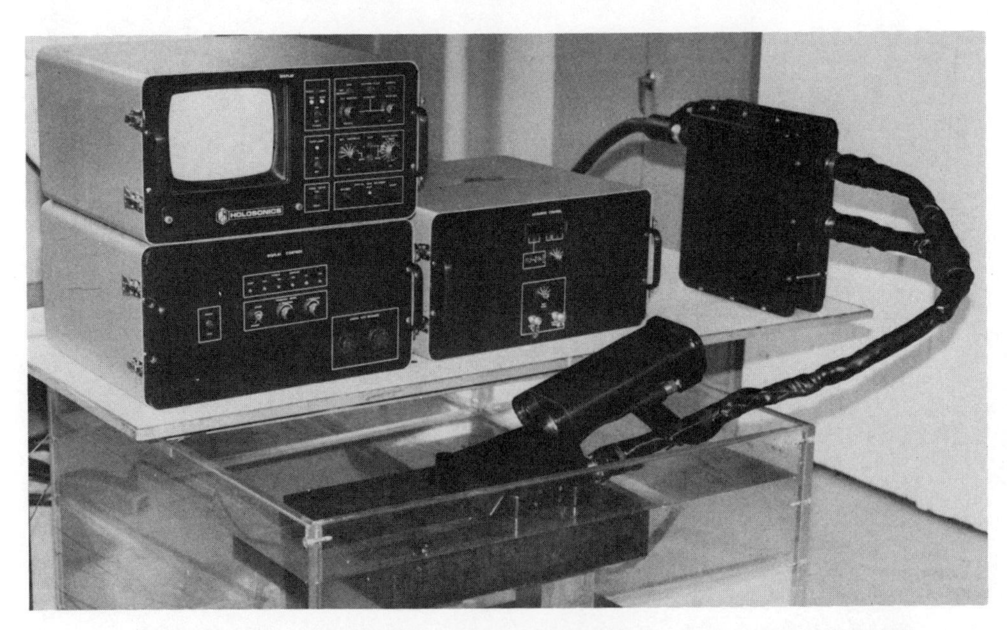

FIGURE 3. ACOUSTICAL HOLOGRAPHY IMAGING SYSTEM FOR UNDERWATER OFFSHORE OIL PLATFORM WELDMENTS

Multiple shear wave focused holograms are constructed at 32 preselected planes in the weld volume orthogonal to the acoustic propagation path.

The defect images are presented in side, plan and pseudo three-dimensional views with the options of rotation, tilt and zoom magnification. Figure 4 shows the two display modes of the weld information, as viewed by the submersible operator. The acoustical side and plan views or three-dimensional view with rotation, tilt, etc. can be selected by the operator.

The two display modes provide optimum defect identification, etc. for assessing weld integrity.

PERFORMANCE REQUIREMENTS

For an underwater weld inspection system, the following factors were considered: maximum and minimum materials thickness range, array flexibility, radius of curvature, lateral resolution,

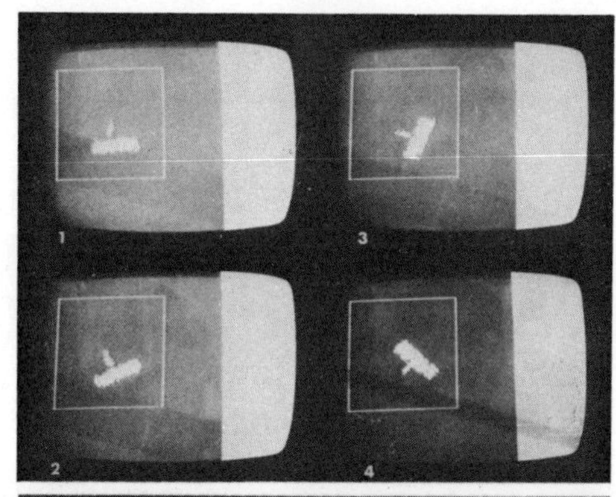

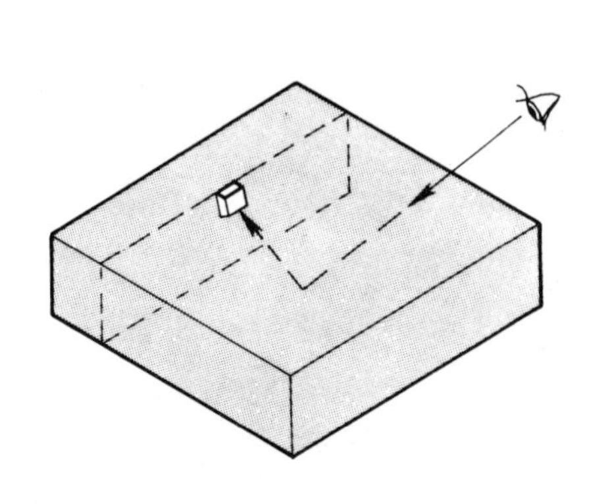

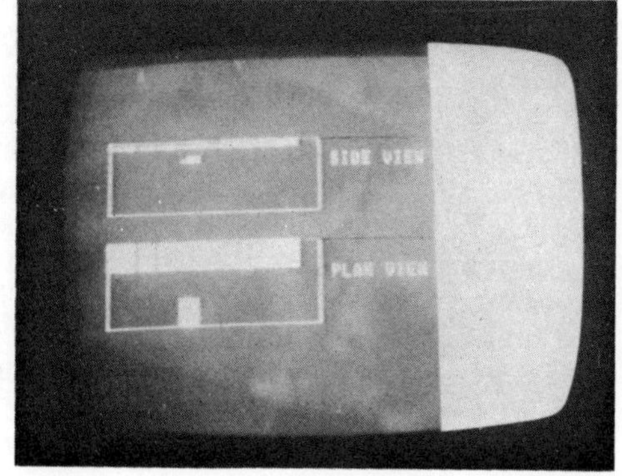

FIGURE 4. ACOUSTIC DISPLAY MODES: 3-D, SIDE AND PLAN VIEWS

defect detection sensitivity, image generation time, power consumption, weight, size, environmental constraints, etc.

The system specifications are given for the North Sea offshore platform weldment inspection task:

- Material thickness range: 25mm to 76mm
- Array flexibility radius curvature range: 500mm to infinity
- Lateral resolution: 6mm - 8mm
- Defect detection sensitivity @ 230mm metal path: 3mm
- Image range in steel: 250mm
- Image generation time single frame: 50 milliseconds
- Television Display: acoustical information - side, plan and 3-D images integrated with optical (external weld surface) information
- Data recording: digital and videotape systems for the acoustical and optical image information
- Acoustic array geometry: 150mm X 111mm X 20mm
- Operational depth: 200 meters
- Operational temperature: 0°C to 35°C
- Power requirements: 325 watts (maximum)

SYSTEM DESCRIPTION AND OPERATION

Figure 5 is a simplified block diagram of the acoustical holography weld inspection system. The system consists of three basic sub-sections: 1) acoustic array/video probe, 2) diver pack and 3) submarine based control unit.

Acoustic Array/Video Probe

The acoustic array contained in the gun-like device consists of 160 individual transducers in five rows of 32 elements each, mounted in a unique, flexible, elastomer matrix to allow contouring of the array to the variable curvature of the platform structure. Figure 6 shows the device positioned on the platform structure inspecting a segment of the weld. The electronic scanning consists of sequentially activating, simultaneously, (6, 7, 8, 9 .... 12, 12, ... 9, 8, 7, 6) transducers along each row. The complete scan of one focal plane requires approximately 53 milliseconds (i.e., employing a 3kHz pulse repetition frequency) and is capable of scanning up to 32 focal planes. The acoustic signal gate automatically sweeps through the entire weld volume in sequence with the focal plane digital addressing. The control unit also permits manual selection of any desired focal plane in the weld volume. The weld volume is inspected by a skip-bounce technique that requires reflection from the lower surface. The central imaging angle is approximately 40° with respect to the bracing surface. The probe also contains a tungsten halogen illumination light and

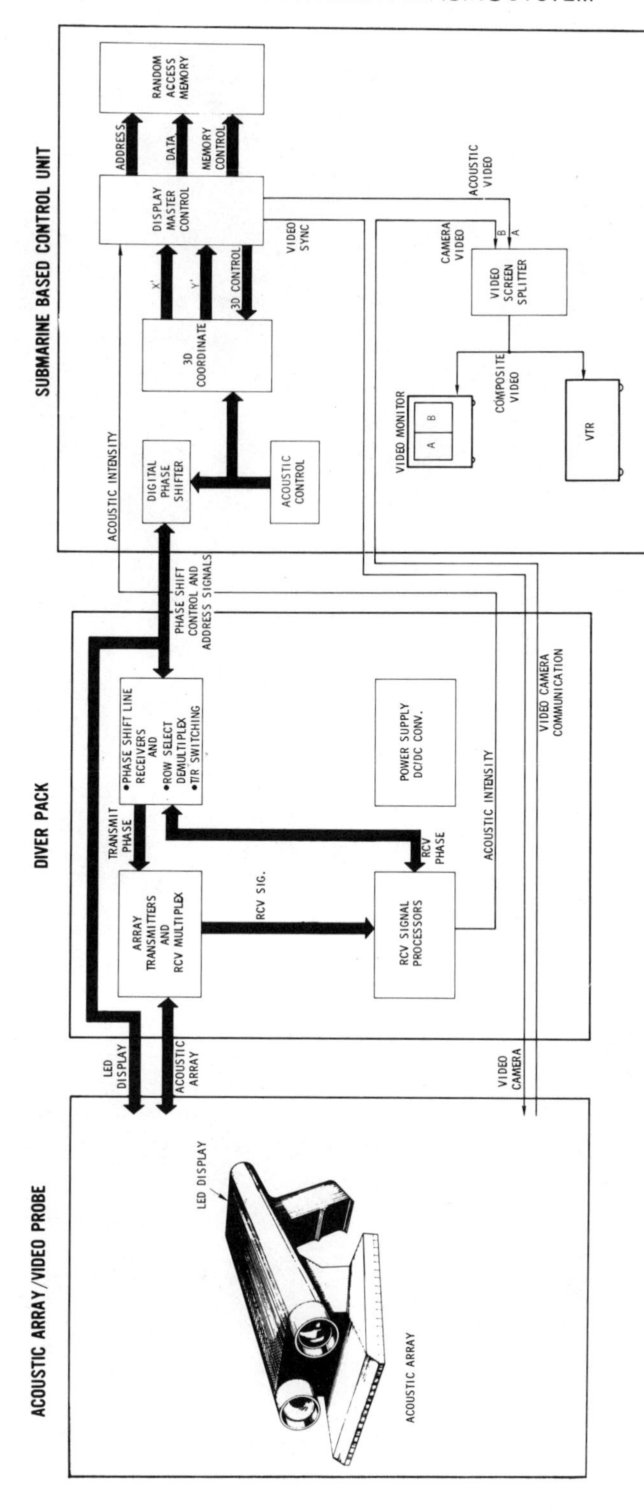

FIGURE 5. ACOUSTIC ARRAY/VIDEO WELD IMAGING/INSPECTION SYSTEM - BLOCK DIAGRAM

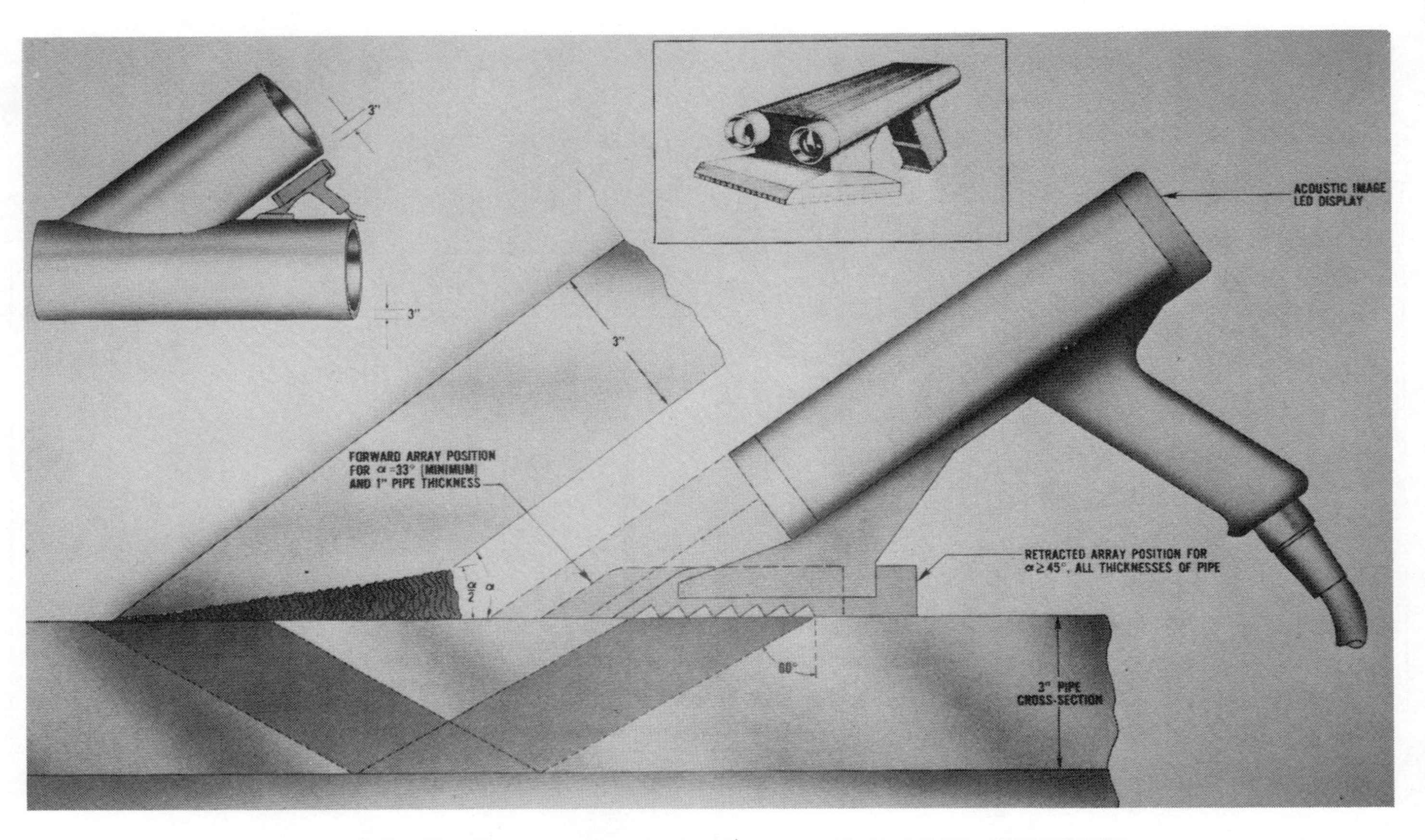

FIGURE 6. FULL SCALE WELD INSPECTION GEOMETRY

a miniature T.V. camera which provides an optical view of the external weld surface. The camera focusses from 7cm to infinity in the ocean environment.

The VLED array located behind the T.V. camera housing provides the diver with a C-Scan image of the inspected weld volume.

Diver Pack Unit

The diver pack unit is contained in an unique pressure vessel capable of withstanding pressures of up to 500 psi (i.e., 343 meter depths). The diver pack has internal honeycomb-type structure to withstand the hydrostatic pressure and allow neutral buoyancy. One-hundred-sixty power amplifiers, 32 receiver channels, signal processors, multiplexing circuits, etc. are mounted on printed circuit cards, heat sunk to the aluminum compartment walls of the diver pack. The compartment walls are in direct contact with two large 28cm x 33cm anodized aluminum covers which also serve as the main heat dissipating surfaces. (Approximately 80 watts of power are dissipated from the diver pack.) Two covers are used for ease of assembly and maintenance of the compact electronics within the diver pack.

Close-up photographs of the diver pack's internal components are shown in Figure 7. The diver transports the neutrally buoyant diver pack with the array/probe to the inspection site. A 15 meter cable is connected between this unit and the submarine.

Submarine Based Control Unit

The submarine based control equipment consists of three separate units: (1) acoustic control unit, (2) digital memory/3-D unit, and (3) image display unit.

Packaged in an environmental case, 48cm X 25cm X 44cm, the acoustic control unit generates addressing, sequencing, and phase data required for acoustic operation. Located here is the 64MHz coherent oscillator. All sequencing, addressing, and phase referencing are derived from this oscillator. The transmit rate is determined by a separate oscillator (3kHz) slaved to the 64MHz reference. By dividing the reference down and driving a counter-decoder, the timing sequences for the address counters and phase shifters are obtained. Fourteen address bits are required to activate the desired sequenced aperture of twelve transducers from 160. Listing the least significant to most significant bits, one address line gives transmit or receive information, five bits determine which of the 32 different scan positions in a row of transducers is active, three bits choose one of five rows being scanned, and the most significant five bits of the address counter determine one of the 32 focal depths that is programmed.

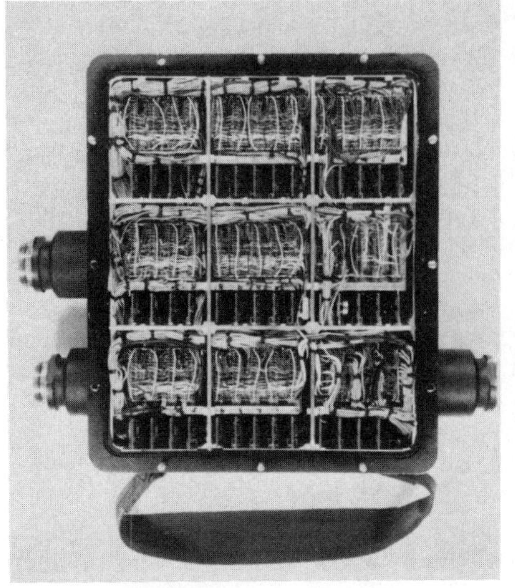

a) DIVER PACK BACK VIEW SHOWING WIRING

b) DIVER PACK FRONT VIEW SHOWING INSERTED P/C BOARDS

FIGURE 7. DIVER PACK INTERNAL CONSTRUCTION

Fifty ohm line drivers located in the acoustic control case are used on all signal and address lines supplied to the diver pack. This provides impedance matching with the 15 meter cable interfacing the two systems.

The display unit, housed in a 48cm X 25cm X 44cm environmental case, is designed to provide essential operator-system interaction. In split screen format, acoustic and video information of the scanned area is displayed on a 23cm video monitor. With a real-time display, the operator has the capability of interacting with the inspection process, based upon his observations. These interactions can be placed into two categories, image enhancement and information recording.

Acoustic image enhancement controls include interlace, metal thickness, and manual focal depth. The interlace switch when activated doubles the sampling points for the scan. This scan is then retained and displayed on the video monitor for operator study.

In the automatic scan mode, the metal thickness control allows the operator to vary the focal scan according to metal thicknesses that may be encountered. This provides for a focal scan of the weldment volume in question. In manual mode, the operator can choose in 8mm steps the metal depth to be viewed. If it is desired to limit the scan to one depth for closer flaw analysis, this provides that option. Video controls provided are the focus switch and lamp switch.

Information record and display controls consist of a data select switch, and a view format select switch. The data select switch chooses between three possible data inputs. In normal operation this is in the gun position. It is here that real-time data from the scanned array are being recorded on digital tape and displayed. When in the internal memory position, data from the last complete scan are stored, recorded and displayed until the switch position is changed. In the external memory position, data stored on digital tape are input, processed, and displayed, in the same way as the scanned array data.

The view format switch in the section/plan position provides a top and side view of the scanned area. In the isometric position, a three-dimensional projected view is recorded and displayed. By adjusting the rotate and tilt controls, the projected view can be altered to the aspect desired.

Electronic Scanned Source-Receiver Matrix Array

Consider a single row of the 160 matrix array consisting of 32 piezoelectric transducer elements. N elements of the 32 elements are addressed digitally by read only memories (ROMS) and erasable program read only memories (EPROMS) provide the proper phasing for focusing in both the transmit and receive modes. The sequential switching and shifting of the basic aperture of twelve elements occurs at the pulse repetition frequency and simulates scanning a focused transducer over the complete matrix array with 32 different focal lengths.

The transmitter receive group of twelve elements concentrates and receives energy from a finite region ΔX, ΔY given by

$$\Delta x = \frac{\lambda f}{a} \quad \text{and} \tag{1}$$

$$\Delta y = \frac{\lambda f}{b} \tag{2}$$

where λ is the acoustic wavelength, f the focal length and (a & b) the effective apertures in the X and Y directions.

The transducers are line focused in the Y direction, by axicon shaping the piezoelectric elements. The five rows of 32 transducers are slanted at the proper central angle to produce shear wave illumination at 50° with respect to the normal. The lateral image resolutions are given by Equations (1) and (2).

If the external surface of the volume to be inspected has a shape requiring the array to flex and conform into some non-flat configuration, the relative phasing of the transmit and receive EPROMS can be programmed to eliminate this distortion and produce the desired focus.

Radiation Pattern

The basic simulated transducer aperture of twelve uniformly excited elements introduces the usual problems of sidelobes. Figure 8 shows the normalized response of the twelve element array unapodized. The grating sidelobes occur at approximately 22° from the central beam in steel when imaging with shear waves. The first sidelobes, as predicted by theory, are typically -13db below the central beam without apodization.

Imaging flaws in cylindrical pipe geometry using skip/bounce techniques minimizes the sidelobe problem by channel divergence. The first test array because of the stringent power and space requirements was devoid of sidelobe reduction techniques.

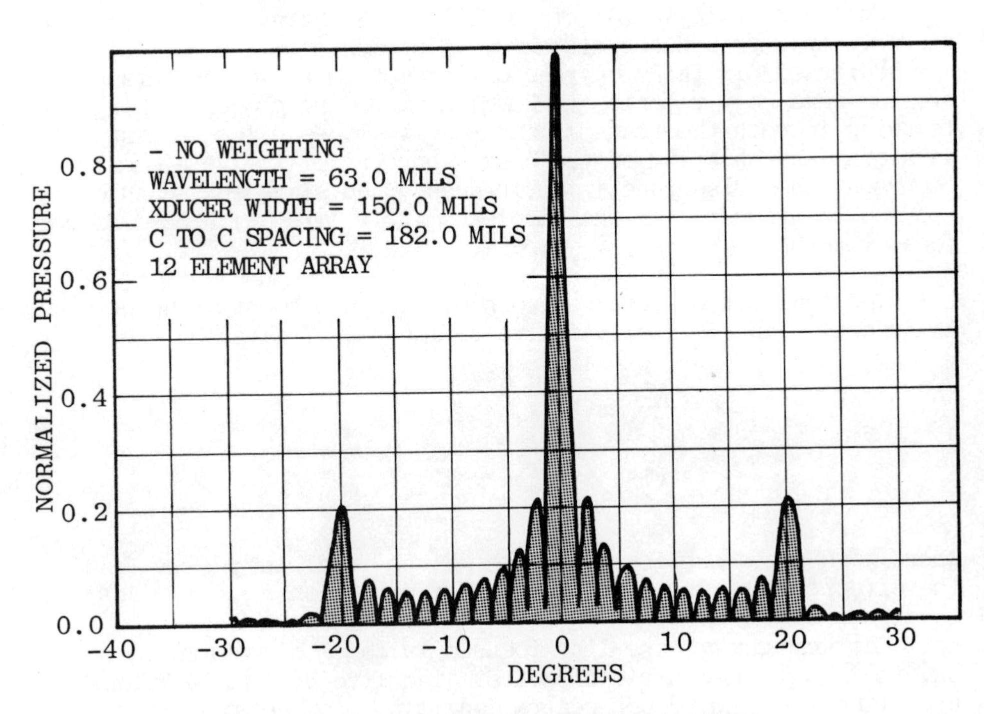

FIGURE 8. ARRAY RADIATION PATTERN

Programmable Simulated Holographic Electronic Reference

The manner in which the simultaneous phasing of each element in the basic aperture is accomplished may be understood from Figure 9. A 64MHz oscillator serves as the clock or coherent source. The digital programmable reference is derived from a series of divider circuits capable of generating 32 discrete phase shift increments 11.25¼ apart. The proper phase selection and sequence for each focal length, surface correction and reference are programmed into each of the 32 EPROMS, one for each row element. ROMS provide the multiplexing by enabling each of the EPROMS as to the sequential element selection of the basic aperture along the row. The EPROMS then generate the phase address to each of the 32 digital phase shifters.

Receiver Signal Processing

Figure 9 illustrates the signal processing of the fundamental receiver aperture of 12 elements. The output of the signal summing amplifier is given by the following expression:

$$S_o(x,y) = \sum_{i=1}^{12} \left\{ A_i B e^{j(\phi_{oi} - \phi_{ri})} \right\}$$

where the phase of the object is

$$\phi_{oi} = \phi'_o - \frac{2\pi}{\lambda} r_i$$

and the reference phase ϕ_{ri}.

If we operate in the focused holographic mode, the reference phase will be composed of two parts; (1) focusing and (2) the simulated off-axis reference. The reference is then defined by the following expression:

$$\phi_{ri} = \phi'_r + \phi_{fi}$$

where $\phi'_r = \frac{\omega_x x}{V_x}$ (off-axis phase)

and ϕ_{fi} (focusing phase)

The phases of the focused holographic signals at the input of the summing amplifier are given by the following expression:

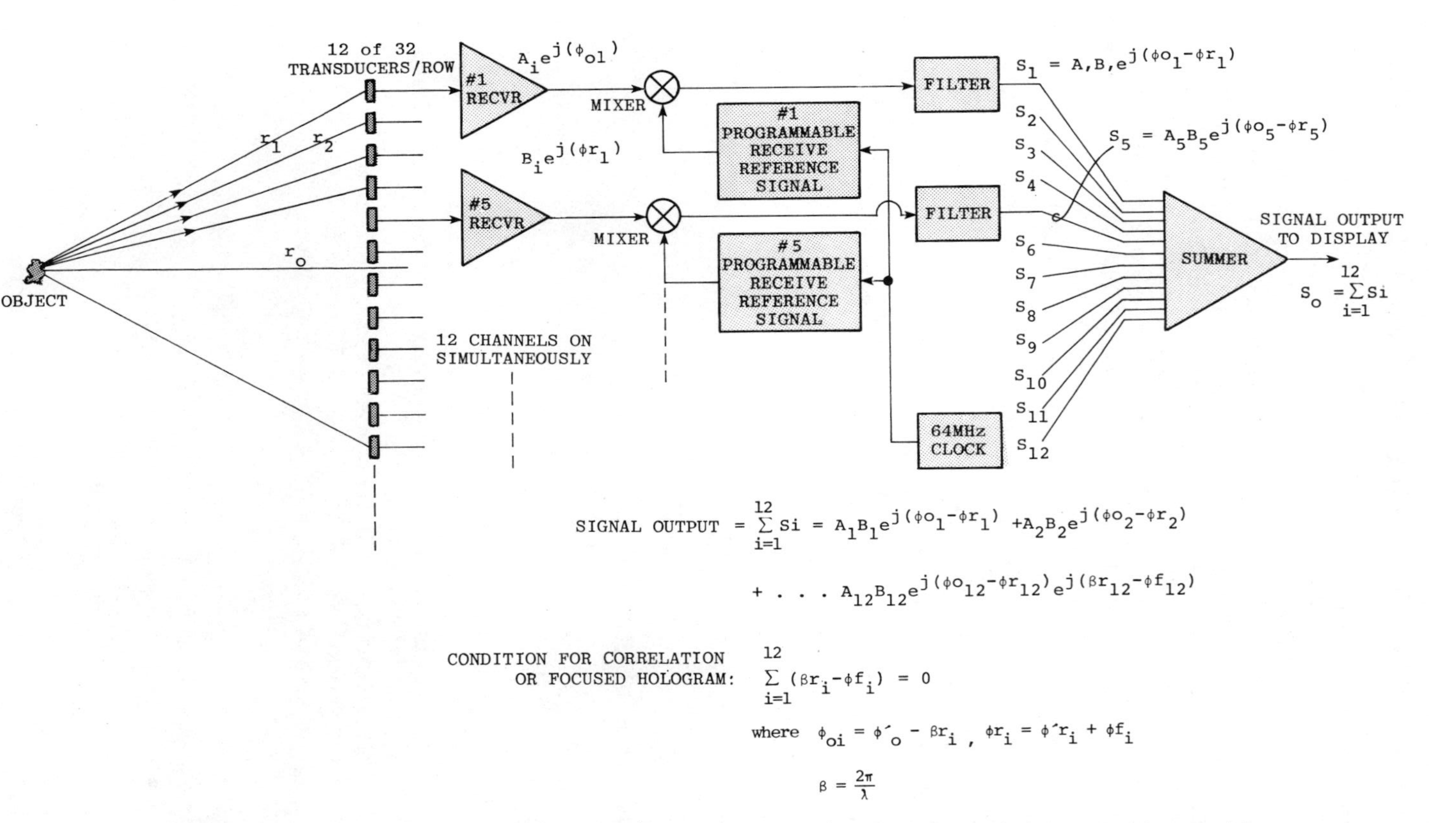

$$\text{SIGNAL OUTPUT} = \sum_{i=1}^{12} Si = A_1 B_1 e^{j(\phi o_1 - \phi r_1)} + A_2 B_2 e^{j(\phi o_2 - \phi r_2)}$$

$$+ \ldots A_{12} B_{12} e^{j(\phi o_{12} - \phi r_{12})} e^{j(\beta r_{12} - \phi f_{12})}$$

CONDITION FOR CORRELATION OR FOCUSED HOLOGRAM: $\sum_{i=1}^{12} (\beta r_i - \phi f_i) = 0$

where $\phi_{oi} = \phi'_o - \beta r_i$, $\phi r_i = \phi' r_i + \phi f_i$

$$\beta = \frac{2\pi}{\lambda}$$

FIGURE 9. HOLOGRAPHIC ARRAY RECEIVER SIGNAL PROCESSING WITH PROGRAMMABLE REFERENCE (12 of 32 channels)

$$\phi_{hi} = \left[\phi'_o(x,y) - \frac{\omega_x x}{V_x} - \frac{2\pi}{\lambda} r_i + \phi_{fi}\right]$$

where the condition for correlation or focusing is given by:

$$\sum_{i=1}^{12} (\phi_{fi} + \frac{2\pi}{\lambda} r_i) = 0.$$

The phase function ϕ_{fi} is programmed into the receive references for all 32 focal depths.

Naturally, non-focused holography is easily accomplished by deleting the focusing part of the reference in all EPROMS.

Transmit Signal Processing

The holographic signal processing with the programmable transmit reference is shown in Figure 10. The procedure or method is very similar to the receive technique.

The received signal at the point object from the 12 transducers is:

$$P_o = \sum_{i=1}^{12} P_i$$

where

$$P_i = A_{ti} e^{j(\phi_i + \frac{2\pi}{\lambda} r_i)} \quad \text{for } i=1, 2...12$$

The condition for source focusing is given by the following expression:

$$\sum_{i=1}^{12} (\phi_{fi} - \frac{2\pi}{\lambda} r_i) = 0.$$

EXPERIMENTAL RESULTS

Focused Array Beam Profiles

Figure 11 shows the basic or fundamental aperture of 12 elements electronically scanning the letter "H" in the X direction.

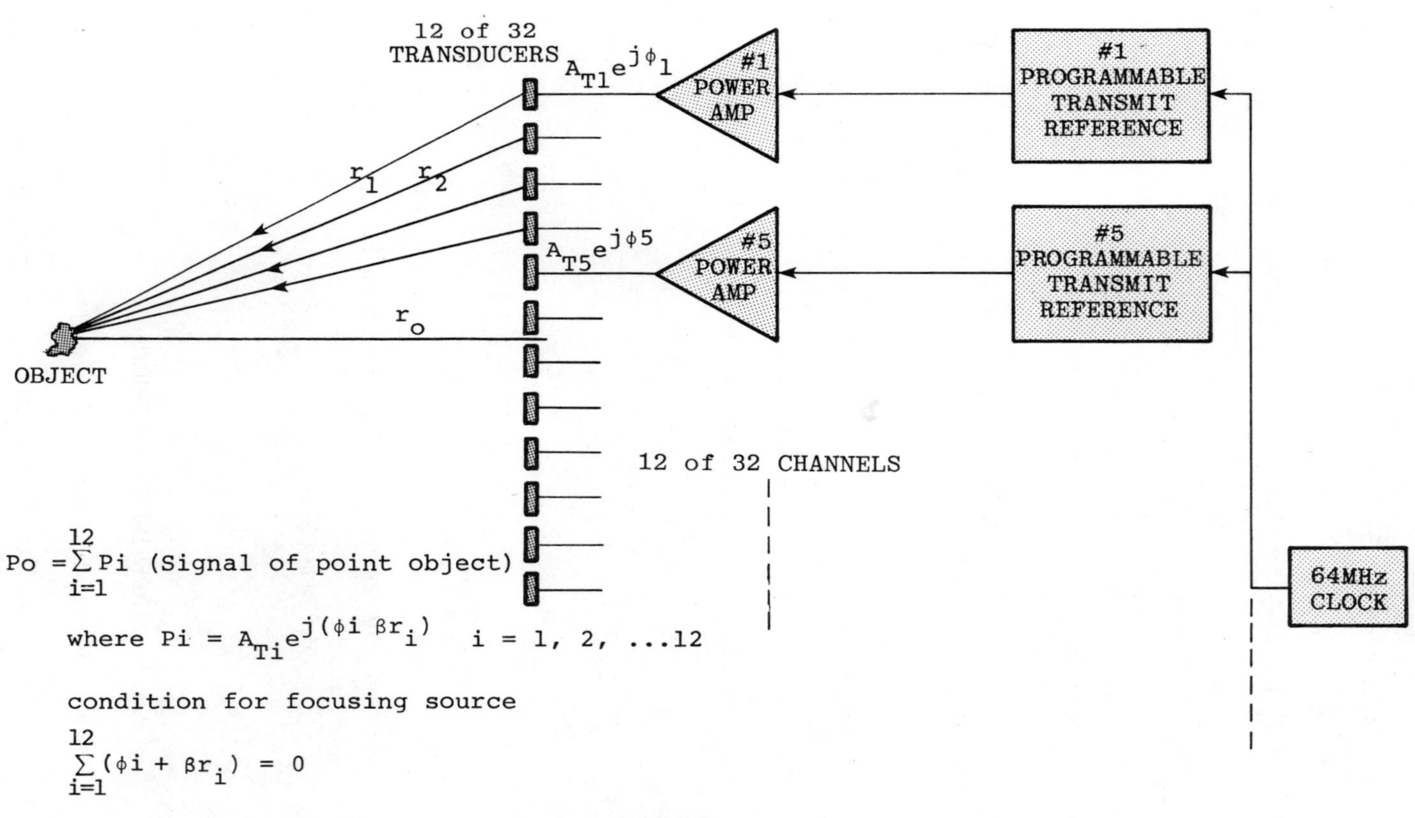

FIGURE 10. HOLOGRAPHIC ARRAY TRANSMIT SIGNAL PROCESSING WITH PROGRAMMABLE REFERENCE (12 of 32 channels)

The target consists of a metal object 5cm x 5cm located 51cm from the array. The lower left picture represents the image without electronic focusing. The lower right picture is the image with source-receiver focusing. The image of the "H" is of excellent quality exhibiting 4mm to 7mm lateral resolution in the X direction. The Y direction resolution is determined by the axicon characteristics.

The upper left picture is the electronically scanned acoustical array beam profile in the X direction illustrating the source focusing capabilities of the system. The upper middle picture illustrates the beam profile in the Y direction (i.e., axicon). The upper right picture is the central beam profile in the X direction.

Shear Wave Image of a 3mm Flaw in Steel

Figure 12 shows the 3mm flaw geometry and the focused hologram image in side and plan view on the T.V. display monitor. The skip/bounce flaw distance is 237mm in steel illustrating the unique sensitivity capabilities of the system. The plan view as a result of unequal lateral resolutions displays the square as a rectangle with the long dimension in the Y direction.

Shear Wave Images of Variable Depth Flaws in Aluminum

Figure 13 shows the variable flaw depth geometry consisting of three flat bottom holes separated 25mm in depth and the associated focused holographic images. These images demonstrate the system's unique ability to electronically scan 32 focal depths throughout the entire weld volume. The three holes at different depths are easily identified in the side view. The plan view shows all three holes at different depths in focus.

Figure 14 shows the letter "H" target geometry and the optical/acoustical display. The left side of the T.V. screen shows the acoustical side and plan views. The letter "H" is slightly tilted as shown in the side view. The optical view is shown on the right side of the T.V. display screen.

The letter "H" illustrates the imaging resolution capabilities of the device in the water at distance of 40cm.

These laboratory test results illustrate the unique capabilities of the focused holographic array system to rapidly detect, locate and size flaws in underwater steel structures. This system is now undergoing tests in the North Sea by International Submarine Services.

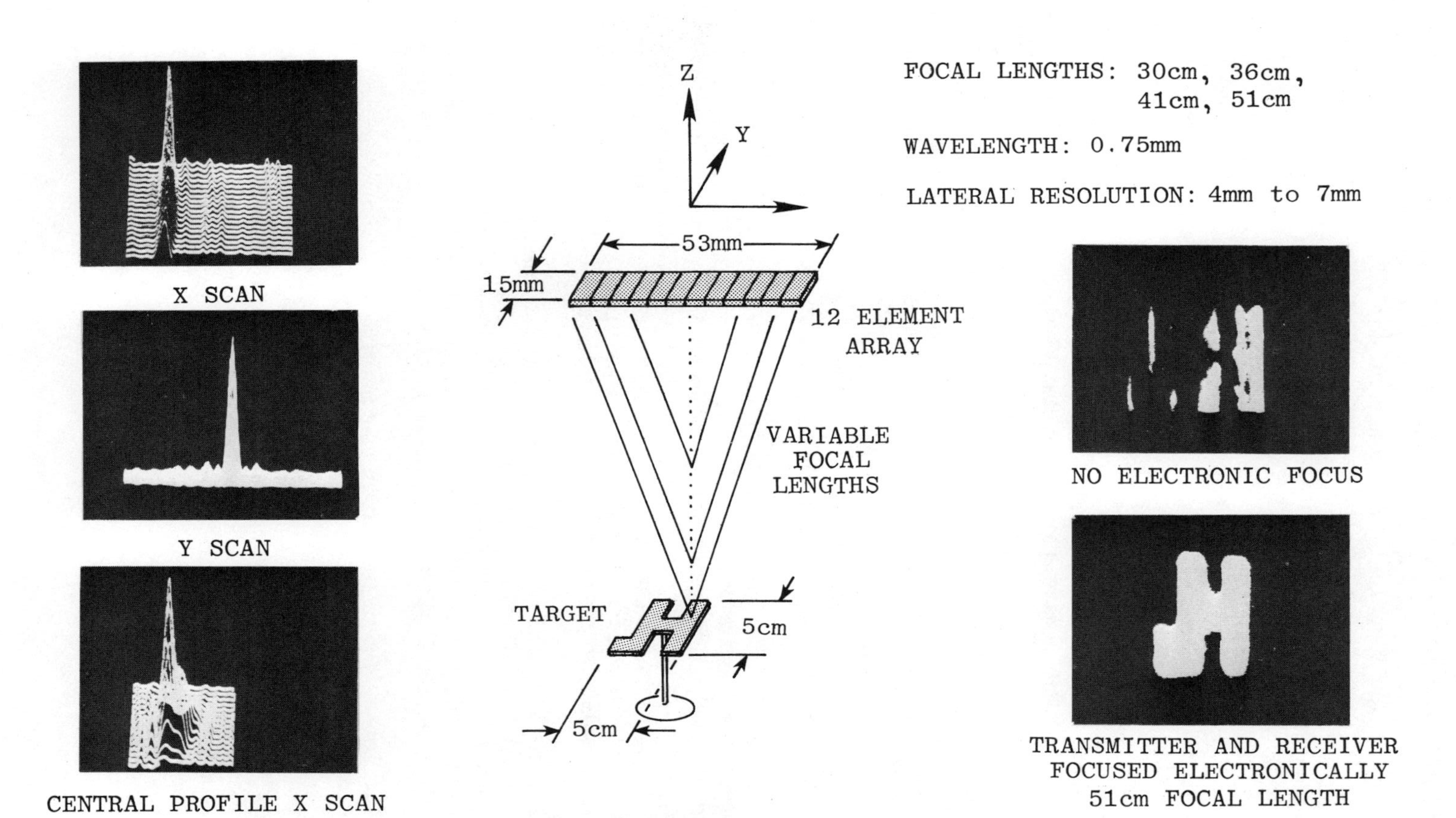

FIGURE 11. ACOUSTIC BEAM PROFILES AND HOLOGRAPHIC IMAGES OF THE TARGET "H"

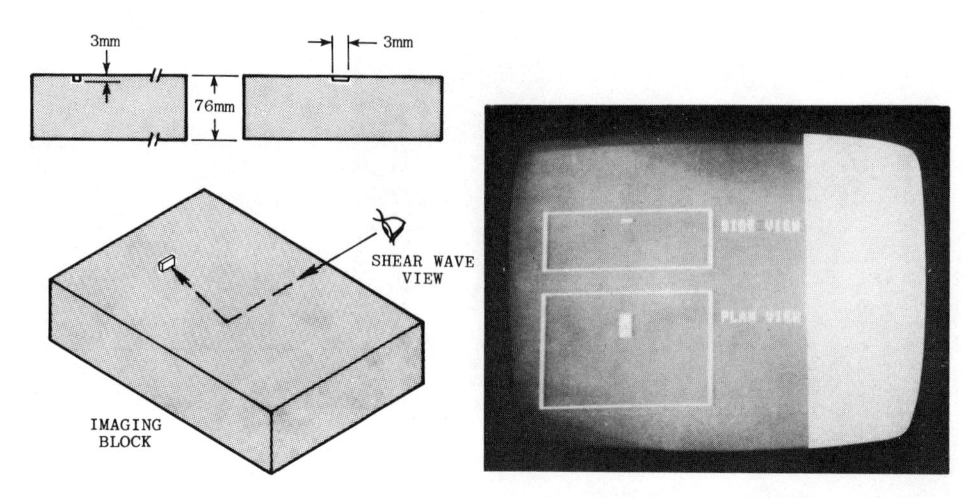

FIGURE 12. HOLOGRAPHIC SHEAR WAVE IMAGES OF A FLAW IN STEEL

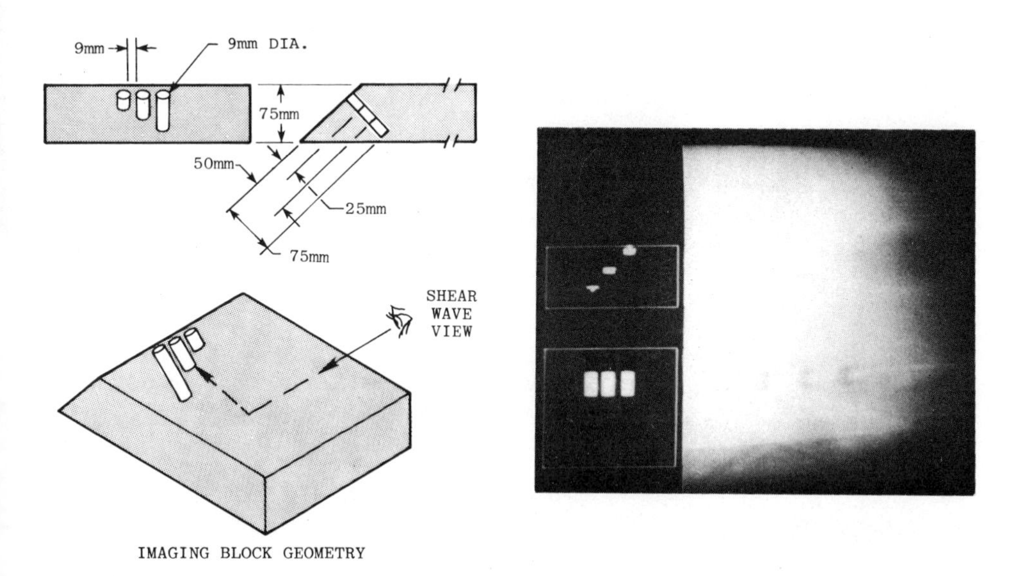

FIGURE 13. HOLOGRAPHIC SHEAR WAVE IMAGES OF VARIABLE DEPTH FLAWS IN ALUMINUM

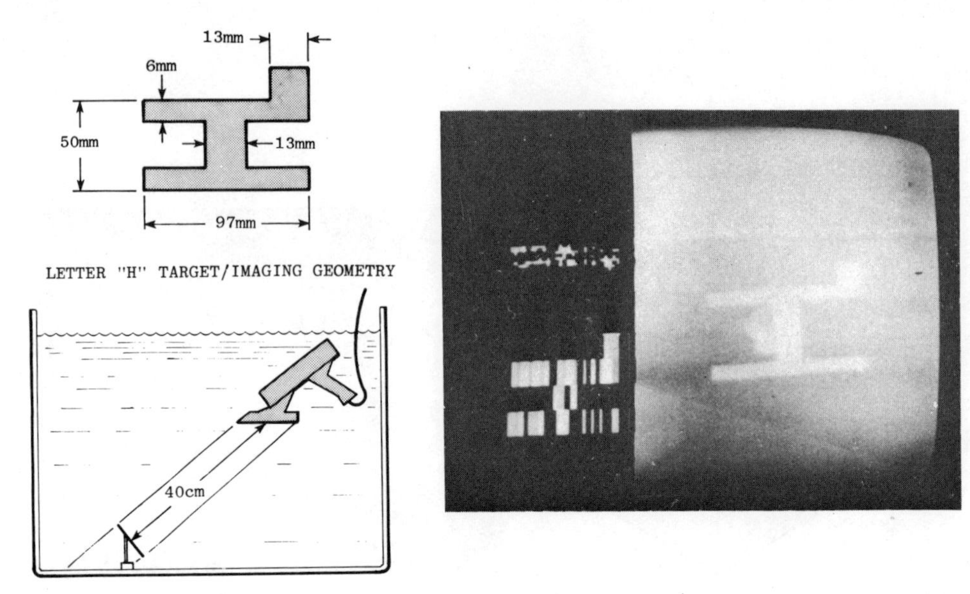

FIGURE 14. OPTICAL ACOUSTICAL IMAGES OF LETTER"H" IN WATER

REFERENCES

1. Kraus, John D., "Antennas," Chapter 4, McGraw Hill, New York (1950).

2. Camp, Leon, "Underwater Acoustics," Chapter 7, Wiley-Interscience (1970).

3. Wintermark, H., "Les Materiaux et le Soudage dans les Constructions," Offshore Conference in Portevin, 1975.

4. Havlice, J., Kino, G., Koful, J., and Quate, C.; "An Electronically Focused Acoustic Imaging Device," Acoustical Holography, Volume 5, Plenum Press, New York, New York, (1974).

5. Burkhardt, C., Grandchamp, P., Hoffman, H.; "Methods for Increasing the Lateral Resolution in B-Scan," Acoustical Holography, Volume 5, Plenum Press, New York, New York (1974).

6. Collins, H.D., "Acoustical Interferometry Using Electrically Simulated Variable Reference and Multiple Paths," Acoustical Holography, Volume 6, Plenum Press, New York, New York (1975).

7. Cook, C.E. and Bernfield, M., "Radar Signals,"; An Introduction to Theory and Application, Academic Press, New York (1967).

8. Collins, H.D. and Brenden, B.B.; "Acoustical Holographic Transverse Wave Scanning Techniques for Imaging Flaws in Thick-Walled Pressure Vessels," Acoustical Holography, Volume 5, Plenum Press, New York, New York (1975).

9. Harvey, J.F., "Theory and Design of Modern Pressure Vessels; Van Nostrand Reinhold, (1974).

10. Timoshenko, S., and Woinowsky-Krieger, S., "Theory of Plates and Shells," McGraw-Hill, Second Ed., (1959).

ULTRASONIC IMAGING USING MONOLITHIC MOSAIC TRANSDUCER UTILIZING TRAPPED ENERGY MODES*

P. Das, G.A. White, B.K. Sinha, C. Lanzl, H.F. Tiersten, and J.F. McDonald

Electrical and Systems Engineering Department
Rensselaer Polytechnic Institute
Troy, New York 12181

ABSTRACT

A transducer array utilizing trapped energy modes can be fabricated simply by plating a large number of separate electrodes on a single piezoelectric plate of uniform thickness. Several two dimensional arrays operating in the frequency range 2-5 MHz have been fabricated using PZT-7A. Experimental results showing the three dimensional radiation pattern of these transducers will be presented. The radiation pattern was determined from a computer aided reconstruction of data obtained from an acousto-optic diffraction measurement. The transducer elements are found to act independently over a higher bandwidth if electronic matching is employed. Preliminary results of ultrasonic imaging using a two dimensional mosaic transducer will also be presented.

INTRODUCTION

The use of ultrasonic imaging in the field of biomedical engineering and nondestructive testing is well established. To obtain images rapidly, the use of electronic rather than mechanical scanning is needed which in turn, requires a large mosaic transducer. A major difficulty in the fabrication of a large mosaic transducer is the achievement of adequate acoustic isolation of the small transducer elements making up the array. In order to obtain the isolation some researchers have combined completely separate

*Partially supported by NASA under Grant No. NGL33-018-003.

individual transducer elements[1] while others have used a large piezoelectric plate with grooves.[2,3] The latter procedure is somewhat less cumbersome but still difficult for very small element sizes. However, for the former case, because of the coupling through the backing plate, the interelement isolation is somewhat restricted. Some attention has been directed towards acoustic isolation schemes which do not require grooves. One of these techniques involves matched terminator backing for the plate.[4] In this way the internal plate reflections that produce coupling are reduced.

Recently, a monolithic mosaic transducer utilizing trapped energy modes was proposed.[5,6,7] The transducer consists of a large number of separated sets of electrodes plated on a single piezoelectric plate of uniform thickness. Each set of electrodes is made to act essentially independently by placing them sufficiently far apart and employing a trapped energy mode, which results from the natural mass loading and electrical shorting of the electrodes.

The energy trapping is achieved by operating in a frequency range in between the frequencies of thickness vibration for the electroded and unelectroded plate. This constraint sets the ultimate limit on the operating bandwidth of the device, which however, can be quite large. This type of transducer array has the advantages of small size, less element positioning error and smaller fabrication cost because neither plate grooving nor bonding are involved, since the electrodes can be metallized using the well known process of photolithography. The earlier work on a linear array consisting of strip electrodes and a two dimensional array consisting of circular electrodes on PZT-7A indicated that the isolation was very good over a 3 dB bandwidth of about 10%, and that the uniformity of response was excellent.[6]

A broadening of the operating bandwidth over the limit set by the material properties is possible with an inductor in the transducer circuit. This is the subject of the next section which is followed by some experimental results of three dimensional radiation patterns obtained by probing with a He-Ne laser and detecting the acousto-optically diffracted light. The three dimensional radiation pattern is obtained by scanning from different angles and making use of a tomographic reconstruction algorithm. Finally, some preliminary images obtained using the trapped energy mode transducers are presented and discussed.

BANDWIDTH LIMITATIONS OF THE TRAPPED ENERGY MODE TRANSDUCER

Achievement of good interelement acoustic isolation in mosaic array transducers often involves a bandwidth limitation on the range of frequencies over which this effect can be obtained. This is true for the trapped energy mode transducer. A monolithic mosaic transducer utilizing trapped energy modes operates between the thickness-extensional frequencies of the electroded and unelectroded portions of ferroelectric plate. Therefore, certain critical design information and insight can be obtained from a pure thickness vibration analysis of an infinitely long plate in conjunction with a qualitative knowledge of the dispersion curves for two dimensional waves in a thin plate. In particular, the one dimensional thickness vibration solution can be used to obtain the limiting bandwidth information as well as requirements on the materials constants of the transducer plate for it to be suitable for such applications. Although the limiting bandwidth information is determined from the homogeneous free vibration analysis, the transduction efficiency and actual realizable bandwidth of the trapped energy mosaic transducer are to be obtained from an analysis of the transducer under loaded conditions and for electrodes of finite dimensions. The difference between the thickness-resonance frequencies of unelectroded and electroded portions of the plate yields the bandwidth limit of the trapped energy mosaic transducer. However, information concerning isolation and the detailed radiation characteristics as a function of frequency require a complete two dimensional analysis of waves in thin plates which is beyond the scope of this work.

Here we present the one dimensional thickness-vibration solution of the electroded and unelectroded infinite plate and indicate a means of increasing the operating bandwidth by introducing an inductance in series with the transducer. This is achieved due to the lowering of the resonant frequency of the electroded ceramic plate from the instance where there is no inductance in the transducer drive circuit. In view of the foregoing discussion, a thickness solution for the admittance of a driven ferroelectric plate is derived and an inductance is incorporated into the driving circuitry so that its influence on the resonance frequency for the total system (including the voltage source, plate and inductor) can be studied. A tradeoff between bandwidth and sensitivity of a loaded thickness-extensional transducer is indicated. The optimum value of inductance could be obtained from an analysis of a loaded transducer which will be discussed in a forthcoming work.

We consider the piezoelectric plate shown in Figure 1. The plate is assumed to be composed of a hexagonal crystalline material in class C_{6v}. Polarized ferroelectric ceramics are in this symmetry class. The x_3-axis is normal to the plane of the plate and coincides with the hexagonal axis of symmetry. The relevant equations[8] for thickness extensional motions in the ferroelectric plate are

$$C_{33}U_{3,33} + e_{33}\,\phi_{,33} = \rho\ddot{u}_3, \qquad (1)$$

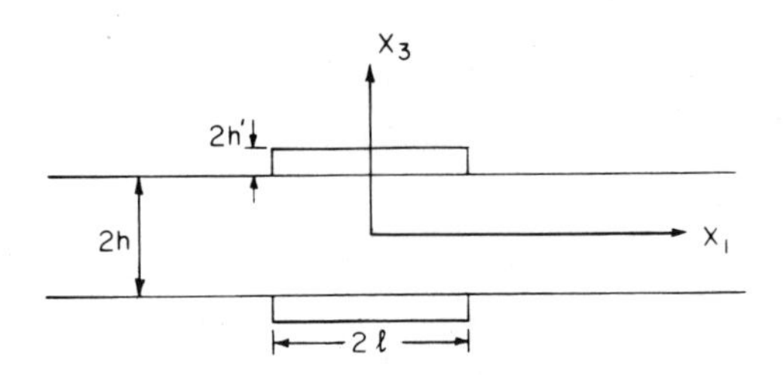

Figure 1 Basic Trapped Energy Mode Transducer Geometry

$$e_{33}U_{3,33} - \epsilon_{33}\phi_{,33} = 0,$$

where U_3, ϕ, and ρ are the x_3-axis component of mechanical displacement, the electric potential and the plate mass density respectively. Also, C_{33}, e_{33} and ϵ_{33} are the appropriate elastic, piezoelectric and dielectric constants respectively. Inasmuch as the dielectric constant of ferroelectric ceramics is much larger than that of air it is both accurate and convenient to ignore the electric field in air and set the normal component of electric displacement to zero at the free surface. Thus the boundary conditions for an unelectroded ceramic plate are

$$C_{33}U_{3,3} + e_{33}\,\phi_{,3} = 0 \text{ at } x_3 = \pm h, \tag{2}$$

$$e_{33}U_{3,3} - \epsilon_{33}\,\phi_{,3} = 0 \text{ at } x_3 = \pm h.$$

A thickness-extensional solution for an unelectroded plate which satisfies the differential equations of motion (1) and boundary conditions (2) is given by

$$U_3 = A \sin \eta x_3\, e^{iwt}, \tag{3}$$

$$\phi = \frac{e_{33}}{\epsilon_{33}} A \sin \eta x_3\, e^{iwt} + L_2,$$

where A and L_2 are arbitrary constants, and provided

$$\omega = \eta_n \sqrt{\frac{\bar{C}_{33}}{\rho}}, \quad \bar{C}_{33} = C_{33} + \frac{e_{33}^2}{\epsilon_{33}}. \tag{4}$$

and the resonant frequencies of free vibration are

$$\omega = \frac{n\pi}{2h} \sqrt{\frac{\bar{C}_{33}}{\rho}} \qquad n = 1, 3, 5, \ldots, \tag{5}$$

where 2h is the thickness of the ferroelectric plate.

On the electroded surfaces the boundary conditions are

$$C_{33}U_{3,3} + e_{33}\,\phi_{,3} = \mp 2\rho^1 h^1 U_3, \quad \phi = 0, \text{ at } x_3 = \pm h, \tag{6}$$

when the electrodes are shorted and where ρ^1 and $2h^1$ are the mass density and thickness of the metal electrodes. A thickness-extensional solution for an electroded plate which satisfies the differential equations of motion (1) and the boundary conditions (6) is given by

$$U_3 = A \sin \eta x_3 \, e^{iwt}, \tag{7}$$

$$\phi = \frac{e_{33}}{\epsilon_{33}} A[\sin \eta x_3 - \frac{x_3}{h} \sin \eta h] e^{iwt},$$

and the eigenfrequencies under the shorted conditions are obtained from the roots of the equation

$$\tan \eta_n h = \frac{\eta_n h}{(k_{33}^2 + R\eta_n^2 h^2)}, \tag{8}$$

where

$$k_{33}^2 = \frac{e_{33}^2}{\bar{C}_{33}\,\epsilon_{33}}, \qquad R = \frac{2\rho^1 h^1}{\rho h}, \tag{9}$$

in conjunction with equation (4).

If the ferroelectric plate is driven by an electric field ϕ_0/h,

$$\phi = \pm\phi_0 \, e^{iwt} \text{ at } x_3 = \pm h, \tag{10}$$

then the amplitude A in equation (7) is given by

$$A = \frac{-e_{33}\left(\frac{\phi_0}{h}\right)}{\bar{C}_{33}\eta_n\left[\cos \eta_n h - (k_{33}^2 + R\eta_n^2 h^2)\frac{\sin \eta_n h}{\eta_n h}\right]} \tag{11}$$

Although we do not provide full details of the computations involved in obtaining dispersion curves for symmetrical thickness-extensional waves in unelectroded and electroded infinite plates, such curves are shown in Figure 2. As noted before, the critical thickness frequencies in Figure 2 are given by

$$\omega_A = \frac{\pi}{2h}\sqrt{\frac{\bar{C}_{33}}{\rho}}, \tag{12}$$

$$\omega_B = \frac{\pi}{h}\sqrt{\frac{C_{55}}{\rho}}, \tag{13}$$

$$\omega_C = \eta_1 h\sqrt{\frac{C_{33}}{\rho}}, \tag{14}$$

where $\eta_1 h$ is the lowest root of equation (8). Furthermore,

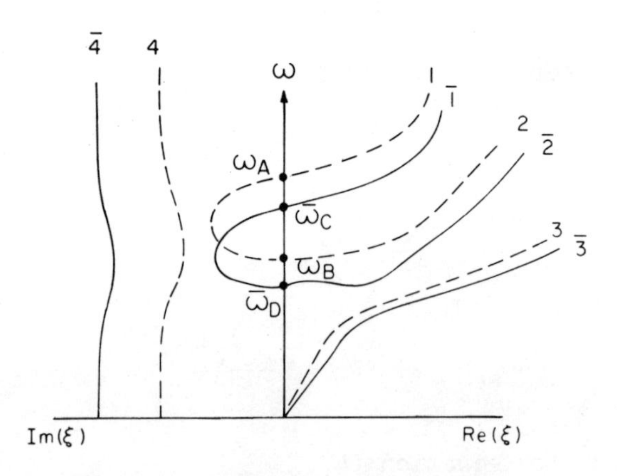

Figure 2 Dispersion Curves for Extensional Waves in an Infinite Plate. The solid curves are for fully electroded and the dotted, unelectroded plates, respectively.

$$\bar{\omega}_D = \eta_1 \sqrt{\frac{C_{55}}{\rho}}, \tag{15}$$

where $\eta_1 h$ is the lowest nontrivial root of

$$\operatorname{Tan} \eta_1 h = -R\eta_1 h, \tag{16}$$

where R is defined by equation (9). Since the mass loading parameter R is on the order of one tenth of a precent, equation (15) can be rewritten in a simpler, more convenient form as follows

$$\omega_D = \frac{\pi}{h}\sqrt{\frac{C_{55}}{\rho}}\ (1 - R). \tag{17}$$

Therefore, we have identified four critical thickness resonance frequencies in the dispersion curves for extensional modes in thin plates. The bandwidth for a trapped energy mode device is given by the difference

$$\Delta\omega = \omega_A - \bar{\omega}_C. \tag{18}$$

It is also clear that the condition for the existence of trapped energy extensional modes is that

$$\bar{C}_{33} > 4C_{55}. \tag{19}$$

For low modes of large coupling materials, the mass loading term $R\eta_1^2 h^2$ in equation (8) is much smaller than the piezoelectric coupling term k_{33}^2, which we will ignore in obtaining the numerical results presented in this section.

In order to develop the circuit terminal behavior for the electroded infinite plate, we calculate its admittance. A transducer element along with an inductor and a generator is shown in Figure 3. The current through the transducer element is given by

$$I = \int_\alpha \dot{D}_3(x_3 = h)d\alpha, \tag{20}$$

where α is the area of the electrode. Substituting from equations (7) and (11) into (20), the total current I through the ceramic plate takes the form

$$I = j\omega\epsilon_{33}\frac{V}{2h}\left[1 + k_{33}^2 \frac{\sin \eta h}{(\eta h \cos \eta h - k_{33}^2 \sin \eta h)}\right], \tag{21}$$

yielding the admittance for the transducer given by

$$Y = j\omega\epsilon_{33} \frac{\alpha}{2h} [1 + k_{33}^2 \frac{\sin \eta h}{(\eta h \cos \eta h - k_{33}^2 \sin \eta h)}]. \tag{22}$$

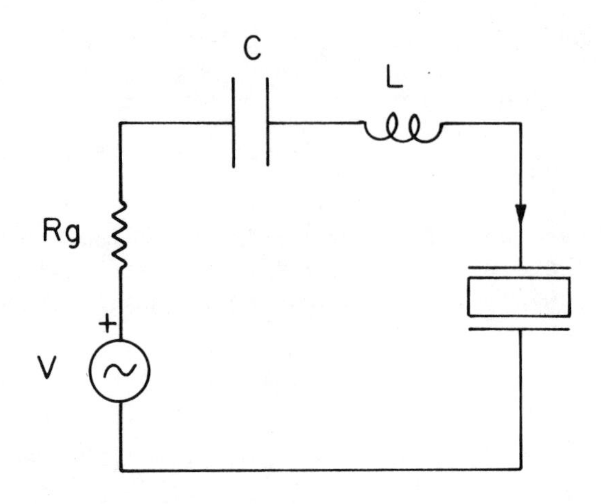

Figure 3 Schematic Diagram of the Driving Circuit

Incorporating the transducer element in the driving circuit of Figure 3 where V_g is the driving voltage, R_g the generator resistance, and L and C are inductance and capacitance in the tuning circuit. Application of Kirchhoff's voltage equation to the circuit yields

$$-\dot{V}_g + R_g\dot{I} + L\ddot{I} + \frac{I}{C} + \frac{\dot{I}}{Y} = 0, \tag{23}$$

where the dot over a quantity represents its time derivative. Substituting for the admittance Y of the transducer element from equation (22) into (23), the current I through the circuit is given by

$$I = \frac{V_g}{R_g + j[\omega L - \frac{1}{\omega c} - \frac{2h(\eta h \cos \eta h - k_{33}^2 \sin \eta h)}{\omega\epsilon_{33}\, \alpha\, \eta h \cos \eta h}} \tag{24}$$

At resonance the reactive part of the impedance vanishes and the resonant frequency is given by

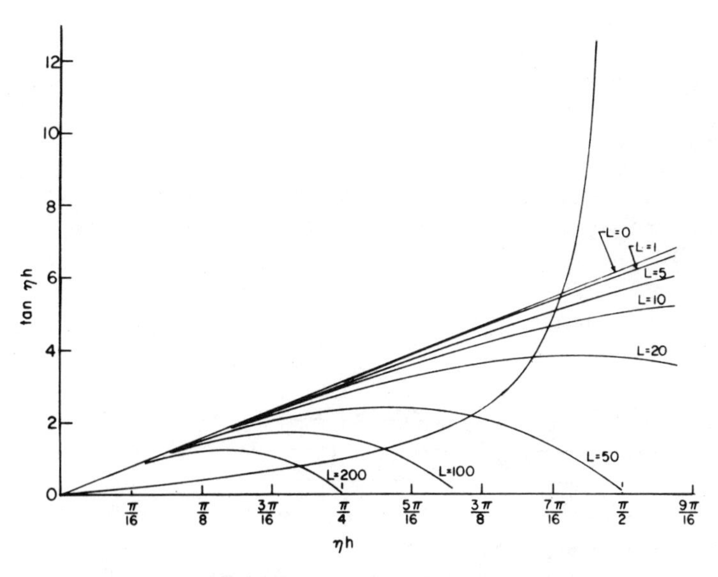

Figure 4 Graphical Procedure for Finding the Lowest Root of Equation (25) for Different Series Inductances L in μH. The intersection of the straight line L = 0, with the tangent function corresponds to a fundamental frequency of 3.54 MHz for a shorted plate of 0.6 mm thickness.

$$\tan \eta h = \frac{\eta h}{k_{33}^2} - L \frac{\alpha \epsilon_{33} \bar{C}_{33}}{2h^3 \rho k_{33}^2} \eta^3 h^3 + \frac{\epsilon_{33}}{2h k_{33}^2} \frac{\alpha \eta h}{C} \qquad (25)$$

where we have substituted for ω from equation (4), equation (25) clearly reveals that if there is no capacitance in the tuning circuit, introduction of an inductance serves to lower the resonant frequency of an electroded ceramic plate. Figure 4 shows the influence of inductance on the resonant frequency of an electroded plate of thickness 0.6 mm, and electrode area 3 sq. mm. Some preliminary measurements on large electrode transducers indicate the lowering of the resonant frequency with series inductances at the expense of sensitivity which falls rapidly with increasing inductance.

SPATIAL RADIATION PATTERN OF THE MOSAIC ARRAY

The three dimensional radiation pattern of the two dimensional mosaic array has been measured in order to verify the isolation between the individual radiating elements. Radiation strength is determined by probing the acoustic waves with a laser beam and measuring the resultant interaction. Specifically, the acoustic wave produces a periodic modulation of the refractive index of the surrounding water, resembling a diffraction grating. The strength of the diffracted laser beam is linearly proportional to acoustic power provided that the acoustic power and the acousto-optic interaction length are held within an upper bound.[9] If these conditions are met, the intensity of the zeroth order diffracted beam gives a direct measure of the integral of acoustic power along the line of the laser beam.

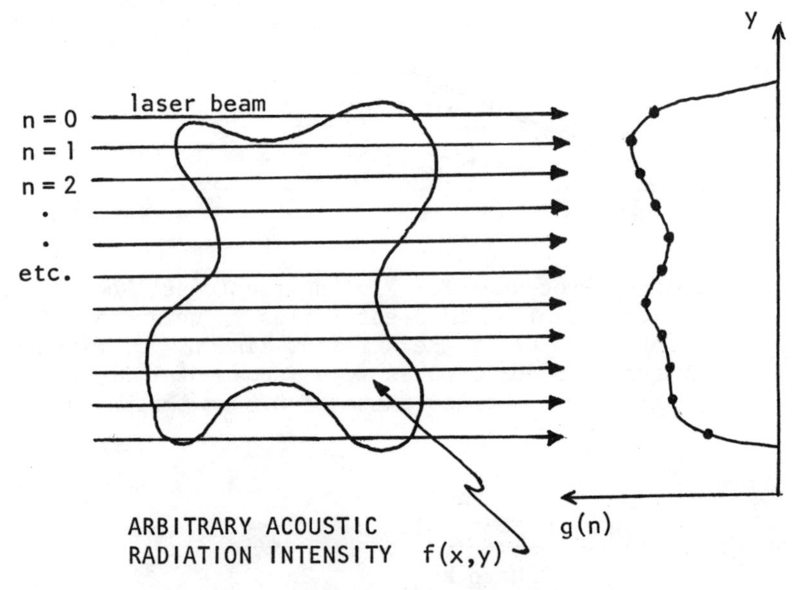

Figure 5 Laser Beam Scanning of an Arbitrary Plane of Radiation

Since the intensity of the diffracted laser beam only gives a measure of a line integral of radiation strength, information about the radiation strength at a particular point cannot be obtained by a single diffraction measurement. The problem of determining the radiation strength in a plane perpendicular to the direction of propagation of the acoustic waves requires diffraction measurements with the laser beam at numerous angles and positions over the plane of radiation. As shown in Figure 5 many parallel line integrals must be made with the laser beam across the plane of radiation under examination; where $f(x,y)$ corresponds to the radiation strength in the plane. Subsequently, the scanning process should be repeated at many different angles $t\Theta_0$ (for $t = 1$ to M and $\Theta_0 = \pi/M$) from the horizontal, and at many distances 'z' from the origin of the radiation. Each scan angle produces a vector $g(n)$ whose values should sum to a constant equal to the total radiation power passing through the scan plane. All vectors $g(n)$, for each angle of scan, taken together form a two dimensional array $g(n;\ t\theta_0)$. In order to compute $f(x,y)$, a two dimensional reconstruction technique using convolution was employed.[10] The computation results are similar to that produced in any form of computer-aided tomography.

First, the array $g(n;\ t\theta_0)$ was transformed to another array $g'(n;\ t\theta_0)$ via a one dimensional filter which operates on all values of n for a particular value of t. This transformation is given by

$$g'(n;\ t\theta_0) = \frac{g(n;\ t\theta_0)}{4} - \frac{1}{\pi^2} \sum_{p\ \mathrm{odd}} \frac{g[(n+p);\ t\theta_0]}{p^2} \tag{26}$$

The transformed array was then used to reconstruct $f(x,y)$ from the following equation

$$f(x,y) = \sum_{t=1}^{M} g'[(x \cos t\theta_0 + y \sin t\theta_0);\ t\theta_0] \tag{27}$$

Interpolation between values of $g'(n)$ may be required for evaluation of equation (27).

Experimentally, the scanning process was performed by holding the laser beam fixed and displacing the transducer under examination in the vertical direction. A rotation of the two dimensional array on an axis parallel to the direction of propagation of the acoustic waves provides the numerous angles of scan necessary. A block diagram of the data acquisition system is shown in Figure 6. The scanning process and acquisition of data is completely automated through use of a Z/80 microcomputer, two high resolution stepping motors and an analog to digital converter. The drop in intensity of the zeroth order diffracted beam is amplified, filtered, peak detected and level shifted before sampling by the computer. Sampled data was stored on a floppy disk for later transmission via phone lines to a Varian 620/i minicomputer where the reconstruction computation was performed. The results were displayed and photographed on a high resolution grey scale monitor. A photograph of the experimental facility is shown in Figure 7.

The two dimensional data array $g(n;\ t\theta_0)$ scanned from a 16 element transducer with only 3 elements excited is shown in Figure 8a where horizontal raster lines represent $g(n)$ for a particular $t\theta_0$. The intensity at each point in the array shown corresponds to the value of a particular element. Figure 8b illustrates the application of equation (26) to Figure 8a. Equation (27) would then be applied to Figure 8b, in order to obtain the two dimensional

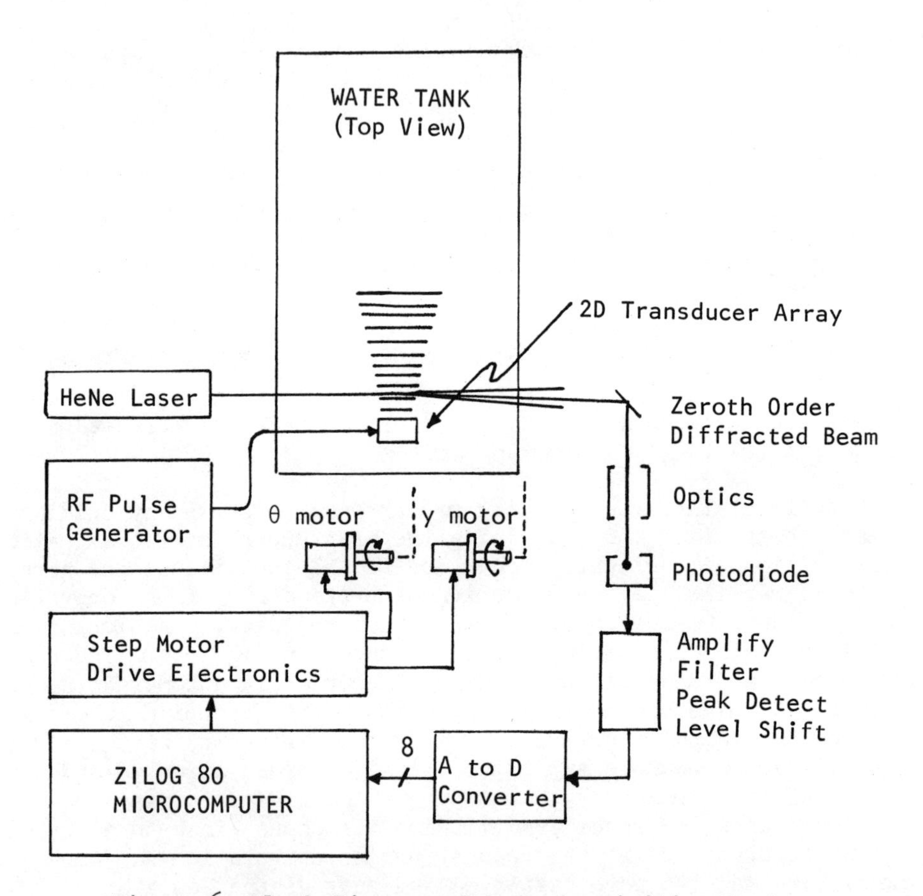

Figure 6 Block Diagram of Data Acquisition System

Figure 7 Experimental Facility Showing Water Tank, He-Ne Laser and Data Acquisition Apparatus

radiation pattern f(x,y). This procedure can be repeated at many different distances 'z' from the transducer face in order to obtain a three dimensional radiation pattern. Photographs of several mosaic transducers with different element geometries and their associated radiation patterns are shown in Figure 9, indicating excellent interelement isolation for the trapped energy mode transducer.

MECHANICALLY SCANNED IMAGES WITH A FOCUSED MOSAIC ARRAY

In order to determine the suitability of the mosaic array for imaging, a lens was fabricated from plexiglass and suspended in the water just in front of the mosaic transducer. An object was then moved mechanically past the focal point of the beam and a receiver measured the transmitted radiation.

Figure 10 is a photograph of the X-Y positioning system which moves the object under examination by means of two stepping motors. Figure 11 shows the results obtained with a piece of machined aluminum. A photograph of the actual objects, the acoustic shadowgraph and a computer edge detection of the shadowgraph are provided.

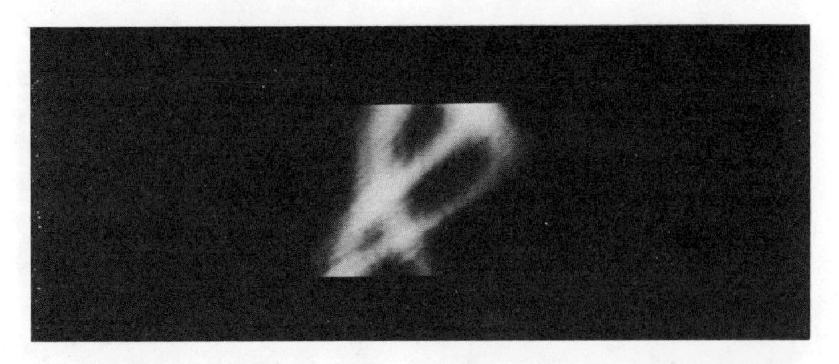

Figure 8a Laser Scan Data; $g(n;\ t\theta_0)$ for a 16 Element Array with only 3 Elements Excited

Figure 8b Filtered Scan Data; $g'(n;\ t\theta_0)$ from the Data on Figure 8

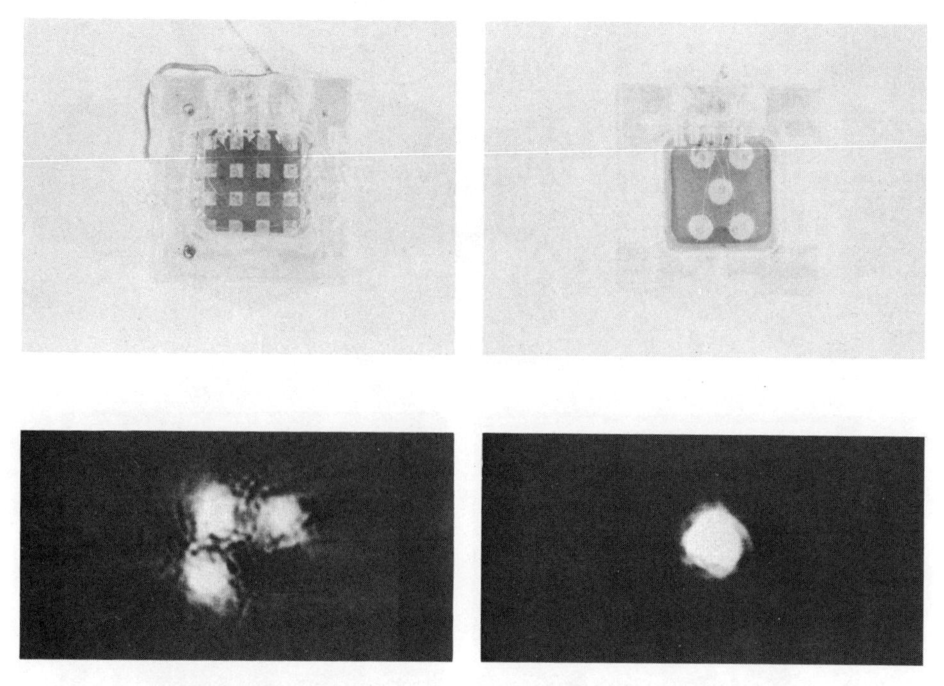

Figure 9 Mosaic Transducers and Associated Two Dimensional Radiation Patterns

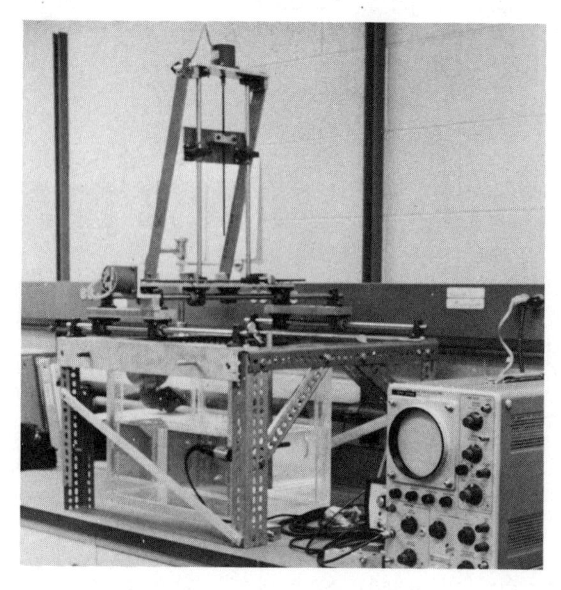

Figure 10 X-Y Positioning System Apparatus

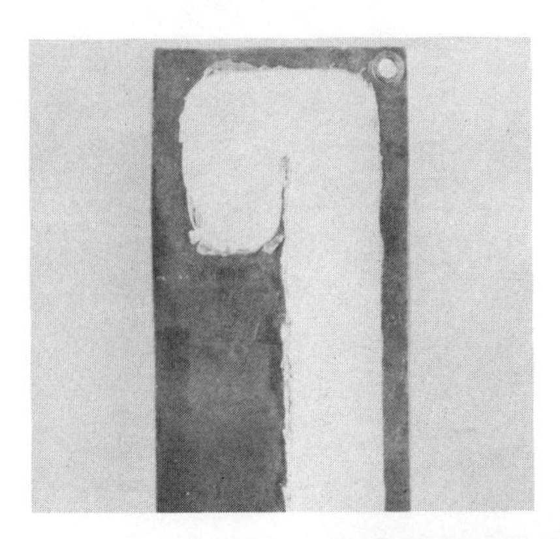

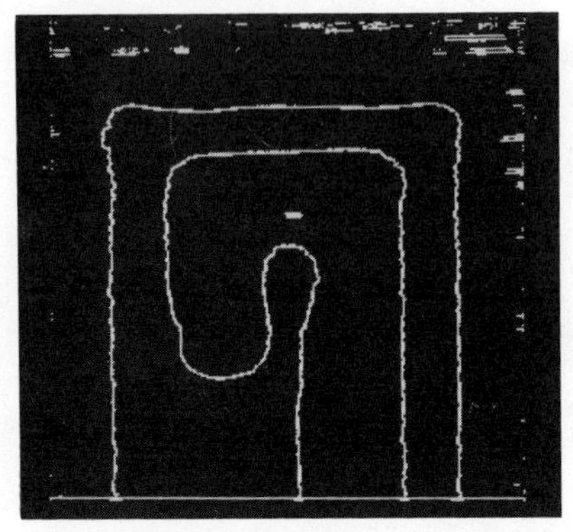

Figure 11 Shadowgraph Data Obtained from a Piece of Machined Aluminum

DISCUSSION

It has been shown that it may be possible to increase the operating bandwidth of the trapped energy mode transducer by placing an inductor in series with the transducer elements. A three dimensional radiation pattern of the monolithic mosaic transducer has verified the isolation between adjacent radiating elements. Future work will be directed towards electronic steering of the two dimensional array.

ACKNOWLEDGEMENT

Special thanks are due to F. M. Mohammed Ayub, Robert Fries and Dominic Scordo for assistance with the experimental apparatus, data reduction and filtering.

REFERENCES

1. V. G. Prokhorov, "Piezoelectric Matrices for the Reception of Acoustic Images and Holograms:, Sov. Phys. Acoust., 18, 408 (1973).

2. M. G. Maginess, J. D. Plummer and J. D. Meindl, "An Acoustic Image Sensor using a Transmit-Receive Array", in Acoustical Holography, R. G. Green, editor (Plenum, New York, 1973), vol. 5, p. 619.

3. N. Takagi, T. Kawaghima, T. Ogura and T. Yamada, "Solid State Acoustic Image Sensor", in Acoustical Holography, G. Wade, editor, (Plenum, New York, 1972), vol. 4, p. 215.

4. B. A. Auld, C. DeSilets and G. S. Kino, "A New Acoustic Array for Acoustic Imaging:, 1974 Ultrasonics Symposium Proceedings, IEEE Cat. No. CH0894-1SU, Institute of Electrical and Electronics Engineers, New York, 1974, p. 24.

5. H. F. Tiersten, J. F. McDonald and P. K. Das, "Monolithic Mosaic Transducer Utilizing Trapped Energy Modes", Appl. Phys. Lett., 29, 761 (1976).

6. H. F. Tiersten, J. F. McDonald, M. F. Tse and P. K. Das, "Monolithic Mosaic Transducer Utilizing Trapped Energy Modes in Piezoelectric Plates", in Acoustical Holography, L. W. Kessler, editor, (Plenum, New York, 1977), vol. 7.

7. H. F. Tiersten, J. F. McDonald and P. Das, "Two Dimensional Monolithic Mosaic Array:, Ultrasonics Symposium Proceedings, IEEE Cat. No. 77CH1264-1SU, Institute of Electrical and Electronics Engineers, New York, 1977, pp. 408-412.

8. H. F. Tiersten, Linear Piezoelectric Plate Vibrations, Plenum Press, New York, Chapter 11, 1969.

9. R. Adler, "Interaction Between Light and Sound", IEEE Spectrum, May 1967, pp. 42-54.

10. G. N. Ramachandran and A. V. Lakshminaryanan, "Three Dimensional Reconstruction from Radiographs and Electron Micrographs: Applications of Convolutions instead of Fourier Transforms", Proc. Nat. Acad. of Science, vol. 68, no. 9, Sept. 1971, pp. 2236-2240.

PROGRESS IN FRESNEL IMAGING;

CLINICAL EVALUATION

P. ALAIS, Laboratoire de Mécanique Physique, Université Pierre et Marie Curie, PARIS, France
B. RICHARD, Service de Biophysique, C.H.U. Cochin, PARIS, France.

INTRODUCTION

This is a progress report on the Fresnel imaging research that we presented at the 7 th symposium 1,2,3 . At this time we had developed both B and C echographic experiments. During these last eighteen months we have built a few machines duplicated from our real time B echographic array and adapted to a permanent clinical use so that we have a better knowledge of the clinical potentiality of the Fresnel focusing technique. These 120 mm arrays have 160 elements and are acted at a recurrent frequency of 4 kHz to furnish a 160 line image in 40 ms. We shall develop here both experimental results obtained in the laboratory and clinical results, trying to correlate these results and quantify the medical requirements of B echographic real time imaging.

These results led us recently to develop a 2nd generation device with new refinements. This device exhibits new possibilities ;

- a depth of field of either 15 or 30 cms instead of 15 only

- 4 zones of focusing depth

- an electronic frame memory permiting to freeze the image and also to operate an efficient tracking focusing technique.

PHYSICAL CONSIDERATIONS ABOUT THE FRESNEL FOCUSING TECHNIQUE

The most important feature of the Fresnel focusing technique is certainly its inherent simplicity. Without entering theoretical details which are developed in the same proceedings[4] , we must recall with the figure 1 that the focusing operation may be reduced, either at the emission or at the reception, to a simple 3 level amplitude modulation of the signal which may be easily carried out by an electronic inverter associated to each element of the array. In addition, Fresnel lenses exhibit, with the wide band signals involved in B echography, focusing properties completely different of what is obtained with classical lenses. Instead of a delay correction, they impose the same phase correction to each spectral component of the signal, which involves focusing at a depth proportional to the involved frequency. A short signal is then focused according a focal caustic rather than a focal spot and realizes the thin acoustical beam insuring a good lateral resolution in a depth of field varying like the band width authorized by the transducers and the electronic circuitry.

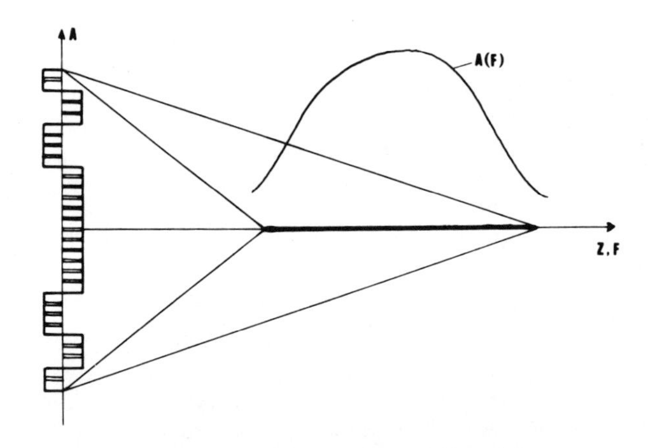

Fig.1 - The Fresnel lens focusing a wide band signal

To check accurately this focusing technique in terms of resolution and secondary lobes we use a very simple technique. We may test a Fresnel profile at any location of the array, the electronic scanning being stopped as for an A echographic investigation. The array is then mechanically displaced looking at a single nylon wire set in a good anechoic aquarium, free from parasitic reflections. The mechanical displacement is then recorded through the X displacement of the spot of an oscilloscope while the Y signal is the detected ultrasonic

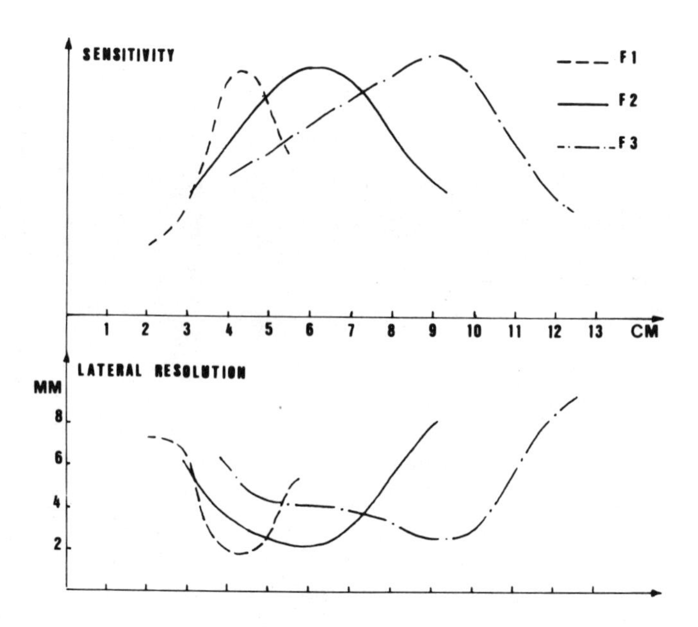

Fig.2 - Axial sensitivity and lateral resolution of the 1st generation imaging device.

response. The time sampling of the signal is then obtained by acting the brightness of the spot. The figure 3 shows examples of results obtained in a few seconds through that way. In these conditions, we are able to check the properties of different Fresnel profiles associated with different types of signals. We had mentioned at the previous symposium that the F.E.R. technique (i.e : focusing both at emission and at reception with the same large aperture Fresnel profile) presents the highest lateral resolution but very unhappily also high level secondary lobes which afford severe artefacts in some clinical situations. A very simple apodizing technique consists in emitting only through the Fresnel central zone and focusing with the entire profile at reception only, what we call the F.R. technique. The figure 2 shows the typical results in terms of lateral resolution and axial sensitivity obtained from a thin nylon wire with the 3 different Fresnel profiles available in the machines used for clinical tests. It may be noted that the F.R. technique does not degrade the lateral resolution too much at the nominal focusing depth and ensures much larger depths of field. But the most important effect of this technique is to reduce the side lobe level in a drastic way as is shown in the Figure 3 which permits to compare the results obtained at the nominal depth for the Fresnel profile F2 with the FER technique and with the FR technique (fig.3). The other results show how the central peak and side lobes vary when reducing the aperture of the same Fresnel profile to 4 and 2 zones and also when using the central zone only i.e a non focusing technique.

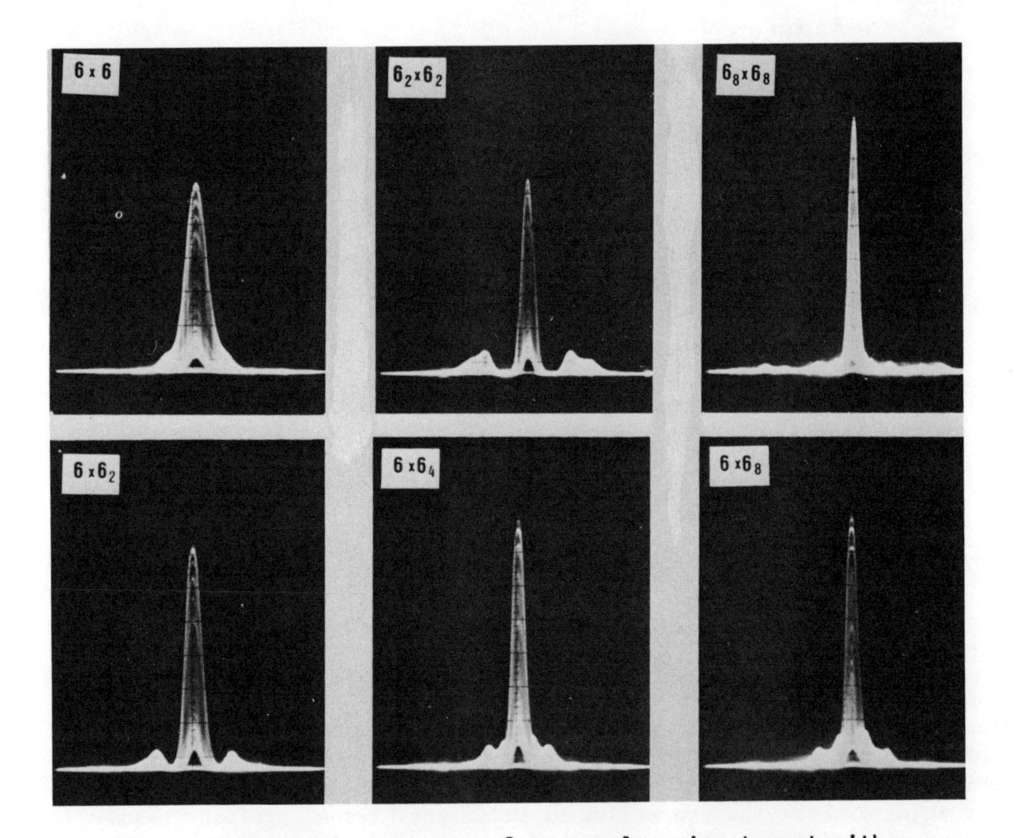

Fig.3 – Echographic responses from a nylon wire target with a Fresnel profile with 6 transducers in the half central zone.
6p x 6q means : emitting with p Fresnel zones and receiving with q zones. The FER technique = $6_8 \times 6_8$ and the FR technique = $6 \times 6_8$. The classical non focusing technique = 6 x 6.

As mentioned in [4] it is possible also to reduce the side lobes by using a Fresnel phase sempling with more than 2 levels. In that case the parasitic waves involved in the creation of side lobes and specially the diverging wave, symmetrical of the focused wave, may be reduced or even disappear. For that reason we built a 4 phase-machine where each transducer is associated to an analog multiplexer instead of a simple inverter. But the results in terms of side lobes were not more convincing that what we already obtained with the simple FR technique and, at least up to now, we have not gone further in that direction.

For insuring a better visualization and tape recording possibility we have built for our clinical machines a video converter permitting the time compression of each 250 μs ultrasonic line of information in videolines of 62,5 μs duration. These video lines are duplicated in the display to give a frame of 320 lines from 160 ultrasonic lines. The whole frame is obtained like in TV from the interlacing of an odd frame and an even frame, each one being obtained in 20 ms. These values are compatible with monitors and tape recorders adapted to the 625 lines european TV standard. Each videoline is sampled in 512 words of 4 bits which permits to obtain pictures very competitive with the oscilloscope one, as it may be seen in the following section.

CLINICAL EVALUATION

The successive prototypes of real time B mode Fresnel focusing devices have been tested in routine clinical work during these past two years.

We first checked the efficiency of this type of large aperture focusing in vivo with the FER technique. The first results confirmed that the lateral resolution obtained in vivo was very similar to the theoretical one as demonstrated by the pictures of fetal spines for example. The pictures looked quite different from other ultrasonic images as they are made of elementary points with similar axial and lateral dimensions instead of the small segments usually seen when axial resolution is far better than lateral resolution.

Unfortunately, despite its excellent resolution, this FER technique was not sufficient to lead to a versatile device : in some difficult cases with very strong reflections, sidelobes may give unacceptable artefacts. Then, the absence of these defects becomes more important than resolution so that pictures could be better without focusing (essentially for heart imaging). With the use of the FR technique, this difficulty was overcome. The little loss in resolution power is compensated by the complete suppression of sidelobes. Images are less sharp than with the FER technique, but the focusing effect remains quite obvious and the same FR focusing may be used in all kind of clinical work. Moreover, with the use of a small aperture in transmission, the depth of field is increased so that, for an exploration depth of 15 cm, the improvement in resolution is apparent at least from 4 to 10 cms in depth with only one Fresnel profile. As a consequence we are now using always this FR technique which looks superior to the other modes in nearly all cases.

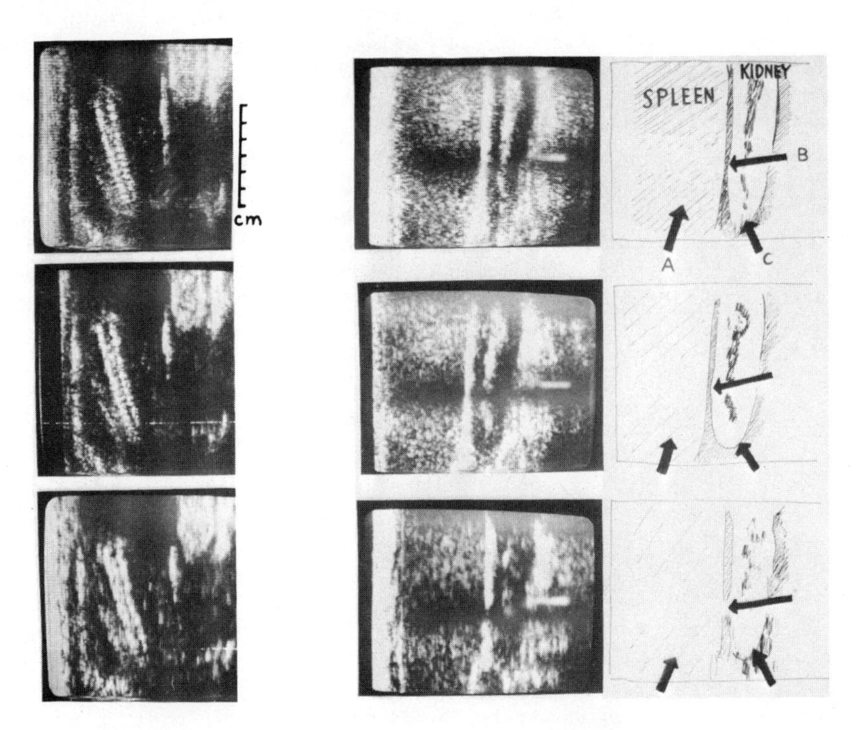

Fig.4 : Fetal Spine

Fig.5

Fig.4 - The top picture corresponds to the FER technique and displays the best resolution. When focusing only at reception, vertebrae are not as well separated but still distinguishable. The last one is non focused.
With little pratice, such freeze frame pictures are easily recognized in the FER or FR case whereas the NF one is meaningful only when observed in motion.

Fig.5 - On the FER picture following points can be noted :
- very rich tissue information coming from parenchyma of the spleen (A)
- limited effect of acoustic shadow of the rib (B)
- sidelobes artifacts at the lower pole of the kidney coming from strong reflections on bowel gas (C)

on the second picture (FR) the contour of the lower pole is much better seen and there is only little loss in tissue detail; acoustic rib shadow is not too strong.
The non focused picture has the typical fuzzy aspect with rough tissue information, suppression of echoes behind the rib, bad outline of the lower pole.

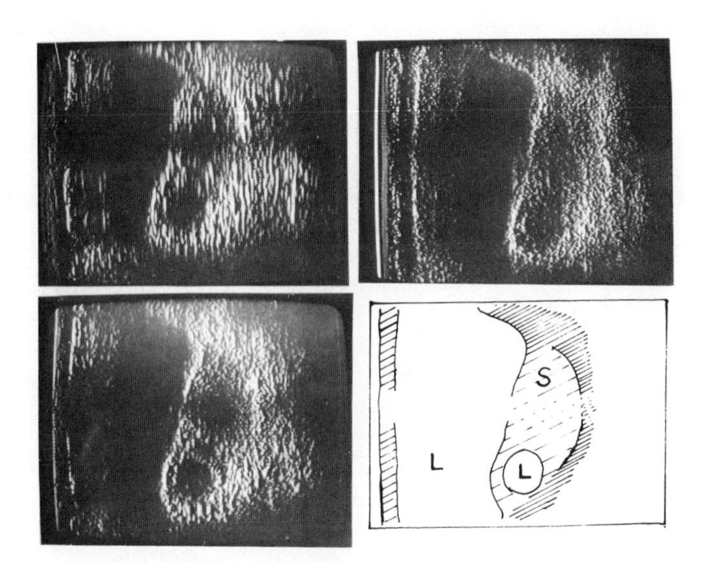

Fig.6 - BLADDER, UTERUS AND LIQUID PARAUTERINE MASS.
A comparison between the different focusing modes shows how the FR technique is the only one which permits to display at the same time a very fine contour of structures, in particular for the small liquid zone beside the uterus (lower part of the picture) without any side-lobe artefact inside this liquid zone.

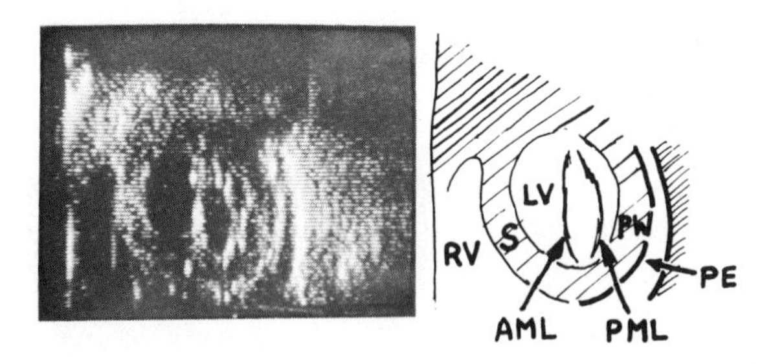

Fig.7 - CARDIAC IMAGING : ventricular cross section
RV : right ventricle, S interventricular septum
AML : anterior mitral leaflet,
PML : posterior mitral leaflet
LV : left ventricle
PW : posterior wall
PE : pericardial effusion

Another important advantage of high resolution in real time machines is the possibility to recognize the structures even on frozen images. As the explored region is necessarily limited, a fine contour of organs is most helpful for recognition on still frame pictures without the surrounding information obtained in manual B mode cross sections. Besides we checked that a high line density such as our 160 line frame is much higher in quality than the 80 line frame that we get when supressing the interlacement. That implies the obvious necessity of higher line densities with higher resolutions and in the same time the fact that the frame lines are almost indistinguishable (fig.8)

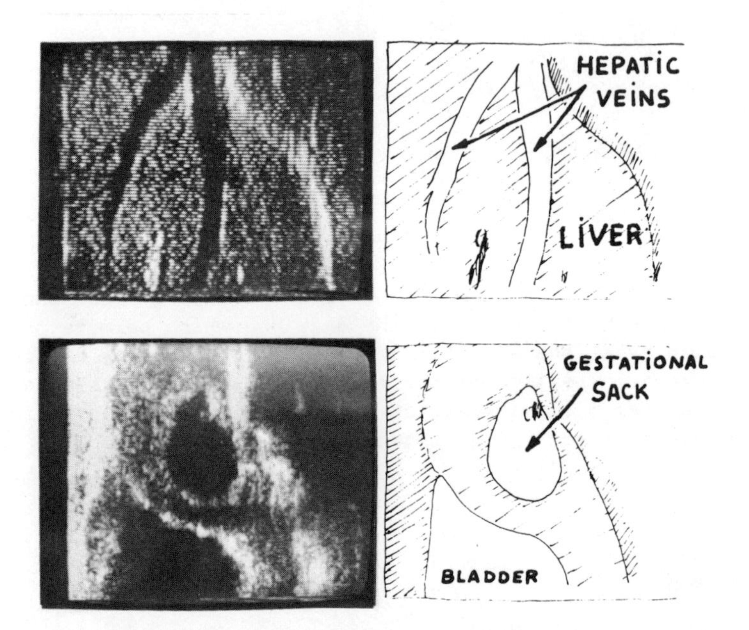

Fig. 8 - This picture shows the difference between the evidence of lineage obtained with only 80 lines (top picture) and the " continuous " aspect of the picture when using 160 lines (below).

To conclude these remarks upon Fresnel focusing in real time imaging, it can be said that, in addition to the interest of real time, the improvement in resolution does not afford any complication to the ultrasonographer; the quality of pictures shortens the learning period, permits faster and more accurate diagnosis. Taking in account the quality of still frame pictures, we don't find very often any more advantage in using our classical manual B scan equipment.

NEW DEVELOPMENTS

Recently we conceived a new device well adapted to a tracking focusing technique. We had shown in 1976 the possibility when using the FR technique to commute sequentially the 3 Fresnel profiles available in our first generation device at proper times so that the highest lateral resolution and sensitivity could be preserved in the whole depth of field. But in the absence of preamplifiers associated to each transducer we must commute very low level signals and must accept to lose some information during a minimum time of a few microseconds after the commutation. For this reason we have developped a new device with 4 zones of focusing depth as it is shown in the figure 9.

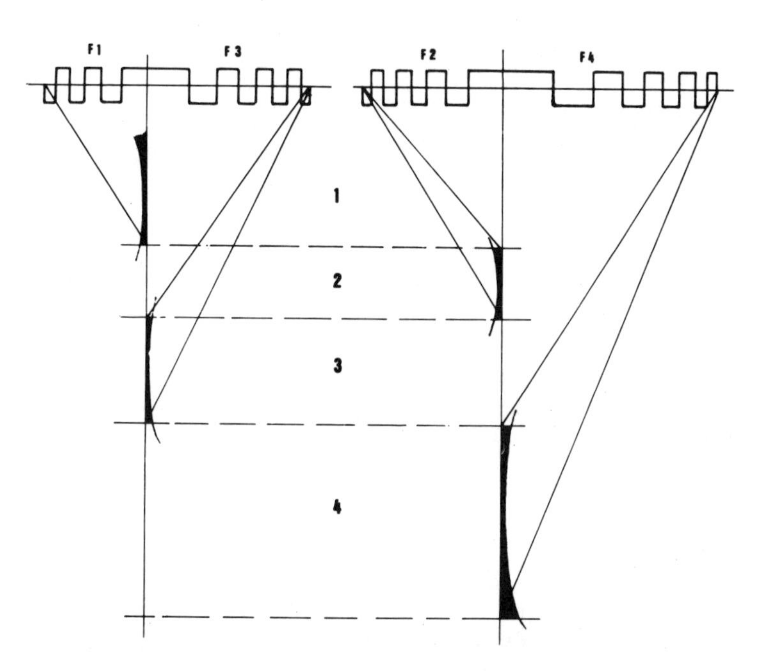

Fig.9 - Tracking focusing with 4 Fresnel profiles.

This device may be acted with any one of the 4 focusing depths and delivers in this case 25 frames per second. But it may be used also with a tracking focusing technique where the commutation is either effected from the zone 1 to the zone 3 or from the zone 2 to the zone 4. The image is viewed on a video monitor through the electronic frame memory and only information delivered in the focusing depth corresponding to a good lateral resolution as indicated on the figure 10 is registered in this memory. This technique gives of course a refreshment of the whole frame in 80 ms instead of 40 ms but the

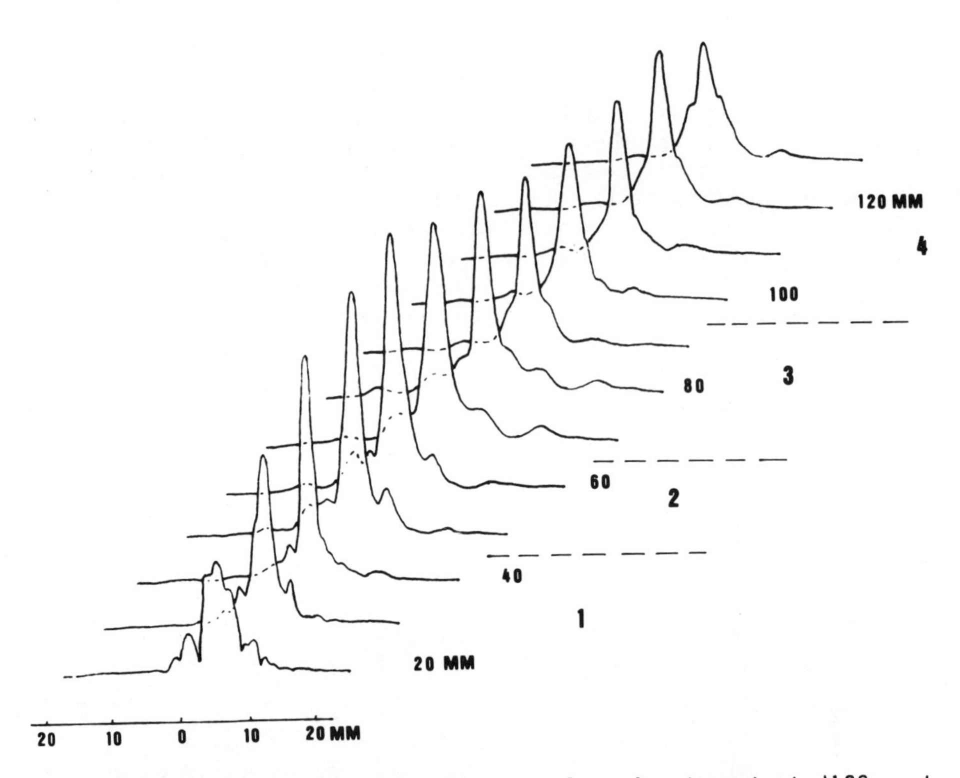

Fig.10 - Echographic responses from a nylon wire target at different depths with the tracking focusing technique

use of the electronic memory avoids flickering effects and commutation bars. The figure 11 permits to compare images obtained from a multiple nylon wire target with each one of the Fresnel profiles and with this tracking focusing technique.

Moreover the same machine permits to use the array alternatively with a 15 cm depth of field or a 30 cm one. In this last case we use a 2 kHz recurrent frequency for the ultrasonic shots and obtain on the oscilloscope a 80 line image of 12 x 30 cms^2. The 4 Fresnel profiles used for this higher depth are of course different and adapted to higher focusing depth. The fig.12 shows an image of the nylon wire target obtained with the zone 4'. This machine is partially achieved and fragmentary clinical tests have been obtained only. The tracking focusing technique seems to be very efficient and the lower refresment rate of the image is not at all annoying for the physician at least in obtetrics and abdominal examinations.

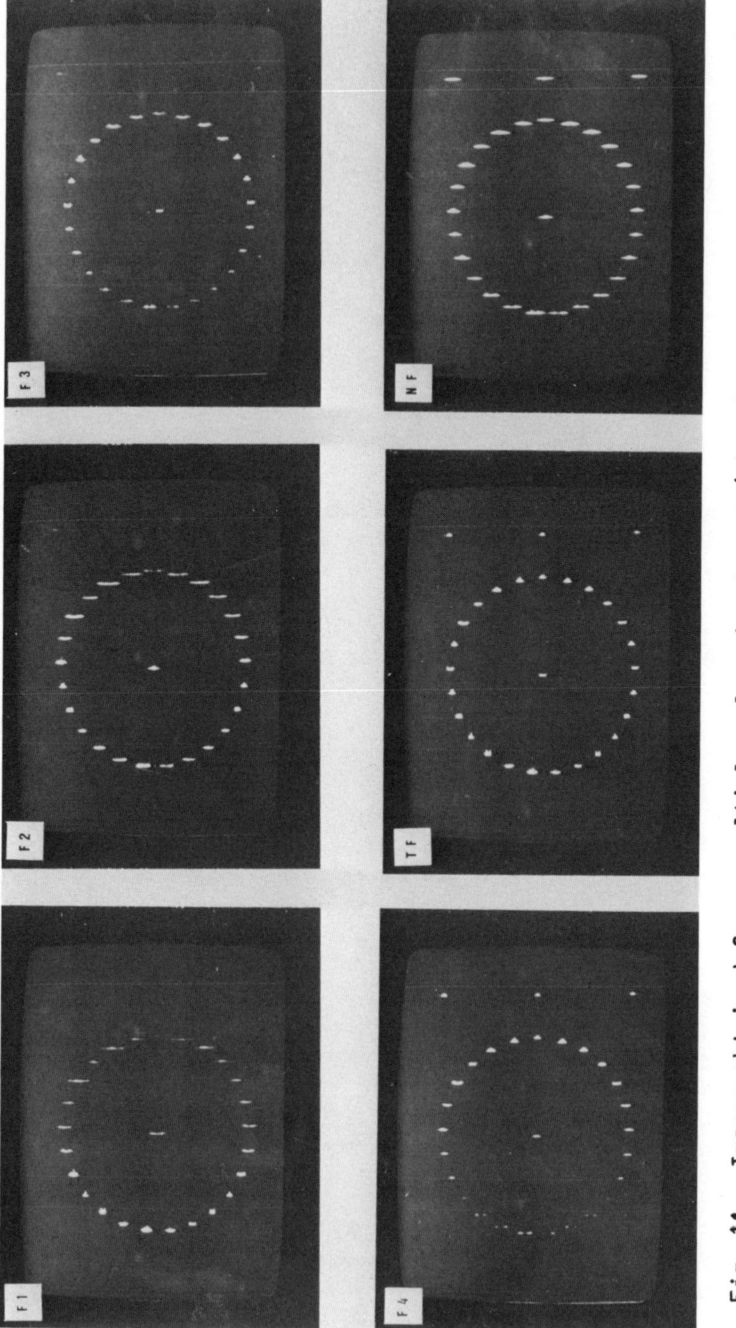

Fig. 11 - Images obtained from a multiple nylon wire target (the circle is 80 mms in diameter) with the different Fresnel profiles (F_i), the tracking focusing technique (TF), the non focusing technique (NF).

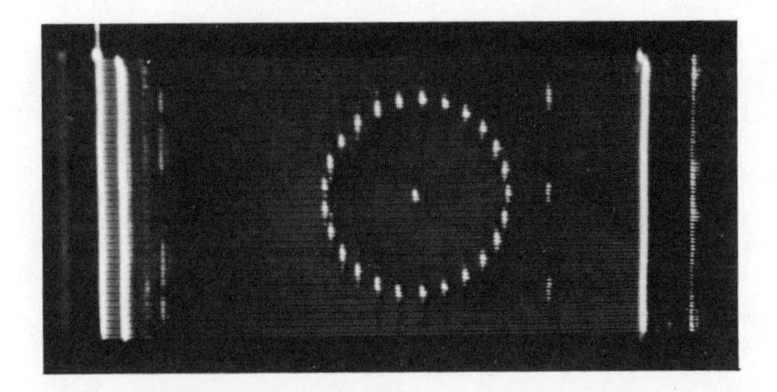

Fig.12 - Image of the same target with the F4 profile associated to the 30 cm depth technique.

CONCLUSION

Our experiments in the laboratory and our exams in the hospital have convinced us that the Fresnel focusing technique constitutes a very interesting compromise for achieving the medical requirements of B echography in image resolution with a simple low cost technique. The new tracking focusing device will be extensively tested in the following months and will probably be an interesting refinement of a technique which is in fact already very efficient, due to its natural depth of field.

ACKNOWLEDGEMENTS

Authors are indebted to T. CHARLEBOIS and M. KAPP for valuable discussions and suggestions. They want also to thank R. LALIMAN, C. SASSIER and J.O. ALQUIE for their efficient technical assistance.

BIBLIOGRAPHY

1 P. ALAIS, M. FINK, " Fresnel zone focusing of linear arrays applied to B and C echography ", Acoustical Holography, Vol.7 pp 509-522.

2 P. ALAIS, M. FINK, B. RICHARD, J. PERRIN, " A holographically focusing array for real time B echography and 3 D investigation" Ultrasound in Medicine, Vol. 3 B (1977), pp 1499-1507.

3 P. ALAIS, M. FINK, B. RICHARD, J. PERRIN, " Applications of a holographically focusing linear array to medical diagnosis " Ultrasound in Medicine, Vol. 3 B (1977), pp. 1509-1518.

4 M. FINK, " Theoretical aspects of the Fresnel focusing technique" Acoustical Holography, Vol. 8.

THEORETICAL ASPECTS OF THE FRESNEL FOCUSING TECHNIQUE

M. FINK

Laboratoire de Mecanique Physique,
Université Pierre et Marie Curie
Paris. France

Introduction

Recent developments of real time echographic device have shown the interest of multi transducer focusing arrays. Large focusing aperture must be synthetised to obtain a good resolution. The use of delay lines, allowing a converging cylindrical lens to be simulated, is one of the classical approach of the focusing problem in the echographic mode. A continuous variation of the focal length of such a lens allows a good focusing on all the echographic range. However, if the dimension of the lens aperture is small enough, the sequential use of a discrete number of focal lengths is sufficient to cover the entire depth of field.

The aim of this paper is to present a theoretical study of another approach of the echographic focusing problem, which may be obtained by the Fresnel focusing technique. It is well known that, for a monochromatic wave, the delay concept may be replaced by a phase shift concept. Thus, in a medium of sound velocity C, the cylindrical delay compensation, adapted to a focal length z

$$(1) \qquad \tau(x) = \frac{z - \sqrt{z^2+x^2}}{C} \simeq \frac{-x^2}{2Cz}$$

may be replaced, in the Fresnel approximation, by a quadratic phase compensation

$$(2) \qquad \varphi(x) = -\alpha x^2 \qquad [2\pi]$$

where α depends of the pulsation ω as

$$\alpha = \frac{\omega}{2cz} \tag{3}$$

Such a Fresnel focusing technique may be very interesting for echographic signal of large frequency bandwidth, when using a phase compensation which is not dependant on the frequency. In this case, the coefficient α is a constant ; this involves a linear dependance of the focal length f_i with the pulsation ω_i.

$$f_i = \frac{1}{2\alpha c}\,\omega_i \tag{4}$$

Therefore, if the signal bandwidth $\Delta\omega$ is large enough, a very large depth of field Δf may be obtained :

$$\Delta f = \frac{1}{2\alpha c}\,\Delta\omega \tag{5}$$

If such a technique appears to be attractive, it must be recognized that non dependant frequency dephasing circuits are not so easy to handle. Therefore, a strong phase sampling of the Fresnel law is necessary (Fig. 1). In fact, on a large frequency bandwidth, only the phase shift of 0 or π can be easily obtained. Each transducer

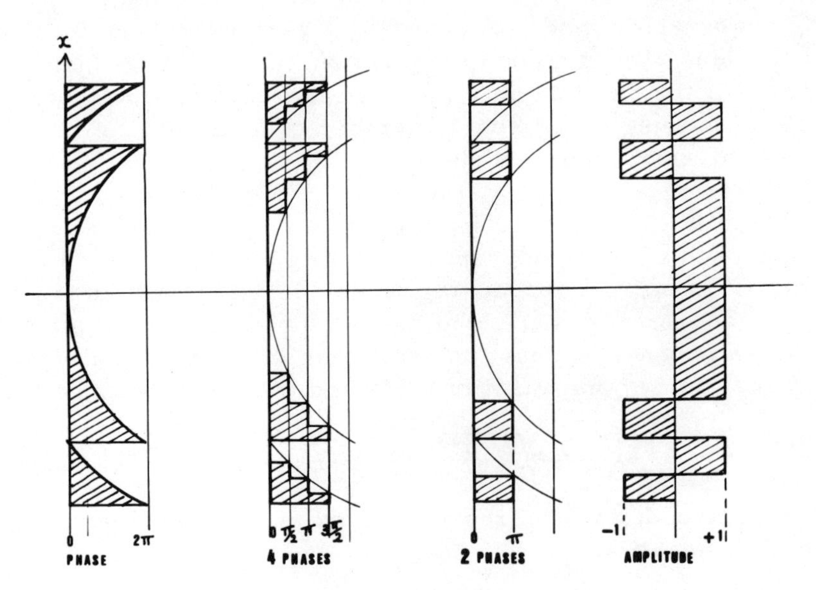

Fig. 1 - Fresnel lens and different phase sampling

is thus connected with a simple electronic inverter, and the Fresnel two state phase compensation $\sqcap_2(\alpha x^2)$ where

$$\sqcap_2(\varphi) = +1 \qquad -\pi/2 < \varphi \leqslant \pi/2 \qquad [2\pi]$$
$$(6) \qquad\qquad = -1 \qquad \pi/2 < \varphi \leqslant 3\pi/2 \qquad [2\pi]$$

can be considered as a simple amplitude modulation of the aperture by a three state law (+1, -1, 0). However ambiguities coming from this rough phase sampling are important. In carrying out a monochromatic diffraction study, the origins of such ambiguities can be easily understood. In the first part of this paper, we shall restrict our investigation to this case.

1 - MONOCHROMATIC RESPONSE

When using a Fourier expansion of the pupil function associated to the two state phase compensation

$$(7) \qquad \sqcap_2(\varphi) \propto \sum_{m=0}^{\infty} \frac{(-1)^m}{2m+1} \left(e^{-i(2m+1)\varphi} + e^{+i(2m+1)\varphi} \right)$$

we may check that in addition to the first term $e^{-i\varphi}$ (U_1^-) corresponding to the principal converging wave of focal length f, different parasitic converging and diverging waves appear (Fig. 2). When dealing with a two dimensional focusing obtained from a one dimensional Fresnel aperture, the contributions of all these secondary waves may be important in the focal plane. This can be explained by the cylindrical nature of these different waves. It is well known, from energy conservation law, that the amplitude of such waves varies slightly as $1/\sqrt{r}$, which replaced the classical ratio $1/r$ observed with spherical waves. In the focal plane, the contribution of the first diverging wave $e^{i\varphi}$ (U_1^+) is the most important. If a simple geometric spreading of this wave is considered, the mean value of the pressure in the focal plane is $\overline{U_1^+} = 1/\sqrt{2}$ For a similar reason, the principal converging wave gives a relatively weak pressure focal spot : Max (U_1^-) = $D/\sqrt{\lambda f}$. Then, the ratio of these two values is

$$(8) \qquad \frac{\overline{U_1^+}}{\mathrm{Max}(U_1^-)} = \frac{\sqrt{\lambda f}}{D\sqrt{2}}$$

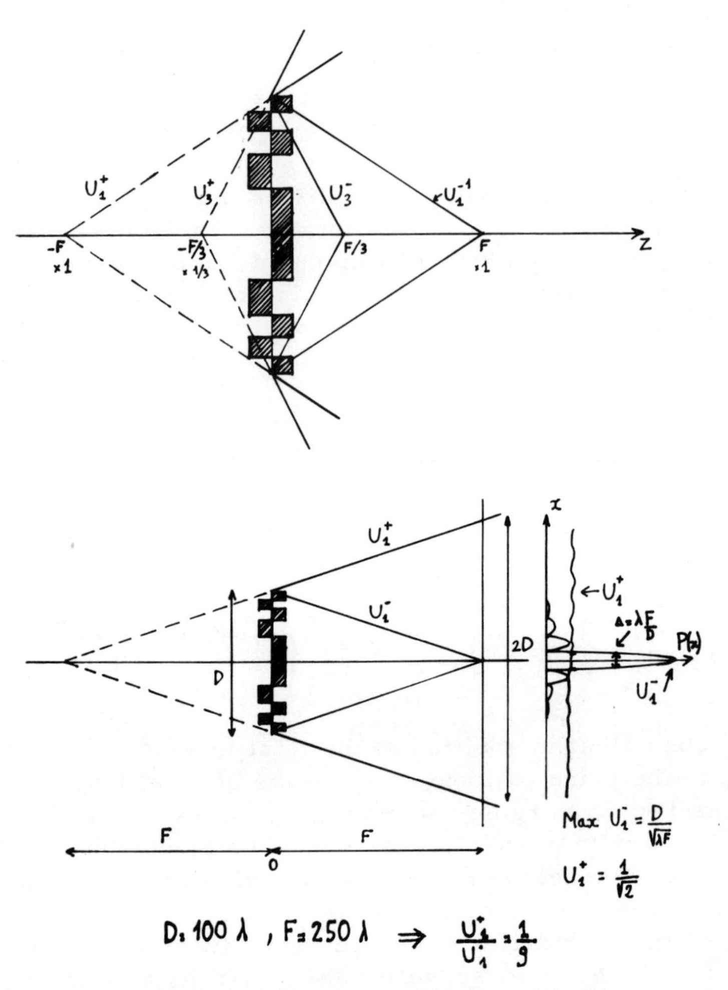

Fig. 2 - Influence of diverging waves with the two state phase focalisation.

and in our experiment ($D=100\lambda, f=250\lambda$), it is about 1/9 (-19 db). In fact, when taking in account all the parasitic waves, a numerical simulation gives a value $\simeq$ 1/5 (-13 db), which may be in some case sufficient to deteriorate echographic pictures, because of the large spreading of this acoustic noise.

Our main study was directed to the minimisation of these parasitic effects. At the last symposium [1] we explained how the use of two different apertures at emission and at reception is able to give a good apodizing effect. In particular, a simple solution is to associate the emission of a flat ordinary transducer, and the two state phase focalisation at reception. The vanishing of diverging waves at emission permits to obtain a good apodization, at a little expense of the lateral resolution.

The ambiguities of the two state phase focalisation process, which is not able to distinguish between a converging and a diverging wave, caused us to study much thinner phase sampling of the Fresnel lenses. When phase compensation is afforded with p different phase states $0, 2\pi/p, 4\pi/p, \ldots$, new pupil functions $\sqcap_p(\alpha x^2)$ are obtained where

$$(9) \qquad \sqcap_p(\varphi) = 1 \qquad -\pi/p < \varphi \leq \pi/p \quad [2\pi]$$
$$= e^{i2\pi/p} \qquad \pi/p < \varphi \leq 3\pi/p \quad [2\pi]$$
$$\vdots$$

The Fourier expansion of these pupil functions and the study of Fourier coefficients give a simple criterium to evaluate the secondary lobes (Fig. 3)
For different values of p we obtain

$$(10) \qquad \sqcap_p(\varphi) \propto \sum_{m=0}^{\infty} \left(\frac{(-1)^m}{pm+1} e^{-i(pm+1)\varphi} + \frac{(-1)^m}{pm+p-1} e^{i(pm+p-1)\varphi} \right)$$

This result must be compared to the one obtained with $\sqcap_2(\alpha x^2)$
For $\sqcap_p(\alpha x^2)$, in addition to the term $e^{-i\varphi}$ (U_1^-), which corresponds to the principal converging wave, we find different terms associated to different focuses located at :

$$(11) \qquad -f/p-1 \;,\; +f/p+1 \;,\; -f/2p-1 \;, \ldots$$

The first diverging wave U_{p-1}^+ is here pondered by $1/(p-1)$, which comes from the value of the Fourier coefficient $(-1)^m/(pm + p - 1)$ when $m = 0$. In addition, the important spreading of this wave, associated to a virtual focus located at $-f/(p-1)$ explains the weak level of the acoustic noise in the focal plane.

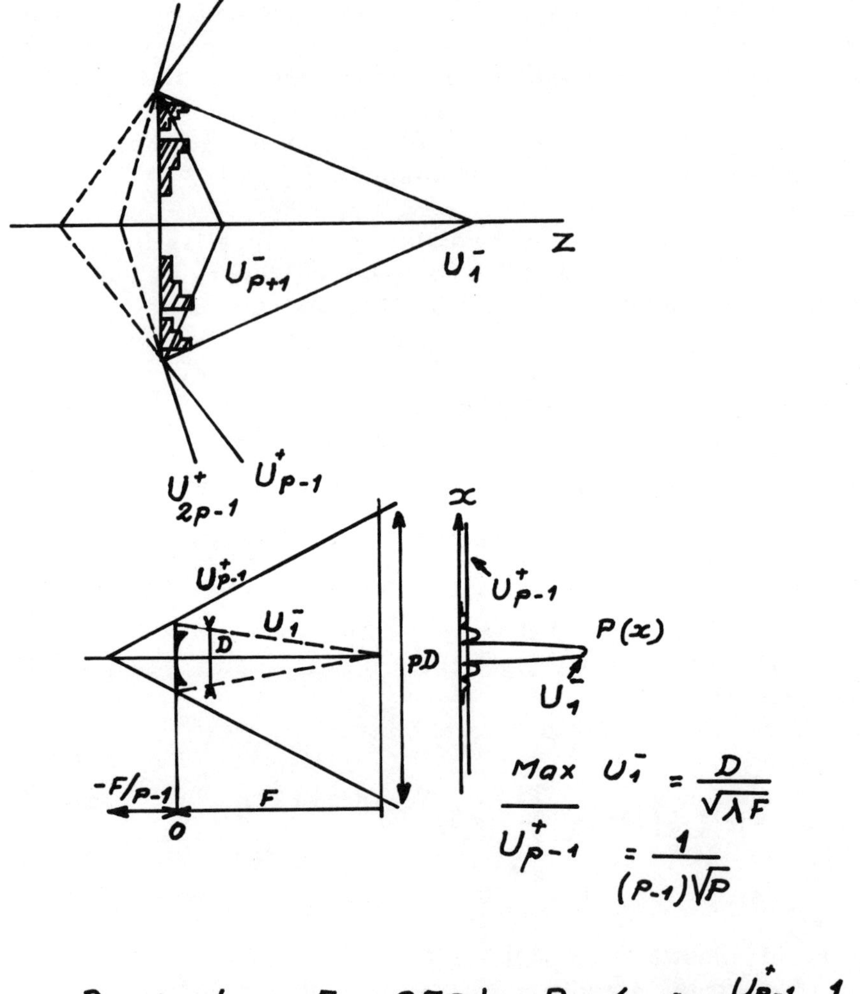

Fig. 3 - Influence of diverging waves with the p state phase focalisation

Thus, the mean value obtained is equal to

$$\overline{U^+_{p-1}} = \frac{1}{(p-1)\sqrt{p}} \tag{12}$$

which must be compared to the focal spot of amplitude $D/\sqrt{\lambda f}$

Then, the ratio of these two values is

$$\frac{U^{+}_{p-1}}{\mathrm{Max}(U^{-}_{1})} = \frac{\sqrt{\lambda f}}{D\sqrt{p}\,(p-1)} \tag{13}$$

and in our experimental case, it is about 1/22 (-27 db) for $p = 3$, and about 1/39 (-32 db) for $p = 4$. For the last case, the secondary lobes are reduced of about 13 db compared to the two state phase law. In fact a numerical simulation (Fig. 4) shows that the apodization is reduced in this case to 8 db.

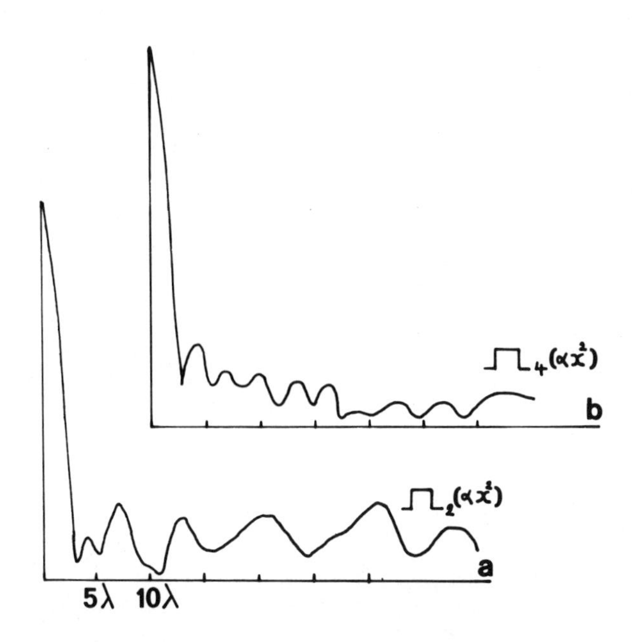

Fig. 4 - Computed pressure at the focal distance.
a : Fresnèl 2 state phase law
b : Fresnel 4 state phase law.

II - IMPULSE ECHOGRAPHIC RESPONSE

In the second part of this paper, the use of Fresnel focusing in the case of short signals is investigated. With short pulses the steady state solution is no longer adequate to describe the structure of the field and a transient solution is required.
After recalling the impulse response function concept, we shall study this function for circular and rectangular Fresnel aperture, and we shall show some very interesting properties of these Fresnel lenses in transient state. In particular, a typical echographic focusing process will be emphasized, in which a strong focusing effect appears, even when using very short emitted pulses, if the same Fresnel profile is used at emission and at reception.

Different authors studied the near field transient radiation pattern for uniform apertures [2, 3], this paper presents an extension of this impulse response approach to non uniform apertures [4]

The impulse response at emission $h_E(\vec{r}, t)$ corresponds to the transient velocity potential field pattern associated with a baffled planar emitter, whose surface is subjected to a velocity impulse $V(t) = \delta(t)$
The solution of the diffraction problem using a Green's function development, permits to write this impulse response at a field point as the sum of contributions from elementary Huygens'sources, each radiating a hemispherical wave

$$h_E(\vec{r}, t) = \int_S \frac{P(\vec{m}_0)\, \delta(t - R/c)}{2\pi R}\, d\vec{m}_0 \tag{14}$$

$$R = |\vec{r} - \vec{r}_0| \quad , \quad \vec{r}_0(\vec{m}_0, 0) \in S \quad , \quad \vec{r}(\vec{m}, z)$$

and where $P(\vec{m}_0)$ is the pupil function associated to the amplitude modulation of Fresnel apertures.
The velocity potential $\varphi(\vec{r}, t)$ and the acoustical pressure $p(\vec{r}, t)$ for an arbitrary velocity function $V(t)$ is then obtained by the use of the convolution integral

$$\varphi(\vec{r}, t) = V(t) \otimes h_E(\vec{r}, t) \tag{15}$$

$$p(\vec{r}, t) = -\rho_0 \frac{\partial \varphi(\vec{r}, t)}{\partial t} \tag{16}$$

where ρ_0 is the medium density.

This theory can be extended to include reflection from a point target using the same aperture in the transmit-receive mode. In fact, with the same modulation law at reception $\mathcal{P}(\vec{m_o})$ we may define a similar impulse response to an impulsive velocity source $S(\vec{r}, t) = \delta(t)\,\delta(\vec{r} - \vec{r_s})$ located at a point $\vec{r_s}$

$$h_R(\vec{r_s}, t) = \int \frac{\mathcal{P}(\vec{m_o})\,\delta(t - R'/c)}{4\pi R'}\, d\vec{m_o} \tag{17}$$

$$R' = |\vec{r} - \vec{r_s}|$$

If the transducers of the aperture are pressure sensitive, the output electrical signal depends on the incident pressure over the aperture

$$E(\vec{r}, t) \propto \frac{\partial S(\vec{r}, t)}{\partial t} \otimes h_R(\vec{r}, t) \tag{18}$$

When using the approximation $S(\vec{r}, t) \propto p(\vec{r}, t)$ we may simply evaluate the echographic response from a point target in the transmit-receive mode

$$E_{ER}(\vec{r}, t) \propto \frac{d^2 V(t)}{dt^2} \otimes h_E(\vec{r}, t) \otimes h_R(\vec{r}, t) \tag{19}$$

In fact, in the echographic mode, when neglecting the different acoustoelèctric transfert function, the impulse echographic response may be assimilated to

$$H_{ER}(\vec{r}, t) = h_E(\vec{r}, t) \otimes h_R(\vec{r}, t) \tag{20}$$

which for an unique aperture is the autoconvolution product of the impulse response.

First we shall evaluate these impulse responses for a circular Fresnel aperture $\sqcap_2(\alpha r^2)$ (Fig. 5) [5, 6]. It must be checked that such Fresnel aperture was conceived for monochromatic use. We have seen that, when using large frequency bandwidth, a large depth of field may be obtained. This is due to the frequency linear dependance on the focal length. Therefore the use of very short signal seems to be interesting. However, when making this assumption, we forget the influence of all parasitic diverging and converging waves associated to each focal spot. In fact, for an impulsive velocity excitation, of infinite spectrum, all these parasitic waves destruct the focusing process. The evaluation of the impulse response is therefore very instructive.

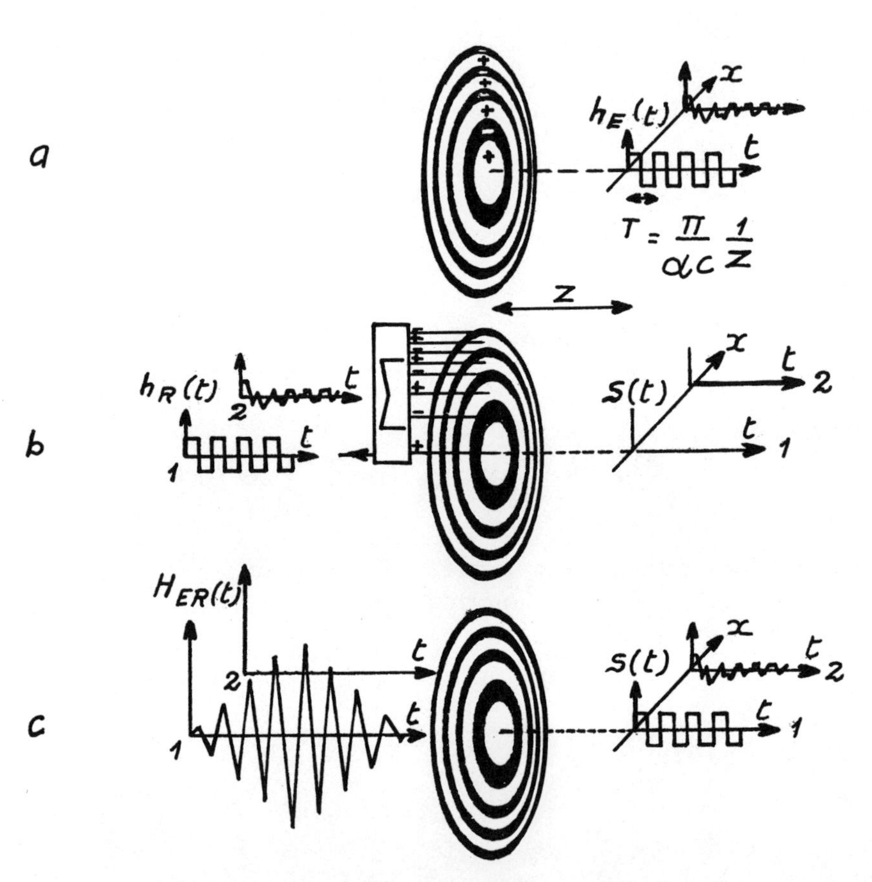

Fig. 5 - Impulse responses for circular Fresnel aperture. (a : transmit mode, b: receive mode, c : transmit-receive mode ; 1 : on axis field point ; 2 : off axis field point).

On the axis of the aperture the response is quasiperiodic

$$h(0,z,t) \propto \sqcap\!\sqcup_2 (2\alpha c z t)\, \mathrm{rect}_D(t) \tag{21}$$

where $D = A^2/2cz$ and $2A$ is the diameter of the aperture.
The quasiperiodicity arises from the fact that the contributions of each Fresnel zone have the same amplitude and the same duration. The shape of this transient response is different for off axis field points. As long as the projection of the field point lies within the diameter of the Fresnel aperture, there will be some part of the velocity potential having a flat topped pattern, the amplitude of which is the same as for the on axis response . On the other hand the remainder of the velocity potential transient after the flat topped initial stage is very erratic ; this is due to the fact that the contributions come from different parts of different Fresnel zones modulated either by + 1 or by - 1 . Therefore, when looking at the amplitude of the impulse response, the focalisation effect does not occur at emission. The same waveform is obtained at reception for the impulse response $h_R(\vec{r}, t)$, and here also, there is no focalisation.
However, a very interesting focusing process occurs when the same Fresnel aperture is used at emission and at reception. Such a process can be explained by the autoconvolution product properties of the impulse response. Due to symmetry the autoconvolution product $H_{ER}(0, z, t)$ rises to high value ; although the erratic shape of the off axis response gives a weak response as can be seen on the figure 5 . In fact, when using Schwarz inequality : $h \otimes h \leqslant \int h^2 dt$, it is easy to show that the maximum value of the echographic response is obtained for high symmetrical response functions (odd or even) , like the on axis response. In this case, the amplitude of the echographic response can be identified with the energy of the impulse response $\int h^2\, dt$, which is proportionnal to the surface of the Fresnel aperture . On the other hand, for a given aperture, the increase of the number of Fresnel zones involves a more erratic off axis response , that is a more efficient focusing process. It can be seen on figure 6 that the autoconvolution process gives very good impulse focusing compared to the use of monochromatic focusing.

Another very interesting property is obtained for this impulse echographic response : whatever the depth value z, the transverse echographic response remains the same. Therefore, we have an infinite depth of field. This is due to the fact that, for a given value of x , the impulse response $h(\vec{r}, t)$ conserves the same shape, but the time scale diminishes as $1/z$

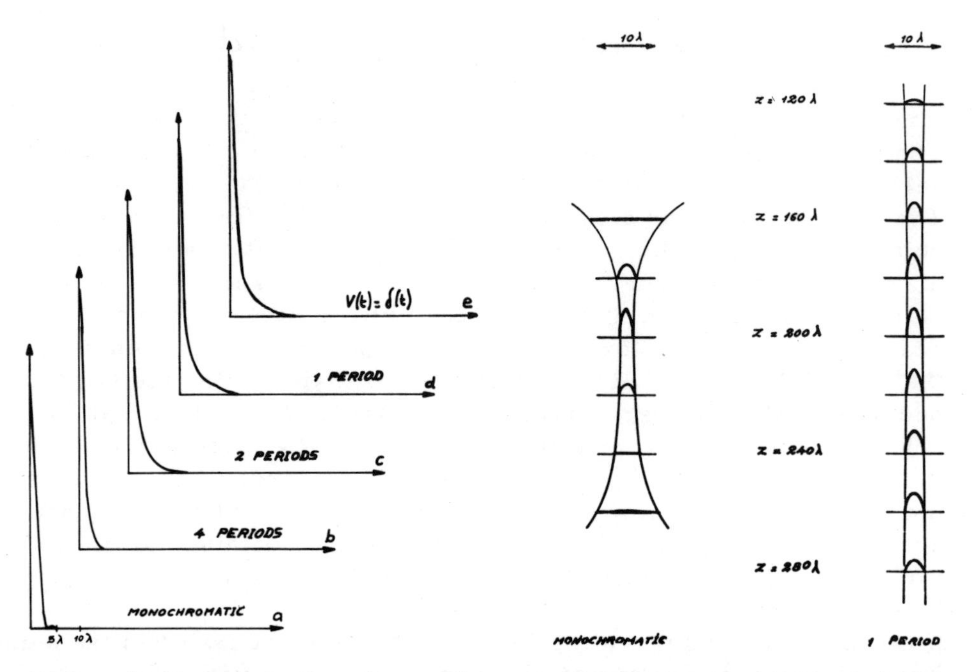

Fig. 6 - Transverse echographic response

Fig. 7 - Beamwidth at 6 db

Circular Fresnel aperture.

$$h(x,z,t) = h(x, z_i, (z/z_i)t) \tag{22}$$

This last property, associated to a Fresnel approximation, arises from the fact that, when looking at the acoustic contributions of different points of the aperture to a point field, the time interval Δt between the arrival of two elementary wave fronts varies as $1/Z$. For the complete echographic response H_{ER}, we find

$$H_{ER}(x,z,t) = z/z_i \; H_{ER}(x, z_i, (z/z_i)t) \tag{23}$$

Thus the transverse echographic response remains the same on an infinite depth. Only the absolute value changes as $1/Z$.

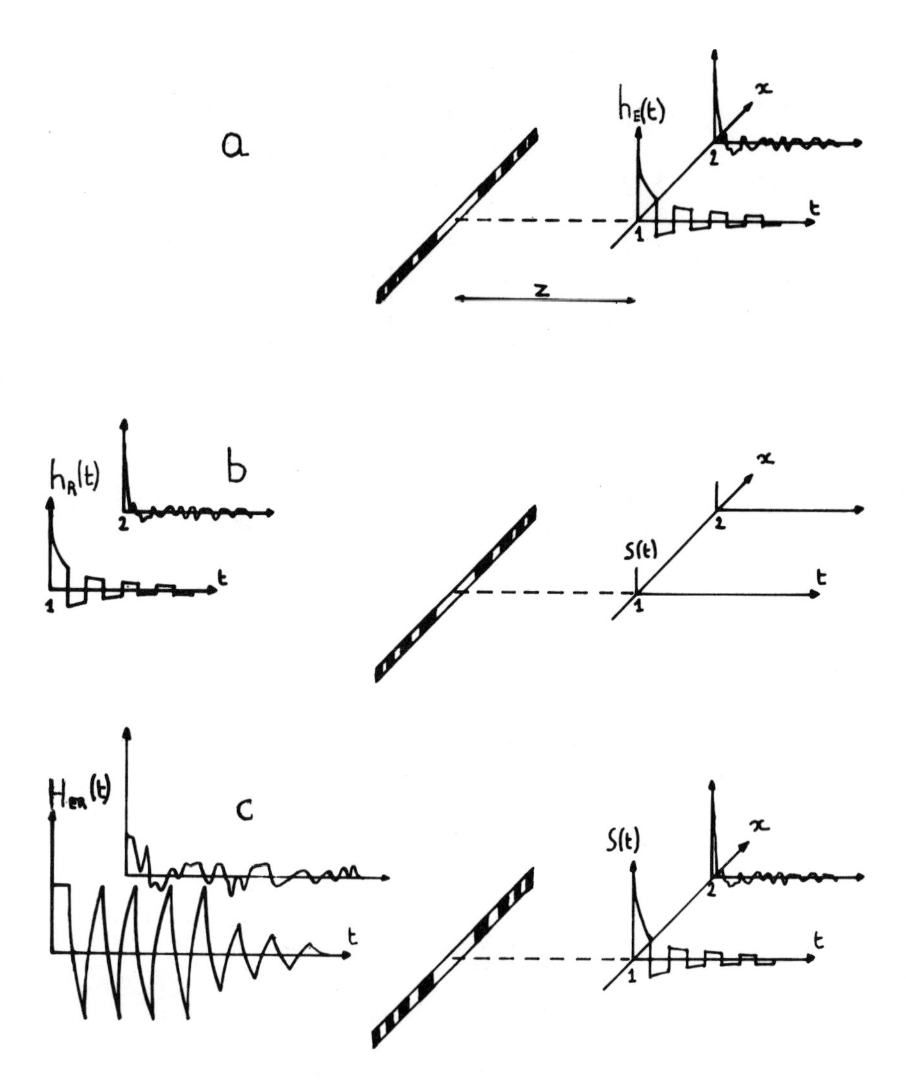

Fig. 8 - Impulse responses for a one dimensional aperture (a : transmit mode ; b : receive mode ; c : transmit-receive mode ; 1 : on axis field point , 2 : off axis field point.)

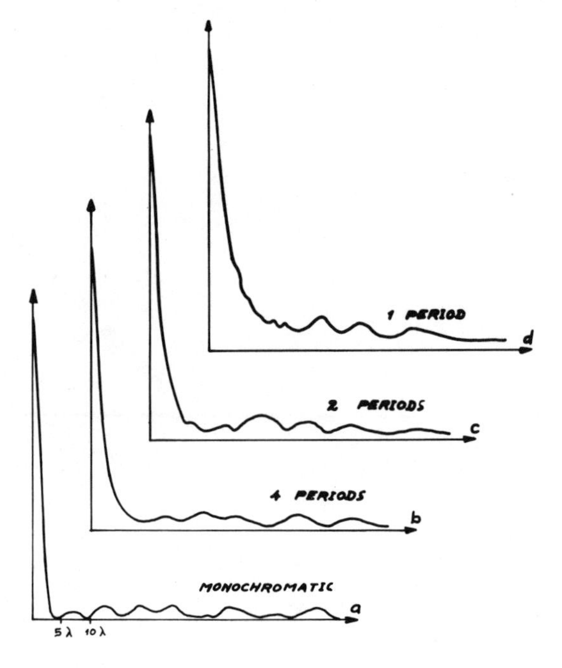

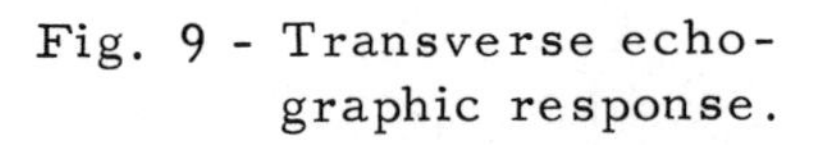

Fig. 9 - Transverse echographic response.

One dimensional Fresnel aperture.

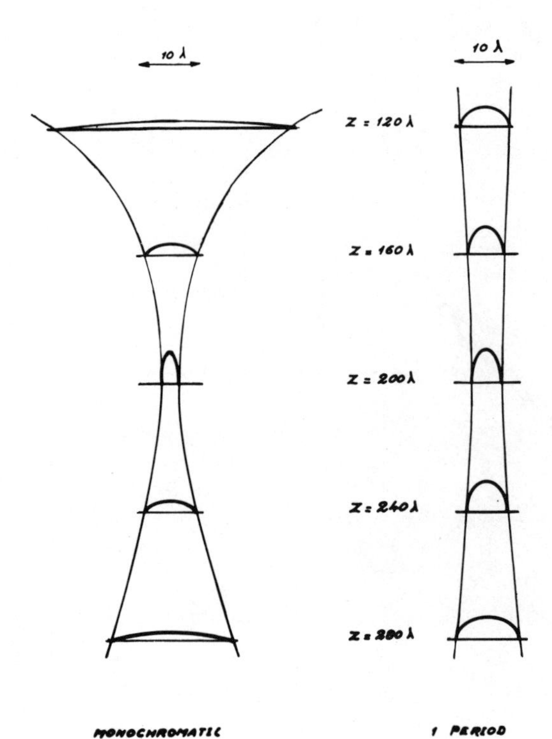

Fig. 10 - Bandwidth at 6 db.

One dimensional Fresnel aperture.

In fact, the ideal impulse excitation must be replaced by a non impulsive excitation $e(t) = \frac{d^2V}{dt^2}$. Thus, to find the echographic response, we must evaluate the convolution integral

$$E_{ER}(\vec{r},t) = H_{ER}(\vec{r},t) \otimes e(t) \qquad (24)$$

For a single sinusoidal signal of period T, the convolution integral emphasizes thus one precise depth Z_0, corresponding to the equality between this period T and the impulse response period $\alpha C Z_0$, that is $Z_0 = \alpha C / T$. In this case, the depth of field is reduced, but as can be seen on figure 7, for one period excitation, it is still deep, compared with the monochromatic case.

After presenting all these interesting properties of circular Fresnel apertures, we shall now study rectangular apertures, which are directly linked to our experiments. For such rectangular apertures, the evaluation of the impulse response appears to be more difficult . In fact, in our case, the transverse dimension of the aperture is small enough to be considered as a one dimensional aperture [4] . The shape of the impulse response (Fig. 8) is then easily obtained. For the on axis response, we still find a quasiperiodic function, but this is now modulated by an hyperbolic law $1/\sqrt{t}$ Such a law comes from the fact that in the unidimensional case , the surface of each Fresnel zone varies as $1/\sqrt{n}$ where n is the number of the zone . A similar effect occurs when looking at the off axis response.
For these apertures, the autoconvolution process is not as efficient as with circular apertures. The symmetry needed for high autoconvolution product is not obtained with the on axis response. The focalisation process occurs only when using short sinusoidal excitations . However, the numerical results (Fig. 9) show that the side lobes level is thus very important. In fact, the lowest level is obtained for monochromatic excitation . However great depth of field may be obtained for short excitations (Fig. 10) as with circular apertures. These results show that, for echographic investigation, apodizing techniques are required, and actually the technique described at the last symposium [1] is the one found to be the most easily obtained. The use of two different apertures at emission and at reception is able to give good apodizing effect.

BIBLIOGRAPHY

1 - P. ALAIS , M. FINK
Acoustical Holography ; Vol. 7 (1976). p. 509, 522.

2 - D. ROBINSON , S. LEES , L. BESS
IEEE . Vol. ASPP - 22 N° 6 - Dec 1974

3 - P.R. STEPANISHEN
JASA - 49 - 283 (1971) , JASA - 49 - 841 (1971)

4 - M. FINK
1978 - Thèse d'Etat, Université Pierre et Marie Curie. Paris

5 - B.A. AULD , S.A. FARNOW,
Acoustical Holography. Vol. 6 (1975) p. 259 -274.

6 - M. LAGREVE
1976 - Thèse de 3° cycle, Université Pierre et Marie Curie
Paris.

AN ANALYSIS OF PULSED ULTRASONIC ARRAYS*

B. P. Hildebrand

Battelle-Northwest
Richland, WA 99352

ABSTRACT

Array theory was developed on the foundation of steady-state analysis. In the electromagnetic regime this has, until now, been acceptable since even short pulses (in time) were physically long (in space). In the ultrasonic regime, however, physically short pulses are common (due to slow velocity of propagation). In this case steady-state theory is no longer adequate. In this paper I develop a simple method of analysis yielding the space-time impulse response of arrays driven under transient conditions. The analysis indicates that time, as well as space apodization, may be used to control the three-dimensional resolution capability of such arrays.

INTRODUCTION

Array theory has attained a relatively high level of maturity. There is a large body of literature pertaining to the effects of element spacing, spacing error and amplitude weighting on the array pattern. However, theory and practice are predicated upon the steady state, so that interference theory can be used; grating and side lobes being manifestations of destructive and constructive interference.

Arrays are common in the electromagnetic and acoustic regimes. Since the velocity of propagation in the electromagnetic regime is high, even short pulses are physically long. Hence steady state

*This work was sponsored by the Xerox Research Center, Palo Alto, California.

theory is adequate for most purposes. In ultrasonics, however, physically short pulses are common. For this reason it is imperative that a theory incorporating transient signals be developed.

The simplest physical picture for explaining how a radiation pattern is formed is by Huygens principle. As shown in Figure 1, the pattern is formed due to the fact that waves originating at different points on the array reinforce each other at some points in space and cancel at others. This behavior requires that a single frequency, or well defined phase fronts are propagated by the elements. In other words, steady state exists.

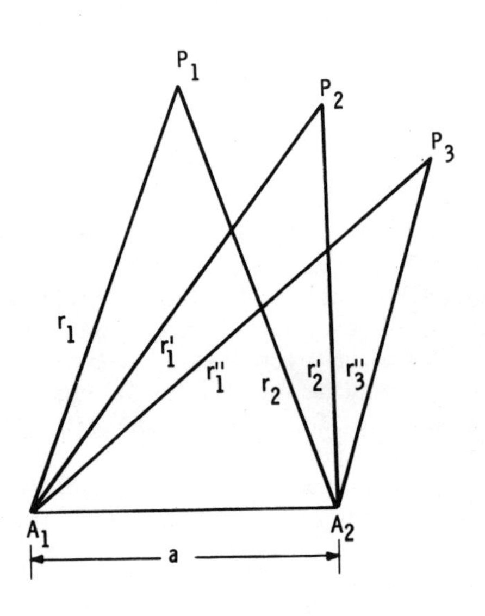

P_1: $r_1 = r_2$ ADDITION

P_2: $r_1' = r_2' + \lambda/2$ CANCELLATION

P_3: $r_1'' = r_3'' + \lambda$ ADDITION

FIGURE 1. This Diagram Illustrates the Effect of Steady State Driving Signals on the Behavior of an Array

This single phenomenon (interference) accounts for all of the well-known properties of arrays and transducers. Figures 2 and 3 show some classic examples of resulting radiation patterns. Figure 2 is the far field pattern of a 16-element ultrasonic array operating at 2.25 MHz in water. The array elements are 1 mm wide, spaced by 2 mm. This figure illustrates the formation of sidelobes and the first order grating lobes. Figure 3 illustrates the effects of interference between elementary areas of a circular flat piston radiator, resulting in dramatic differences in the near and far field. Note that in the near field complete cancellation occurs at certain points on axis.

One of the earliest references discussing transient effects is concerned with the time it takes to establish a radiation pattern.(1) This can be illustrated by again referring to Figure 1. Suppose that all elements of the array are turned on at the same time and we monitor the field at P_2. Since the energy from element A_1 must travel further than that from A_2, it is obvious the cancellation will not occur until both signals are present. This interval of time is proportional to a/c, where a is the element spacing and c the velocity of propagation. Thus, the steady state

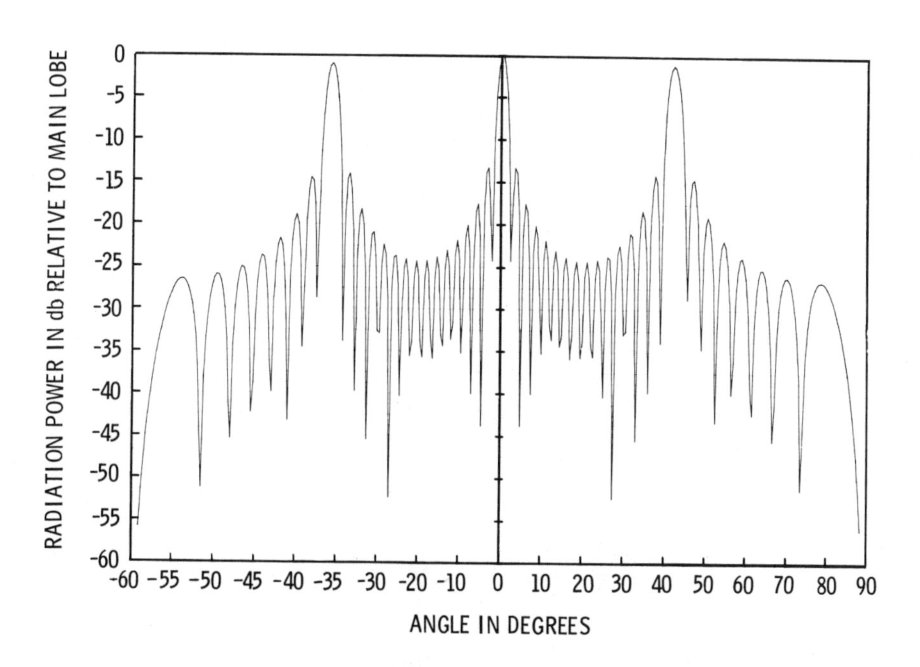

FIGURE 2. Far Field Radiation Pattern for a 16-Element Ultrasonic Array Driven in the Steady State

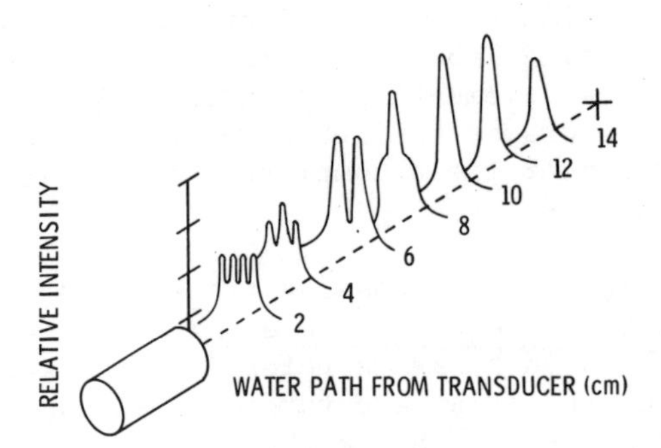

FIGURE 3. Radiation Pattern for a Flat Circular Piston Driven in the Steady State

pattern will not be built up for this short interval of time. In the cited reference, the author considers this a limitation on the scan rate of the antenna. However, he does point out that during the transient period, the side-lobes are no higher than during steady state.

Another point of view regarding non-classical array theory is to consider the array's operation with wide band signals.(2) In this reference the authors assume a signal waveform with a specific frequency spectrum of finite width. Rigorously, of course, this is still a steady state system in the sense that in the time domain, the signal must have infinite extent. The authors of the cited reference use this type of signal to synthesize array patterns with fewer unequally spaced elements than is classically required for equal spacing. I will not attempt to repeat their argument here since it is not intuitively obvious and does not pertain to the present paper. I mention it here for completeness.

In ultrasonic applications, the velocity of propagation is very small compared to that of electromagnetic energy. As an example, a 1 μsec pulse of sound in water occupies only 1.5 mm in contrast to 3×10^4 m for a radio wave pulse. Consequently, it is highly unlikely that an arbitrary point in space will receive energy from all array elements, or even from all points of a transducer, at the same instant in time. This problem has been considered to some extent in the acoustic literature. A short overview of the field prior to 1969 is given by Freedman.(3) The papers he summarizes concern themselves primarily with the performance of piston radiators under transient excitation. These, and more recent analyses showed that the array patterns were considerably modified, especially in the near field.(4,5,6,7) The latter papers provide detailed accounts of the impulse response of piston radiators and make the observation that in the near field the response is radically different than for the steady state. Specifically, the nulls occurring in the near field as a function of distance are no longer present.

The latter papers analysed particular radiators, principally circular pistons, by digital computer techniques. In this paper, I use analytical methods to solve for the transient response of arrays, and uncover some interesting possibilities for shaping the three-dimensional impulse response.

ANALYSIS

Consider Figure 4 as a two-element interferometer with element spacing X. There are two elements radiating the complex signals s(t), and $s[t - (R_o - r_1)/c]$, respectively. This means that the two elements are time delayed so that the signals will arrive at point P_o simultaneously. We wish to know what the field will be like at some other point, P_1. Neglecting spherical spreading, this is simply the summation

$$E(P_1,t) = s\left(t - \frac{R_o - r_o}{c} - \frac{r_1}{c}\right) + s\left(t - \frac{R_1}{c}\right). \tag{1}$$

The intensity of the field at this point becomes

$$I(P_1,t) = \left|s\left(t - \frac{R_o - r_o + r_1}{c}\right)\right|^2 + \left|s\left(t - \frac{R_1}{c}\right)\right|^2$$

$$+ 2\mathrm{Re}\left\{s\left(t - \frac{R_o - r_o + r_1}{c}\right)s^*\left(t - \frac{R_1}{c}\right)\right\}. \tag{2}$$

As a simple specific example we assume

$$s(t) = \mathrm{rect}\left(\frac{t}{T}\right), \tag{3}$$

where T = length of the unit amplitude pulse.

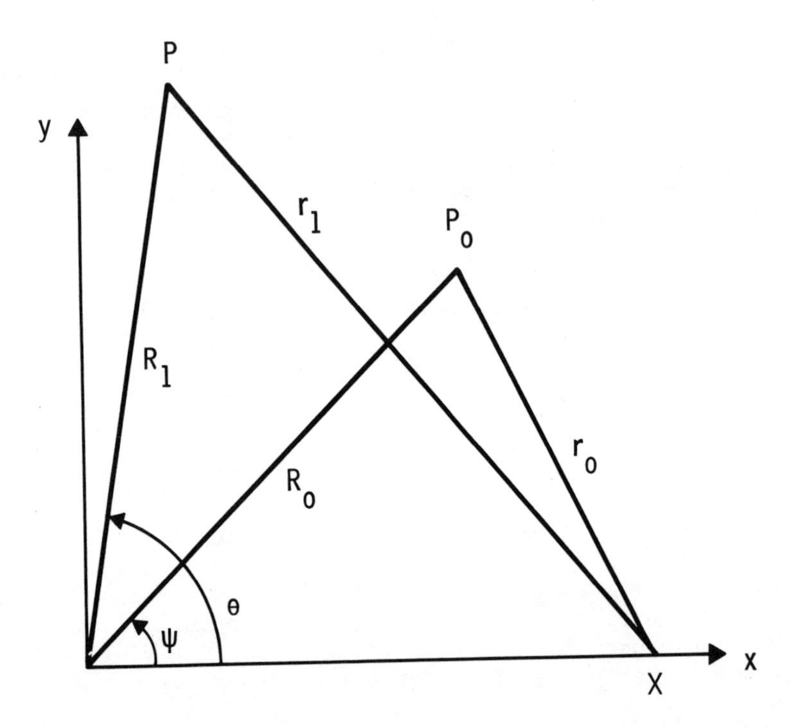

FIGURE 4. Geometry for the Analysis of a Focused Array

Then

$$I(P_1,t) = \text{rect}\left(\frac{t - \frac{R_0 - r_0 + r_1}{c}}{T}\right) + \text{rect}\left(\frac{t - \frac{R_1}{c}}{T}\right)$$

$$+ 2\,\text{rect}\left(\frac{t - \frac{R_0 - r_0 + r_1}{c}}{T}\right)\text{rect}\left(\frac{t - \frac{R_1}{c}}{T}\right). \tag{4}$$

This equation predicts that the time history of the field intensity at point P_1 consists of two separate unit pulses unless P_1 lies close enough to P_0 to provide some pulse overlap, in which case the third term of Equation 4 comes into play. The region in space over which this occurs can be found by the following geometric argument.

By inspection of the third term we find that overlap occurs when

$$- T < \frac{(R_o - r_o)}{c} - \frac{(R_1 - r_1)}{c} < T. \tag{5}$$

The boundaries of the region

$$- L < (R_o - r_o) - (R_1 - r_1) < L, \tag{6}$$

where L = cT, the physical length of the pulse, may be described by the hyperboloids of revolution

$$R_1 - r_1 = R_o - r_o \pm L. \tag{7}$$

If we write this in two dimensions we can use textbook formulas to help sketch this case. That is,

$$\frac{(x - X/2)^2}{a^2} - \frac{y^2}{b^2} = 1, \tag{8}$$

where $a = \dfrac{R_o - r_o \pm L}{2}$,

$b^2 = \dfrac{X^2}{4} - a^2$,

X = separation of the elements, and the asymptotes to the hyperbolas are

$m = \dfrac{b}{a}$,

If we use an example where the interferometer is designed for overlap at $\Psi = 45°$ and X = 10 cm, T = 5 μsec, R_o = 20 cm and $c = 1.5 \times 10^5$ cm/sec we find that the aymptotes to the two hyperbolas lie at the angles $\Psi_1 = 63.26°$, $\Psi_2 = 53.13°$, and the asymptote to the focal line is at $\Psi_o = 58.26°$. Figure 5 sketches the situation and provides a simple method for determining the overlap region. If we wish to find the angular extent of the region at a particular range, we simply draw a circle of the desired radius centered on the origin and measure the intercept angles. Similarly, a straight line at the desired angle can be used to find range overlap. In Figure 5 this is illustrated for an angle of Ψ=40° and a range of 14 cm.

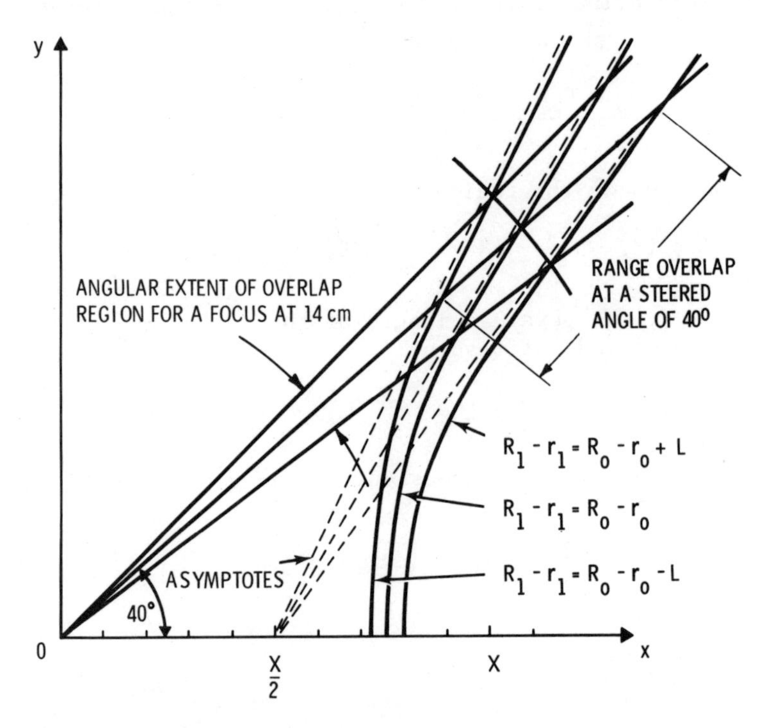

FIGURE 5. Geometrical Approach to Obtaining the Region in Space Where Interference Can Occur in a Pulsed Array

The preceding analysis provides a convenient way to estimate the region in space where the pulses from the interferometer elements coexist. However, for a multi-element array or continuous aperture, it can only provide an overlap region and not the desired intensity distribution. For this we need to provide a more general analysis.

Consider Figure 4 again, now as a one-dimensional aperture, each elementary area of which is driven with an impulse in time. Hence, at each point, x, of the aperture, the signal applied is

$$f(x)\,\delta\left(t - \frac{R_o - r_o(x)}{c}\right) \tag{9}$$

where $f(x)$ = aperture weighting function, and $[R_o - r_o(x)]/c$ denotes a time delay such that the impulse from each element will arrive at P_o simultaneously.

The field at any point P_1, will be proportional to

$$h(P_1,t) = \int_{-\infty}^{\infty} \frac{f(x)}{r_1(x)} \delta\left(t - \frac{R_0 - r_0(x) + r_1(x)}{c}\right) dx, \tag{10}$$

where we now include a spherical spreading factor $1/r_1(x)$. In order to evaluate this integral, the delta-function must be rewritten to reflect its dependence on the space variable x. According to Papoulis,[8]

$$\delta[\alpha(x)] = \sum \frac{\delta(x - x_n)}{|\alpha'(x_n)|} \tag{11}$$

where x_n are the roots of $\alpha(x) = 0$, and $\alpha'(x_n)$ is the derivative of $\alpha(x)$ evaluated at x_n.

In this particular case

$$\alpha(x) = \left(t - \frac{R_0}{c} + \frac{r_0(x)}{c} - \frac{r_1(x)}{c}\right) \tag{12}$$

$$\alpha'(x) = \frac{1}{c}\left[x\left(\frac{1}{r_1(x)} - \frac{1}{r_0(x)}\right) - \left(\frac{R_1}{r_1(x)} \cos\theta - \frac{R_0}{r_0(x)} \cos\Psi\right)\right] \text{ where}$$

$$r_0(x) = \left[R_0^2 + x^2 - 2x R_0 \cos\Psi\right]^{1/2} \text{ and}$$

$$r_1(x) = \left[R_1^2 + x^2 - 2x R_1 \cos\theta\right]^{1/2}$$

After substitution of the expressions for r_0 and r_1 and considerable algebraic manipulation the result is the quadratic equation

$$\alpha(x) = Ax^2 + Bx + C, \tag{13}$$

$$\alpha'(x) = 2Ax + B,$$

$$x_1 = \frac{B}{2A}\left[-1 + \sqrt{1 - \frac{4AC}{B^2}},\right]$$

$$x_2 = \frac{B}{2A}\left[-1 - \sqrt{1 - \frac{4AC}{B^2}},\right]$$

$$\alpha'(x_1) = \sqrt{B^2 - 4AC},$$

$$\alpha'(x_2) = -\sqrt{B^2 - 4AC}$$

where $A = (R_o \cos \Psi - R_1 \cos \theta)^2 - K_o^{\,2}$,

$$B = K_1(R_o \cos \Psi - R_1 \cos \theta) - 2K_o^{\,2} R_1 \cos \theta,$$

$$C = \left(\frac{K_1}{2}\right)^2 - (K_o R_1)^2,$$

$$K_o = ct - R_o,$$

$$K_1 = R_o^{\,2} - R_1^{\,2} - K_o^{\,2}.$$

Hence, we have the impulse response

$$h(P_1,t) = \sum_{n=1}^{2} \frac{f(x_n)}{|\alpha'(x_n)|r_1(x_n)} \tag{14}$$

This becomes a messy equation; consequently we make the Fresnel approximation by retaining only second order terms in the binomial expansion of $r_o(x)$ and $r_1(x)$ with the result that for uniform weighting, $f(x) = \text{rect}\,(x/X)$,

$$h(P_1,t) = \left| \frac{c}{R_1\sqrt{(\cos\theta - \cos\psi)^2 + 2\left(\frac{1}{R_1} - \frac{1}{R_o}\right)\left(ct - R_1\right)}} \right| \text{rect}\left[\frac{ct - R_1}{\frac{X^2}{2}\left(\frac{1}{R_1} - \frac{1}{R_o}\right) + X(\cos\theta - \cos\psi)}\right]. \tag{15}$$

Figures 6 and 7 show profiles of this function for $R_1 = R_o$ and $\theta = \psi$, respectively. Figure 8 attempts to show the complete impulse response as a function of angle for several values of distance R_1.

With the impulse response known, it is possible to obtain the response to other input driving functions by convolving the impulse response, $h(P_1,t)$, with the input function. For example, if the driving function is a pulse, rect (t/T), and we look on the focal sphere, $(R_1 = R_o)$ we have

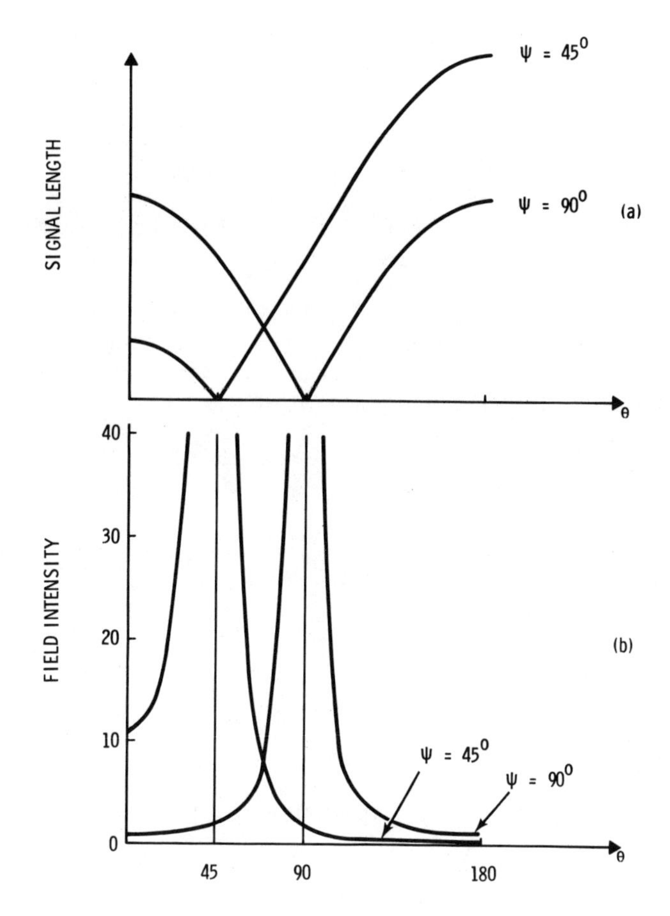

FIGURE 6. Intensity (a), and Signal Length (b), as a Function of Angle for an Aperture Driven with an Impulse

$$E(P_1,t) = h(P_1,t) * \text{rect}\,\frac{t}{T} \tag{16}$$

$$= \frac{c}{R_0\,|\cos\theta - \cos\psi|}\int \text{rect}\left[\frac{u}{\frac{X}{c}(\cos\theta - \cos\psi)}\right] \text{rect}\left[\frac{(t - R_0/c) - u}{T}\right] du$$

or $$\frac{c}{|x_1 - x_0|} \int \text{rect}\left[\frac{u}{\frac{X}{cR_0}(x_1 - x_0)}\right] \text{rect}\left[\frac{(t - R_0/c) - u}{T}\right] du$$

where x_1 = x-coordinate of P_1.

Cross-sectional plots of this function are shown in Figure 9 with a full view sketch in Figure 10.

The cases just analyzed allow us to infer a number of interesting properties of pulsed arrays. Grating lobe suppression may be inferred by a small extension of the interferometer analysis. If the pulse used to drive the elements is a gated sine wave; i.e., the signal is coherent, then only in the region of overlap will interference effects occur. Then the third term of equation 4 carries the added multiplier

$$\text{Cos}\left(\frac{\omega}{c}\left[(R_0 - x_0) - (R_1 - r_1)\right]\right) \tag{17}$$

where ω = radian frequency.

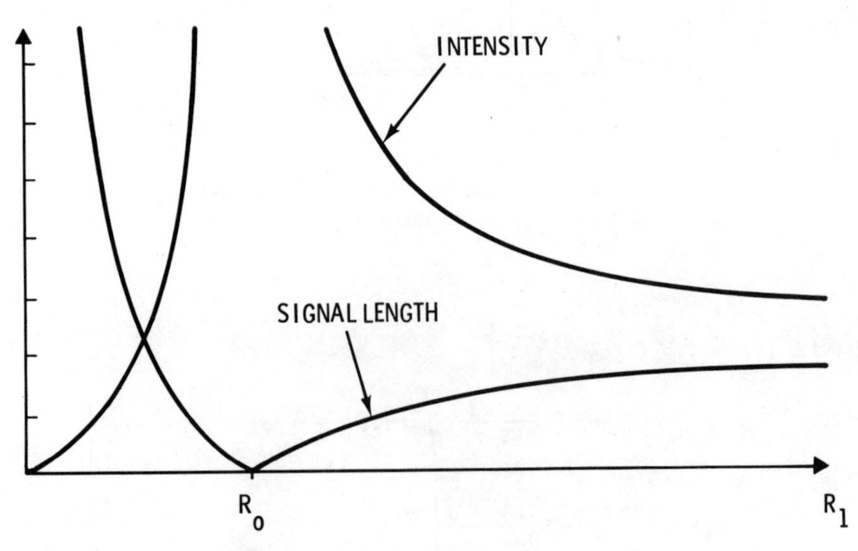

FIGURE 7. Intensity and Signal Length as a Function of Range for an Aperture Driven with an Impulse

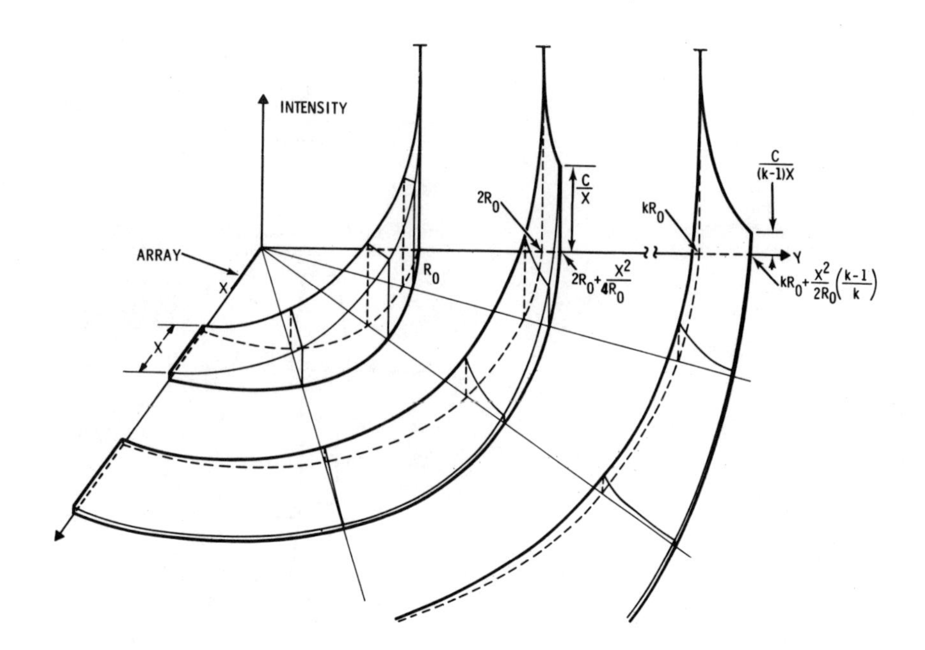

FIGURE 8. Impulse Response of a Continuous Aperture Focused at R_o

This represents a system of fringes existing in space wherever the pulses from the two sources overlap. If the Fresnel conditions are imposed and the function inside the cosine set to $2n\pi$, we obtain the result that the maxima occur at the angles given by ($R_1 = R_o$),

$$\cos \theta_n = \cos \psi \pm \frac{n\lambda}{X} \tag{18}$$

Hence, for our example, with λ = 0.75 mm (2MHz) the grating lobes are spaced by about 0.60° in the vicinity of the main lobe at Ψ = 45°. Since overlap occurs over an angle of approximately ±5.0° we would have 17 grating lobes within this region. However, if more elements, spaced by D, are added within the length X, the grating lobes spread further apart according to

$$\cos \theta_n = \cos \Psi \pm n \frac{\lambda}{D} \tag{19}$$

For our example, if we want only one lobe, we must choose D so that

$$\left| \frac{\lambda}{2D} \right| > \frac{L}{X} \quad \text{or} \quad D \leq \frac{\lambda X}{2L} = 6.67\lambda. \tag{20}$$

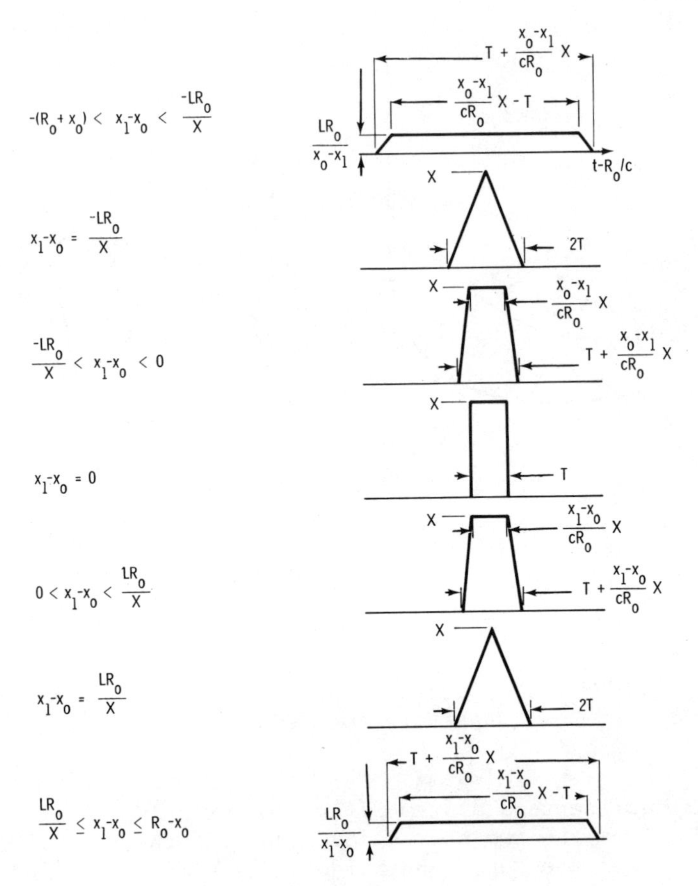

FIGURE 9. Azimuth Cross-Section of the Field on the Focal Sphere of an Array Driven by a Square Pulse

The number of array elements is

$$N = \frac{X}{D} .$$

Substituting $D = (\lambda X)/2L$ from equation 20 we find

$$N = \frac{2L}{\lambda} = \frac{2cT}{\lambda} . \tag{21}$$

Figure 11 shows a graph of element spacing, D, and number of elements, N, versus pulse length, T, which will yield a power pattern unencumbered by grating lobes. This shows that even a two-element interferometer can be unambiguous if the pulse is short enough.

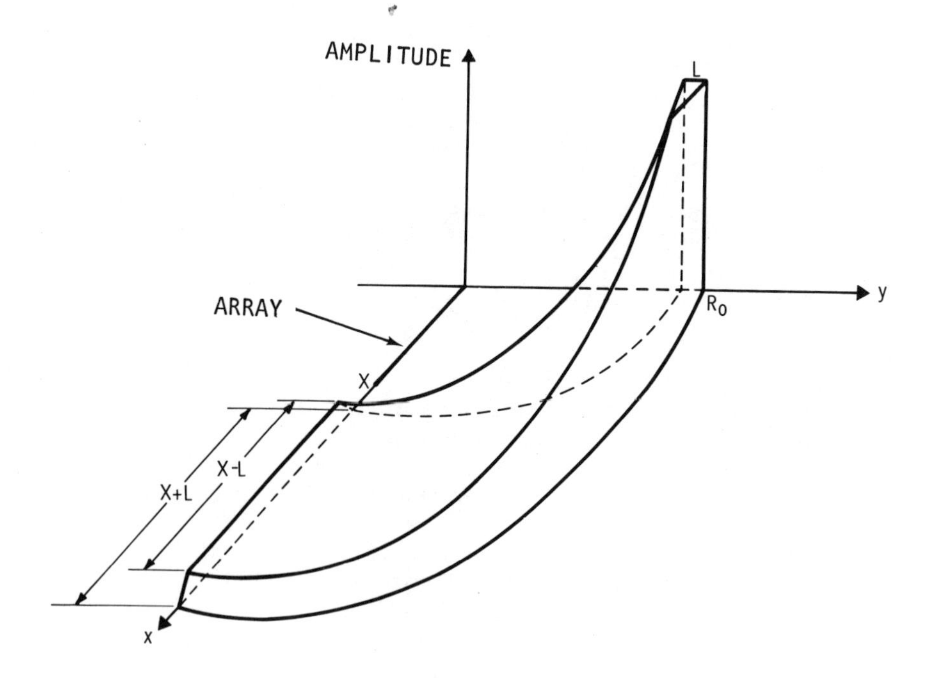

FIGURE 10. Field on the Focal Sphere of an Array Driven by a Square Pulse

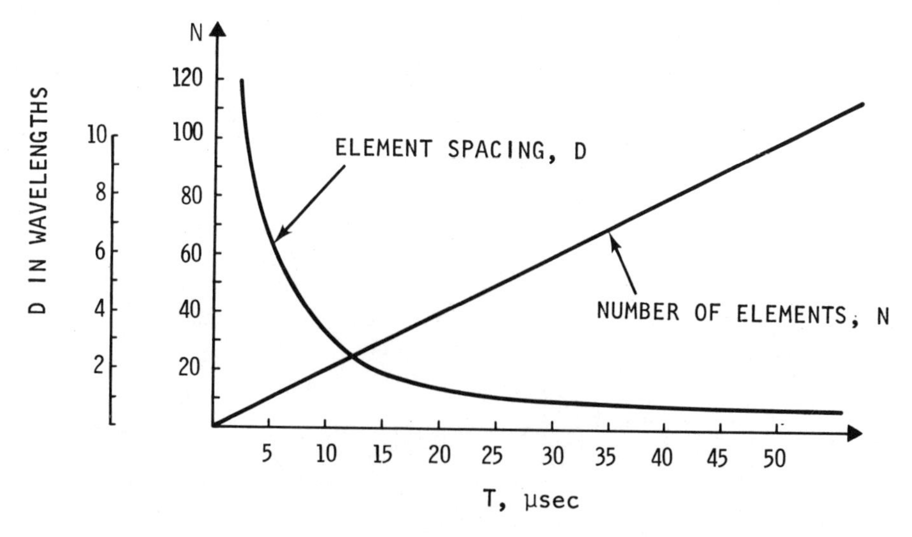

Figure 11. Plot of Element Spacing and Number for an Array with no Grating Lobes

We realize, of course, that two elements are not practical since the signal-to-noise ratio will be rather low. The amplitude at focus will only be twice that in nonoverlapping regions of the wave. When more elements are added, this situation improves at the cost of a larger region in space in which at least some of the signals overlap. Figure 10 illustrates this region for the filled aperture. Thus, grating lobes will exist, but be suppressed according to the inverse law

$$\left|\frac{1}{R_0} \cdot \frac{1}{x_1 - x_0}\right|^2 \text{ or } \left|\frac{1}{R_0} \cdot \frac{1}{\cos\theta - \cos\Psi}\right|^2$$

Focal sharpening or beam shaping is another exciting possibility. For example, equation 16 was identified as the convolution between a space and a time weighted function. This means that time apodization as well as space apodization will affect the beam pattern. A particularly simple example will serve to illustrate. Suppose that the elements are pulsed with a triangular pulse having the same width, T, as did the rectangular pulse used to drive Figures 9 and 10. Then equation 16 becomes

$$E(P_1,t) = \frac{c}{x_1 - x_0} \int \text{rect}\left[\frac{u}{\frac{(x_1 - x_0)X}{cR_0}}\right] \text{tri}\left[\frac{(t - R_0/c) - u}{T}\right] du, \tag{22}$$

where tri (t/T) describes an isosceles triangle of unit height and base T.

Figure 12 describes this function for various regions of x_1 and Figure 13 sketches the field on the focal sphere. In order to make the beam sharpening more apparent Figure 14 shows the two beam profiles along the focal sphere. Note the significantly sharper beam cross-section for the triangular pulse. We have apparently been able to use time apodization in place of the normal space apodization to suppress side-lobes and grating lobes.

The analysis is easily extended to two-dimensional arrays. The resulting impulse response on the focal sphere becomes

$$\frac{c}{|(x_1 - x_0)(y_1 - y_0)|} \iint f\left[\frac{u}{\frac{x_1 - x_0}{cR_0}}, \frac{v}{\frac{y_1 - y_0}{cR_0}}\right] \delta\left[u + v + \left(t - \frac{R_0}{c}\right)\right] du dv, \tag{23}$$

where $u = \frac{X}{cR_0}(x_1 - x_0)$, $v = \frac{Y}{cR_0}(y_1 - y_0)$.

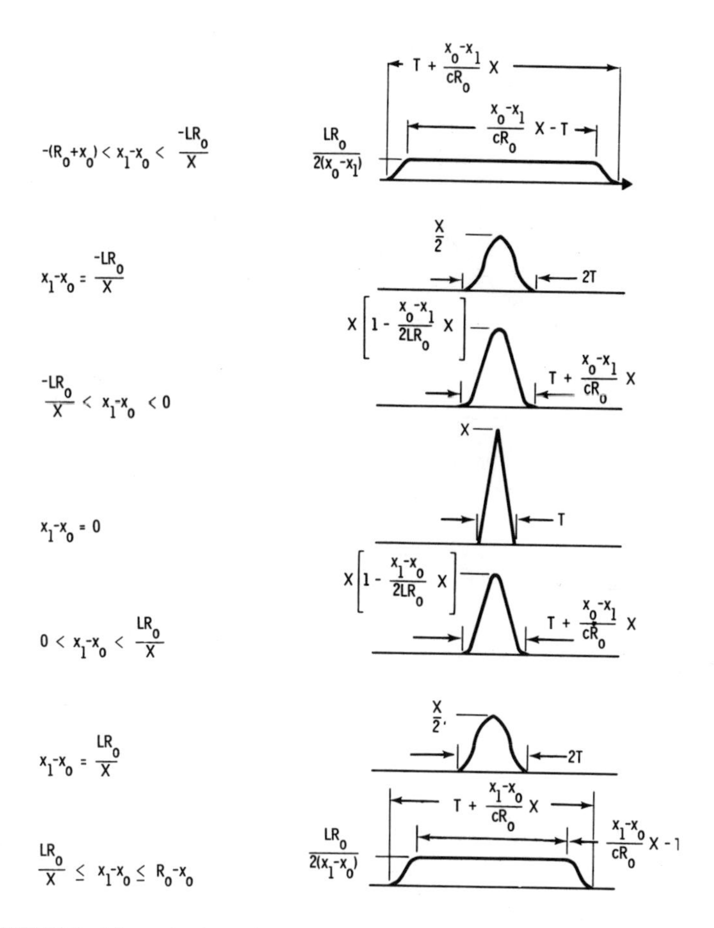

FIGURE 12. Azimuth Cross-Section of the Field on the Focal Sphere of an Array Driven by a Triangular Pulse

$\delta[\alpha(x,y)]$ represents a line mass on the curve $\alpha(x,y) = 0$ with density $\lambda(x,y) = \dfrac{1}{\sqrt{\alpha_x^2 + \alpha_y^2}}$, where $\alpha_x = \dfrac{\partial\alpha}{\partial x}$ and $\alpha_y = \dfrac{\partial\alpha}{\partial y}$.

Consequently, the integral represents the common area as the line mass moves through the two dimensional function, f, as shown in Figure 15. If the driving function is something other than an impulse, the same idea holds except that the line mass becomes a line volume whose profile is the driving function. Since the

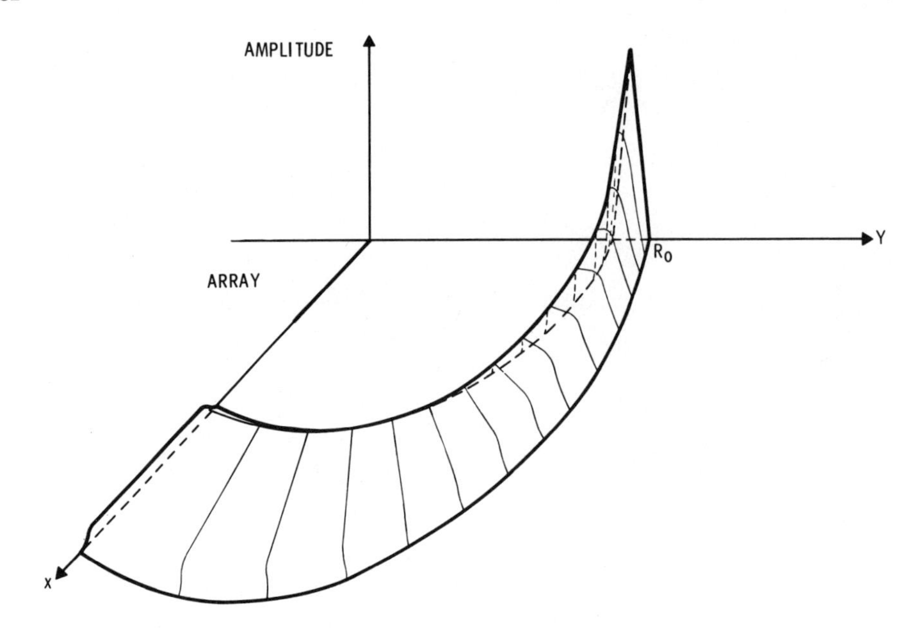

FIGURE 13. Field on the Focal Sphere of an Array Driven by a Triangular Pulse

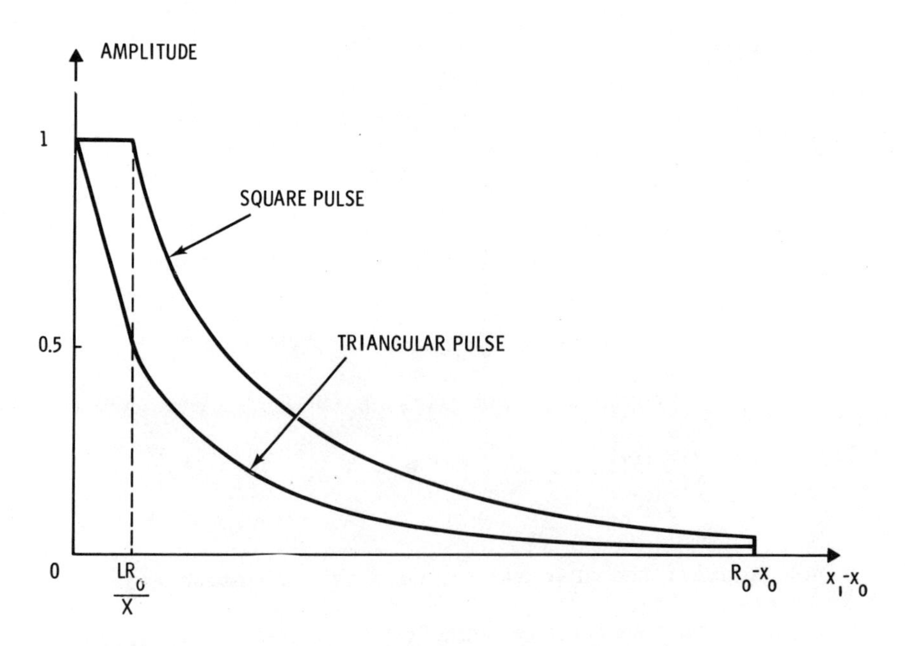

FIGURE 14. Focal Sphere Cross-Section of the Field of an Array Driven by a Rectangular and a Triangular Pulse

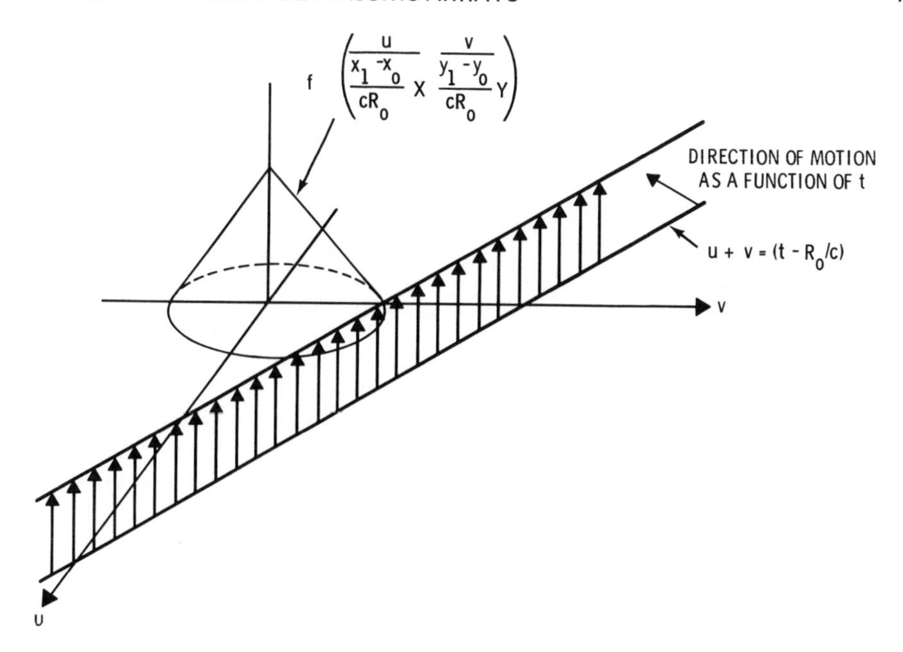

FIGURE 15. Physical Interpretation of the Expression for the Impulse Response of a Two-Dimensional Array (Equation 23)

aperture function, f, changes shape as the field point is changed, the value of the integral changes as a function of viewing angle. When the field point coincides with the focal point, f becomes a delta function and so does the field.

IMPLEMENTATION

The implementation of dynamically steered and focused arrays using pulsed driving signals is conceptually simpler than standard phased array steering since timing is substituted for phasing. Modern digital techniques allow very accurate pulse and gate control with the possibility that very fast steering and focusing can be obtained.

Work done at Battelle Northwest in the past has resulted in a method for producing a unipolar triangular ultrasonic pulse.(9,10) The base of the triangular pulse is equal to one period of the transducers natural frequency. Thus, a 2MHz transducer would yield a pulse length of 0.77 mm in water. This would provide an extremely sharp focus; a 10 cm long array would yield a 6 db intensity focal spot of 0.375 mm in range and 3 mm in azimuth for an F-2 focus. This contrasts with a spot of 15 mm in range and 3.5 mm in azimuth for a coherently focused array. Thus, pulsed operation improves range resolution without sacrificing azimuth resolution.

CONCLUSION

I have presented a simple method for analyzing the behavior of arrays in pulsed operation. This analysis indicates that significantly different, and perhaps superior, properties manifest themselves. When an array is driven with a pure pulse, side-lobes and grating lobes are non-existent. However, these manifestations of interference theory are replaced by a smoothly decreasing function radiating outward from the focus. If the driving function is a gated sine wave, side-lobes and grating-lobes reestablish themselves under this envelope function.

A second interesting observation is that the three-dimensional array pattern can be substantially improved by shaping the driving pulse. Thus, time weighting, in addition to array weighting, can be used to improve array performance.

Some practical advantages for operating arrays in the pulsed mode are, simpler beam steering and focusing using digital electronics, and a tradeoff between pulse length and number of elements.

REFERENCES

1. Charles Polk, "Transient Behaviour of Aperture Antennas," IRE Proceedings, 1281-1288, July 1960.

2. F. I. Tseng and D. K. Chang, "A Synthesis Technique for Linear Arrays with Wide-Band Elements," IEEE Proceedings, 51, 1679-1681, 1963.

3. A. Freedman, "Transient Fields of Acoustic Radiators," JASA, vol. 48, no. 1 (Part 2). 135-139, 1970.

4. P. R. Stepanishen, "Impulse Response and Radiation Impedance of an Annular Piston," JASA, vol. 56, no. 2, 305-313, 1974.

5. W. L. Beaver, "Sonic Nearfields of a Pulsed Piston Radiator," JASA, vol. 56, no. 4, 1043-1049, 1974.

6. D. E. Robinson, S. Lees, and L. Bess, "Near Field Transient Radiation Patterns for Circular Pistons," IEEE Transactions ASSP-22, 395-403, 1974.

7. A. Penttinen and M. Lenekkala, "The Impulse Response and Pressure Nearfield of a Curved Ultrasonic Radiator," J. Phys. D: Appl. Phys., 9, 1547-1557, 1976.

8. A. Papoulis, "Systems and Transforms with Applications in Optics," McGraw-Hill, New York, NY, (1968) 36-38.

9. N. E. Dixon and T. J. Davis, "A New Triangular Acoustic Pulse--Its Generation and Unique Properties for NDT Applications," BNWL-1526, 1971.

10. N. E. Dixon, "Method of Generating Unipolar and Bipolar Pulses," U.S. Patent 3,656,012, 1972.

THREE DIMENSIONAL PASSIVE ACOUSTICAL IMAGING SYSTEM USING HEMISPHERICAL ARRAY DETECTORS

Takuso Sato, Yoichi Nakamura, Kimio Sasaki and Kazuho Uemura
Faculty of Science and Engineering, Tokyo Institute of Technology,
4259 Nagatsuda, Midori-ku, Yokohama-shi, Japan

ABSTRACT

A High resolution three dimensional passive acoustical imaging system is realized by using array detectors arranged hemispherically covering the object. Auto and cross power or bispectral analyses of the detected signals are carried out so that images of the distributions of power of single frequency component or coherence among three frequency components are obtained. One of the most significant features of the system is the ability to display high quality images of any desired cross sections of three dimensional noise emitting objects simply by the choice of parameters at the stage of image reconstruction. Three dimensionally high resolutions, for instance resolution volume of 3cm × 3cm × 5cm for waves at 10 kHz over the observation volume of 60cm×60cm ×60cm, are achieved. An image is obtained within five minutes and it is displayed on a color TV monitor or printed out graphycally. Moreover, to obtain structures which are much smaller than the wave length a new image reconstruction process which uses the matrix correspondence between the detected hologram and the complex amplitudes of mesh points on the object's surface is proposed. The system is a powerful means for precise analysis and control of noise sources.

I. INTRODUCTION

Passive acoustical imaging system is required for the imaging of spatial distribution of acoustical sources in many fields. Several holographic and other imaging methods have been developed as useful means for this purpose[1,2,3,4]. One of the special features of these methods is the detection of wave fields at a plane separated far from the object resulting in the reconstruction of images of the undisturbed wave field of the object.

The detection of a wave field at the plane separated far from the object, however, usually restricts the detectable region of information emitted from the object, as the spatial angle which is covered by a fixed detecting area of the plane decreases as the distance is increased.

This fact affects the reduction of the resolving powers of the imaging system, especially since the resolution in range direction is drastically lost and consequently only the image of accumulated wave field in this direction is imaged. This is the fatal drawback for the precise imaging of a three dimensional wave field of an object.

Of course the spatial angle can be increased by the increase of the scales of the detecting plane. However, as much as the plane is used the scales of the plane must be increased to as large as the distance between the object and the plane.

In this paper we propose a method which uses the hemispherical arrangement of detectors so that as much as possible information from the object can be received.

Thus all information about the wave fields scattered from the object in the front direction, that is to the hemisphere, is received by the detectors. The proper processing of the detected signals gives the range resolution which is about one half of that of the azimuth resolution.

Moreover, the wave field on the surface of the hemisphere is related with the wave field of the object if it is placed near the center of the hemisphere by the Fourier transform. This property can be used effectively in the processes of image reconstruction. The construction, fundamental properties, and some experimental results, as well as, an effective means of signal processing, are shown in the following.

II. PRINCIPLE AND CONSTRUCTION OF THE SYSTEM

The fundamental construction of the system is shown schematically in Fig.1. To realize a hemispherical detection of the wave field a set of detectors (microphones) are arranged along a half circle and rotated. (see Fig.1) The signals are amplified and fed to a computer through a multiplexer and an A/D convertor.

One of the signals, usually the signal detected by the detector placed at the center of the circle, is used as the reference. After calculating the hologram on the hemisphere from these signals through spectral analyses, the image is reconstructed by using the conventional calculation of the back propagation of waves from the hologram.

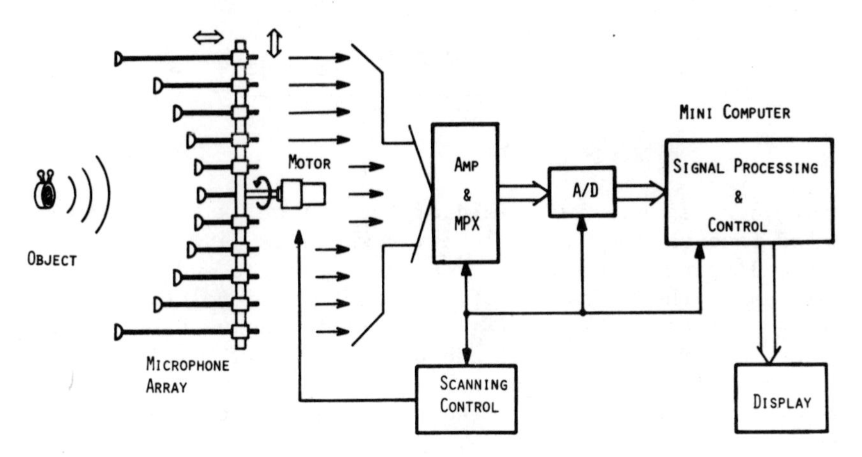

Fig. 1. Schematic Construction of the System

Now let us explain more closely the processes by using formula. Let us use the system of coordinates shown in Fig.2. where π is the detecting surface and a point Q on it is represented by $Q(\alpha,\beta,D)$ and a point P on the object is represented by $P(r,\theta,z)$.

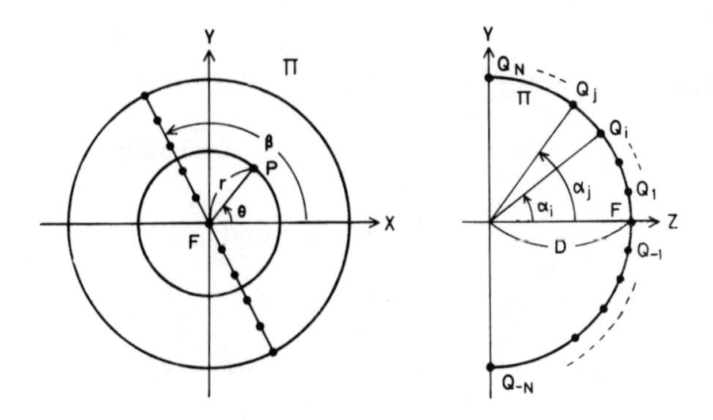

Fig. 2. Coordinates of Object and Detecting Hemisphere

Let us assume that the object is emitting strictly stationary random signals. And let us use the signal $u_1(t)$ detected at a fixed point F(0,0,D) as the reference. Then the hologram $H(\alpha,\beta,D;k)$ at a point $Q(\alpha,\beta,D)$ on the detecting surface of wavenumber k is derived from the auto and cross-polyspectral analyses of the signal $u_1(t)$ and the signal $u_2(t)$ detected at Q as follows[2]:

$$H(\alpha,\beta,D;k) = \frac{S_{u;11\cdots2}(\omega_1,\omega_2,\cdots,\omega_{n-1})}{S_{u;11\cdots1}(\omega_1,\omega_2,\cdots,\omega_{n-1})} \tag{1}$$

$$= \frac{\iiint \frac{A(r,\theta,z)}{d_2} e^{-jkd_2}\, rdrd\theta dz}{\iiint \frac{A(r,\theta,z)}{d_1} e^{-jkd_1}\, rdrd\theta dz} \tag{2}$$

where $A(r,\theta,z)$ is the wave field of the object, k is the wavenumber of frequency ω_n and velocity v which is given by $k=\omega_n/v$, d_1 is the distance between point P on the object and the fixed point F, d_2 is the distance between point P and point Q on the detecting surface, $S_{u;11\cdots1}(\omega_1,\cdots,\omega_{n-1})$ is the n-th order autopolyspectrum[2] of signal $u_1(t)$ and $S_{u;11\cdots12}(\omega_1,\cdots,\omega_{n-1})$ is the n-th order cross-polyspectrum between $u_1(t)$ and $u_2(t)$ where n-1 frequency components of $u_1(t)$ and one frequency component of $u_2(t)$ are used.

If the point P is located around the center of the hemisphere d_2 in the numerator is approximated as follows:

$$d_2 = \sqrt{D^2 + r^2 + z^2 - 2zD\cos\alpha - 2rD\sin\alpha\cos(\theta-\beta)}$$
$$\cong D - z\cos\alpha - r\sin\alpha\cos(\theta-\beta) \tag{3}$$

On the other hand the denominator is a function of only k for a fixed object, hence let us express it by F(k). Then Eq.(2) is reduced as follows:

$$H(\alpha,\beta,D;k) = \frac{e^{-jkD}}{F(k)} \iiint A(r,\theta,z) \times e^{j\{(k\sin\alpha\cos\beta)r\cos\theta + (k\sin\alpha\sin\beta)r\sin\theta + (k\cos\alpha)z\}} \times rdrd\theta dz \tag{4}$$

or

$$= \frac{e^{-jkD}}{F(k)} \iiint A(x,y,z) \times e^{j\{(k\sin\alpha\cos\beta)x + (k\sin\alpha\sin\beta)y + (k\cos\alpha)z\}} \times dxdydz \tag{5}$$

in rectangular coordinates. Eq.(5) shows clearly that the hologram signal $H(\alpha,\beta,D;k)$ is a Fourier transform of the object wave field $A(r,\theta,z)$, this fact can be applied when the operations of image restoration is required.

As for the image reconstruction the method of back propagation of waves is used. That is, for the image point $P'(r',\theta',z')$ to be reconstructed, first the distance $d'_{P;Q}$ between P' and a hologram point $Q_{n,m}(\alpha_n,\beta_m,D)$ is calculated and the image as the wave field is reconstructed by

$$I(r',\theta',z') = \sum_{\alpha_n}\sum_{\beta_m} H(\alpha_n,\beta_m,D;k)e^{jkd'_{P;Q}} \tag{6}$$

If the same approximation as Eq.(3) is used it can expressed as follows:

$$I(r',\theta',z') = \iiint A(r,\theta,z)g(r,\theta,z;r',\theta',z')rdrd\theta dz \tag{7}$$

where $g(r,\theta,z;r',\theta',z')$ is the corresponding point spread function of the imaging system which is given by

$$\begin{aligned} g(r,\theta,z;r',\theta',z') = \sum_{\alpha_n}\sum_{\beta_m} \exp[j\{&(k\sin\alpha_n\cos\beta_m)(r\cos\theta - r'\cos\theta') \\ &+ (k\sin\alpha_n\sin\beta_m)(r\sin\theta - r'\sin\theta') \\ &+ (k\cos\alpha_n)(z - z')\}] \end{aligned} \tag{8}$$

$$\begin{aligned} g(x,y,z;x',y',z') &= g(x - x',y - y',z - z') \\ &= \sum_{\alpha_n}\sum_{\beta_m} \exp[j\{(k\sin\alpha_n\cos\beta_m)(x - x') \\ &\quad + (k\sin\alpha_n\sin\beta_m)(y - y') \\ &\quad + (k\cos\beta_m)(z - z')\}] \end{aligned} \tag{9}$$

in rectangular coordinates. The approximate expression for distance can be used for the objects near the center, but the exact expression must be used for the objects placed away from the center for precise reconstruction of image.

III. RESOLUTIONS OF THE SYSTEM

Resolutions of the system can be evaluated by observing the shape of the point spread function of Eqs.(8) or (9). To simplify the analysis let us take an arrangement of detectors as follows:

$$\sin\alpha_n = \Delta\cdot n \quad (n=0,\pm1,\cdots,\pm N) \tag{10}$$

where Δ is a positive constant and $\Delta N \leqq 1$.
Then the point spread function as for azimuth resolution is given by

$$g_a(r,\theta;r',\theta') = \sum_{m=1}^{M} \frac{\sin\{(N+\frac{1}{2})\Theta\}}{\sin(\frac{1}{2}\Theta)} \tag{11}$$

where

$$\Theta = (k\cdot\Delta\cdot\cos\beta_m)(r\cos\theta - r'\cos\theta') + (k\cdot\Delta\cdot\sin\beta_m)(r\sin\theta - r'\sin\theta') \tag{12}$$

On the other hand the point spread function as for the range resolution is given by

$$g_r(z;z') = \sum_{\alpha_n} e^{j(k\cos\alpha_n)(z - z')} \tag{13}$$

A numerical example of these functions, which corresponds to the images for a point object, were calculated under the following conditions; frequency:10kHz (λ=3.4cm), β_m=10·m (m=1,2,···,18),

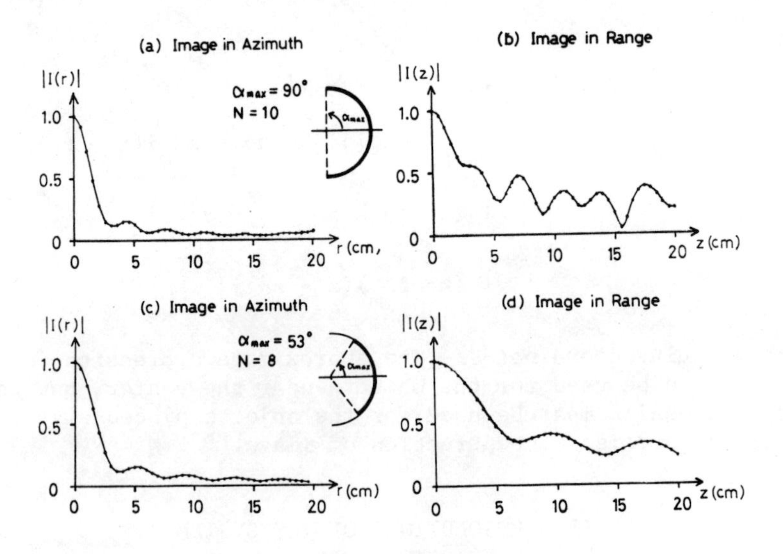

Fig. 3. Reconstructed Images a Point Object (I)
(Point Spread Function)

Δ=0.1 and N=8 or 10. The results are as shown in Fig.3.

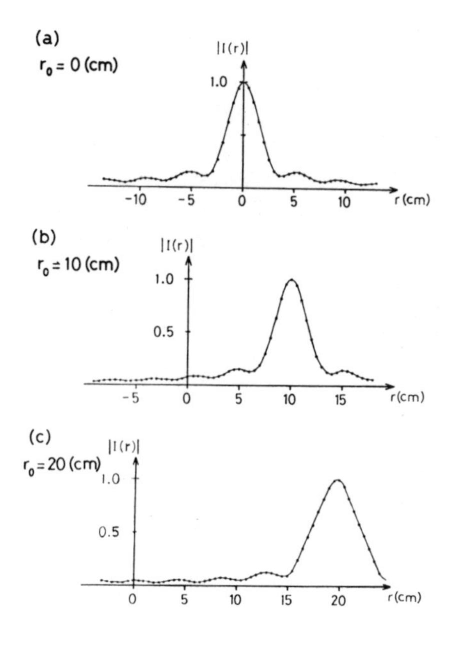

Fig. 4.
Reconstructed Images for a Point Object (II)

About one half of the azimuth resolving power is obtained in range direction. Although the resolutions are increased with the increase of N, N=8 gives fairly good resolutions. Hence this value is adopted in the following practical system. The reconstructed images for a point object placed away from the center of the hemisphere are shown in Fig.4.

The approximation of Eq.(3) was used for the image reconstruction. These results show that object over about 40cm can be observed without suffering from large distortions even when the approximation is used. Practically a region of 60cm × 60cm × 60cm may be used.

The effects of finiteness of the number of detectors and the special arrangement of them can be compensated by using suitable weighting for the hologram since the Fourier transform of the object corresponds to the hologram and the reconstructed image is improved.

Examples of images obtained by using compensations are shown in Fig.5 and Fig.6.

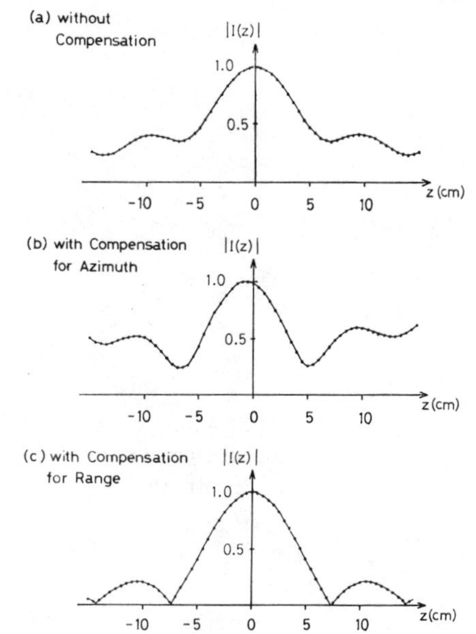

Fig. 5.
Effect of Compensation for Range Resolution

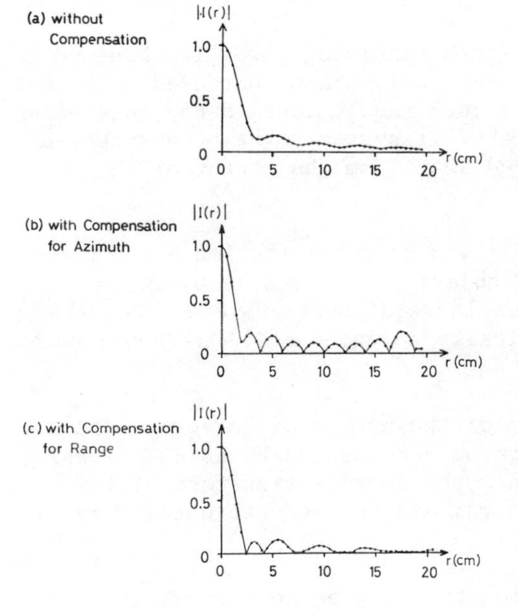

Fig. 6.
Effect of Compensation for Azimuth Resolution

They are compensated in two ways, that is (b) shows the effect of compensation such that the azimuth resolution is maximized, while (c) shows the effects of compensation such that the range resolution is maximized.

From these results the compensation used for (c) is adopted in the following practical system, since it gives good resolutions both in azimuth and range.

IV. THE CONSTRUCTED SYSTEM

A concrete system was constructed based on the principles and discussions in the previous chapters.

Typical specifications of the system are as follows: maximum span of the microphone array:165cm, number of microphones:17, rotating speed of the array:0.5 r.p.m. , used frequency range 200Hz ~15kHz. The whole system is controlled by a computer as shown in Fig.1.

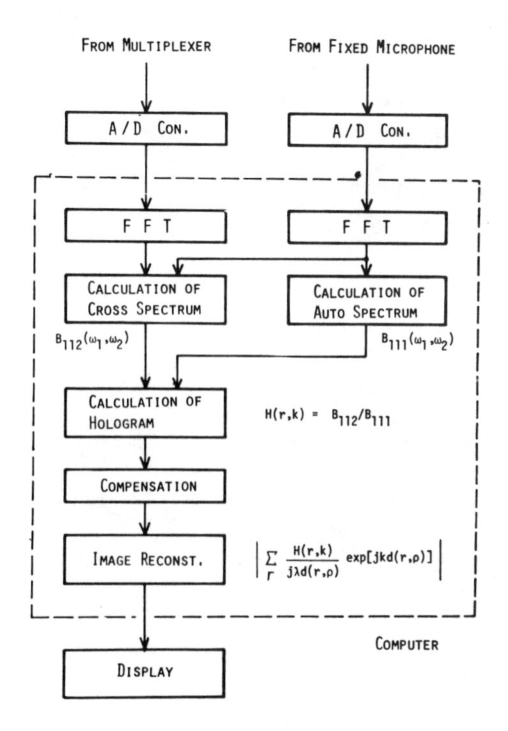

Fig. 7. Signal Processings in Computer

The details of the signal processings in the computer are shown in Fig.7. Power spectral analyses or higher order spectral analyses are chosen according to the desired image, that is, if the distribution of power is required, power spectral analysis is used, while if the distribution of the degree of coherence among frequency components is required, then higher order spectral analysis must be used.

The compensation is carried out by using suitable weight for the raw hologram. Once the compensated hologram is obtained, the reconstruction plane can be chosen at any desired one by the choice of the parameters at the stage of image reconstruction. Thus image of object wave field of any desired three dimensional cross section is obtained. The obtained results are graphically printed out or displayed on a graphic display by indicating the difference of amplitude by distinct letters.

V. EXPERIMENTAL RESULTS

A few results of the experiments are shown in this chapter. As a typical example, the following parameters were adopted: frequency used for image reconstruction ; 10kHz, span of microphone array; 80 cm, number of microphone; 17, Δ=0.1 (see Eq.10), the array was rotated by the step of 10°, the detected signals were sampled at 30kHz and 512 sampled data were fed for the FFT analyses. It takes about 10 minutes to obtain the whole hologram.

First, the point spread function of the system was measured by using a plane driven acoustical transformer, which is a kind of

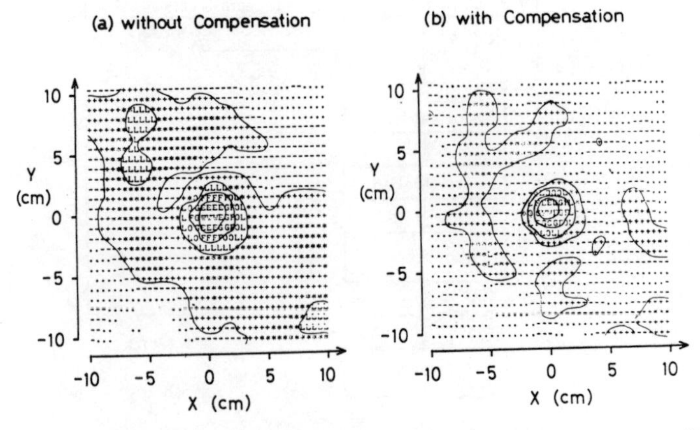

Fig. 8. Reconstructed Images of Point-like Object(I)
- X-Y Cross-Section -

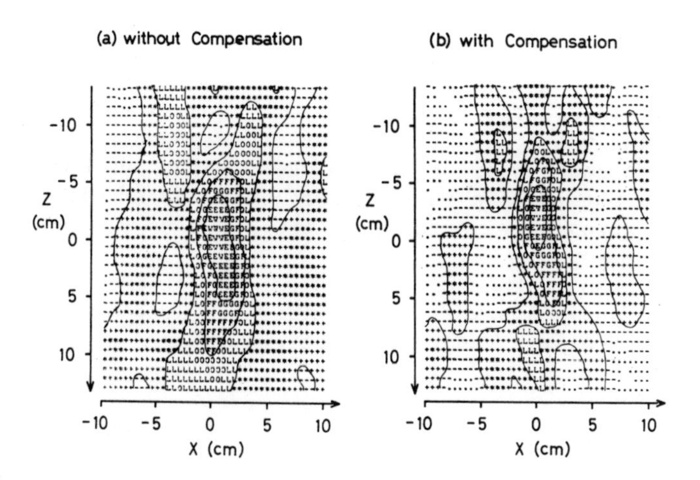

Fig. 9. Reconstructed Images of a Point-like Object
- X-Z Cross-Section -

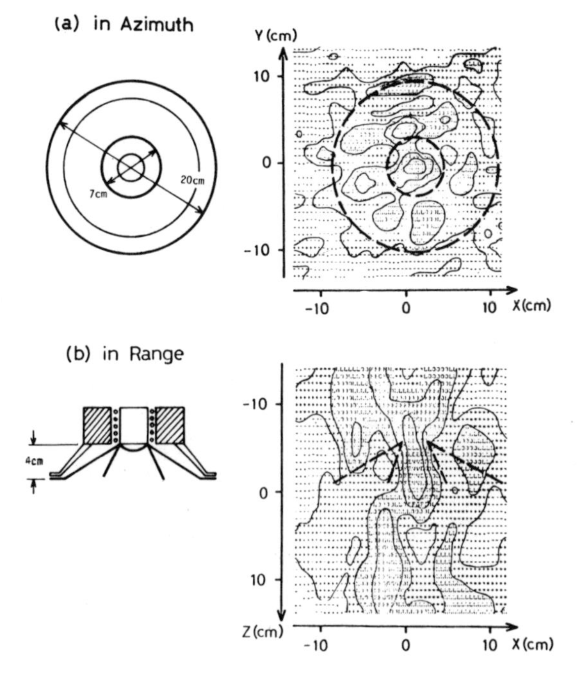

Fig. 10. Reconstructed Images of an Audio Speaker with Double Cones

speaker of 1cm diameter, as the object. It was excited by a white noise from a noise generator. The results are shown in Figs.8 and 9.

We can see clearly the effects of hologram compensation both in azimuth and range directions where the compensation which maximizes the range resolution was used. About one half of resolving power is obtained in range direction compared with that of the azimuth direction as expected from the previous discussions.

Next, as an example of a practical object the wave field of a double coned audio speaker was examined. It was excited by white noise. The reconstructed cross-sectional images of the wave field are shown in Figs.10 and 11.

We can see clearly the effect of the supplimentary small cone for the improvement of resulting acoustical wave field by comparing these results. The special feature that any desired cross-sectional image can be reconstructed and displayed from the same hologram may offer a useful means for the design of desired audio systems and other many applications.

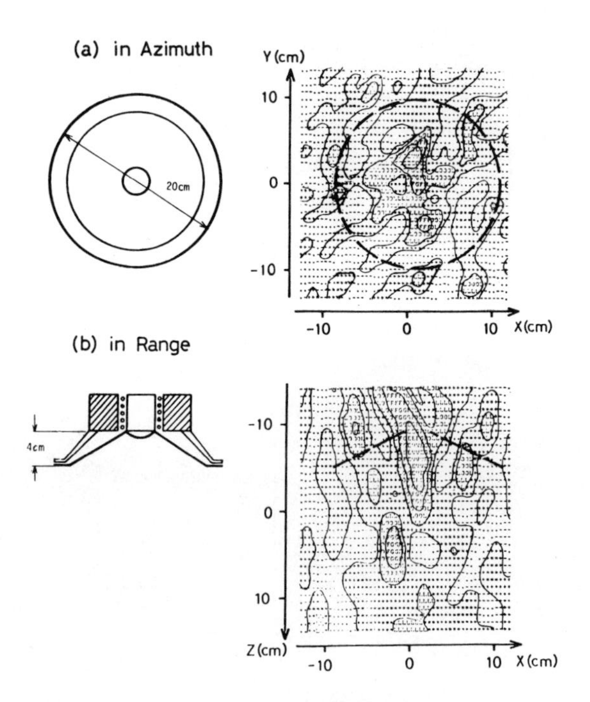

Fig. 11. Reconstructed Images of the Audio Speaker after Taking off the Supplimentary Cone

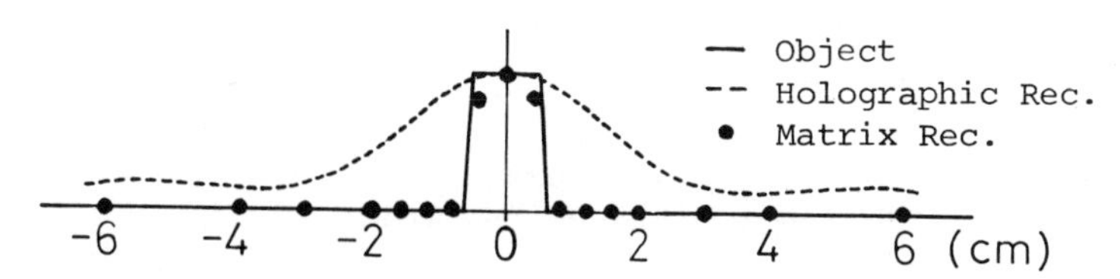

Fig. 12. Superresolution by Matrix Correspondence

VI. SUPERRESOLUTION IMAGING USING MATRIX CORRESPONDENCE BETWEEN DETECTED HOLOGRAM AND OBJECT'S WAVE FIELD

In some cases, the structures of noise emitting objects are much smaller than the wave length in audio frequency range. For the imaging of the object of this class we may use the direct correspondence between the detected signals on the hologram surface and the complex amplitudes of the mesh points on the surface of the object.[5] Usually the object's surface can be meshed properly, since their physical constructions can be known beforehand. Moreover the wave propagation can be considered to follow the conventional linear wave equation, hence the correspondence can be realized through proper matrix. The mesh points may be chosen with arbitrarily small intervals resulting in a superresolution imaging.

A numerical example is shown in Fig.12, where even the fine structure of which scale is about one third of the wave length is reconstructed.

VII. CONCLUSIONS

The application of a hemispherical detection of an acoustical wave field made possible the three dimensional image reconstruction of the wave field of a noise emitting object. The experimental results obtained by the constructed system showed the usefulness of the method.

Further application for the imaging of distribution of coherence among frequency components may offer a powerful means for more precise analysis of audio system and diagnosis of noise emitting machine systems.

Moreover the superresolution imaging which uses the inverse matrix calculation may open another fields of application of the system.

REFERENCES

1. T.Sato, K.Sasaki, Y.Nakamura and M.Nonaka, "Bispectral Passive Holographic Imaging System", Acoustical Holography Vol.7 L.W.Kessler Ed.179, Plenum Press, New York, (1977).

2. K.Sasaki, T.Sato and Y.Nakamura, "Holographic Passive Sonar", IEEE Trans. Sonics & Ultrasonics, Vol.SU-24, No.3, May, 193 (1977).

3. S.Ueha, M.Fujinami, K.Umezawa and J.Tsujiuchi, "Mapping of Noise-like Sound Sources with Acoustical Holography", Appl. Opt., Vol.4, 14(1975).

4. J.Brillingsley and R.Kinns, "The Acoustic Telescope", J.Sound & Vib. Vol.48, 485(1976).

5. D.P.Vasholz, "General Formulation of the Source Extraction Problem", J.Acoust.Soc.Am. Vol.61, No.6, June,1550 (1977).

PERIODIC SAMPLING ERRORS IN SCANNED ULTRASONIC HOLOGRAPHY

M I J Beale
I & AP Division, Building 347.1, AERE Harwell,
Oxfordshire OX11 ORA.

ABSTRACT

This paper describes an investigation into the effect of periodic sampling errors arising from systematic errors in the scanning mechanism which are not followed by the rest of the system. The effect of such errors on the images of both point and planar reflectors is analysed and spurious image detail is shown to be generated. A method for quantifying periodic sampling errors is outlined.

LIST OF SYMBOLS

A	Amplitude of holographic signal
B	Amplitude of recorded bias
F	Focal length of Fourier lens
$Jn(e\Omega)$	Bessel function of order n of the first kind, argument $e\Omega$
e	Amplitude of sampling error in the scan direction
f	Ultrasonic frequency
n	Integer
S	Distance from back focal plane to screen
t	Time
w	Space frequency of error = $2\pi \div$ wavelength of error
x	Coordinate in scan direction
x_f	Satellite to main carrier distance in back focal plane
x_s	Coordinate on image screen
z	Coordinate of depth below scan line
α	Angle from normal of equivalent phase reference
λ	Ultrasound wavelength
λ_ℓ	Reconstruction light wavelength
θ	Angle of sound path to scan normal
ν	Reconstruction light frequency
$\emptyset$	Phase constant
ψ	Phase constant
Ω	Space frequency = $\frac{4\pi}{\lambda}$ Sin θ

INTRODUCTION

Ultrasonic holography offers a method of converting an ultrasonic image into an optical image whilst retaining three dimensional position information. A number of different implementations are possible but we are here concerned only with systems in which the aperture is sampled at intervals by a mechanically scanned transducer. This paper describes an investigation into the effect of periodic sampling errors arising from systematic errors in the scanning mechanism.

Attention has been concentrated on periodic rather than random errors, the former are likely to occur in any mechanically scanned system and have been found to introduce spurious image detail. The effect of random errors is simply to degrade image resolution (Ref 1).

A theoretical treatment is given of the effect of periodic sampling errors on the images of both point and planar reflectors. The image of a point reflector has a weak spurious point on either side of the true image and the image of a planar reflector is modulated by a spurious grating pattern. The focussing properties of the grating pattern are analysed and comparison made with imaging a simple grating with a screen and lens.

Measurements in the back focal plane of the reconstruction lens are described which enable the amplitude and space frequency of the errors to be determined.

The principles of scanned ultrasonic holography are not discussed in detail as reference is made to the literature (Ref 2, 3). However, a brief description is given of the holography system developed at AERE Harwell as the performance of the scan and its subsequent improvement are described.

DETAILS OF THE SYSTEM

A scanned ultrasonic holography system was developed at Harwell for the Non Destructive Testing Centre there (Ref 2). The design objectives included flexibility in use, a wide dynamic range and freedom from spurious image detail. In this system an ultrasonic transducer, operated in pulse-echo, is mechanically scanned over a rectangular aperture in a raster pattern. The received signal is multiplied by an electronically generated reference and the holographic output recorded on a facsimile recorder. The recording is photographically reduced to a 35mm transparency and the image is then reconstructed optically.

The aperture is scanned by driving the transducer in a fast reciprocating motion along one axis while slowly moving it along the other. The reciprocating motion is achieved by mounting the transducer on a carriage connected to a loop of tape which passes over two drums, one at each end of the scan line. The carriage runs on support rails and is driven by a motor coupled, via a train of gears, to one of the drums. The slow scan motion is achieved by driving the fast scan assembly on support rails by a nut which bears onto a rotating screwed thread.

Periodic scanning errors may be generated in either scan direction by errors in the gears or the screwed thread. A further source of periodic errors is mechanical vibration of the scan. It will be shown that errors in the transducer position as small as 10μm may be significant.

The motion of the transducer must be synchronised to the processing electronics and the recorder. This is done using stepping motors to drive both scan and recorder. In the ideal situation the transducer samples the aperture at equal increments in time while scanning with a uniform velocity, hence the sampling is of uniform spacing. With the assumption of uniform sampling spacing the electronics and recorder form the hologram. Thus any perturbation of the scan velocity which is not followed by the rest of the system, results in an error in the hologram. It is such errors that are investigated here. Although in principle velocity perturbations on the recorder could introduce the same type of error in the hologram, the dominant errors were found in practice to be due to the scan.

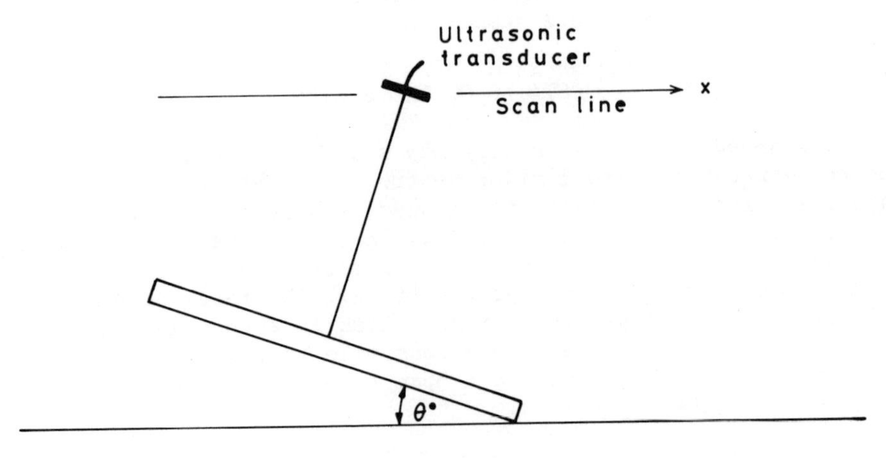

Fig 1 Scanning an inclined planar reflector

HOLOGRAM FORMATION

To determine the effect of a periodic sampling error on the reconstructed images we shall first investigate the imaging of an inclined planar reflector. A planar reflector is used as it facilitates the analysis and measurement of the error, the imaging of a point reflector will be considered later.

Error Free Hologram of a Planar Reflector

A hologram is recorded using a transducer oriented to receive a specular reflection from a planar reflector inclined to the plane of the scan as in Fig 1. Consideration of the sound path lengths shows the received signal for a pulse-echo system to be of the form:-

$$A \cos\left[\frac{4\pi}{\lambda} x \sin\theta + 2\pi ft + \Psi\right] \qquad \text{Eq 1}$$

x = displacement in the scan direction

θ = angle of inclination of reflector from scan line

λ, f = wavelength and frequency of ultrasound

Ψ = arbitrary phase constant

The two dimensional case has been adopted for simplicity of notation, the extension to three dimensions can be done on inspection as there is no interaction between the two dimensions in the scan plane. Further simplification is achieved by neglecting the details of the sampling nature of the system, the sampling rate is assumed to be adequate (Ref 2, 3), and also by assuming the phase reference corresponds to a beam at normal incidence to the aperture.

After electronic processing the recorded hologram is a sinusoidal grating of the form:-

$$B + A \cos\left[\Omega x + \Psi\right] \qquad \text{Eq 2}$$

$\Omega = \frac{4\pi}{\lambda} \sin\theta$

B = Bias necessary to record the signal on the unipolar recorder

Hologram of a Planar Reflector With a Periodic Sampling Error

If there is a periodic error in the scan then the form of the recorded hologram will be modified. In general an error will have a resolved component along the scan line and another perpendicular to it. Attention will be concentrated on the former type as it is likely to be dominant in practice, the latter type may be reduced to a very low level by increasing the rigidity of the static components of the scan. Either type of error may be analysed in the following manner but it will be assumed here that the error is confined to the scan line and is of the form

$$e \, \mathrm{Sin} \, (wx + \emptyset) \qquad \text{Eq 3}$$

e = amplitude of error

w = space frequency of error

$\emptyset$ = arbitrary phase constant

Consideration of the sound path lengths perturbed by the error shows the recorded hologram becomes (cf Eq 2):-

$$B + A \, \mathrm{Cos} \left[\Omega \left(x + e \, \mathrm{Sin} \, (wx + \emptyset) \right) + \Psi \right] \qquad \text{Eq 4}$$

For simplicity it is assumed that the hologram is not demagnified.

The effect of the error is analogous to phase modulating a carrier wave in communications theory. Interpretation of this expression is facilitated by its series expansion (Ref 4):-

$$\begin{aligned} \mathrm{Cos} \left[\Omega \left(x + e \, \mathrm{Sin} \, (wx + \emptyset) \right) + \Psi \right] = \; & J_0(e\Omega) \, \mathrm{Cos} \left[\Omega x + \Psi \right] \\ & + J_1(e\Omega) \, \mathrm{Cos} \left[(\Omega + w) \, x + \Psi + \emptyset \right] \\ & - J_1(e\Omega) \, \mathrm{Cos} \left[(\Omega - w) \, x + \Psi - \emptyset \right] \\ & + J_2(e\Omega) \, \ldots\ldots\ldots \end{aligned} \qquad \text{Eq 5}$$

For small errors ($e\Omega \ll 1$) terms in $J_2(e\Omega)$ and higher may be neglected, $J_0(e\Omega) \simeq 1$ and $J_1(e\Omega) \simeq \frac{1}{2}e\Omega$. The recorded hologram may be expressed as

$$\begin{aligned} B + A \, \mathrm{Cos} \left[\Omega x + \Psi \right] & + \tfrac{1}{2}Ae\Omega \, \mathrm{Cos} \left[(\Omega + w)x + \Psi + \emptyset \right] \\ & + \tfrac{1}{2}Ae\Omega \, \mathrm{Cos} \left[(\Omega - w)x + \Psi - \emptyset + \pi \right] \end{aligned} \qquad \text{Eq 6}$$

The hologram consists of a bias term and three superimposed gratings, the principal grating being identical to that recorded by an error free system. The amplitudes of the secondary gratings are a function of the reflector inclination angle and are zero for a reflector parallel to the scan line. Hence a scan error confined to the scan line is of more importance if an inclined transducer is used such as in shear wave holography (Ref 5), than if a normal transducer is used as in compressional wave holography.

HOLOGRAM RECONSTRUCTION

The error free hologram and the sampling error hologram are reconstructed in the normal manner using parallel coherent light (Fig 2 and Ref 2). The transmitted light forms an image on a screen which may be moved into the back focal plane of the lens or positioned to focus at any chosen depth in the image. If the screen is in the back focal plane of the lens, the Fourier Transform of the space frequency content of the hologram is displayed. The exact Fourier Transform of the hologram is observed only if the hologram lies in the front focal plane, otherwise the Fourier Transform is modified by a quadratic phase term (Ref 6). From the position and intensity of a point of light in the back focal plane the corresponding space frequency and its amplitude present in the hologram may be calculated.

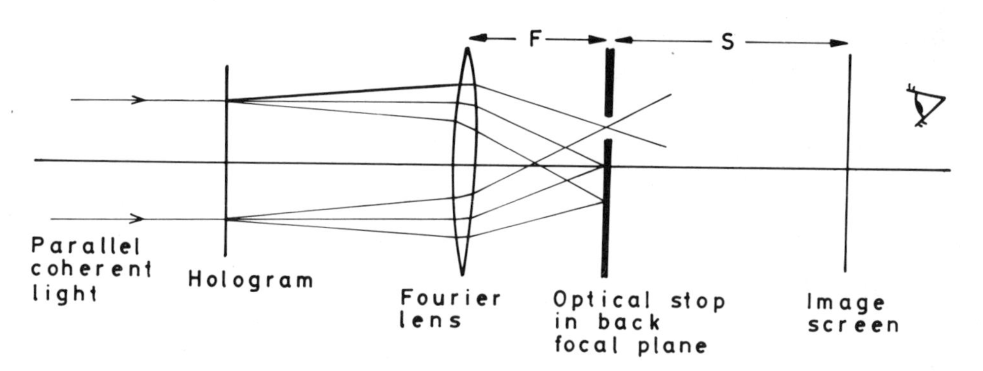

Fig 2 Optical reconstruction system

Space Frequency Content

The back focal plane contains three points of light when the error free hologram is reconstructed. A central point and two weaker points symmetrically placed about it (Fig 3). The central point corresponds to the Fourier Transform of the bias term and the two weaker points to that of the sinusoidal grating. The latter points reconstruct the true and pseudoscopic images of the planar reflector. Due to the rectangular aperture the points have in fact a $\left(\frac{\mathrm{Sin}\ x}{x}\right)^2$ intensity distribution (Ref 6). Pursuing the communications analogy, we refer to these points as the 'main carriers' as it is these that are phase modulated by the error frequency.

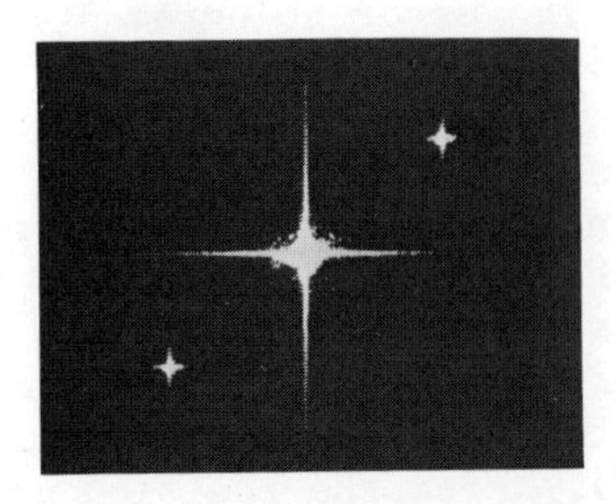

Fig 3 Photograph of back focal plane of error free hologram

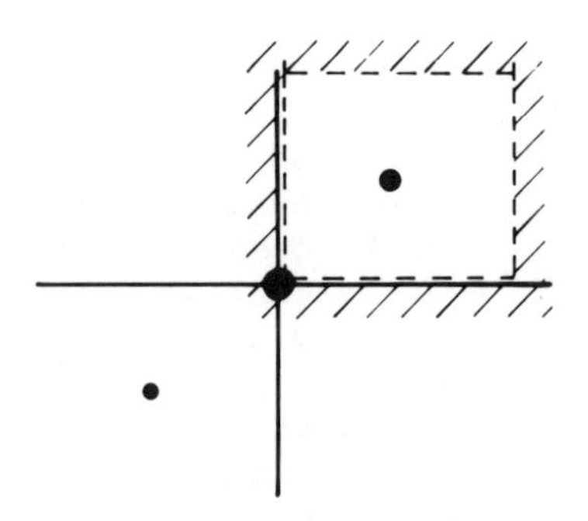

Fig 4 Diagram to show spatial filter position on Figure 3

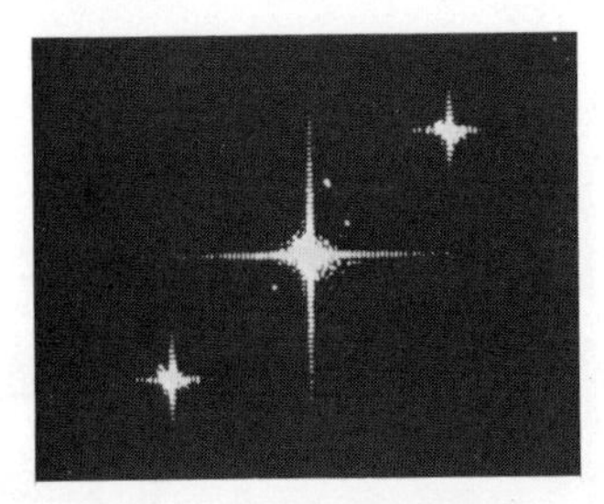

Fig 5 Photograph of back focal plane of hologram recorded with a single frequency sampling error

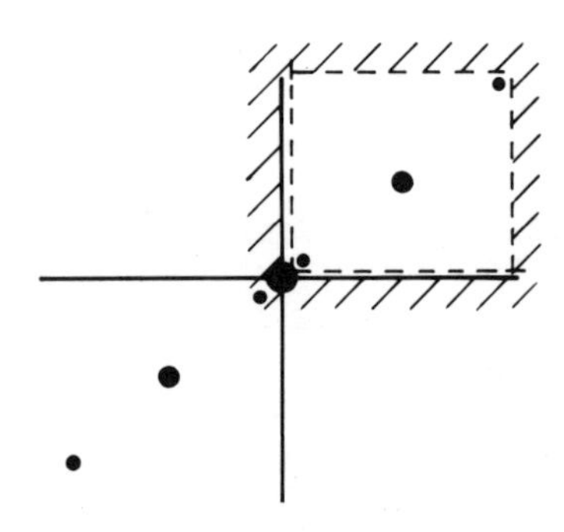

Fig 6 Diagram to show spatial filter position on Figure 5

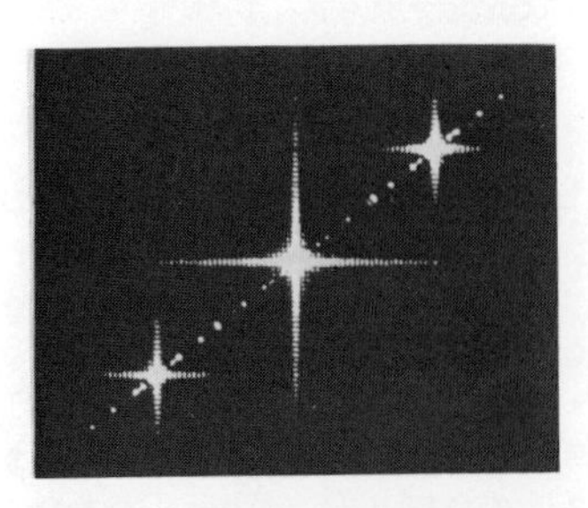

Fig 7 Photograph of back focal plane of hologram recorded with multiple sampling errors

When the sampling error hologram is reconstructed the back focal plane contains three points of light with positions identical to those of the error free hologram. But, four additional weak points are observed; these correspond to the Fourier Transform of the two additional space frequencies present in this hologram (Eq 5). Each space frequency gives two points symmetrical about the central point. The extra points will be referred to as 'satellites' and, due to the relationship between the space frequencies in the hologram, they are symmetrically placed about the main carriers (Fig 5).

Reconstructed Image

The normal condition of reconstruction is with a spatial filter positioned in the Fourier plane to stop out both the central point and the distribution corresponding to the spurious image. In this way the true image is isolated for viewing (Fig 8). Under this condition a single point, the main carrier, forms the image from the error free hologram (Fig 4), from the sampling error hologram the main carrier and its two satellites form the image (Fig 6).

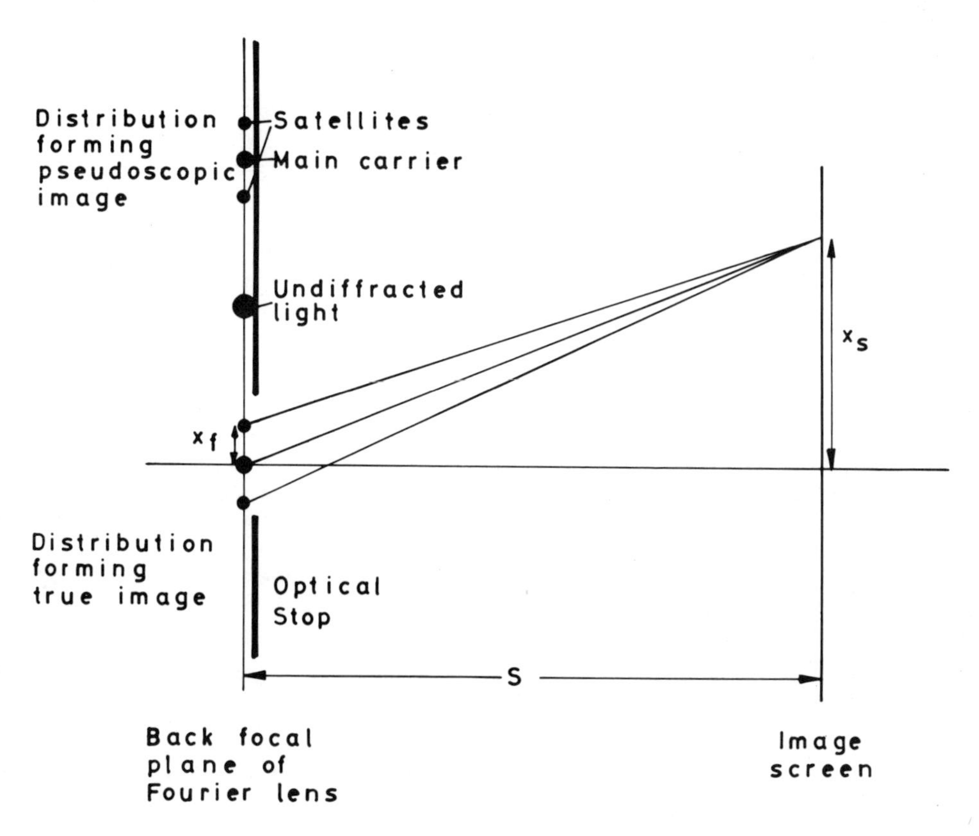

Fig 8 Reconstruction of sampling error hologram

In the case of a planar reflector an image as such cannot be located as the two diffracted beams are plane waves bounded by the aperture. Thus the main carrier alone reconstructs a uniformly illuminated rectangle while the satellites also reconstruct similar but fainter images. The three images overlap and interfere to form a grating pattern (Fig 9). The effect of the scan error has been to introduce spurious image detail.

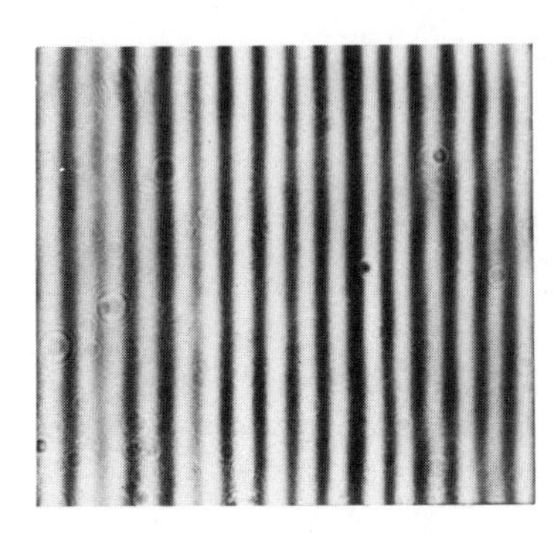

Fig 9 Reconstructed image of planar reflector recorded with a periodic sampling error showing the spurious grating pattern

The focussing properties of the grating pattern are not those predicted by simple geometrical optics and will now be analysed.

Focussing Properties of the Image

We now determine the nature of the image formed by the three points of light from the back focal plane of the reconstructed sampling error hologram. To simplify the analysis the hologram will be assumed to be in the front focal plane of the lens, thus the back focal plane contains the exact Fourier Transform of the hologram.

The image intensity at any point on the screen is obtained by the vector addition of the light arriving there from each of the three points. The phase of each contribution at the screen is dependent on its phase in the back focal plane and its path length to the screen. The phase of each point in the back focal plane is given by the Fourier Transform of the hologram (Eq 5), either of the two symmetrical distribution can be taken, both lead to the same result. Taking phases of $\Psi, \Psi + \emptyset$ and $\Psi - \emptyset + \pi$ for the main carrier and satellites respectively and using a parabolic approximation to the path length from back focal plane to screen, valid under the imaging conditions used, the wave front in the plane of the screen takes the form:-

$$\tfrac{1}{2} J_0(e\Omega) \cos\left[\nu t - \frac{2\pi}{\lambda_\ell} S\left(1 + \frac{x_s^2}{2S^2}\right) + \Psi\right]$$
$$+ \tfrac{1}{2} J_1(e\Omega) \cos\left[\nu t - \frac{2\pi}{\lambda_\ell} S\left(1 + \frac{(x_s - x_f)^2}{2S^2}\right) + \Psi + \emptyset\right]$$
$$+ \tfrac{1}{2} J_1(e\Omega) \cos\left[\nu t - \frac{2\pi}{\lambda_\ell} S\left(1 + \frac{(x_s + x_f)^2}{2S^2}\right) + \Psi - \emptyset + \pi\right]$$

Eq 7

x_s = Position on screen

$x_f = \dfrac{\lambda_\ell w F}{2\pi}$ = point separation in back focal plane

F = Focal length of Fourier lens

S = Screen to back focal plane distance

If the screen position S is chosen such that the following relationship is held

$$\frac{2\pi}{\lambda_\ell} \cdot \frac{x_f^2}{2S} = (2n-1)\frac{\pi}{2}$$

Eq 8

n = integer

Then Eq 7 reduces to a:-

$$\tfrac{1}{2} J_0(e\Omega)\left[1 \pm 2\frac{J_1(e\Omega)}{J_0(e\Omega)} \sin\left[\frac{2\pi}{\lambda_\ell} x_s \frac{x_f}{S} + \emptyset\right]\right] \cdot \cos\left[\nu t - \frac{2\pi}{\lambda_\ell} S\left(1 + \frac{x_s^2}{2S^2}\right) + \Psi\right]$$

Eq 9

Thus a grating pattern is observed on the screen with an intensity ratio of

$$\left[\frac{1 + 2\left(\dfrac{J_1(e\Omega)}{J_0(e\Omega)}\right)}{1 - 2\left(\dfrac{J_1(e\Omega)}{J_0(e\Omega)}\right)}\right]^2$$

Eq 10

and space frequency

$$\frac{2\pi}{\lambda \ell} \frac{x_f}{S} = \frac{wF}{S} \qquad \text{Eq 11}$$

For screen positions not obeying Eq 8 an out of focus grating pattern results. Thus as the screen position is varied the grating pattern goes through periodic focussed positions. With the hologram positioned in the front focal plane of the lens the geometrical optics focus position is with the screen at S = infinity. Under this condition the grating pattern is at its worst focus as the equality of Eq 8 is in error by $\frac{\pi}{2}$. This rather surprising result arises from the phase relationships of the three space frequencies constituting the hologram, the resultant vector of the sidebands is $\frac{\pi}{2}$ out of phase with the main carrier.

Imaging a simple grating such as the error free hologram with a lens forms an in-focus image on the screen under the conditions defined by geometrical optics. In this case, following the procedure above, the resultant of the main carriers is found to be in phase with the central point.

Though not pursued here it may be shown that for the sampling error hologram the grating pattern is always at its worst focus when the screen is positioned by geometrical optics to focus on the plane of the hologram irrespective of the position of the hologram in front of the lens.

A METHOD FOR THE MEASUREMENT OF SAMPLING ERRORS

To quantify a scan error in terms of space frequency and amplitude, the position and intensity of the points in the back focal plane are measured and Eq 7 used. The errors found in the Harwell scan and their subsequent reduction are now described.

Initially the direction in which the errors lay was identified by measuring the intensity of the satellite spots as a function of the inclination of the reflector. In both slow and fast scan directions the errors were confined to the scan plane, no out of plane components were detected.

In the slow scan direction a single frequency error was present due to a bend on the screwed drive thread, its space frequency corresponded to the pitch of the thread (20 tpi = 0.787mm^{-1}) (Fig 5). Replacement of the brass nut, bearing directly onto the screwed thread, with a doubly gimballed cork-lined nut reduced the error amplitude from $\simeq$30μm to $\simeq$10μm (Fig 3).

The errors in the fast direction consisted of a number of space frequencies some of which were dependent on the scan speed (Fig 7). The speed dependent errors were due to mechanical resonances in the scan, driven by the forces developed at the stroke reversal point. The scan was modified to minimise these forces, to increase its rigidity and to reduce slackness in the drive system. The speed independent errors were due to cutting errors in the drive gears and noise from the carriage bearings. Prior to modification the error amplitudes were of the order of 40μm and were reduced to $\simeq$5μm.

The severity of the grating pattern caused by such errors depends on the wavelength of ultrasound used and the inclination of the reflector. The calculated variation of satellite to carrier intensity is plotted against reflector inclination for a number of error amplitudes (Fig 10). The intensity ratio is $\left[\frac{J_1(e\Omega)}{J_0(e\Omega)}\right]^2$ from Eq 7 and the error amplitude is expressed as a fraction of the ultrasound wavelength. The variation of the contrast ratio of the resultant grating pattern with the satellite to carrier intensity ratio is shown in Fig 11. The contrast ratio is expressed as $\frac{I_{max} - I_{min}}{I_{max} + I_{min}}$ where Imax and Imin are the extremes of intensity found in the grating pattern at its optimum focus. An error of amplitude $\lambda/100$ when scanning a reflector inclined at $\theta=24^\circ$ results in a satellite to carrier intensity ratio of $\simeq$1:1500 and a contrast ratio of 0.1. Though the satellite intensity is relatively weak the resultant grating pattern is easily visible, ie the system is very sensitive to small periodic perturbations when imaging a planar reflector. The very strong grating of Fig 9 resulted from an error of $\simeq\lambda/20$ with an inclination of 38°.

HOLOGRAM OF A POINT REFLECTOR WITH A SAMPLING ERROR

We now extend the analysis of periodic sampling errors to the imaging of a point reflector. Consideration of the sound path lengths shows the error free hologram to be of the form (Fig 12):-

$$B + \frac{A}{\sqrt{x^2 + z^2}} \cos\left[\frac{4\pi}{\lambda}\sqrt{x^2 + z^2}\right] \qquad \text{Eq 12}$$

x Scan coordinate

z Range perpendicular to scan

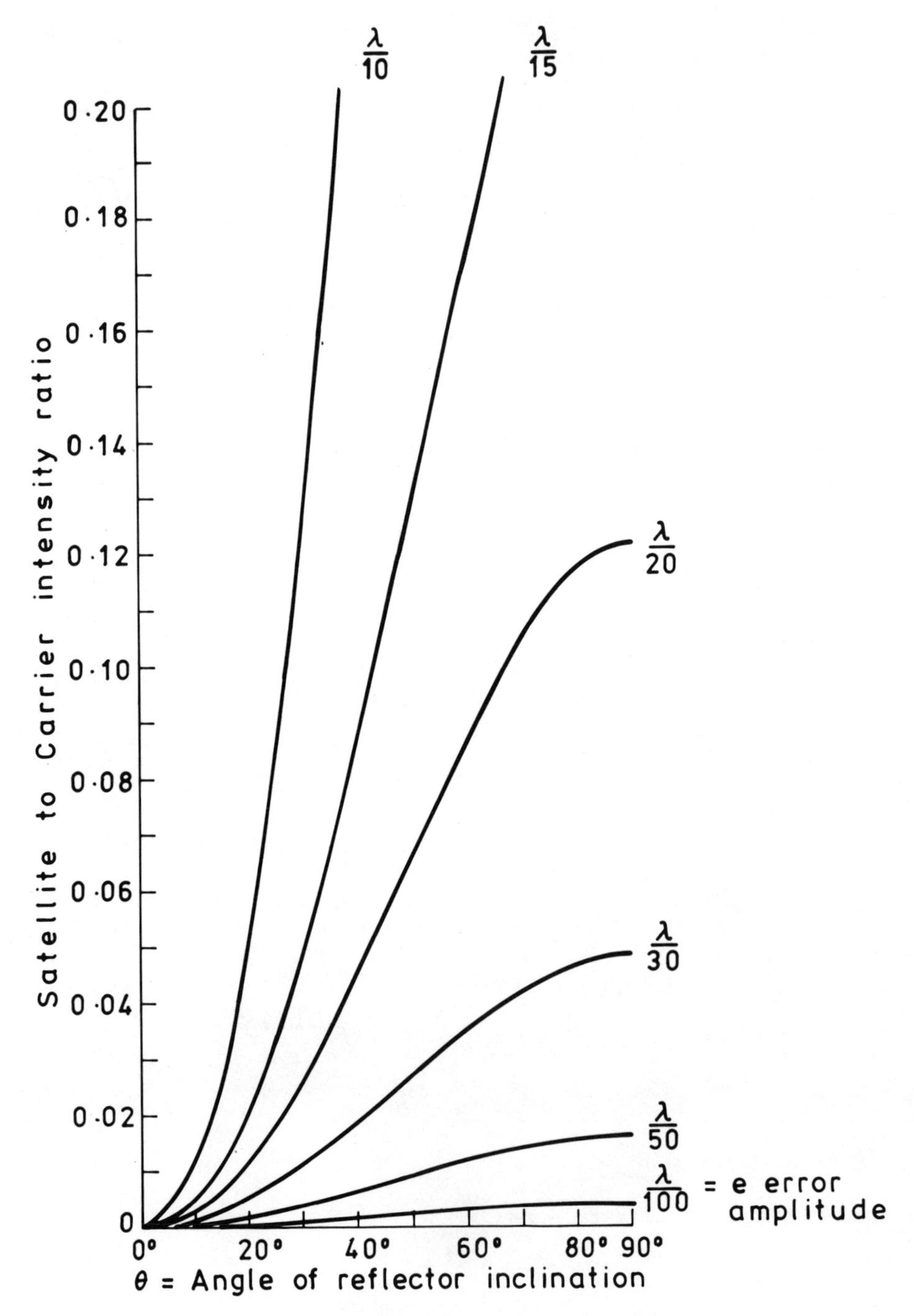

Fig 10 Graph of Satellite to Carrier intensity ratio against Angle of reflector inclination

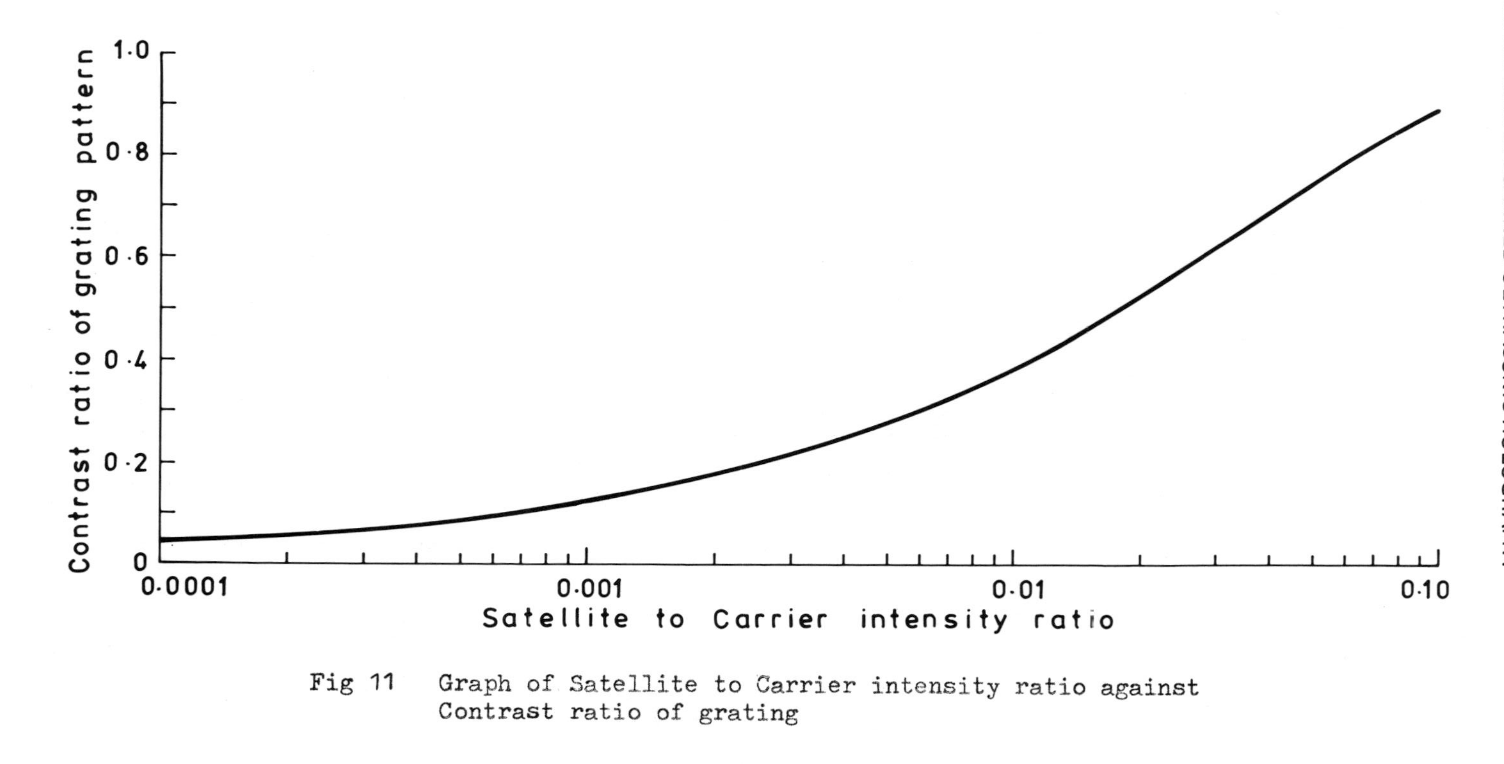

Fig 11 Graph of Satellite to Carrier intensity ratio against Contrast ratio of grating

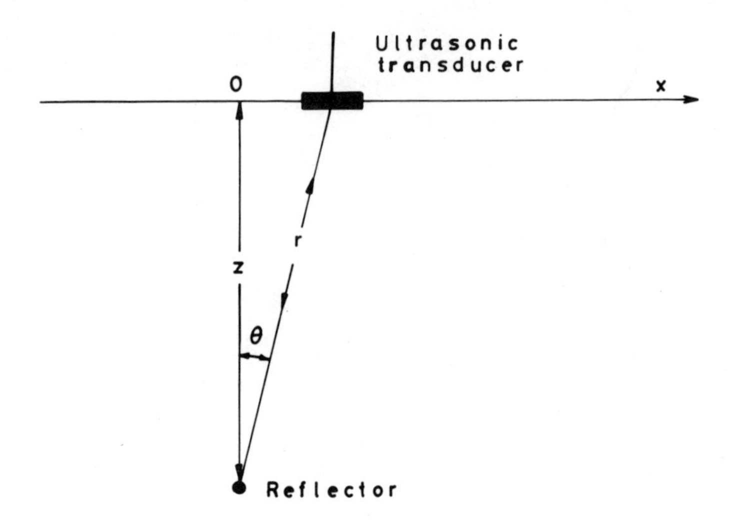

Fig 12 Scanning a point reflector

On reconstructing this hologram the true image would be superimposed on the pseudoscopic image and the undiffracted light. The image could be separated by using an inclined reference, however for simplicity of notation a normal reference will be used, the results deduced apply equally for the case of an inclined reference.

With an error in the scan direction as previously of

$$e \text{ Sin } wx$$

The recorded hologram, supressing the amplitude term, becomes:-

$$B + A \text{ Cos} \left[\frac{4\pi}{\lambda} \sqrt{(x + e \text{ Sin } wx)^2 + z^2} \right] \qquad \text{Eq 13}$$

For small amplitude errors $e \ll \lambda$ and $e \ll z$ substituting

$$\text{Sin } \theta = \frac{x}{\sqrt{x^2 + z^2}} \text{ and } \Omega = \frac{4\pi}{\lambda} \text{ Sin } \theta$$

Equation 13 becomes

$$B + A \text{ Cos} \left[\frac{4\pi}{\lambda} \sqrt{x^2 + z^2} + e\Omega \text{ Sin } wx \right] \qquad \text{Eq 14}$$

Note Ω not now a constant as it was when imaging the planar reflector.

With further series expansion (Ref 4)

$$\text{Cos}\left[\frac{4\pi}{\lambda}\sqrt{x^2+z^2} + e\,\Omega\,\text{Sin}\,wx\right] = J_0(e\Omega)\,\text{Cos}\left[\frac{4\pi}{\lambda}\sqrt{x^2+z^2}\right]$$

$$+J_1(e\Omega)\,\text{Cos}\left[\frac{4\pi}{\lambda}\sqrt{x^2+z^2} + wx\right]$$

$$-J_1(e\Omega)\,\text{Cos}\left[\frac{4\pi}{\lambda}\sqrt{x^2+z^2} - wx\right]$$

$$+J_2(e\Omega)\ \ldots\ldots\ldots\ldots \qquad \text{Eq 15}$$

For small errors $e\Omega \ll 1$, $J_0(e\Omega) \simeq 1$ and $J_1(e\Omega) \simeq \frac{1}{2}e\Omega$ and terms in $J_2(e\Omega)$ and higher may be neglected. Thus the recorded hologram becomes

$$B + A\,\text{Cos}\left[\frac{4\pi}{\lambda}\sqrt{x^2+z^2}\right] + A\,\tfrac{1}{2}e\Omega\,\text{Cos}\left[\frac{4\pi}{\lambda}\sqrt{x^2+z^2} + wx\right]$$

$$- A\,\tfrac{1}{2}e\Omega\,\text{Cos}\left[\frac{4\pi}{\lambda}\sqrt{x^2+z^2} - wx\right] \qquad \text{Eq 16}$$

The hologram consists of three terms, the first is identical to that of the error free hologram and thus images to a point. The other terms are identical in phase to those that would be recorded by an error free system if two phase references inclined at $\pm\alpha^o$ were used during the recording, $\alpha^o = \text{Sin}^{-1}\left[\frac{w\lambda\ell}{4\pi}\right]$. Hence the reconstructed image consists of the true image with a spurious point at α^o on either side of it. The spurious terms have an amplitude weighting term that is dependent on the scan position, this results in the attenuation of low space frequency components of the image relative to high space frequencies. Though this must affect the spurious images it will not seriously disturb their character. The spurious point images will in any case be of very low intensity for small amplitude errors and are unlikely to be significant. Under typical scan conditions the intensity ratio of true to spurious images will be of the order of $\left[\frac{2e}{\lambda}\right]^2$, (see Eq 16).

CONCLUSION

Periodic sampling errors have been shown to introduce spurious image detail into the reconstructions of holograms of both planar and point reflectors. The image of a point reflector has a spurious point on either side of it, however these spurious images are very weak and unlikely to be significant. The image of a planar reflector has superimposed upon it a spurious grating pattern. The system has been shown to be particularly sensitive to periodic errors when imaging planar objects and an error of amplitude $\lambda/100$ shown to affect the image significantly. The severity of the grating pattern increases with the angle of inclination of the reflector to the scan plane and hence is likely to be more significant when using an inclined transducer as in shear wave holography.

The sensitivity of the system to periodic errors is in contrast to its sensitivity to random errors, the effect of which were investigated by Hildebrand and Collins (Ref 1). They concluded that random errors of the type investigated here do not affect the image significantly provided they are of amplitude less than $\lambda/4$.

ACKNOWLEDGEMENTS

Many thanks to E E Aldridge, A B Clare and D A Shepherd for their guidance and support in this work. Also to the European Coal and Steel Community and the Mechanical Engineering and Machine Tools Requirements Board (DoI) UK for their financial assistance.

REFERENCES

1. H D Collins and B P Hildebrand. The effects of scanning position and motion errors on hologram resolution. Acoustical Holography, Vol 4, Plenum Press 1972.

2. E E Aldridge. Acoustical Holography, Merrow 1971.

3. B P Hildebrand and B B Brenden. An Introduction to Acoustical Holography, Hilger 1972.

4. N W McLachlan. Bessel Functions for Engineers. OUP 1934.

5. H D Collins and B B Brenden. Acoustical Holographic Transverse Wave Scanning Technique for Imaging Flaws in Thick Walled Pressure Vessels. Acoustical Holography, Vol 5, Plenum Press 1973.

6. J W Goodman. Introduction to Fourier Optics. McGraw-Hill 1968.

DESCRIPTION OF A NAVY HOLOGRAPHIC UNDERWATER ACOUSTIC IMAGING SYSTEM (AIS)

J. L. Sutton, J. V. Thorn, N. O. Booth, B. A. Saltzer

Naval Ocean Systems Center, code 5213

San Diego, California 92152

BACKGROUND

The Acoustic Imaging System (AIS) was developed by the Naval Ocean Systems Center (NOSC), San Diego, for use aboard deep diving Navy submersibles and also as a research tool. The operational specifications of the AIS were chosen to lie between those of closed circuit television (short range, high resolution) and most sonars (long range, low resolution) to provide moderate range (5-100 feet) and moderate resolution (0.3 degree) (figure 1) for classification of objects. Holography was selected as the most promising technique for this underwater acoustic imaging application over focused or beamformed techniques (figure 2). The primary reasons for selecting holography were for (1) compactness and lack of any cable-snagging protrusions, (2) failure tolerance, and (3) simplicity of hardware. Submersible pilots are generally very particular about external equipment getting entangled in cables. Since holography requires no lens, only the hydrophone array and its immediate processing electronics need to be in the water. Failure tolerance comes from the fact that the most unreliable components in an acoustic imaging system appear to be the hydrophone elements and their associated circuits. In holography, these components are in the Fourier transform domain. Hence, any errors or failures in them contribute only to the overall noise background of the reconstructed image. In focused systems, these components are in the image plane, and failures give rise to holes in the image.

The above advantages apply equally well to beamformed systems. The major difference between holography and beamforming, which are mathematically identical (reference 1), is in the implementing hardware. In general, beamformed systems manipulate the high fre-

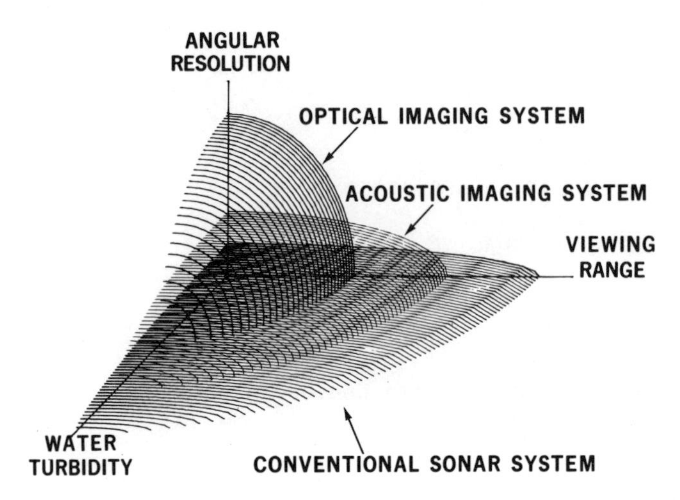

Figure 1. Relative performance of optical, acoustical and sonar viewing systems versus range, resolution, and water turbidity.

quency signal directly in many parallel processing channels. If the entire image is to be collected and processed all at once this involves a lot of complex electronics. Holography, on the other hand, converts the acoustic signal to a baseband hologram as soon as possible. This hologram can then be processed (reconstructed) in a single channel. This is important for size, power, complexity, cost and adaptability. A digital minicomputer was used as a high quality reconstructor for the AIS. Once the hologram is acquired, the computer can perform many tasks with the data and images before display of the image. These include variable focusing, aperture shading, error correction, image smoothing and dynamic range compression. Hence, acoustic holography was selected for the AIS. A filled square array of hydrophones was implemented in the AIS in order to gather the entire hologram in one snapshot. This was necessitated by the instability of both the platform and the scenes that are likely underwater. At a moderate range of say, 100 ft, the two way travel time of a fan beam would be 40 msec. The time for 50 such beams to be scanned over the field of view, one at a time, would be 2 seconds or more allowing for spurious echoes to die down. A vehicle can move a significant amount in 2 seconds, so the acoustic image would probably be smeared and useless. In light of this result, a raster scanned pencil beam is clearly out of the question unless more sophisticated techniques are employed.

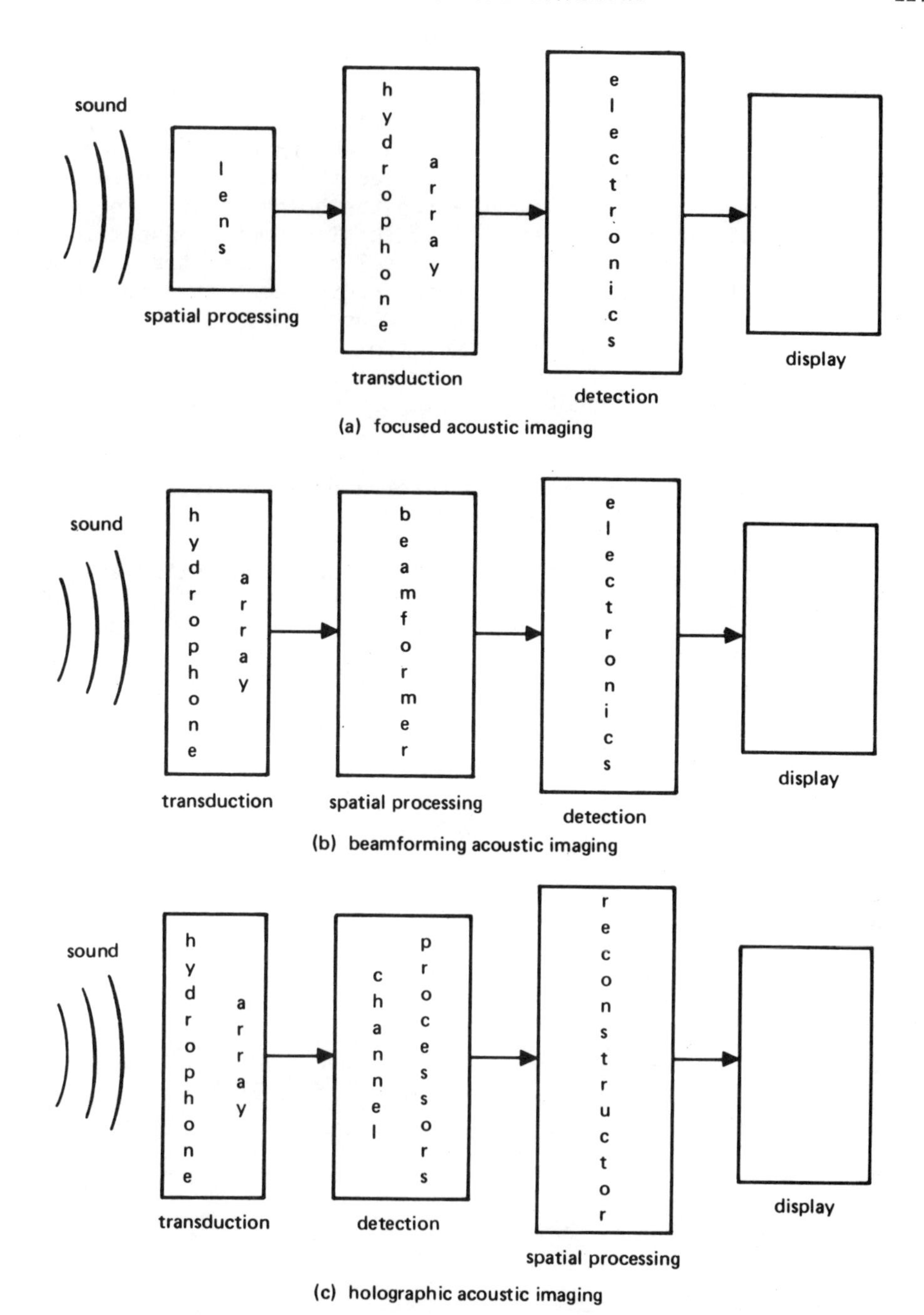

Figure 2. The basic types of undersea acoustic imaging systems.

OPERATION

The operation of the AIS aboard a small submersible is illustrated in Figure 3. The transmit/receive (T/R) unit sends ultrasonic sound into the water, receives the return echo by means of a square array of hydrophones, and processes the signal to create an acoustic hologram. The T/R unit sends the holographic information through the hull of the submersible to the T/R control panel in the form of digital data. The T/R control panel then transfers the holographic data to the computer reconstructor control panel, which in turn sends it to the computer reconstructor. The computer reconstructor reconstructs the image from the acoustic hologram under the direction of the computer reconstructor control panel. The hologram and the reconstructed image are displayed on separate cathode ray tube displays.

The T/R unit, shown mounted on a pan-and-tilt carriage in Figure 4, houses a 250-watt, 642-KHz transmitter as well as a 48-by-96 PZT hydrophone array, only half of which is used to form a uniform 48-by-48 array. In the array circuitry are 2304 channel processors (one for each hydrophone) mounted on printed circuit cards with 48 channel processors on each card. One card with the channel processors in place appears in Figure 5. The channel processors function in parallel to create and scan out

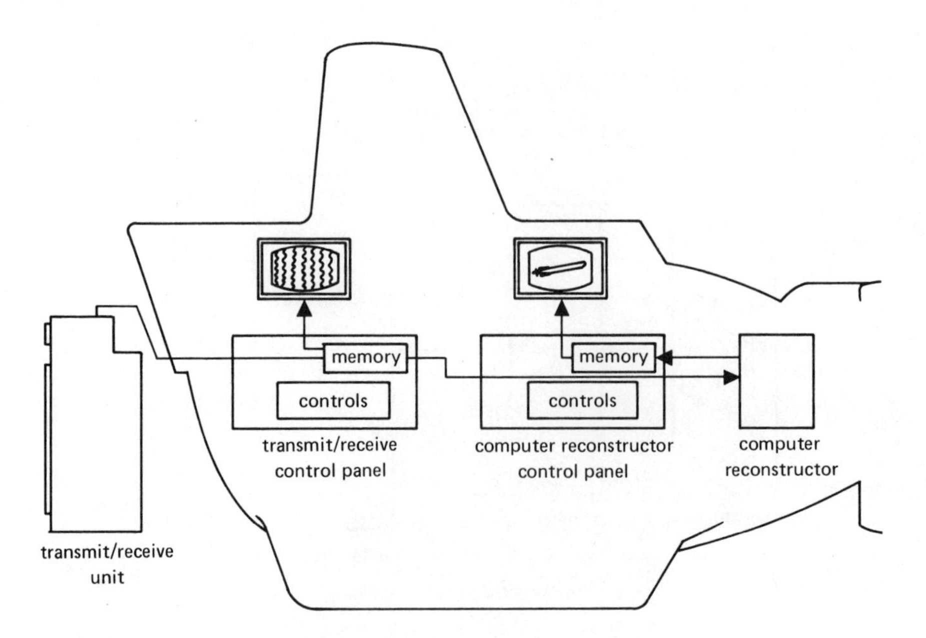

Figure 3. The operation of the AIS aboard a small manned submersible. The system would work equally well on a remotely operated vehicle such as CURV III.

Figure 4. The AIS transmit/receive unit mounted on a pan-and-tilt carriage for preliminary evaluation. The transmitter is at the top above the square hydrophone array. The unit is oil-filled and pressure-compensated for operation at depths to 12,000 feet (3,658 meters).

Figure 5. An AIS channel processor card. There are 48 such cards in the transmit/receive unit; each carries 48 channel processors.

acoustic holograms in a fraction of a second. It is this technique that enables the system to provide near real-time viewing. Figure 6 shows a diagram of a single channel processor. Each processor amplifies the electronic signals from a hydrophone, mixes them with a reference signal to determine their phase and amplitude, integrates them over a period of time, and multiplexes them to an analog-to-digital converter. Note that two mixing signals in quadrature are actually used in each channel. Arbitrarily named, one reference produces the sine term A sin θ where A is the amplitude and θ is the phase. The other gives the term A cos θ. The resulting hologram representation is known variously as complex, I.F., sine/cosine, or single sideband. The result is that there is no conjugate image in the reconstruction, even when the spatial aperture is undersampled.

The 48 channel processor cards and the acoustic transmitter electronics are all packaged inside the T/R unit using pressure tolerant electronics (PTE) techniques. That is, the thin-walled housing is filled with a pressure compensation oil which transmits the ambient hydrostatic pressure directly to the electronic component packages. Semiconductors and many passive components operate with no apparent degredation under these conditions at pressures as high as 15K psi. The PTE techniques were required by AIS for use aboard deep diving (12K ft) submersibles in order to connect the 2304 transducers to their corresponding electronic channel processors. The processing electronics were fabricated with integrated and discrete components mounted on multi-layer printed circuit boards.

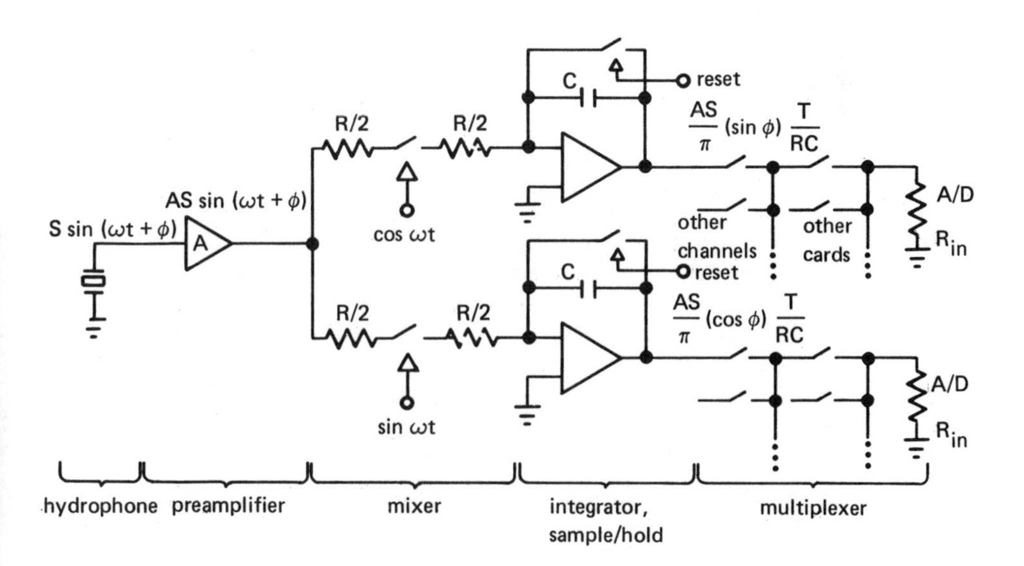

Figure 6. Diagram of AIS channel processor.

All system controls and the computer reconstructor are mounted in a single control rack as shown in Figure 7. The T/R control panel includes controls regulating the transmit, range, and receive gates. These determine the length of time the system transmits (transmit gate), waits to receive (range gate), and receives (receive gate), respectively. They enable the operator to select a target at some range of interest, blank out returns from other ranges, and then adjust the system for optimum viewing of the target. Figure 8 shows a simplified diagram of the transmit/receive cycle. The transmit gate length may be varied by a thumbswitch on the control panel from 1 usec to 1 sec in steps of approximately 1 percent. Shorter gate lengths are used to reduce average transmitted power, decrease reverberation noise by decreasing the ensonified water volume or by decreasing interference from targets at other ranges. Longer gate lengths are used to increase the depth of field by increasing the volume of ensonified water, or to improve the effective signal-to-noise ratio by allowing the channel processors more time to respond.

Similarly, the range gate may be varied from 1 usec to 1 sec in steps of approximately 1 percent. The speed of sound in water is about 5 feet/msec, so the range gate allows the operator to control the range of the system to an accuracy of approximately 0.005 feet.

The receive gate is variable from 1 usec to 1 sec in steps of approximately 1 percent. Short receive gates are used for separating returns from targets at slightly different ranges. Longer receive gates are used for looking at greater depths of field, or again, for improving the signal-to-noise ratio by allowing a longer integration time.

The computer reconstructor control panel governs the operation of the computer reconstructor and image display. The range and speed of sound are used to focus the system on the target in the same manner as a camera is focused. The computer processes this information in the same way as an acoustic lens does in a focused system. Data from portions of the array can be electronically blanked out to approximate smaller array aperture. As in a camera, this increases the depth of focus at short ranges. Other controls regulate the modes of operation and the display intensity and contrast.

DESIGN GOALS

Table 1 summarizes the design goals established for AIS. They include important performance parameters and some of the component characteristics necessary to achieve this level of performance. Resolution, image dynamic range, and image frame rate are particu-

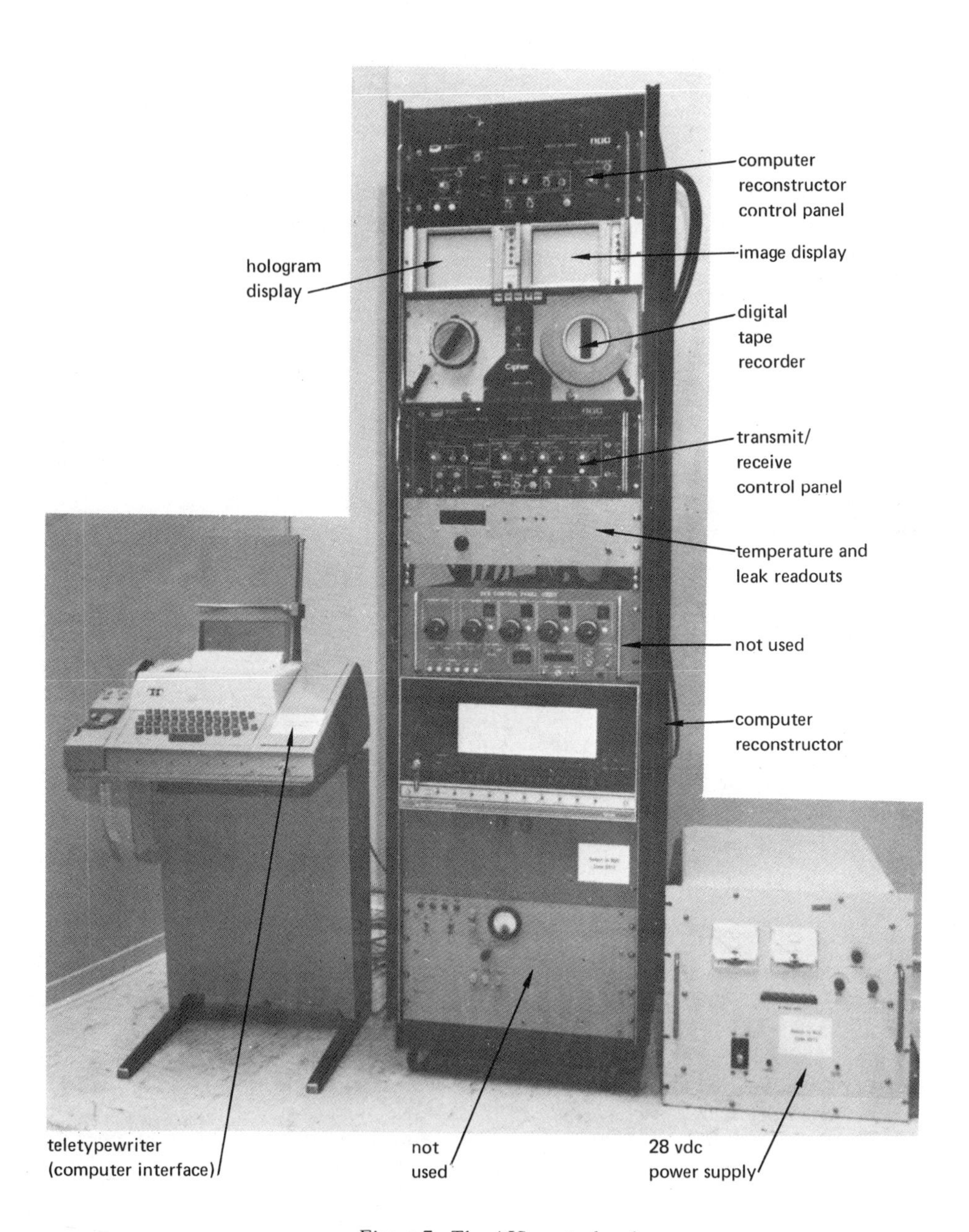

Figure 7. The AIS control rack.

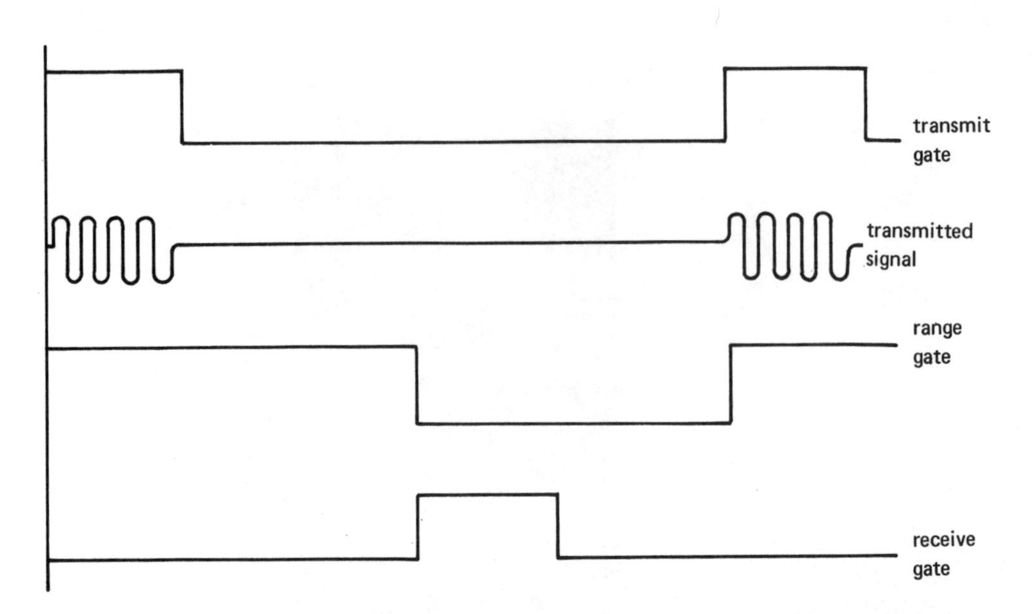

Figure 8. AIS control panel timing. This figure illustrates the control sequence in a single transmit/receive cycle. The transmitted signal is controlled by the transmit gate. The receive gate activates the receive circuitry following the end of the range gate. Hence, the depth of field of the AIS is controlled primarily by the gating circuits.

larly useful indexes to the AIS's performance. They illustrate some of the options that were open in the design of the system and reflect some of the decisions that had to be made.

Resolution is a measure of the system's ability to detect detail. It depends in part upon the aperture of the hydrophone array (that is, upon its dimensions) and upon the acoustic frequency used to ensonify the target. Increasing either of these would increase the system's performance, but not without cost. A larger array would be unwieldy and expensive. Higher frequency sound would be more sharply attenuated in seawater, so power requirements would soar or, alternatively, range would decrease.

The image dynamic range (IDR) is a measure of the signal-to-noise ratio in the displayed image and is defined for our purposes as the ratio of peak power to average background noise when observing a point target. Earlier work (Refs 2-5) had shown that the IDR must be as high as possible when using a single acou-

Table 1. AIS DESIGN GOALS.

Acoustic Frequency	642 kHz
Maximum Acoustic Power	250 W (pulsed)
Angular Resolution	1/4 deg (2 1/2 in. at 50 ft; 63 mm at 15 m)
Field of View	11 deg vertical X 11 deg horizontal
Number of Resolution Elements in Field of View	48 vertical X 48 horizontal
Image Dynamic Range	53 dB (32 dB achieved)
Hologram Frame Rate	30 frames per second
Image Frame Rate	1 image per 2 sec
Range to Reverberation Limit	100 ft (30.5 m)
Weight (in water)	900 lb (408 kg)
Weight (in air)	2,000 lb (907 kg)
Hydrophone Array	48 X 96 elements in 2-ft (0.6-m) square
Depth Capability	12,000 ft (3,658 m)

stic projector because most targets are specular reflectors of sound. They are smooth compared to the wavelength of sound used to ensonify them and reflect it as a mirror would reflect light. Figure 9 illustrates the concept of specularity and its opposite, diffusivity. In order to determine the shape of a target it is necessary that the diffuse reflections not be buried in the noise of the system. The reflected highlights may be as much as 60 dB above the diffuse returns.

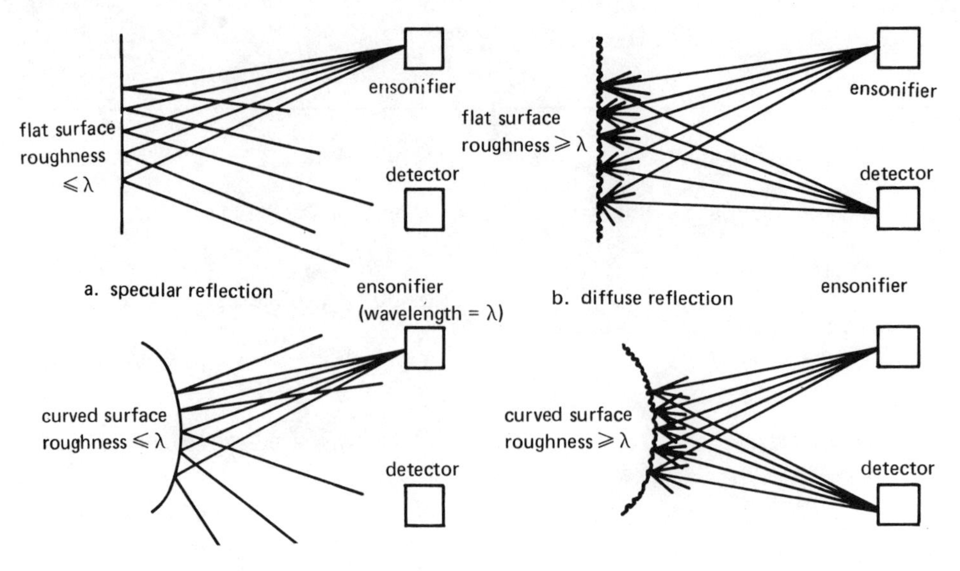

Figure 9. Specular and diffuse reflection.

The greatest obstacles to achieving a high IDR are noise sources in the channel processing electronics and amplitude and phase nonuniformities in the hydrophone array.

Some of these nonuniformities between channel processor circuits can be removed to some degree by means of a correction hologram--a matrix of amplitude, phase, and DC offset correction terms. Current results with an attempt at a correction hologram indicate some success, but further work is required to determine the useful limits of this technique.

The image frame rate of the system is a measure of its speed. The AIS can display 1 image every 2 seconds. The image frame rate is limited by the speed of the computer. Faster optical reconstruction methods may exist, but none offers the image dynamic range and flexibility of digital processing. Faster digital reconstructors are available and may be adapted to the AIS in the future.

INITIAL EVALUATION

The AIS has undergone initial evaluation in San Diego Bay off the NOSC pier. The objectives of the first tests were to verify the operation of the T/R unit in the water, obtain preliminary measurements of system performance parameters, evaluate hologram and image quality, and evaluate the operator's response to the performance of the system. Figure 10 shows the test assembly. The

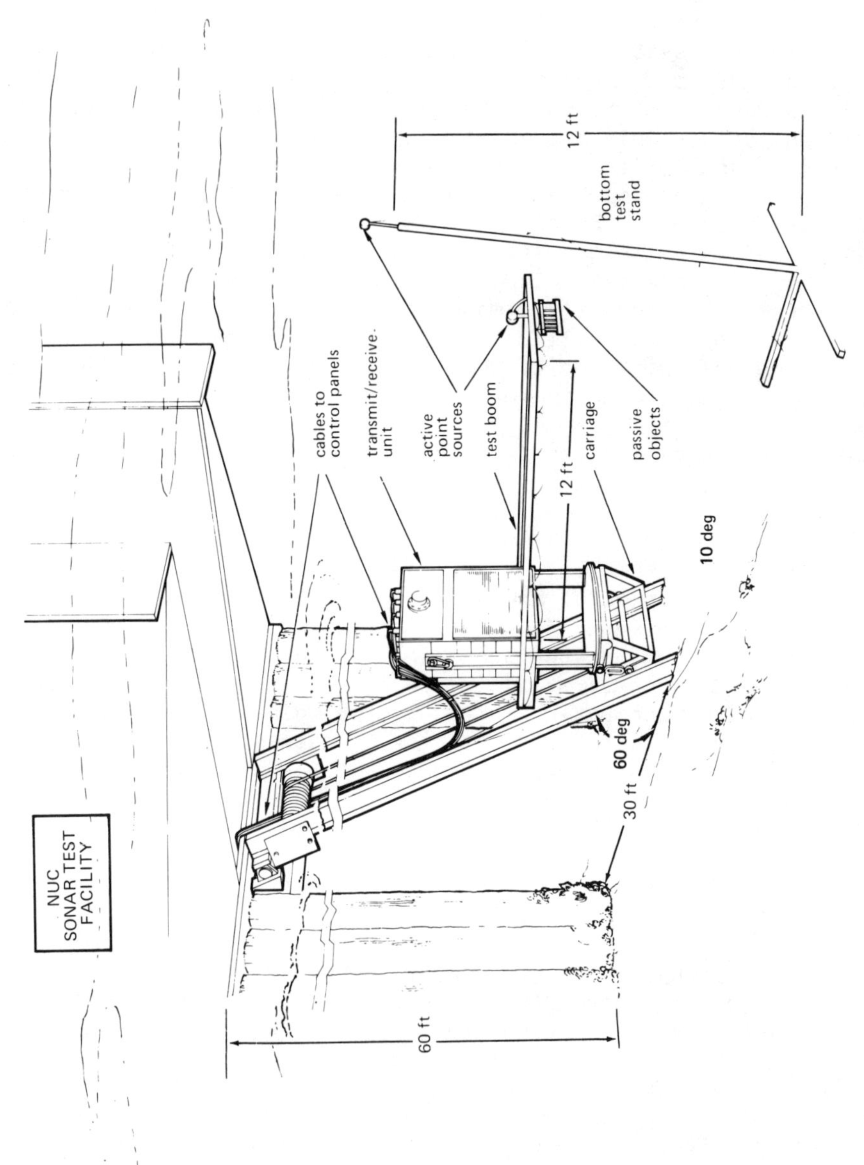

Figure 10. The test assembly used for preliminary evaluation of the AIS.

T/R unit, mounted on a pan-and-tilt carriage, was lowered along the tracks to a depth of approximately 10 feet (3 meters). The bay is about 50 feet (15 meters) deep at this point. Five targets of varying shapes, sizes, and materials were ensonified at ranges of 6 to 12 feet (1.8 to 3.7 meters). Additionally, point sources were used to test the system in the passive mode. Measurements were made of the system's angular resolution, image dynamic range, field of view, and depth of focus. The results generally coincided with the design goals. Only the image dynamic range fell short of expectations; it was limited by the display cathode ray tube to 16 dB. Subsequently, a dynamic range compression algorithm was incorporated in the computer. IDR's of 32 dB were then observed and measured. The overall quality of the images collected was good. Figure 11 presents a metal target and the image reconstructed from its hologram. The target was made by bolting together two 12-by-12 inch (300- by 300-millimeter) plates, leaving a 6-inch (150-millimeter) distance between them. The target appears end-on in this photograph. The image was formed with the target at a range of 12 feet (3.7 meters). Control and adjustment of the system proved to be straightforward and presented no problems for the operator.

CONCLUSIONS

The AIS was conceived and designed in 1971. It was intended to be a brute force approach to building an acoustic imaging system for investigating underwater acoustic imaging using a filled square aperture of hydrophones, a single projector, and a slow but capable and available minicomputer. The system did prove its point. Underwater, holographic acoustic imaging does work, and the system has given us valuable insights into how to better engineer more sophisticated acoustic imaging systems in the future. A few such observations:

1. From a cost-effectiveness standpoint, synthetic aperture systems appear to be desirable over filled arrays.

2. Even though vehicle geometry severely constrains acoustic projector placement, more projectors spaced farther from the array appear desirable. This would improve image quality by reducing specularity.

3. With limited geometries, specularity will continue to be a problem, and so wide IDR will probably still be necessary. This means engineering tolerances and error budgets will still be paramount.

4. Research efforts are required into what aspects of objects give rise to acoustic image details, and which of those are of use in human perception of the images.

a. Metal target 12 by 12 inches (300 by 300 millimeters)

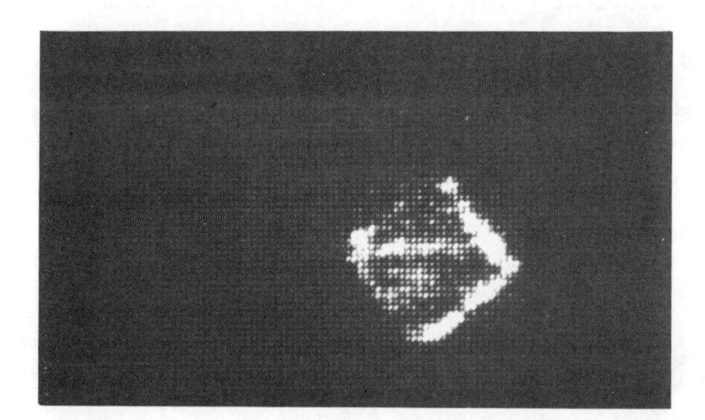

b. Image reconstructed with the target at a range of 12 feet (3.7 meters)

Figure 11. Preliminary testing of the AIS.

5. More use should be made of image enhancement techniques.

6. Most near-term applications of underwater acoustic imaging will probably be in the short range (1-10m), high resolution (<0.1 degree) regime, with moderately wide fields of view (30-60 degrees). That is, TV replacements in murky water.

SUMMARY

The Acoustic Imaging System (AIS) was built as an experimental tool to explore the usefulness of moderate range (5-100 feet), moderate resolution (0.3 degree) underwater acoustic imaging aboard deep diving Navy submersibles. The holographic technique was used with a filled square array of 48X48 hydrophone/electronic channels at 642 KHz, with a digital minicomputer reconstructor. Testing has shown the AIS to meet nearly all of its design goals. The design and testing of the AIS has also provided us with much valuable insight into the design, fabrication and use of underwater holographic acoustic imaging systems that may be required in the future.

REFERENCES

1. "Acoustical Holography - A Comparison with Phased Array Sonar", by P.N. Keating, Acoustical Holography, Vol 5, Edited by P.S. Green, Plenum Press, New York, p. 231.

2. "An Experimental Focused Acoustic Imaging System", by J. L. Sutton, Acoustical Holography, Vol 4, Edited by G. Wade, Plenum Press, New York, pp. 351-369.

3. "Gain and Phase Variations in Holographic Acoustic Imaging Systems", by J. V. Thorn, Acoustical Holography, Vol 4, edited by G. Wade, Plenum Press, New York, pp. 569-581.

4. "The Effects of Circuit Parameters on Image Quality in a Holographic Acoustic Imaging System", by J. L. Sutton, Acoustical Holography, Vol 5, edited by P. S. Green, Plenum Press, New York, pp. 573-590.

5. TEST AND EVALUATION OF AN EXPERIMENTAL HOLOGRAPHIC ACOUSTIC IMAGING SYSTEM by J. V. Thorn, N. O. Booth, J. L. Sutton, B. A. Saltzer, Naval Undersea Center Technical Publication 398, November 1974. (available through the Defense Documentation Center)

A COMPUTER-AIDED ULTRASONIC IMAGING SYSTEM

J.P. POWERS, CAPT. J.Y.R. DE BLOIS*,
LT. R.T. O'BRYON† AND LT. J.W. PATTON†

Department of Electrical Engineering, Naval
Postgraduate School, Monterey, California 93940

ABSTRACT

A preliminary version of an ultrasonic imaging system with capability of two dimensional coherent data processing and computer image processing has been built and tested. The system consists of two parts: data acquisition, and computer processing and display. The data acquisition system detects and records the coherent ultrasonic diffraction pattern emitted from or reflected by an object. This pattern is then quantized and entered into the computer memory. The computer portion processes the data and displays the resulting image on a high resolution computer controlled graphics terminal. The use of coherent detection offers advantages in that coherent processing techniques as well as incoherent techniques can be used to obtain an optimal image. Such operations as matched filtering, removal of the receiver transducer's directivity pattern, edge enhancement, etc. are easily performed. Applications of the system to ultrasonic imaging, acoustic transducer calibration and studies of scalar wave propagation will be discussed.

INTRODUCTION

Inverse diffraction[1-3] was suggested as an imaging technique about ten years ago. In this technique, an

*Canadian Forces
†United States Navy

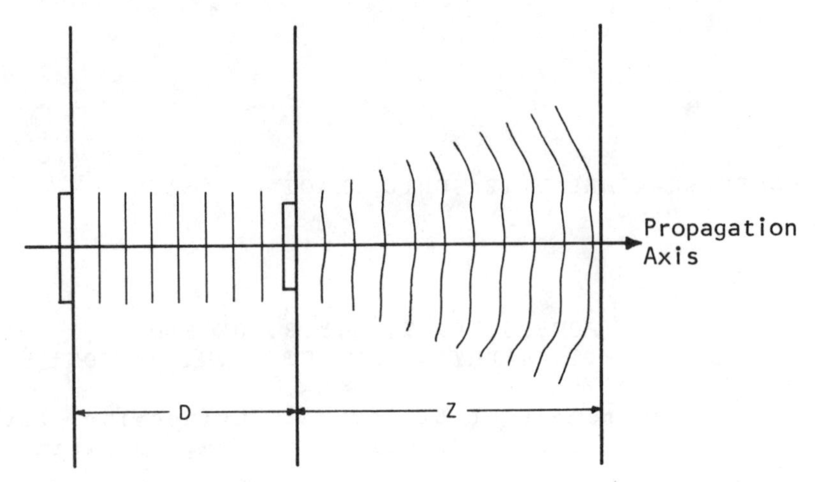

Fig. 1 Relative locations of the transducer, object and diffraction planes for a transmission system.

object (Fig. 1) is insonified by a transducer and the transmitted (or reflected) wave is detected coherently (i.e., both amplitude and phase) in some portion of a diffraction plane located an arbitrary distance Z away from the object. Detection of the complex wave by a linear detector can be accomplished in a variety of ways: a two-dimensional matrix of detector elements, a sweep of a linear array of elements, or a raster scan of a single receiver. Piezoelectric receiving elements offer an ideal combination of linearity and sensitivity[4] for this application. After wave detection, the amplitude and phase of the wave can be sampled, digitized and placed in a computer memory. By programming the inverse diffraction equations[5], the computer can 'propagate' the waves backward in space and at the distance Z from input plane an image of the object will be found. Such a computer aided technique combines the advantages of acoustic linear detection with the rapidly expanding technology of computer memory, manipulation (such as array processors) and display. Because of the relatively long acoustic wavelengths, a meaningful image can be produced from an amount of data that is relatively manageable in terms of storage and processing. Recent advances in solid state memory and fast processing hardware and software also will further alleviate this data handling requirement. Computer aided acoustic imaging also allows one to simultaneously apply many of the recent advances in image processing to the acoustic images. Such techniques can improve signal-to-noise ratios, improve resolution, or help recognize features, for example.

Having previously demonstrated some of the concepts of inverse diffraction imaging with computer simulated objects[6], we have built a system that can provide real acoustic data to test the technique. It is the purpose of this paper to present this system and our first results. The scope of the imaging system work at the Naval Postgraduate School (NPS) is to provide a research vehicle to investigate a variety of computer aided imaging techniques rather than building a prototype of a working system. The NPS system consists of two parts: a data acquisition section and the computer processing and display section. The computer section uses the large scale computing facilities at NPS as well as some of the graphic display systems of the Computer Science Laboratory. No special attempt has been made to maximize the speed or efficiency of the system or to build a dedicated system. Emphasis instead has been placed on operational flexibility to investigate various geometries and processing techniques.

DATA ACQUISITION

The purpose of the data acquisition section is to record (on an instrumentation tape recorder) the amplitude, phase and receiver X-Y position information for subsequent digitization and formatting for entry in a computer. Figure 2 shows a block diagram of the system. An insonifying transducer, object and receiving probe are placed in an echo-free test tank. Figure 3 shows a typical source (and mounting) and a receiving point transducer. The source is a 2" x 2" (5.08 cm x 5.08 cm) gold coated quartz transducer operating at 1 MHz; the receiver is a .04" (1.01 mm) diameter PZT-5 ceramic receiver with a fundamental frequency of 1.014 MHz. The position of the receiving element is controlled by a precision X-Y screw system under electronic control that generates the X-Y raster sweep. (Modification of this control allows variation in the sweep geometry.) The position of the receiver in the raster is recorded by the generation of position marks every half wavelength. Electronics have been designed and added to system to count the number of

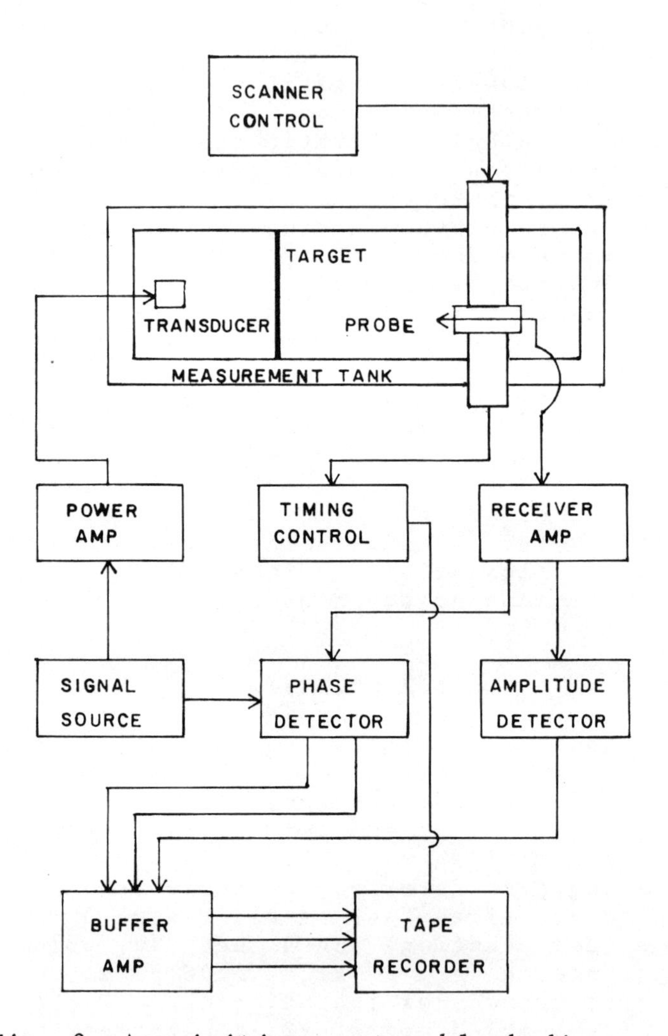

Fig. 2 Acquisition system block diagram.

Fig. 3 Source (A) and receiving transducer (B)

samples per row, the number of rows, and automatically ensure precise vertical alignment of the position locations. The position pulses are later used in the digitizing process to trigger the sampling of this amplitude and phase channels so precise alignment of the position locations is crucial to the success of the technique.

The complex signal received by the linear detector is detected, preamplified and split into two: one portion for the amplitude detection circuitry and the other for the phase detection circuits. Following detection, the analog amplitude and phase information and the position location pulses are recorded on an instrumentation tape recorder. The amplitude detector (Figure 4) is primarily two 30 dBV log amps whose combined characteristics cover the 60 dBV dynamic range observed in representative amplitude data. Figure 5 shows the overall characteristics of the amplitude detector with a slight discontinuity in the slope observed between -20 and -30 dBV where the log amp stages transition into each other. Software processing can easily correct for this discontinuity once the data is in the computer.

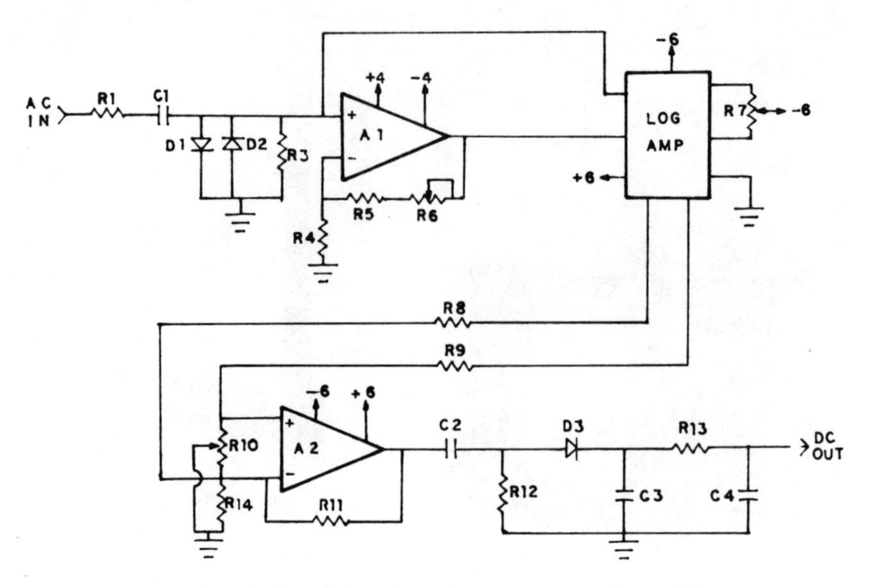

Fig. 4 Amplitude detector circuit

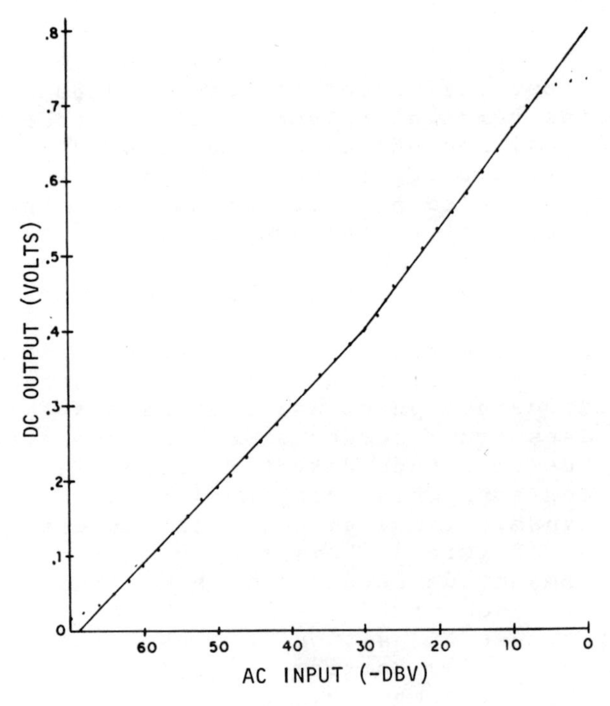

Fig. 5 Transfer characteristic of the amplitude detector

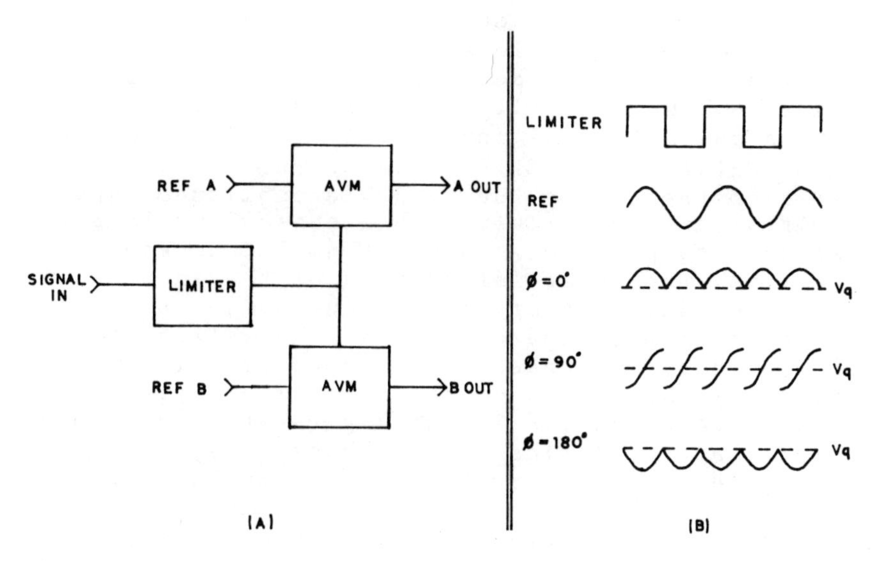

Fig. 6 Phase detector: (A) block diagram; (B) output for various phase relationships

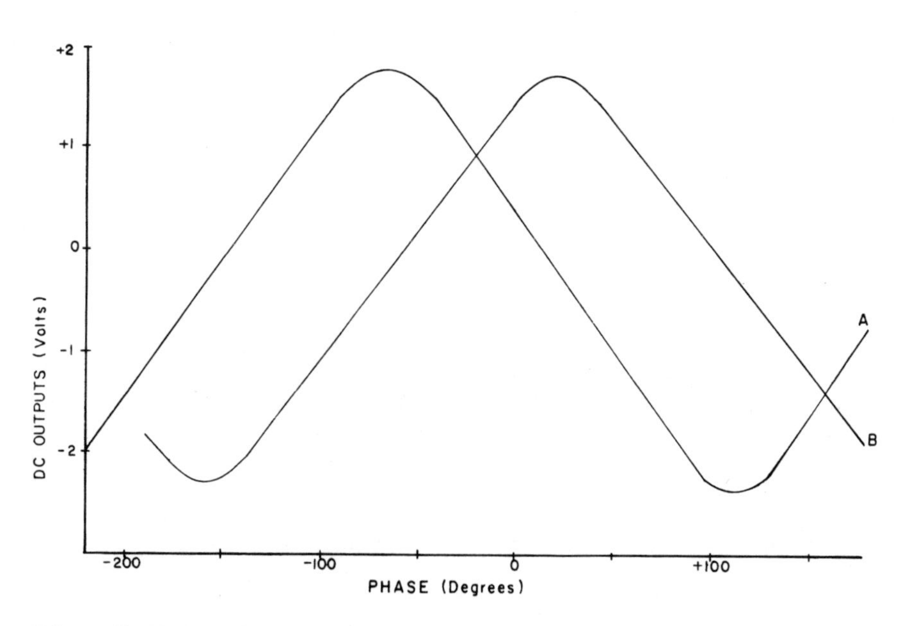

Fig. 7 Experimental phase detector characteristics (channels A and B)

Two forms of phase detector have been built and tested. The first (of lesser accuracy) uses analog voltage multipliers to determine the phase. As shown in Figure 6, two multipliers are required to obtain unambiguous phase information over a full 0 - 360° range. Signals 'REF A' and 'REF B' are quadrature signals obtained from the source.The variable amplitude input signal is limited (in our circuit with a phase lock loop and comparator) to produce a constant amplitude square wave that carries the phase information of the input signal. As shown in the right of Fig. 6, the dc component of the product of the square wave and the reference indicates the relative phase of the input signal. Low pass filtering of the output of the multipliers provides a phase detected signal. Figure 7 gives the experimental transfer characteristics of the phase detectors. Since curves A and B are not quite sin and cos curves (especially in the peaks and troughs and in the fact that they are not exactly 90° out of phase), some phase errors (±3°) were encountered near the peaks and troughs. Consequently, a commercial integrated circuit phase detector with linear characteristics as shown in Figure 8a was used for our second phase detector. Two quadrature phase detector circuits having characteristics as shown in Figure 8b can be combined to produce an overall phase transfer function of Figure 9. This circuitry is presently being implemented. The phase voltage is then also recorded in analog form for subsequent digitizing.

COMPUTER PROCESSING AND DISPLAY

The sampling, A/D conversion, and formatting of the computer compatible tape was done on a XDS 9300 computer combined with a CI 5000 analog computer. Large blocks of data would be processed on the large IBM 360 computer while image display and simple processing are done on a dual PDP-11/50 computer system driving an interactive Ramtek CX-100A color display terminal (Figure 10). Early work has generated a basic interactive display capability, while further work is required to develop more software for processing and display. Additionally, the present display has a limitation of 16 gray scales (although many pseudocolor schemes are possible).

TEST RESULTS

The data acquisition system has been completely electronically tested and calibrated producing the

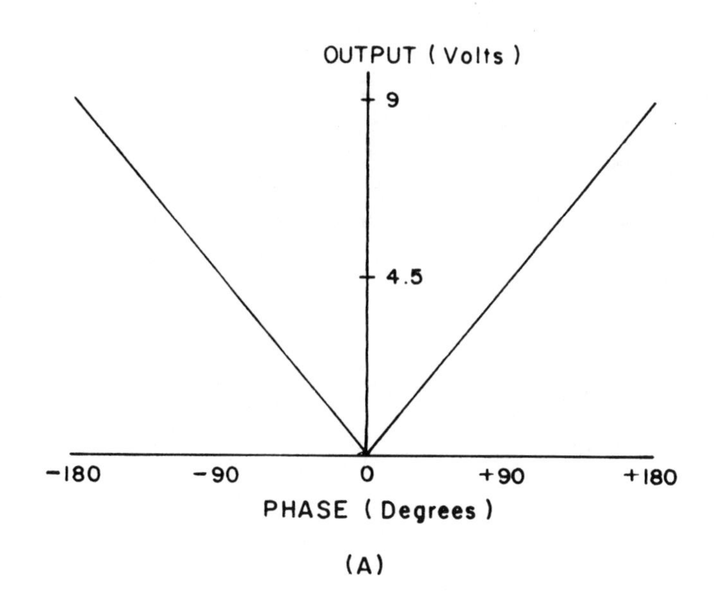

(A)

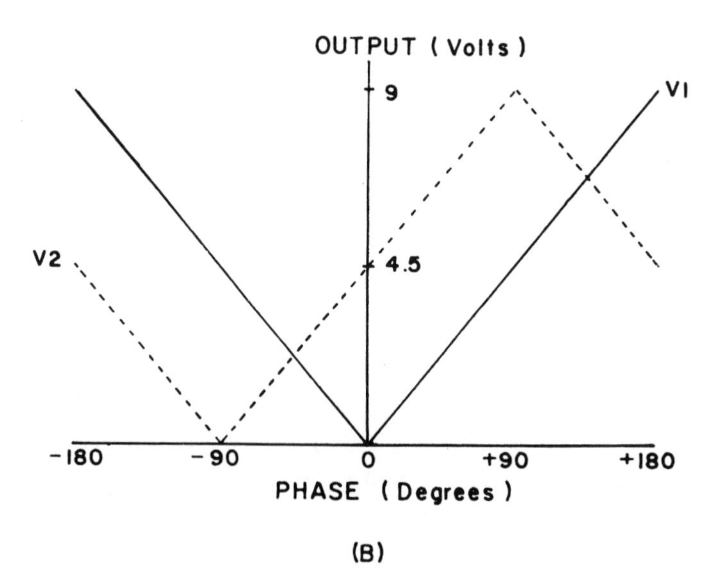

(B)

Fig. 8 (A) Transfer characteristics of commercial integrated circuit phase detector;
(B) Quadrature outputs from two channels

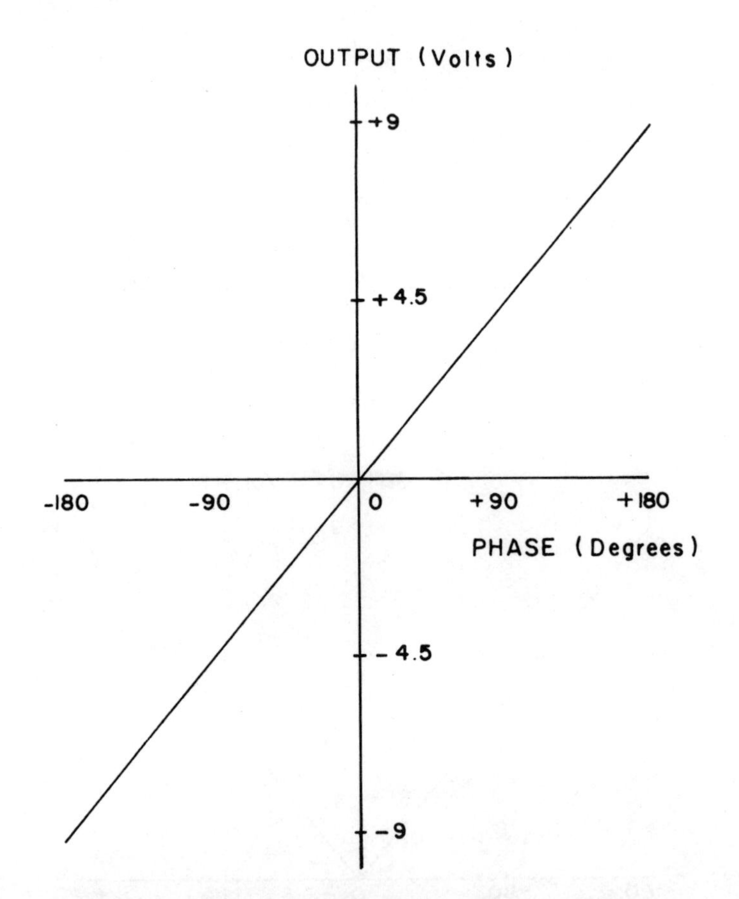

Fig. 9 Phase transfer function after combination of quadrature outputs

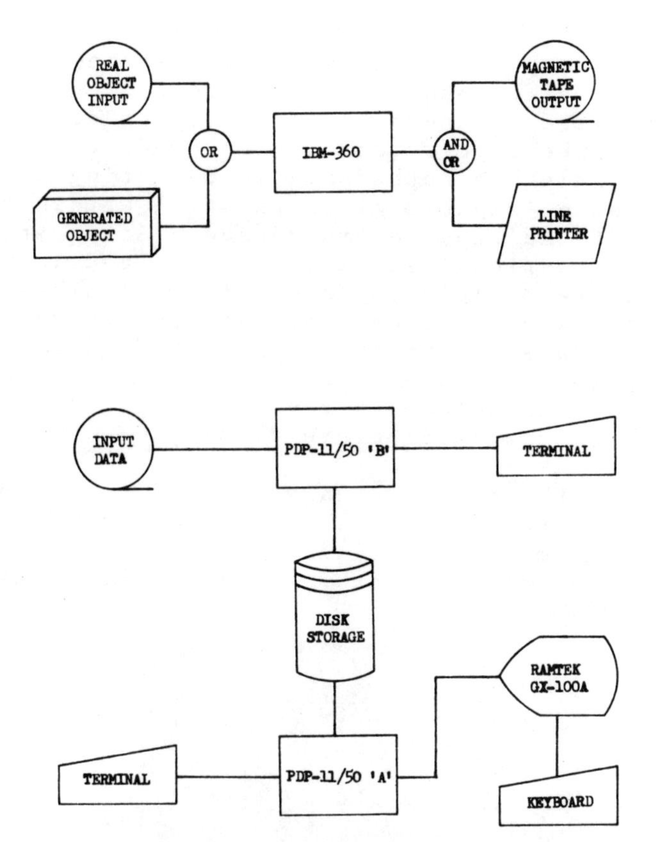

Fig. 10 Block diagram of computer processing system

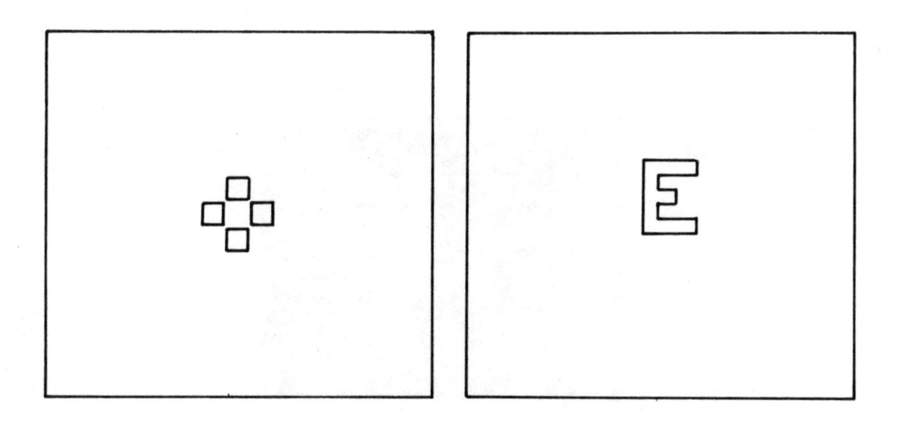

Fig. 11 Test object shapes

transfer characteristics of Figures 5 and 7. Calibration points were provided by HP 3575A gain-phase meter. The data was also found to be reproducible. Building vibrations were found to be an important source of phase noise while the tape recorder also produced noise (20-50 mV) which is bothersome in detecting some of the weaker value of signals although still within the specifications of the tape recorder used. Attenuation measurements of various materials have helped in the design of the test tank lining; phase information has helped in aligning the source transducer parallel to the data plane. The data acquisition system has also proven useful in finding dead elements in transducer arrays or phasing reversal between array elements. These applications use the raw data before display.

To test the display characteristics, two objects (Figure 11) were used: a cruciform of four rectangles holes (5 x 6 mm on a side) and a letter 'E' (total size: 15 mm x 20 mm, stroke width: 5 mm). The forms were cut

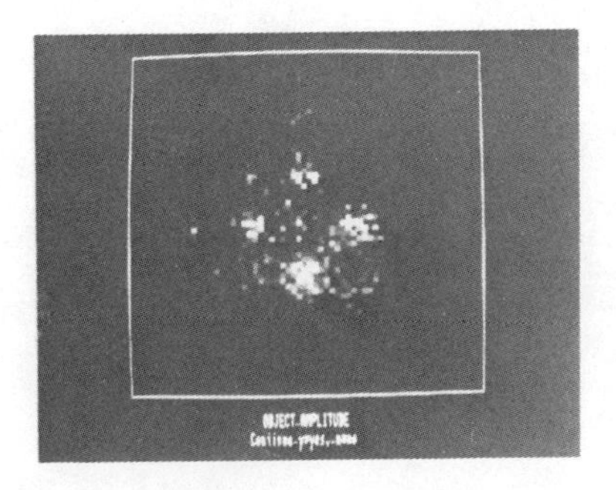

(A)

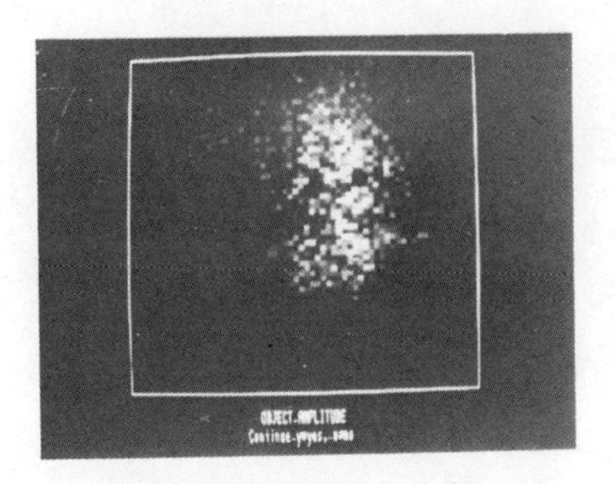

(B)

Fig. 12 Computer displayed images of test objects

out of 5 mm thick cork. Data was sampled (64 x 64 points) at half wavelength vertical and horizontal spacings. The object-receiver spacings were only 4 and 6 wavelengths respectively making the images 'shadowgrams' and requiring no inverse diffraction. The first images of these objects from the computer display are shown in Figure 12. Sixteen gray scales are used with 64 x 64 data cells. The letter 'E' image exhibits a large amount of noise (of unknown origin) in the upper left corner. Further refinements and experimentation are obviously required.

SUMMARY

The primitive images shown here are the first obtained with the system. Further areas of interest for future work include refinement of the data acquisition system to remove some of the 'bugs' observed, much more work on the computer display system to enhance the interactive capability between user and machine, more test examples while increasing the data field to 128 x 128, and finally implementation of a broad range of image processing procedures to enhance the images.

ACKNOWLEDGEMENTS

The authors acknowledge support of this work by the Naval Postgraduate School Foundation Research Program and the National Science Foundation.

REFERENCES

1. T. R. Shewell and E. Wolf, "Inverse diffraction and a new reciprocity theorem," J. Opt. Soc. Am., 58(12): 1596-1603, 1968.

2. M. M. Sondhi, "Reconstruction of objects from their sound diffraction patterns," J. Ac. Soc. Am., 46(5): 1558-1164, 1969.

3. A. L. Boyer, et al., "Reconstruction of ultrasonic images by backward propagation," Acoustical Holography, Vol. 3(A. F. Metherell, Ed.), Plenum Press, pp. 333-348, 1971.

4. H. Berger, "A survey of ultrasonic image detection methods," Acoustical Holography, Vol. 1, (A.F. Metherell, et al., Eds.), Plenum Press, pp. 27-48, 1969.

5. J. P. Powers, "Computer simulation of linear acoustic diffraction," Acoustical Holography, Vol. 7, (L. W. Kessler, Ed.), Plenum Press, pp. 193-205, 1977.

6. J. P. Powers and D. E. Mueller, "A computerized acoustic imaging technique incorporating automatic object recognition," Acoustical Holography, Vol. 5, (P. S. Green, Ed.), Plenum Press, pp. 527-539, 1974.

AN EXPERIMENTAL UNDERWATER ACOUSTIC IMAGING SYSTEM USING MULTI-BEAM SCANNING

Kazuhiko Nitadori, Kunihiko Mano, and Hiroshi Kamata

OKI Electric Industry Co., Ltd.
Tokyo, 108 Japan

ABSTRACT

An experimental, high resolution, real-time underwater acoustic imaging system using multi-beam scanning method has been built and tested successfully. The system uses a 4x4-element sparsely arranged planar projector array and a 32x32-element densely arranged hydrophone array and obtains images with equivalent resolution to that obtained by a 128x128-element hydrophone array. The image is reconstructed numerically by a dedicated FFT processor. Partial picture elements of 32x32 are obtained by a single transmission of sound pulses and the whole picture elements of an image of 256x256 are obtained by 64 transmissions within two seconds.

INTRODUCTION

Underwater viewing is one of the main application areas of acoustic imaging. For the purpose of applying to the underwater use, acoustic lens type [1]-[3] and holographic type [4], [5] imaging apparatus have been developed. The acoustic lens type imaging does not require signal processing because an acoustic lens performs image formation, but the main performance of the image, such as resolution, is limited by that of the lens. The use of this type of imaging apparatus for underwater viewing is also limited because a mechanical movement is required for adjusting a range and manufacturing and handling of large aperture lenses are not easy.

In the holographic acoustic imaging apparatus, acoustic portion is simplified because the image forming operation is performed by signal processing; hence, it becomes possible to realize a large scale transducer array for a high resolution use or a large sized transducer array for a low frequency use. By the use of basic holographic imaging, however, a high resolution imaging apparatus can hardly be realized because of an economic constraint since the number of resolution cells is almost equal to the number of hydrophone elements used and a large scale hydrophone array and accompanying electronics are required for the high resolution use.

The transducer elements can be saved in holographic imaging by the use of synthetic aperture technique. But, the conventional synthetic aperture method in which transducer elements are moved mechanically or electronically will not be effective in the underwater applications where the environment is largely varying. One of the authors proposed a new imaging scheme called multi-beam scanning method which is based on the synthetic aperture principle, but does not require the severe stability of the environment [6].

We have built an experimental, high resolution, real-time underwater acoustic imaging system operating with this scheme and tested it in an ocean environment successfully. The system has been operating as theoretically expected and taking high resolution images during the test.

The system uses a 4x4-element sparsely arranged planar projector array and a 32x32-element densely arranged planar hydrophone array and obtains real-time, orthographic images with equivalent resolution to that obtained by a 128x128-element hydrophone array. In this paper, we describe the principle of operation, the configuration, and the performance of the system.

MULTI-BEAM SCANNING METHOD

Suppose acoustic projectors and hydrophones are arranged on a T/R plane and an object to be observed is on an object plane that is parallel to and distant by z from the T/R plane, as dipicted in Fig. 1. Each of the projectors and the hydrophones are assumed to have the identical pupil function $g_t(x, y)$ and $g_r(x, y)$, respectively, and to be centered at points (u_t, v_t) and (u_r, v_r), respectively, where

$$\begin{aligned} u_t &= u_{to} + mL_x \quad (m = 0, 1, \ldots, M_x-1) \\ v_t &= v_{to} + nL_y \quad (n = 0, 1, \ldots, M_y-1) \end{aligned} \tag{1}$$

and

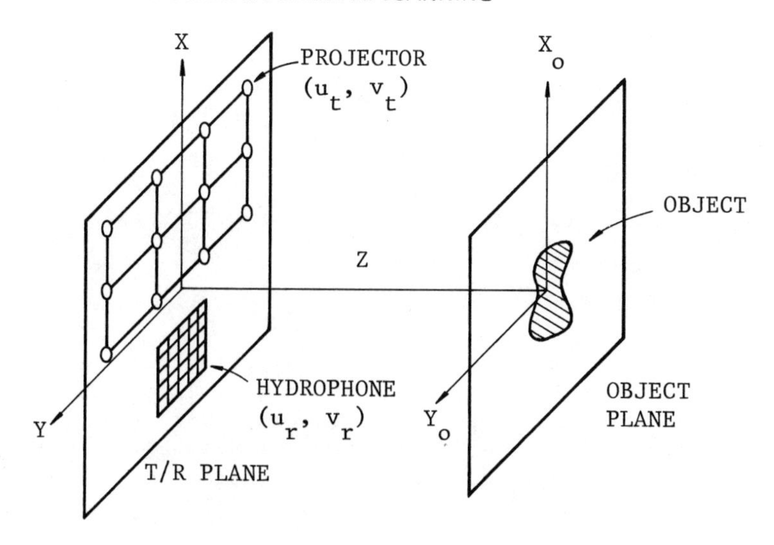

Fig. 1 Geometry of imaging.

$$u_r = u_{ro} + mL_x/N_x \quad (m = 0, 1, \ldots, N_x-1)$$
$$v_r = v_{ro} + nL_y/N_y \quad (n = 0, 1, \ldots, N_y-1) \tag{2}$$

First, the transmitting signal with the complex amplitude

$$S_t(u_t, v_t; x_i, y_i) = \exp\{-ik/2z\cdot(u_t^2 + v_t^2 - 2x_iu_t - 2y_iv_t)\} \tag{3}$$

is radiated simultaneously from each projector located at (u_t, v_t), where $k = 2\pi/\lambda$ is a wave number and λ is a wave length of sound. The radiated sound is focused at a point (x_i, y_i) on the object plane and its complex amplitude at a point (x_o, y_o) on the object plane is represented by

$$U_o(x_o, y_o; x_i, y_i) = A(z))\iint_{-\infty}^{\infty}\sum_{(u_t, v_t)} g_t(x - u_t, y - v_t)\cdot$$
$$S_t(u_t, v_t; x_i, y_i)\exp\{ik/2z\cdot[(x - x_o)^2 + (y - y_o)^2]\}dxdy$$
$$= A(z)\ h_t(x_i, y_i; x_o, y_o) \tag{4}$$

assuming a Fresnel approximation, where

$$A(z) = \exp(ikz)/ikz \tag{5}$$

$$h_t(x_i, y_i; x_o, y_o) = \iint_{-\infty}^{\infty} g_t(x, y) \exp\{ik/2z.[(x - x_o)^2 + (y - y_o)^2]\} P_t((x_o - x_i - x)/\lambda z, (y_o - y_i - y)/\lambda z)\, dxdy \tag{6}$$

$$P_t(f_x, f_y) = \sum_{(u_t, v_t)} \exp\{-i2\pi(u_t f_x + v_t f_y)\} \tag{7}$$

The function $h_t(\ )$ in (6) is a transmitting beam pattern and $P_t(\ )$ in (7) is a spatial Fourier transform of the transmitting aperture. When the pupil of the projector is sufficiently small, the transmitting beam pattern is approximated by $P_t((x_o - x_i)/\lambda z, (y_o - y_i)/\lambda z)$. For the rectangular array consisting of $M_x x M_y$ projectors located at the point (u_t, v_t) in (1), the amplitude response of the beam is represented by

$$|P_t(x/\lambda z, y/\lambda z)| = \left| \frac{\sin M_x \pi x/\Delta x}{\sin \pi x/\Delta_x} \cdot \frac{\sin M_y \pi y/\Delta_y}{\sin \pi y/\Delta y} \right| \tag{8}$$

where

$$\Delta x = \lambda z/L_x, \qquad \Delta_y = \lambda y/L_y \tag{9}$$

The cross section along x-axis of the amplitude response of the transmitting beam pattern is shown in Fig. 2(a).

The transmitting beam has many periodic peaks (grating lobes) with periods Δ_x and Δ_y along x and y-axis, respectively. That is, the transmitting beam focused at a point (x_i, y_i) is also focused at the points (x_i', y_i') represented by

$$x_i' = x + p\Delta_x, \qquad y_i' = y_i + q\Delta_y \tag{10}$$

where p and q are any integers.

Let $S_r(u_r, v_r; x_i, y_i)$ denote the complex amplitude of the received signal received by the hydrophone located at a point (u_r, v_r) for the transmitting signal in (3) focused at a point (x_i, y_i). Perform the following operation for the received signals $\{S_r(u_r, v_r; x_i, y_i)\}$ for image reconstruction:

$$f(x_i + p\Delta_x, y_i + q\Delta_y) = \sum_{m=o}^{N_x - 1} \sum_{n=o}^{N_y - 1} S_p(m, n; x_i, y_i).$$

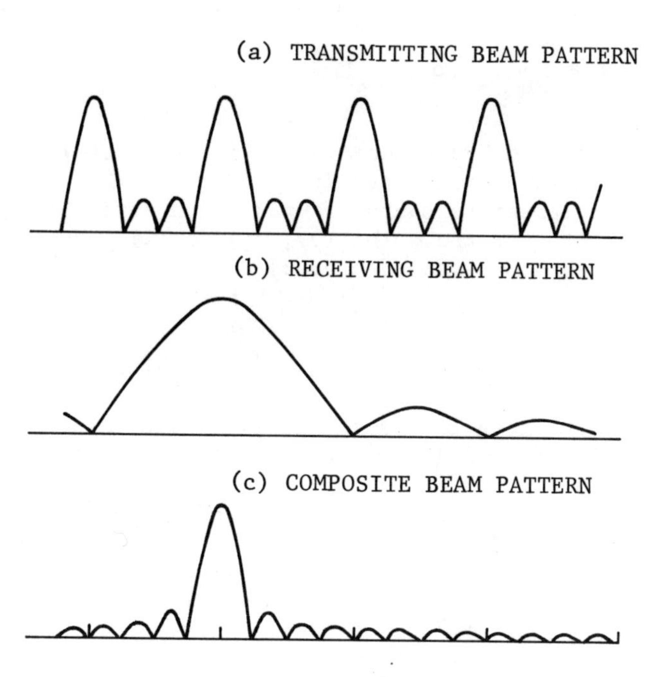

Fig. 2 Cross sections of transmitting, receiving, and composite beam pattern (M_x= 4, N_x= 32).

$$\exp\{i2\pi(pm/Nx+ qn/N_y)\} \tag{11}$$

$$S_p(m, n; x_i, y_i) = S_r(u_r, v_r; x_i, y_i)$$
$$\exp\{-ik/2z\cdot(u_r{}^2+ v_r{}^2-2x_iu_r-2y_iv_r)\} \tag{12}$$

where (u_r, v_r) is in (2). The operation in (11) is the two-dimensional DFT (discrete Fourier transform) of $S_p(m, n; x_i, y_i)$ and can be efficiently calculated by the use of an FFT algorithm. By this operation, $N_x x N_y$ receiving beams focused at the points (x_i+ $p\Delta_x$, y_i+ $q\Delta_y$) on the object plane are generated and the same number of picture elements corresponding to these points are resolved.

One of these receiving beams has a beam pattern

$$h_r(x_i', y_i'; x_o, y_o) = \iint_{-\infty}^{\infty} g_r(x, y) \exp\{ik/2z\cdot[(x - x_o)^2 + (y - y_o)^2]\} P_r((x_o - x_i' - x)/\lambda z, (y_o - y_i' - y)/\lambda z)\, dxdy \quad (13)$$

$$P_r(f_x, f_x) = \sum_{(u_r, v_r)} \exp\{-i2\pi(u_r f_x + v_r f_y)\} \quad (14)$$

When the pupil of the hydrophone is sufficiently small, the receiving beam pattern is approximated by $P_r((x_o - x_i')/\lambda z, (y_o - y_i')/\lambda z)$. For the N_xx N_y-element rectangular hydrophone array each element of which is located at the point (u_r, v_r) in (2), the amplitude response of the beam is represented by

$$\left|P_r(x/\lambda z, y/\lambda z)\right| = \left|\frac{\sin \pi x/\Delta x}{\sin \pi x/N_x\Delta_x} \cdot \frac{\sin \pi y/\Delta y}{\sin \pi y/N_y\Delta_y}\right| \quad (15)$$

The cross section along x-axis of the amplitude response of the receiving beam pattern is shown in Fig. 2(b).

The receiving beam pattern has a broad peak and periodic zeros along x and y-axis with periods Δ_x and Δ_y, respectively. The composite beam $h_t(x_i, y_i; x_o, y_o)\cdot h_r(x_i', y_i'; x_o, y_o)$ of the transmitting beam $h_t(x_i, y_i; x_o, y_o)$ and one of the receiving beams $h_r(x_i', y_i'; x_o, y_o)$ has a pencil beam pattern with a sharp peak at the point (x_i', y_i'), where (x_i, y_i) and (x_i', y_i') are related by (10). This composite beam pattern is a point spread function of the imaging system. The cross section of the composite beam along x-axis is shown in Fig. 2(c). These N_xx N_y pencil beams are generated by a single transmission of the transmitting signal in (3) and the image reconstruction operation in (11) and (12); consequently, the same number of picture elements are resolved.

Since these N_xx N_y picture elements are separated each other by Δ_x and Δ_y along x and y-axis, respectively, other picture elements must be obtained to fill the space. In order to obtain these picture elements, the steering (x_i, y_i) of the transmitting beam is steped by a distance corresponding to one picture element and the above procedure is repeated. In general, the distance of this scanning step is chosen to a half of the Rayleigh distance, hence, $2M_x$x $2M_y$ transmissions are sufficient to obtain the whole picture elements of an image.

As the imaging process is thus completed by scanning the whole object plane with the transmitting and receiving multi-beam, this scheme is refered to as multi-beam scanning method.

In Fig. 3, the block diagram of a typical imaging apparatus

using this scheme is shown. A scanner generates the steering of a transmitting and receiving multi-beam and a beamformer generates a steered transmitting signal, which is radiated from a rectangular planar projector array and focused at the steering point on the object plane. The sounds reflected from an object are received by a rectangular planar hydrophone array, amplified by amplifiers, and detected by quadrature detectors. The quadrature detector consists of a pair of product detectors, whose reference signals are the in-phase and quadrature component of the transmitting signal. The complex amplitudes of the received signals detected by the quadrature detectors are multiplexed by a multiplexer and converted to digital codes by an A/D converter. For these digital codes a signal processor performs the image reconstruction operation and numerically reconstructs the acoustic image of the object, whose intensity is calculated and displayed on a CRT display.

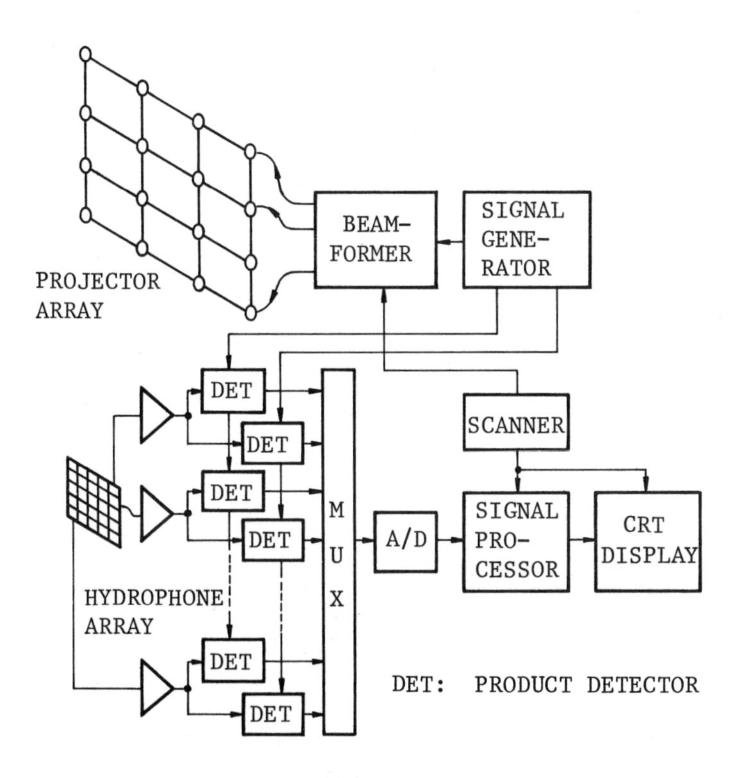

Fig. 3 Block diagram of an imaging apparatus using multi-beam scanning method.

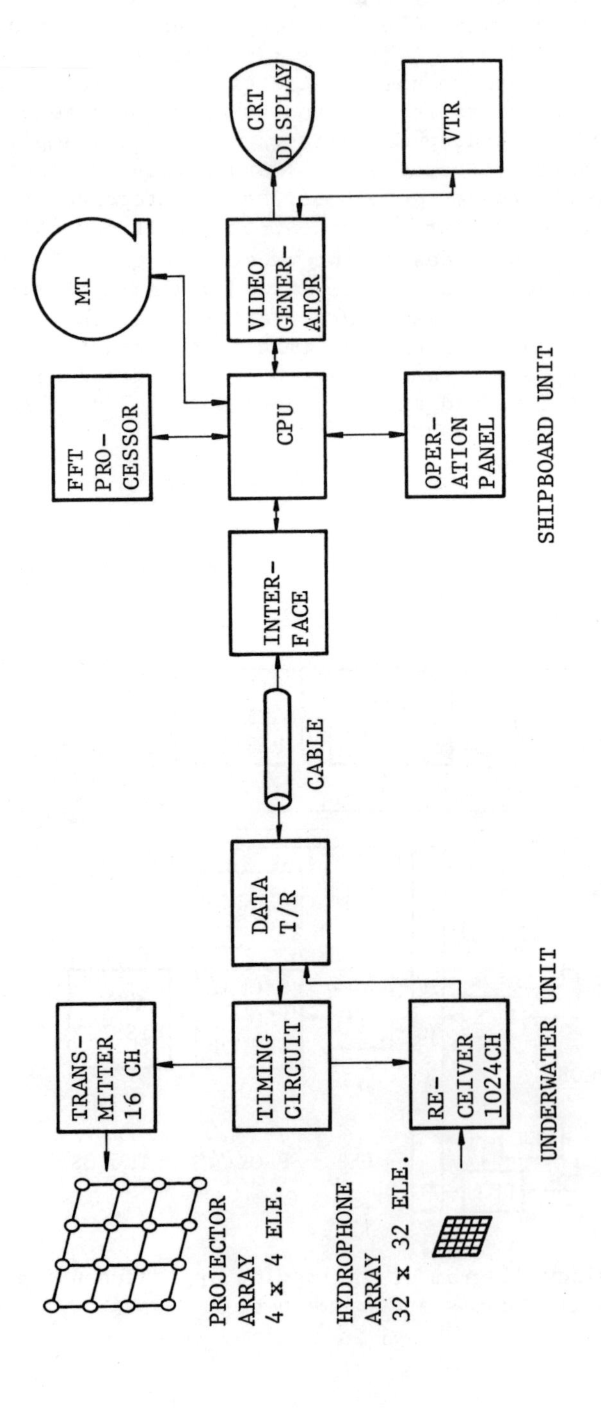

Fig. 4 Block diagram of the experimental underwater acoustic imaging system.

SYSTEM CONFIGURATION [7]

An experimental, real-time, high resolution underwater acoustic imaging system has been built which operates with multi-beam scanning method. The system consists of an underwater unit and a shipboard unit with a cable connecting between them. The block diagram of the system is shown in Fig. 4, the appearence of the underwater unit is shown in Fig. 5, and the system parameters are summarized in Table 1.

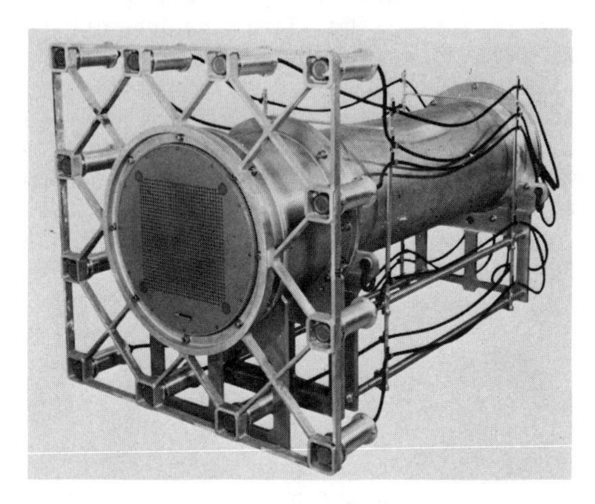

Fig. 5 Underwater unit of the acoustic imaging system.

Table 1 System parameters of the experimental underwater acoustic imaging system

Imaging scheme:	Multi-beam scanning method
Image reconstruction:	Digital signal processing
Acoustic frequency:	200 kHz (wave length: 7.5 mm)
Field of view:	40 degrees (both horizontal and vertical)
Angular resolution:	0.33 degree (theoretical) 0.40 degree (measured)
Picture elements displayed:	256x256
Range:	2 - 100 m
Image reconstruction time:	2.0 seconds
Projector array:	4x4-element planar array
Hydrophone array:	32x32-element planar array
Transmission power:	200 watts max.

The underwater unit consists of a 4x4-element projector array, a 32x32-element hydrophone array, a 16-channel transmitter, a 1024-channel receiver, a data transceiver, and a timing circuit. All of them are contained in a cylindrical container. The shipboard unit consists of a CPU (a mini-computer), an FFT processor, an operation panel, an interface with the underwater unit, a video generator, a CRT display, a magnetic tape handler, and a VTR. They are mounted in two racks.

An acoustic projector is composed of a lead zirconate titanate ceramic disk, 29 mm in diameter and 5 mm thick. The ceramic-dicing technique is applied to eliminate unwanted vibration modes and to suppress sidelobes with an amplitude shading. In order to reduce the deviation of the amplitude and phase response at the operating frequency of 200 kHz, the ceramic disk is used in non-resonance. The appearence and the directivity pattern of a typical projector are shown in Figs. 6 and 7, respectively. The 4x4-element projector array is constructed from these projectors arranged with a regular spacing of 320 mm. The standard deviation of the amplitude response of these projectors is 1.0 dB.

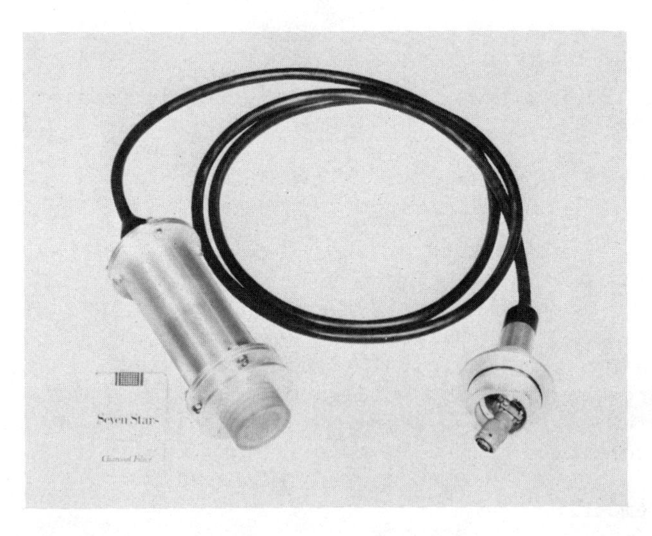

Fig. 6 Projector

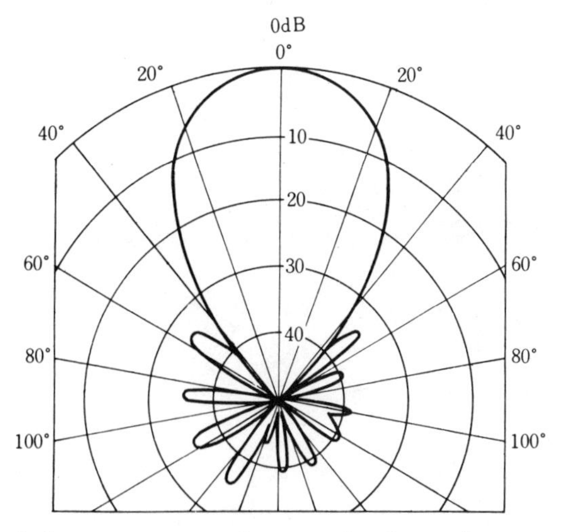

Fig. 7 Directivity pattern of a typical projector at 200 kHz.

A hydrophone is composed of a short-column of lead zirconate titanate ceramic, 4.5 mm in diameter and 3.5 mm thick, cemented on an epoxy resin socket. The circumference of the column is shielded from sound by rubber corks. It also is used in non-resonance. The appearence and the directivity pattern of a typical hydrophone are shown in Figs. 8 and 9, respectively. The hydrophone array is constructed from these 32x32 hydrophones arranged with a regular spacing of 10 mm and molded by rubber in front of them. High reliability, high uniformity, and low cross-talk level are attained by this construction. The standard deviation of their sensitivity is 1.2 dB.

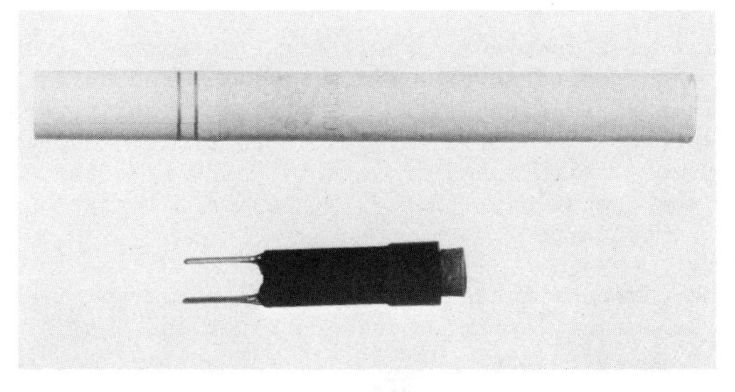

Fig. 8 Hydrophone (bottom) compared with a cigarette (top).

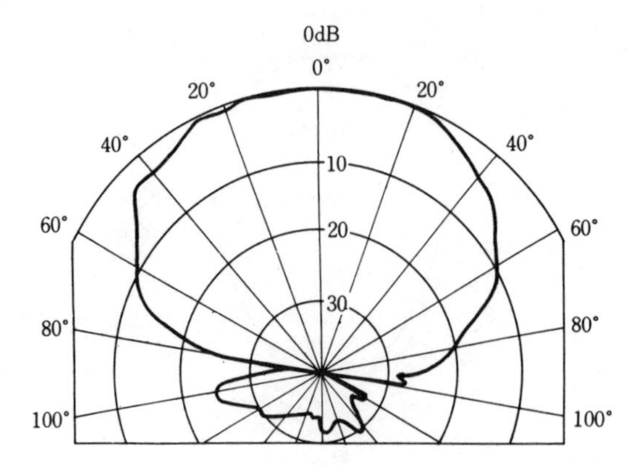

Fig. 9 Directivity pattern of a typical hydrophone at 200 kHz.

The transmitter generates 16-channel transmitting signal with the common frequency, pulse width, and amplitude and the independent phase and sends it to each of the projectors. Phase data are given from the CPU so that the transmitting multi-beam focusing at a specified point is generated.

The receiver contains 1024 channel processors and 8 A/D converters. Each channel processor consists of a preamplifier and a quadrature detector and detects the complex amplitude of a received signal [7]. The output signals of the quadrature detectors are serially read by two-stage multiplexers and converted to digital codes by the A/D converters.

The channel processor is made from standard integrated circuits and passive components. One three-layer printed circuit board contains 32 channel processors. The receiver consists of 32 sheets of these boards.

The digitized received signals are transmitted through the cable to the shipboard unit. Then, the image reconstruction operation is performed in a signal processor consisting of the FFT processor and the CPU. The FFT processor is a dedicated, micro-programmed, high speed processor, which is efficient in performing the operations related to the image formation, such as DFT, phase compensation.

The image reconstructed by the signal processor is written into a refresh memory in the video generator and read from it periodically to refresh the CRT display (a color TV monitor). The image data read from the refresh memory are compressed in dynamic range linearly or logarithmically, transformed to (R, G, B) signals by a RAM table, converted to video signals by D/A converters, then displayed on the color TV monitor in pseudo-color or monochrome. Images can be integrated by means of the refresh memory to improve signal-to-noise ratio for stationary objects.

The image data and the received signals are recorded digitally in a magnetic tape. The video signal is also recorded by the VTR.

An image consisting of 256x256 picture elements is reconstructed in this system within two seconds, during which 64 acoustic pulses are transmitted and received.

SYSTEM PERFORMANCE

The maximum field of view where no alias occurs is represented by

$$\theta = 2 \sin^{-1}(\lambda/\Delta_r) \tag{16}$$

where Δ_r is the spacing of hydrophones in the hydrophone array. In our system, $\theta = 44^o$ since $\Delta_r = 10$ mm and $\lambda = 7.5$ mm. In order to suppress the aliases from the outside of the field of view, the directivity pattern of the projector is controlled so as to insonify the field of view only. (see Fig. 7) Effective field of view is somewhat narrower than this value due to the roll-off characteristic of the beam pattern.

Suppose the resolution is defined as the Rayleigh distance, the corresponding angular resolution is expressed by

$$\Delta\theta = \sin^{-1}(\lambda/L) \tag{17}$$

Where L is the length of the synthesized aperture. In our system L = 4x320 mm, hence we have $\Delta\theta = 0.33^o$. On the other hand, a resolution test was made in a water tank in which a V-shaped metal rods target was used. Two rods are observed separately at the center of the target where the two rods are separated by the distance corresponding to an angle of view of 0.4^o [7].

The maximum range of the system is estimated as the following. Suppose SL, NL, and DI denote the source level, the noise level, and the receiving directivity index of the system, respectively, TL denote the transmission loss of sound between the system and an object, and TS denote the target strength of the object. Let these parameters be expressed in dB at a unit distance of 1 m, then the signal-to-noise ratio of the image is represented by

$$SNR = SL - 2TL + TS - NL + DI \tag{18}$$

Hence, if the minimum required SNR is expressed by DT (detection threshold), the maximum allowable transmission loss (two-way) is expressed as

$$2TL = TS + FM \tag{19}$$

where FM (figure of merit) is defined by [8]

$$FM = SL - (NL - DI + DT) \tag{20}$$

In our system, SL = 204 dB (0 dB = 1 μP/ref 1 m), NL = 83 dB (0 dB = 1 V/1 μP) for a receiving gate width of 1 mS, and DI = 30 dB, hence we have FM = 141 dB, assuming DT = 10 dB.

Suppose the transmission loss is the sum of a spherical spreading loss and an absorption loss. The former is expressed by

$$TL' = 20 \log z \tag{21}$$

where z is a distance in meter. The latter is proportional to the distance z and the absorption coefficient is 0.1 and 0.4 dB/m for turbid water of particle concentration of 100 and 1000 ppm, respectively, at 200 kHz [9].

In Fig. 10, the two-way transmission loss 2TL versus a distance z is shown. From this graph it is estimated that the object of a target strength of 0 dB and -40 dB (e.g., a rigid sphere of radius 2 m and 2 cm, respectively [10]) will be visible at a distance of 200 m and 100 m, respectively, in turbid water of particle concentration of 100 ppm. The image of sea bottom was actually taken at a range of 100 m, the maximum range of the system.

TEST RESULTS

Acoustic images of many kinds of objects were taken by this imaging system in the ocean. Typical examples of these are shown in Figs. 11 - 14. High resolution, high quality images were obtained for the objects with complex structures. However, background noise was visible in the image of the objects with specular potion due to the deficiency of image dynamic range.

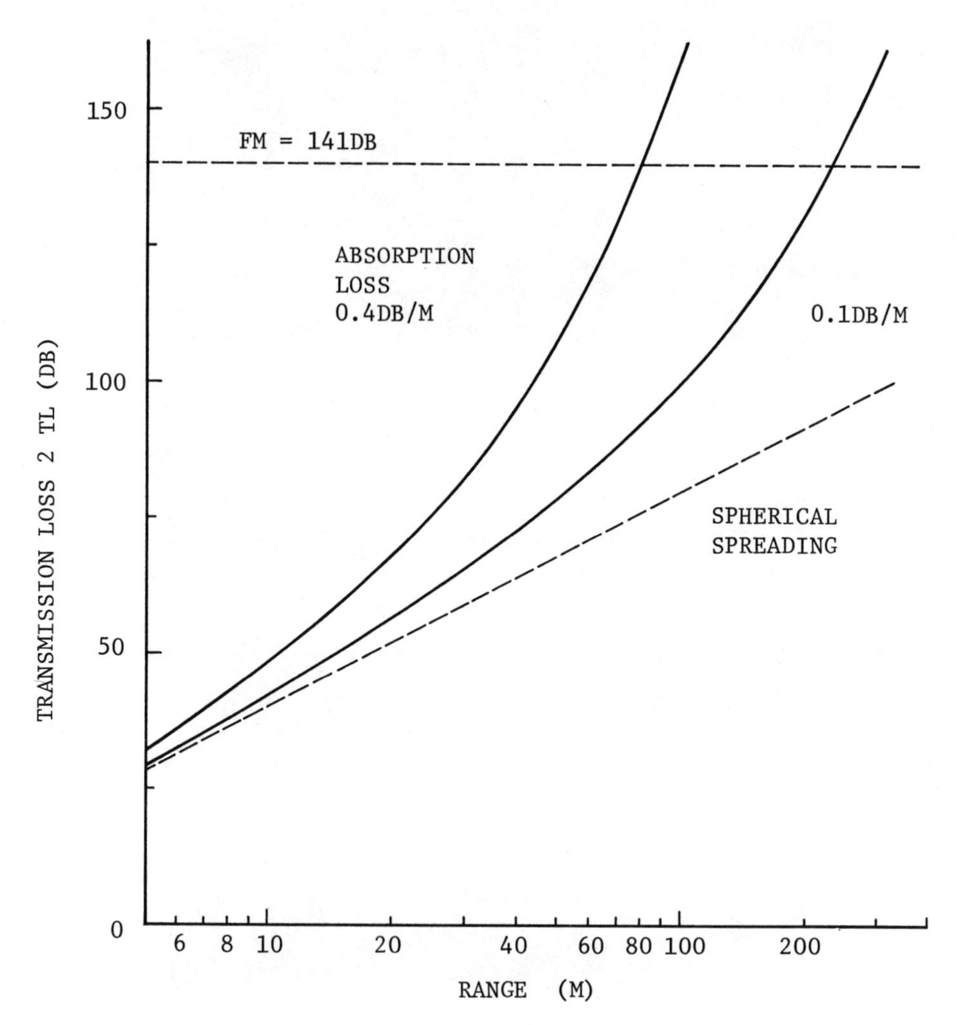

Fig. 10 Two-way transmission loss of sound vs. range.

Fig. 11 Acoustic image of a bicycle taken at a range of 3.6 m.

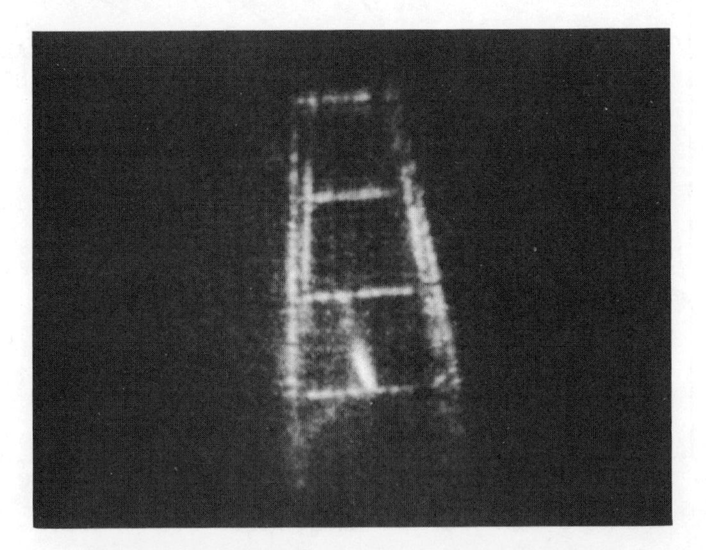

Fig. 12 Acoustic image of a stepladder taken at a range of 3.5 m.

CONCLUSIONS

An experimental, high resolution, real time underwater acoustic imaging system has been built and tested successfully which operates with a new imaging scheme called multi-beam scanning method. This

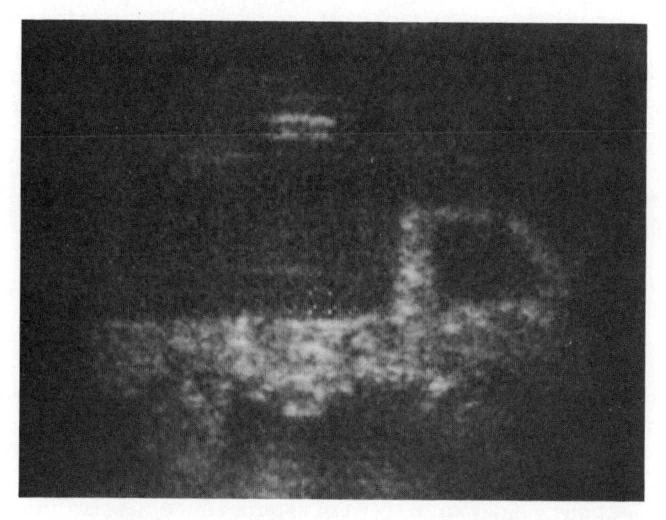

Fig. 13 Acoustic image of a car taken at a range of 5 m (logarithmically compressed).

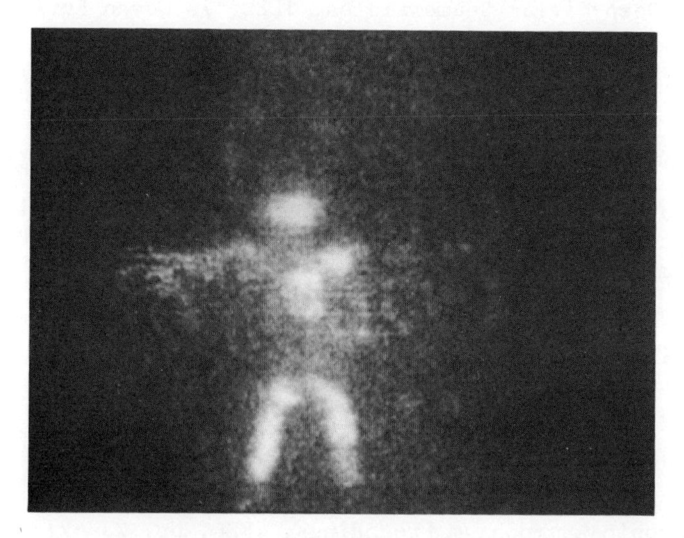

Fig. 14 Acoustic image of a diver with a diving suit taken at a range of 3.3 m (logarithmically compressed).

method has been proved to be effective for obtaining high resolution acoustic images economically. Also, techniques for implementing a large scale transducer array and transmitter/receiver electronics required for obtaining high resolution images have been developed. By these technicques it becomes possible to implement many kinds of acoustic imaging apparatus suitable to each application area.

In order to improve the performance of the imaging system, the followings should be investigated further: (1) Extending image dynamic range by suppressing the sidelobe level of the point spread function by means of shading. (2) Realizing 3D imaging function by adopting pulse echo technique.

This work was partly supported by Japan Ship's Machinery Association.

REFERENCES

[1] N. Takagi, et al.: Solid-State Acoustic Image Sensor, in Acoustical Holography, Vol. 4, Plenum Press (1972), pp. 215-236.

[2] J. L. Sutton: An Experimental Focused Acoustic Imaging System, ibid., pp. 351-369.

[3] A.L. Rolle: Focussed or Lens-Type Ultrasonic Imaging for Small Deep-Diving Submersibles. IEEE '73 Ocean Environment (1973) pp. 284-300.

[4] E. Maron, et al.: Design and Preliminary Test of an Underwater Viewing System Using Sound Holography, in Acoustical Holography, Vol. 3, Plenum Press (1971), pp. 191-209.

[5] N. Booth and B. Saltzer: An Experimental Holographic Acoustic Imaging System, in Acoustical Holography, Vol. 4, Plenum Press (1972), pp. 371-380.

[6] K. Nitadori: Synthetic Aperture Approach to Multi-Beam Scanning Acoustical Imaging, in Acoustical Holography, Vol. 6, Plenum Press (1975), pp. 507-523.

[7] K. Mano and K. Nitadori: An Experimental Underwater Viewing System Using Acoustical Holography, 1977 Ultrasonics Symposium Procedings, IEEE, pp. 272-277.

[8] R.J. Urick: Principles of Underwater Sound, McGraw-Hill (1975), pp. 18.

[9] P. S. Green, et al.: Acoustic Imaging in a Turbid Underwater Environment, J. Acoust. Soc. Amer., 44, 6 (1968), pp. 1719-1730.

[10] R. J. Urick: ibid., pp. 243.

THREE-DIMENSIONAL ACOUSTIC IMAGING

R. F. Koppelmann and P. N. Keating

Bendix Research Laboratories

Southfield, Michigan 48076

ABSTRACT

The implementation of a sonar system exhibiting high resolution in three dimensions has been investigated. This investigation involved the design of a sonar system capable of gathering data in a single "ping" sufficient to generate a 40 x 40 x 54 point image volume. Experimental verification of the system's performance demonstrated that three-dimensional data can be acquired and processed on a single-"ping" basis and that the resulting image field data can be rotated (under software control) so that the operator can view extended objects within the volume from different aspects.

1. INTRODUCTION

The application of holographic techniques to high resolution sonar provides a system capable of gathering large quantities of information. There has been a considerable effort to improve the accuracy with which this information is measured; however, little work has been done in the area of acquiring and processing the available three-dimensional data on a single-"ping" (system cycle) basis.

This paper describes a sonar system with high resolution in three dimensions and presents an approach to the problem of handling large volumes of three-dimensional data on a single-"ping" basis.

In this paper we discuss the design of the data acquisition hardware and also briefly outline the processing software. The display of the three-dimensional image field is also discussed and experimental results are presented to demonstrate the system's capability.

2. SYSTEM DESIGN

Our previous system design was capable of providing high resolution in three dimensions(1); however, the signal processing hardware was entirely analog and as such would only provide high range descrimination for a single range slot. The purpose of the new design was to provide continuous range coverage (on a single-"ping" basis) while retaining the high resolution exhibited by the earlier system.

Figure 1 shows a conventional, analog homodyne section where the received signal is multiplied by quadrature replicas of the transmitted signal. The resulting products are filtered to extract the D.C. components which represent the complex holographic data for that channel.

The transmitted signal is normally a C.W. burst, the duration of which determines the signal-to-noise ratio of the received signal. The range accuracy under these conditions is approximately equal to the transmitted pulse length and, as the pulse length is increased to improve the signal-to-noise ratio, there is an accompanying loss in this accuracy. It has been shown(1), however, that if pulses with large time/bandwidth products are used it is possible to increase the transmitted energy and still maintain good range resolution in a holographic system.

If coded transmissions are used, the temporal processing can still be implemented using conventional homodyne hardware; however, the number of range slots which can be interrogated is normally

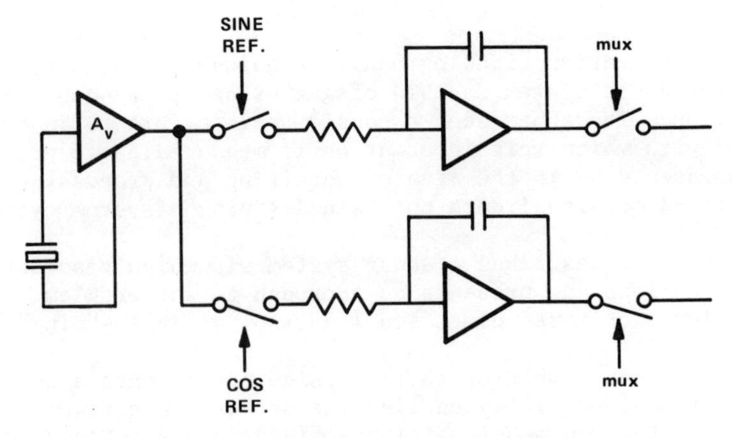

Figure 1--Conventional Homodyne Processor

limited to one per system cycle or at best one per code duration. Overlapping codes can be decoded using hardware if a separate homodyne section is assigned to each range slot over the length of the codes. This would not be cost-effective, however, since each channel would have 31 homodyne sections if, for example, a 31-bit code is used and if the range slots were set at 1 bit intervals.

One solution to this problem is to use a minicomputer to perform the decoding operation, thereby replacing the decoding hardware with software. This, of course, requires either that the computer accept the data as it is received or that the data be stored in each channel until there is sufficient time to process it. Since the data is received in parallel and the minicomputer can only accept it in a serial fashion, the data rate to the computer for a large system would be in excess of the computer's capability. Therefore, only the second option is realizable.

Our design approach uses biphase-coded transmissions for improved range resolution and also employs digitization and buffer storage of the received signals at the channel level so that the data can be supplied to the minicomputer at a rate which is within its capability. Figure 2 shows a block diagram of the channel electronics. Included is a heterodyne section, an analog-to-digital conversion section, and a storage section. The heterodyne section reduces the received signals frequency range to the information bandwidth, which has been set at 50 kHz. The sampling frequency is ideally set at 4 times (or a multiple of 4 times) the information bandwidth so that quadrature decoding can be easily implemented.

The homodyne processing and the decoding of the coded signals are performed in the minicomputer, leaving a system which has hardware dedicated to general-purpose data collection and where any special purpose processing is performed in the software.

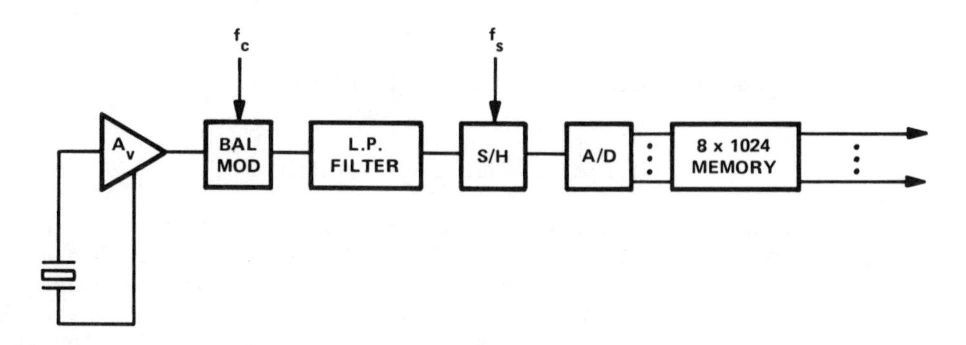

Figure 2--Block Diagram of Signal Digitization and Storage Section

Since the frequency of the received signal and the sampling frequency are known, the data count, as the data is transferred to the minicomputer, can be used to index the software waveform for the homodyne processing. In a similar manner, after each set of complex bit values is calculated in the homodyne section, the bit count can be used to index the code masks associated with each range slot.

The temporal processing results in a set of complex data for each element which corresponds to two words for each of the range slots being interrogated. After this processing is completed for the entire array, the data are sorted into sets corresponding to the same range which are processed in the normal manner[2] to yield image information for each range slot.

The channel storage memory for this design is an 8 x 1024 static random-access memory which is capable (assuming a sample frequency of 200 kHz) of storing a time-series of length equivalent to 25 ft.

The experimental sonar system includes a 20 x 20 element receive array. In a production system each of the 400 channels would require a separate processor; however, as a cost saving measure, only a single row of 20 processors was built. In order to acquire data from all 400 channels, switching was provided between the 20 rows of hydrophone elements and the row of preprocessing channels. Since only 20 channels can be processed at a time, the experimental system had to undergo 20 system cycles (consisting of transmission, ranging, data acquisition, and temporal processing) to acquire the data required to fill the 20 x 20 array. After each system cycle the data set resulting from the temporal processing was stored in the computer memory. The element switching network was then stepped to its next position and a new system cycle was initiated until the data from all 20 element rows had been gathered and processed.

The system software required for operation consists of five major sections: (a) system cycling and row-switching section, (b) homodyne section, (c) a parallel decoding section, (d) a reconstruction section, and (e) a display section.

The system is schematically shown in Figure 3.

3. THREE-DIMENSIONAL IMAGE DISPLAY

The spatial processing of the holographic data obtained using the 20 x 20 array yields an image volume 40 elements wide, 40 elements high, and having a depth equal to the number of range bins selected[1]. With the memory used, the maximum number of range bins was limited to 10. In this implementation, the range resolution was 15 cm, chosen appreciably larger than the 3.5 cm obtained earlier[1], so that the total range resolved was 150 cm or about 5 ft.

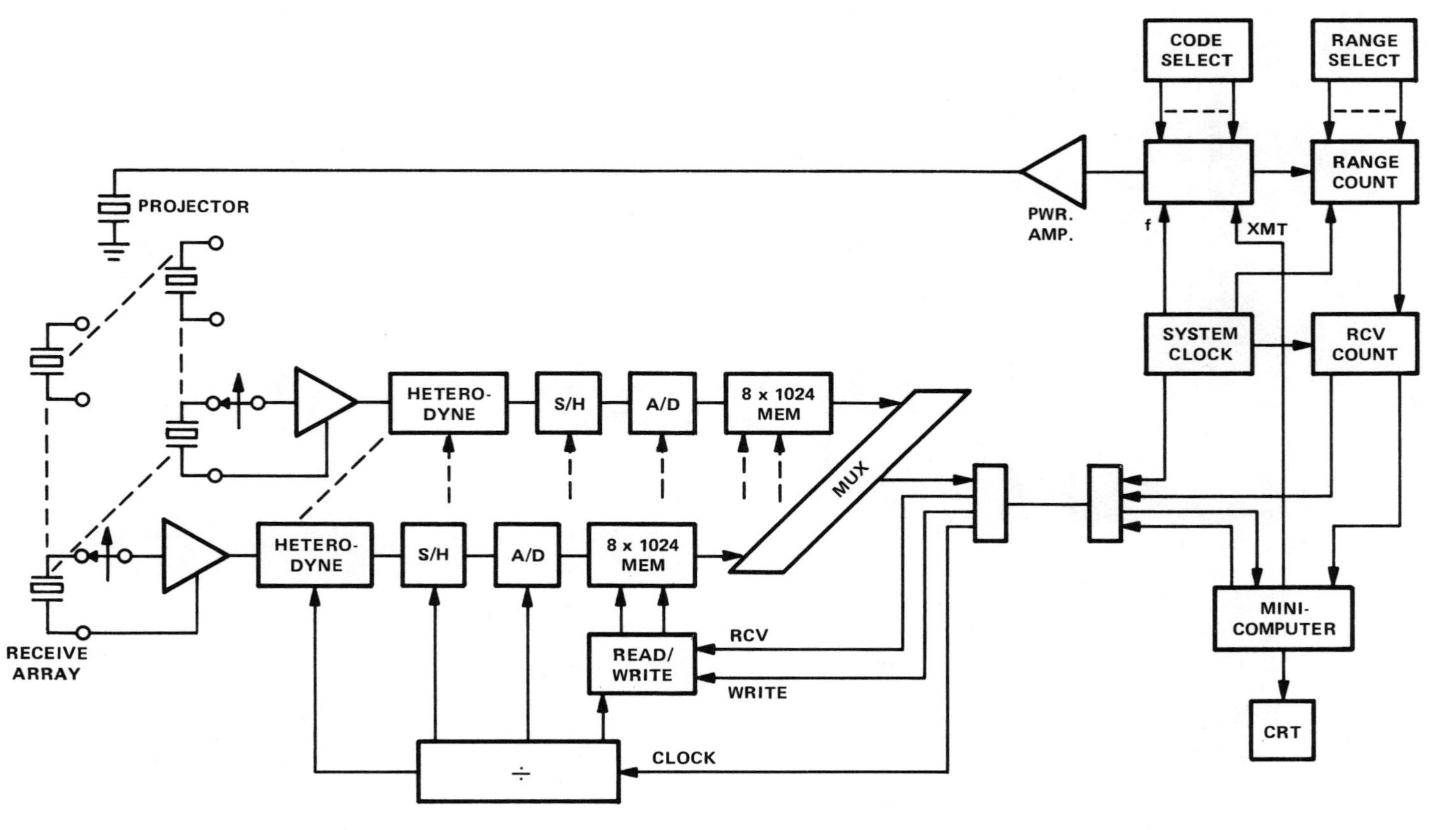

Figure 3--Block Diagram - High Resolution Sonar System

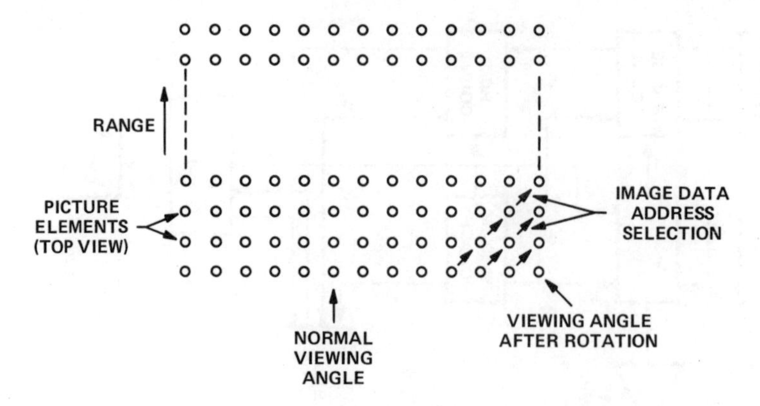

Figure 4--Picture Element Selection for Rotation of Image Volume

The resulting image information was displayed on a CRT; therefore, in order to provide a three-dimensional effect, the first image frame was displayed normally and all subsequent frames were stepped in both the x and y direction. Image rotation was accomplished by selecting picture elements from different range slots and displaying them in the same picture frame. Figure 4 shows the picture element selection for a 45° image rotation.

4. EXPERIMENTAL RESULTS

In order to demonstrate the system's three-dimensional capabilities, two 3" x 3" targets were placed on a line perpendicular to the face of the receive array and were spaced approximately 34". The target furthest from the array was moved to a lateral position 2" to the right and 2" up from the lateral position of the closest target to insure that it would not be completely shadowed (see Figure 5). The transmissions were coded using a 13-bit Barker code with a bit length set to the equivalent of 15 cm (6 inches). The range slots were set at one bit intervals. Figure 6 shows the image of the two targets.

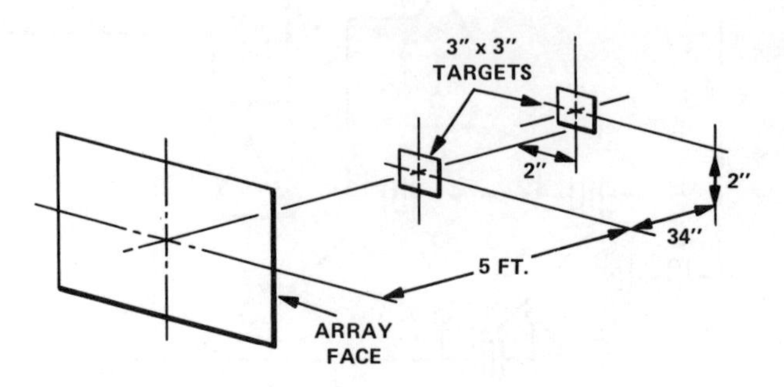

Figure 5--Test Setup

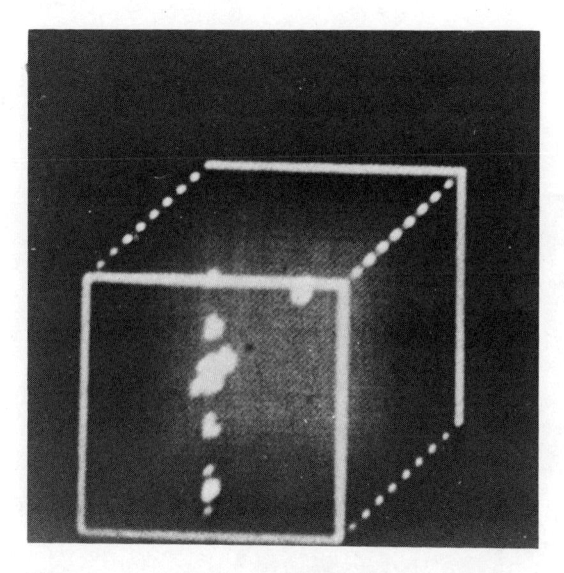

Figure 6--Volume Image of Two Targets

The objects of interest are usually centered in the display volume and the intensity of the peripheral picture elements is consequently low, making it difficult to distinguish the volume outline; therefore, the points outlining the volume have been deliberately brightened in the figures to indicate the angle through which the display had been rotated.

Although precautions were taken to hold the targets in place during the data acquisition cycle, there was always some target movement which resulted in vertical smearing of the image. This (in Figure 6 and most subsequent pictures) results in image points being brightened above and below the target image. This, of course, would not occur if the processing hardware was applied to all of the receive elements as would be the case in a production system.

Figures 7 a and b show that with a slight software modification the three-dimensional volume can be sliced in either the horizontal or vertical direction (or for that matter, any direction). Since the range spacing can be determined from the code bit-length and the image point spacing can be calculated once the range is known, it is possible to determine the spatial relationships of targets within the volume being viewed.

Figure 8 shows the targets as would be seen looking directly into the volume from the viewing angles indicated.

Figure 9 shows the same volume rotated 45° in the vertical direction and then rotated in the horizontal direction.

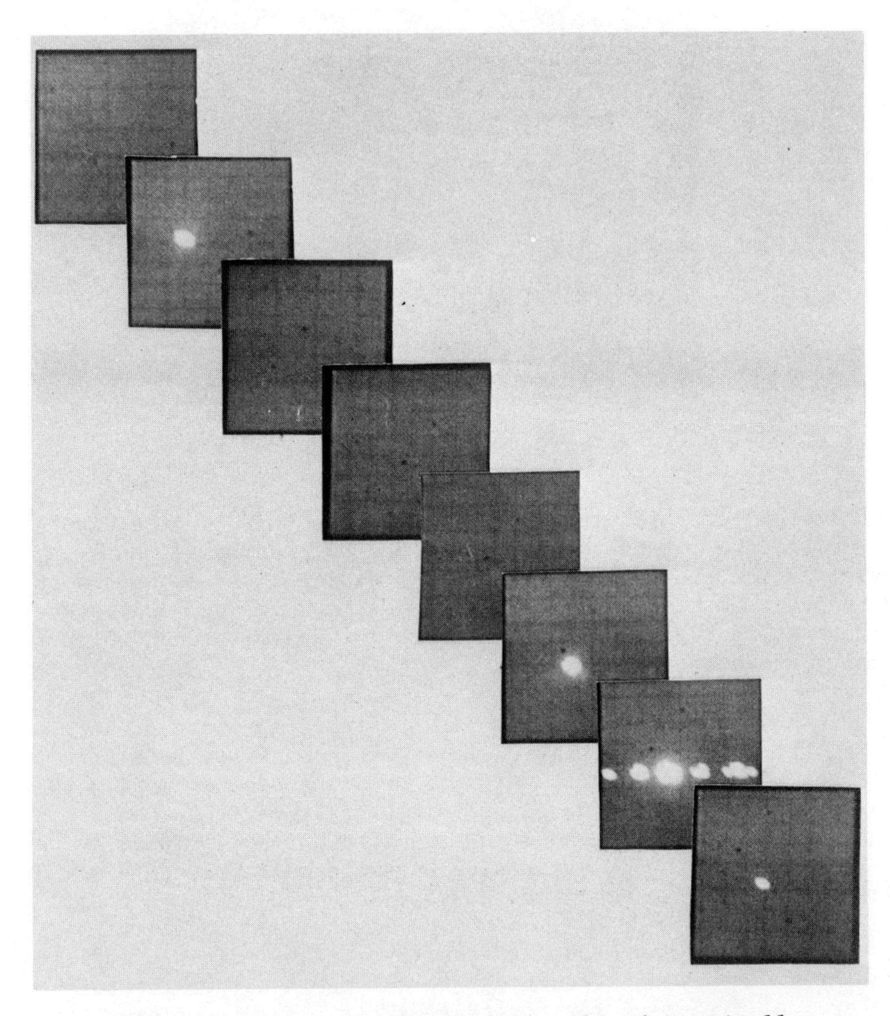

Figure 7a--Two Target Volume Image Sliced Vertically

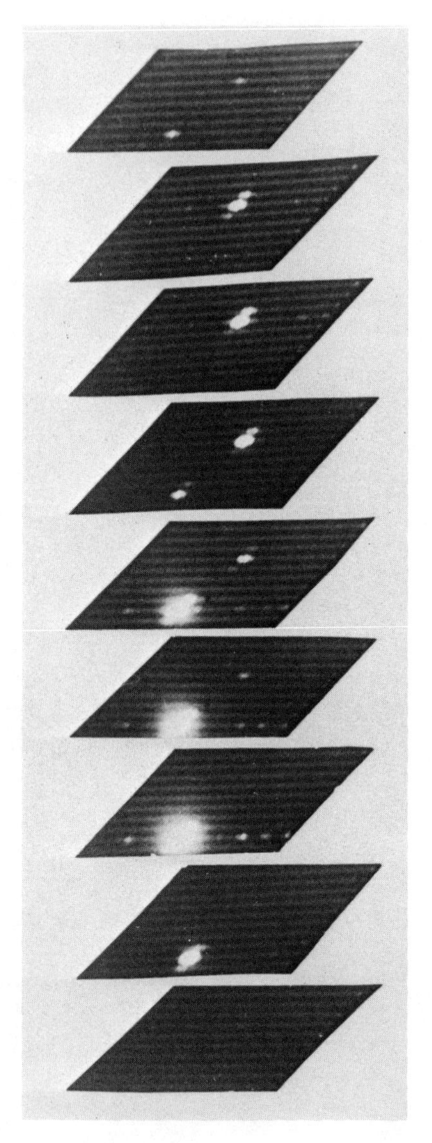

Figure 7b--Two Target Volume Image Sliced Horizontally

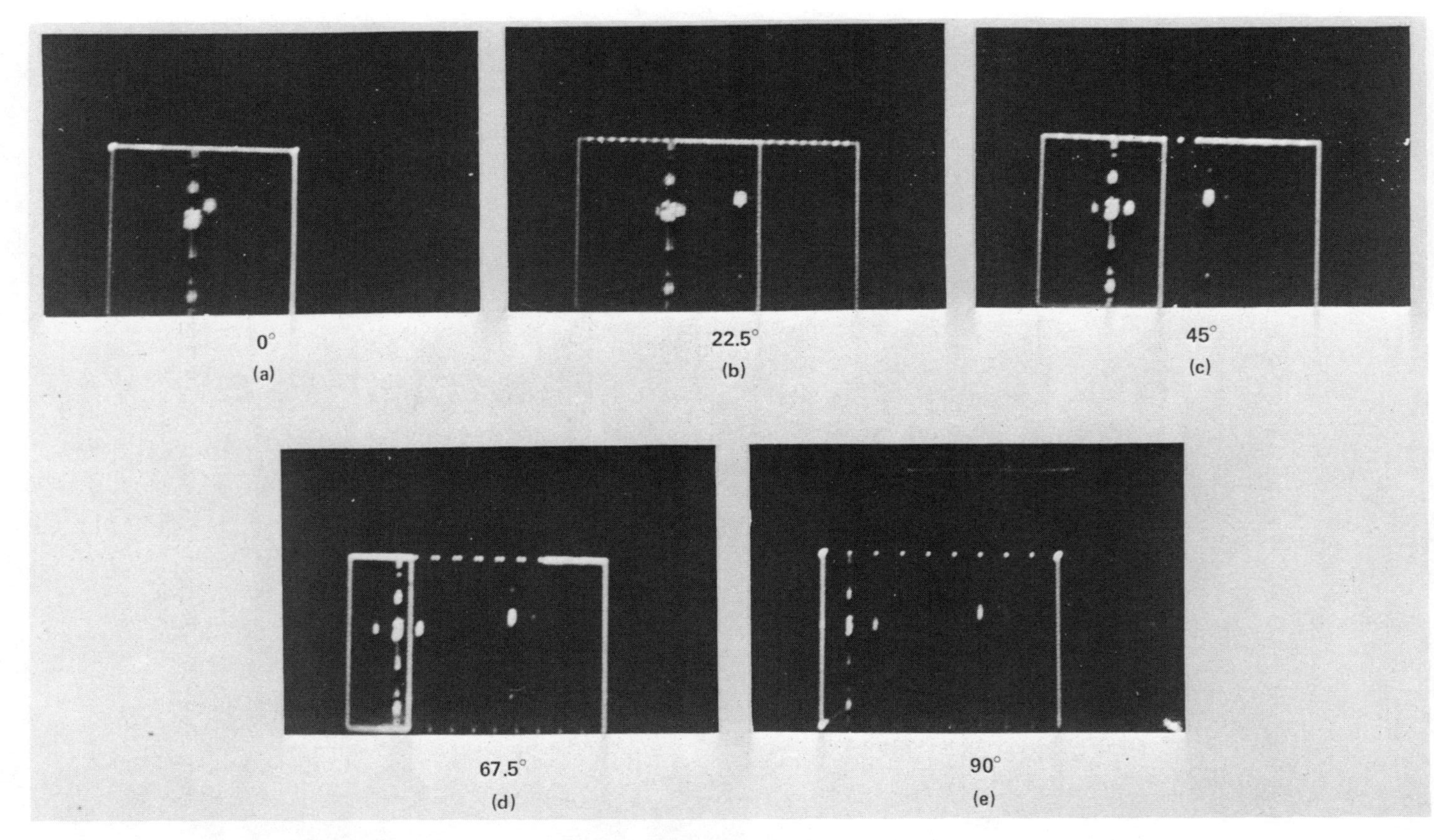

Figure 8--Two Target Volume Image at 22.5° Horizontal Aspect Intervals

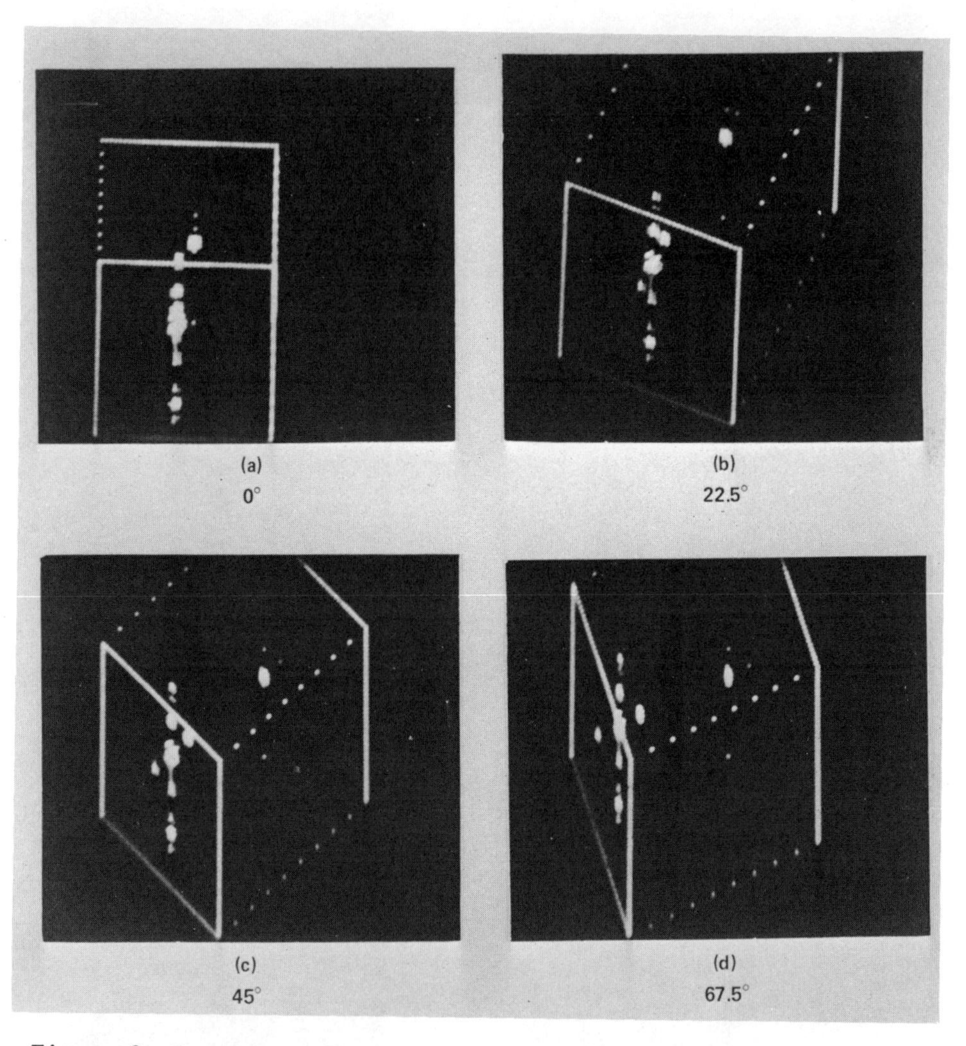

Figure 9--Two Target Volume Image from 45° Vertical Aspect and at 22.5° Horizontal Aspect Intervals

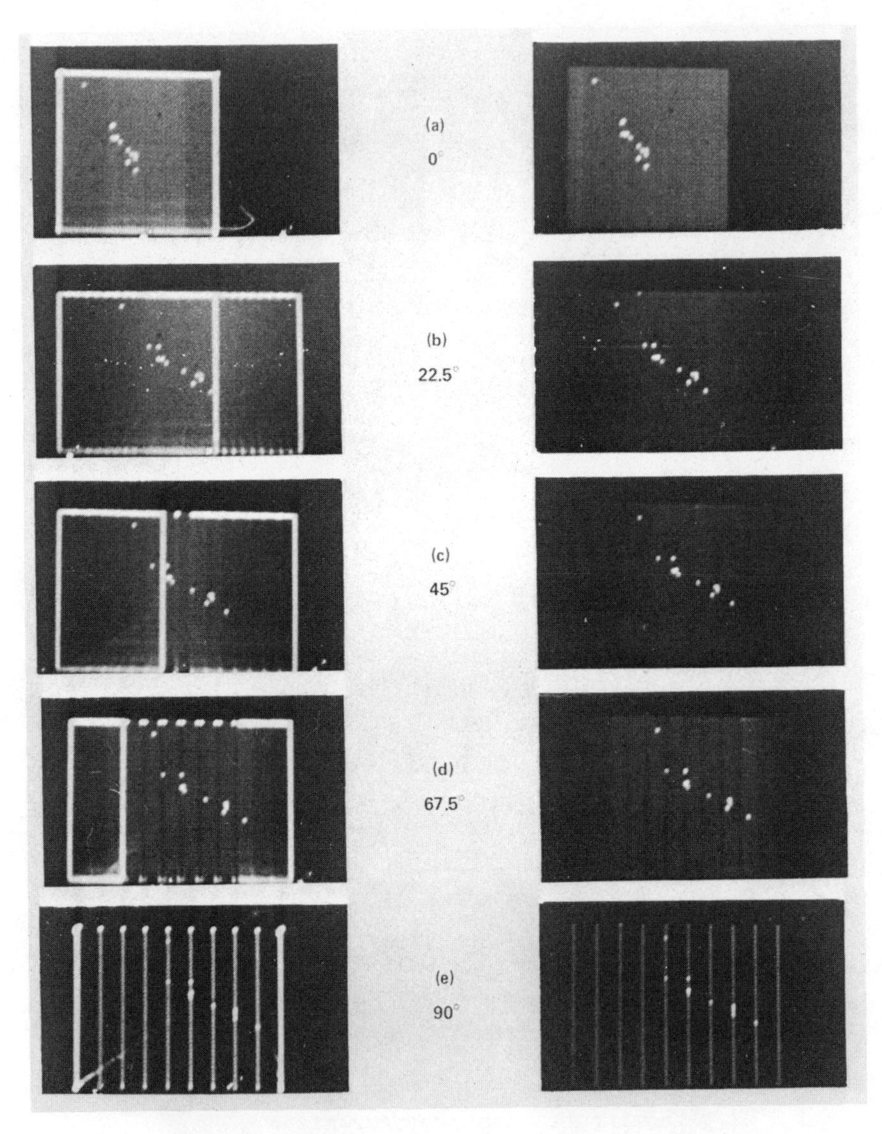

Figure 10--Tube Image Vs. Aspect Angle

A comparison of Figures 8 and 9 demonstrates that considerably more information about a target scenario can be obtained if the objects are viewed from different angles.

To further illustrate this capability, a tube 3" in diameter and 18" in length was placed at an angle approximately 22.5° with the horizontal plane and rotated approximately 67.5° with respect to the array face. It can be seen in Figure 10 that the tube image is foreshortened when viewed from most angles and is seen in its full extension as the volume is rotated through 67.5°.

The foregoing demonstrates that even though all of the available image information is being displayed, it may be presented in a format which obscures the true characteristics of the object being viewed. However, if the information can be presented from several different viewpoints, the operator will then be in a better position to make a decision as to the nature of the object being imaged.

5. SUMMARY

The implementation of a sonar system exhibiting high resolution in three dimensions has been described. The system has the capability of acquiring and processing three-dimensional data on a single-"ping" basis. It has been additionally demonstrated that this capability when combined with operator-controlled rotation of the scene can provide a valuable tool for target classification in that the operator can view objects within the field of view from several aspects. This is especially important when viewing extended objects.

REFERENCES

1. P. Keating, R. Koppelmann, and R. Mueller, "Maximization of Resolution in Three Dimensions," Acoustic Holography, Vol. 6, ed. N. Booth, p. 525 (Plenum Press, New York, 1975).

2. P. Keating, R. Koppelmann, R. Mueller, and R. Steinberg, "Complex On-Axis Holograms and Reconstruction without Conjugate Images," Acoustic Holography, Vol. 5, ed. P. S. Green, p. 515, (Plenum Press, New York, 1975).

ANGLE-LOOK SONAR

G.A. GILMOUR

Westinghouse Electric Corporation, Oceanic Division

P.O. Box 1488, Annapolis, Maryland 21404

ABSTRACT

The Angle-Look Sonar (ALS) is a new technique to take a swath of high-resolution sonar shadowgraph data in the path of a vehicle. The technique was developed to fill the "gap" of side-look sonars, but can also be used as the only imaging sonar for some applications.

The ALS technique consists of mounting a side look sonar type transducer in a geometry such that the ping intersects the bottom on a diagonal line in front of the vehicle. The image is built up one line at a time, as in a side look sonar, but in a herringbone pattern.

Results are shown taken with a 0.1° electronically focusing side look sonar system with the transducer mounted in the ALS geometry.

INTRODUCTION

The Angle-Look Sonar (ALS) is a new technique to take a swath of high-resolution sonar shadowgraph data in the path of a vehicle. The technique was developed to complement side look sonar (SLS) and fill the SLS "gap" with a comparable quality record.

The idea consists of mounting a side-look sonar type transducer in a geometry such that the ping intersects the bottom as shown in Figure 1. Each ping takes data on a line which extends from point 1 to point 3 crossing the path of the vehicle at point 2, the midpoint of the line. Time resolves the position along the line as shown by the intersection of the range arcs with the ping track. The angular resolution of the sonar determines the width of the ping track.

Displaying the data from this sonar is nearly as easy as displaying the data from a side-look sonar. The record, i.e., film or paper, is moved beneath the cathode ray tube or stylus at an angle of β instead of 90°. This provides a hard copy of the "gap" filling record, which has not been accomplished satisfactorily for previous forward-looking systems. Alternatively, the record can be scan-converted and shown on a television monitor. The TV can be modified to change the horizontal sweep such that it has a vertical component, thus correcting the distortion inherent in the Angle-Look geometry.

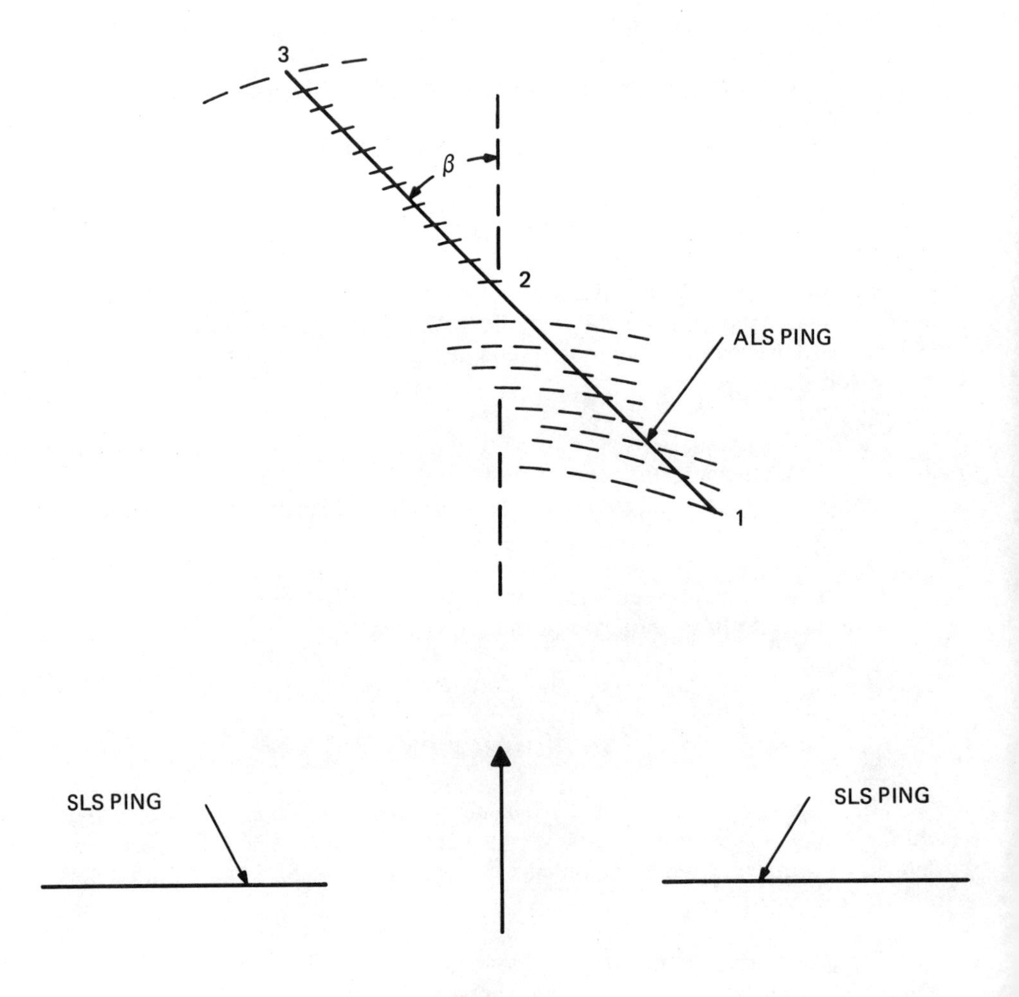

Figure 1. Sonar Ping Geometry

BACKGROUND

Side-looking sonars have a coverage gap underneath the vehicle as shown in Figure 2. This gap may be a true gap if the transducer mounting prevents sound from being transmitted into, or received from, the gap. More often, the gap is simply a region of poor echo characteristics. Since time is used to resolve the range dimension in an SLS, the return from point B arrives at nearly the same time as the return from point A. The bottom range resolution is therefore poor in the "gap". Also, in SLS, shadows are more important for target interpretation than highlights. If the angle of attack, α, is close to 90° there are no shadows. Most workers in the field would consider data from angles of attack of less than $\alpha_{min} = 60°$ acceptable. One could argue that by reducing the ping length, data into angles of attack of 70° would be acceptable. However, at this point the shadows are getting very short. Therefore, the gap to be covered is inside angles of attack of 60-70° which correspond to widths of 1.15 h and 0.73 h, respectively.

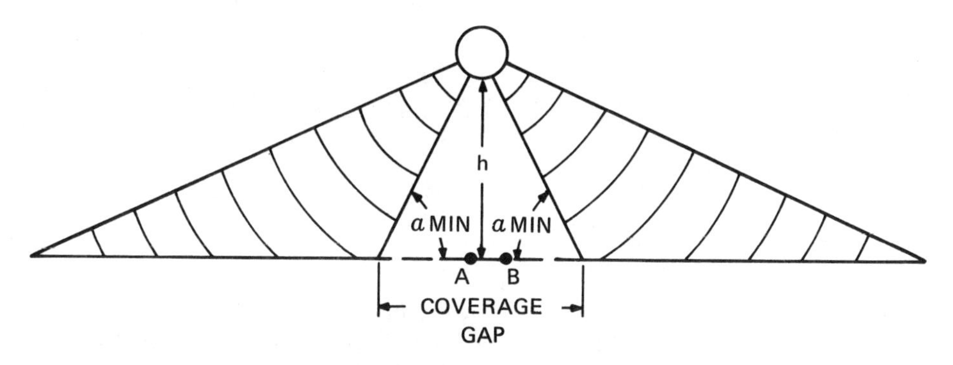

Figure 2. SLS Geometry (Vertical Plane)

ANGLE-LOOK SONAR THEORY

The theory of the Angle-Look Sonar is similar to that of side-look sonar. However, the geometry is difficult to visualize.

System Geometry

Consider the sonar geometries shown in Figures 3 through 5. The standard SLS geometry is shown in Figure 3. In Figure 4, the transducer has been rotated through 60° in the horizontal plane. This causes the beam to intersect the bottom at an angle of 30° relative to the direction of travel instead of the 90° angle of the SLS. Next, the transducer is lowered by 45° in the vertical plane, as shown in Figure 5. This moves the plane of sound under the vehicle to the ALS position.

System Theory

Selection of the best Angle-Look Sonar geometry requires a series of compromises. The "along-track" resolution and the swath width are the most important considerations. The "along-track" resolution is very similar to that of a side-look sonar. The plane of the ping just intersects the bottom at a different angle. Consider Figure 6. The plane of the ping intersects the bottom at an angle ϕ. The resolution along-track at point **d** is bad. This is where the plane of sound is at 90° to the track. The along-track resolution becomes acceptable at some point **c**. The considerations are identical to those of a side-look sonar. Point **d** corresponds to the point immediately under the SLS where there is no range resolution. Point **c** corresponds to the arbitrary point where the SLS range resolution is considered to be acceptable. This is about 60°. θp is the angle in the plane of sound. There is some minimum range point with an angle θp_{min} that represents the start of acceptable data.

Now consider the maximum range point, **a**. The important consideration at maximum range is the angle of attack against the bottom. Historically, ratios of maximum range to altitude of 5 to 10 have been used in side-look sonar. These correspond to angles of attack at maximum range of 11.5° to 5.7°. In side-look sonar the desire to minimize the gap has led to these shallow angles of attack. A system designed with an Angle-Look Sonar to fill the gap should probably fly at a higher altitude to improve the angle of attack to approximately 11° to 15° at maximum range.

There appears to be no reason to require a shallow angle of attack at maximum range for the Angle-Look Sonar. In fact, if the maximum range required to fill the gap can be minimized, the length of the transducer wing can be kept shorter. Therefore, something around 20° appears to be a reasonable value for θa_{max}, the angle of attack at maximum range for the ALS. In side-look sonar the angle of attack is the same as the angle in the plane of the ping. However, in the Angle-Look Sonar the angle in the plane of the ping (θp_{max} in Figure 6) must be greater. The relationship between the angle of attack and the angle in the plane of the ping is

$$\theta a = \theta p \sin \phi. \tag{1}$$

Figure 3. SLS Geometry

Figure 4. Rotated Forward By 60°

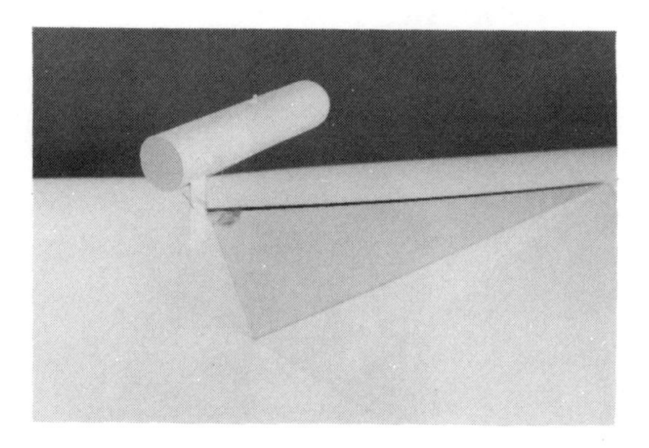

Figure 5. Rotated Under By 45° To ALS Geometry

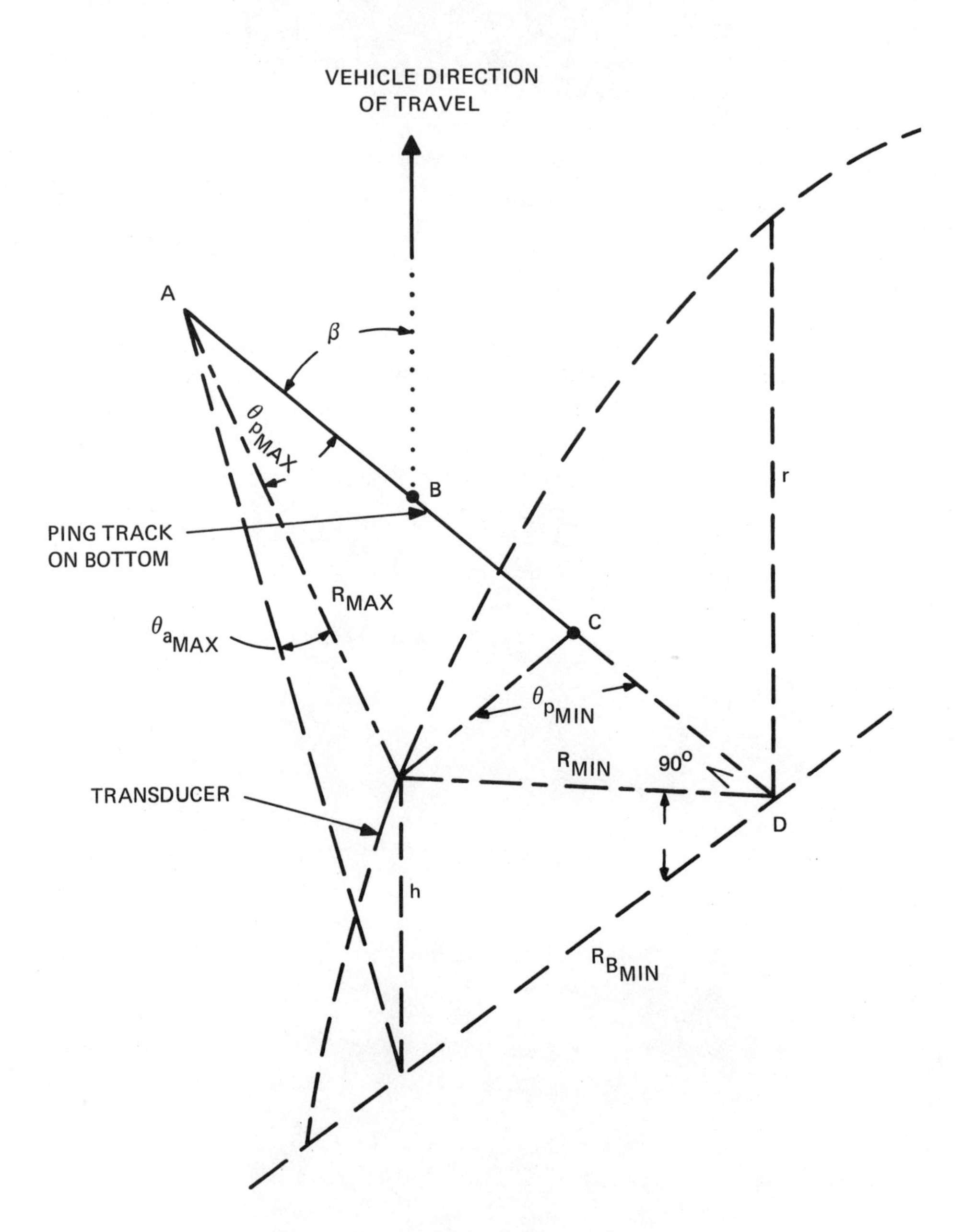

Figure 6. Angle-Look Sonar Geometry

The maximum slant range is related to the altitude (h) and the angle of attack at maximum range by

$$R_{max} = (h/\sin \theta a_{max}). \quad (2)$$

Our selection of $\theta a_{max} = 20°$ gives

$$R_{max} = 2.92h \approx 3\ h. \quad (3)$$

Let us now consider the dimensions of the track on the bottom. The distance along-track from C to A is the useful region. Since it is desirable to cover the gap equally on both sides, the vehicle should travel over the midpoint of the track, i.e., point B. If we call β the angle the track makes with the direction of travel, then the ALS swath width is $T_{ca} \sin \beta$. In order to optimize the width we need the equations for T_{ca} and β.

$$\tan \theta p_{max} = (R_{min}/T_{da}) \quad (4)$$

$$\tan \theta p_{min} = (R_{min}/T_{dc}) \quad (5)$$

$$T_{ca} = T_{da} - T_{dc} \quad (6)$$

$$T_{ca} = (R_{min}/\tan \theta p_{max}) - (R_{min}/\tan \theta p_{min}) \quad (7)$$

$$T_{ca} = R_{min} (\cot \theta p_{max} - \cot \theta p_{min}) \quad (8)$$

The minimum slant range R_{min} is related to the altitude by

$$\sin\phi = (h/R_{min}). \quad (9)$$

θp_{min} is typically around 60° but we will retain the more general term for the time being. θp_{max} is related to θa_{max}, the angle of attack at maximum range through equation (1). Since θa_{max} is the limitation, we convert equation (8) to equation (10) by use of (1) and (9).

$$T_{ca} = (h/\sin \phi) [\cot (\theta a_{max}/\sin \phi) - \cot \theta p_{min}] \quad (10)$$

Now we turn our attention to β.

$$\tan \beta = (R_{bmin}/T_{db}) \quad (11)$$

The minimum bottom range R_{bmin} is related to the minimum slant range by

$$R_{bmin} = R_{min} \cos \phi. \quad (12)$$

Since b is by definition the middle of line ca,

$$T_{cb} = (T_{ca}/2) \tag{13}$$

and

$$T_{db} = T_{dc} + (T_{ca}/2). \tag{14}$$

Using (6)

$$T_{db} = (T_{da}/2) + (T_{dc}/2). \tag{15}$$

Using (15), (4), (5), and (12) in (11) we get

$$\tan \beta = [(2\ R_{min} \cos \phi)/(R_{min} \cot \theta p_{max} + R_{min} \cot \theta p_{min})]. \tag{16}$$

Cancelling out the R_{mins} and again changing the θp_{max} to θa_{max} by (1) gives

$$\tan \beta = (2 \cos \phi)/[\cot (\theta a_{max}/\sin \phi) + \cot \theta p_{min}]. \tag{17}$$

This is the desired expression for β.

The swath width is

$$W_s = T_{ca} \sin \beta. \tag{18}$$

Using (10) and (17) in (18) we get

$$W_s = \frac{h}{\sin \phi}\left[\cot\left(\frac{\theta a_{max}}{\sin \phi}\right) - \cot \theta p_{min}\right] \sin \tan^{-1}\left[\frac{2 \cos \phi}{\cot (\theta a_{max}/\sin \phi) + \cot \theta p_{min}}\right] \tag{19}$$

This is the desired expression for the swath width.

The whole purpose of this derivation was to select desirable angles for β and ϕ. Let us take $\theta p_{min} = 60°$, i.e., the angle in the plane of the ping where the along-track resolution becomes acceptable. We previously suggested $\theta a_{max} = 20°$, the angle of attack at maximum range. Equation (17) has been plotted in Figure 7 and shows the relationship between ϕ and β under these constraints.

It is desirable to have the widest swath width possible under these constraints. However, we need one more constraint. As the angle ϕ becomes very shallow, the width of the track becomes very large. Some tradeoff between the width of the track and the swath width must exist. The width of the track is related to the width of the beam by

$$\omega_t = (\omega_b/\sin \phi). \tag{20}$$

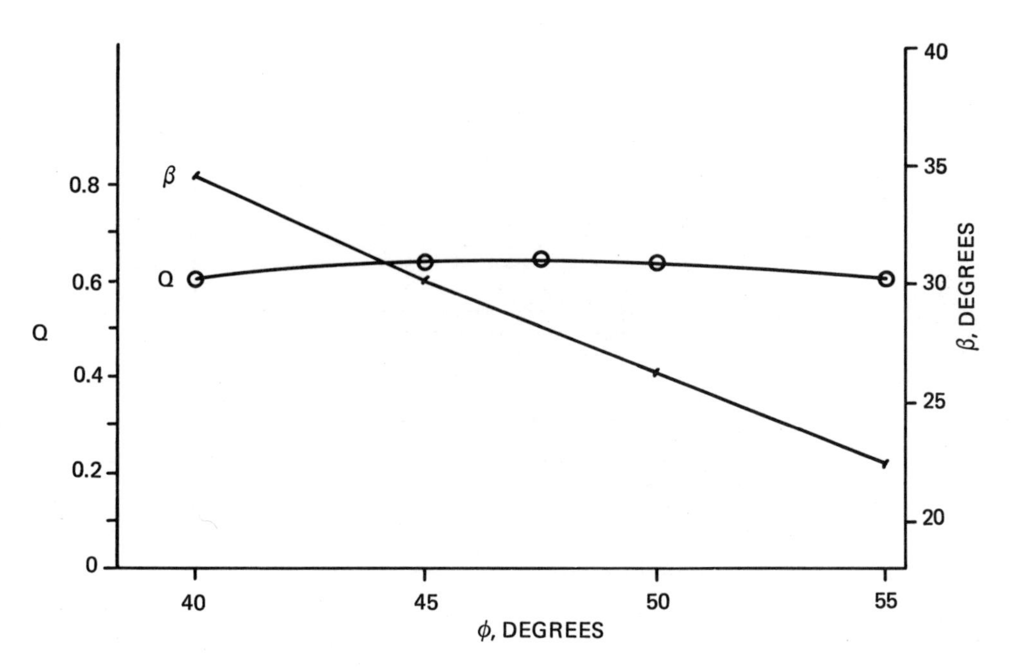

Figure 7. Angle Relationships for ALS

Let us consider a quality factor that maximizes the swath width and minimizes the width of the beam, i.e., we would like W_s/ω_t to be maximum. We must hold the altitude (h) constant since the SLS gap to be filled is proportional to h. Therefore, we want to maximize the quantity

$$Q = (W_s \; \omega_b/h \; \omega_t) = (W_s \sin \phi/h) \tag{21}$$

which we will call a quality factor. W_s/h can be considered to be a normalized swath width and ω_t/ω_b a normalized beam width.

Using (20) and (19) we get:

$$Q = \left[\cot\left(\frac{\theta a_{max}}{\sin \phi}\right) - \cot \theta p_{max}\right] \sin \tan^{-1}\left[\frac{2 \cos \phi}{\cot (\theta a_{max}/\sin \phi) + \cot \theta p_{min}}\right] \tag{22}$$

This expression has been plotted as a function of ϕ under the $\theta a_{max} = 20°$ and $\theta p_{min} = 60°$ constraints. The peak is at about $\phi = 47.5°$ but the function is very insensitive to angle. Angles of $\phi = 45°$ and $\beta = 30°$ are within 1% of the quality of the optimum point, and were selected for convenience. The swath width is 0.91 h which fills the SLS gap inside about 65°. This is not far from the desirable value of 60°. Something around 62 or 63° on both the ALS and the SLS probably matches things up as well as possible. In

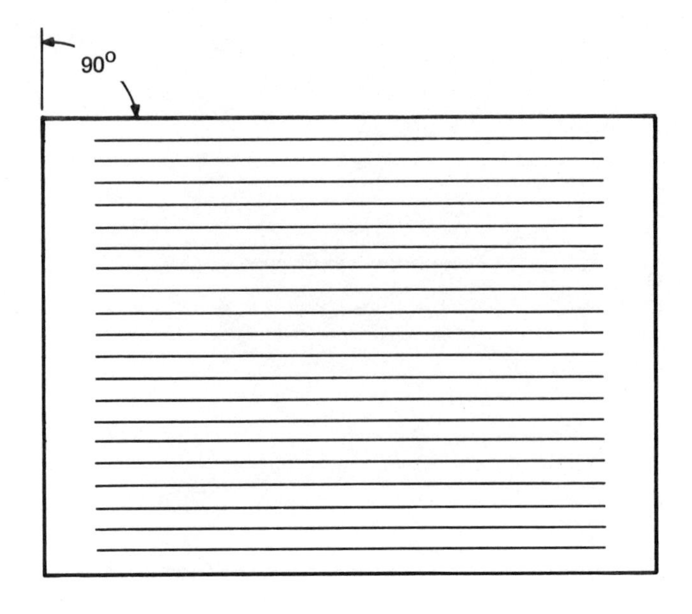

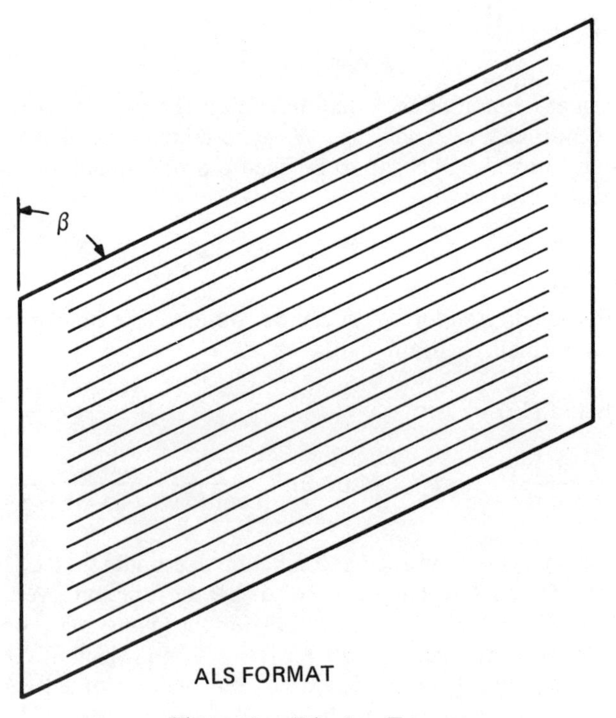

Figure 8. Display Formats

practice the data will be there from 90° on and the operator will have overlapping records from which to choose. Therefore, further exercises in mathematics to optimize the transition point are not justified. It should be noted that the long-range point of the ALS which was limited by our angle of attack selection of 20° only fills the SLS gap on the other side to 65°. The range should be extended to give some overlap at this point also. The amount figures out to about 22%, making

$$R_{max} \approx 3.5 \text{ h}. \quad (23)$$

This makes the angle of attack at maximum range about 16°.

DISPLAY

Side-look sonar displays can be placed into two categories: hard-copy and scan-converted for TV. An Angle-Look Sonar display could be developed using either principle. Data were taken during our at-sea tests on magnetic tape and fiber optic SLS film recorder. The SLS presentation distorts the ALS record, as will be shown. A fiber optic film recorder could be built to correctly display ALS records by rotating the CRT relative to the film. This is simple in principle but would require a major mechanical redesign of present recorders.

The correct format was obtained in the laboratory by playing the tape recordings through a SLS scan converter onto an electrostatic TV monitor. The vertical drive circuitry for the electrostatic TV could easily be modified to write the raster in ALS format rather than the SLS format as shown in Figure 8.

The film recorder writes a negative image as shown in Figure 9C. The SLS display scan converted to TV is a positive image as shown in 9B. The correct representation is shown in Figure 9A.

CHESAPEAKE BAY TEST RESULTS

Angle-Look Sonar tests were run in the Chesapeake Bay using a SLS transducer mounted on the side of the Westinghouse RV NORTHSTAR in an ALS configuration.

Test Configuration

A rather extensive structure was built so that the transducer could be lowered and locked into position. The structure is shown in the up position in Figure 10. When lowered into position the pipes that pivot from the side of the boat were locked into position. This held the transducer at the desired angle of 45° down and 30° back. A faired wing (such as shown on the model in the Frontispiece) would be a more desirable configuration. However, our goal was simply to get an existing transducer into the water such that feasibility could be proven. Therefore, we used no fairings, ran all ALS tests at slow speed (1 beam), and pulled the transducer structure out of the water for high-speed transit.

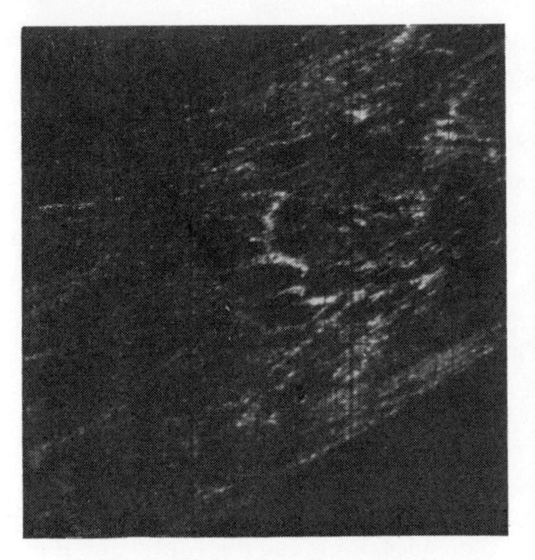

A. ALS Display on Electrostatic TV (with distortion corrected)

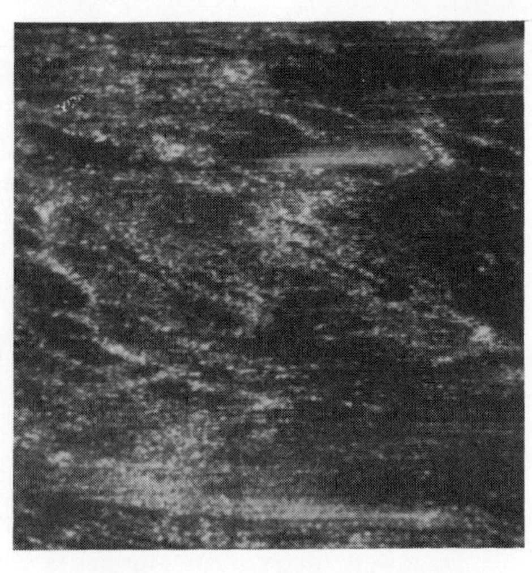

B. Display on Magnetic TV (written as an SLS)

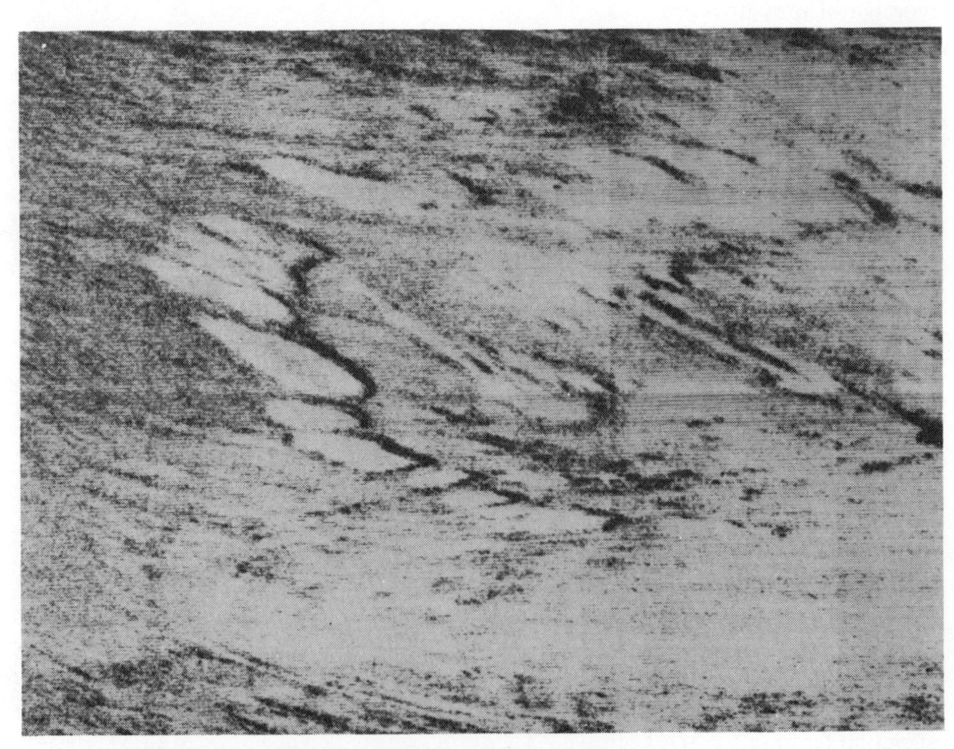

C. Display From Film Recorder (written as an SLS)

Figure 9. ALS Displays

Figure 10. ALS Structure and Transducer in Up Position

Figure 11. Model Showing ALS Beam Emanating From A Wing (SLS Beams Also Shown)

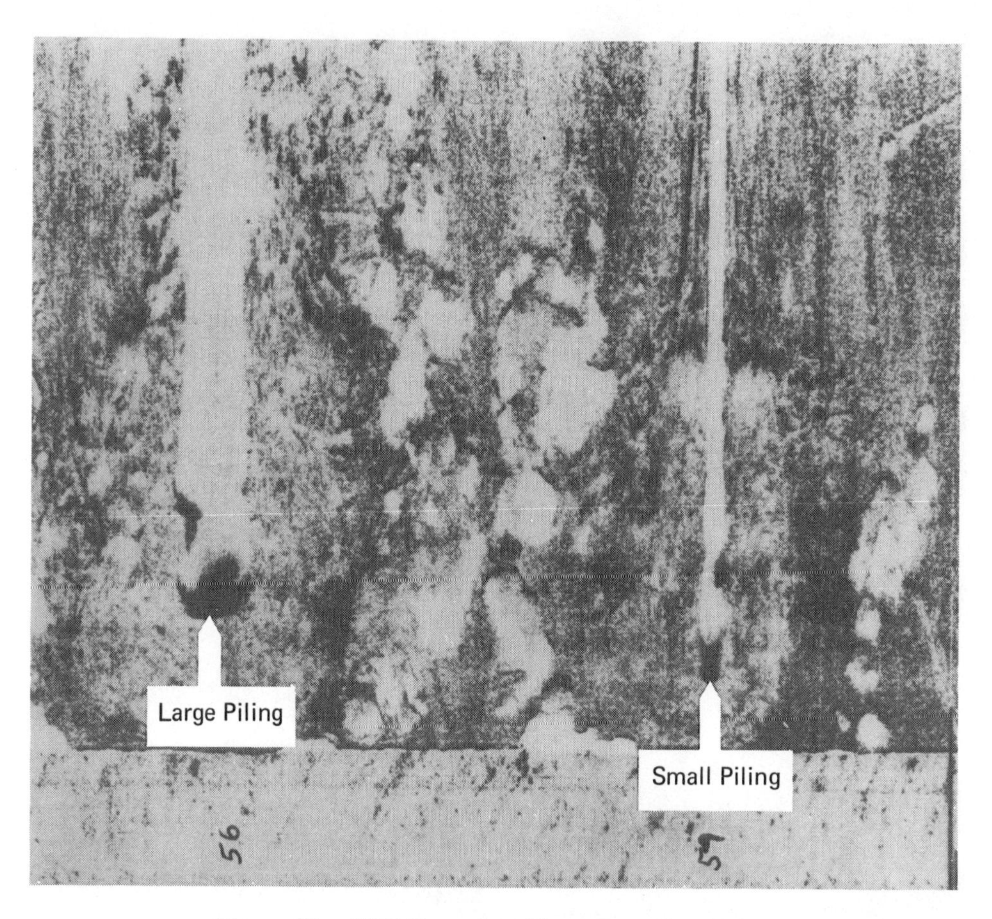

Figure 12. SLS Record: Chesapeake Bay Bridge

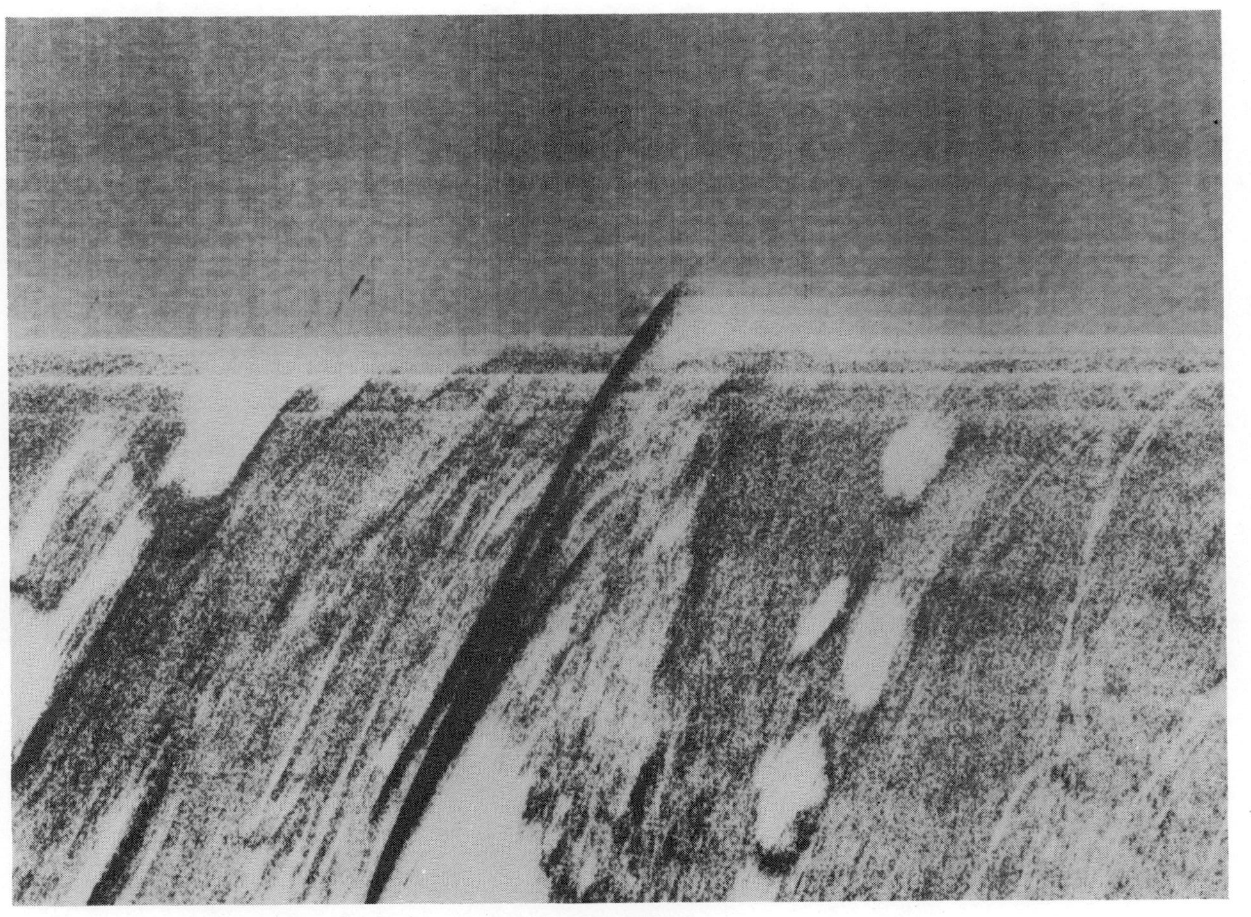

Figure 13A. Chesapeake Bay Bridge: ALS Written as SLS

Figure 13B. ALS Written as SLS on Standard TV

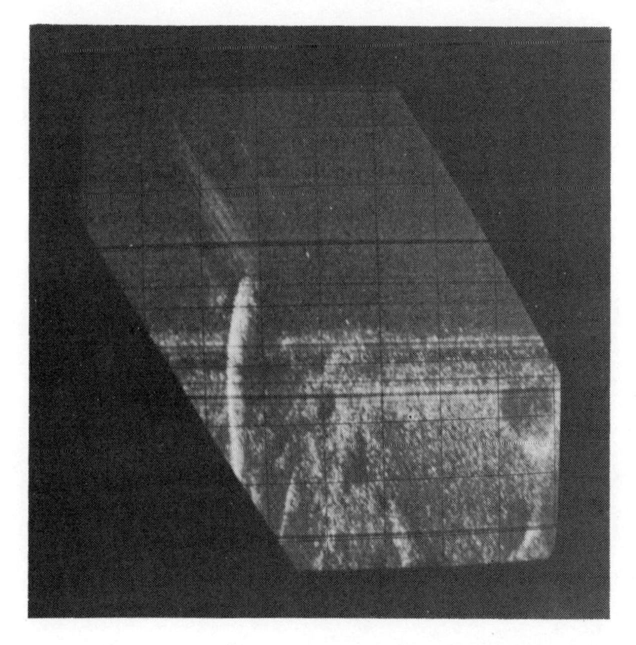

Figure 13C. ALS Corrected Angle on Electrostatic TV

The use of a surface ship for the ALS test platform was a mixed blessing. It was convenient not to have to deploy a towed vehicle. However, we had some problems from reflections off the surface (not as bad as in the SLS tests). A more serious problem was ship roll. Roll is in the vertical plane of a side-look sonar and hardly affects the results at all. Roll changes the intersection line of the ALS relative to the bottom. Also, the way we had the transducer mounted resulted in quite a lever arm relative to the centerline of the ship. Therefore, the transducer amplified the rolling of the ship. Most of our good results were taken on calm days. The importance of stabilizing the vehicle led to the idea of mounting the ALS transducer in a stabilizing wing and adding a second wing on the opposite side as shown on the model. (See Figure 11.)

The electronic all-range focusing feature of the experimental SLS meant that altitude off the bottom was not critical. However, the distance the beam reaches under the vehicle is directly proportional to the altitude off the bottom.

Test Results

An SLS record taken under the new span of the Chesapeake Bay Bridge at Annapolis is shown in Figure 12. The pilings have shadows behind them and there are many depressions in the bottom which give weak returns (indicated by lighter areas). An ALS pass near the same small piling is shown in Figure 13. The piling gives a distorted geometry since it comes up through the water column. However, bottom features hold the same geometry between SLS and ALS (corrected) displays.

Conclusions

The results from the tests prove conclusively that the Angle-Look Sonar works. The perspective on targets in the water column is distorted. However, since most targets of interest will be laying on the bottom, the ALS should be at least as effective as the SLS it complements in detecting them.

SEISMIC HOLOGRAPHY IN A NORWEGIAN FIORD

S. Ljunggren, O. Løvhaugen and E. Mehlum

Central Institute for Industrial Research

Forskningsvn 1, Blindern - Oslo 3, Norway

ABSTRACT

The classical methods used in seismic surveys are based on measuring the time taken for a pulse of sound to travel from a sound source to a receiver via the various geological reflectors. Another use of sound in seismic surveys would be to sample the returning wavefront with phase and amplitude over an area. The measured wavefront can be used to reconstruct the geological reflectors. In this paper most emphasis is put on an experiment in seismic holography carried out in a Norwegian fiord to determine the feasibility of the method. Both the registration and the processing techniques employed are discussed. Examples of reconstructions of the geological structure together with a comparison with results obtained by classical methods are presented. The possibility of using optical methods to reconstruct acoustical fields is also discussed.

INTRODUCTION

Seismic holography [1,2] is a method, whereby one can reconstruct the acoustical properties of a geological structure from a two-dimensionally recorded wavefront. It differs from the classical seismic methods [3], by which reconstructions are obtained from pulse-echo recordings with focused beams of sound with coincident source and receiver focused at the same spot.

Experiments in seismic holography have been carried out previously [1,2]. The objective of the experiment described in this paper was to determine the feasibility of a scheme in which a pulsed source in a fixed position was fired repeatedly, and signals were recorded patch by patch by a hydrophone array to cover an area of 200 x 200 m^2. This scheme was used to ensure identical insonification of the geological structure in each recording. An equally important objective was to observe the quality of the reconstructions obtained from the recorded wavefront.

The recorded time variations of the pressure was processed to give the complex Fourier coefficients at a certain temporal frequency at each sampling point. The resulting monochromatic wavefront could then be used either as input to a digital reconstruction method based on the theory of inverse propagation, or as input to make a synthetic hologram to get an optical reconstruction. Reconstructions of simulated data have been made by both methods. Because of the extreme near field geometry in the experiment (the depth of the sea was 30 m, whereas the extent of the recording area was 200 m) the digital method was chosen to reconstruct the geological structure. We feel, however, that the method we have developed for obtaining optical reconstructions is promising for other geometries.

RECORDING THE WAVEFRONT

The soundfield incident on the geological structure was produced by a 1 $inch^3$ airgun resting in a fixed position. See Fig. 1.

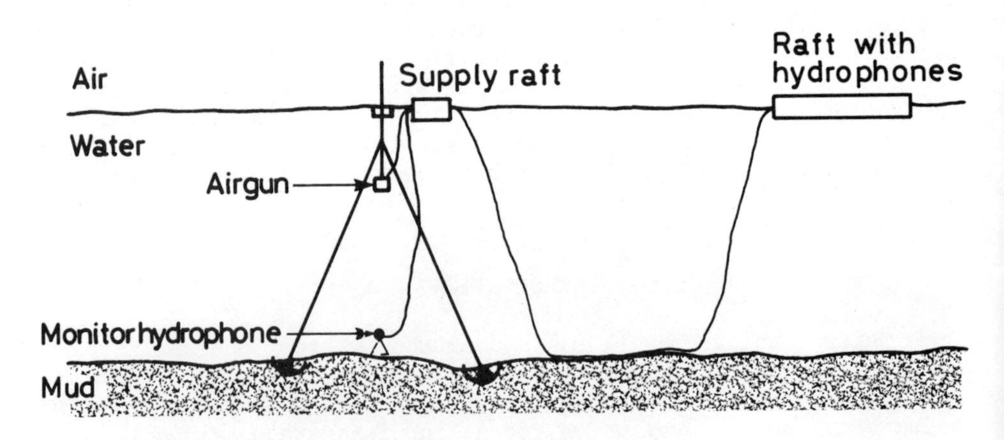

Fig. 1. The layout of the experiment. The 10 x 10 m^2 hydrophone array with the recording equipment was moved around to cover the 200 x 200 m^2 recording area.

The airgun signal was monitored by a hydrophone in a fixed position at the sea floor below the airgun. After being scattered by the geological structure the signal was recorded by a hydrophone array of 10 by 10 elements, each 1 m apart and mounted on a movable raft. The monitor hydrophone made it possible during the processing to compensate for variations in the airgun output from one shot to another. The monitor signal was also used as a time reference for the signals received by the array in different recordings. 23 of the hydrophone signals plus the monitor signal could be recorded at one time by a digital recorder. Four recordings had to be done at each position of the raft to cover the 92 array hydrophones in use. Each recording lasted 1 s after the shot was fired. The recording procedure was repeated until an area of 200 x 200 m^2 was covered, thereby producing an effective array of 40 000 sampling points.

The position of the raft, and thus the position of the sampling points, was determined to within 0.01 m using ordinary surveying equipment positioned on shore. Due to boating and other activities the overall signal-to-noise ratio was at times as low as 2:1.

SIGNAL PROCESSING

From spectral analysis of the monitor signals it was decided to make the monochromatic wavefront at a frequency of 240 Hz. Typical spectra of the monitor signals and the recorded signals are shown in Fig. 2. The monochromatic wave field was constructed by calculating the complex Fourier component $F_{240}(x,y)$ at 240 Hz in each sampling point from the recorded signal $s(x,y,t)$,

$$F_{240}(x,y) = \int_{\infty}^{\infty} s(x,y,t)\cdot e^{-i2\pi\cdot 240\cdot t}dt.$$

A section of the wavefront put together of signals from four different raft positions, with four recordings made at each position, is shown in Fig. 3. In this figure the length of each phasor is proportional to $|F_{240}(x,y)|$, while the angle between each phasor and the horisontal is equal to the difference $\mathrm{Arg}(F_{240}(x,y))-\mathrm{Arg}(F_{240}(0,0))$. Fig. 4a shows a binary representation of the whole 200 x 200 sampling points wavefront. A black spot means that the difference $\mathrm{Arg}(F_{240}(x,y))-\mathrm{Arg}(F_{240}(0,0))$ is positive, a white spot means the difference is negative.

From the same data that gave the wavefront at 240 Hz it is possible to achieve wavefronts at any other frequency at which the airgun has an output of acoustical energy. Differences in reconstructions at different temporal frequencies will reveal areas with dispersion and wavelength dependent reflections and absorptions.

One of the main interests in the experiment was to investigate the quality of the imaging of buried diffractors. The strong re-

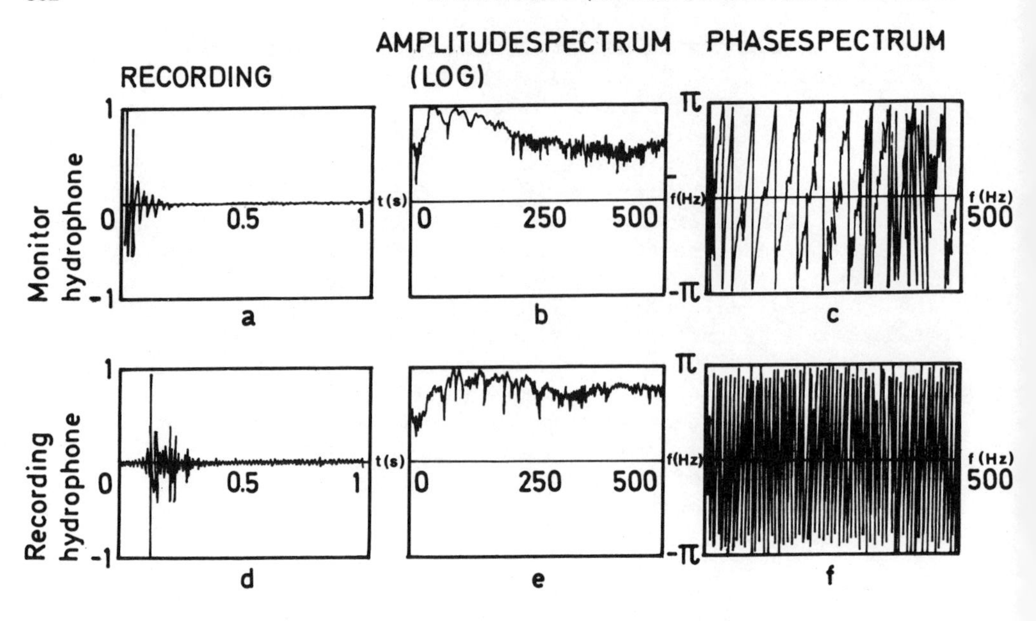

Fig. 2. Typical recordings are shown for the monitor hydrophone in a, and for an array hydrophone in d. The record covers 1 s of the airgun shot. In b and e amplitude spectra (0-500 Hz) are shown for respectively the monitor hydrophone and an array hydrophone, whereas phase spectra (0-500 Hz) for the two recordings are shown in c and f.

flections of the airgun in the sea floor were undesirable. The first 250 milliseconds of the recording were therefore suppressed before the processing. The suppressed part of the signal lies in the shaded area in Fig. 5. The resulting wavefront is shown in Fig. 4b. The ringlike structure in the wavefront centered about the xy-coordinate of the source, has disappeared.

To reduce noise coming from sources within the water or at the water surface, we lowpass-filtered the angular spectrum of the monochromatic wavefront. The angular spectrum was calculated using a two-dimensional FFT. Fig. 4c and 4d show the wavefronts in 4a and 4b respectively, after removal of the plane waves incident at an angle greater than 45°.

RECONSTRUCTING THE ACOUSTICAL FIELD USING INVERSE DIFFRACTION

To reconstruct the acoustical field below the recording area the method of inverse diffraction was used. The method is described

Fig. 3. A section of the monochromatic wavefront represented by phasors, i.e. the length of the phasor is proportional to the modulus of the Fourier coefficient, the relative angles give the difference in phase angles between recordings.

by Shewell and Wolf [4] and Lalor [5], and have been used by Demetrakopoulous [6] for digital reconstructions of acoustical holograms. The reconstructions were made under the assumption that the medium is homogeneous, although we feel that the angular spectrum approach would be very powerful for reconstructing layered media.

The angular spectrum method is identical to solving the inverse diffraction integral, given by

$$p(x,y,z_o) = \int_{-\infty}^{\infty}\int_{-\infty}^{\infty} p(x',y',0)\cdot \left\{\frac{-1}{2\pi}\cdot\frac{d}{dz_o}\left[\frac{\exp\left(-i\frac{2\pi}{\lambda}\sqrt{(x-x')^2+(y-y')^2+z_o^2}\right)}{\sqrt{(x-x')^2+(y-y')^2+z_o^2}}\right]\right\}\cdot dx'dy' ,$$

using the convolution theorem.

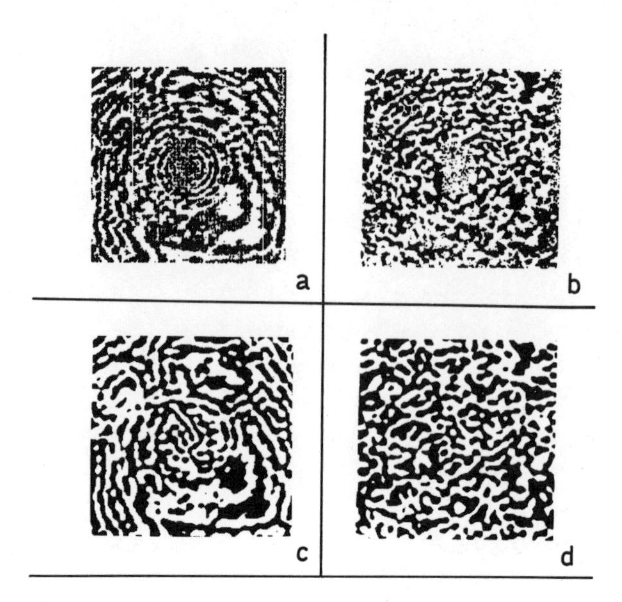

Fig. 4. Monochromatic wavefronts in binary representation.
(a) The wavefront of the 0-1 s recording.
(b) The wavefront of the recording after suppression of the first 250 ms.
(c) The wavefront in (a) after lowpass filtering of the angular spectrum. Max. angle of incidence 45 deg.
(d) The wavefront in (b) after lowpass filtering as in (c).

Here $p(x,y,z)$ is the pressure at the coordinate x,y,z. x',y' is the coordinate in the registration plane. z_o is the distance between the registration plane and the plane where the field is reconstructed and λ is the wavelength.

The convolution theorem states that if a function $h(x,y)$ can be expressed in the form

$$h(x,y) = \int_{-\infty}^{\infty}\int_{-\infty}^{\infty} f(x',y')\cdot g(x-x',y-y')dx'dy,$$

then $h(x,y)$ can be written in terms of the two-dimensional Fourier transform $F\{\ \}$ and the two-dimensional inverse Fourier transform $F^{-1}\{\ \}$ as

$$h(x,y) = F^{-1}\Big\{F\{f(x,y)\}\cdot F\{g(x,y)\}\Big\}.$$

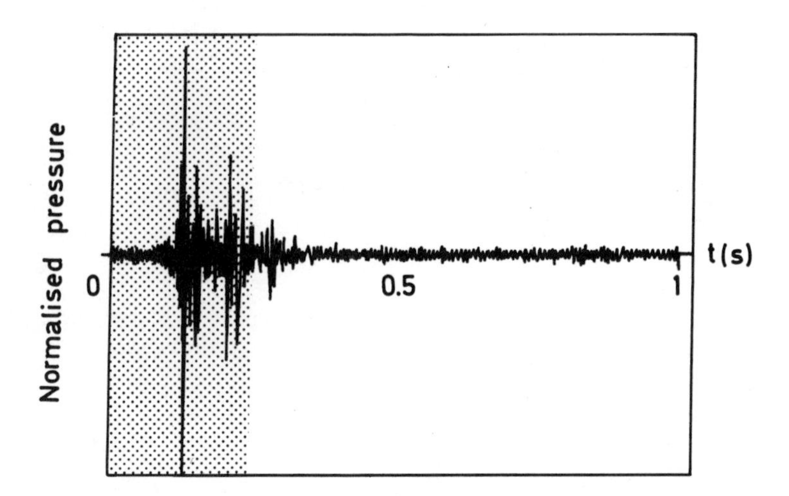

Fig. 5. A recording where the shaded area shows the 250 ms of the signals that were suppressed.

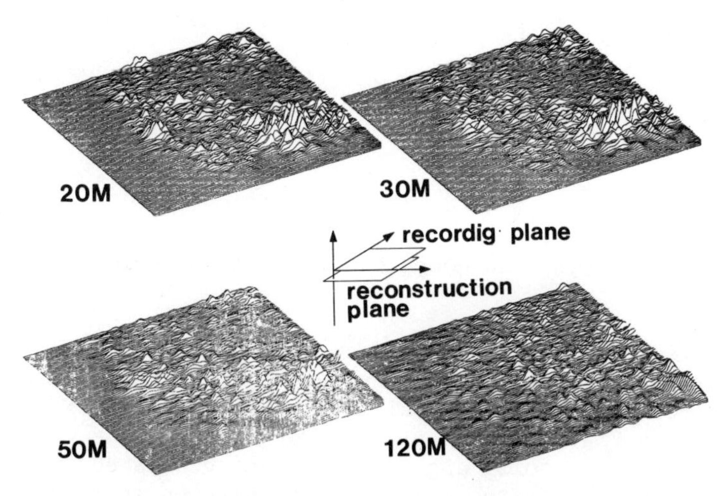

Fig. 6. Digital reconstructions of the wavefront shown in Fig. 4d at 20 m, 30 m, 50 m and 120 m depths as imaged through water.

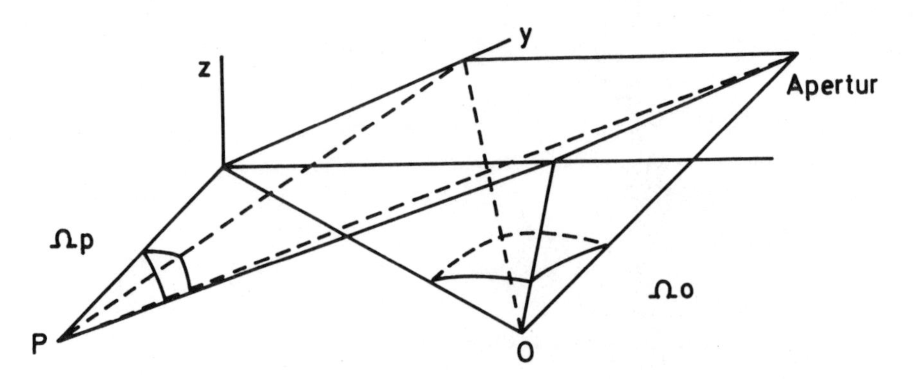

Fig. 7. The relative strength in reconstruction of two homogeneous scatterers O and P is determined by the ratio of solid angles Ω_o/Ω_p.

If one makes the identification

$$h(x,y) = p(x,y,z_o),$$

$$f(x,y) = p(x,y,0),$$

and $$g(x,y) = \frac{-1}{2\pi}\frac{d}{dz_o}\left[\frac{e^{-i\frac{2\pi}{\lambda}\sqrt{x^2+y^2+z_o^{\,2}}}}{\sqrt{x^2+y^2+z_o^{\,2}}}\right],$$

one gets the expression

$$p(x,y,z_o) = F^{-1}\left\{F\{p(x,y,0)\} \cdot F\{g(x,y)\}\right\},$$

from which $p(x,y,z_o)$ is computed using the two-dimensional Fast Fourier Transform.

By repeating this process for various depths z_o we get a reconstruction of the three-dimensional acoustical field below the recording area. Acoustical intensity plots at different depths are shown in Fig. 6. These plots were given a crude compensation for the difference in solid angle the recording area covers seen from the two reconstructed points. The compensation is made under the assumption that the diffractors that gave rise to the recorded wavefront are isotropic scatterers. Without this compensation the scatterer P in Fig. 7 would receive much less energy in reconstruction than the scatterer O, even if both scatterers were of equal strength.

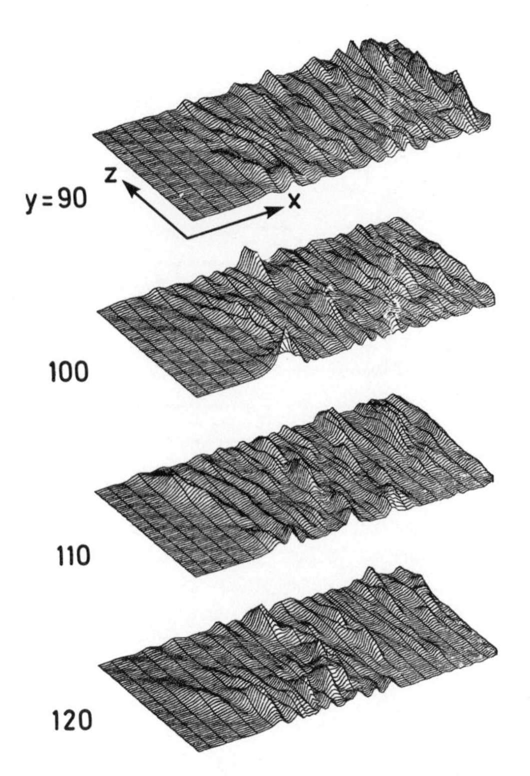

Fig. 8. Vertical cross sections of the reconstructed acoustical intensity field. z is the vertical axis, x is along the shore. The plots are produced for the distances 90 m, 100 m, 110 m, 120 m from the edge of the recording area nearest shore.

INTERPRETATION OF THE RECONSTRUCTED ACOUSTICAL FIELD

The final aim of our work is not to reconstruct the acoustical field, but to find diffractors and reflecting layers. The diffractors show up as points of focused energy in the reconstructed field. To find reflectors one has to determine the position and the strength of the mirror image of the source. Fig. 8 shows vertical x-z-plots of the acoustical field. Together with x-y-plots as shown in Fig. 6, these plots make it possible to single out the focused points. The diffractors shown in Fig. 8 are much too small to show up in traditional pulse-echo methods. To obtain some sort of a comparison between the present mapping of diffractors and the traditional seismic sections we have plotted histograms of the number of diffractors versus depth, as seen in water. These histograms are shown in Fig. 9a, b, and c. Pulse-echo data from a section in the recording area appears in Fig. 10. These data are collected using an

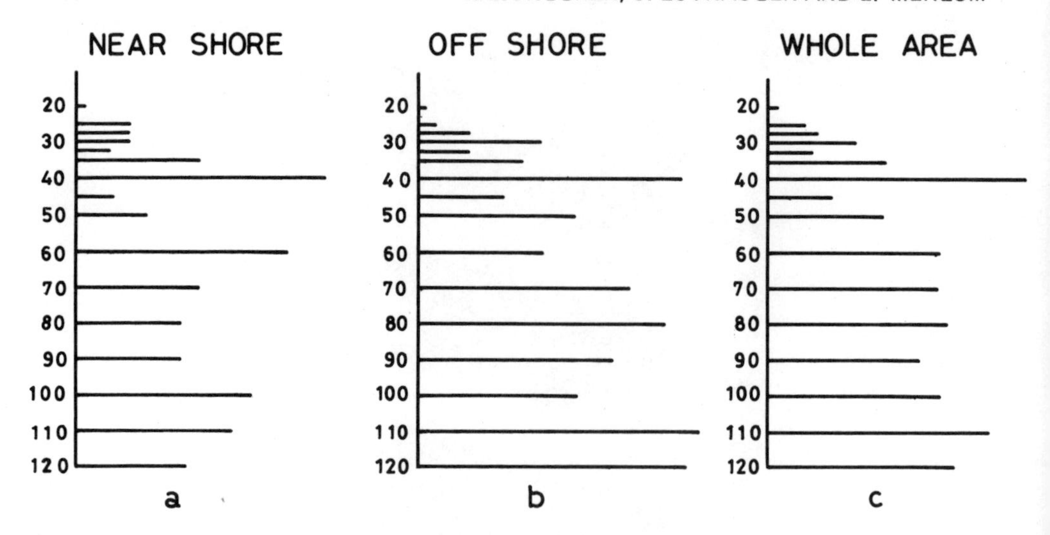

Fig. 9. Histogram of the numbers of diffractors as function of depth.
(a) Relative number of diffractors in the half of the reconstructed volume near shore.
(b) Relative number of diffractors in the half of the reconstructed volume farthest from the shore.
(c) Relative number of diffractors in the whole volume.

electromagnetic boomer having a maximum frequency of 2.5 kHz. An interpreted version of the section is shown in Fig. 11.

Comparing the histograms in Fig. 9 and the section in Fig. 11 we see that at a depth of 40 m where there is an interface between soft mud and clay, there is also a much greater number of diffractors than at other depths. We conclude from this that this interface scatters sound, partly diffusely, partly specularly, and that there is much more diffuse scattering taking place at the interface than at neighboring depths. Under the assumption that this scattering property is characteristic for other interfaces as well, we should expect interfaces in Fig. 9 at depths where the number of scatterers has a maximum value. From Fig. 9a and Fig. 9b we are thus led to believe that there is an interface at a depth of 60 m near the shore going downwards towards 70-80 m farther from the shore. If we take into account that the depts in Fig. 9 are imaged through water, whereas the depts in Fig. 11 are the pulse-echo times multiplied by a constant sound speed, we may guess that the interface at 50-52 m depth in Fig. 11 corresponds to the interface at 60-80 m in Fig. 9. If our guess is right, then the sound speed in the layer of clay between the two interfaces is about 1.2-1.3 times greater than the sound speed in water.

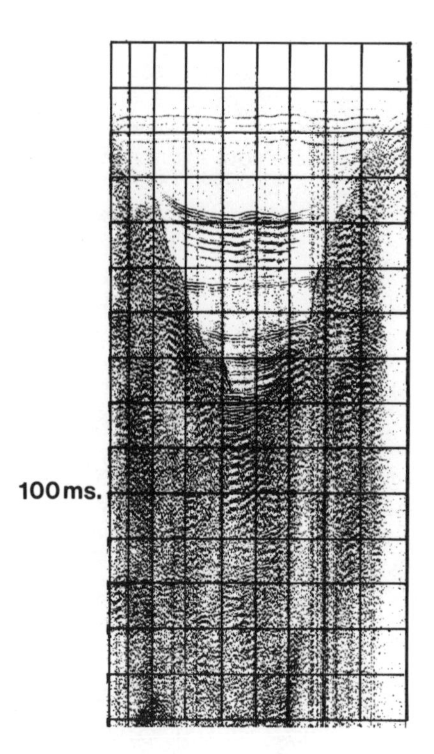

Fig. 10. Vertical cross section of the recording area recorded with a boomer in pulse-echo mode. From sea to the shore is left to right.

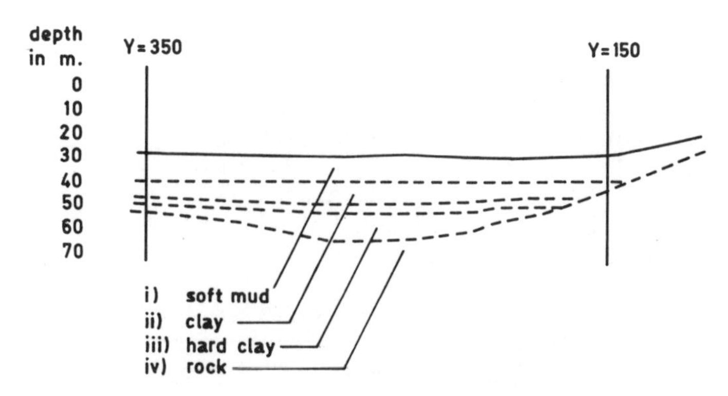

Fig. 11. Interpreted version of the cross section in Fig. 10.

OPTICAL RECONSTRUCTIONS OF THE ACOUSTICAL FIELD

The digital representation of the scatterers as shown in Fig. 6 and Fig. 8 clearly shows the need of a method to present the three-dimensional structure under reconstruction. We will present some optical reconstructions obtained from simulated acoustical data. To obtain high quality optical reconstructions we make use of our equipment for producing computer generated optical holograms. This equipment consists of a numerically controlled scanning electron beam machine which exposes electron resist. Examples of high quality optical reconstructions obtained by this equipment are given in [7]. Since the simulations were done in an early stage of the project, the simulated data are produced for a geometry different from the geometry in the previously described experiment. We therefore start with a description of the simulation model.

The model consists of a plane object emitting waves at a long wavelength. The wavefront from the object is calculated in a rectangular array of 400 by 400 points lying 0.2 m apart in a plane parallel to the object 400 m away from it. The object is 10 x 10 m^2. In the reconstruction we want the object reduced to 1 x 1 cm^2 at a distance of 40 cm away from the synthetic hologram. The wavelength in recording is 6.3 cm, which corresponds to a frequency of about 20 kHz in water. We want to reconstruct the hologram at 0.63 μm with He-Ne laser light. The important fact here is that the wavelength ratio, 6.3 cm / 0.63 μm, is 100 times the ratio between the object in recording and the object in reconstruction, 10 m / 1 cm.

The method for transforming the long wavelength wavefront to an optical hologram will be described in [8]. Here we will only give a brief scetch. Suppose the long wavelength hologram data, represented by the function $f(x_1,y_1)$, are recorded in the geometry shown in Fig. 12a. Next, suppose that the hologram coordinates (x_1,y_1), the wavelength λ_R, and consequently also the reconstruction, are scaled down by the same factor α. This scaling is performed to achieve a desired reconstruction size. From this conceptual, intermediate stage the hologram coordinates (now $(x_1/\alpha,\ y_1/\alpha)$) are scaled down by a factor m, and the resulting hologram is to be reconstructed by light of wavelength λ_2. The wavelength in the intermediate stage is $\lambda_1 = \lambda_R/\alpha$. The hologram values, $f(mx',my')$, as function of the coordinates in reconstruction (x',y'), are then multiplied with a complex function, $C(x',y')$, given by

$$(1)\qquad C(x',y') = \left[\frac{(x_s-mx')^2 + (y_s-my')^2 + z_s^2}{(x_s-x')^2 + (y_s-y')^2 + z_s^2}\right]^{1/2} \cdot$$

$$\exp\left\{-ik_1[(x_s-mx')^2+(y_s-my')^2+z_s^2]^{1/2} + ik_2[(x_s-x')^2+(y_s-y')^2+z_s^2]^{1/2}\right\}$$

where

x_s, y_s, z_s	is a coordinate near the object center
$\left.\begin{matrix} mx' = x_1/\alpha \\ my' = y_1/\alpha \end{matrix}\right\}$	is the hologram coordinate in the intermediate stage
x', y'	is the final hologram coordinate
$k_1 = \frac{2\pi}{\lambda_1}$	is the wavenumber in the intermediate stage
$k = \frac{2\pi}{\lambda_2}$	is the wavenumber during reconstruction.

The function $C(x',y')$ in (1) is found by an optimization procedure in which the ideal demand is that any object point shall be reconstructed with constant magnification in all directions. The object coordinates have the point x_s, y_s, z_s as origin, and the volume in the object space over which the optimization is done should be symmetrically situated round the point x_s, y_s, z_s along all three axis.

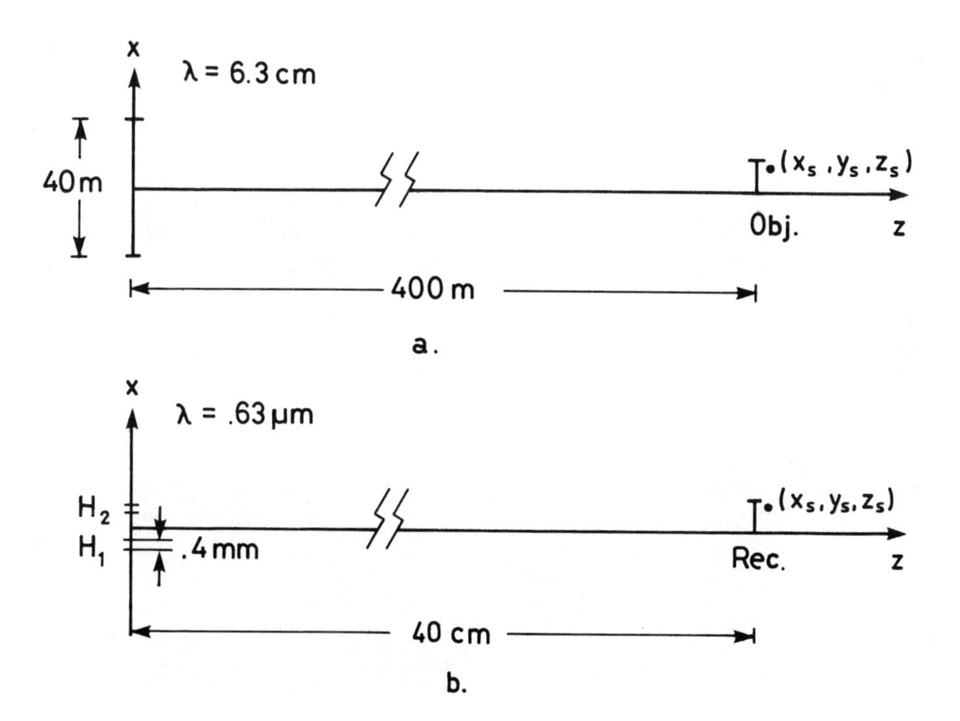

Fig. 12. Geometry for transforming a long wavelength hologram to an optical hologram.
(a) In recording.
(b) In reconstruction.

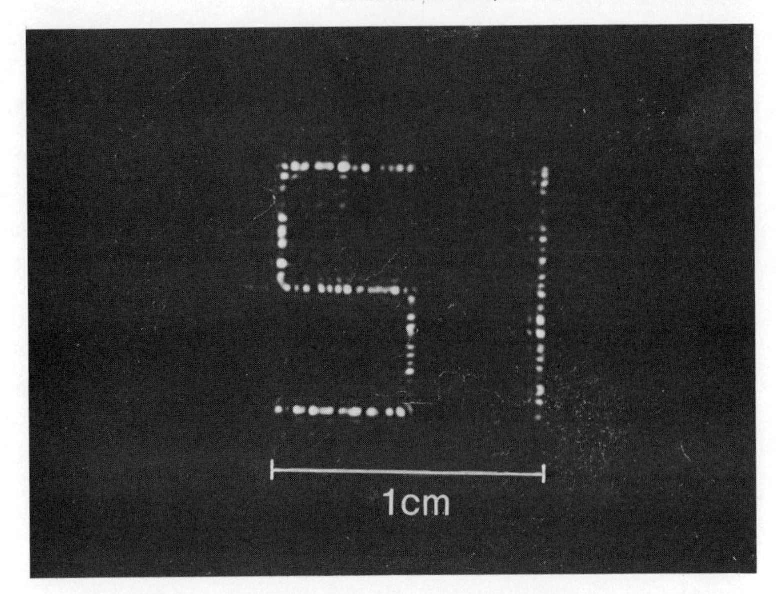

Fig. 13. Optical reconstruction of a simulated long wavelength hologram. λ recording / λ reconstruction = 10^5 / 1. Object size / Image size = 10^3 / 1.

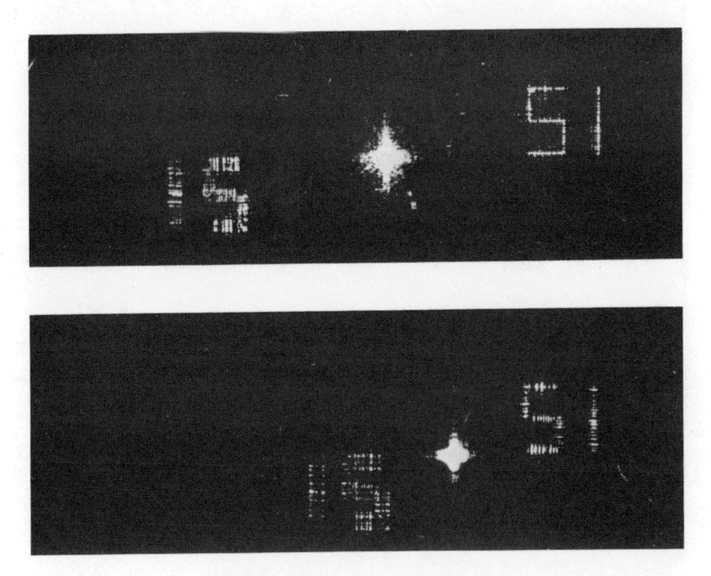

Fig. 14. (a) Reconstruction in Fig. 13 with central beam showing.
(b) Reconstruction of the same object with a new optical hologram shifted 1 cm to the right. Both optical holograms are made from the same long wavelength data.

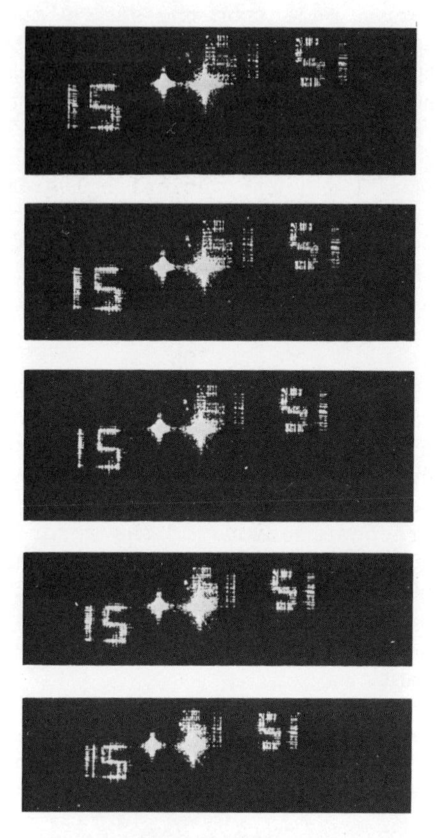

Fig. 15. Cross sections of the optical fields reconstructed by the optical holograms. The distances from the holograms to the reconstruction planes are from top to bottom 36 cm, 38 cm, 40 cm, 42 cm, and 44 cm, respectively.

Fig. 13 shows the reconstruction of our object, which consists of the letters SI. The method is an approximate one, but it seems to have the desired property that it scales down the object by equal factors transversely and longitudinally.

Another possibility offered by the method is that one can extend the optical hologram by repeating the process with a slightly different function C(x',y'). Fig. 14 shows the reconstruction from another optical hologram made from the same simulated long wavelength hologram. The new complex function gives an optical hologram which is placed 1 cm to the right of the first hologram. If the two holograms are reconstructed simultaneously, the depth resolution in the resulting optical reconstruction will increase. This effect is illustrated in Fig. 15, where cross sections of the optical field reconstructed by the two 1 x 1 mm^2 optical holograms are shown at increasing distance from the hologram plane. The distance along the z-axis between each recording is 2 cm. The overall distance from the hologram plane to the focused image is 40 cm.

CONCLUSION

From the experiment described in this paper we conclude that it is possible to obtain a monochromatic seismic wavefront by the use of a pulsed, repeatable source and sequential recordings with a movable hydrophone array. The processing we have carried out shows that digital reconstructions of the recorded wavefront provide information about buried diffractors that are much too small to be detected by pulse-echo recordings. Our optical reconstructions based on simulated acoustical data, indicate that a three-dimensional visualization of the geological structure may be obtained by the use of synthetic holograms.

We conclude that seismic holography has great potentials for detailed inspection of geological structures, whereas large scale surveys should be conducted using pulse-echo methods.

This work has been founded by Saga Petroleum A/S & Co., Norsk Hydro a.s, A/S Kongsberg Våpenfabrikk, Geophysical Company of Norway A.S., the Aker Group, Royal Norwegian Council for Scientific and Industrial Research, and Central Institute for Industrial Research.

REFERENCES

1. J.B. Farr:
Earth Holography as a Method to Delineate Buried Structures.
Acoustical Holography, Vol. 6, Plenum Press, 435 (1975).

2. G.L. Fitzpatrick, H.R. Nicholls, R.D. Munson:
An Experiment in Seismic Holography.
Bureau of Mines Report of Investigation No. 7607, 1972.

3. K.H. Waters:
Reflection Seismology.
(John Wiley & Sons, 1978).

4. J.R. Shewell and E. Wolf:
Inverse Diffraction and a New Reciprocity Theorem.
J. Opt. Soc. Am., 58, 1596 (1968).

5. E. Lalor:
Inverse Wave Propagator.
J. Math. Phys., 9, 2001 (1968).

6. T.H. Demetrakopoulous:
Synthetic Holograms and Image Reconstructions from Suboptical Diffraction Patterns.
Thesis, University of Illinois at Urbana, 73-A.542,
University Microfilms, Ann Arbor, Michigan, USA.

7. G. Hutton:
Fast-Fourier-Transform Holography: Recent Results.
Optics Letters, 3, 30 (1978).

8. O. Løvhaugen, E. Mehlum:
Transformation of Long Wavelength Wavefronts to Optical Holograms.
To be submitted for publication.

IMAGING CHARACTERISTICS OF CLINICAL ULTRASONIC TRANSDUCERS IN TISSUE-EQUIVALENT MATERIAL*

R.A. Banjavic, J.A. Zagzebski, E.L. Madsen
and M.M. Goodsitt
Medical Physics Section, Departments of Radiology and
Human Oncology, University of Wisconsin, Madison,
Wisconsin, 53706

ABSTRACT

Our group has been involved in the comparison of ultrasonic pulse-echo transducer characteristics in water versus attenuating and scattering mediums. For this purpose we are using a gelatin-based material with a speed of sound of 1570 m/sec, a variable attenuation coefficient between 0.3 - 1.8 dB/cm/MHz, and a linear dependence of attenuation on frequency over the diagnostic range. Thus we can duplicate under controlled conditions similar interactions that a pulsed ultrasound beam encounters when traversing soft tissue structures. The specific findings regarding clinical pulse-echo equipment and transducers include an increased beam divergence, as seen in tissue samples, in the tissue-equivalent gel as compared to water only. Also, the focal plane and, in fact, the entire focal region whose boundaries are defined by the beam area being twice that of the focal plane are shifted axially closer to the transducer face for the attenuating path. These differences are being studied as a function of transducer frequency, diameter, and focal properties as well as the pulser-receiver combination which is used.

INTRODUCTION

The work to be described here is being carried out to better understand the nature of medical ultrasonic transducer beams. the ultimate goal of this work is to characterize ultrasonic transducer pulse-echo profiles in the complex field made up of human soft tissue. The characterization of transducers, as well as their quality assurance testing and selection for specific clinical imaging procedures, are routinely carried out using measurements in a nonattenuating medium such as water. The results of earlier observations in our laboratory when a water path was partially replaced with fresh mammalian soft tissues such as liver and kidney showed the introduction of definite and significant changes in the ultrasonic beam shape and size by the presence of soft tissue.[1] Although in the water field alone the beam profiles were smooth and well-defined, the profiles were significantly distorted after traversing soft tissue. The isoecho response contours or curves representing equal echo amplitudes for various levels of sensitivity determined in planes parallel to the transducer face became more irregular and the widths of specific contours increased, indicating in some cases a significant loss of resolution after traversing soft tissue. This effect was consistently noted for the choice of transducer frequency, diameter, and focal properties over the range of clinical probes which we studied.

In this study, we have concentrated on one aspect of these pulse-echo profile changes, namely, that of the increase in the isoecho response contours exhibited when the beam traverses soft tissue samples. Our purpose was to measure pulse-echo profiles in an attenuating medium, under controlled geometry, representing some of the same sources of profile variations observed in excised tissues. We hope to demonstrate that estimates of lateral resolution based on such profile measurements would be more realistic than estimates obtained for nonattenuating paths where sources of beam distortion and contour enlargement are absent.

MATERIALS AND METHOD

The set of transducers shown in Table 1 were used for our transducer response measurements. They represent a typical group of general purpose, pulse-echo, clinical scanning transducers. As can be

TABLE 1 Description of the transducers used in this study.

SERIAL NO.	NOMINAL FREQ. (MHz)	CENTER FREQ. (MHz)	FOCUS (RADIUS OF CURVATURE) (CM)	CRYSTAL DIAM. (MM)	BANDWIDTH (%)
012689H	2.25	2.30	Long Internal (10.0)	13	62
019784H	2.25	3.09	Long Internal (10.0)	19	41
34793H	3.50	3.80	Long Internal (10.0)	13	61
30296H	3.50	3.30	Long Internal (10.0)	19	41
35645H (#1)	3.50	3.95	Long Internal (10.0)	19	46
5545 (#2)	3.50	3.30	Long Internal (10.0)	19	55

seen, all are long internally focused transducers with the piezoelectric element having a radius of curvature of 10 cm. The other important items to note are the range of center frequencies or the maximum point of the frequency spectra of the RF pulses for a given nominal frequency and the half amplitude response bandwidth as a percentage of the center frequency. This data was either obtained directly from the specification sheets supplied by the manufacturer or from a swept frequency pulse-echo response technique in our lab. The results using both techniques were in good agreement. The final 3.5 MHz transducer listed as #2 represents a different manufacturer than the rest. The transducers marked as #1 and #2 will be referred to in later discussions and used as examples of representative data.

The equipment used in the collection of the quantitative transducer information was designed and built in our laboratory. It consists of an aluminum frame assembly which is free to slide on a steel supporting structure. One of two identical stepping motors is rigidly attached to this aluminum frame and moves a sliding cart via a rack-and-pinion drive. On the cart itself is the second stepping motor which also has a rack-and-pinion drive in a direction orthogonal to both the sliding cart's motion as well as the aluminum frame assembly's motion. Hence, for a fixed position of the aluminum frame assembly on the steel supporting structure, there are two degrees of freedom. The rack which is driven by the motor fixed to the sliding cart is designed to either allow a chosen target, such as a spherical reflector, or a transducer to be attached. In this manner, the apparatus can be used to either scan a transducer beam with a target or move a transducer over a region of interest in a phantom or water tank in a two-dimensional fashion. The minimum size of the step increment in both directions is 0.45 mm. The translator control for the stepping motors is shared with an isodensitometer used for scanning radiographic films. This had already been interfaced to a PDP8/I computer with 32 K of memory so that a program could be adapted for the transducer beam measurements. The accuracy of the position coordinates in the scanning plane was better than 0.5 mm over 30 cm of motion.

The data that was analyzed in the first part of this work on transducer performance was derived from the A-mode signals obtained from commercial pulse-echo units with no swept gain applied. The signal was digitized using a transient recorder. Pulse-echo response beam profiles obtained from both the RF wave voltage amplitude read directly from an oscilloscope and the values in dB of an internal attenuator for a fixed A-mode signal deflection amplitude were in good agreement when a spherical reflector was used as a target.

Beam characteristics of the six transducers were studied both in a tank filled with a nonattenuating water/ethanol mixture with a speed of sound of ∿ 1570 m/sec at room temperature and through a suitable ultrasonically tissue-equivalent material having the same speed of sound as the water/ethanol path but with an attenuation value in the range of human soft tissue.[2,3] The equality of the speeds of sound for both the nonattenuating and tissue-equivalent paths helped avoid effects in the transducer beam profile and directivity function due to differences in the indices of refraction. The latter material consists of an animal-hide based gelatin mixture containing water, alcohol, preservatives, and appropriate concentrations of additives such as graphite or talc. Figure 1 shows a graph of the attenuation coefficient as a function of frequency for various talc/gel concentrations.[2] Similar curves for graphite have been obtained.[3] By appropriate concentration selection, any desired attenuation within the soft tissue range can be obtained while maintaining the same approximate proportionality between the attenuation coefficient and frequency as exhibited by most human soft tissues. The speed of sound in the gelatin materials is representative of the speed of sound in most human soft tissues other that fat. The attenuation value chosen for the material in this study was 0.7 dB/cm/MHz. This value appears to be representative of recent attenuation data for liver.[4]

Using the talc/gel material a profiling phantom was constructed. It consisted of an array of steel spherical reflectors, 4.8 mm in diameter, positioned laterally at 2 cm intervals and at

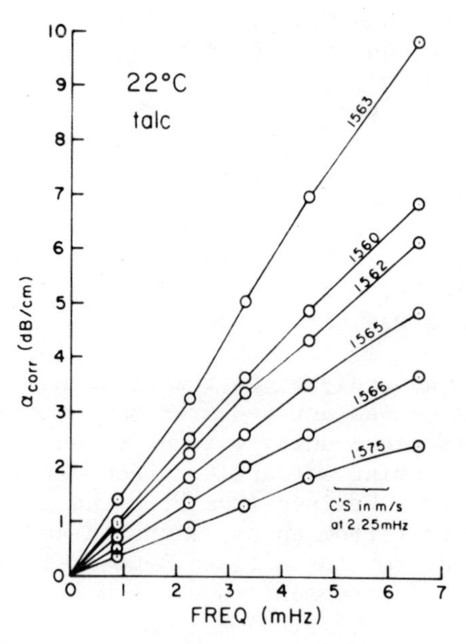

FIGURE 1 The acoustic attenuation coefficient versus ultrasonic frequency for six different concentrations of talc powder in gel. The speed of sound is shown for one frequency.

distances from the top of the phantom ranging from 2 to 18 cm. The top of the phantom was designed to include a water-coupling trough for ease of measurement. A steel spherical reflector as a target for both the nonattenuating as well as attenuating paths was chosen on the grounds set forth by Lypacewicz and Hill of comparability with human tissue targets, appropriate absolute target strength and general convenience in mounting and alignment.[5]

RESULTS

The quantitative data, which we assembled, is in the form of pulse-echo response isocontour plots and isocontour widths. The isocontours represented are in terms of dB relative to the maximum signal from the steel spherical reflector. This resulted in pulse-echo isoresponse curves in any given plane parallel to the transducer face at various axial distances from the transducer. The distances examined ranged from 2 to 18 cm from the transducer face. The isocontours were determined and plotted directly by the computer. Three sample depths were chosen--one before the focal region; one in the focal region; and one beyond the focal region. Figure 2 shows the isocontour responses at depths of 5 cm, 10 cm, and 15 cm in the water/ethanol mixture for transducers #1 and #2 in Table 1. The grid spaces in these and all future plots represent a 1 mm distance. The innermost contour is 6 dB below the maximum signal level in each plane. The other contours represent the -10 dB, -20 dB, -30 dB and -40 dB levels. (The actual computer plots are color-coded for ease of identifying various contour levels.) The small regions of asymmetry are probably real and appeared for several other transducers tested. In Figure 2a, transducer #2 does not appear to have the uniform radial decrease in signal response as does #1. The response is apparently peaked along the beam axis for the stronger signals. Figure 2b shows the same two transducers at 10 cm depth. Here the differences between the two transducers are even more obvious. The outermost, or -40 dB contour, is clearly larger for #2 even though the overall response appears more symmetric than #1. The asymmetry of #1 is even more pronounced than at 5 cm, especially in the -30 dB and -40 dB contours. Figure 2c shows the isocontour plots for these transducers at 15 cm depth in the water/ethanol mixture. Here, the increased beam size of #2 is quite evident. Also, the asymmetry of #1 is consistent with that noted at previous depths. As a check of both the reproducibility of these plots and the invariance of the relative response on the size of the spherical reflector, the transducer was rotated 180°and the target was replaced by a 12.7 mm diameter steel sphere. Figure 3 shows this isocontour response alongside the previous response for transducer #1. The size and shape of the contours are in good agreement.

In addition to obtaining isocontour response plots at specified

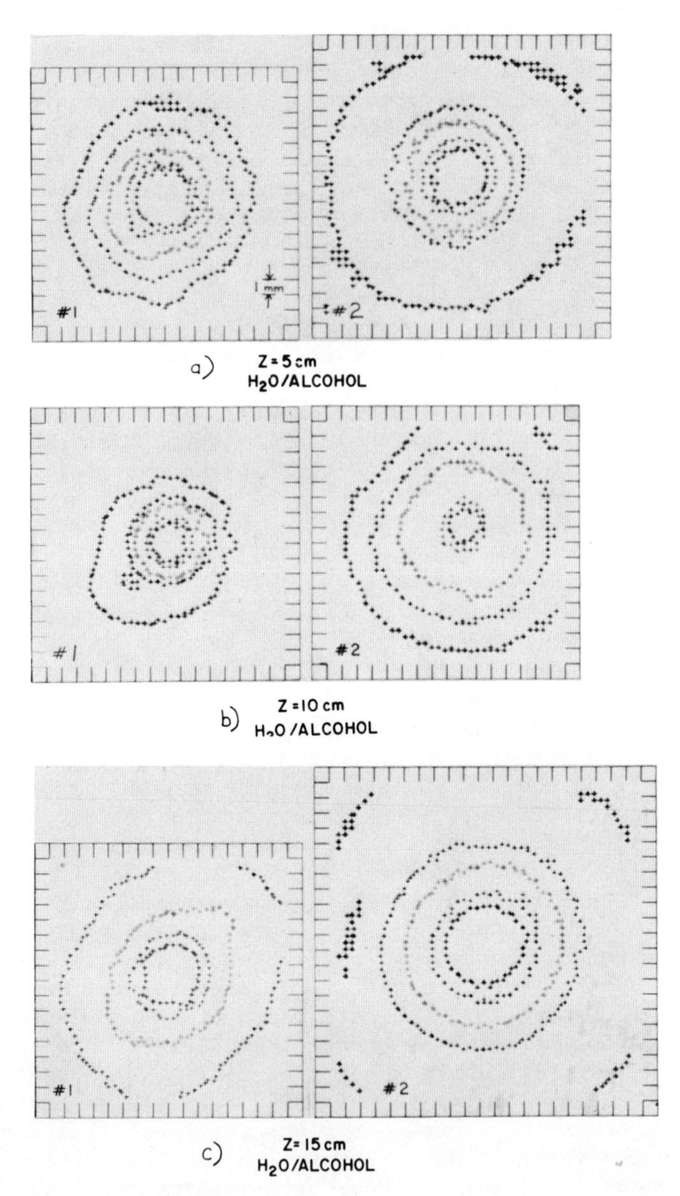

FIGURE 2 Isocontour plots for the two 3.5 Mhz, 19 mm diameter, long internally focused transducers listed in Table 1 as #1 and #2. The figure shows measurements in a water/ethanol bath with a spped of sound of 1570 m/sec for three different depths:a) 5 cm, b) 10 cm, and c) 15 cm. The innermost contour points represent -6dB with respect to the maximum signal level at that depth. The other levels are, in order from the center of the plots outward, the -10 dB, the -20 dB, the -30 dB and the -40 dB levels.

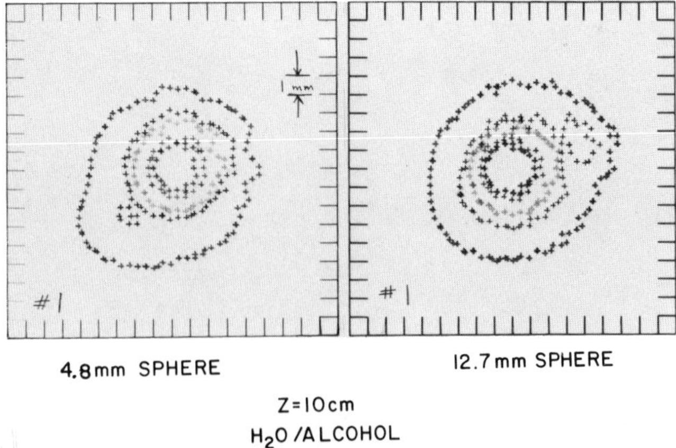

FIGURE 3 Isocontour plots for transducer #1 in Table 1 at 10 cm depth in water/ethanol bath where a 4.8 mm diameter spherical target was used on the left and a 12.7 mm diameter spherical target was used on the right after a 180° rotation of the transducer about its central axis.

depths, the width of each isoecho contour was determined as a function of distance in the water tank as well as in the tissue-equivalent material. This is done by using the apparatus described earlier to scan the beam in two perpendicular directions through the maximum response and to then average the appropriate contour widths. As an example of these measurements for the two transducers in Figure 2, a graph of the width of the group of dB contours as a function of distance in the water/ethanol mixture is shown in Figure 4a for transducer #1 and Figure 4b for transducer #2. Graphs such as these also demonstrate the position of the focal plane or region of minimum beam isocontour area. For Figure 4a, the minimum beam area (or width) can be seen to be about 8.0 cm for all the contour levels. Also, the rate of change in the width of the pulse-echo responses is less for the higher level or lower negative-numbered dB contours. If we define the focal region as the distance over which the ultrasonic beam area is no greater than twice the minimum beam area, or where the beam width changes by no more than a factor of $\sqrt{2}$, and if we estimate this beam area by averaging the areas of the isocontour levels; the appropriate boundaries for the focal region of this transducer would be from about 6.2 cm out to 10.5 cm. The implications drawn from the isocontour plots shown earlier are clearly evident when examining Figure 4b where the lower level contour widths or larger negative-numbered dB values are much larger than for transducer #1 and the slope of the lines for all contours below -10 dB are significantly

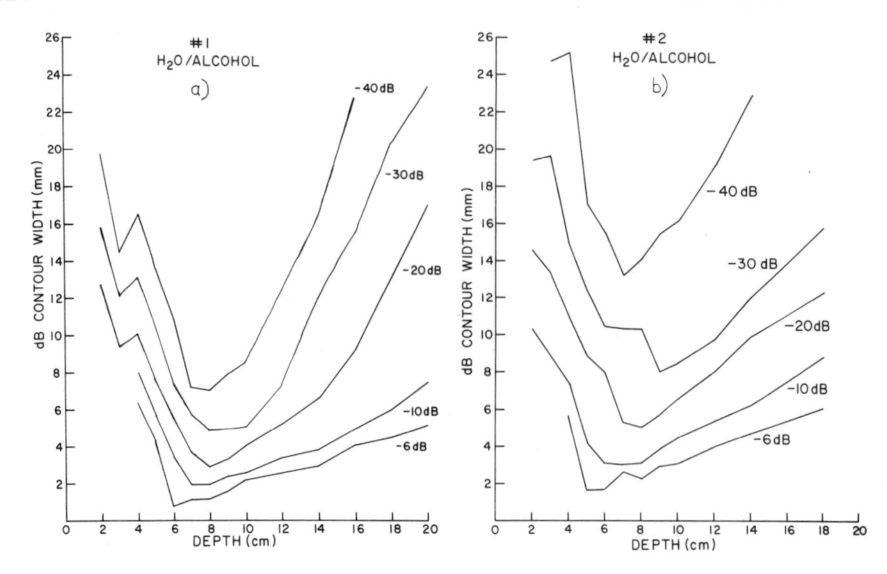

FIGURE 4 Graphs of isodB contour widths versus depth in the water/ethanol bath. The two 3.5 MHz, 19 mm diameter, long focused transducers represented are, as labelled in Table 1, a) #1 and b) #2. The widths are in mm and the depths in cm.

less. The focal plane shows a much greater variation in position as a function of signal strength with the -10 dB contour width minimal at 7.5 cm depth. The average focal region in this case extends from 5.4 cm to 11.5 cm or a total distance of 6.1 cm. Thus it would appear that for our "supposedly" identical transducers, #2 is less sharply focused than #1.

In order to better relate this type of data to realistic estimates of lateral resolution in soft tissue, measurements were also taken with the profiling phantom described earlier. Figure 5 is a graph of the dB contour widths vs. depth curves for measurements in this gel phantom. The important observations are that the contour widths in the tissue-equivalent material are both larger and exhibit a greater rate of change in size as a function of depth than in the nonattenuating bath. For example, for transducer #1, the -20 dB contour width at 8 cm depth in gel was 6 mm as compared to 3 mm in the nonattenuating mixture; while for transducer #2 the corresponding dimensions were 8 mm and 5 mm. As for the beam divergence, transducer #1 showed a variation of 2 to 3 mm in the -10 dB contour width between 8 and 12 cm depth in the water/ethanol bath, whereas the corresponding variation over the same depth range was from 4 to 8 mm in the talc/gel phantom. This would imply a degradation in the lateral resolution from that estimated on the basis of measurements in the nonattenuating medium. Another observation from examining the curves in Figure 5 is that there is

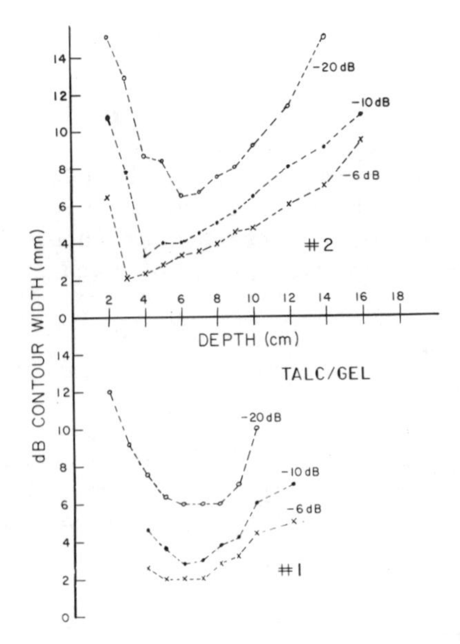

FIGURE 5 Graphs of isodB contour widths versus depth in the talc/gel phantom. Transducers #1 (lower curves) and #2 (upper curves) are shown. These transducers are listed in Table 1.

significantly less difference between the characteristics of the two transducers in the talc/gel phantom as far as the size of the dB contours as well as the position of the focal plane and length of focal region are concerned. Also, the position of the minimum beam size or focal plane is closer to the transducer face for the attenuating path than in the water/alcohol mixture! Isocontour widths in the focal region of these transducers were also determined using a second clinical pulser-receiver unit. The results of these two measurements were in complete agreement. The apparent shift in both the focal plane and entire focal region closer to the transducer face for soft tissue like material was also evidenced for other transducers tested. Table 2 shows a summary of these measurements for a variety of transducers. It can be seen that the wider diameter transducers seem to show a larger effect as far as the shift in the position of the focal plane is concerned.

The isoresponse contours for the two 3.5 MHz transducers which exhibited marked differences when scanned in the nonattenuating bath were plotted for the talc/gel path. Figure 6 is the result at 5 cm depth in the talc/gel phantom for these two transducers. As was indicated by the curves in Figure 5, the differences between transducers noted in the water/alcohol plots of Figure 2 are less pronounced here. Both beams exhibit the increased isocontour widths through the attenuating material relative to the water path as mentioned earlier. The other feature of note is that the back-

TABLE 2 The position of the focal plane or point of minimum beam area and the positions of the boundaries of the focal region where the beam area equals twice the minimum area in terms of mm from the transducer face. The values are given for both the water/ethanol path and the talc/gel path.

			Water/ethanol		Talc/gel	
Serial No.	Nominal Freq. (mHz)	Diameter (mm)	Focal Plane	Focal Region	Focal Plane	Focal Region
012689 H	2.25	13	55	90-33	50	69-31
019784 H	2.25	19	80	120-60	60	90-42
34793 H	3.50	13	60	95-40	40	70-30
30296 H	3.50	19	75	95-63	58	80-46
35545 H (#1)	3.50	19	80	105-62	60	86-40
5545 H (#2)	3.50	19	75	115-54	55	82-34
019701 H	3.50	13	40	65-30	30	55-20

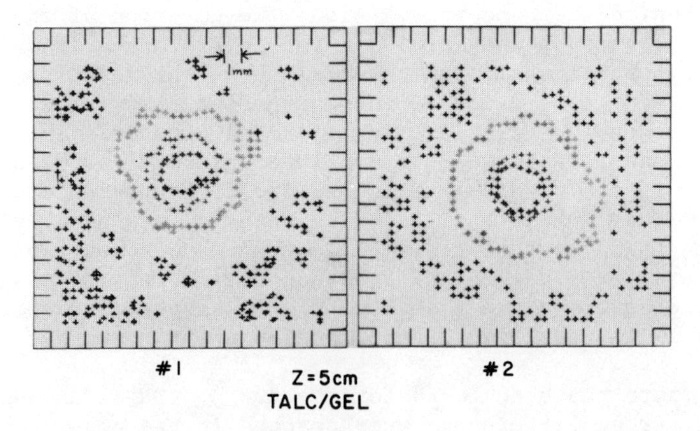

FIGURE 6 Isocontour plots for the two 3.5 MHz, 19 mm diameter, long focused transducers labelled in Table 1 as #1 and #2 at 5 cm depth in the talc/gel phantom. #1 is on the right.

ground texture scatter from the gel occurs at about 30 dB below the maximum response for these particular transducers. The differences between the paths through the two choices of materials is more evident in Figure 7 which compares the isocontour plots for transducer #2 at 10 cm depth in both water/alcohol and tissue-equivalent gel. Here the increased isoecho contour size is obvious as well as the maintenance of the axial symmetry in both cases. This isocontour enlargement is similar to that found for the beam passing through fresh mammalian liver tissue samples of the same path length; even though the distortion observed in that earlier experiment is absent.[1] A further interesting observation was that the asymmetries for the beam emitted by transducer #1 which were present for the water/alcohol path are smoothed out when transmitting through the gel material. Presumably, this also is a manifestation of the preferential attenuation of higher frequency components of the beam in the talc/gel path.

SUMMARY

We have duplicated under a controlled laboratory geometry using a spherical reflector one of the major effects exhibited by a pulsed-echo beam transmitting through soft tissue, namely, increased relative beam divergence and accompanying loss of lateral resolution. This "spreading" can probably be attributed to attenuation based effects and most likely is not accompanied by an increase in intensity at larger radial distances from the beam's central axis. The broad band clinical ultrasonic pulses do not show the characteristic "divergence" in nonattenuating paths due

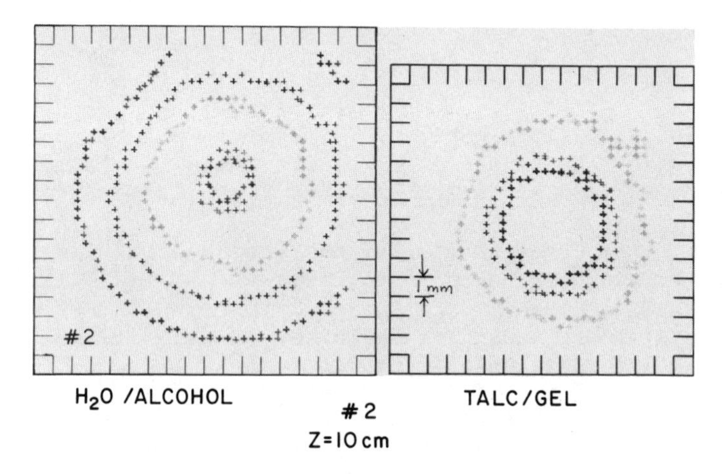

FIGURE 7 Isocontour plots for the 3.5 MHz, 19 mm diameter, long focused transducer labelled as #2 in Table 1 at 10 cm depth in both the water/ethanol path and the talc/gel phantom. The -6, -10, -20, -30, and -40 dB isocontours are shown.

to insufficient frequency - dependent attenuation alteration of the frequency content of the RF waveform. In addition, we have seen that transducers which exhibited marked differences in isocontour shape and size in a nonattenuating path were much more equivalent in a path through tissue-equivalent material. Also, a significant shift in the position of the focal plane was found when the ultrasound beam traversed the attenuating path as compared to when it traversed a path in water/ethanol alone. A contributing factor to this effect could be the resultant shift of the center frequency of the broad band pulses forming the beam to lower values due to the frequency dependence of the attenuation. In the nonfocused, continuous-wave approximation, this would result in a decrease in the distance to the last axial maxima in the beam which is usually given by $Yo \cong a^2/\lambda = fa^2/c$, where a is the radius of the transducer, λ is the assumed wavelength of the beam, f is the assumed frequency and c is the speed of sound in the material through which the beam is propagating. The conclusions drawn here regarding estimation of lateral resolution can hopefully be extended to clinical scanning procedures and B-mode echo signal display. Understanding the nature of the transducer beam under uniform scattering conditions and controlled geometric configurations in attenuating materials is a first step in understanding the results for inhomogenous distributions of scatterers of various sizes and strengths such as is found in human soft tissue.

*Work supported in part by the National Cancer Institute Wisconsin Clinical Cancer Center Grant 5-P01-CA-19278-02.

REFERENCES

1. Banjavic, R.A., Zagzebski, J.A., Madsen, E.L., and Jutila, R.E., "Ultrasonic Beam Sensitivity Profile Changes in Mammalian Tissue", ULTRASOUND IN MEDICINE, 4, 515-518 (1978).

2. Madsen, E.L., Zagzebski, J.A., Banjavic, R.A., Burlew, M.M., "Further Developments in Soft-Tissue-Equivalent, Gelatin-Based Materials", Abstract, 23rd Annual Conference of American Institute of Ultrasound in Medicine, October 19-23, 1978, San Diego, California (under review).

3. Madsen, E.L., Zagzebski, J.A., Banjavic, R.A., and Jutila, R.E., "Tissue Mimicking Material for Ultrasound Phantoms", MEDICAL PHYSICS, 1978 (in press).

4. Kuc, R., Schwartz, M., "Statistical Estimation of the Acoustic Attenuation Coefficient Slope for Liver Tissue from Reflected Ultrasonic Signals", PROCEEDINGS OF THE SECOND INTERNATIONAL SYMPOSIUM ON ULTRASONIC TISSUE CHARACTERIZATION, Gaithersburg, Maryland, June, 1977 (in press).

5. Lypacewicz, G., Hill, C.R., "Choice of Standard Target for Medical Pulse-Echo Equipment Evaluation", ULTRASOUND IN MEDICINE AND BIOLOGY, 1, 287-289 (1974).

ACKNOWLEDGMENTS

The authors wish to thank Mr. Orlando Canto for his assistance in the preparation of the figures and tables for this publication.

MAGNETOSTRICTIVE IMAGING SYSTEMS

David Dameron and Albert Macovski

Department of Electrical Engineering

Stanford University, Stanford, CA 94305

ABSTRACT

Many ferromagnetic materials, especially nickel and certain ferrites, demonstrate magnetostrictive properties, i.e., they undergo a change in length along the direction of an applied magnetic field. This phenomenon is used to form an imaging system, using both electromagnetic and ultrasonic waves, with many characteristics different from the normal pulse-echo ultrasonic system.

Magnetostrictive particles are placed in the region of the body to be imaged, for example, a colloidal suspension in the G.I. tract. Instead of transmitting an ultrasonic wave into the body, an electromagnetic wave is used at low ultrasonic frequencies to propagate a magnetic field. This magnetic field passes through the body at the speed of light with interactions at the administered particles. The magnetostrictive particles vibrate with the magnetic field, generating an ultrasonic wave. This ultrasonic wave can be received as any pulse-echo signal by any ultrasonic receiving system.

Several advantages of this technique are apparent. First, since the electromagnetic waves travel to the magnetostrictive sites at the speed of light, round trip transit time is cut in half versus pulse-echo systems. Second, the higher energy electromagnetic transmitted wave does not interact appreciably with body tissues, only with the magnetostrictive particles. Finally, and most important, this type of imaging system is an emissive one, only those regions having the administered material produce an ultrasonic return wave.

INTRODUCTION

We will analyse in this paper a new modality of ultrasonic imaging system; the magnetostrictive imaging system. Many ferromagnetic materials, most notably among them nickel and certain ferrites, demonstrate magnetostrictive propoerties. That is, they experience a change in length along the direction of an applied magnetic field. This phenomenon is used to advantage to form an imaging system with many characteristics different from the normal pulse-echo ultrasonic system.

A magnetostrictive system, shown schematically in Fig. 1, works basically as follows. Magnetostrictive particles are placed in the region of the body to be imaged, a suspension of particles in the G.I. tract is one example. Instead of transmitting an ultrasonic wave into the body, an electromagnetic pulse is used; more specifically, a magnetic field in the low MHz region. This magnetic field passes through the body at the speed of light with essentially no interaction, except at the magnetostrictive particles. The magnetostrictive particles vibrate with the magnetic field, either at the same or twice the frequency, depending on the use, generating an ultrasonic wave. This ultrasonic wave can be received as any pulse-echo reflected wave by any ultrasonic receiving system, such as a B-scanner.

Several advantages of this technique are immediate. First, since the electromagnetic wave travels to the magnetostrictive sites at the speed of light, several orders of magnitude faster than the speed of sound, round trip transit time is cut in half vs. pulse-echo systems. This means that twice the number of lines or beams can be used in a given frame time. Second, the higher energy transmitted electromagnetic wave does not interact sighificantly with body tissue, only with the magnetostrictive particles*. The ultrasound generated has its highest intensity at the imaged sites, so one does not have to worry about high ultrasonic power levels at the skin or absorption and attenuation of the transmitted beam. Finally, and most important, this imaging system is an emissive one. Only those regions containing magnetostrictive material actually produce an ultrasonic return wave. Adjacent tissues are

---

* The major compound in the body that would interact with magnetic fields is hemoglobin in the blood. At one time it was thought that hemoglobin could be used as the magnetic-ultrasonic transducer, enabling direct non-invasive imaging of blood and blood vessels. At this time this technique does not appear to have enough sensitivity to be useful.

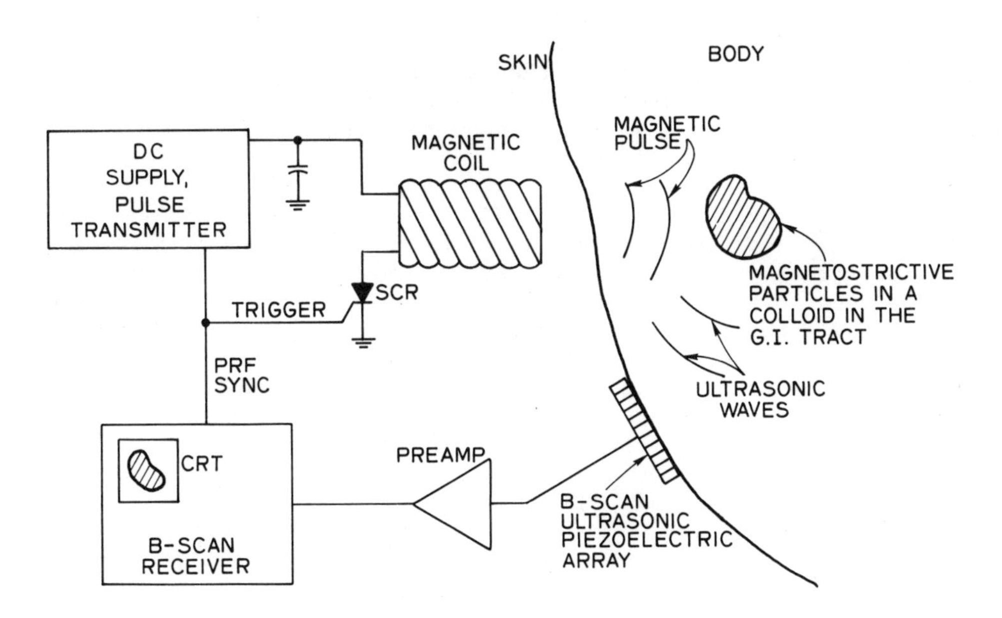

Fig. 1. Schematic diagram of a magnetostrictive imaging system. The magnetic coil transmits a magnetic pulse which is converted to ultrasonic waves by the magnetostrictive particles and received by the piezoelectric array.

not sources and thus do not contribute clutter to the image, although the ultrasound does need a path to the receiving array. Perturbations of the received ultrasound due to reflections and scattering will always come after the desired signal, since the desired signal propagates in a direct line to the receiver. This is unlike pulse-echo signals which can be masked by multiple reflections from a reflector closer to the receiver.

It should be noted that in some studies, the density of magnetostrictive particles which produce the level of the ultrasonic return, are of no direct significance. Rather the boundaries of the magnetostrictive particles, which define the surface of the enclosing volume, is the desired information. Because in this case we are looking only for the presence or absence of magnetostrictive generated ultrasound, a binary decision to locate the surface boundaries, a smaller signal to noise ratio can be tolerated than for gray-scale imaging.

In imaging the G.I. tract, visualization of the surface area may be most important. Tumors, polyps, and other growths are all perturbations of the G.I. tract surface and are what is desired to be located. The G.I. tract is almost impossible to image or visualize with normal X-ray radiography because the small changes in the X-ray mass absorption coefficient from the interior to the walls of the tract produce a low contrast image. To increase the contrast, barium is injected into the tract in order to increase the mass absorption coefficient of the interior; this procedure is called a barium X-ray study.

There are still two major drawbacks in these X-ray studies. The first is that a conventional X-ray radiograph is a projection image, all planes transverse to the X-ray beams are summed together (with different magnifications). An object in one plane cannot be distinguished from a similar object along the same X-ray path in another plane; this fact makes location difficult. Also, since a protrusion which is small compared to the G.I. tract diameter will produce an absorption change along the projection much smaller than the total absorption, the net contrast will again be small. The second drawback is the invasive character of X-rays. The physician would like to examine the images in real time for an extended period of time, investigating suspicious areas with alternate views from different angles. Also, he would like to observe real time phenomena such as peristalsis. The radiation dose from X-rays limits the time the physician can search through views.

Pulse-echo B-scan ultrasound, which can also be used to image a volume, has neither of these two drawbacks. It is not a projection technique, the range or depth is given directly by the transit time of the echo pulses. Good lateral resolution can be achieved by focusing the ultrasonic beams dynamically on receive. Ultrasonic imaging is also non-invasive. There are no time constraints in which to collect all the image frames.

However, there are problems with pulse-echo ultrasonic images of the G.I. tract. Each wall encountered produces a strong echo with both specular and diffuse components. A large number of them will also produce reverberations. Reverberations between walls or interfaces close to the skin, where the incident ultrasonic beam is strong, will produce echo's later in time (depth) which can obscure weaker genuine echo's at greater depths. Regions outside the receiver beamwidth, but within the transmitted pattern or its sidelobes, can scatter specular or diffuse echo's into the receiver beamwidth, introducing more clutter in the image, the specular echo's being more concentrated in time (space). Gas pockets which partially block the transmit beam introduce even more artifacts, being essentially perfect reflectors. What obviously is needed is an emissive imaging system, with its stated advantages. Magnetostrictive imaging is such a system, keeping all the advantages both of pulse-echo ultrasound and emissive type imaging.

MATERIALS

Many magnetic materials exhibit a small but definite change in length along the direction of an applied magnetic field. This effect, which is independent of the sign of the applied field, is called magnetostriction. Although in some geometries materials contract in magnetic fields due to forces from induced eddy currents, magnetostriction is a different phenomenon caused by domain crystal anisotropy, and can produce both extensions and contractions in materials.[1] The fact that the change in length is a function of only the absolute value of the applied magnetic field can be used to advantage in some cases, producing a frequency doubling of the magnetic field in the length vibrations. This effect can be controlled by the amount of bias magnetic field applied.

The change in length of magnetostrictive materials limits at higher magnetic fields, the limiting value. $(\Delta\ell/\ell)_{sat}$ is called the saturation magnetostrictive coefficient. The amount of ultrasonic power which can be produced is also limited for the same reason. Table 1 gives the saturation magnetostrictive coefficient for various materials. Note that the values are quite small. Recently, extremely large room temperature magnetostrictive coefficients have been found in Laves phase RFe_2 alloys, where R is a rare earth[2,3]. These values are also included in Table 1 and may enable much easier ultrasonic generation via magnetostriction.

TABLE 1

SATURATION MAGNETOSTRICTIVE COEFFICIENTS FOR VARIOUS MATERIALS

Material	$(\Delta\ell/\ell)$saturation
Annealed nickel, 99.92% Ni	-33×10^{-6}
Permalloy, 45% Ni, 55% Fe	$+27 \times 10^{-6}$
Alfer, 13% Az, 87% Fe	$+40 \times 10^{-6}$
Ferrous ferrite $Fe^{2+} + Fe_2{}^{3+} + O_4$	$+40 \times 10^{-6}$
Ferrite 7A2, Ni-Cu-Co ferrite	-28×10^{-6}
$SmFe_2$	-2340×10^{-6}
$TbFe_2$	-2600×10^{-6}
$DyFe_2$	$+640 \times 10^{-6}$
$ErFe_2$	-340×10^{-6}
$Tb_{1-x}Dy_xFe_2$	$+1500 \times 10$

We have chosen nickel for experimental tests. It is readily available in powder form, important for low eddy current losses at high frequencies, and which enables it to be mixed in a colloid. Although its toxicity is somewhat unknown, it can be encapsulated with an inert material for biological use. Fig. 2 shows the magnetostriction of various materials, including nickel, as a function of field strength[4]. Nickel is saturated at a field strength of 1.6×10^4 amp/m (200 Oersteds), producing half that strain at approximately 800 amp/m (10 Oersteds).

APPLIED MAGNETIC FIELDS

For the model of the magnetostrictive particles, we will use ferromagnetic spheres in an uniform external field, as shown in Fig. 3. It can be shown in this case that[5]

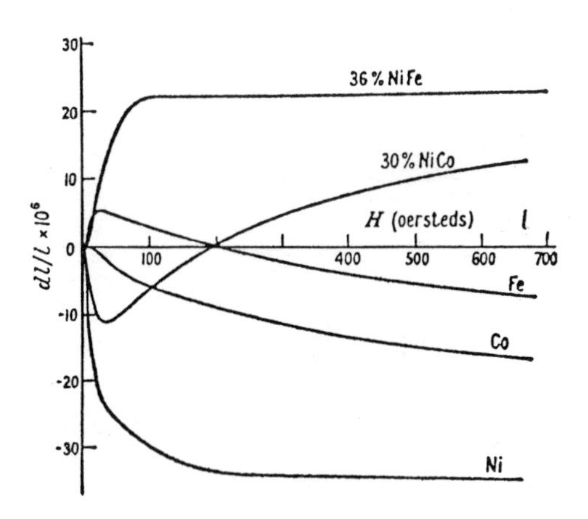

Fig. 2. The magnetostriction of some common substances as a function of magnetic field strength H. Of these substances, nickel, Ni, exhibits the strongest magnetostriction of $\Delta\ell/\ell$, -33×10^{-6}.

$$H_{int} = \frac{3}{2 + \mu} H_{applied} \quad (1)$$

$H_{applied}$ is assumed to be constant throughout the region.

A solenoid is the usual method of producing AC magnetic fields, as shown in Fig. 4(a). The solenoid is of length L and radius a. Current I flows through the windings with n turns/unit length, or N = nL turns. For this case and r = 0, the magnetic intensity, H_z is calculated to be

$$H_z = \frac{I_n}{2} \left[\frac{z + L}{\sqrt{(z + L)^2 + a^2}} - \frac{z}{\sqrt{z^2 + a^2}} \right] \quad (2)$$

If a rod with permittivity μ is placed inside the solenoid, the magnetic intensity just beyond the rod will be increased by an amount proportional to μ for relatively small increases in μ, as will the inductance of the coil.

A second scheme to produce larger fields is a magnetic "C" structure, as shown in Fig. 4(b). The fields are assumed to be completely contained inside the magnetic conductor, outside the air gap; fringing fields in the gap are neglected. For this case, using Ampere's law

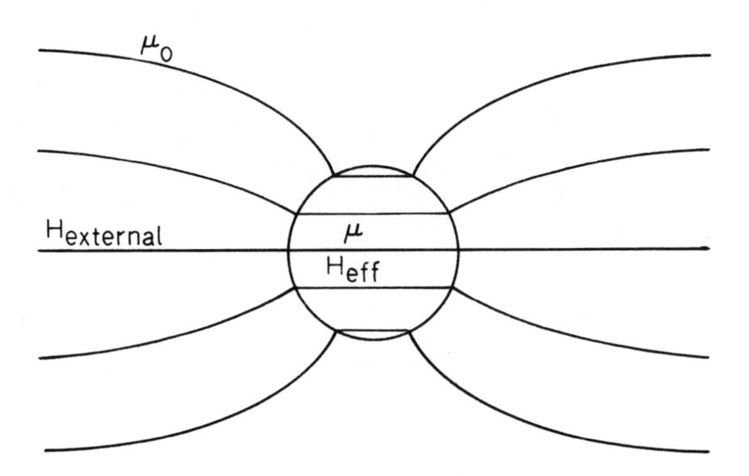

Fig. 3. Magnetic field lines in a ferromagnetic sphere. The magnetostrictive particles were modelled as spheres in order to calculate their internal magnetic fields.

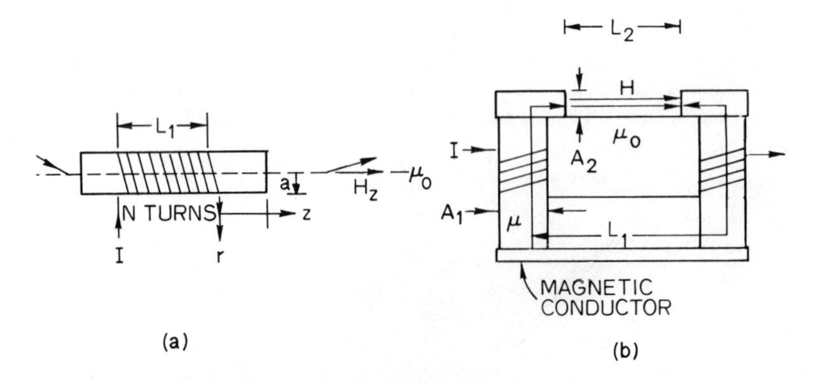

Fig. 4. Experimental magnetic sources. (a) shows a solenoidal source, while (b) shows a magnetic conductor formed into a "C" structure to concentrate the field.

$$NI = \frac{B}{\mu} L_1 + \frac{B}{\mu_o} L_2$$

and,

$$H = \frac{NI}{(L_2 + L_1/\mu_r)} \frac{A_1}{A_2} \tag{3}$$

where H is the magnetic intensity in the gap, and μ_r is the relative permittivity of the magnetic material. Usually μ_r is chosen so that L_1/μ_r can be neglected with comparison to L_2.

ACOUSTICS

In this section, the ultrasonic power generated by a vibrating (magnetostrictive) particle is derived. We use a one dimensional model. Assume that the surface of the particle is vibrating with displacement.

$$y = y_m \exp i(kx - \omega t) = \Delta \ell$$

Although this is not generally a good approximation for magnetostrictive materials with a magnetic excitation frequency of ω, for a magnetic pulse of 1/2 cycle of a sine wave at frequency ω, the magnetic fields are always of one sine, so that the absolute value response does not matter. ω is also chosen to be the center frequency of the ultrasonic receiver for maximum response. The fundamental frequency will be the only one which will significantly pass through the receiver's bandpass - for most current ultrasonic B-scan receivers. Harmonic reception due to the pulse shape and nonlinear magnetostriction may increase the received power and depth resolution, but this single-frequency model is adequate for determining sensitivity.

The power in a one dimensional acoustic wave is given by

$$I = \frac{R_a}{2} |\upsilon|^2 = \frac{R_a}{2} \omega^2 (y_m)^2 \tag{4}$$

where is the particle velocity in the acoustic wave. Solving for y_m:

$$y_m = \left(\frac{2I}{R_a \omega^2}\right)^{1/2} \tag{5}$$

We use the magnetostrictive relation

$$\ell = \Delta\ell/(\Delta\ell/\ell)$$

in Eq. 5, where $(\Delta\ell/\ell)$ is the operating magnetostrictive coefficient at the given applied magnetic field, not necessarily at saturation. $(\Delta\ell/\ell)$ is less than or equal to the saturation magnetostrictive coefficient. This gives an expression for ℓ, the needed length of magnetostrictive material to produce an ultrasonic power level, I, where all the magnetostrictive material is excited in phase so all of the $\Delta\ell$'s add in phase, as given by

$$\ell = \frac{1}{(\Delta\ell/\ell)} \left(\frac{2I}{R_a \omega^2}\right)^{1/2} \tag{6}$$

Because the magnetic fields travel several orders of magnitude faster than the acoustic fields, (the ratio being 2.22×10^4 for water) all the material is excited essentially instantaneously.

EXPERIMENTAL TESTS

Experiments were performed with various magnetostrictive phantoms to determine the feasibility of this type of imaging system. We used frequencies from 0.5 - 1.0 MHz having an acoustic wavelength of 1.5 - 3.0 mm. A frequency much lower than these values would yield a (depth) resolution (approximately one wavelength) which would be inadequate for imaging the biological structures of interest. The SCR transmit circuits described in the next section limited the maximum frequency which could be used.

A small lead metaniobate transducer, 4 by 4 mm, which was resonant at 1 MHz was used to receive the ultrasound. Ultrasonic imaging systems using ceramic transducers generally have a lower sensitivity level of 10^{-11} watts/cm^2, determined by the signal to noise ratio. They also have a required sensitivity range of 100 dB. Realizing that the magnetostrictively generated ultrasound has to travel only one way in the body, from the source to receiver, with only half the corresponding attenuation of that of a normal B-scan system, and that there are no small reflection coefficients to further decrease the ultrasonic amplitudes, we picked a power level of 10^{-6} watts/cm^2 for the magnetostrictive source, in the center of the receiver dynamic range*.

Referring next to Fig. 2, we picked a magnetostrictive operating point in nickel of $(\Delta\ell/\ell)$ of -16×10^{-6}, approximately half that of the saturation value. The corresponding magnetic field in nickel is 7.95×10^{2} amp/m (10 Oersteds). Substituting these values into Eq. 1 yields:

$$H_{applied} = 1.59 \times 10^{5} \text{ amp/m}$$

using a μ of 600 for nickel. The value of $(\Delta\ell)$ in water ($R_a = 1.5 \times 10^5$ gm cm^{-2} sec^{-1}) is 1.838×10^{-9} cm, yielding a total length or thickness of magnetostrictive material of (using Eq. 6)

$$\ell > \frac{1.838 \times 10^{-9}}{|-16 \times 10^{-6}|} \text{ cm} = 1.15 \times 10^{-4} \text{ cm}$$

The nickel powder used has a diameter less than 2.54×10^{-2} cm. Even with this size, each particle is sufficiently large to produce the required ultrasonic fields. Although every particle is excited in phase, different ultrasonic path lengths to the receiver will cause coherent cnacelling. This cancellation can be written as a convolution of the magnetostrictive sources in space with the incident magnetic pulse shape. This convolution is in addition to that of the receiver pattern. For this reason, the magnetic pulse (and current pulse through the transmit coil) should be short.

For a transmitter, we used a magnetic "C" structure with 3 SCR capacitive discharge circuits. Each circuit contained a 3-turn coil around the "C" and produced a peak current of 28 amps at 0.5 MHz. Although this structure is unsuitable for many clinical uses, it can be used for experimental studies. For this structure

$$N = 3 \times 3 = 9 \text{ turns}$$

$$L_2 = 2 \text{ cm}$$

$$A_1/A_2 = 3$$

and,

$$H = (28)(9)(3)/(3 \times 10^{-2}) = 2.52 \times 10^{4} \text{ amp-turns/m} \quad \text{(peak)} \tag{7}$$

---

*This power level results in a received voltage of 9 mV RMS, assuming 50% transducer efficiency.

EXPERIMENTAL RESULTS

The phantom we used to produce magnetostrictive signals is a disk composed of nickel powder and epoxy in the volumetric ratio of 0.89:1. Although this ratio represents a concentration of nickel much too large for clinical use, it is acceptable for a feasibility study. The disk has a diameter of 26 mm and a thickness of 7 mm. It has an acoustic velocity of 2.7 mm/microsecond, close to that of the epoxy alone.

A diagram of the experimental set-up is shown in Fig. 5. The magnetostrictive disk is placed face up in a small container of water placed between the poles of the magnetizing coil and "C" structure. We measured the magnetic field between the poles to be 5% of the field at the coils*. This high field loss is due to the large air gap, so that most of the magnetic flux escapes the ferrite before it reaches the poles. Using the fields calculated

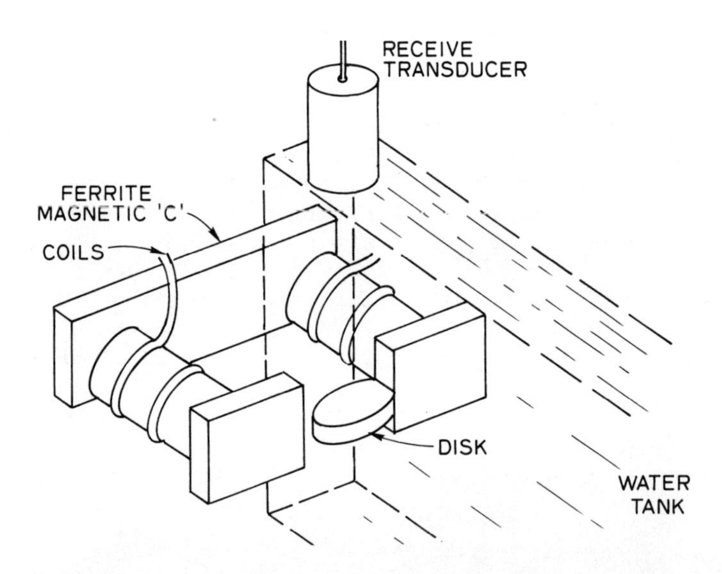

Fig. 5. Experimental magnetostrictive setup. A magnetostrictive disk is placed in a water tank between the poles of a magnetic "C" structure. The ultrasonic receive transducer picks up the waves generated by the disk.

---

*In light of this fact, we feel that a single short solenoid will work better for clinical applications. This requires a higher voltage SCR.

in Eq. 7, an applied field of 1.26×10^3 amp/m is produced, or a field internal to the nickel of 6.28 amp/m, less than 0.5% of its saturation value.

Making the following assumptions:

1) For small field strengths, the displacement is proportional to the square of the magnetic field.
2) A half cycle is transmitted, so that 1/2 acoustic wavelength (2.7 mm at 0.5 MHz) length of material produces a pulse $2/\pi$ times its peak value.
3) Only 0.89/1.89 = 47% of the disk is nickel (active).

The received voltage is calculated to be 39.4 μv, RMS, withing the range of the measured voltages.

Using the same setup as in Fig. 5, we made a B-scan image of the disk with a Picker "Echoview" scanner, which uses a single element transducer and a storage screen. The image is shown in Fig. 6. The top thin line is transmitter pickup, the bottom segment is the disk image. Note that the large divisions each correspond to 10 mm (for a pulse-echo system). The disk was actually 60 mm away from the receiver transducer, demonstrating the one-way path propagation of magnetostrictive imaging systems.

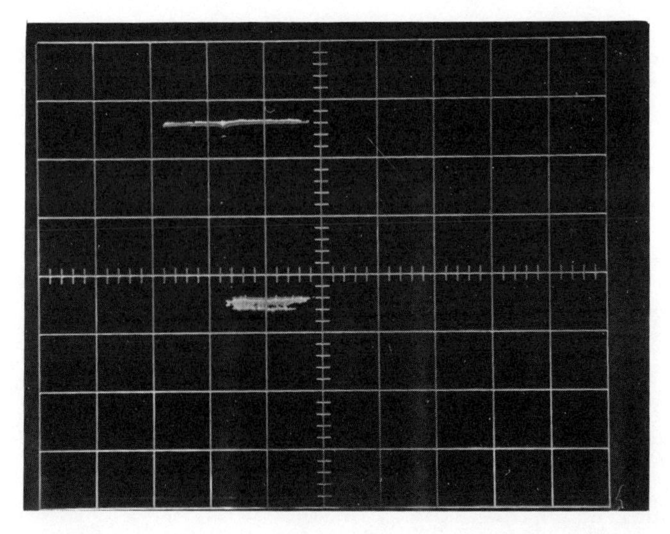

Fig. 6. B-scan photograph of the magnetostrictive nickel powder disk as in Fig. 5. The top line is direct transmitter pickup, the bottom line is the image of the disk 60 mm from the receive transducer. The scale is 10 mm/div, but is calibrated for 2-way ultrasound so that the vertical scale is actually 20 mm/div one way.

CONCLUSIONS

An imaging technique using magnetostrictive particles as the emissive sources for an organ-specific system appears possible using today's technologies. Visualization studies of the G.I. tract using X-rays are invasive, while conventional ultrasonic studies are hindered by two-way wave travel and multiple reflection interfaces. For the maximum benefits of a magnetostrictive system, a (dynamically) focussed B-scan or acoustic lens receiving system should be used, capable of imaging throughout a volume of interest.

Ultrasonic power levels do not pose any hazards in these systems. Since the magnetostrictive elements operate well below their saturation limits, we can obtain larger received signals by just increasing the transmitted fields. Power absorption at these RF frequencies by biological tissues is slight, but will be a future topic of study.

Several possibilities exist for improving these systems. Circuits where the current levels are not limited by the SCR switch would enable larger currents and, therefore, received signals. A single solenoidal transmitter coil may prove optimum for clinical use. Magnetostrictive systems could be operated in conjuction with normal ultrasonic imaging systems (on alternate scan lines), in order to place the magnetostrictive returns in perspective, anatomically.

The new rare earth magnetostrictive alloys are promising in that they have saturation magnetostrictive coefficients over two magnitudes greater than nickel, while operating with similar field strengths. With these alloys, a magnetostrictive imaging system for the G.I. tract is practicable today.

REFERENCES

1. Carl Heck, Magnetic Materials and Their Applications, Crane Russak and Co., New York, 1974, p. 666.

2. A.E. Clark, "Highly Magnetostrictive Rare Earth Alloys," in Magnetism and Magnetic Materials, 1976, AIP Conf. Proc. 34, p. 13 (1976).

3. R. Abbundi and A.E. Clark, "Anomalous Thermal Expansion and Magnetostriction of Single Crystal $Tb_{.27}Dy_{.73}Fe_2$," IEEE Transactions on Magnetics, Vol. MAG-13, No.5, September 1977, pp. 1519-20.

4. E.W. Lee, "Magnetostriction and Magnetomechanical Effects," Reports on Progress in Physics, Vol. 18, 1955, pp. 184-229.

5. J.D. Jackson, Classical Electrodynamics, John Wiley and Sons, New York, 1962, pp. 115, 156-165.

MAGNETOSONIC IMAGE CONVERTER

N. Collings, N. Patil, L.M. Lambert

Depts. of Physics and Electrical Engineering

University of Vermont, Burlington, Vt. 05401

1. INTRODUCTION

One of the more recent works concerned with the storage of stress information by means of magnetic thin films was that of Weinstein et al [1] and their Sonic Film Memory. In this memory, strain pulses were used in conjunction with current pulses to store digital information in a magnetostrictive thin film evaporated onto a glass cylinder. The easy axis of magnetization was circumferential around the cylinder. A conductor ran down the center of the glass cylinder and a transducer for generating sonic waves was mounted on the end of the cylinder. The transducer generated a 10 nanosecond pulse of ultrasound which propagated down the cylinder. A current pulse was also sent down the conductor. The magnetization of a particular location or band could be changed if the strain pulse and current pulse arrived at that location at the same instant of time.

The relevance of this earlier work to the present research is that it demonstrates, in a workable system, one of the two ways in which ultrasound can effect the magnetic state of a magnetostrictive material. In the Sonic Memory, the current pulse provided a magnetic bias and the ultrasound caused the switching or change of magnetic flux density.

Another method by which ultrasound can effect the magnetic state of a magnetostrictive material, is to use stress as a bias and use a current pulse for the switching. This method was demonstrated successfully in bulk (not thin films) nickel and cobalt by Gobram and Youssef[2].

The use of a current bias is particularly suited to the storage of digital information; in as much as this technique is characterized by a threshold amplitude for switching by means of a strain pulse as shown by Weinstein et al. The use of a stress bias, on the other hand, is more suited to the storage of analogue information because there appears to be a dynamic range of magnetic states for a given current pulse.

2. THEORY

2.1 Introduction. In any system which is characterized by a change in magnetic flux density due to ultrasonic stressing there is the problem of measuring this change. One optical method is by use of the magneto-optic Kerr effect. This effect depends on the rotation of the plane of polarization of linearly polarized light when it is reflected from a magnetic surface. The plane of polarization is rotated an amount proportional to the absolute value of the magnetic flux density of the reflecting surface as well as the direction of magnetization. By use of a Glan-Thompson or Nicols prism set near extinction (i.e., not transmitting light of the original polarization), this rotation of the plane of polarization is converted into a change in transmitted light intensity.

In the present Magnetosonic Image Converter, the light passing through the so-called analyzer prism is detected by a use of a photomultiplier tube. The light source itself is a He-Ne laser with a Brewster window for polarizing the light followed by a Glan-Thompson "polarizing" prism. Figures 1 and 2 show the laser and the associated optics. The light, in this figure, reflects off the magnetic thin film mounted at the top of the Lucite tank. An ultrasonic transducer is mounted below the film under water, and generates 4.9 mHz ultrasound for stressing the film and changing the magnetic flux density of the film and hence the light passing through the analyzer to the photomultiplier tube.

2.2 Magneto-optic effects. It is important to describe, at least briefly, the various possible magneto-

Figure 1. Photograph of the experimental arrangement.

optic effects that one must consider in the Magnetosonic Image Converter system. In addition, one must be able to optimize the selected magneto-optic effect in order to obtain the greatest change in light intensity for a given change in magnetic flux density. A further enhancement of certain of the magneto-optic effects can be achieved by the use of thin dielectric layer of high refractive index material such as zinc sulphide (e.g., 510Å of ZnS for 6328Å light).

Clearly, there are two possible choices for the polarization of incident light reflecting from a surface. One polarization would be parallel to the plane of incidence, called a p-wave, and another would be normal to this plane, called an s-wave. One can now relate the magnetization, M, of the magnetic thin film to these light waves. If M is in an arbitrary direction relative to the plane of incidence, the component of M parallel to the plane of incidence gives rise to the longitudinal Kerr effect. On the other hand, the component of M perpendicular to the plane of incidence gives rise to the transverse Kerr effect. Actually there is a third possibility. The magnetization M could be normal to the reflecting surface. This is called the polar Kerr effect.

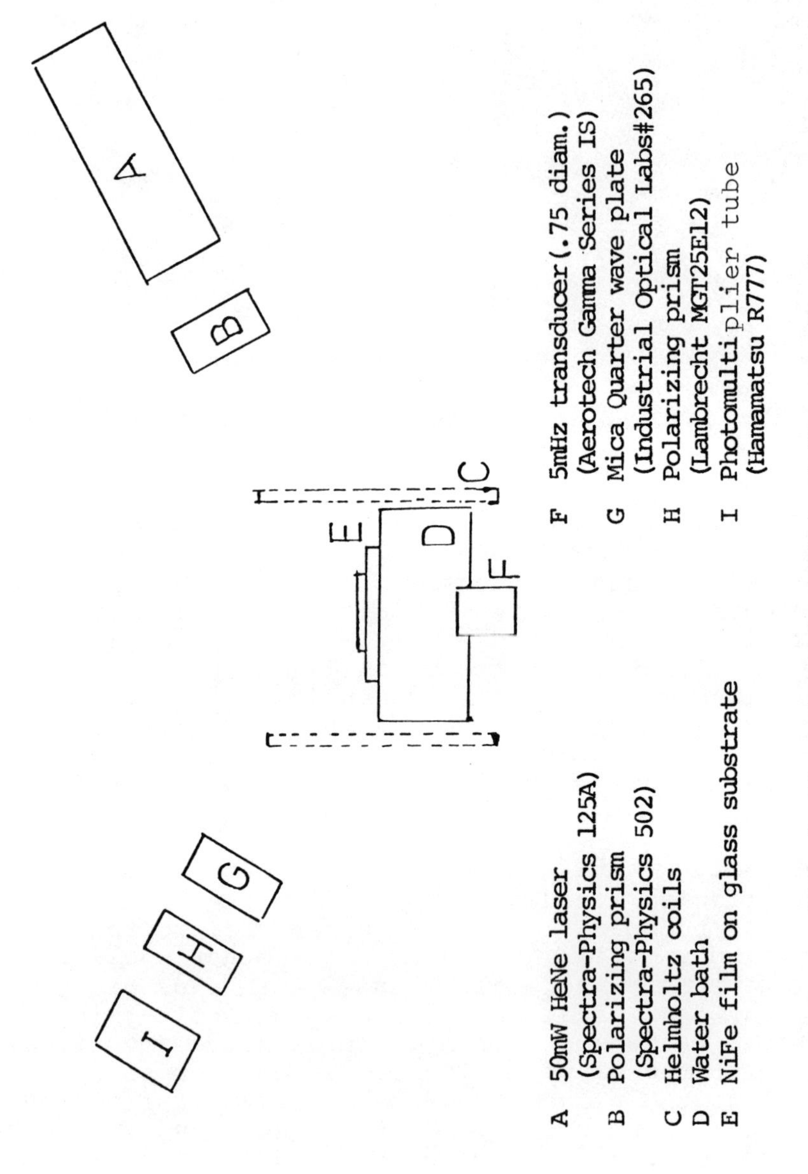

FIG. 2. The schematic layout of the experiment.

There would be an admixture of the longitudinal and transverse Kerr effects when M is in or normal to the plane of incidence except for the fact that the transverse Kerr effect corresponds to a direct change in reflectivity rather than a rotation of the plane of polarization thus can be removed with the analyzer prism set near extinction. The polar Kerr effect, even though the largest of the three effects, is not possible with thin films due to the shape anisotropy which constrains M to be in the plane of the thin film.

Finally, the longitudinal effect causes a rotation of the plane of polarization because the magnetic film is slightly birefringent. Neither the p or s-wave of the incident light corresponds to a principal axis for this birefringence. Thus the reflected light can be represented as the sum of two components of a reflected wave, R, and a small orthogonal wave, K, called the Kerr coefficient. The Kerr coefficient (orthogonal wave) is not in phase with the reflected wave and the sum of the two components has a resulting ellipticity. This ellipticity is a dominant effect and must be removed by use of a retardation plate (see Appendix A).

The Kerr coefficient, K, can be enhanced by use of an overcoating of the magnetic film with a high dielectric constant material such as zinc sulphide [3]. The dielectric layer causes multiple reflections; each reflection causes an additional rotation of the plane of polarization. The theoretical improvement in the modulus of K versus the angle of incidence of the light beam in shown in Figure 3. This result was obtained by use of the formulae derived by Lissberger [3] and Robinson [4] and by the use of the optical constants appropriate for 80-20 nickel-iron obtained from Tanaka [5]. As noted here, an improvement of approximately 2.5 is obtainable.

2.3 Contrast. One of the major feasibility concerns of the proposed Magnetosonic Image Converter was the maximum contrast obtainable between neighboring magnetic areas for which the direction of M is different. Since the sensitivity of the eye and photographic film depends on the logarithm of the light intensity, the lower light levels offer the best contrast for a given intensity difference between adjacent areas. However, the low light levels provide small magneto-optical signals. The optimum light level can be computed, for a given optical arrangement, in terms of the analyzer

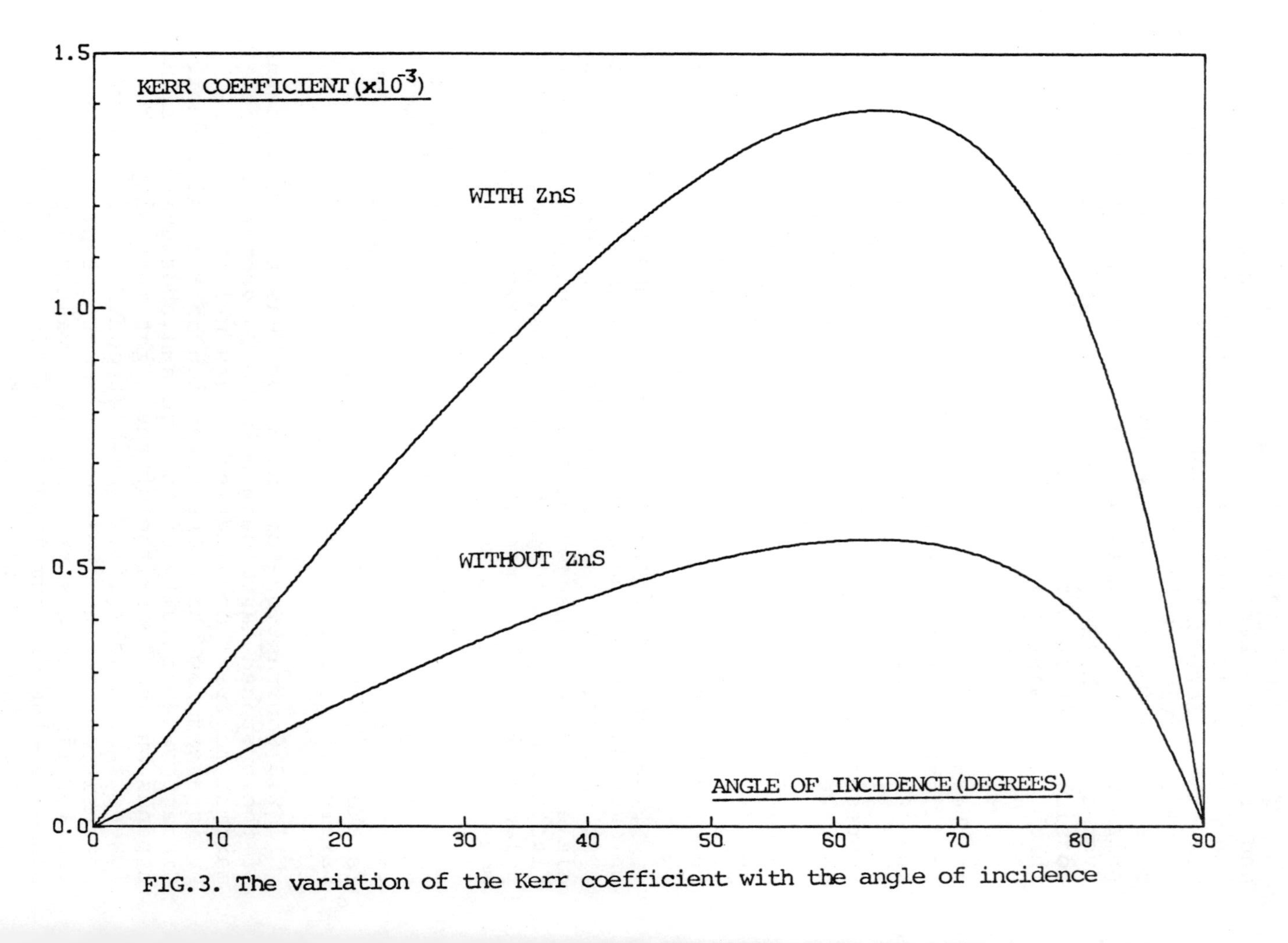

FIG.3. The variation of the Kerr coefficient with the angle of incidence

angle, ε . ε is usually very small, less than 0.15 radians, and thus it is important to include the imperfections of the polarizing prisms in the analysis. For example, if T_1 is the amplitude of the electric field vector transmitted by the polarizing prism, and T_2 is the small orthogonal component associated with T_1, then the matrix of the incident s-wave becomes $[T_2', T_1']$. Correspondingly, the Jones matrix for the analyzing prism, inclined at an angle ε to the principal axis becomes $\begin{bmatrix} T_1 & T_1\varepsilon \\ T_1\varepsilon & T_2 \end{bmatrix}$. One can define the extinction coefficients of the polarizing and analyzing prisms as $|C^1|^2$ and $|C|^2$ respectively and

$$|C^1|^2 \equiv \left(\frac{T_2'}{T_1'}\right)^2 \quad ; \quad |C_2|^2 \equiv \left(\frac{T_1}{T_2}\right)^2 \tag{1}$$

The best quality prisms have the lowest extinction coefficients.

For purposes of analysis we will assume the retardation plate to be perfect and to introduce no losses in the system.

An analytic equation for the final light intensity, in terms of the Kerr coefficient, K, and the analyzer angle, ε, can now be derived. Since K is proportional to the component of $\underline{M}$ in the plane of incidence for the longitudinal Kerr effect, neighboring magnetic domains with different components of $\underline{M}$ will generate different light intensities. Three specific cases are to be examined: (1), when the $\underline{M}$ vectors in adjacent areas are anti-parallel and lie in the plane of incidence; (2), when the $\underline{M}$ vectors are anti-parallel and inclined at an angle of 45° to the plane of incidence; and (3), when one $\underline{M}$ vector lies in the plane of incidence and the other is perpendicular to this plane.

The contrast between adjacent domains, which is the logarithm of the intensity ratio, is plotted as a function of analyzer angle, ε , in Figures 4 and 5. In Figure 4, the magnitudes of the extinction coefficients, $|C^1|^2$ and $|C|^2$ is taken as 10^{-6}. This value is an optimum value for available prisms. In Figure 5, the magnitudes of the extinction coefficients are taken at more realistic values of $|C^1|^2 = 2.5\text{x}10^{-5}$ and $|C|^2 = 8.5\text{x}10^{-5}$ (see Appendix B).

To obtain the best contrast, one should employ an incident s-wave and a retardation plate. This optical arrangement was used in the computations for

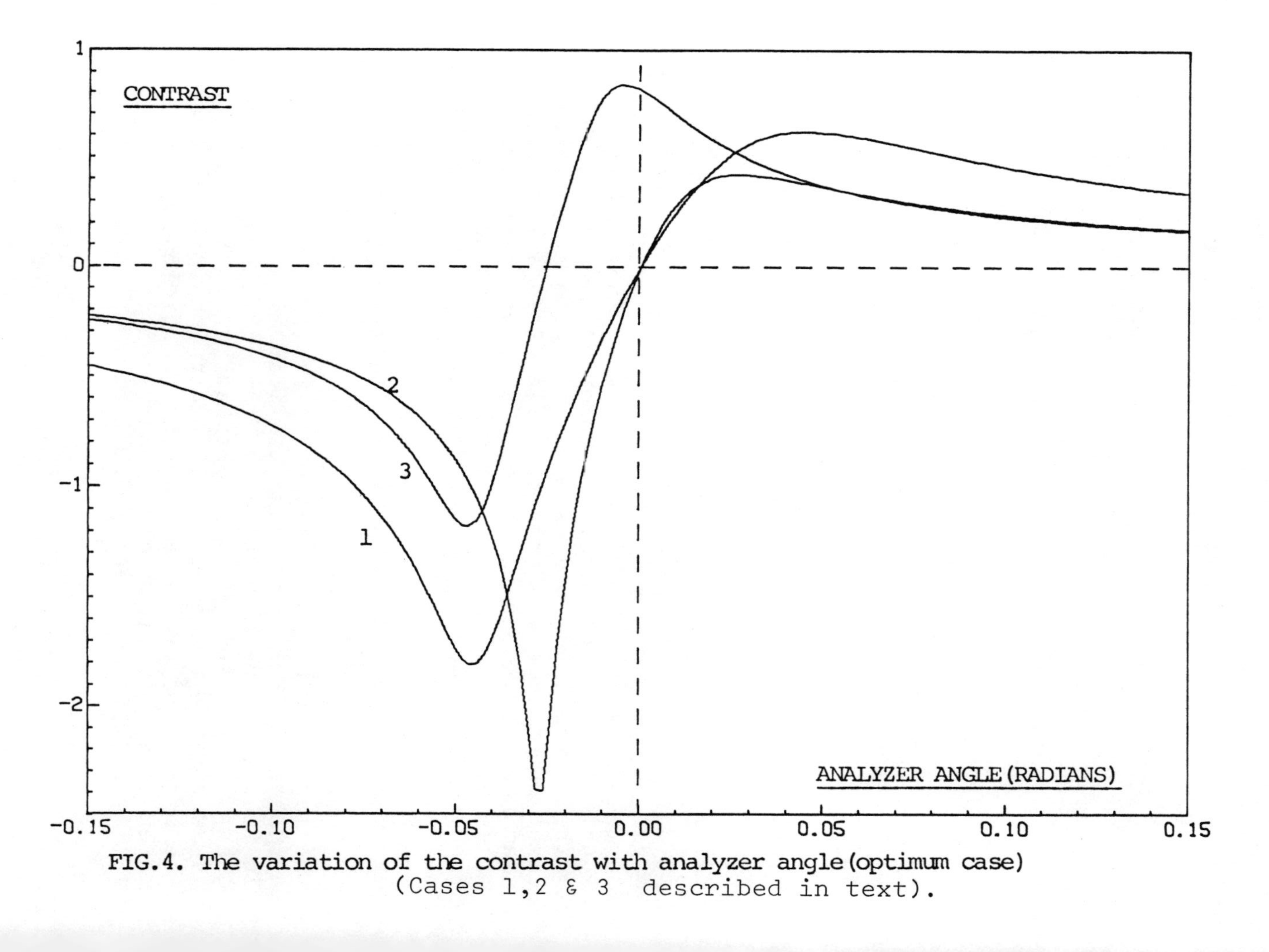

FIG.4. The variation of the contrast with analyzer angle (optimum case)
(Cases 1, 2 & 3 described in text).

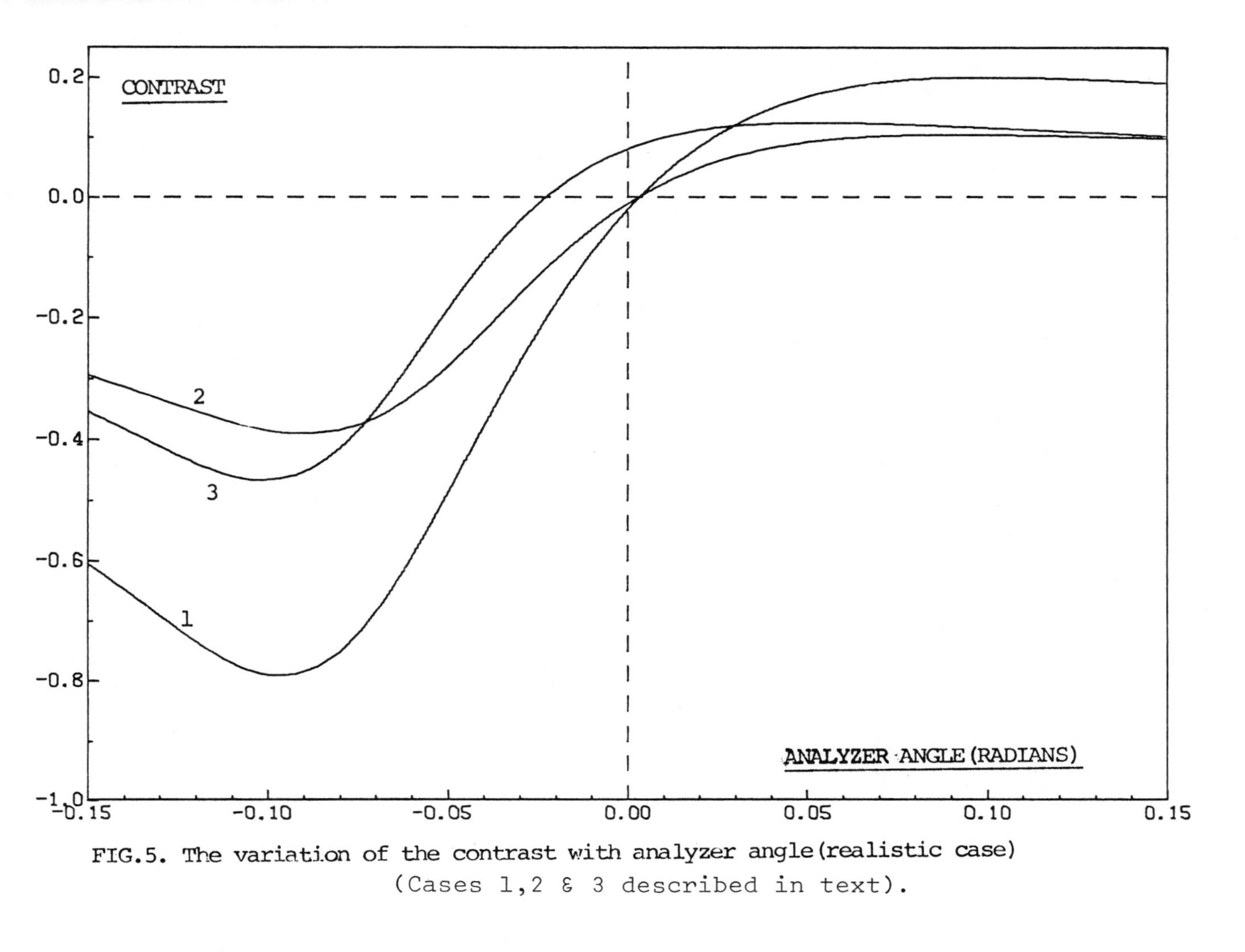

FIG.5. The variation of the contrast with analyzer angle (realistic case) (Cases 1,2 & 3 described in text).

Figures 4 and 5. Also for the best contrast, the analyzer should be set at angle of $\varepsilon = -0.05$ radians for example in Figure 4 and $\varepsilon = -0.1$ radians for example in Figure 5. If one uses these settings, the brightness of the image would be low and the use of an image intensifier tube, such as the RCA 4550, would be necessary. The use of an image intensifier does not alter the contrast calculations, however, since both intensities (from adjacent magnetic domains) will be multiplied by the same magnification factor.

The s-wave is the better arrangement for the following reason. The contrast is directly proportional to the Kerr rotation, R. Furthermore, the zinc sulphide dielectric layer acts as an anti-reflection coating for the s-wave, i.e. R_s is small. Therefore, from (A.1) it is evident that R, the Kerr rotation, will be large and, hence, the contrast will be large.

2.4 Noise in optical systems. The two major sources of noise in any optical system are the light source and the light detector. In the present Magnetosonic Image Converter system these sources correspond to the intensity stability of the laser and the shot noise of the phototube. In situations in which the laser noise dominates, the signal to noise ratio, S/N, is proportional to the Kerr rotation; therefore, the setting for maximum contrast is also the setting for the best S/N ratio. In cases in which shot noise dominates, the S/N ratio becomes proportional to the difference of the two intensities, involved in the contrast ratio, divided by the square root of their sum. One must then compute this S/N ratio and choose the analyzer angle which gives the best compromise between good contrast and a good S/N ratio.

The shot noise contribution to the total noise, in the S/N ratio is increased when the number of photons striking the detector from a given area of magnetic film is decreased. Thus if large areas of magnetic film are illuminated from a laser by use of a beam expander, the shot noise increases. An alternate technique for illuminating the magnetic film is to scan a high intensity laser beam across the film analogous to a TV raster, but since the number of photons, from a given area, is similarly reduced, there are no noise advantages from this system.

3. FABRICATION

3.1 Introduction. The magnetic material used in this research was primarily 60-40 NiFe. This alloy combines high magnetostrictive constants with low anisotropy constants (7); and, consequently, it has a high stress sensitivity. For consistent results, it was found important to use a vacuum-melted alloy because it is more homogeneous. The NiFe alloy used was purchased from Materials Research Corporation and Ventron in the form of 1/2" diameter slugs, with a starting purity of: Ni (99.99) and Fe (99.95).

Glass was the most convenient substrate for mounting the thin films. The 18x18x1 mm glass slides were undistorted by the high substrate temperatures used in the vacuum system. They were also not distorted by the radiation pressure of the ultrasonic wavefront - a condition for keeping the light reflected from the film in a constant direction. The use of a thick glass substrate was also a convenient way to apply a large in-plane stress to the film. For example the stress in a 1000Å thin film would be about 10^4 times the stress in the 1 mm glass slide. The ultrasonic wave passes through the glass slide to the glass-metal film interface. It is not known at present to what degree the glass distorts the ultrasonic plane wave.

3.2 Preparation of magnetic thin films. The films are made in a 25" diameter stainless steel bell jar vacuum system 32" high. The system has a 6" diffusion pump and a 1500 ℓ/min pump which enables the chamber to be pumped down to about 8×10^{-6} torr. The pressure in the chamber is initially measured by a thermocouple gauge and then by an ion gauge. Most evaporation runs were made between 2×10^{-5} and 4×10^{-5} torr. A uniform magnetic field of 27 Oe is provided by two rectangular 63"x32" water cooled field coils. The magnetic field is in the plane of the film.

The thin films are deposited onto glass slides 2"x1.75"x.04". The slides are placed in a copper substrate holder and clamped down. The substrate holder is placed in a stainless steel tray which is located about 1.75" from heater coils and 15" from the evaporation source. The heater is capable of heating the substrates up to 325°C. The temperature is monitored by a chromel-alumel thermocouple in contact with the substrate holder and controlled by a temperature control unit.

The thickness of the NiFe film is monitored by a quartz crystal deposition control, Sloan Omni II. The frequency shift between the time the shutter is opened and closed is proportional to the thickness of the deposition. The unit also indicates the rate of deposition. The thickness is measured once again by use of a Tolansky interferometer. The thickness of the ZnS coating is monitored by means of a laser-photodiode set up as shown in Figure 6. The output of the photodiode is recorded on a chart. The ZnS is deposited until a minimum is obtained on the recorder. ZnS is evaporated by resistance heating a tungsten basket inside a crucible containing the ZnS powder.

The NiFe slugs used were placed in a 5/8" diameter alumina crucible which was placed inside another 1" diameter alumina crucible. The NiFe is evaporated by rf induction heating of the slug. The rate of deposition of the NiFe can be controlled by adjusting the plate and grid current of the rf generator.

The glass slides used are washed in an ultrasonic bath, boiled in distilled water,and then isopropyl alcohol. They are dried and placed in the substrate holder.

In a typical run, the slides are cleaned and placed in the substrate holder in the tray and the tray is then placed in the chamber. The slides are heated to the required substrate temperature about 300°C,while the system is pumping down. The magnetic field is switched on and the NiFe slugs are rf heated. The rate of deposition is observed on the evaporation monitor unit. When a satisfactory rate is obtained, around 15Å/sec, the shutter is opened and left open until the required thickness is obtained. The rf generator and the heater is then shut off. The ZnS powder is next heated and the shutter is again opened after a satisfactory rate of deposition is observed. When a minimum is obtained in the reflected laser light on the recorder, the shutter is closed and the ZnS heater is shut off. The magnetic field is switched off after the slides have cooled to below 100°C. The chamber is opened to air after the slides have cooled to below 30°C.

4. EXPERIMENTAL RESULTS

4.1 Test equipment.The Magnetosonic Image Converter

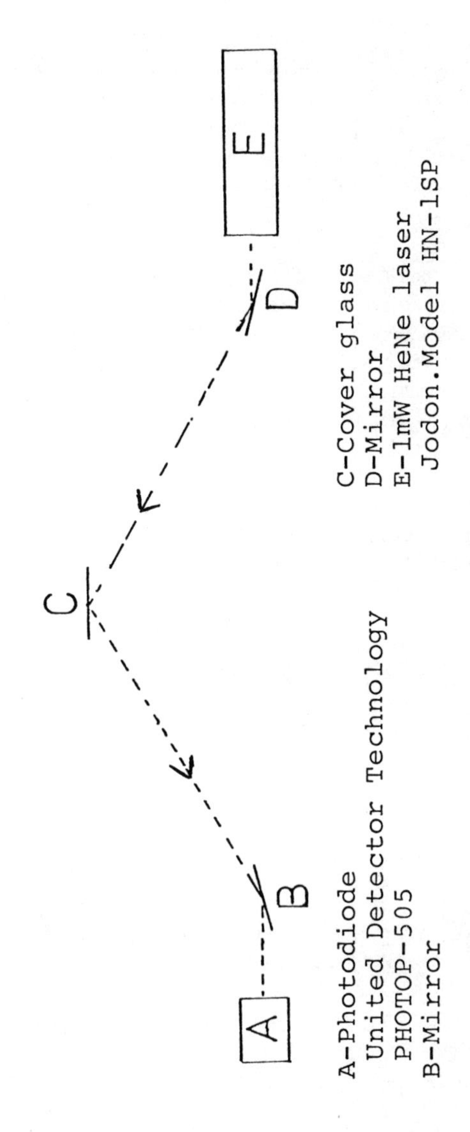

Fig.6 Arrangement for monitoring the ZnS deposition.

test system is shown in Figure 7. Seen in this view is the ultrasonic tank with a 1" diameter O-ring for making a water tight seal between the magnetic film and the water bath. A similar O-ring seal is used with a hold-down cover which is held in place by four screws as shown. The magnetic thin film is on a glass substrate with the film on the top surface. The film is clamped between the O-rings giving a stress, due to flexing of the film, perpendicular to the plane of incidence. An ultrasonic transducer in the tank generates waves, which travel upward striking the substrate and dynamically stressing the magnetic film.

4.2 Test procedures. Hysteresis loops are routinely plotted for every magnetic thin film sample made. To obtain these loops, an alternating current (0.3 Hz) is used to drive a Helmholtz coil generating a magnetic field intensity sufficient to saturate the magnetic flux density. A small resistor in series with the coils supplies a voltage proportional to the magnetic field to the X axis of a chart recorder. The voltage generated by a photomultiplier tube, proportional to the magnetic flux density, drives the Y axis of the recorder. Figure 8a shows a typical easy axis hysteresis loop and the change that occurs in the loop under stressing. As seen here, the coercivity (the magnetic field intensity required to reduce the magnetic flux density to zero) decreases and the remanent magnetization (magnetic flux density at zero magnetic field) is considerably less than for an unstressed loop. This substantial change in the stressed hysteresis loop is a prerequisite for the film to be sensitive to ultrasound; although, not all films exhibiting this change in hysteresis loop are sensitive to ultrasound. The sensitivity of the various magnetic thin films to ultrasound varies greatly and is dependent on the amount of stress bias used (flexing of the film in the test tank holder). The control of sensitivity by stress bias is useful for varying the dynamic range of stress sensitivity of the magnetic thin films.

The effect of ultrasound on the hysteresis loop can be seen clearly in Figure 8b. In this figure, the abscissa has been expanded for greater clarity. As seen here, the coercivity and remanent magnetic flux density both decrease. The ultrasonic transducer (black circle in Figure 7.) is mounted 1" below the film. The transducer was calibrated by the radiation force technique. At 4.9 mHz, 12V applied to the transducer gave a sound intensity of 160 mW/cm^2.

Fig. 7. Photograph of the top surface of the water tank

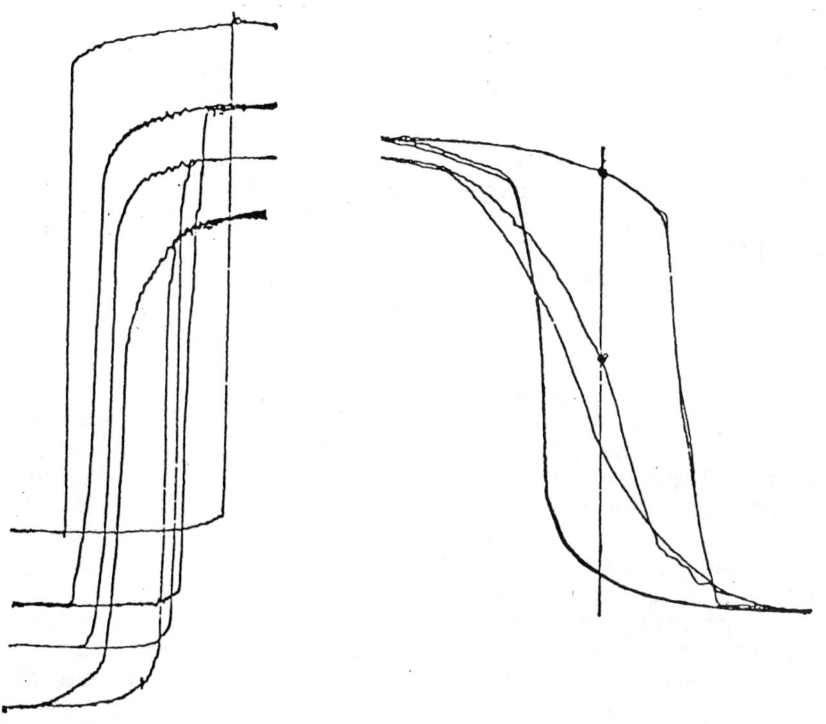

Fig. 8a. The change in the hysteresis loop with increasing stress.

Fig. 8b. The change in the hysteresis loop due to ultrasound (30V).

The more relevant property of the magnetic film for our application is the decrease in remanent magnetization with irradiation by ultrasound as portrayed in Figure 9. to the right of the hysteresis loop. The ratio of the change in $\underline{M}$ for 10V, 20V, and 30V ultrasonic transducer voltage is 1:5:16. This ratio differs from the 1:4:9 ratio expected if the change in $\underline{M}$ is proportional to the intensity of the ultrasound; however since the overall change in $\underline{M}$ was close to the dynamic range of the thin film (30V caused $\underline{M}$ to go approximately to zero), non-linearities are perhaps to be expected. One of the important features of magnetic thin film ultrasonic sensors is that after irradiation the resulting change in $\underline{M}$ is permanently stored as a magnetic flux condition until erased by remagnetization. The steps in this figure occur during a three to four second burst of 4.9 mHz ultrasonic irradiation.

In these figures, measurements were made with a laser beam 2 mm in diameter. Because changes in sensitivity arise from domain wall motion between adjacent domains when irradiated with ultrasound, smaller diameter light beams were used to probe the interior of these domains. In Figure 10, the hysteresis loop is measured with a 0.5 mm diameter light beam. It is uncertain whether all domain wall behavior has been eliminated (so that only rotational processes are observed) but the less abrupt changes of magnetization with each pulse of ultrasound would support this conjecture. The flux changes are proportional to the amplitude of the ultrasound rather than the intensity.

The 0.5 mm diameter beam was used to measure the change in magnetic flux density between various areas of the magnetic film with and without irradiation by ultrasound. Five areas with a 4 mm spacing between them were selected. The change in magnetic flux density for each area undergoing irradiation was measured, the results of which are shown for the top line of Figure 11.

One notes that the sensitivity of various areas is not the same. There are several reasons why this may have occurred. First, and perhaps most likely, we are operating in the near field of the ultrasonic transducer and the amplitude of the wavefronts may not be the same a at all five areas. Secondly, areas 1 and 5 were near the edges of the test film and demagnetizing fields can become important in determining the behavior of such areas. Finally, there may be inhomogeneity in either the film composition or the stress biasing.

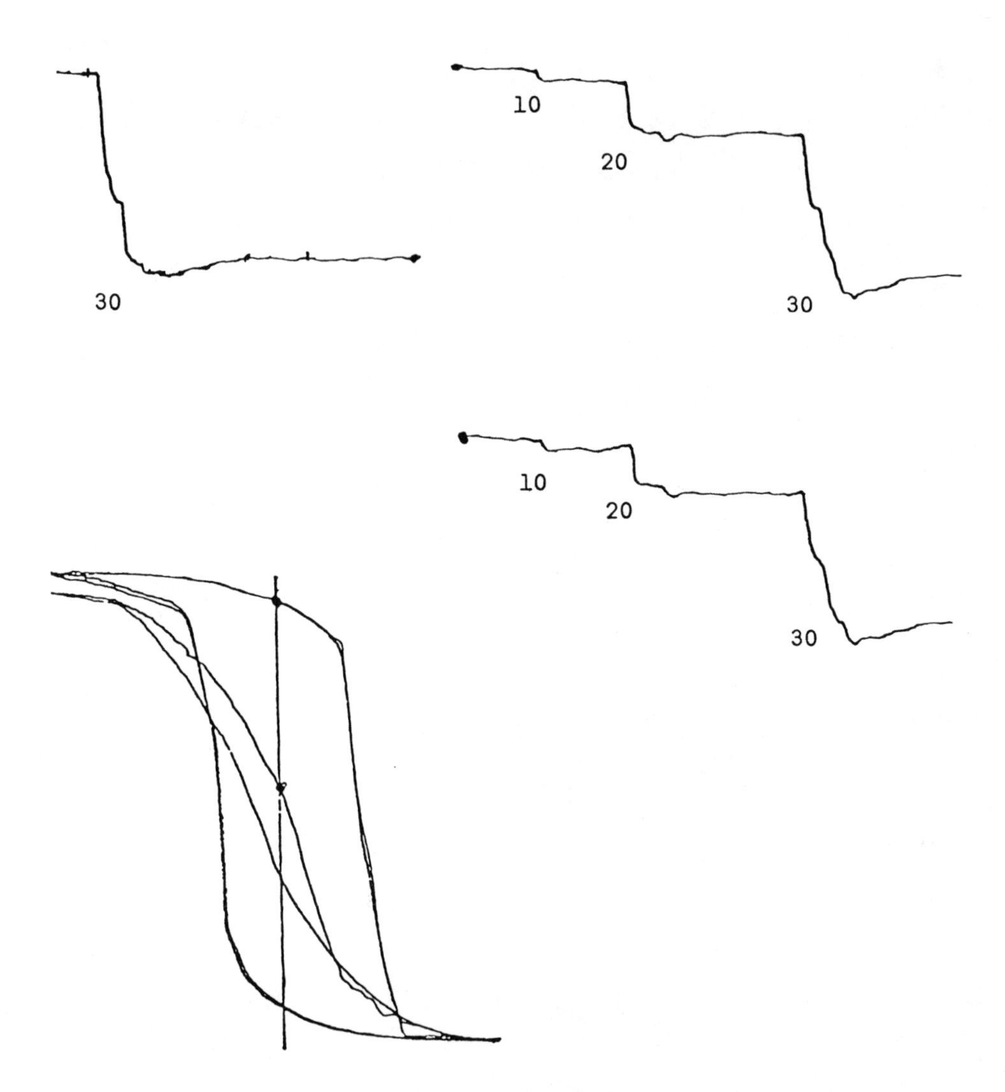

FIG. 9. The decrease in the remanent magnetization due to ultrasound. (Numbers refer to the voltage applied to the transducer).

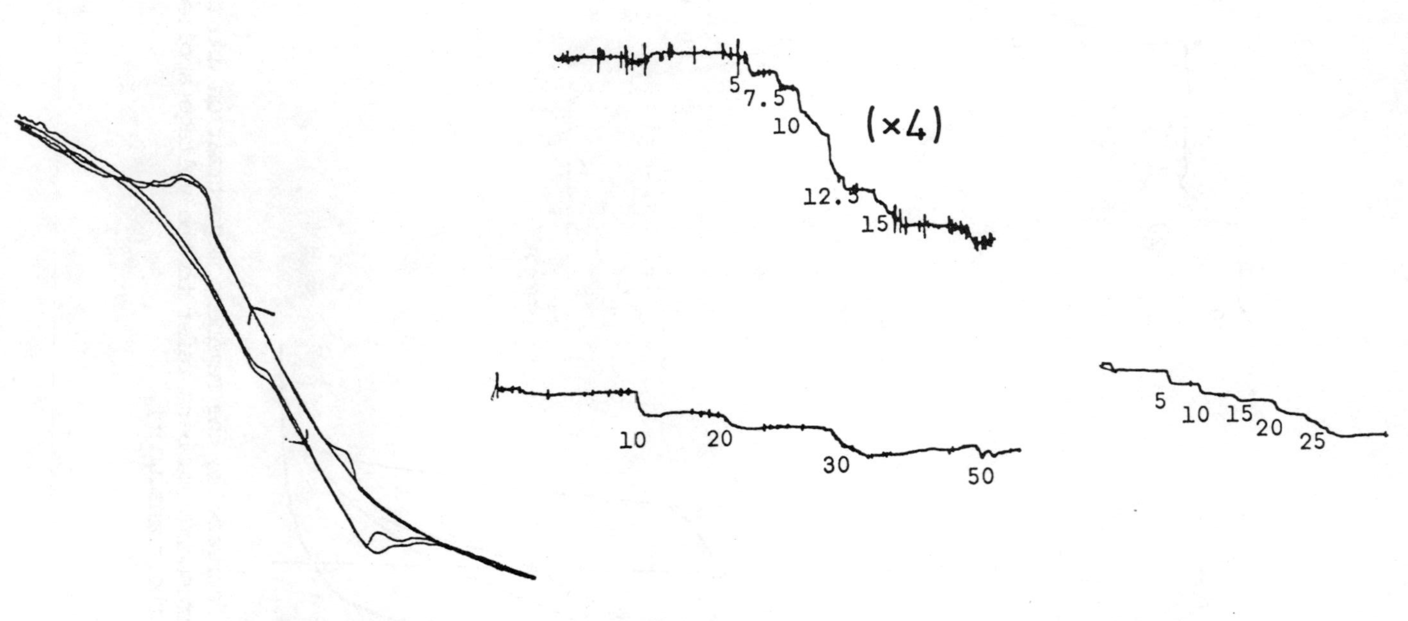

FIG. 10. The decrease in the remanent magnetization using a focussed laser beam.

The lower lines of Figure 11 shows the corresponding magnetization changes which occur when a polystyrene attenuator is placed between the transducer and areas 1 and 2 and at the edge of 3. The polystyrene is 1/8" thick and is located approximately 1/4" below the magnetic film. As seen, the amplitude of the ultrasound is severely reduced upon passing through the polystyrene resulting in small changes in magnetic flux density of areas 1 and 2. The attenuation at spot 4 is perhaps due to diffraction and reflection at the edge of the polystyrene attenuator.

5. CONCLUSIONS

Research on the Magnetosonic Image Converter has demonstrated that magnetic thin films can be used to make a permanent record of the amplitude of ultrasonic waves. The sensitivity of these ultrasonic wave detectors can be varied by the use of stress bias.

The next stage in the research program will be to record the images of simple objects. In order to enhance the light level, an image intensifier tube (RCA 4550) would be used. In order to improve the contrast, we propose to use a high power Argon laser in place of the present low power He-Ne laser.

This next test system will record and view film areas of around 3 square inches. The proposed system would employ an Argon laser; the polarizing prism and beam expander described in Appendix B; a 2" diameter mica retardation plate; a 4 sq. in. sheet of HN-22 polaroid; and the RCA 4550 image intensifier tube.

The use of magnetic films with areas of 3 square inches poses the problem of the uniformity of the sensitivity across the film suggested by Figure 11. At the present time, 1/8" diameter thin film test samples are being prepared and their sensitivity measured. It is possible that the uniformity problem could be solved by using a matrix of these or smaller elements instead of a continuous film.

In the final system a concave mirror will focus the ultrasound on the magnetic film. This will allow greater resolution and permit the specimen and trans-

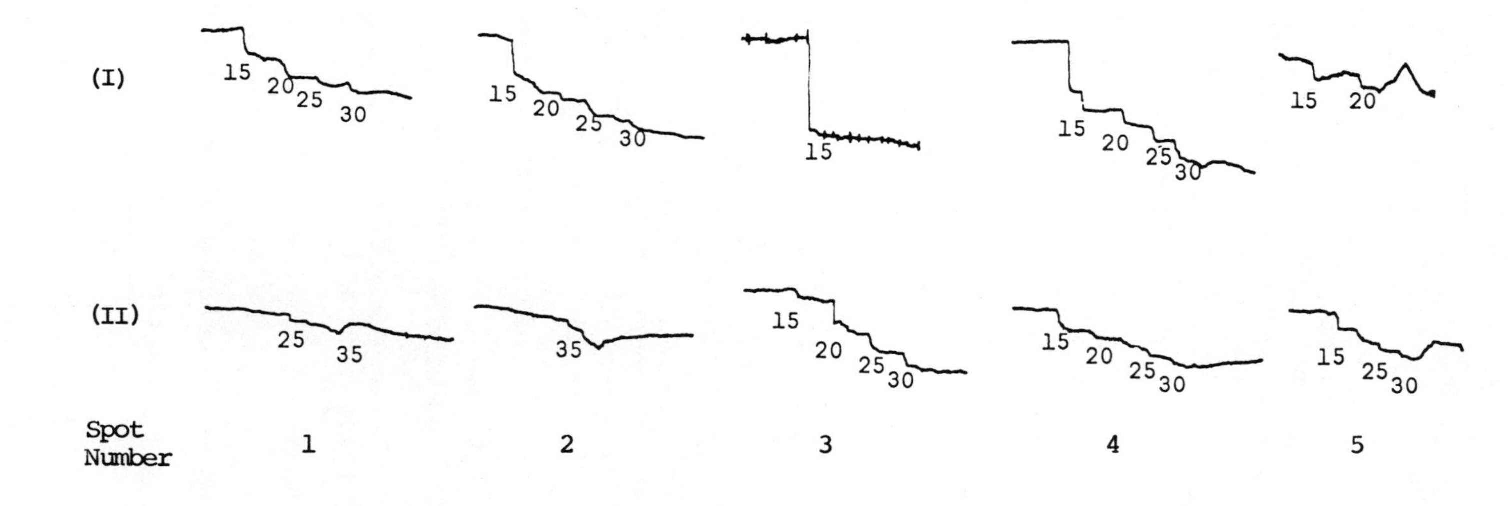

FIG. 11. The decrease in the remanent magnetization at five spots on the magnetic film.
Line (I)-- all spots exposed to ultrasound.
Line (II)--spots 1&2 shaded from ultrasound.

ducer to be placed in a line parallel to the plane of the magnetic film.

If the images obtained are of good quality, then a motion picture facility can be adapted to the above system by including small magnetic field coils to erase the magnetic information for each frame. The frame speed will be limited only by the persistence of the phosphor on the image intensifier tube (about 10 msec.).

Improvement of the magnetic films, with regard to ultrasonic stress sensitivity is, perhaps, the most difficult problem whose solution will require an adequate theoretical understanding of the interaction between the ultrasound and the local magnetization in the film.

APPENDIX A

In order to represent each optical component mathematically, a Jones matrix is constructed which, when it operates on a vector (E_p, E_s), gives the same result as the effect of the component on an electric field vector (E_p, E_s) (where E_p and E_s are the component amplitudes of the light wave parallel and orthogonal to the plane of incidence).

In particular, the metal film is represented by

$$\begin{bmatrix} R_p & K \\ -K & R_s \end{bmatrix}$$

when the magnetization lies in the plane of incidence. (R_p and R_s are the reflected wave amplitudes for the p- and s-waves; and K is the Kerr coefficient.)

A quarter wave retardation plate, with its optic axes in the principal axes of the light wave is represented by

$$\begin{bmatrix} 1 \pm i & 0 \\ 0 & 1 \mp i \end{bmatrix}$$

(the choice of signs depends on whether the retarding axis is parallel or perpendicular to the plane of incidence).

An analyzing prism, inclined at a small angle, ε , to the principal axes, is represented by

$$\begin{bmatrix} 1 & \varepsilon \\ \varepsilon & 0 \end{bmatrix}$$

For the arrangement of Figure 2, with an incident s-wave the electric field vector incident on the photomultiplier tube will be

$$\begin{bmatrix} E_{pf} \\ E_{sf} \end{bmatrix} = \begin{bmatrix} 1 & \varepsilon \\ \varepsilon & 0 \end{bmatrix} \times \begin{bmatrix} 1-i & \\ & 1+i \end{bmatrix} \times \begin{bmatrix} R_p & K \\ -K & R_s \end{bmatrix} \times \begin{bmatrix} 0 \\ 1 \end{bmatrix}$$

and the light intensity $I = |E_{pf}|^2 + |E_{sf}|^2$
$= 2i\varepsilon(R_s K^* - R_s^* K) + 2\varepsilon^2 |R_s|^2$

When the magnetization in the film is reversed, K changes sign, and the intensity is

$$I' = -2i\varepsilon(R_s K^* - R_s^* K) + 2\varepsilon^2 |R_s|^2$$

The change in light intensity, $\Delta I = 4i\varepsilon(R_s K^* - R_s^* K)$
The reflected light is characterized by two parameters, the rotation (R), and the ellipticity (E) (the ellipticity is numerically equal to the ratio of the minor to major axes of the ellipse) and in terms of the above formalism:

$$R = 1/2 \text{ Real } \left(\frac{K}{R_s} + \frac{K^*}{R_s^*} \right) \qquad (A.1)$$

$$E = \frac{1}{2i} \text{ Imag } \left(\frac{K}{R_s} - \frac{K^*}{R_s^*} \right) \qquad (A.2)$$

$$\Delta I = 8\varepsilon E |R_s|^2$$

If the phase plate is removed; then, repeating the above analysis

$$\Delta I = 8\varepsilon R |R_s|^2$$

If the ellipticity is greater than the rotation a retardation plate is required to optimize the signal.

APPENDIX B

The improvement of the extinction coefficients, $|C^1|^2$ and $|C|^2$, is of primary importance for optimum contrast.

Unfortunately, when larger areas of magnetic film are to be viewed, such as by beam expansion or scanning, there is a deterioration of both coefficients. With a beam expander there is scattering at the pinhole and also a deterioration if the lenses are not strain free. With beam scanning systems there is, in some cases, a complete loss of the polarization quality of the incident beam. Hence, $|C^1|^2$ is increased. $|C|^2$ increases if it is necessary to use a sheet polarizer, such as the Polaroid HN-22, as the analyzer because of the limited aperture of Glan-Thompson prisms.

The realistic values for $|C^1|^2$ and $|C|^2$, quoted in 2.3, correspond to such a system and Figure 12 shows the present arrangement. The polarization of the laser beam is improved by use of a Lambrecht polarizing prism (MGT 3S5) with a measured extinction coefficient of 3.3×10^{-6}. A beam expander (Spectra-Physics Models 332 and 336 with L6 lens and A6 aperture) is attached to this and the extinction coefficient is degraded to 2.5×10^{-5}. The HN-22 Polariod sheet, used as an analyzer, has an extinction coefficient of 8.5×10^{-5}.

The optimum values, quoted in 2.3, correspond to the use of high quality Glan-Thompson prisms, which could accept reasonably large beam diameters. This would be expensive. Most of the improvement can be obtained, however, if such a high quality prism is used solely as the polarizer.

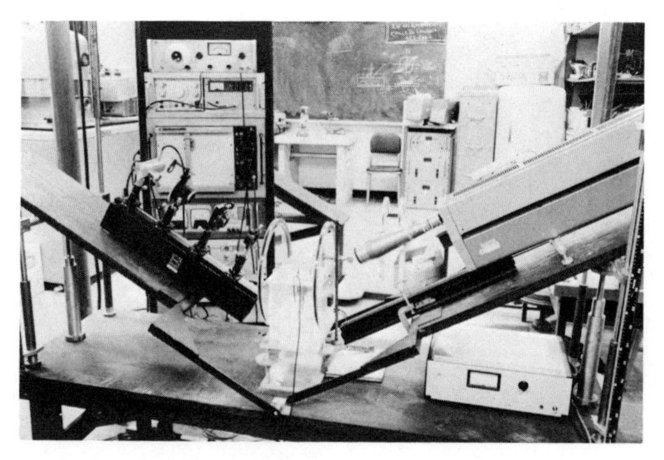

Fig. 12. Photograph of the experimental arrangement, including the beam expander.

ACKNOWLEDGEMENTS

The authors wish to thank and acknowledge the calibration of the transducer employed in this research by Dr. D. Miller of the University of Vermont.

This research was supported by the National Institute of Health through grant GM20226.

REFERENCES

1. Weinstein, H., Onyshkevych, L., Karstad, K., and Shahbender, R., RCA Review, 317, (June, 1967).

2. Gobram, N.K., and Youssef, H., J. Phys. C, 1, S24, (1970).

3. Lissberger, P.H., J. Opt. Soc. Am., 51, 957, (1961).

4. Robinson, C.C., J. Opt. Soc. Am., 54, 1220, (1964).

5. Tanaka, S., Jap. J. App. Phys. 2, 548, (1963).

6. Lissberger, P.H., J. Opt. Soc. Am., 54, 804, (1964).

7. Bozorth, R.M., Rev. of Mod. Phys. 25, 42 (1953).

ACOUSTIC IMAGING VIA THE ACOUSTO-PHOTOREFRACTIVE EFFECT

N. J. BERG, J. N. LEE, and B. J. UDELSON

U.S. Army Electronics Research and Development Command
Harry Diamond Laboratories
2800 Powder Mill Road   Adelphi, MD 20783

A new technique has been discovered for storing acoustic images using the interaction between a traveling surface acoustic wave (SAW) and intense short-duration laser pulses. The laser irradiation results in a semipermanent index-of-refraction pattern proportional to the SAW signal in the lithium-niobate ($LiNbO_3$) storage crystal. It is conjectured that the wavelengths in the laser beam excite charge carriers across the bandgap of $LiNbO_3$, and that these carriers drift in the intense piezoelectric field accompanying the SAW until they are retrapped. The displacement of the retrapped carriers from their original positions causes local variations in the electric field which manifest themselves as index-of-refraction changes caused by the electro-optic nature of the $LiNbO_3$. The stored image can be read out either by scanning the image with a narrow light beam or by illuminating the stored image uniformly and generating an acoustic impulse to interact with the light diffracted by the image. The stored image can be erased by exposure to incoherent ultraviolet radiation, and a new image can then be stored. The storage time in pure $LiNbO_3$ is approximately two months. Iron-doped $LiNbO_3$ was found to be about five times as sensitive as the pure material.

I. INTRODUCTION

A new storage mechanism for acoustic signals has been discovered wherein an index-of-refraction change (δn) in lithium niobate ($LiNbO_3$) results from the acousto-photorefractive interaction of high-intensity, short-duration, laser pulses with propagating surface acoustic waves (SAW's). This acousto-photorefractive effect can be used to store acoustic signals for long periods of time.

Storage for as long as two months has already been demonstrated. This storage effect is phenomenologically similar to a nonlinear "two-photon" photorefractive effect previously reported,[1] but is not a simple extension of that effect. Section II describes the experimental results of an investigation of this storage phenomenon. A possible application of this storage mechanism to acoustic imaging for medical diagnostic purposes is discussed in section III.

II. ACOUSTO-PHOTOREFRACTIVE EFFECT

A. Experimental Results

Our experiments to date have concentrated on the storage of acoustic waves located at and just below the surface of a slab of $LiNbO_3$. This storage results from the interaction of a pulsed high-intensity laser beam with SAW's traversing the surface of the $LiNbO_3$. The storage of bulk acoustic waves traveling in $LiNbO_3$ can in principle also be achieved with the same interaction mechanism.

Figure 1 shows the setup used for our experiments. During the writing (storage) phase, illustrated at the left of the figure, a Nd:YAG laser was used in which the output was a single-mode (TEM_{oo}) Gaussian pulse having a 12-ns width and an energy of 100 mJ. The 1060-nm infrared (IR) output was converted to green (530 nm) by using either a CD*A temperature-tuned or a KDP angle-tuned doubler. By means of appropriate lenses, the resultant light was converted into a sheet beam about 1 cm wide and 250 μm thick. This sheet beam was then focused along the top edge of the side of the $LiNbO_3$ crystal, so that it passed through the region of SAW propagation. Since many pulses were often used for storing a signal, the writing (Nd:YAG) laser pulse was synchronized with the launching of the acoustic waves to within 3 ns of uncertainty; this was done to assure that the acoustic wave was in the same position during each laser pulse. The light beams used for both writing and reading were incident along the x-axis, perpendicular to the z-direction (optical c-axis) of acoustic propagation.

The reading phase, illustrated at the right of Fig. 1, was accomplished with a 10-mW cw He-Ne laser. Since the index-of-refraction pattern stored by the acousto-photorefractive effect was only a small ac component riding on top of a much larger ($\sim 10^3$) dc index change, the ac signal could not be measured by the means used conventionally for observing index-of-refraction changes. Two methods were used for observing these changes. The presence of an ac index-of-refraction pattern could be verified qualitatively by observing the diffraction of light transmitted through this newly formed diffraction grating. Quantitative information was obtained by propagating a live acoustic wave in the crystal during illumination with the low-intensity cw reading laser. Both the stored

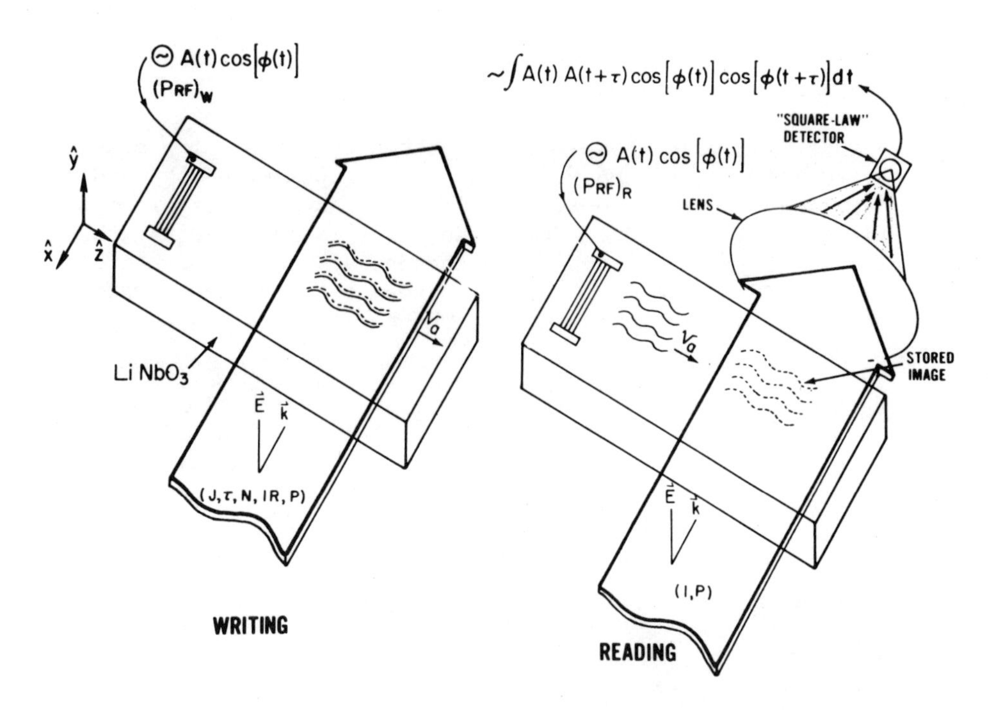

Fig. 1. Acousto-Optic Correlator in Both Writing (left) and Reading (right) Modes.

and live signals modulated the light beam via the Bragg acousto-optic interaction. The doubly modulated light was focused onto a "square-law" detector diode. The detector output is proportional to the correlation of the two acoustic signals.(2)

There are two practical methods of interacting a live signal with a stored signal. First, the live signal introduced into the crystal may be identical to the stored signal. In this case, the output signal from the detector diode will be the self-correlation of the stored signal. The self-correlation output usually has the shape of a high-amplitude, narrow spike, with several much smaller amplitude spikes on either side of the principal spike. This type of interaction is of particular importance in the communications field and for use by the military, where it is of interest to compare incoming live signals with previously stored signals. Most of our experiments have used this approach, and our data and figures in this paper are expressed in terms of the self-correlation output. The other method of interacting with the stored signal is to introduce a live impulse function into the crystal. The resulting correlation output of any signal with an impulse function is the signal itself, so that the live impulse function acts to recall the stored signal. It is this method which would be used to read out stored acoustic images.

Figure 2 shows the experimental setup in the writing mode. The Nd:YAG laser is at the right end of the optical bench, and the focusing lenses are at the left. Figure 3 shows the experimental setup in the reading mode. The He-Ne laser is just behind the second element from the right; the element at the far right, a plane mirror for deflecting the He-Ne beam into the lens train, is in place only in the reading mode. The photodetector diode is at the far left.

In the experiments described herein, the refraction index changes were obtained in plates of Y-cut Z-propagating single-crystal $LiNbO_3$ which had been fabricated into SAW delay lines by

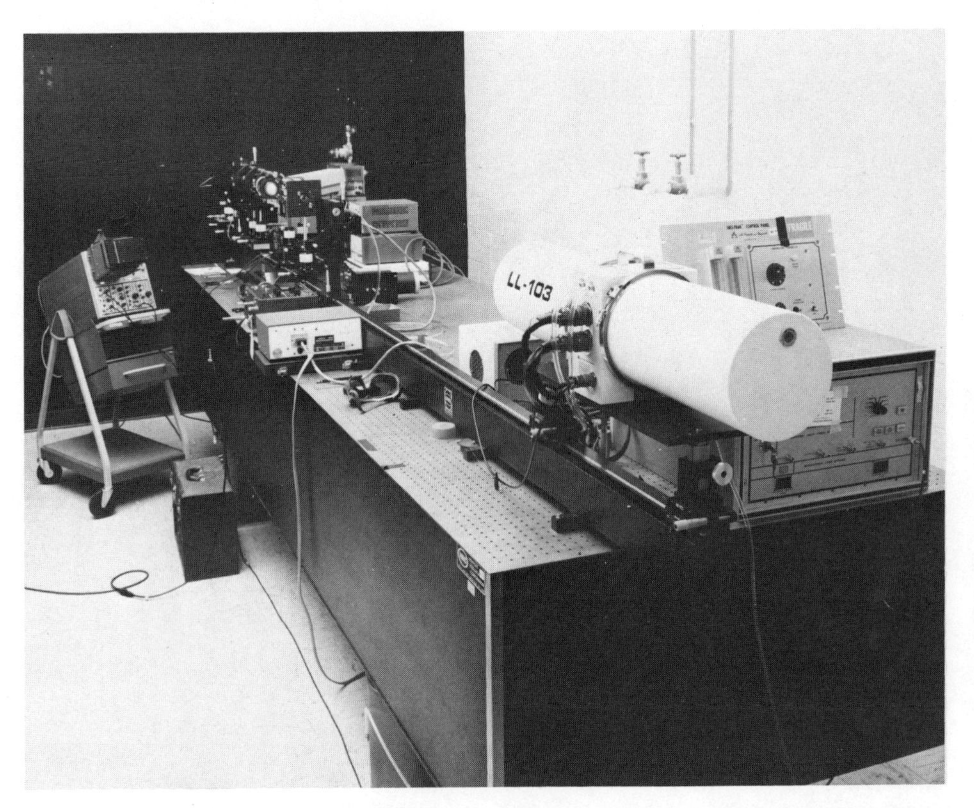

Fig. 2. Acousto-Optic Correlator Operating in the Writing Mode.

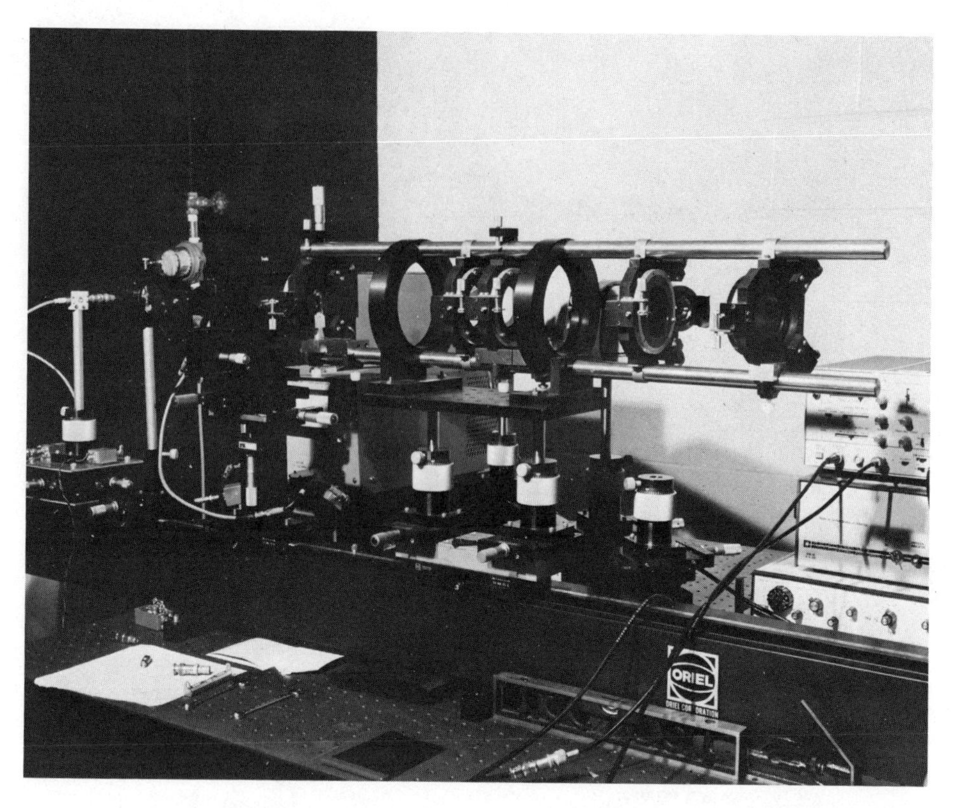

Fig. 3. Acousto-Optic Correlator Operating in the Reading Mode.

the deposition of interdigitated-finger transducers at each end. The spacing between transducers was 7 cm, corresponding to an interaction time of about 20 μs. The acoustic aperture of the transducers (ie; the length of the transducer fingers) was 1.5 cm, and the transducer center frequency was 10 MHz with a 1-MHz bandwidth. Figure 4 shows one of these devices.

The correlation output power ($\sim\delta n^2$)[3] is plotted as a function of several parameters in Figs. 5 and 6. Figure 5 shows curves of the power output plotted against both the number of pulses (N) and the incident laser energy density per pulse (J) of the writing beam. The data as a function of N were obtained by normalizing the correlation output power of each measured point to the same writing laser energy, using the proportionality factor of $J^{2.5}$ obtained empirically from the curve at the right of Fig. 5. The plot in Fig. 6 of output power as a function of the rf input power during storage shows that the curve is almost linear over three orders of magnitude of input power. The apparent saturation at high input power is

Fig. 4. Close-Up of $LiNbO_3$ Crystal in Which Acousto-Optic Interaction Takes Place.

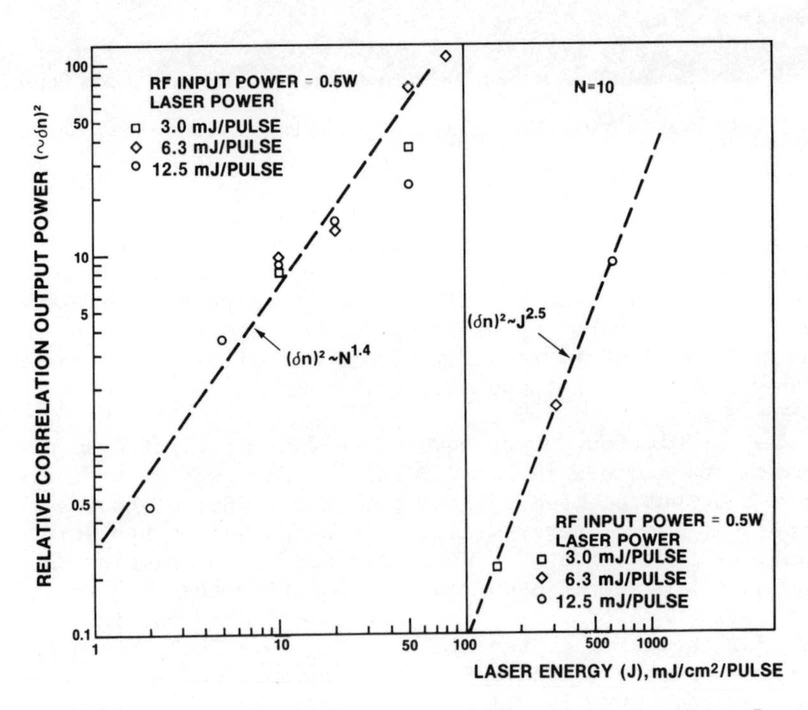

Fig. 5. Relative Correlation Output Power Plotted versus Laser Energy Density.

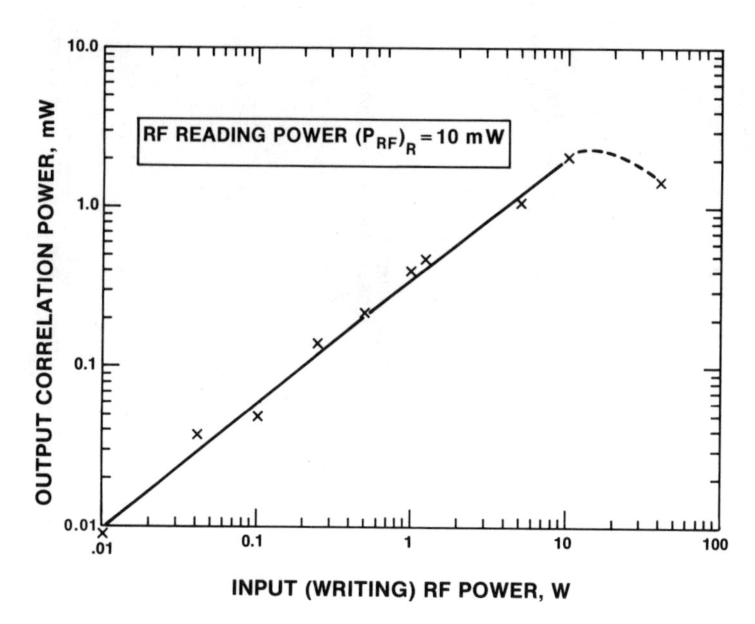

Fig. 6. Correlator Output Power Plotted versus Input rf Writing Power.

attributed to the Hooke's law regime of operation being exceeded, with power being lost to nonlinear modes.

The information in Figs. 5 and 6 has been summarized into an engineering-type nomograph, and is presented in Fig. 7. The axes are the number of laser pulses (N) and laser energy density per pulse (J), with lines of constant insertion loss plotted in the figure. Insertion loss is defined here as the ratio (in dB) of the power output obtained from the "live" convolution of two 10-mW signals in an acousto-optic convolver to the power in the self-correlation signal. It should be observed from Fig. 7 that for a single laser storage pulse of 10^3mJ/cm^2, which is still below the damage threshold for $LiNbO_3$, the insertion loss is ~30 dB.

Figure 8 shows a plot of the decay of δn at room temperature as a function of time after storage for two cases: (1) green

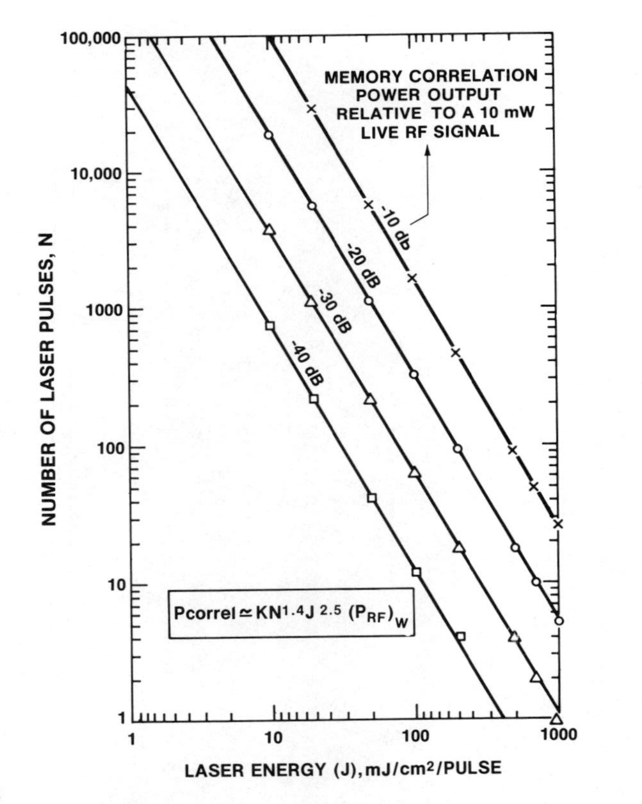

Fig. 7. Curves of Constant Output Power from Memory Correlator as a Function of the Number of Laser Pulses and of Laser Energy Density.

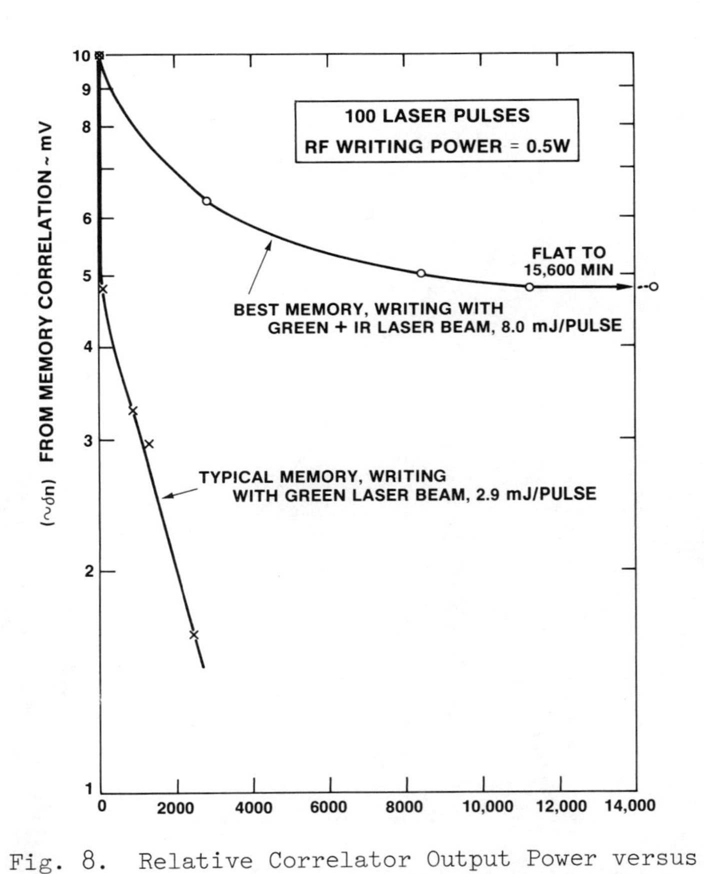

Fig. 8. Relative Correlator Output Power versus Storage Time for Two Cases: Green Writing Beam Alone and Combined Green and IR Writing Beam.

illumination only, and (2) green together with IR. The two cases show different time constants, the time constants for the second case being several orders of magnitude larger than those of the first. We feel that the relatively flat regime obtained with the "green with IR" is in reality the long-term dielectric relaxation time for charge in a relatively pure dielectric.[4]

Figure 9 demonstrates the long-term storage of a surface acoustic signal. The particular signal stored was a seven-bit Barker code. This signal consists of a pulse divided into seven equal segments, where a 180-degree phase change is introduced after the third, fifth, and sixth segments. Barker codes are used in radar applications because of the unique self-correlation output that results. The slow degradation of the stored pattern as a function of time, observed in Fig. 9, manifested itself in a distortion of the sidelobes; however, the main correlation peak remained relatively unchanged.

The induced acousto-photorefractive index changes were erased by (1) exposure to ^{60}Co-gamma radiation, (2) ultraviolet exposure

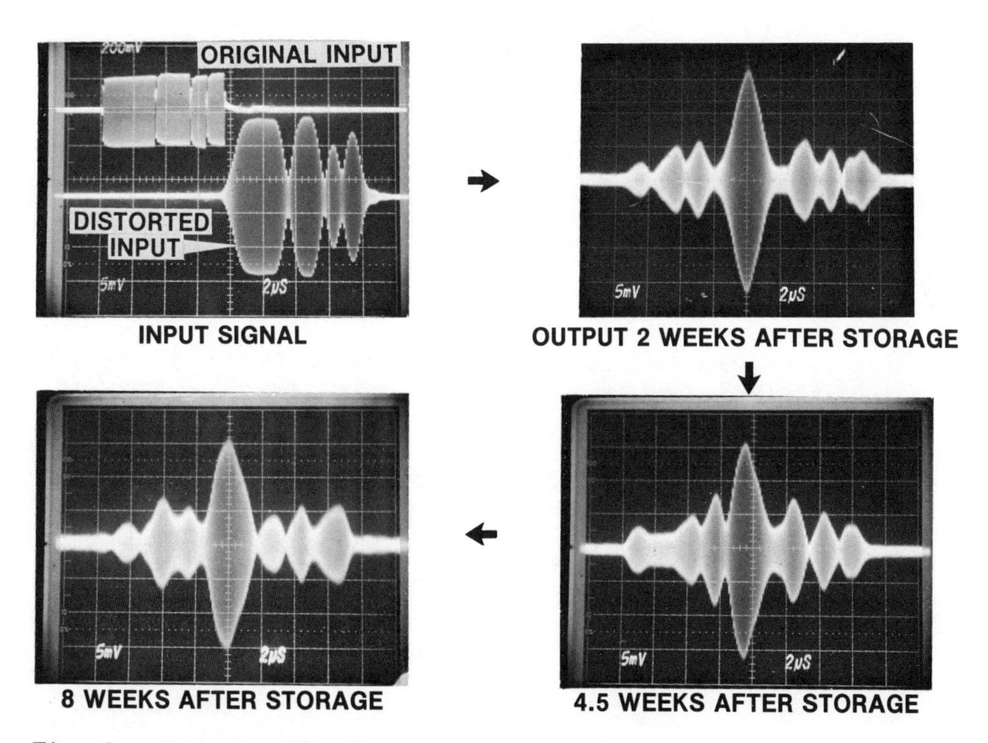

Fig. 9. Changes with Time of the Output Waveforms of Memory Correlator for a Seven-Bit Barker Code.

from a mercury lamp, or (3) a temperature anneal at 250°C. These erasure results are in accord with what has been observed for the photorefractive effect.[5,6,7]

A few storage experiments were also made with 0.005-percent iron-doped $LiNbO_3$ instead of high-purity $LiNbO_3$. These experiments indicated that the power amplitude of the stored SAW signal was five times greater than that for the high-purity $LiNbO_3$. The stored signal in the iron-doped $LiNbO_3$ would be expected to decay more rapidly than in the pure $LiNbO_3$; our limited experimental data appear to confirm this.

B. Discussion of Experimental Results

It has been reported previously[1] that the index change for the nonlinear photorefractive effect varies linearly with N and J^2. According to the data of Fig. 5, the acousto-photorefractive effect (δn) varies as $N^{0.7}$ and as $J^{1.3}$. It is felt nevertheless that this new effect is based on the same physical principles as the previously reported photorefractive effect. An exact model for the nonlinear photorefractive effect has not been derived. It is thought, however,[1] that photo-carriers are generated by a two-photon process, involving either two 530-nm photons or one photon at 530 nm and one at 1060 nm. In any case, the generated photocarriers move from their original position by some transport process, and then become retrapped. The resultant change in the local electric field manifests itself, through the electro-optic effect, as a change in the index of refraction. The same general effect is believed to occur in the case of the acousto-photorefractive effect. In this case, the acoustic wave provides the means for moving the generated photocarriers, as discussed in the next paragraph.

The magnitude of δn_{ac} (normalized with respect to n) produced by the acousto-photorefractive effect can be estimated by comparing it with the index-of-refraction change produced by the elasto-optic effect, and is found to be $\sim 10^{-7}$. This is to be compared with a value of $\delta n_{dc} \approx 10^{-4}$ based on von der Linde's data for the nonlinear photorefractive effect.[1] If the acousto-photorefractive effect were to originate merely from the acoustically induced strain, s, producing a modulation of the nonlinear photorefractive effect, then (since, according Uchida and Niizeki,[8] $s \simeq 10^{-6}$) a value of $\delta n_{ac} \simeq 10^{-10}$ would be expected. This value is almost three orders of magnitude smaller than that observed. This discrepancy leads us to speculate that the electric field accompanying the acoustic wave in the piezoelectric material (rather than density changes alone) may be responsible for the movement of charge in the acousto-photorefractive effect, as pictured in Fig. 10. This speculation is reinforced by a simple derivation of the magnitude of δn_{ac} resulting from a change in the local electric field. If we assume

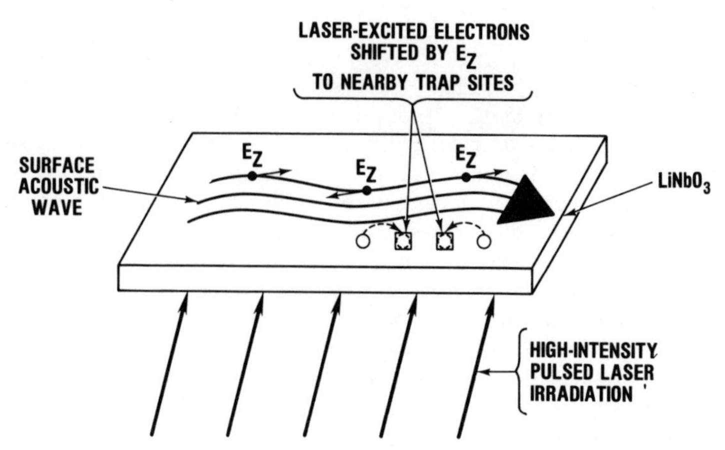

Fig. 10. Proposed Storage Mechanism for Acousto-Photorefractive Effect.

that the motion of the photo-carriers results from the electric field accompanying the acoustic wave, then these charges would eventually build up a pattern which should cancel out this field, i.e., be of the same magnitude but of the opposite polarity. The δn_{ac} resulting from this charge pattern would thus be directly proportional to the electric field accompanying the acoustic wave, where the proportionality constant is the appropriate electro-optic coefficient ($n_o^3 r_{22} = 41 \times 10^{-12}$ m/V, according to Turner(9)). Taking into account the insertion loss of the transducers, and their geometry, the relationship between δn_{ac} and input rf power (P) is simply

$$\delta n_{ac} \approx 2 \times 10^{-6} \sqrt{P} \ . \tag{1}$$

For the 10-mW input of our experiments, this yields a predicted δn_{ac} of 2×10^{-7}, in good agreement with the experimentally measured value of 10^{-7}.

III. POSSIBLE APPLICATION OF ACOUSTO-PHOTOREFRACTIVE EFFECT TO ACOUSTIC IMAGING

As mentioned earlier, an acoustic signal stored in $LiNbO_3$ may be recalled by introducing a live impulse function into the $LiNbO_3$. The interaction of the low-level cw laser beam with both the live and stored signals results in an output proportional to the correlation of the two signals. When one of the input signals is an impulse function, the output signal is proportional to the other signal: i.e., the stored signal. This recall mechanism was verified experimentally in one of our early experiments, the results of which are shown in Fig. 11. The left trace shows a five-bit Barker code that was transmitted through a $LiNbO_3$ delay line via a SAW. The right trace shows the signal that resulted when a 1.5-μs-wide pulse (in fact a poor simulation of an impulse function) was used to read out the stored Barker code. The 1.5-μs width of the impulse function was determined by the transducer geometry available at that time. Signals that more closely approximate an impulse function have been generated subsequently by using an interdigitated transducer having only three fingers. For Fig. 11, the left trace has been time-inverted to facilitate comparison of the left and right traces. The inversion of the output signal resulted from the fact that the impulse function was propagated in the same direction as the original signal, and therefore interacted first with the back edge of the stored signal. The recall of a signal in the correct time sequence may be achieved simply by introducing the impulse function at the other end of the $LiNbO_3$ crystal.

A possible application of the acousto-photorefractive effect, as described earlier, to acoustic imaging for medical diagnostics is outlined in concept in Figs. 12 and 13. To simplify the model, the application as pictured in those figures uses the interaction of a laser beam with a bulk acoustic wave. However, SAW devices could be used in a manner similar to that shown in Figs. 12 and 13

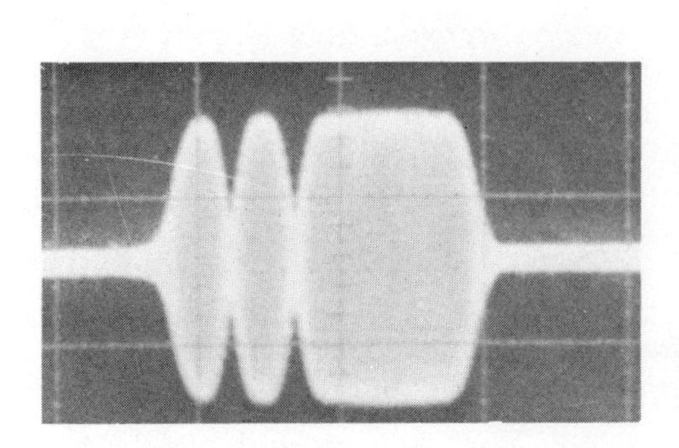

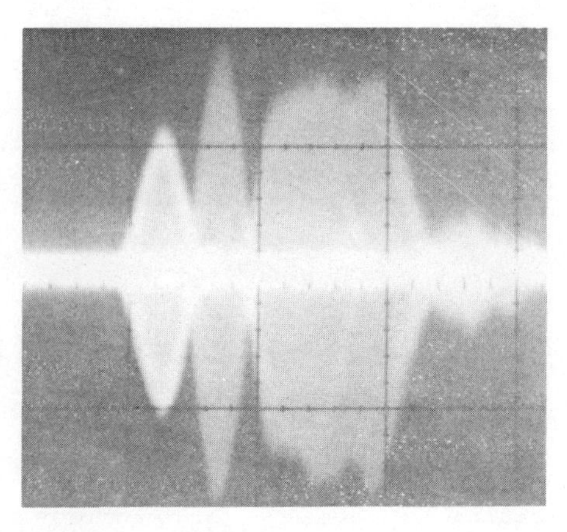

Fig. 11. Oscilloscope Traces of a Five-Bit Barker Code (left), and the Same Signal as Recalled from Storage Using an Impulse Function (right).

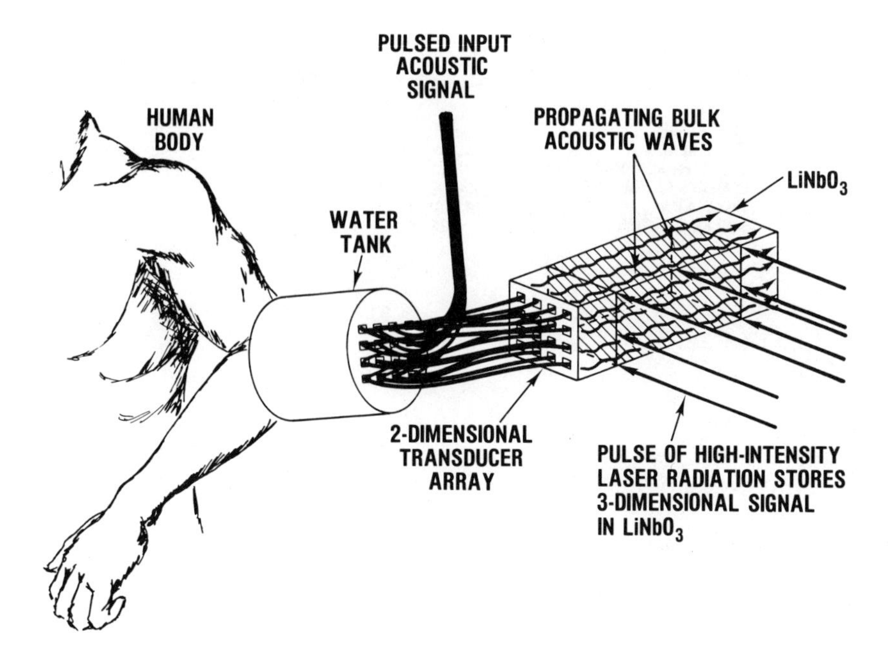

Fig. 12. Storage Process for a Possible Application of Acousto-Photorefractive Effect to Acoustic Imaging.

to store and to recall the acoustic imaging data. In Fig. 12, a pulsed input acoustic signal is introduced into the top of a water tank via an appropriate transducer or array of transducers. The three-dimensional acoustic image reflected by the human body at the bottom of the tank is picked up by a two-dimensional transducer array at the top of the tank and is transferred electrically to an identical array of two-dimensional transducers at the end of a piece of bulk $LiNbO_3$. The signal introduced onto each transducer causes a bulk wave to be propagated along the length of the $LiNbO_3$. During simultaneous transit of these bulk waves, the entire $LiNbO_3$ crystal is irradiated by a high-intensity pulsed beam of green laser light in a direction perpendicular to that of the acoustic wave propagation. In the manner described earlier, this laser pulse causes each propagating acoustic wave to be stored as small local changes of the index of refraction of the $LiNbO_3$. The result is that the entire three-dimensional acoustic image is stored within the $LiNbO_3$.

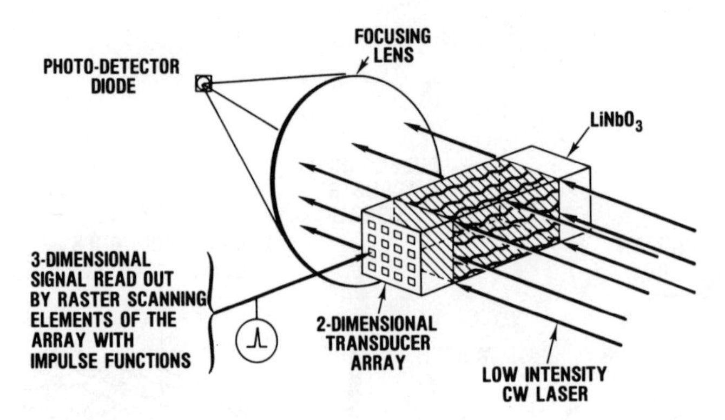

Fig. 13. Read Out Process for Application to Acoustic Imaging Shown in Fig. 12.

A means of reading out the stored three-dimensional signal is shown in Fig. 13. A low-intensity cw He-Ne laser beam is introduced onto one side of the $LiNbO_3$ crystal, so as to flood the entire crystal. The laser light transmitted through the $LiNbO_3$ is then focused by means of a simple converging optical lens onto a small photo-detector diode. A particular stored bulk acoustic wave may be recalled (in a time-reversed manner) by introducing an impulse function onto the same transducer element that originally propagated that bulk wave. By raster scanning all the elements of the transducer array using impulse functions, the entire three-dimensional signal may be read out.

SUMMARY

An acousto-photorefractive effect has been found in $LiNbO_3$ which is qualitatively similar to the nonlinear photorefractive effect reported previously. When combined green and infrared light are used to store the signals, storage times as long as two months have been obtained. The stored signal may be read out by causing a low-intensity cw laser to interact with both the stored signal and a live impulse function introduced onto the $LiNbO_3$. A possible application of this storage phenomenon to acoustic imaging for diagnostic purposes is suggested.

REFERENCES

1) D. von der Linde et al, Appl. Phys. Lett. 25, 155 (1974).
2) N. J. Berg and B. J. Udelson, 1976 Ultrasonics Symp. Proc., 183 (1976).

3) R. Adler, IEEE Spectrum 4, 42 (May 1967).
4) J. P. Huignard et al, "Optical Systems for Photosensitive Materials for Information Storage" (North-Holland, Amsterdam, 1976), Chapt. 16.
5) J. J. Amodei et al, Appl. Optica 11, 390 (1972).
6) Y. Ohmori et al, Japan J. Appl. Phys. 16, 181 (1977).
7) D. L. Staebler et al, Appl. Phys. Lett. 26, 182 (1975).
8) N. Uchida and N. Niizeki, Proc. IEEE 61, 1073 (1973).
9) E. H. Turner, Appl. Phys. Lett. 8, 303 (1966).
10) A. Bers and J. H. Cafarella, Appl. Phys. Lett. 25, 133 (1974).
11) C. E. Cook and M. Bernfeld, "Radar Signals" (Academic Press, NY, 1967), Chapt. 8.

A POLYVINYLIDENE FLUORIDE BOW-TIE IMAGING ELEMENT

David H. Dameron and John G. Linvill

Stanford Electronics Laboratories
Stanford University
Stanford, California 94305

ABSTRACT

Polyvinylidene fluoride is a piezoelectric polymer film useful for imaging elements. It has many ultrasonic applications and is well suited to array technology as a replacement to ceramic piezoelectric materials. In a bow-tie configuration, this film effectively couples ultrasonic energy to and from air, producing a radiation pattern matched to the imaging requirements of array systems. We have designed and built a bow-tie array useful at frequencies up to 100 kHz. This frequency limit is imposed by element dimensions as well as ultrasonic attenuation in air. Experimental measurements confirm our model predictions. Typical applications of this device are intrusion detectors, tactile imagers, phased array sector scanners, and any other requirement where efficient coupling to air is a major objective.

INTRODUCTION

Polyvinylidene fluoride (PVF_2) is a piezoelectric polymer useful in mnay applications where conventional piezoelectric materials such as PZT are inappropriate. PVF_2 is normally manufactured in thin flexible sheets from 5 to 100 microns in thickness. The sheets have a stiffness and density much less than ceramic

---

The authors gratefully acknowledge the support of the National Science Foundation under grant ENG75 22329.

transducers [1]. In addition, it is well suited for non-planar structures such as a bow-tie element [2].

A bow-tie imaging element is fabricated from two sheets of PVF_2 bonded together at their centers to form an 'X' or bow-tie shaped structure. This structure, in addition to the operation of the PVF_2 in the transverse or 31 mode, permits the bow-tie to operate into low impedance acoustic loads, such as air. The element shape also permits a certain control over the element's radiation pattern. This element can easily be incorporated into arrays for acoustic imaging of objects in air.

ELECTROMECHANICAL MODEL

Fig. 1 is an illustration of a bow-tie element. Each arm of PVF_2 is length H, thickness τ, and width W. The 4 arms formed from two sheets are mechanically fixed at their ends spaced a distance 2D, where $(D/H) < 1$, and bonded together at the common apex. The polarization sense of the 4 arms is shown by the arrows. In operation, the 4 arms are under net tension at all times. Elec-

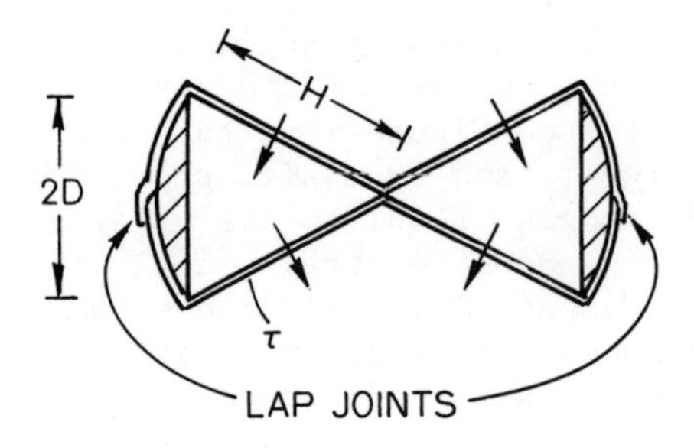

Sheets cemented together at the vertex.

Electrodes are etched off over the frame.

Frame holds sheets in tension.

Inner surfaces are energized by voltage V.

Outer surfaces are grounded.

Polarization senses are shown by arrows.

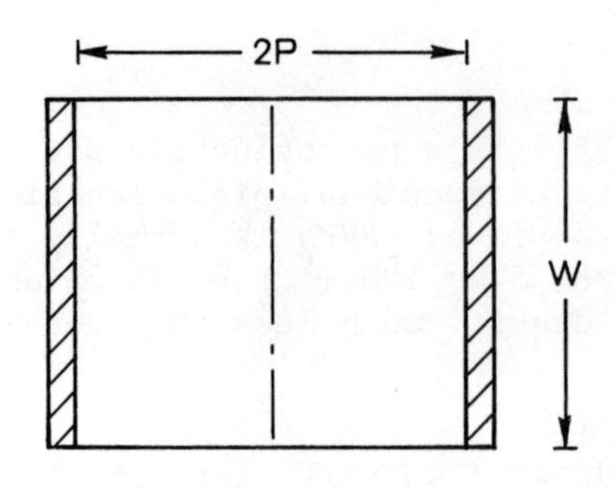

Fig. 1. THE BOW-TIE TRANSDUCER CONSTRUCTION

trodes are deposited on both sides of the 4 arms, the outer electrodes are grounded and the inner electrodes are connected to a voltage source or receiver, as shown in Fig. 2. In the transverse mode of operation, when a voltage is placed across the sheets, the two top or bottom arms tend to shorten while the other two tend to lengthen due to the polarizations of the sheets. The net result is that the apex is moved sideways with the impressed voltage.

Piezoelectric equations can be written for PVF_2 as [3]

$$D = \varepsilon^S \mathcal{E} + e_{31} S \tag{1}$$

$$T = - e_{31} \mathcal{E} + c^E S \tag{2}$$

where D is electric displacement, $\mathcal{E}$ is electric field, T is mechanical tension, and S is mechanical strain. ε^S is the permittivity under constant strain, c^E is the stiffness under constant electric field strenght, and e_{31} is the transverse charge density-strain coefficient for PVF_2. Using similar triangles to balance the apex force in Fig. 1, one gets

$$f_v = 4 \left(\frac{D}{H}\right) f_i \tag{3}$$

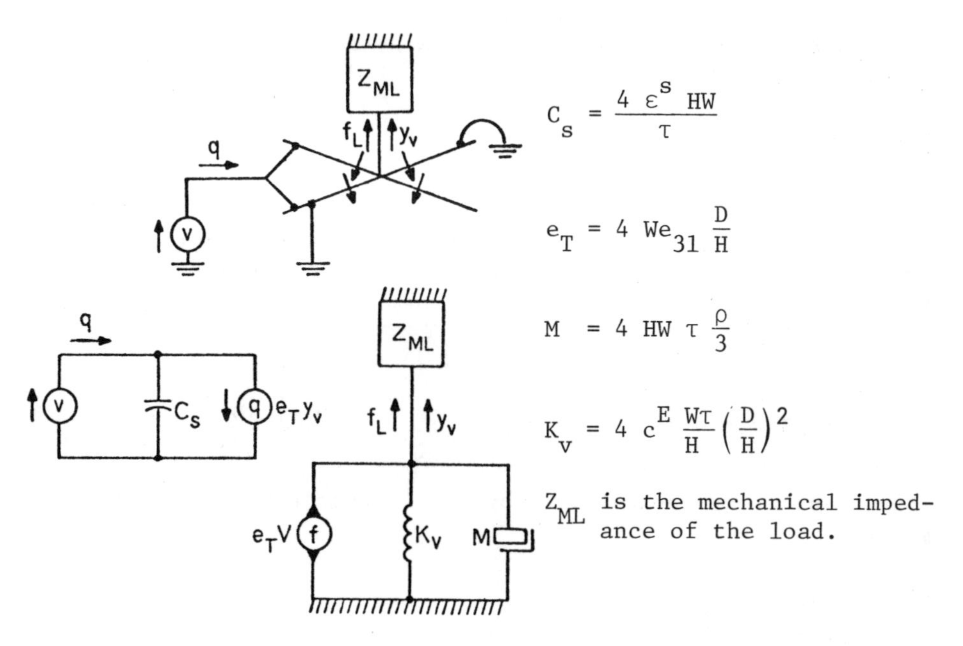

$$C_s = \frac{4\ \varepsilon^S\ HW}{\tau}$$

$$e_T = 4\ We_{31} \frac{D}{H}$$

$$M = 4\ HW\ \tau\ \frac{\rho}{3}$$

$$K_v = 4\ c^E \frac{W\tau}{H} \left(\frac{D}{H}\right)^2$$

Z_{ML} is the mechanical impedance of the load.

Fig. 2. LUMPED MODEL OF BOW-TIE TRANSDUCER AND LOAD

where $\pm f_i$ is the incremental tension in each of the 4 arms. To simplify the steady state analysis, the bow-tie will be considered as a lumped model with a lumped load. The sheets in the arms are assumed planar so that the transverse motion is a linear function of distance toward the apex. Using these results, and summing forces, a lumped model is derived in a straightforward manner and is shown in Fig. 2. For linear motion, the effective mass of each arm is 1/3 of its total mass, ρ is the mass density of the PVF_2. The internal (viscous) damping of the PVF_2 is neglected with respect to the damping component (the real part of the characteristic impedance) of the load.

The resonant frequency of the bow-tie element is ω_o, given in terms of the parameters as follows.

$$\omega_o = \sqrt{\frac{K_v}{M}} = \sqrt{\frac{c^E}{\rho}} \frac{\sqrt{3}}{H} \left(\frac{D}{H}\right) \tag{4}$$

With air as a load on both bow-tie faces, we use the characteristic impedance of air,

$$Z_A = \sqrt{\rho c} = (415 \ \mathrm{Nsm}^{-3}) \tag{5}$$

where ρ is the density of air and c is its effective stiffness. Since the mean displacement across the bow-tie sheets is half the maximum displacement at the apex, for linear motion one can write the mechanical load impedance as half that with a constant displacement.

$$Z_{ML} = 2\left(\frac{1}{2}\right) 2HWZ_A = 2\ HWZ_A \tag{6}$$

Using this lumped model, one finds the ratio of the (complex) velocity at the apex to the applied voltage for a radian frequency ω to be:

$$\frac{\mathcal{V}_v}{V} = \frac{\left[4\ e_{31} W(D/H)\right] j\omega}{\left(\frac{4\ HW\tau\rho}{3}\right)(j\omega)^2 + (2\ HWZ_A)(j\omega) + \frac{4\ c^E\ W\tau}{H}\left(\frac{D}{H}\right)^2} \tag{7}$$

At the resonant frequency given by Eq. 4, this expression reduces to

$$\frac{2\ e_{31}(D/H)}{HZ_A} \tag{8}$$

The 3 dB bandwidth of the bow-tie is, from Eq. 7,

$$\Delta\omega_{3dB} = \frac{3Z_A}{\tau\rho} \tag{9}$$

It is seen that it is inversely proportional to the sheet thickness τ. This is due to the fact that the radiated power is proportional to area, while stored power is proportional to volume = τ area.

MATCHING TO LOW IMPEDANCE LOADS

The characteristic acoustic impedance for PVF_2 in the 31 mode is

$$Z_{PVF_2} = \sqrt{c^E \rho} = 2.55 \times 10^6 \ \text{kgm}^{-2}\text{s}^{-1}$$

which is much larger than that for air, 4.15×10^2 kgm^{-2}s^{-1}. For this reason, a structure such as the bow-tie is needed to transform the PVF_2 impedance to the air.[4] As before, the displacement of the 4 arms is assumed a linear function of distance toward the apex, so that the bow-tie drives an equivalent amount of air with a mean displacement one half that of the maximum at the vertex. A transformation of impedances occurs because each arm of the bow-tie, with cross sectional area of $W\tau$, drives a much larger column of air of $PW = \sqrt{H^2 - D^2}\, W$, as in Fig. 1. Due to the mechanical advantage of the triangular configuration of the bow-tie, the velocity at the apex into the load is the velocity along each of the PVF_2 arms multiplied by the ratio H/D. The force is transformed in a similar manner by the factor 4(D/H). These two factors mean that the PVF_2 impedance is transformed due to mechanical advantage by the additional factor $4(D/H)^2$. Using these factors to match the PVF_2 impedance to that of an air load yields:

$$Z_{load} = Z_{PVF_2}\, 4\left(\frac{D}{H}\right)^2 \frac{W\tau}{W\sqrt{H^2 - D^2}}\, \frac{1}{2} = 2\left(\frac{D}{H}\right)^2 \frac{\tau}{\sqrt{H^2 - D^2}}\, Z_{PVF_2} \tag{10}$$

Using the values for air and a 30 μ thick PVF_2 thickness gives a ratio (H/D) = 8.8, an easily obtainable value.

In order to match the electrical side of the bow-tie transducer, damping internal to the PVF_2 must be considered. Changing the ratio (H/D) changes the transformed load impedance seen by the electrical side of the bow-tie so that a given power transfer or efficiency is obtained with a given electrical source impedance.

RADIATION PATTERNS

The radiation pattern as measured in Fig. 3 for one side of a bow-tie element is also a function of θ or the ratio (D/H) and the length, H of a side in terms of wavelengths in the load media (air), as well as the linear weighting of the velocity along a side. Since, for a fixed θ or ratio (D/H), Eq. 4 determines H as a function of ω_o at resonance and the speed of sound in air (331 m/s) determines the wavelength, this angle or ratio also determines the length H in wavelengths in air. A smaller ratio (D/H) means a smaller H and a wider effective radiation pattern. As the ratio (D/H) goes to zero, assuming again a linear displacement along the arms, the transducer degenerates to a planar surface with a triangular amplitude function. This function has a far field amplitude of $\text{sinc}^2(\frac{H}{\lambda} \sin \phi)$. Fig. 4 shows the simulated radiation pattern of a bow-tie into air with $\theta = 30°$, or $(D/H) = \frac{1}{2}$. The two radiating arms have produced an on-axis response less than the maximum. This behavior is characteristic of non-planar radiating structures.

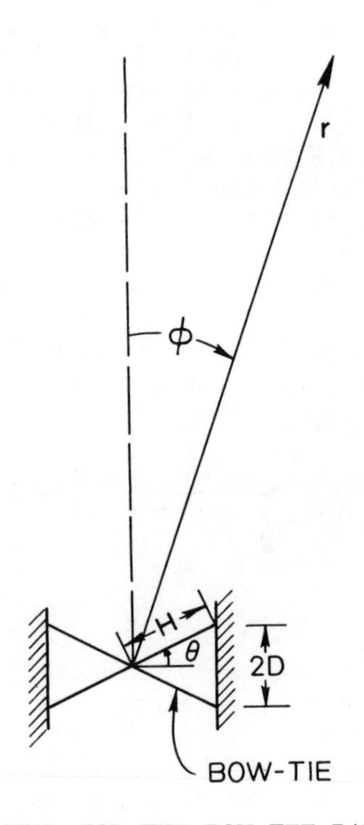

Fig. 3. GEOMETRY FOR THE BOW-TIE RADIATION PATTERN

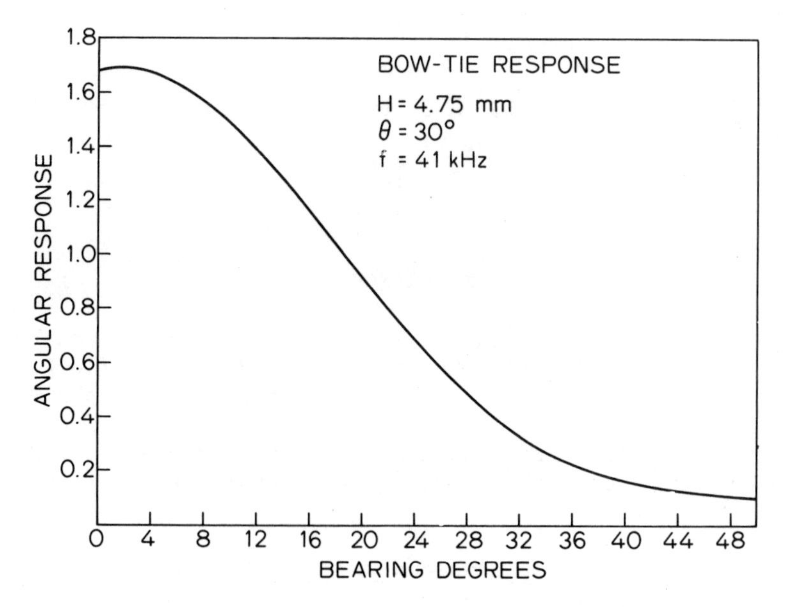

Fig. 4. RADIATION PATTERN FOR A BOW-TIE WITH (D/H) = 1

CONCLUSIONS

The bow-tie element can be an effective transducer for coupling into low impedance loads. By changing the ratio (D/H), different sized elements are obtained for a given frequency, producing different radiation patterns. Changing this ratio also changes the value of load impedances seen by the PVF_2. A large range of load impedances can be matched in this fashion.

APPENDIX A

Parameter Values for Commercial Kureha PVF_2 Film

$$c^E = 3.6 \times 10^9 \ N/m^2$$

$$\varepsilon^S = 12 \cdot \varepsilon_o = 1.04 \times 10^{-10} \ F/m$$

$$d_{31} = 27 \times 10^{-12} \ c/N \text{ or } m/V$$

$$d_{31} c^E = e_{31} = 0.0972$$

$$\frac{d_{31}^2 c^E}{\varepsilon^T} = k_{31}^2 = 0.0249$$

$$\rho = 1.8 \times 10^3 \ kg/m^3$$

$$\sqrt{c^E/\rho} = 1.414 \times 10^3 \ m/s \quad \text{transverse acoustic velocity}$$

$$Z_{PVF_2} = \sqrt{c^E \rho} = 2.55 \times 10^6 \ kgm^{-2}s^{-1}$$

REFERENCES

1. H. Ohigashi, "Electromechanical Properties of Polarized Polyvinylidene Fluoride Films Studied by the Piezoelectric Resonance Method," J. Appl. Phys, 47, No.3, March 1978, pp.949-55.

2. John G. Linvill, "PVF_2 - Models, Measurements, Device Ideas," Technical Report 4834-3, Stanford Electronics Laboratories, March 1978.

3. B. A. Auld, Acoustic Fields and Waves in Solids, Vol I, Wiley, New York, 1973, pp.271-5.

4. M. Tamura, T. Yamaguchi, T. Oyaba, and T. Yoshimi, "Electroacoustic Transducers with Piezoelectric High Polymer Films," Journal of the Audio Engineering Society," Vol 23, No. 1, Jan/Feb 1975, pp. 21-26.

A MATRIX TECHNIQUE FOR ANALYZING THE PERFORMANCE OF MULTILAYERED FRONT MATCHED AND BACKED PIEZOELECTRIC CERAMIC TRANSDUCERS

George K. Lewis
Group Research
Searle Diagnostics Inc.
2000 Nuclear Drive
Des Plaines, IL 60018

INTRODUCTION

Transmission line analogies have been utilized by many authors to predict the frequency response of piezoelectric transducers.[2,4,5,6] Although these analogies are extremely useful, the model grows in complexity as the number of matching sections increase. Also, when lumped transmission line analogies are utilized, care must be exercised in calculating transfer functions since the element values in the model tend to blow up at certain frequencies. Thus, Martin et.al.[3] have found it necessary to use a modified numerical form of L'Hospital's rule when calculating the performance of transducers at these frequencies. Furthermore, most authors address the problem of transducer analysis in the frequency domain, and little reference has been made to the actual transient performance.[7] To bridge this gap, a matrix formularization has been developed which simplifies the frequency analysis of multiple front matched and backed piezoelectric ceramics and avoids the problem of indeterminancy at certain frequencies. By utilization of complex matrix inversion, the frequency response of the transducer is found and by using a fast Fourier transform routine the actual transient response is obtained. To show the utility of this technique it has been employed in analyzing the pressure response of a piezoelectric ceramic having ten matching layers with a stepped approximation in an exponential impedance profile.

DEVELOPMENT OF MATRIX MODEL

The development of the present matrix equation is based upon

Redwood's[7] characterization of the force and particle displacement in a piezoelectric transducer in terms of their Laplace transforms. This analysis can be extended to model the effect of multiple matching and backing by analyzing the force and particle displacement in each layer and setting up the appropriate boundary conditions to solve for the unknown quantities.

With reference to figure 1, a transducer with a double layer front matching plate and backing is illustrated. This transducer is driven by a voltage source in series with an equivalent impedance. Although this is a relatively simple model, it will be instructive in illustrating the development of the matrix representation. Furthermore, once the matrix is developed, it becomes a relatively simple task to analyze multiple acoustic layers as might occur with finite bonding thicknesses or as more matching sections are added to the transducer, and it becomes relatively easy to change electrical driving conditions by using Thevenin's Theorem.

With respect to figure (2), the Laplace transform with respect to time which relates the particle displacement and force existing in a piezoelectric ceramic to the electrical conditions has been derived by Redwood to be:

$$\overline{\xi}_c = A_c \exp\left(\frac{-px}{v_c}\right) + B_c \exp\left(\frac{px}{v_c}\right) \quad \text{and} \tag{1}$$

$$\overline{F}_c = pZ_c \left\{ -A_c \exp\left(\frac{-px}{v_c}\right) + B_c \exp\left(\frac{px}{v_c}\right)\right\} - h\overline{Q} \tag{2}$$

where $\overline{\xi}_c$ = Laplace transform of particle displacement

$\overline{F}_c$ = Laplace transform of particle force

$\overline{Q}$ = Laplace transform of charge on ceramic

Z_c = Characteristic mechanical impedance of the ceramic (Area x density x velocity of sound)

h = A piezoelectric constant relating electric field to mechanical strain

$A_c \exp\left(\frac{-px}{v_c}\right)$ represents a particle displacement wave traveling in the positive x-direction with velocity v_c and

$B_c \exp\left(\frac{px}{v_c}\right)$ represents a particle displacement wave traveling in the negative x-direction

p = Laplace Variable

Taking into account the electrical terminating conditions, one can write an equation relating the variables: voltage, charge, and particle displacement on the two faces of the ceramic, namely,

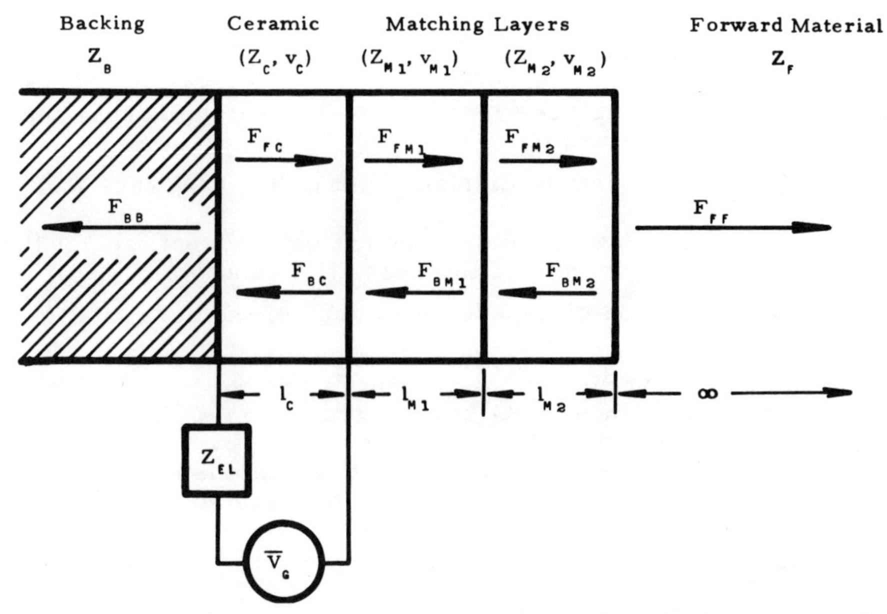

Figure 1: A double layer front matched and backed ultrasound transducer having an arbitrary input voltage and source impedance.

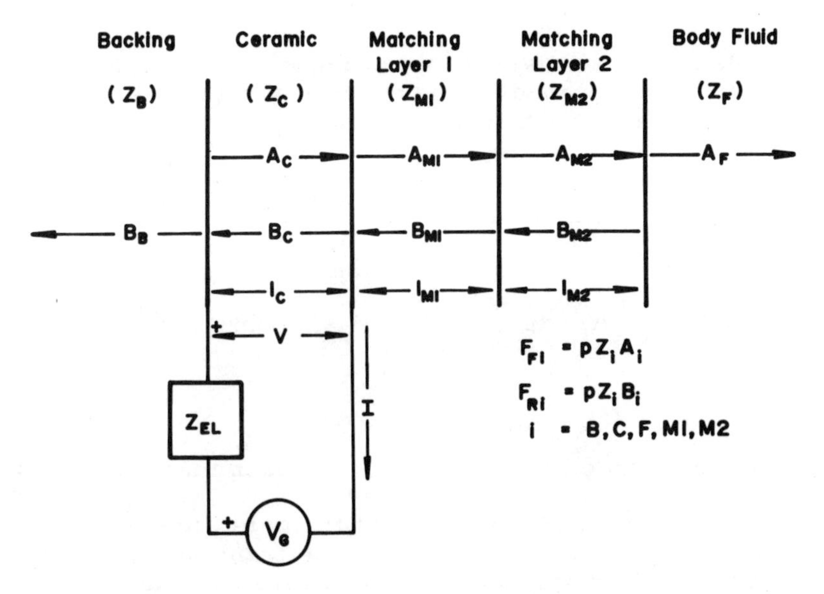

Figure 2: Relation between the force and particle displacement in the various transducer layers to the electrical terminating voltage and source impedance.

$$\overline{V} = -h\left\{\left(\overline{\xi}_c\right)_{lc} - \left(\overline{\xi}_c\right)_{lc}\right\} + \frac{\overline{Q}}{C_o} \tag{3}$$

where $\overline{V}$ = Laplace transform of the voltage across the ceramic,

$\left(\overline{\xi}_c\right)_{lc}$ = Particle displacement on the back face and

$\left(\overline{\xi}_c\right)_o$ = Particle displacement on the front face of the ceramic of thickness (lc), and

C_o = the clamped capacitance of the ceramic.

Furthermore, the transformed current is related to the charge across the ceramic by $\overline{I} = p\overline{Q}$ and the transformed voltage across the caramic is related to the generator voltage by $\overline{V} = \overline{V}_G - \overline{I}\,\overline{Z}_{el}$. By utilizing these relations together with equations (2) and (3), the force existing in the ceramic is related back to the electrical driving condition by

$$\begin{aligned} F_c = {} & pZ_c\left\{-A_c \exp\left(\frac{-px_c}{v_c}\right) + B_c \exp\left(\frac{px_c}{v_c}\right)\right\} \\ & - \frac{hC_o}{pZ_{el}C_o + 1}\quad V_G + h\left\{\left(\overline{\xi}_c\right)_{lc} - \left(\overline{\xi}_c\right)_o\right\} \end{aligned} \tag{4}$$

For plane acoustic waves travelling in homogoneous front matching and backing media, the equations which describe the particle displacement and force are given by:

$$\overline{\xi}_i = A_i \exp\left(\frac{-px_i}{v_i}\right) + B_i \exp\left(\frac{px_i}{v_i}\right) \tag{5}$$

and $$\overline{\overline{F}}_i = pZ_i\left\{-A_i \exp\left(\frac{-px_i}{v_i}\right) + B_i \exp\left(\frac{px_i}{v_i}\right)\right\} \tag{6}$$

for i = M1, M2, F and B

and x_i = a defined coordinate axis for each material.

If the piezoelectric properties of the ceramic are known and the mechanical properties of the acoustic transmission media are also known, then the only unknown quantities are the values for A_i and B_i in each of the acoustic media. Given the driving voltage source and electrical impedance, these unknown quantities can be found from the boundary conditions such that the particle displacement and force across a boundary must be continuous, thus

$$\bar{\xi}_C\Big|_{x_C = 0} = \bar{\xi}_B\Big|_{x_B = 0} \qquad \bar{F}_C\Big|_{x_C = 0} = \bar{F}_B\Big|_{x_B = 0} \tag{7}$$

$$\bar{\xi}_C\Big|_{x_C = lc} = \bar{\xi}_{M1}\Big|_{x_{M1} = 0} \qquad \bar{F}_C\Big|_{x_C = lc} = \bar{F}_{M1}\Big|_{x_{M1} = 0} \tag{8}$$

$$\bar{\xi}_{M1}\Big|_{x_{M1} = l_{M1}} = \bar{\xi}_{M2}\Big|_{x_{M2} = 0} \qquad \bar{F}_{M1}\Big|_{x_{M1} = l_{M1}} = \bar{F}_{M2}\Big|_{x_{M2} = 0} \tag{9}$$

$$\bar{\xi}_{M2}\Big|_{x_{M2} = l_{M2}} = \bar{\xi}_F\Big|_{x_F = 0} \qquad \bar{F}_{M2}\Big|_{x_{M2} = l_{M2}} = \bar{F}_F\Big|_{x_F = 0} \tag{10}$$

From these boundary conditions and equations (1) through (5) and noting that only a forward wave exists in the fluid and only a backward wave in the backing there results

$$B_B - A_C - B_C = 0 \tag{11}$$

$$pZ_B B_B + \left\{pZ_C + hh_G C_O \left[R_C^{-1} - 1\right]\right\} A_C + \left\{-pZ_C + hh_G C_O \left[R_C - 1\right]\right\} B_C = -h_G C_O V_G \tag{12}$$

$$A_C R_C^{-1} + B_C R_C = A_{M1} + B_{M1} \tag{13}$$

$$\left\{-pZ_C R_C^{-1} + hh_G C_O \left[1 - R_C^{-1}\right]\right\} A_C + \left\{pZ_C R_C + hh_G C_O \left[1 - R_C\right]\right\} B_C + pZ_{M1} A_{M1} - pZ_{M1} B_{M1} = h_G C_O V_G \tag{14}$$

$$A_{M1} R_{M1}^{-1} + B_{M1} R_{M1} = A_{M2} + B_{M2} \tag{15}$$

$$-pZ_{M1} A_{M1} R_{M1}^{-1} + pZ_{M1} B_{M1} R_{M1} = -pZ_{M2} A_{M2} + pZ_{M2} B_{M2} \tag{16}$$

$$A_{M2} R_{M2}^{-1} + B_{M2} R_{M2} = A_F \tag{17}$$

$$-pZ_{M2} R_{M2}^{-1} + pZ_{M2} B_{M2} R_{M2} = -pZ_F A_F \tag{18}$$

where $h_G = h/(1 + pC_O Z_{EL})$

and $R_i = e^{pt_i}$ where $t_i = l_i / v_i$ is a term which relates to the propagation delay through the ith material.

After some simplification, these eight equations can be put into a matrix form which relates the forward and backward acoustic pressure wave in each material to the input voltage source (Equation 19).

To find the impulse frequency response for the transmitted pressure wave, the Laplace variable (p) is changed to the complex frequency variable (jw) and the matrix is solved by a complex matrix inversion routine for $\overline{F}_{FF}(jw)$ with $\overline{V}_G(jw)$ set equal to unity. By taking the inverse Fourier transform of this function, the time domain pressure response is found.

It should be noted that most of the element values in the matrix are zero and aside from the four terms which relate to the electromechanical conversion factor (the terms containing $\overline{G}$ in the matrix equation) the magnitudes of the remaining terms are constant for all frequencies, being either unity or ratios of adjacent impedance levels. Because of the simplicity of the matrix, it can be easily extended by iterative techniques to handle any number of front matching layers. Each additional layer, however, increases the side of the matrix by a factor of two, to account for the forward and backward travelling pressure wave in each layer.

RESULTS

To determine the optimum performance to be expected from a piezoelectric ceramic transmitting into body fluid it seemed reasonable that one should try to duplicate as nearly as possible the pulse response that is obtained for a ceramic transmitting into its own characteristic impedance. For a ceramic transmitting into a load equal to its characteristic impedance, one would expect nearly maximum energy transfer with the least amount of reverberation. With this in mind and using the necessary parameters for a 2.25 MHz, 2.54 cm diameter PZT-7A piezoelectric ceramic, the transmitted pressure frequency response was calculated from the matrix formularization by complex matrix inversion. This was done by setting all the matching layers in equation (19) equal to the characterisitc impedance of the ceramic. Once the frequency response was obtained, a fast Fourier transform routine was performed on this data to derive the time domain pressure signal.

Figure (3a,b) illustrates the impulse pressure frequency response function obtained for an air backed ceramic driven by a

Equation 19 Matrix equation relating the forward and backward travelling acoustic pressure waves to the input driving voltage.

$$\begin{vmatrix} 1 & -1 & -1 & 0 & 0 & 0 & 0 & 0 \\ Z_B/Z_C & 1 + G(R_C^{-1} - 1) & -1 + G(R_C - 1) & 0 & 0 & 0 & 0 & 0 \\ 0 & R_C^{-1} & R_C & -Z_C/Z_{M1} & -Z_C/Z_{M1} & 0 & 0 & 0 \\ 0 & R_C^{-1} + G(R_C^{-1} - 1) & -R_C + G(R_C - 1) & -1 & 1 & 0 & 0 & 0 \\ 0 & 0 & 0 & R_{M1}^{-1} & R_{M1} & -Z_{M1}/Z_{M2} & -Z_{M1}/Z_{M2} & 0 \\ 0 & 0 & 0 & R_{M1}^{-1} & -R_{M1} & -1 & 1 & 0 \\ 0 & 0 & 0 & 0 & 0 & R_{M2}^{-1} & R_{M2} & -Z_{M2}/Z_F \\ 0 & 0 & 0 & 0 & 0 & R_{M2}^{-1} & -R_{M2} & -1 \end{vmatrix} \times \begin{vmatrix} \overline{F}_{BB} \\ -\overline{F}_{FC} \\ \overline{F}_{BC} \\ -\overline{F}_{FM1} \\ \overline{F}_{BM1} \\ -\overline{F}_{FM2} \\ \overline{F}_{BM2} \\ -\overline{F}_{FF} \end{vmatrix} = \begin{vmatrix} 0 \\ -1 \\ 0 \\ -1 \\ 0 \\ 0 \\ 0 \\ 0 \end{vmatrix} \times h_G C_O \overline{V}_G$$

where $h_G = h/(1 + pC_O Z_{EL})$

$G = hh_G C_O / pZ_C$

F_{Bi} = Fourier transform of backward travelling acoustic pressure wave in i^{th} material = $pZ_i B_i$

F_{Fi} = Fourier transform of forward travelling acoustic pressure wave in i^{th} material = $pZ_i A_i$

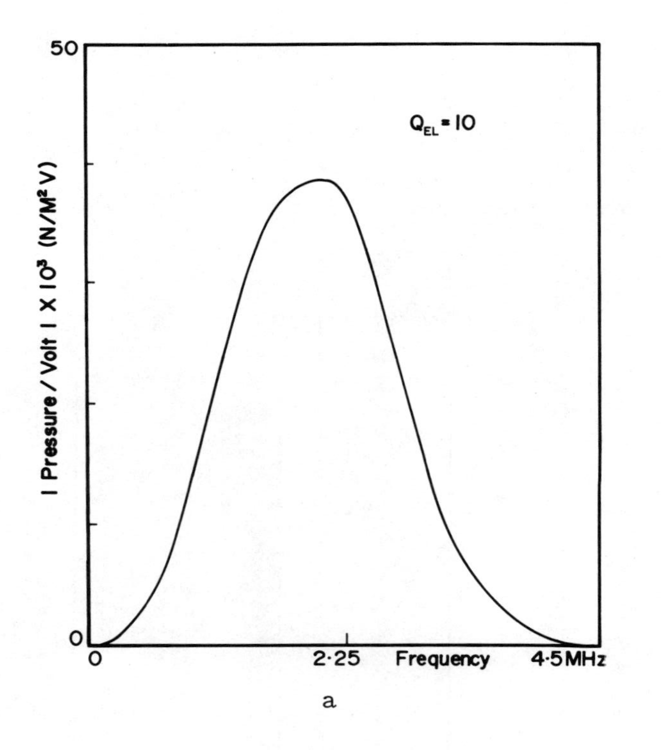

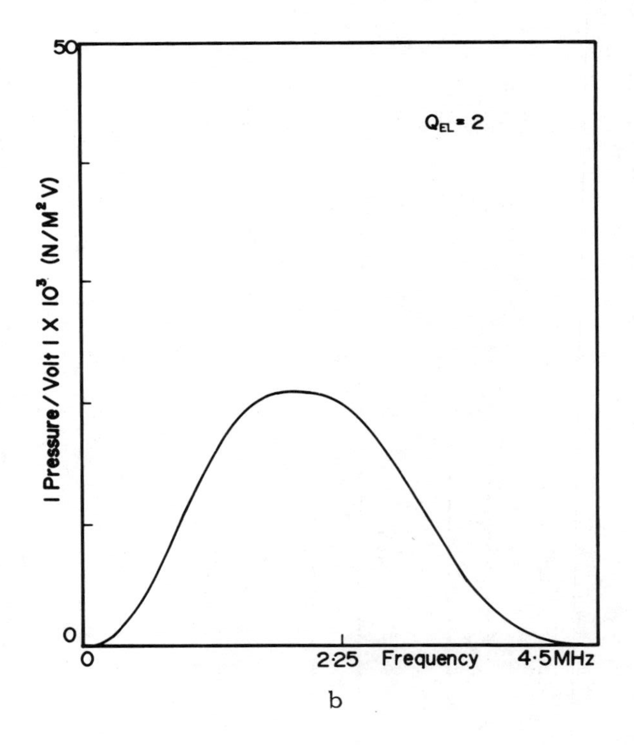

Figure 3: Output pressure frequency response curves for an air backed piezoelectric ceramic transmitting into a load having an impedance equal to the characteristic impedance of the ceramic and driven by series resonant R-L circuits having electrical Q_{EL}'s of 10 and 2 respectively.

series R-L circuit for two different electrical Q's while figure (4a,b) illustrates the corresponding time domain function. When the ceramic is then backed in its own characteristic impedance the frequency and time response curves are as illustrated in figures (5a,b) and (6a,b) respectively.

From these results it is observed that with increased electrical Q, the frequency response curves become more peaked with a corresponding increase in the number of half-cycles that occur in the time domain. For an electrical Q of two, the time domain signal approaches 3 half-cycles for the air backed case and 2 half-cycles for the backed ceramic. It should also be observed that the resonant frequency is shifted down from the 2.25 MHz nominal frequency of the ceramic.

Using the preceding result as a standard, the problem remaining is to be able to generate in body tissue a pulse which mimicks this result. That is, to maximize the gain-bandwidth product by transmitting the maximum energy in the shortest time. This becomes a difficult problem since there is an approximate 22 to 1 impedance mismatch between the ceramic and body tissue. To decrease this mismatch, several research groups have proposed various combinations of single and double quarter-wave front matching layers in conjunction with various electrical terminating networks (1,2,9). These have resulted in increased transmitted signal, but have fallen short of the standard with respect to magnitude and decreased pulse length. This mainly results from the fact that the bandpass characterisitic of the overall transducer response is being limited by the frequency response of the matching layers.

By using a modified version of the matrix formularization for transducers an investigation was conducted into the pressure transfer function for the matching layers acting alone. Figure (7) illustrates the transmitted frequency impulse response curves for one, two, five, and ten quarter-wave matching layers. The impedances of the various layers were picked to equalize the reflection coefficient across each boundary. In essence, this becomes a stepped approximation to an exponential impedance variation. As can be observed from these results, as more layers are added, the bandpass characteristic becomes much broader and there is less peak-to-peak ripple variation in the pass-band region. This is a significant result in that as more layers are added there is less tendency for the matching plate to modulate the frequency response of the ceramic and its terminating electrical network. In fact, it can be shown (Appendix A) that in the limit, an exponential variation from a high to a low impedance level, or vice versa, will transmit a pressure impulse as a single impulse containing all the energy in the original pulse.

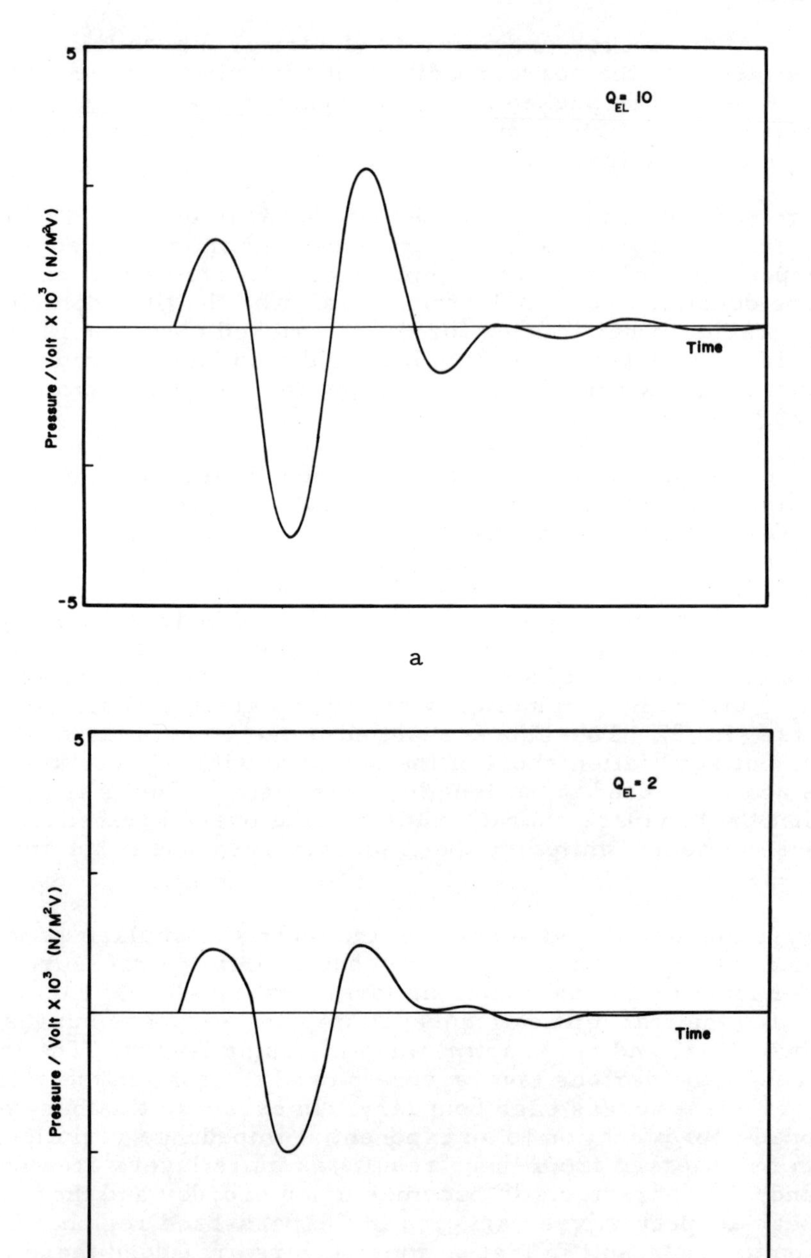

Figure 4: Time domain output pressure for frequency response curves in figure 3.

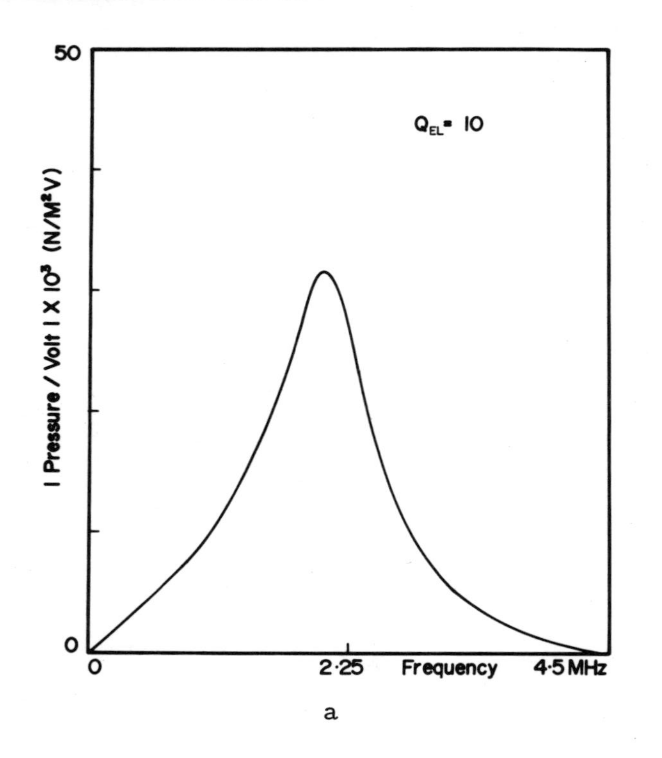

a

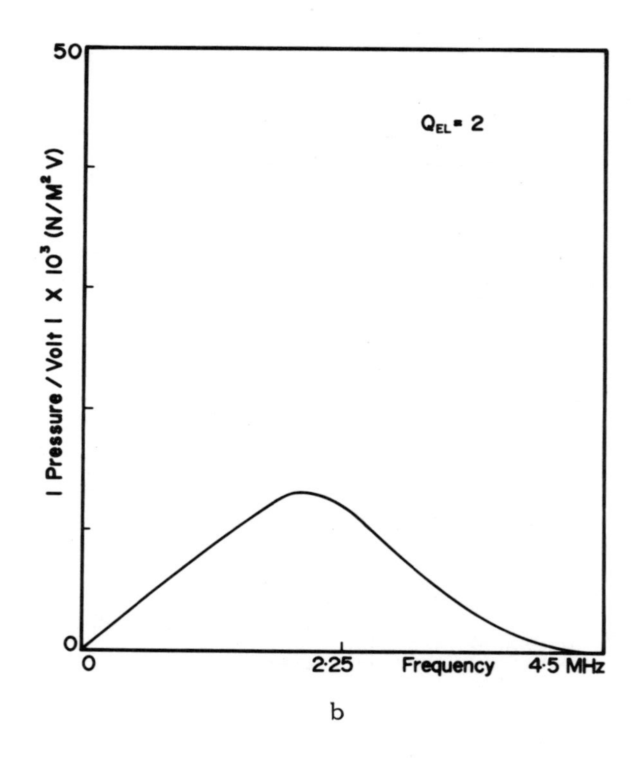

b

Figure 5: Output pressure frequency response curves for a ceramic backed by its own characteristic impedance and transmitting into a load having an impedance equal to its characteristic impedance and driven by a series tuned resonant R-L circuits having electrical Q_{EL}'s of 10 and 2 respectively.

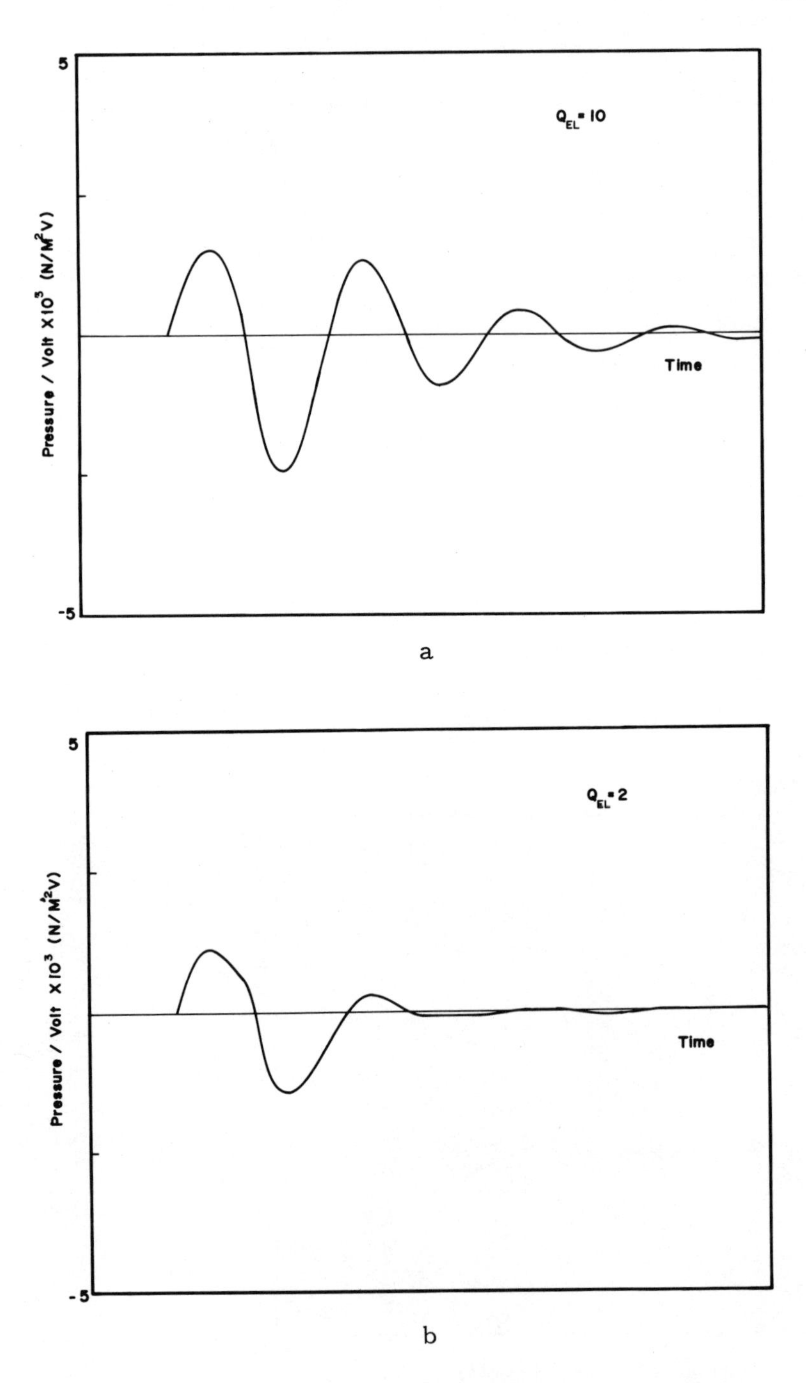

Figure 6: Time domain output pressure for frequency response curves in figure 5.

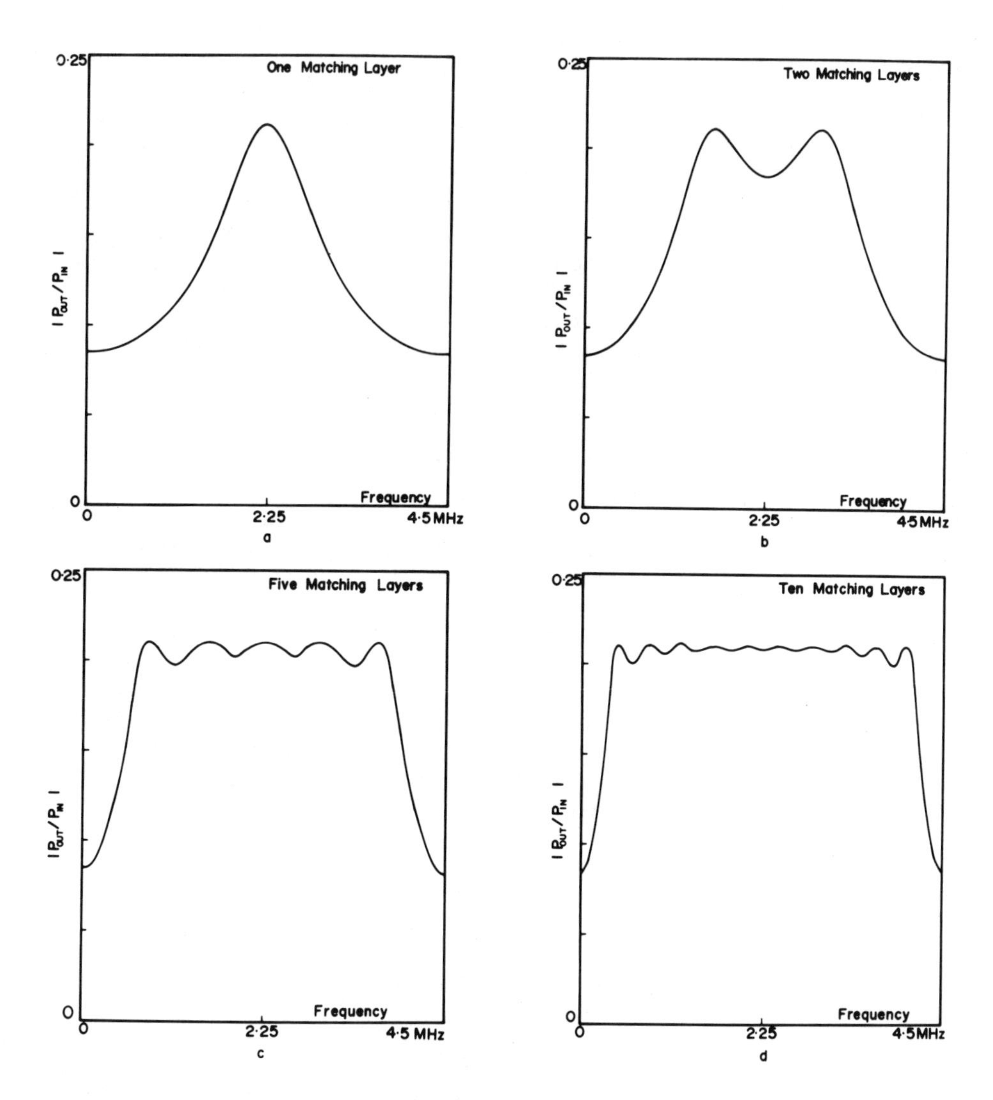

Figure 7: Pressure wave transfer function for one, two, five and ten quarter wave matching layers having impedance profiles which equalize the reflection coefficient across each boundary. Impedance values were chosen according to equations (A1-A5) to match impedance of ceramic (35.7 M Rayls for PZT-7A) to that of water (1.6 M Rayls).

Considering this property for an exponentially graded matching plate, one would expect that the acoustic impedance presented to the ceramic would then be real and frequency independent and could be made equal to the characteristic impedance of the transducer. In which case, one would expect a frequency and time response similar to that obtained for the standard result. To test this hypothesis, the matrix equation was iterated to take into account ten front matching layers having a stepped exponential impedance taper. Figure (8a,b) illustrates the impulse pressure frequency response obtained for an air backed ceramic under the same electrical terminating conditions as used in the reference standard while figure (9a,b) illustrates the corresponding time domain signal. When the ceramic is then backed in its own characteristic impedance, the frequency and time response curves are illustrated in figures (10a,b) and (11a,b) respectively.

When these results are compared to those obtained for the reference case, it is noted that the overall shapes of the frequency response curves are similar except for a scale factor and the fact that the multilayered system introduces high time ripple which reflects itself in the time domain as a small delayed reverberant component. Obviously, this effect would be worsened by increasing the electrical Q and improved by adding more matching layers.

In comparing the time domain signals, it is observed that they too are similar except for the scale factor. This factor is related to the transformer ratio for the matching layers. For an ideal transformer, the pressure amplitude is decreased in travelling from a high to low impedance material by a ratio related to the square root of the terminating impedance levels (Appendix A). Thus, in going from the ceramic impedance level (35.7 M Rayls for PZT-7A) to that of body tissue (1.6 M Rayls) an ideal transformer would reduce the pressure amplitude by approximately 21%. Looking at the results for the 10 layer matching scheme, this ideal value is nearly achieved. Thus, by using an exponentially graded matching plate, one should be theoretically able to maximize the overall gain-bandwidth product.

CONCLUSIONS

In conclusion, a matrix model has been developed which facilitates the analysis of multilayered front matched and backed piezoelectric ceramics in the frequency domain.

By utilization of the fast Fourier transform, the time domain pressure response function can be found.

Furthermore, it was shown that the number of quarter wave sections in the front matching plate can have a significant influence

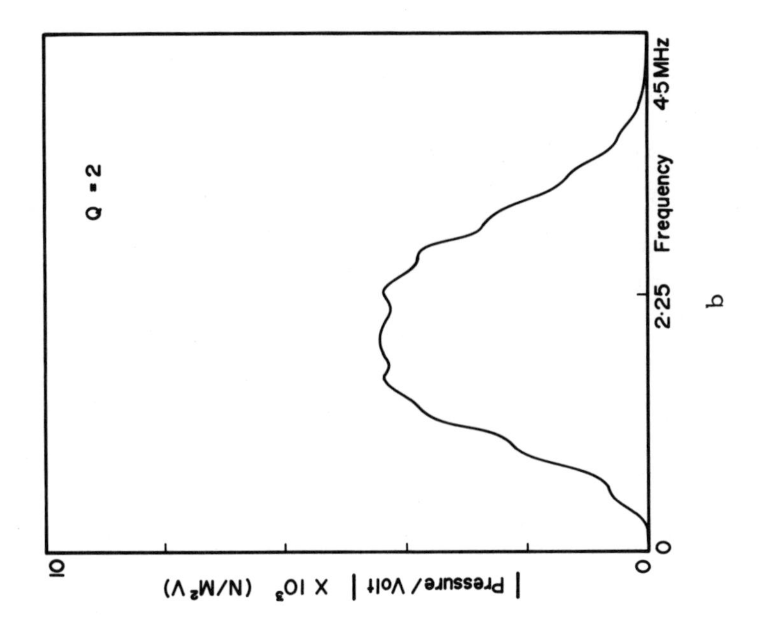

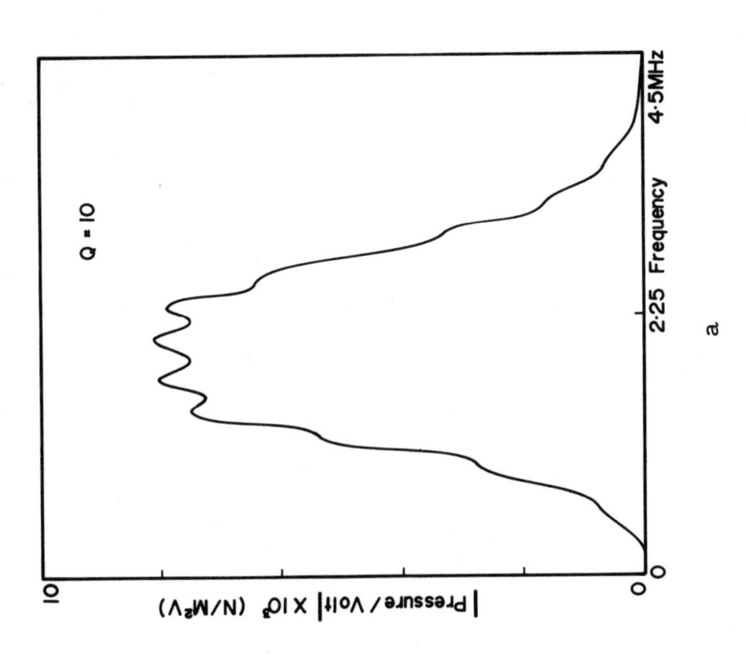

Figure 8: Output pressure frequency response curves for an air backed, PZT-7A piezoelectric ceramic having ten matching layers and transmitting into a load having an impedance equivalent to body tissue and driven by series resonant R-L circuits having electrical Q_{EL}'s of 10 and 2 respectively.

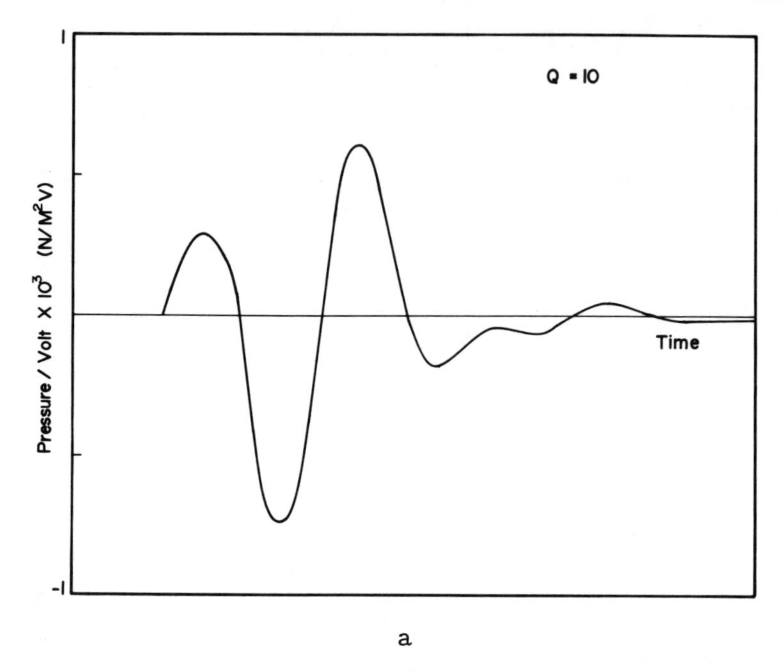

a

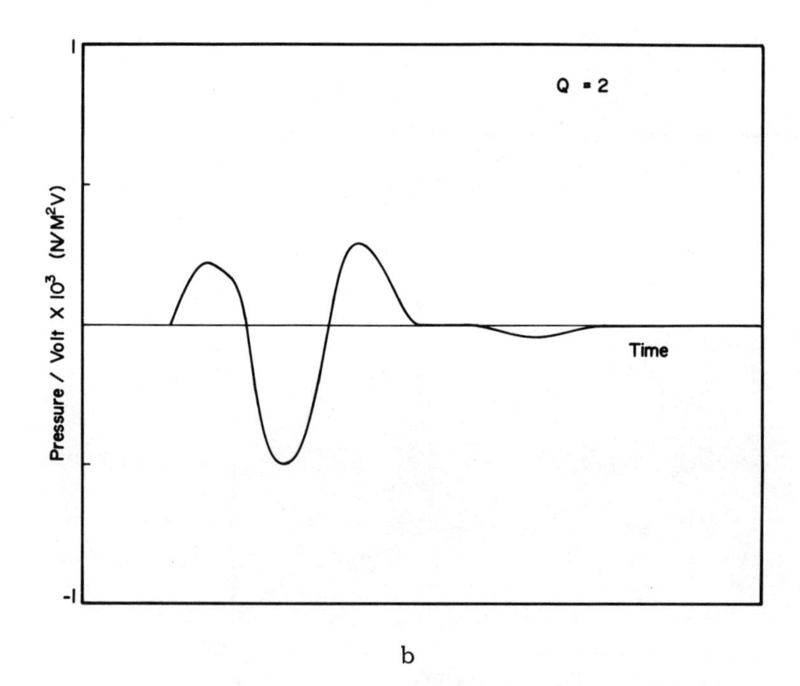

b

Figure 9: Time domain output pressure signal for frequency response curves in figure 8.

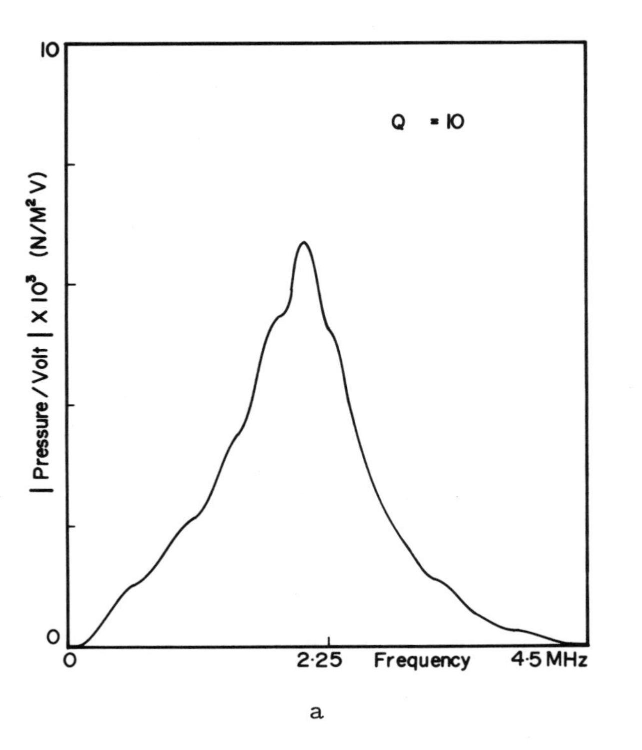

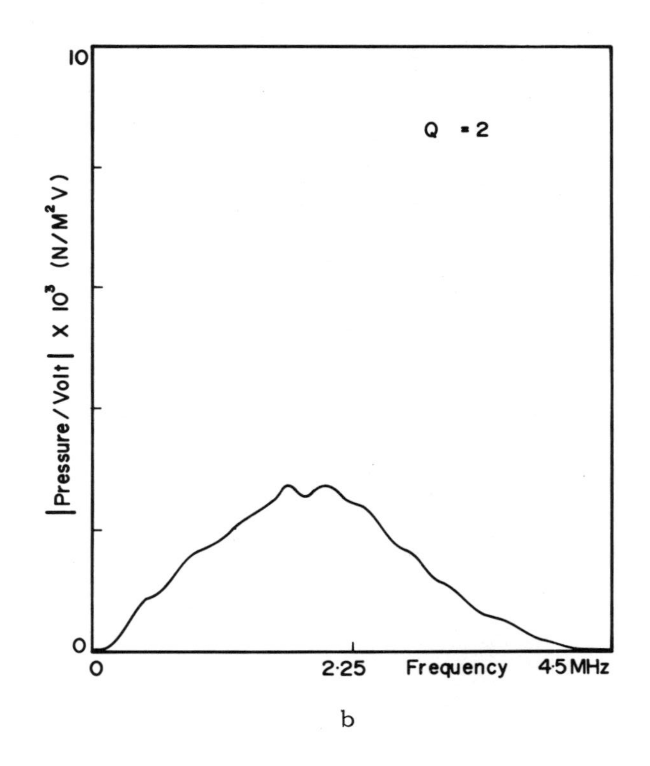

Figure 10: Output pressure frequency response curves for a PZT-7A piezoelectric ceramic backed in its own characteristic impedance and having ten matching layers and transmitting into a load having an impedance equivalent to body tissue and driven by series resonant R-L circuits having electrical Q_{EL}'s of 10 and 2.

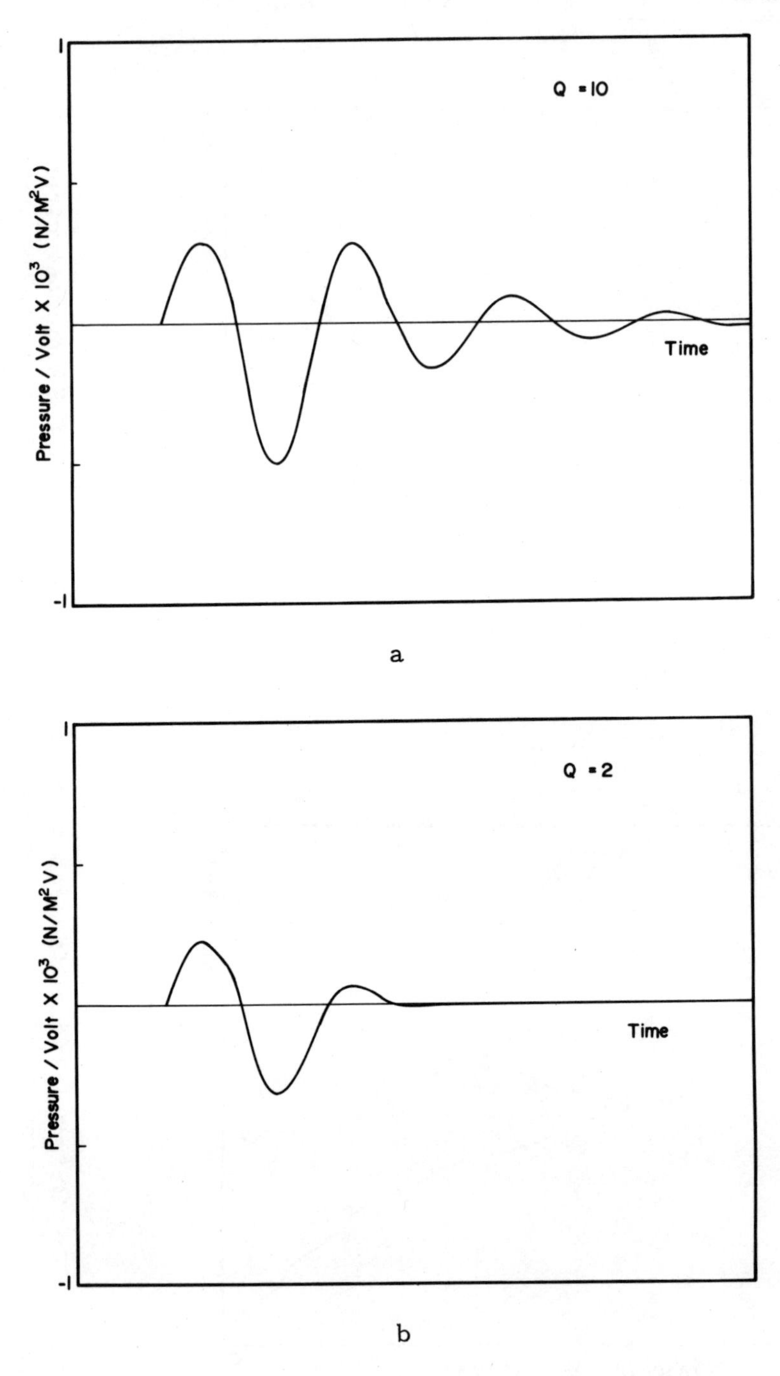

Figure 11: Time domain output pressure signal for frequency response curves in figure 10.

on the overall pressure transfer characteristic of the transducer and that the use of only one or two matching layers is the limiting factor in achieving broadband operation.

It was also shown that the use of an exponentially graded impedance plate transmits a pressure impulse as a single impulse containing all the energy in the initial pulse.

Because of this, by utilization of the matrix technique, it was shown that if one approximates an exponential taper by ten quarter wave matching layers, then the transmitted pressure pulse approaches the ideal three half-cycle pulse for an air backed transducer and a two half-cycle pulse for a backed ceramic.

APPENDIX A

Consider the (n) matching layers in figure (A1). An impulse in pressure originating in the material with characteristic acoustic impedance Z_0 propagates through the N layers and emerges from layer Z_N. Obviously, as the pulse enters the various layers, it will set up a series of transmitted as well as reflected pulses. However, the magnitude of the first transmitted pulse will have a value given by

Magnitude of First Transmitted Pulse =

$$\frac{2Z_1}{Z_1 + Z_0} \cdot \frac{2Z_2}{Z_1 + Z_2} \cdots\cdots \frac{2Z_{N+1}}{Z_{N+1} + Z_N} \qquad \text{(A1)}$$

If each layer has an impedance which is a fixed percent lower or higher than the preceding layer such that

$$Z_1 = aZ_0, \; Z_2 = aZ_1, \; \cdots\cdots \; Z_{N+1} = aZ_N$$

then equation (A1) can be rewritten as:

$$\left(\frac{2a}{a+1}\right)^{n+1} \qquad \text{(A2)}$$

If one considers a piezoelectric ceramic having an impedance (Z_C) transmitting a pulse into the human body with an impedance (Z_B) and one wants to place (n) matching layers between the two extremes in impedance, then one would pick the parameter (a) in equation (A2) to be

$$a = \sqrt[n+1]{Z_B/Z_C} \qquad \text{(A3)}$$

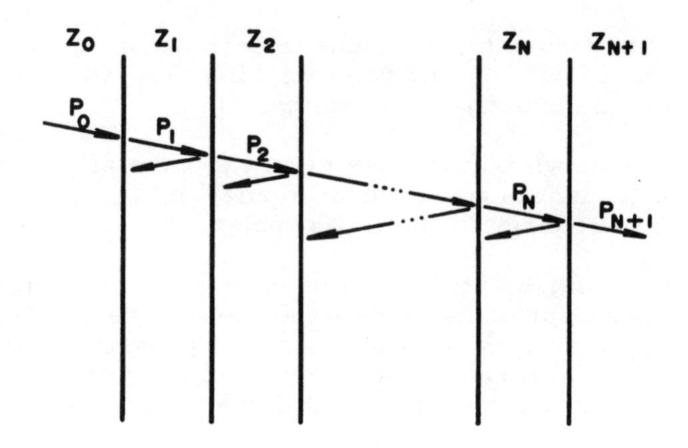

$$P_{N+1} = \frac{2\,Z_{N+1}}{Z_N + Z_{N+1}}\, P_N$$

Figure A1: Relation between the magnitude of the first transmitted pulse to the input pulse for a series of (N) acoustic matching layers.

If equation (A3) is substituted into equation (A2) one finds that the magnitude of the first transmitted pulse is:

$$\frac{Z_B}{Z_C}\;\frac{2^{n+1}}{\left(\sqrt[n+1]{Z_B/Z_C}+1\right)^{n+1}} \tag{A4}$$

By picking a large number for (n) it can be shown that

$$\lim_{n\to\infty}\;\frac{Z_B}{Z_C}\left(\frac{2}{1+\sqrt[n+1]{Z_B/Z_C}}\right)^{n+1} \longrightarrow \sqrt{\frac{Z_B}{Z_C}} \tag{A5}$$

This is an extremely useful result since it states that all of the energy in the initial input pulse is transmitted to the first output pulse. This can be shown when one considers that the input power is given by:

$$\text{Power in} = \frac{(\text{Input Pressure})^2}{Z_C} \tag{A6}$$

and the output power by

$$\text{Power out} = \frac{(\text{Output Pressure})^2}{Z_B} \quad \text{(A7)}$$

Since the magnitude of the first transmitted pulse is

$$\sqrt{\frac{Z_B}{Z_C}} \times \text{Input Pressure Magnitude} \quad \text{(A8)}$$

when this is substituted into (A7) one finds that the power in the first transmitted pulse equals the initial power.

Since all the input power is transmitted to the first output pulse, no energy is available to cause afterringing or reflections back into the ceramic.

ACKNOWLEDGMENTS

The author would like to gratefully acknowledge the work of L. Walker, A. Smudde and J. Domnanovich for aid in preparation of this manuscript and for their expert photographic reproductions.

REFERENCES

1. Goll, J.H., Auld, B.A., Multilayer Impedance Matching Schemes for Broadbanding of Water Loaded Piezoelectric Transducers and High Q Electrical Resonators, IEEE Transactions on Sonics and Ultrasonics, Vol. SU-22, No. 1, January 1975

2. Kossoff, G., The Effects of Backing and Matching on the Performance of Piezoelectric Ceramic Transducers, IEEE Transactions on Sonics and Ultrasonics, Vol. SU-13, No. 1, March 1966

3. Martin, R.W., and Sigelmann, R.A., Force and Electrical Thevenin Equivalent Circuits and Simulations for Thickness Mode Piezoelectric Transducers, J. Acoust, Soc. Am., Vol. 58, No. 2, August 1975

4. Mason, W.P., Electromechanical Transducers and Wave Filters (Von Nostrand, New York 1948)

5. Meeker, T.R., Thickness Mode Piezoelectric Transducers, Ultrasonics, January 1972

6. Reeder, T.M. and Sperry, W.R., Broad-Band Coupling to High-Q Resonant Loads, IEEE Transactions on Microwave Theory and Technique, Vol. MTT-20, No. 7, July 1972

7. Redwood, M., Transient Performance of a Piezoelectric Transducer, The Journal of the Acoustical Society of America, Vol. 33, No. 4, April 1961

8. Riblet, H.J., General Synthesis of Quarter-Wave Impedance Transformers, IEEE Transactions on Microwave Theory and Techniques, January 1957

9. Sittig, E.K., Effects of Bonding and Electrode Layers on the Transmission Parameters of Piezoelectric Transducers Used in Ultrasonic Digital Delay Lines, IEEE Transactions on Sonics and Ultrasonics, Vol. SU-16, No. 1, January 1969

ON OBTAINING MAXIMUM PERFORMANCE FROM LIQUID SURFACE LEVITATION HOLOGRAPHY

Alfred V. Clark, Jr.

Naval Research Laboratory

Washington, D.C. 20375

INTRODUCTION

The concept of liquid surface levitation holography as a tool for acoustical imaging has come to fruition with the availability of commercial systems. The analysis of such systems has been carried out by a number of authors, but perhaps the most thorough analysis has been done by Pille [1] and by Pille and Hildebrand [2]. These authors decomposed the levitation holography system into several major components and derived transfer functions for each of them.

Part of the present work is a direct outgrowth of the treatment of Pille and Hildebrand, in that it is based upon and extends their analysis. Their treatment can be used to predict the best image quality theoretically possible with levitation holography, provided certain constraints are satisfied. The nature of the constraints identified by them is explored and quantified here for a typical system. There is an additional constraint which is also derived in this paper.

The satisfaction of these constraints will lead to maximum performance of the levitation holography system, when the usual method of optical reconstruction of the acoustic hologram is employed. The system is also capable of producing enhanced images when a new method of reconstruction is used. This simple method can, in many cases, remove unwanted portions of the image which are caused by diffraction and interference effects in the acoustic object being examined. It can also remove some background "noise" for easier target identification. In addition, the method preserves phase information about the acoustic target, which in the usual method of reconstruction is lost. Hence, this new method has the capability of rendering the outline of phase objects visible.

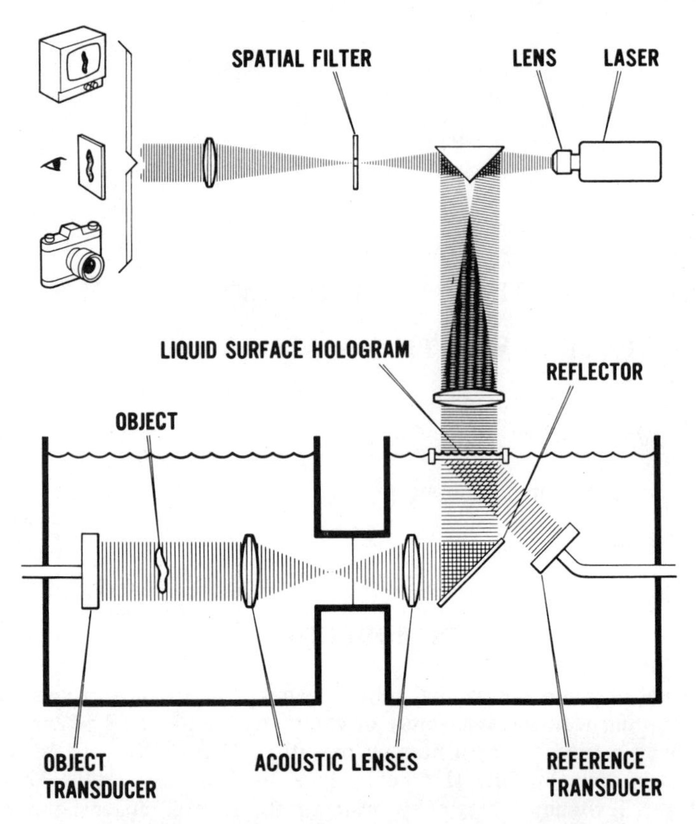

Fig. 1. A typical liquid surface levitation holography system.

DERIVATION OF CONSTRAINTS FOR MAXIMUM PERFORMANCE

A typical liquid surface levitation holography system is shown in Fig. 1. An object is placed in a water bath and insonified by the object transducer. The sound field is modified by the presence of the object, due to changes in transmission, internal reflection, mode conversion, attenuation, etc. The modified sound field contains information about the acoustic nature of the object, and is focused by an acoustic lens system onto the liquid surface of a thin film of imaging fluid. The imaging fluid is contained in a small tank, called a minitank.

A second (reference) transducer also insonifies the surface. The liquid surface levitates in response to the radiation pressure in the acoustic field. At the surface, this pressure is proportional to the square of the acoustic particle velocity [1]. Part of the radiation pressure results from the interference between the velocity fields in the focused image and in the reference beam. The corresponding levitation constitutes a phase hologram and contains information about the acoustic object.

To extract this information, the surface is illuminated with laser light. The surface levitation phase modulates the reflected light. The light is reflected as beams (orders) traveling with a multiplicity of spatial frequencies. Assuming perfect fidelity, there are two beams whose optical intensities are proportional to the intensity in the sound field emerging from the acoustic object. Either beam can be selected to form the optical reconstruction of the acoustic object. This is done by the optical lens system and spatial filter shown in Fig. 1. The optical image is formed on a viewing screen for visual display.

The levitation system can be broken down into several subsystems, namely,

1. the acoustic lens system,
2. the minitank,
3. the liquid surface, and
4. the optical system.

It is assumed that the object transducer emits a plane wave which passes through a thin acoustic object. The sound field emerging from the object has velocity potential ϕ_o, which can be thought of as a superposition of plane waves travelling in different directions. Because of the finite size of the acoustic lens system, waves which propagate at an angle greater than the cutoff angle (determined by lens system geometry; see e.g., Ref. [3]) will not reach the minitank (which contains the film of imaging fluid).

It will be shown in this paper that each plane wave incident on the minitank may be diffracted into a multiplicity of plane waves traveling through the minitank. This is undesirable, as it can degrade the acoustic image. It is necesary to avoid this condition in order to obtain maximum system performance.

At the liquid surface, the velocity potential ϕ_o in the focused acoustic image causes a corresponding vertical particle velocity, V_o. V_o is in general a complex quantity; i.e. $V_o = |V_o| e^{iX_o}$. The reference transducer also generates a beam idealized as a single plane wave with velocity potential $\phi_r = A_r e^{2\pi i \eta_r y}$ at the surface. This wave enters the minitank and causes a vertical partical velocity V_r at the liquid surface.

Due to nonlinear effects in the acoustic field, there will be a steady momentum flux/area in the vertical direction at the liquid surface; this is the radiation pressure, P_r. For a perfectly reflecting surface, P_r is given by [3]

$$P_r = \frac{1}{4} \rho_u \Big[|V_o|^2 + |V_r|^2 + 2|V_o|\,|V_r| \cos(2\pi\eta_r y - X_o) \Big] \tag{1}$$

where ρ_u is the density of the imaging fluid. The temporal variation of P_r has been suppressed.

The radiation pressure causes the liquid surface to levitate. It is convenient to let

$$P_r = P_{rb} + P_{ri} + P_{ri}^* \tag{2}$$

where

$$P_{rb} = \frac{1}{4} \rho_u \Big[|V_o|^2 + |V_r|^2 \Big] \tag{3}$$

$$P_{ri} = \frac{1}{4} \rho_u |V_o|\,|V_r|\, e^{iX_o} e^{-2\pi i \eta_r y} \tag{4}$$

and P_{ri}^* is the complex conjugate of P_{ri}. P_{rb} is the sum of radiation pressures in the reference beam and focused image, as if each were separately present at the liquid surface. P_{ri} is proportional to V_o, and P_{ri}^* is proportional to V_o^*. P_{rb} can be thought of as noise, in that it causes a low spatial frequency bulge of the surface which carries no useful information about V_o. P_{ri} and P_{ri}^* can be considered signal, as they carry information about V_o, and hence about the nature of the acoustic object.

In order to faithfully record the acoustic image, it is necessary that the liquid surface levitation be independent of the spatial frequency content of the radiation pressure. In Ref. [1] and [2], it is shown that this is approximately true at a certain time after a pulse of radiation pressure excites the surface. Due to viscosity, the surface response to P_{ri}, the information-bearing term in the radiation pressure, will decay. Then the surface is again excited with another pulse of radiation pressure, and sampled by the laser after the appropriate time delay. This process is repeated at a repetition rate high enough form a flicker-free optical image.

Another condition for faithful recording of the acoustic image by the surface is that the surface respond in a linear fashion to the radiation pressure. The behavior of the imaging fluid is governed by the momentum conservation equation of fluid mechanics, which is, in general, nonlinear. However, in the limit of small displacements of the fluid particles, the equation can be linearized [4]. This places another constraint on operation of the system.

The response of the liquid surface due to P_{ri} will be called h_i. In addition to h_i, the liquid surface levitation also has a low spatial frequency component, called h_b. This is the response to the noise component, P_{rb}, of the total radiation pressure, P_r. If the "bulge", h_b, is not small over the illuminated area of the phase hologram, it can superpose a low spatial frequency noise on the reflected laser light [1]. This can, in some cases, cause degradation of the optical signal. Consequently, it is necessary to place a constraint on the magnitude of the bulge for best performance of the system.

The information-bearing components P_{ri} and P_{ri}^* in the total radiation pressure field cause a ripple pattern to be imposed on the liquid surface. When the laser light reflects from this ripple pattern, it will be in the form of beams travelling in discrete directions. In the limit of small displacements there are two beams which have optical intensities proportional to the displacement of the surface. Either beam can be extracted by the spatial filtering scheme shown in Fig. 1, and used to form the optical image. It will be shown that if surface levitations are not constrained to be small, then the relation between optical and acoustical intensitites become ambigous, and it is difficult to interpret the optical image. The requirement of small levitation is one which must be met for maximum system performance.

Effect of Membrane Displacement

The focused acoustic image can be represented by a superposition of plane waves in the region between the acoustic lens system and the liquid surface. Assume that the acoustic object can be represented by a transmission function, S_o, and is insonified by a quasi-monochromatic wave of duration $2\Delta t$ and temporal frequency ω_a. Let the particle velocity be derived from a velocity potential, ϕ_o. In the incident wave,

$\phi_o = \phi_{inc}\, e^{-i\omega_a t}\, Rect(\hat{t}/\Delta t)$ where $\hat{t} = t - \Delta t$ and $Rect(\hat{t}/\Delta t) = 1$ for $|\hat{t}| < \Delta t$ and zero otherwise. In what follows, a two-dimensional problem is assumed for simplicity.

The velocity potential in the wave emerging from the object is represented by $S_o(y_o)\, \phi_{inc}(y_o)\, e^{-i\omega_a t}\, Rect(\hat{t}/\Delta t)$ where y_o identifies a point in the acoustic object. Letting $\psi_o = S_o\, \phi_{inc}$, the field between the lens system and the minitank is given by [3]:

$$\phi_o = \int_{-\eta_M}^{\eta_M} \overline{\psi}_o(\eta)\, e^{2\pi i(\eta y + \zeta z)}\, e^{-i\omega_a t} \times Rect\left[\frac{y \sin\theta_l + z\cos\theta_l - c\hat{t}}{\Delta t}\right] d\eta. \tag{5}$$

The spatial frequency η is defined by

$$\eta = \frac{\sin\theta_l}{\lambda_l} \tag{6}$$

where θ_l is the angle between the direction of propagation of the wave

$$\overline{\psi}_o(\eta)\, e^{2\pi i(\eta y + \zeta z)}\, e^{-i\omega_a t}$$

and the acoustic axis; λ_l is the acoustic wavelength in the water bath. $\overline{\psi}_o(\eta)$ is the Fourier transform of $\psi_o(y_o)$. Waves propagating at angles $|\theta_l| > \theta_M$ will be cut off by the acoustic lens system; $\eta_M = \sin\theta_M/\lambda_l$ is the cutoff spatial frequency. ϕ_o reaches the minitank at $t = 0$.

Since in general, the minitank fluid will be other than water, a thin membrane at the bottom of the minitank separates the imaging fluid from the water bath in which the object is immersed. The membrane also serves to isolate the liquid surface from environmental disturbances. The set of plane waves in (5) is incident on the membrane; since the acoustic properties of the imaging fluid are normally different from those of water, displacement of the membrane can lead to degradation of the acoustic image.

To see how this can come about, consider the case of the membrane deformed into a sinusoidal pattern. The question of how such a deformation occurs will be dealt with later. Then, letting w be the membrane displacement, it is assumed that $w = W \cos Ky$. Let the membrane displacement be such that $KW \ll 1$, so that the membrane displacement has a small slope. It is also assumed that $K \ll 2\pi/\lambda_u$, with λ_u the acoustic wavelength in the minitank.

Consider the incident wave $\overline{\psi}_o(\eta)\, e^{2\pi i(\eta y + \zeta z)}$; i.e., one plane wave component of (5). The temporal variation has been suppressed. Upon emerging from the membrane, the velocity potential ϕ_{tr} becomes $\overline{T}_{12}(\theta')\, \phi_o(z = w)$. Here θ' is the local angle of incidence, measured from the normal **n** to the membrane, as shown in Fig. 2; $\overline{T}_{12}$ is the transmission coefficient for the membrane.

Now $\theta' = \theta_l + \dfrac{\partial w}{\partial y}$; since small slopes are assumed, it can be shown that $\overline{T}_{12}(\theta') \cong \overline{T}_{12}(\theta_l)$, and that Snell's Law simplifies to

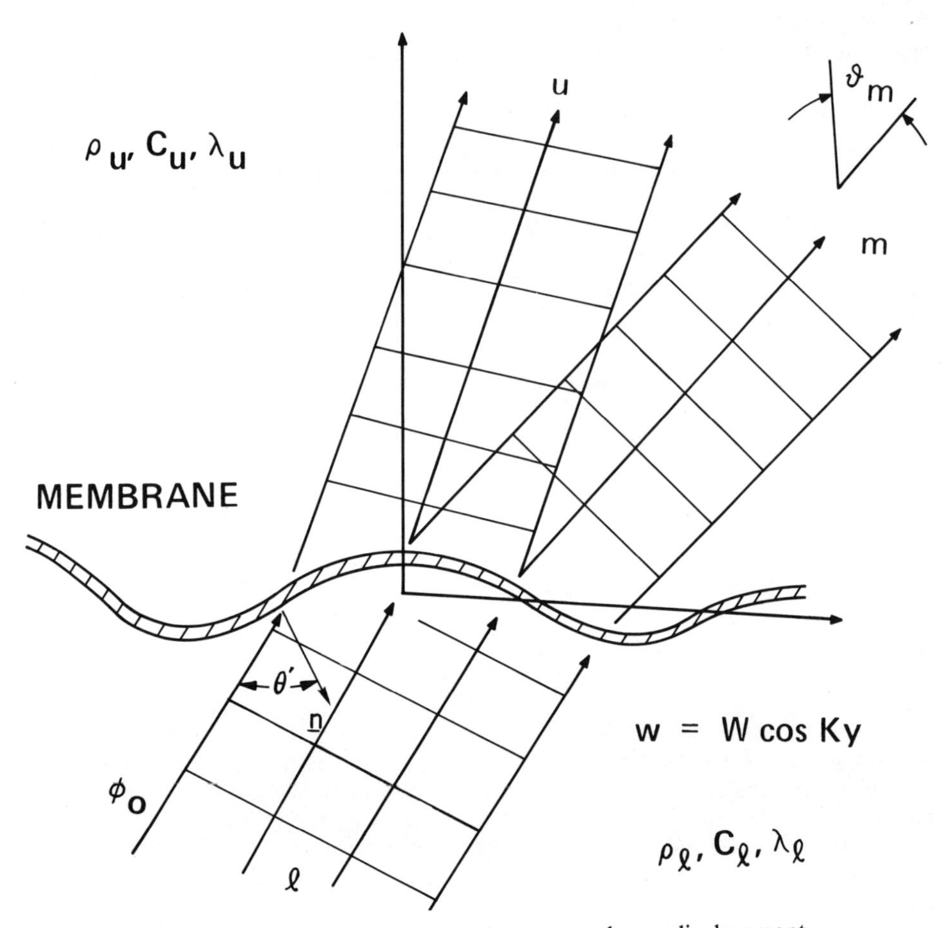

Fig. 2. Diffraction of plane wave due to membrane displacement.

$$\frac{\sin \theta_u}{c_u} = \frac{\sin \theta_l}{c_l} \tag{7}$$

where c_l is the sound speed in the lower fluid and c_u the sound speed in the upper (imaging) fluid. These results allow the expression for ϕ_{tr} to be written as

$$\phi_{tr} = \overline{\psi}_o(\eta)\ \overline{T}_{12}(\theta_l)\ e^{2\pi i\eta y}\ e^{2\pi i(\zeta' z + (\zeta - \zeta')w)} \tag{8}$$

Here $\eta = \sin \theta_u/c_u$ (Snell's Law) and $2\pi\zeta' = \sqrt{(\omega_a/c_u)^2 - 4\pi^2\eta^2}$ to satisfy the wave equation. Note the appearance of the phase factor $\exp[2\pi i(\zeta - \zeta')w]$, which appears due to: 1) the displacement of the membrane, and 2) the difference in sound speed between the upper and lower fluids.

Since $w = W \cos Ky$, the identity [5]

$$\exp\left[2\pi i(\zeta - \zeta')\ W \cos Ky\right] = \sum_{-\infty}^{\infty} i^n J_n\left(2\pi[\zeta - \zeta']\ W\right) e^{inKy} \tag{9}$$

can be used. Here J_n is the nth order Bessel function of the first kind. Let $\sin \vartheta_n = \sin \theta_u + nK/k_u$ and $k_u = \omega_a/c_u$. Then for small n, and $k' = 2\pi(\zeta - \zeta')$

$$\phi_{tr} = \bar{\psi}_o(\eta)\ \bar{T}_{12}(\theta_l) \sum_n i^n J_n(k'W)\ e^{ik_u(y \sin \vartheta_n + z \cos \vartheta_n)} \tag{10}$$

Hence, ϕ_{tr} is a superposition of diffracted orders with direction cosines $\cos \vartheta_n$, and amplitude $J_n(k'W)$. Note that the above satisfies the wave equation in the minitank, and the boundary condition

$$\phi_{tr}(z = w) \cong \bar{T}_{12}(\theta_l)\ \phi_o(z = w) \tag{11}$$

is approximately satisfied for $n << k_u/K$.

If $k'w$ is $<< 1$, then only the $n = 0$ order is significant. The 0 order is merely the wave which would be diffracted in accordance with Snell's Law. Hence, if w is small ($<< 1/k'$), or if $c_u \cong c_l$, then the higher order diffracted waves are negligible, and every plane wave in the focused image (5) will be incident on the liquid surface with its spatial frequency preserved.

If $k'w$ is of order unity, then the amplitudes of the higher orders can be the same size as that of the zero order. In this case, every plane wave component $\bar{\psi}_o(\eta)\ e^{2\pi i(\eta y + \zeta z)}$ of (5) will give rise to several diffracted orders, all of which combine at the liquid surface.

To see what effect this has on the process of recording the acoustic image, recall that the liquid surface levitates in response to the radiation pressure there. That part of the radiation pressure which contains information about $\psi_o(y_o)$ will be called P_{ri}, and the corresponding levitation, h_i.

For a pulse of radiation pressure of duration $2\Delta t$, it can be shown that [3]

$$P_{ri} = \frac{1}{4}\ \rho_u < Re[V_o]\ Re[V_r^*] > Rect\ (\hat{t}/\Delta t). \tag{12}$$

The symbol $Re[V_o]$ denotes the real part of $V_o(y,t)$, and $< >$ denotes a time-average over the pulse duration; i.e.,

$$<Re[V_o]Re[V_r^*]> = \frac{1}{2\Delta t}\int_o^{2\Delta t} Re[V_o]Re[V_r^*]dt. \tag{13}$$

For ease of notation, $V_o V_r^*$ will be used in what follows to denote $<Re[V_o]\ Re[V_r^*]>$.

Furthermore, the velocity potential ϕ_{tr} is related to V_o by

$$V_o = \frac{\partial\phi_{tr}(z = d)}{\partial z} \tag{14}$$

where d is the minitank depth. Pille and Hildebrand have shown [2] that multiple internal reflections can occur in the minitank when $c_u \neq c_l$. For simplicity, this effect is ignored, since it is desired to study only the effect of the membrane on the imaging process.

From (10) it is seen that, ignoring constant phase factors,

$$V_o \propto \bar{\psi}_o(\eta)\ \bar{T}_{12}(\theta_i) e^{-i\omega_a t} \mathrm{Rect}(\hat{t}/\Delta t) \times \sum_n i^n \cos\ \vartheta_n J_n(k'W)\, e^{ik_u y \sin\ \vartheta_n} \tag{15}$$

where $\propto$ denotes proportionality. Now $t = 0$ when the waves reach the liquid surface.

Furthermore, if the velocity potential in the reference beam is given by

$$\phi_r \propto A_r\, e^{2\pi i \eta_r y}\, e^{2\pi i \zeta_r z}\, e^{i\omega_a t} \tag{16}$$

then

$$V_r \propto A_r \bar{T}_{12}(\theta_r) \sum_m i^m \cos\ \vartheta_m\ J_m(k'W(t))\ e^{ik_u y \sin\ \vartheta_m} \times e^{-i\omega_a t}\, \mathrm{Rect}(\hat{t}/\Delta t) \tag{17}$$

with

$$\sin\ \vartheta_m = \left(\frac{c_u}{c_l}\right) \sin\ \theta_r + mK/k_u. \tag{18}$$

Hence, the reference beam is also diffracted into a number of orders by the membrane displacement.

The radiation pressure, P_{ri}, will have a component due to interference of the zero orders of the image field and of the reference beam. In addition, there will be interference between all the higher (non-zero) orders, which can give rise to degradation of the recording process at the surface.

For simplicity, suppose that $k'w$ is small enough so that only the 0 and ± 1 orders in both fields are significant. Let the Bessel functions be replaced by their small argument approximations

$$J_o(k'w) \cong 1 \qquad J_1(k'w) \cong \frac{k'w}{2} \qquad J_{-1}(k'w) \cong \frac{-k'w}{2}$$

Then $V_o V_r^*$ has a component proportional to

$$A_r \bar{\psi}_o(\eta) e^{2\pi i(\eta - \eta_r)y}\, \mathrm{Rect}(\hat{t}/\Delta t)$$

which is the interference of the zero orders. This is termed the (0, 0) component; note that the spatial frequency is $\eta - \eta_r$. There will be two components, called the (0, ± 1) components

$$iA_r \bar{\psi}_o(\eta) \left[e^{i(2\pi[\eta-\eta_r]-K)y} - e^{i(2\pi[\eta-\eta_r]+K)y}\right] < k'W(t) > \mathrm{Rect}(\hat{t}/\Delta t)$$

which are due to the interference of the zero order in the acoustic image field with the ± 1 orders in the reference beam field. Note the change in spatial frequency of the interference pattern of amount $\pm K$. Likewise, the ± 1 diffracted orders in the image field interfere with the zero order of the reference beam field to give the (± 1, 0) contribution where the spatial frequency of the interference pattern again changes by $\pm K$. Finally, there will be interference between both sets of ± 1 orders in the two fields. It is easy to see that these (± 1, ± 1) components will cause interference patterns with spatial frequency changes of $\pm K \pm K$.

In Ref. [1] and [2], it is shown that the response of the surface to radiation pressure of the form $P_r = p_r(\hat{\eta})\, e^{i\hat{\eta}y}\, F(t)$ is

$$h = \overline{H}(\hat{\eta}, t) p_r(\hat{\eta}) e^{i\hat{\eta}y} \tag{19}$$

where

$$\overline{H}(\hat{\eta}, t) = \frac{\omega_\eta\, e^{-q\omega_\eta t} \sin \omega_\eta t \overset{t}{*} F(t)}{\rho_u g + \gamma\hat{\eta}^2} \tag{20}$$

The symbol $\overset{t}{*}$ represents convolution in the time domain. Here ω_η is the frequency of free vibration of the surface associated with propagation of a free wave in the form $h \propto e^{i\hat{\eta}y}\, e^{i\omega_\eta t}$. Letting γ be the surface tension, g the gravitational acceleration, and d the minitank depth, ω_η is given by

$$\omega_\eta^2 = (\rho_u g + \gamma\hat{\eta}^2)\ (\hat{\eta}/\rho_u)\ \tanh\, \hat{\eta}d. \tag{21}$$

The authors of [1] and [2] have found that

$$q = 2\nu\rho\hat{\eta}\sqrt{\frac{(\hat{\eta}/\rho_u)}{\rho_u g + \gamma\hat{\eta}^2}}\ (\tanh\hat{\eta}d)^E \tag{22}$$

where E = -2.3 for $\hat{\eta}d < \pi/2$ and zero otherwise. ν is the coefficient of kinematic viscosity, assumed small, i.e., $2\hat{\eta}^2\nu << \omega_\eta$.

The above results can be used to calculate the response, h_{oo}, to the (0, 0) components of $V_o V_r^*$:

$$h_{oo} \propto \frac{A_r\overline{\psi}_o(\eta)\ e^{i\hat{\eta}y}\left[\omega_\eta \sin \omega_\eta t \overset{t}{*} Rect(\hat{t}/\Delta t)\right]}{\rho_u g + \gamma\hat{\eta}^2} \tag{23}$$

where the effect of viscosity is negligible and $\hat{\eta} = 2\pi(\eta - \eta_r)$. In order to continue, it is necessary to assign values to K, and also to the system parameters.

K can range from a minimum of π/R_t to a maximum of $2\pi(\eta_M + \eta_r)$, where R_t is the minitank radius, and η_M is the maximum spatial frequency present in the focused image (5). It is assumed that η_M is $0(\eta_r)$.

Assume the surface tension $\gamma \sim 0$(10 dyne/cm) and $\rho_u \sim 0$(1 gm/cm^3); the notation $A \sim 0(B)$ denotes that A is the same order of magnitude as B. Then, for $\hat{\eta} \sim 0$(100 rad/cm), ω_η is 0(10^3 rad/s). Assume a typical pulse duration, $2\Delta t$, of 10^{-4} seconds. For these typical values, h_{oo} will be proportional to

$$\frac{2\omega_\eta\Delta t \sin \omega_\eta t}{\gamma\hat{\eta}^2}$$

when $t > 2\Delta t$.

The response h_{01} to the (0, +1) component of $V_o V_r^*$ is more complicated, since it depends upon the temporal variation of $k'W(t)$:

$$h_{01} \propto \frac{A_r\overline{\psi}_o(\eta)\ e^{i\hat{\eta}y}\left[\omega_\eta \sin \omega_\eta t \overset{t}{*} <k'W>\ \mathrm{Rect}(\hat{t}/\Delta t)\right]}{\rho_u g + \gamma\hat{\eta}^2} \tag{24}$$

Here $\hat{\eta} = 2\pi(\eta - \eta_r - K)$. Similar expressions can be written for the liquid surface response due to the $(\pm 1, 0)$, and $(\pm 1, \pm 1)$ components $V_o V_r^*$.

The temporal variation of $k' W(t)$ depends upon the response of the membrane to a pulse of radiation pressure. This is calculated in APPENDIX A. The membrane has a characteristic time variation $|W| \sin \omega_M t$ so that the liquid surface response h_{01} to the $(0, +1)$ component is proportional to

$$\frac{k'|W| \, \omega_\eta < \sin \omega_M t > \text{Rect}(\hat{t}/\Delta t) \overset{t}{*} \sin \omega_\eta t}{\rho_u g + \gamma \hat{\eta}^2}$$

If $h_{01} << h_{oo}$, then it can be shown that no degradation of the image will result. If h_{01} is $\geqslant h_{oo}$, then the interference of the various orders may cause some degradation.

Suppose that the membrane tension T, is of order 10^6 dyne/cm. In APPENDIX A it is demonstrated that the natural frequency ω_M of the membrane is given by

$$\omega_M^2 = \left[(\rho_l - \rho_u) \, gK + TK^3\right] [\rho_u + \rho_l]^{-1} \tag{25}$$

where it is assumed that $2\rho_M bK << \rho_l$ for $2b$ = membrane thickness, ρ_M = membrane density. Let $\rho_u - \rho_l \sim 0(\rho_u)$ as it would be for a typical imaging fluid, e.g., Freon E-5. Assuming water to be the lower fluid,

$$\omega_M \sim 0(10^2 \text{ rad/s}) \text{ for } K \sim 0(.1 \text{ rad/cm})$$

$$\omega_M \sim 0(10^3 \text{ rad/s}) \text{ for } K \sim 0(1 \text{ rad/cm})$$

$$\omega_M \sim 0(10^6 \text{ rad/s}) \text{ for } K \sim 0(100 \text{ rad/cm})$$

If $K \sim 0(100 \text{ rad/cm})$, then $< \sin \omega_M t > \to 0$ since $\omega_M \Delta t$ is $>> 1$. Consequently, only values of K of order unity or less need be considered.

Assume for simplicity that the membrane is excited by a single pulse and oscillates with temporal variation $|W| \sin \omega_M t$. Recall that the transducers emit pulses at a given repetition rate, typically 60 Hz. This means that the oscillating membrane is insonified by sound pulses of duration $2\Delta t$ every 1/60 of a second. As a result, the time-average $< \sin \omega_M t >$ may vary from pulse to pulse. For values of K of order unity or less, $\sin \omega_M t$ is approximately constant over each pulse duration, since $\omega_M \Delta t$ is $<< 1$. As an upper bound on h_{01}, a value of unity is assigned to $< \sin \omega_M t >$ for all pulses.

Hence, the maximum value of h_{01} is proportional to

$$\frac{k'|W| \, \omega_\eta \Delta t}{\gamma \hat{\eta}^2}$$

with $\hat{\eta} = 2\pi(\eta - \eta_r) - K$ and $K \sim 0(1 \text{ rad/cm})$. h_{01} will be negligible compared to h_{oo} if

$$k'|W| << 1.$$

In APPENDIX A, it is shown that

$$|W| \sim 0 \left| \frac{2\Delta t\ P_r\ K}{\omega_M(\rho_u + \rho_l)} \right| \tag{26}$$

Since P_{rb} is the component of P_r having spatial frequencies of order 1 rad/cm or less the requirement $k'|W| << 1$ gives

$$P_{rb} << \frac{\omega_M(\rho_u + \rho_l)}{2\Delta t K k'} \tag{27}$$

The radiation pressure, P_{rb}, can be expressed in terms of the intensity, I_o, in the focused image, and I_r, the intensity in the reference beam [5]:

$$P_{rb} = \frac{2(I_o + I_r)}{c_u} \tag{28}$$

This allows the requirement $k'|W| << 1$ to be rewritten in terms of the acoustic intensities:

$$I_o + I_r << \frac{\omega_M(\rho_u + \rho_l)c_u}{4\Delta t K k'} \tag{29}$$

This is the first of four major constraints for maximum performance. For a typical system employing Freon E-5 as the imaging fluid, $\rho_u = 1.8$ gm/cm^3 and $c_u = .7(10^5)$ cm/s. Letting $2\Delta t = 10^{-4}$ s, $\cos\theta_l \cong 1$ and assuming a frequency of 5 MHz gives

$$I_o + I_r << 10^2 \text{ watts/cm}^2. \tag{30}$$

Recall that K is 0(1 rad/cm). Now the center spatial frequency, $\hat{\eta}_r$, in $\bar{P}_{ri}(\hat{\eta})$ is typically much larger than 0(1 rad/cm), so that the effect of h_{01} is to add at most a low spatial frequency distortion to the levitation of the liquid surface. This effect is much the same as that due to the bulge distortion, considered later.

Linearity of Liquid Surface Response

Another requirement to be satisfied for proper operation of the levitation system is that the liquid surface respond in a linear fashion to a pulse of radiation pressure. Nonlinearity of response can occur due to convective acceleration in the imaging fluid, as represented by the term $\mathbf{v} \cdot \nabla \mathbf{v}$ in the Navier-Stokes equation

$$\rho[\dot{\mathbf{v}} + \mathbf{v} \cdot \nabla \mathbf{v}] = -\nabla P + \mu \nabla^2 \mathbf{v} + \rho \mathbf{g} \tag{31}$$

for an incompressible fluid. Here $\mathbf{v}$ is the fluid particle velocity in the minitank caused by levitation of the surface, and dots over a quantity denote a time derivative. If the levitation of the surface is much less than λ_s, the wavelength of a gravity-capillary wave, then the Navier-Stokes equation can be linearized [4].

Recall that if the radiation pressure at the liquid surface is of the form $P_r = p_r(\hat{\eta}) e^{i\hat{\eta}y} \mathrm{Rect}(\hat{t}/\Delta t)$ then

$$h(y, t) = \bar{H}(\hat{\eta}, t)\ p_r(\hat{\eta})\ e^{i\hat{\eta}y} \tag{32}$$

where for negligible viscosity,

$$\overline{H}(\hat{\eta},\ t) = \frac{\omega_\eta \sin \omega_\eta t \overset{t}{*} Rect(\hat{t}/\Delta t)}{\rho_u g + \gamma \hat{\eta}^2} \tag{33}$$

Note that the corresponding wavelength, λ_s, is given by $2\pi/\hat{\eta}$.

Recall also that the total radiation pressure is given by

$$P_r = P_{rb} + P_{ri} + P_{ri}^* \tag{34}$$

where

$$P_{rb} = \frac{1}{4}\rho_u(|V_o|^2 + |V_r|^2)$$

and

$$P_{ri} = \frac{1}{4}\rho_u |V_o||V_r| e^{iX_o} e^{-2\pi i \eta_r y} \tag{35}$$

Consider the radiation pressure due to P_{ri}; this will have a spatial frequency which lies in the bandwidth $-\hat{\eta}_M + \hat{\eta}_r \leqslant \hat{\eta} \leqslant \hat{\eta}_M + \hat{\eta}_r$ where $\hat{\eta}_M$ is the maximum spatial frequency passed by the acoustic lens system. Since P_{ri} is $0(\sqrt{I_o I_r}/c_u)$, the requirement $h_s << \lambda_s$ gives

$$\sqrt{I_o I_r} << \frac{\pi \omega_\eta \rho_u c_u}{\hat{\eta}^2 \Delta t \tanh \hat{\eta} d}. \tag{36}$$

where it will be recalled that d is the minitank depth and it is assumed that $\omega_\eta \Delta t$ is $<<$ 1. Using previously cited values of $\hat{\eta}$, ρ_u, γ etc. and assuming $d \sim 0(1$ mm) or larger, the inquality becomes

$$\sqrt{I_o I_r} << 10^2 \text{ watt/cm}^2 \tag{37}$$

In a similar fashion, it is found that the requirement

$$I_o + I_r << \frac{\pi \omega_\eta \rho_u c_u}{\hat{\eta}^2 \Delta t \tanh \hat{\eta} d} \tag{38}$$

must be imposed so that the bulge levitation, h_b, will be much less than the corresponding wavelength, λ_s. Here $\hat{\eta}$ is of the order of π/R_t, where R_t is the minitank radius. With the values quoted above for system parameters, and assuming $R_t \sim$ 0(10 cm),

$$I_o + I_r << 10^5 \text{ watt/cm}^2. \tag{39}$$

The preceding analysis has been based on the assumption that the radiation pressures P_{ri} and P_{rb} each have one spatial frequency component. It can be shown [3] that inequality (36) must still be satisfied when P_{ri} has many spatial frequency components.

Effect of Bulge Distortion

Assume that the liquid surface is iluminated by laser light of wavelength λ_o at normal incidence; i.e., $\Psi_o \propto Ae^{ik_o z}$ where Ψ_o is the optical disturbance and $k_o = 2\pi/\lambda_o$.

Let the characteristic wavelength of the liquid surface displacement be much longer than an optical wavelength. Assuming small slope of the surface, the reflected light, Ψ, is given by

$$\Psi \propto A\, e^{ik_o z}\, e^{-2ik_o h} \tag{40}$$

The levitation, h, can be represented as $h = h_i + h_i^* + h_o + h_r + h_N$. The quantity $(h_i + h_i^*)$ is the levitation corresponding to the interference between the focused acoustic image and the reference beam. For perfect fidelity of the levitation system, h_i is proportional to $\phi_o \phi_r^*$, where

ϕ_o = velocity potential in focused acoustic image

ϕ_r = velocity potential in reference beam.

Likewise, h_i^*, the complex conjugate of h_i, is proportional to $\phi_o^* \phi_r$. h_o is the levitation due to $\phi_o \phi_o^*$, the self-interference of the image, assuming it contains one spatial frequency component. h_r is the levitation due to $\phi_r \phi_r^*$, the self-interference of the reference beam. h_N is a "noise" term, accounting for the fact that the reference beam and acoustic image are not composed of single plane waves, but are made up of waves of varying spatial frequency and are of finite extent.

When the expression for h is substituted into (40), there results

$$\Psi \propto A\, e^{-2ik_o h_N}\, e^{-2ik_o (h_i + h_i^*)} \tag{41}$$

The constant phase factor $e^{-2ik_o(h_o + h_r)}$ has been dropped, since it has no effect on the imaging process.

If $h_i \propto \phi_o \phi_r^*$ and $\phi_o = |\phi_o|\, e^{iX_o}$, $\phi_r = A_r e^{2\pi i \eta_r y}$ at the liquid surface, then $h_i + h_i^* \propto A_r |\phi_o| \cos(2\pi\eta_r y - X_o)$. The exponential $e^{-2ik_o(h_i + h_i^*)}\, e^{ik_o z}$ has the series expansion [1]

$$\exp\left[-2ik_o\alpha|\phi_o| \cos(2\pi\eta_r y - X_o) + ik_o z\right]$$

$$= \sum_{-\infty}^{\infty} -i^m J_m(2k_o\alpha|\phi_o|)\, e^{imX_o}\, e^{-2\pi i m \eta_r y}\, e^{ik_o z} \tag{42}$$

where α is a factor of proportionality. As pointed out by the authors of [1] and [2], this represents a set of orders of light with direction cosines $\cos\theta_m$, where $\sin\theta_m = 2\pi m \eta_r / k_o$. Substituting the above into the expression (41) for the reflected field shows that a phase distortion $e^{-2ik_o h_N}$ will be present on each of the reflected orders. It is of interest to study the effect of this distortion on the optical reconstruction.

For simplicity, it will be assumed that $k_o h_i$ (or, equivalently, $k_o\alpha|\phi_o|$) is < 1, so that only the 0 and ± 1 orders of reflected light are significant. The 0 order is the specularly reflected light; the $+1$ order has direction cosine $\cos\theta_1$, where $\sin\theta_1 = 2\pi\eta_r/k_o$; the -1 order has direction cosine $\cos\theta_{-1}$, where $\sin\theta_{-1} = -2\pi\eta_r/k_o$.

In the usual method of optical reconstruction, only the $+1$ (or -1) order is used to produce the optical image. This is done by allowing the orders to pass through a lens placed a focal length above the surface. In the back focal plane of the lens, a pinhole passes only the appropriate order. The light field at the pinhole is the Fourier transform of the reflected order. A second lens is placed so that in its back focal plane, where there is a ground glass viewing screen, it produces the transform of the light passed by the pinhole. This transform is merely the scaled, inverted form of the +1 order (or -1 order). For simplicity, the scaling (magnification) will be neglected.

Since $k_o h_i$ (or $k_o\alpha|\phi_o|$) is small, the optical disturbance in the +1 order at the viewing screen is

$$\Psi \propto e^{-2ik_o h_N} k_o\alpha|\phi_o| \, e^{iX_o} \, e^{2\pi i\eta_r y} \tag{43}$$

The small argument form

$$J_1(k_o\alpha|\phi_o|) \cong k_o\alpha|\phi_o| \tag{44}$$

has been used in the above. It is also assumed that the spatial frequency content of h_N is low enough so that all the +1 order light passes through the pinhole.

To consider the effect of the phase distortion $e^{-2ik_o h_N}$, let $k_o h_N < 1$ for simplicity, so that $e^{-2ik_o h_N} \cong 1 - 2ik_o h_N$; then the optical intensity at the viewing screen is

$$|\Psi|^2 \propto (k_o\alpha|\phi_o|)^2 + 4k_o^4\alpha^2|\phi_o|^2 h_N^2 \tag{45}$$

The term $(k_o\alpha|\phi_o|)^2$ is the optical intensity when only the undistorted +1 order is present. The term $4k_o^4\alpha^2|\phi_o|^2h_N^2$ arises because of the low spatial frequency distortion of the liquid surface, which causes the phase distortion $e^{-2ik_o h_N}$.

The bandwidth of spatial frequencies present in the optical image is (ignoring magnification) $-\eta_M \leqslant \eta \leqslant \eta_M$, where η_M is the maximum spatial frequency passed by the acoustic lens system. Typically, the spatial frequencies in h_N are much lower than this. Taking the Fourier transform of $|\phi_o|^2h_N^2$ shows that its spatial frequency components are given by:

$$\overline{|\phi_o|^2h_N^2} = \overline{|\phi_o|^2} \overset{\eta}{*} \overline{h_N^2} \tag{46}$$

where the bars denote the Fourier transform, and $\overset{\eta}{*}$ denotes the convolution in the spatial frequency domain. Let the maximum spatial frequency present in h_N be η_N, and $\eta_N << \eta_M$. The convolution operation causes the bandwidth of spatial frequencies to extend from $-\eta_N - \eta_M$ to $\eta_N + \eta_M$. However, because η_N is $<< \eta_M$, there will be only a slight increase in bandwidth.

The spatial frequencies of the undistorted image are given by $k_o^2\alpha^2\overline{|\phi_o|^2}$. Since the bandwidth of spatial frequencies in (45) is still of order η_M, and $k_o^2h_N^2$ is < 1, the low spatial frequency distortion can usually be ignored.

There is a class of objects whose dominant spatial frequencies can be $0(\eta_N)$. These are low-contrast objects, whose acoustic transmission function, S_o, varies by less than an order of magnitude over the field of view of the acoustic lens system. As an idealized example, suppose that

$$S_o = \begin{cases} A & |y| > y_o \\ A - \Delta A & |y| < y_o \end{cases}$$

where a two-dimensional object is assumed for simplicity, and $(A - \Delta A)/A$ is order unity. If the object is insonified by a plane wave of constant amplitude, then $\phi_o \propto S_o$. The spatial frequency content of $\overline{|\phi_o|^2}$ is approximately

$$\overline{|\phi_o|^2} = A^2\delta(\eta) - 2A(\Delta A)y_o \; sinc(2\pi\eta y_o) \tag{47}$$

with $\delta(\eta)$ the Dirac delta function, representing the d.c. spatial frequency of the background $S_o^2 = A^2$. The sinc function is defined by $sinc(x) \equiv (\sin x)/x$.

A measure of the bandwidth of spatial frequencies in $\overline{|\phi_o|^2}$ is the width of the main lobe of the sinc function; this gives $\eta_M = .5/y_o$. If $\eta_M \leqslant \eta_N$, then the convolution $\overline{h_N^2}^{\eta} * A\delta(\eta)$ in (46) can have spatial frequencies in the bandwidth $(-\eta_M, \eta_M)$ which are the same order of magnitude as $2A(\Delta A)y_o$. In the spatial domain, this implies that the region of low contrast, $|y| < y_o$, is as large or larger than K^{-1}, where K is a characteristic spatial frequency of h_N. The contrast ratio, $\Delta A/A$, is also the same order of magnitude (or less) than the maximum value of $k_o^2 h_N^2$.

The quantities S_o^2 and $S_o^2[1 + 4k_o^2 h_N^2]$ are plotted in Fig. 3 for the idealized case

$$h_N = |h_N'| \cos Ky$$

i.e., the bulge distortion is assumed to have one dominant spatial frequency, K. Recall that the optical intensity is proportional to S_o^2 for perfect fidelity. The effect of bulge distortion is to superpose a fringe pattern of spatial frequency $2K$ on the intensity S_o^2. Since the depth of modulation, $(k_o h_N)^2$, of the fringe pattern is of order $\Delta A/A$, it will be difficult to detect the presence of the low-contrast object. If $(k_o h_N)^2$ is reduced so that it is $<< \Delta A/A$, then the low-contrast object will be detected.

Note that if $k_o h_N$ becomes > 1, then the distortion of the optical image becomes even worse. This occurs because the amplitude of the distortion term $k_o^2 h_N^2$ increases, and also because terms such as $k_o^4 h_N^4$, etc. will then appear in the expression for $|\Psi|^2$ at the viewing screen. These higher-order distortion terms will contain higher spatial frequencies. The result will be a kind of spurious resolution, in which these higher frequencies modulate the intensity of the true image.

Another factor, neglected to this point, is that as $k_o h_N$ increases, it causes the distorted +1 order beam $k_o \alpha |\phi_o| \, e^{ik_o(y \sin\theta_1 + z)} \, e^{-2ik_o h_N}$ to have spatial frequency components which may not pass through the pinhole. This will also cause degradation in the optical image.

Calculation of Bulge Distortion

Recall that the liquid surface is subjected to a train of pulses of radiation pressure. The laser samples the surface at a time t_s after each pulse reaches the surface. If there is a time at which h_N is $<< k_o^{-1}$, sampling the surface then could eliminate the problem of image degradation due to h_N. It will be shown here that for typical system parameters, this cannot be done, as the low spatial frequency "bulge", h_N, reaches a steady-state value.

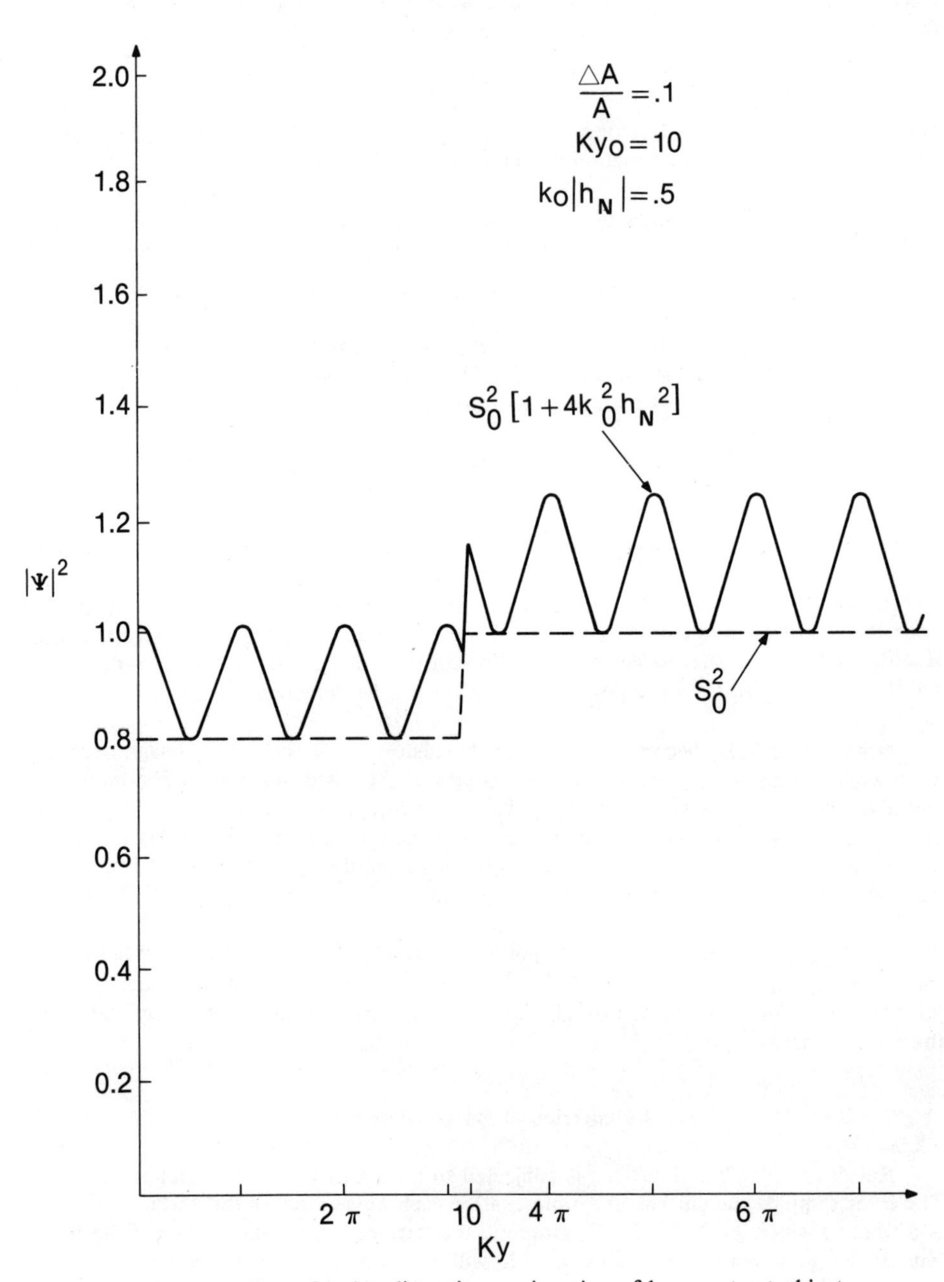

Fig. 3. Effect of bulge distortion on imaging of low-contrast object.

Assume that the sum of radiation pressure in the reference beam and in the acoustic image is of the form

$$P_{rb} = P \; Rect(y/y_b) \sum_{n=0}^{N} Rect\left(\frac{\hat{t} - nt_1}{\Delta t}\right), \qquad t > Nt_1 \tag{48}$$

at the surface. P_{rb} is the radiation pressure which causes the bulge, and is applied over the region $|y| \leqslant y_b$; a two-dimensional problem is assumed for simplicity. Also, the pulses of radiation pressure are applied at times $t = nt_1$, and are of duration $2\Delta t$.

The minitank which contains the imaging fluid is assumed to have rigid walls at $y = \pm R_t$, and a rigid bottom at $z = 0$. This last implies that any displacement of the membrane at the bottom of the minitank is much less than the levitation of the liquid surface.

Following Pille [1], the radiation pressure is put in series form

$$P_{rb} = \sum_{n=0}^{N} \left\{ \sum_{j=0}^{\infty} P_j \cos K_j y \right\} Rect\left(\frac{\hat{t} - nt_1}{\Delta t}\right) \tag{49}$$

with

$$P_j = \begin{cases} Py_b/R_t, & j = 0 \\ (2Py_b/R_t) \; sinc \; K_j y_b, & j > 0 \end{cases} \tag{50}$$

and $K_j = j\pi/R_t$. Since the radiation pressure is symmetric, only the symmetric modes of the liquid surface will be excited. Hence, the levitation h_b can be expanded in the form

$$h_b = \sum_{j=0}^{\infty} h_j(t) \cos K_j y \tag{51}$$

As pointed out by Pille [1], the above expression for P_{rb} repeats periodically at intervals $\pm 2mR_t$, with $m = 1, 2, \ldots$, etc. The same is true of h_b. Hence, the problem can be considered to be the response of a liquid surface of infinite lateral extent ($y \rightarrow \pm\infty$) to pulses of radiation pressure of width $2y_b$ which are centered at $y = \pm 2mR_t$.

The response of a liquid surface of infinite lateral extent has been calculated by the authors of [1] and [2], and is given by (19)-(22). Note that a slightly viscous imaging fluid is assumed, defined by $2\hat{\eta}^2\nu << \omega_\eta$. Since only discrete spatial frequencies K_m occur in the expressions for h_b and P_{rb}, $\hat{\eta}$ is to be replaced by K_m in the expressions for $\overline{H}(\hat{\eta},t)$, ω_η, etc. Then the liquid surface response to a train of pulses is

$$h_b(y,t) = \sum_{m=1}^{\infty} \frac{P_m \omega_m \cos K_m y}{\rho_u g + \gamma K_m^2} \left\{ \sum_{n=0}^{N} e^{-q\omega_m t} \sin \omega_m t \overset{t}{*} \mathrm{Rect}\left(\frac{\hat{t} - nt_1}{\Delta t}\right) \right\} \tag{52}$$

Note that $\omega_m = 0$ for $m = 0$, so that there is no response of the zero spatial frequency mode. Such a mode would cause a uniform levitation of the liquid surface; for an incompressible fluid, this would violate mass conservation. Hence, the modal sum (over m) runs from $m = 1$ to ∞ in the above.

It is useful to classify the modes as low, intermediate- and high frequency modes. Low-frequency modes are those for which $\omega_m t_1$ is $<< 1$. For intermediate-frequency modes, $\omega_m t_1$ is of order unity, and when $\omega_m t_1$ is $>> 1$, the mode is said to be high-frequency. A low-frequency mode responds very little to the initial pulse of radiation pressure, by the time the next pulse arrives at $t = t_1$. It will be shown that the result of this slow response is that low-frequency modal responses build up to a steady-state value. High-frequency modal responses vanish as $t \rightarrow t_1$, due to viscous damping.

The response of a low-spatial frequency mode is proportional to the summation in brackets above:

$$\sum_{n=0}^{N} e^{-q\omega_m t} \sin \omega_m t \overset{t}{*} \mathrm{Rect}\left(\frac{\hat{t}-nt_1}{\Delta t}\right)$$

For typical system parameters, the summation can be approximated by the convolution

$$2\left(\frac{\Delta t}{t_1}\right)[e^{-q\omega_m t} \sin \omega_m t \; G(t) \overset{t}{*} G(t)]$$

where $G(t)$ is the Heaviside step function; this is done in detail in Ref [3]. Each low-frequency mode acts as a single degree of freedom oscillator, driven by a pulse train. There are many pulses applied during the period of oscillation, $2\pi/\omega_m$. Consequently, the oscillator behaves as if driven by a step function with weighting factor $2\Delta t/t_1$.

The convolution is within a few percent of the steady-state value, ω_m^{-1}, when $t =$ .5 sec for the parameters used previously. A first approximation can be obtained for h_b by summing over only the low-frequency modal responses, and neglecting γK_m^2 compared to $\rho_u g$:

$$h_b(y,t) \cong \frac{2P\Delta t}{\rho_u g t_1}\left[\mathrm{Rect}(y/y_b) - \left(\frac{y_b}{R_t}\right)\right]. \tag{53}$$

The surface is elevated in the region $|y| < y_b$ and depressed elsewhere, so that mass conservation is satisfied.

The above equation predicts the presence of discontinuities at $y = \pm\, y_b$, a clearly impossible situation. The effect of surface tension will not permit this to happen. The restoring force/area due to surface tension is represented by the term γK_m^2, neglected above. This term will be significant for intermediate frequency modes, whose contributions must be included in the regions near $y = \pm\, y_b$.

To obtain a better approximation for h_b, the summation (52) has been performed for $N = 60$, using 50 modes of the liquid surface. Since the typical pulse repetition rate is approximately 60 hz, $N = 60$ corresponds to the steady-state bulge value reached 1 second after the surface is initially excited. The result of this computation for system parameters cited previously is shown in Fig. 4, where it is assumed that $y_b/R_t = .7$. Here $h_o + h_r$ is given by the value of h_b at $y = 0$, the center of the mini-

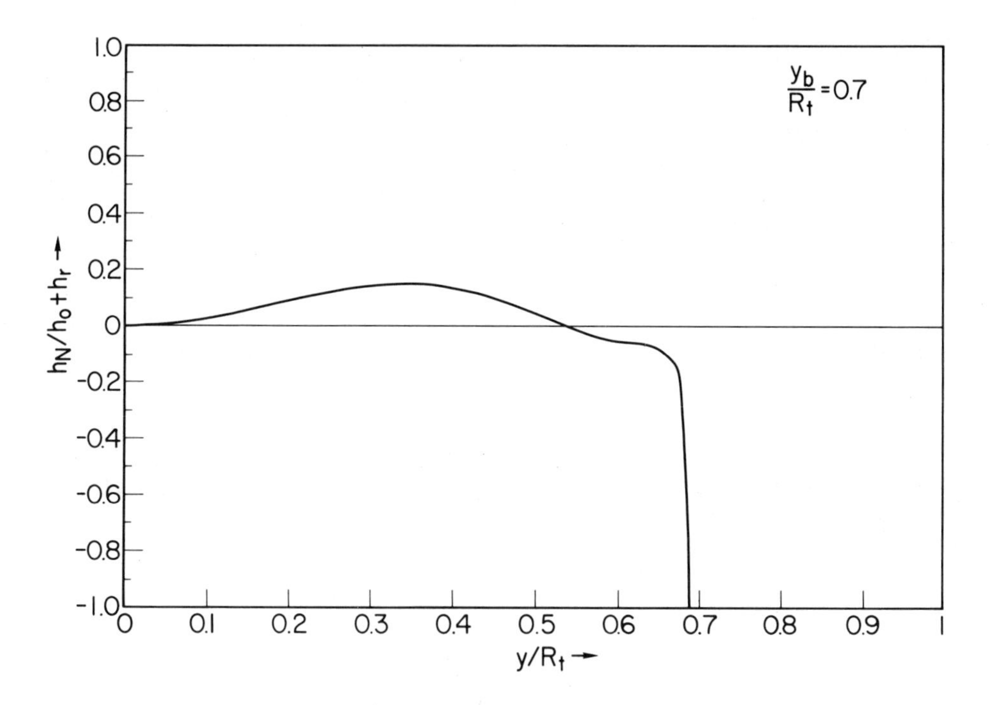

Fig. 4. Calculated bulge distortion for typical system parameters.

tank. It is interesting to note that $h_o + h_r$ is within 10% of the value estimated by (53). Furthermore, h_N is of order $0.1(h_o + h_r)$.

Recall that $k_o^2 h_N^2$ must be $<< \Delta A/A$ for a low-contrast object to be detectable. Assume that ΔA must be at least 0(.1A) for a low contrast object to be detectable even in the absence of bulge distortion; then $k_o h_N$ should be $< .1$.

Since $h_N \sim 0(.1[h_o + h_r])$ and

$$h_o + h_r \cong \frac{2\Delta t \; P_{rb} \; [\mathrm{Rect}(y/y_b) - y_b/R_t]}{\rho_u g t_1} \tag{54}$$

the requirement $k_o h_N < .1$ can be put in the form

$$\frac{2k_o \Delta t \; P_{rb} \; (1 - y_b/R_t)}{\rho_u g t_1} < 1 \tag{56}$$

For $P_{rb} \sim 0 \;\; (2(I_o + I_r)/c_u)$, the inequality can be rewritten to give

$$I_o + I_r < \frac{\rho_u c_u \lambda_o t_1 g}{(8\pi) \; \Delta t \; (1 - y_b/R_t)} \tag{57}$$

For values cited previously, this becomes

$$I_o + I_r \leqslant 0(10^{-2}) \text{ watt/cm}^2 \tag{58}$$

assuming $\lambda_o \cong .6(10^{-4})$ cm.

Contrast Reversal

In the usual method of optical reconstruction, the $+1$ (or -1) order of light reflected from the liquid surface is selected to form the optical image. The reflected light has been shown to be of the form

$$\Psi \propto \sum_{-\infty}^{\infty} -i^m \, J_m(2k_o \, \alpha|\phi_o|) \; e^{imX_o} \; e^{-ik_o y \sin\theta_m} \tag{59}$$

where $\sin\theta_m = 2\pi m \eta_r / k_o$, and $2k_o h_N$ is assumed $<< 1$. When the $+1$ order is allowed to pass through the pinhole (see Fig. 1), the optical intensity at the viewing screen is

$$|\Psi|^2 \propto J_1^2(2k_o \, \alpha \, |\phi_o|). \tag{60}$$

Since ϕ_o is the velocity potential in the focused acoustic image at the liquid surface, it is desirable to have $|\Psi|^2$ proportional to $|\phi_o|^2$. From the behavior of the first order Bessel function, it is seen that $2k_o \, \alpha \, |\phi_o|$ should be less than unity.

In fact, if this quantity is greater than (approximately) 1.8, then a contrast reversal will occur. To illustrate this, suppose that there is an area in the acoustic image where $2k_o \, \alpha \, |\phi_o| \cong 1.8$. If the power transmitted through the acoustic object is increased, then $\alpha|\phi_o|$ will increase. However, the corresponding region in the optical image will become darker; i.e., a contrast reversal has occurred. This is due to the fact that

$J_1(2k_o\,\alpha\,|\phi_o|)$ has its maximum at $2k_o\,\alpha\,|\phi_o| \cong 1.8$; see Fig. 5. Image interpretation becomes difficult, since it is normally assumed that as more acoustic power is transmitted through an area of the acoustic object, the corresponding area in the optical image will become brighter.

As the amount of acoustic power transmitted by the object is increased, the optical image becomes (theoretically) darker until $2k_o\,\alpha\,|\phi_o| \cong 3.8$. Then, since the first order Bessel function again increases as the argument increases, the optical image will become brighter.

Hence, for a linear relation between acoustic intensity and optical intensity, it is necessary that $2k_o\,\alpha\,|\phi_o| < 1$. If contrast reversal is to be avoided, it is necessary to require $2k_o\,\alpha\,|\phi_o| < 1.8$. The latter inequality is chosen, since some nonlinearity can be tolerated.

The coefficient α relates $|\phi_o|$ to h_i, i.e., $|h_i| = \alpha\,|\phi_o|$ for perfect fidelity of the recording system. Furthermore, $h_i = \overline{H}(\hat{\eta}_r,t_s)\,P_{ri}$ where it is assumed that $\overline{H}(\hat{\eta},t_s)$ is (approximately) constant over the bandwidth of spatial frequencies in $\overline{P}_{ri}(\hat{\eta})$. Since $P_{ri} \sim 0(\sqrt{I_o I_r / c_u})$, the necessary condition for no contrast reversal becomes

$$\sqrt{I_o\,I_r} < \frac{1.8\,\lambda_o\,c_u}{4\pi\,\overline{H}(\hat{\eta}_r,\,t_s)} \tag{61}$$

Assume that the optical wavelength, λ_o, is $6(10^{-7})$m. Let a pulse of radiation pressure lasting 10^{-4} second be incident on a minitank filled with Freon E-5. If the spatial frequency $\hat{\eta}_r$ in the reference beam is 100 rad/cm and t_s is $0(10^{-4}$ s) then $\overline{H}(\hat{\eta}_r,\,t_s) \cong (10^3\rho_u g)^{-1}$; see Ref. [1] and [2]. Substituting into the inequality (61) shows that for this typical system,

$$\sqrt{I_o\,I_r} \leqslant 0(10^{-1}\ \text{watt/cm}^2) \tag{62}$$

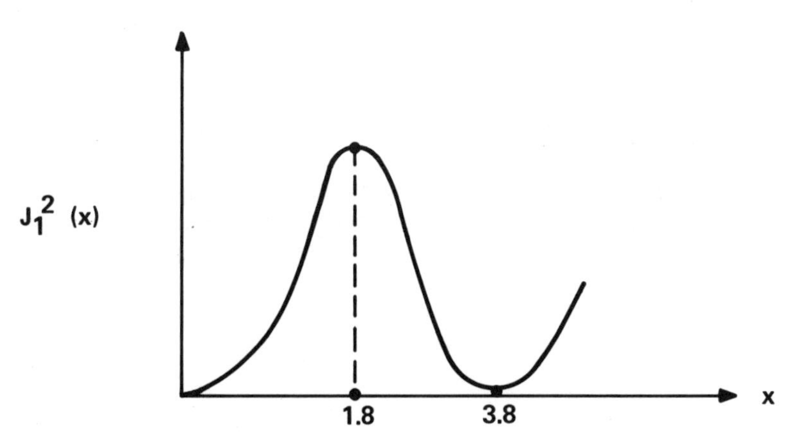

Fig. 5. Plot of J_1^2 (x).

Governing Inequalities-High Contrast Objects

It has been demonstrated that for all imaging applications, care must be taken to insure that the liquid surface acts as a linear recorder, and that contrast reversal does not occur. This necessitates that the acoustic intensity in the reference beam and in the focused acoustic image satisfy the inequalities

$$I << \frac{\pi \omega_\eta \rho_u c_u}{\hat{\eta}^2 \Delta t \tanh(\hat{\eta} d)} \tag{63}$$

and

$$\sqrt{I_o I_r} << \frac{1.8\lambda_o c_u}{4\pi \bar{H}(\hat{\eta}_r, t_s)} \tag{64}$$

In the first of these, I is to be interpreted as the sum $I_o + I_r$ when $\hat{\eta}$ is a characteristic spatial frequency associated with the bulge levitation, h_b. When $\hat{\eta}$ is associated with h_i, then I becomes the product $\sqrt{I_o I_r}$.

The above inequalities have been evaluated for typical system parameters. The first of these gives

$$\sqrt{I_o I_r} \leqslant 0(1 \text{ watt/cm}^2) \tag{65}$$

and

$$I_o + I_r \leqslant 0(10^3 \text{ watt/cm}^2) \tag{66}$$

so that linearity is obtained. In like fashion,

$$\sqrt{I_o I_r} \leqslant 0(10^{-1} \text{ watt/cm}^2) \tag{67}$$

so that contrast reversal does not occur.

For high-contrast acoustic objects, these are the chief constraints which should be satisfied to obtain maximum performance. Violation of these will cause a decrease in system performance from that predicted in [1] and [2]. Recall that high-contrast objects are those which have an acoustic transmission function which varies by more than an order of magnitude over the field of view of the imaging system.

It is interesting to estimate which of the three preceeding inequalities is the most stringent. Let $(I_o)_b$ denote the intensity in the acoustically "bright" part of the focused image, and $(I_o)_d$ the intensity in the "dark" region. In most imaging applications, $(I_o)_b$ will be $\leqslant I_r$, due to insertion loss and attenuation. As long as $(I_o)_b$ is greater than $0(10^{-8} I_r)$, the requirement $I_o + I_r \leqslant 10^3$ watt/cm^2 is automatically satisfied. Assuming this is so, it remains to determine whether

$$\sqrt{(I_o)_b I_r} \leqslant 0(1 \text{ watt/cm}^2) \quad (h >> \lambda_s) \tag{68}$$

or

$$\sqrt{I_o I_r} \leqslant 0(10^{-1} \text{ watt/cm}^2) \qquad \text{(no contrast reversal)} \tag{69}$$

is more restrictive. Obviously, if contrast reversal is not allowed in the acoustically bright area, then the latter inequality governs.

Since $(I_o)_d$ is $<<$ $(I_o)_b$, it may not be possible to identify targets in the acoustically dark region. It may be necessary to tolerate contrast reversal in the bright region in this case. Comparing the inequalities above, it is seen that if $(I_o)_d$ is of order $10^{-2}(I_o)_b$, then both inequalities govern the operation of the levitation system. When $(I_o)_d$ is of order $10^{-3}(I_o)_b$ or less, then the requirement $h << \lambda_s$ is more restrictive.

Governing Inequalities-Low Contrast Objects

In the case of low-contrast objects, there are two additional constraints; namely, that h_N be much less than $1/k_o$, and that the membrane displacement, w, be much less than $1/k'$. These have been previously expressed in terms of the inequalities

$$I_o + I_r < \frac{\rho_u c_u \lambda_o t_1 g}{8\pi \Delta t \; (1 - y_b/R_t)} \tag{70}$$

and

$$I_o + I_r << \frac{\omega_M(\rho_u + \rho_l) \, c_u}{2\Delta t \; K \; k'} \tag{71}$$

For system parameters used here, the above assume the form

$$I_o + I_r \leqslant 0(10^{-2} \text{ watt/cm}^2) \tag{72}$$

and

$$I_o + I_r \leqslant 0(1 \text{ watt/cm}^2) \tag{73}$$

respectively. It is immediately obvious that the requirement of negligible bulge distortion is more restrictive than that of negligible membrane distortion.

Furthermore, assuming that I_o is smaller than $0(I_r)$, it is seen that inequality (72) is even more restrictive than (65) and (67). Hence, for low-contrast objects and system parameters quoted previously, maximum performance is achieved when bulge distortion is negligible.

Suppose that $y_b \rightarrow R_t$, i.e., the entire liquid surface is insonified. If P_{rb} has only a zero spatial frequency component, then the bulge levitation decreases by five orders of magnitude, for the parameters used here. Hence, bulge distortion will no longer be the controlling factor for operation of the system.

Assume I_o is $0(I_r)$ or less, and $y_b \rightarrow R_t$. The nature of the governing constraint will depend upon the ratio I_o/I_r. When I_o/I_r is $0(1)$, then contrast reversal will occur before effects of membrane distortion are noticeable. When the ratio is $0(.01)$, then both effects are present. As I_o/I_r is decreased below $0(.01)$, the effect of membrane distortion will be felt before contrast reversal occurs.

MULTIPLE-ORDER RECONSTRUCTION

The preceding analysis has dealt with ways of maximizing performance from typical levitation holography systems by applying constraints on acoustic intensity. In what follows, a method of extending the capability of levitation holography is explored, by relaxing the requirement that only one order of reflected light be used for visual display. In the usual method of optical reconstruction, only the +1 order (or −1 order) light reflected from the surface is used. It has been found that improved imaging can sometimes be obtained by using several orders of reflected light.

To see why this comes about, recall that the reflected optical disturbance has the form

$$\Psi \propto \sum_{-\infty}^{\infty} -i^m J_m(2k_o|h_i|) e^{imX_o} e^{-im\hat{\eta}_r y} e^{ik_o z}. \tag{74}$$

It is assumed for simplicity that $k_o h_N$ is negligible. Suppose that only the +1 and +2 orders are allowed to pass through the optics. Let the acoustic object be an amplitude object; for perfect fidelity of the levitation system, $X_o = 0$. Then, neglecting any scale factors due to magnification by the lens system, the optical disturbance at the viewing screen is

$$\Psi \propto J_1(2k_o|h_i|)\, e^{-i\hat{\eta}_r y} - i\, J_2(2k_o|h_i|)\, e^{-2i\hat{\eta}_r y} \tag{75}$$

The optical intensity viewed by the observer is $|\Psi|^2$:

$$|\Psi|^2 \propto J_1^2(2k_o|h_i|^2) + 2J_1(2k_o|h_i|)\, J_2(2k_o|h_i|) \sin \hat{\eta}_r y + J_2^2(2k_o|h_i|). \tag{76}$$

The first term above represents the image obtained with the usual single-order method of reconstruction. The second term arises due to the interference of the two orders at the viewing screen. The third term is the optical intensity in the +2 order.

Assume the acoustic object consists of a large area of low transmission (target area) with a "noisy" background. Suppose that the background has a characteristic spatial frequency K which is much less than $\hat{\eta}_r$. For simplicity, the acoustic object will be idealized by the two-dimensional transmission function:

$$S_o^2 = \begin{cases} A - \Delta A, & |y| < y_o \\ A + B \cos Ky, & |y| > y_o \end{cases} \tag{77}$$

It is assumed that A is $\geqslant B$.

For perfect fidelity, $h_i \propto S_o$. Assuming $k_o h_i \leqslant .5$ so that

$$J_1(2k_o h_i) \cong k_o h_i \qquad J_2(2k_o h_i) \cong \frac{1}{2}(k_o h_i)^2,$$

the optical intensiy then becomes

$$|\Psi|^2 \propto (k_o h_i)^2 + (k_o h_i)^3 \sin \hat{\eta}_r y + \frac{1}{4}\, (k_o h_i)^4. \tag{78}$$

The last term is the intensity in the +2 order, and is negligible. In the region $|y| < y_o$, the optical image consisits of a darkened area laced with fringes of spatial frequency $\hat{\eta}_r$ and amplitude $k_o h_i$; see Fig. 6. The image in the region $|y| > y_o$ is more complicated. It consists in part of a fringe pattern with spatial frequency $\hat{\eta}_r$, having amplitude $(k_o h_i)^3$. Since $h_i \propto S_o$, and S_o^2 varies sinusoidally for $|y| > y_o$, the amplitude of the fringe pattern varies. Furthermore, the fringe pattern "rides along" on the term $(k_o h_i)^2$, which is the intensity in the +1 order. This is clearly shown in the figure, since the curve labeled "+1 order" is the average of the fringe pattern.

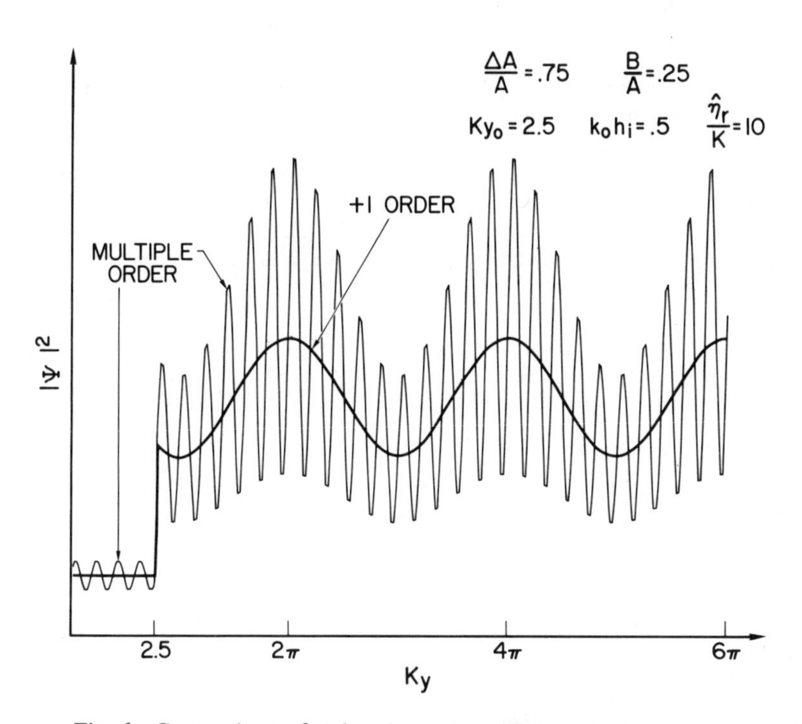

Fig. 6. Comparison of $+1$ order and multiple order images for simple object on noisy background.

When an optical image having the intensity given by (78) is viewed by an observer, it appears as if the "noise", given by $B \cos Ky$, has vanished. The observer is aware of the fringe pattern of spatial frequency $\hat{\eta}_r$. The depth of modulation varies much more slowly than $\hat{\eta}_r$, since it depends upon K, and K is $<< \hat{\eta}_r$. As long as the maxima of the fringe pattern are larger than the adjacent minima, the fringe pattern appears to "wash out" the noise.

A more practical example is that of Fig. 7, which shows a delamination in a carbon composite panel. In the +1 order image, the delamination appears as a darkening in the upper right hand corner. However, there is little to distinguish it from the noisy background. In the multiple-order image, the background has been washed out, and the defect stands out clearly as a darkened area laced with fringes.

In Fig. 8, a carbon bleeder panel was used to provide a noisy background for the targets, which are pieces of Kraft paper in the shapes of letters "*N*" and "*L*". The bleeder panel consists a sheet of fiberglass bonded with epoxy to the back of a carbon composite panel. The +1 order image (left hand picture) was obtained by tipping the panel at about a 30° angle to the insonifying beam. A slight rotation from this orientation would cause the letter "*L*" to be undetectable. The "*N*" could not be detected at all. When multiple orders are used, it is straightforward to obtain the right hand picture. Note that the "*N*" is now clearly visible.

In some cases, the image can be enhanced by including the light specularly reflected from the surface (0 order). Consider for instance the case of using the 0 and +1 orders to form the optical image. The intensity seen by the observer is given by

$$|\Psi|^2 \propto J_o^2(2k_o|h_i|) + 2J_o(2h_o|h_i|)J_1(2k_o|h_i|) \sin \hat{\eta}_r y + J_1^2(2k_o|h_i|). \qquad (79)$$

In regions of little surface levitation, there is a large amount of specular reflection; for small arguments, $J_o(2k_o|h_i|) \cong 1$. Where the levitation is $0(1/k_o)$, there will be significant +1 order reflection. Regions of small levitation correspond to acoustically "dark" areas of the acoustic object; i.e., regions of low transmission.

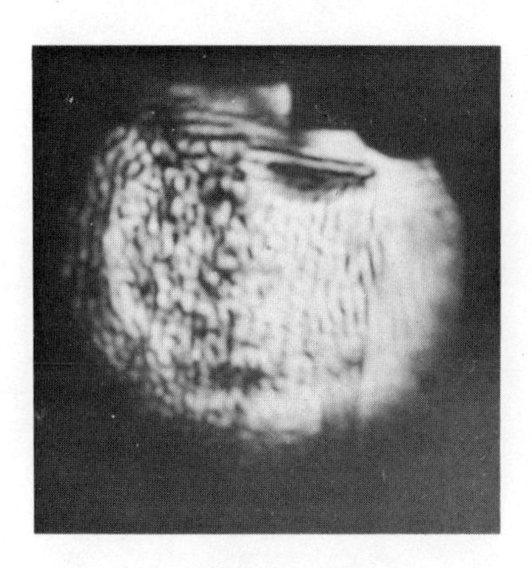

1ST ORDER ONLY

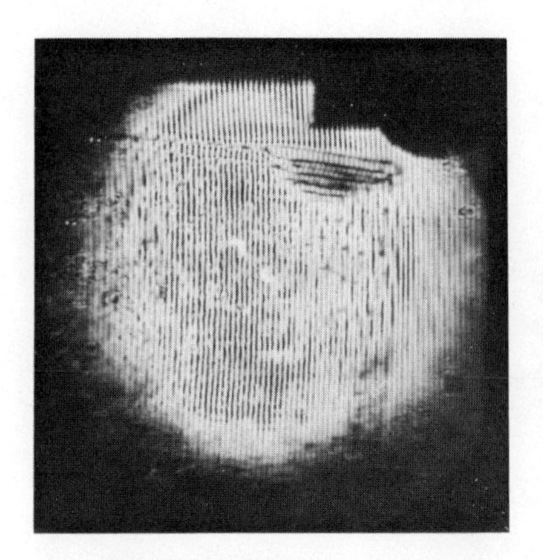

ALL POSITIVE ORDERS

Fig. 7. Delamination in carbon panel.

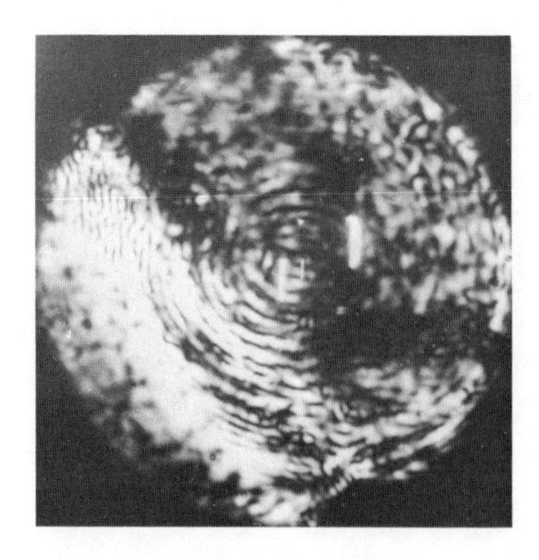

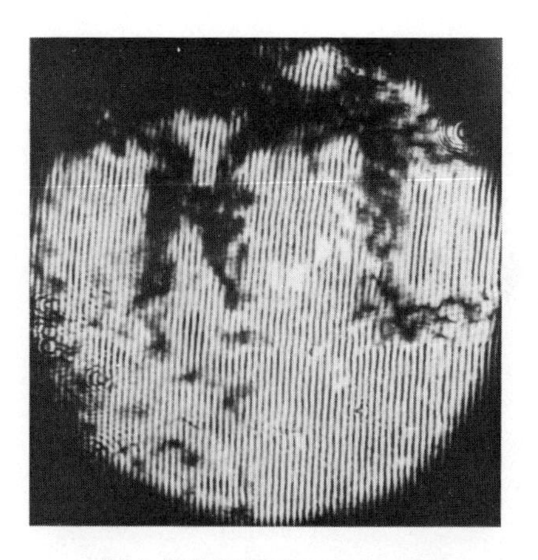

ALL POSITIVE ORDERS

Fig. 8. 0.05 MM Kraft paper on carbon bleeder panel.

Consider the case of a target on a noisy background, represented by (77). If $k_o h_i$ is 0(1), then there will be significant amounts of 0 and +1 order light reflected from the region $|y| > y_o$, the noisy background. When the surface has a local minimum (corresponding to a zero of $B \cos Ky$ in S_o), the amount of 0 order light increases, at the expense of the +1 order light. At local maxima, the converse is true;, i.e., the +1 order increases at the expense of the 0 order. Hence, the noisy background tends to wash out in the optical reconstruction,leaving chiefly the fringe pattern $2J_o(2h_o|h_i|) \times J_1(2k_o)|h_i|) \sin \hat{\eta}_r y$.

An example of this is shown in Fig. 9. The left hand picture is the +1 order image of a plastic wedge having an angle of about 30° at its apex. The dark horizontal lines in the wedge are due to destructive interference caused by multiple internal reflections. The sides and tip of the wedge are difficult to locate, due to diffraction around the wedge. The right hand figure is the result of using multiple orders. The edges of the wedge are now more easily distinguished, since the edge diffraction is washed out by the fringe pattern of spatial frequency $\hat{\eta}_r$. The tip of the wedge is more visible, since the "noisy background" due to diffraction has washed out.

Another case in which the addition of the 0 order light aids in optical reconstruction is that of a highly attenuative acoustic object which causes small surface levitation, i.e., $k_o h_i << 1$. With the direct (+1 order only) imaging technique, the optical intensity is proportional to $(k_o|h_i|)^2$, so that the object may not be visible. When the 0 order light is also used, the intensity is given by

$$|\Psi|^2 \propto 1 + k_o^2 h_i^2 + 2k_o h_i \sin \hat{\eta}_r y. \tag{80}$$

The first term on the right hand side is a bright background due to the intensity in the 0 order. The second term is the direct image (not visible). The last term is the interference between the orders, and is larger than the direct image by a factor of $(k_o h_i)^{-1}$, which is >> 1.

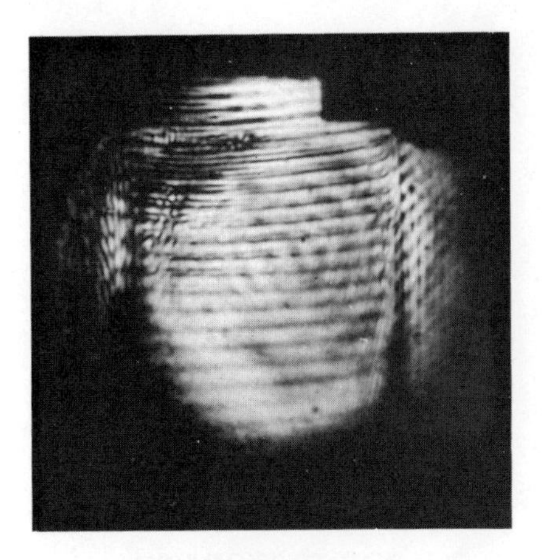

1ST ORDER ONLY (WEDGE)

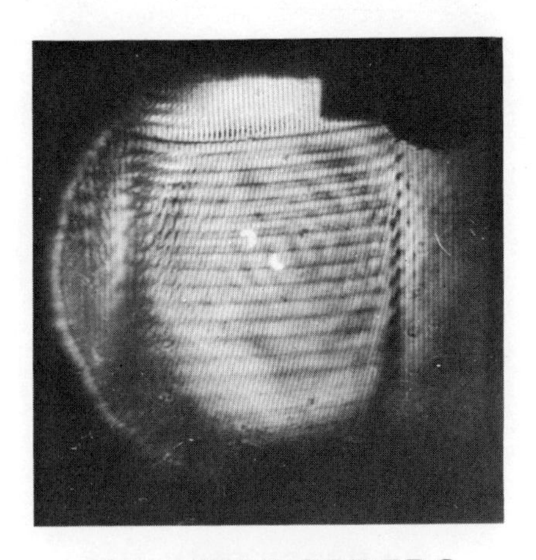

MULTIPLE ORDERS (WEDGE)

Fig. 9. Effect of imaging with orders $\geqslant 0$.

The intensity in the interference term is proportional to S_o. In the direct method, the optical intensity is proportional to S_o^2. However, it is the outline of the object that is detected by eye, and for simple objects, the difference between S_o and S_o^2 can be neglected. Hence, the interference term provides an acceptable image of the object, laced with fringes. It is assumed that the characteristic dimension of the object is $>> \hat{\eta}_r^{-1}$.

An interesting effect can sometimes be obtained by using the -1 order as well as all orders $\geqslant 0$. Areas of the acoustic image which are acoustically dark appear bright in the optical image. To see how this comes about, consider the resulting optical intensity

$$|\Psi|^2 \propto J_o^2(2k_oh_i) + 4J_1^2(2k_oh_i)\cos^2\hat{\eta}_ry. \tag{81}$$

The first term is again the intensity in the specular reflection. Using the identity $\cos^2\hat{\eta}_ry \equiv \frac{1}{2}(1+\cos 2\,\hat{\eta}_ry)$ shows that the second term is due to the intensities in the $+1$ and -1 orders, plus their interference. There is no interference between the specular reflection and higher orders.

In regions where k_oh_i is $<< 1$ (acoustically dark areas), almost all light is specularly reflected, and there are no fringes. This is because there is no interference between the 0 and ± 1 orders, unlike the case where the 0 and $+1$ orders are used. Consequently, if the image is first viewed using 0 and $+1$ orders, fringes will be evident, some of which vanish when the -1 order is added. It appears to the observer as if a constrast reversal has occurred.

This is illustrated in Figs. 10-13. The first of these shows the direct image of a .0625 cm (.025 in.) plastic ruler fixed to the back of a .625 cm (.25 in) fiberglass panel. The second is the result of using all positive orders (chiefly the $+1$ and $+2$ orders). The background has washed out, making the ruler outline more visible. In the next

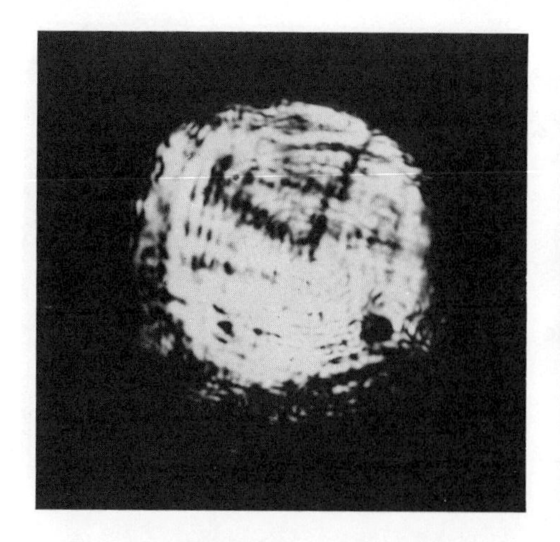

Fig. 10. Direct (+1 order) image of plastic ruler behind fiberglass panel.

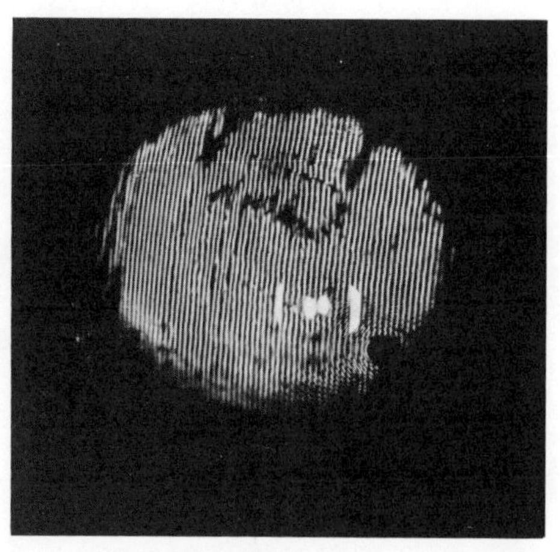

Fig. 11. Image of ruler with all positive orders.

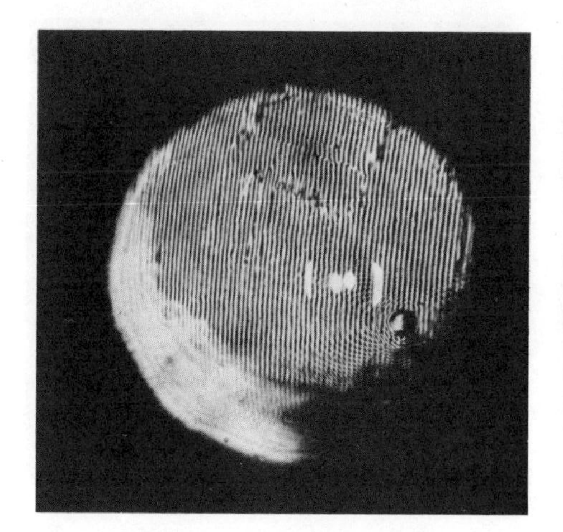

Fig. 12. Image of ruler with all orders $\geqslant 0$.

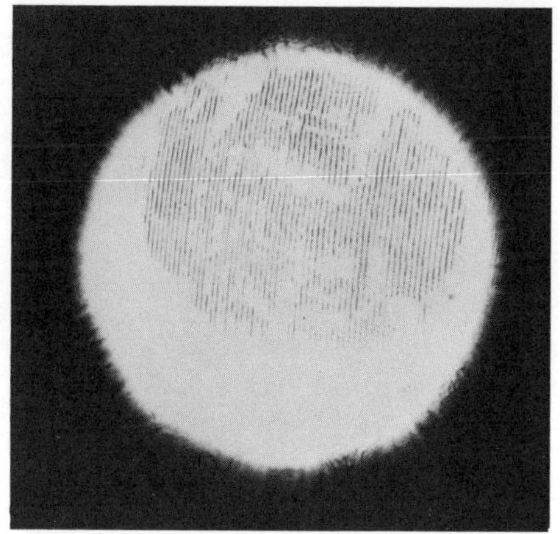

Fig. 13. Image of ruler with all orders $\geqslant 0$, and -1 order.

figure, the addition of the 0 order has caused the ruler outline to become lighter, but still visible. The last figure is made with the addition of the -1 order. The ruler outline now appears bright.

The preceding analysis has shown how enhanced images can be obtained in some instances by multiple-order reconstruction. The experimental results show better image quality when several orders are used, as opposed to the usual method of using only the $+1$ (or -1) order. It is also interesting to compare the image quality obtained with multiple-order levitation holography with that obtained from conventional *C*-scan methods.

Fig. 14. Test object for comparison between C-scan acoustical holography.

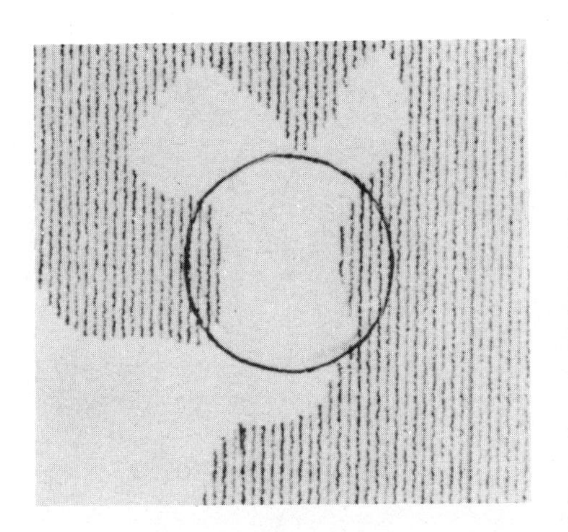

C-SCAN IMAGE (OBJECT CIRCLED)

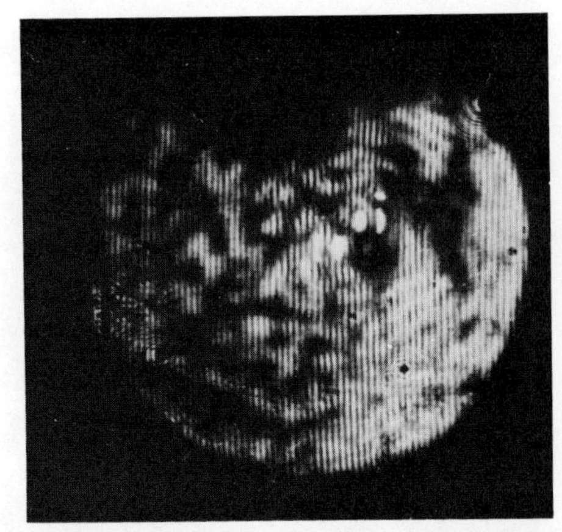

ACOUSTIC HOLOGRAPHIC IMAGE

Fig. 15. Images obtained with C-Scan and acoustical holography.

The result of one such experiment is shown in Figs. 14 and 15. The first of these shows the test object, a piece of .05 mm Teflon tape fixed to the back of a carbon bleeder panel by means of cellulose acetate tape. Fig. 15 shows the best effort *C*-scan and levitation holography images of this object. The object is not resolved in the *C*-scan; it lies in the area circled in the figure. The white area above and below the circle is due to the presence of the cellulose acetate tape which probably has trapped micro-bubbles of air. The object is resolved when levitation holography is used, as indicated by the figure.

To this point, only amplitude objects have been considered. The use of multiple orders in reconstruction also can be used to obtain the outline of phase objects, which cause a change in phase (but not in amplitude) of an acoustic wave passing through them. Note that in the direct method of reconstruction, the optical intensity is proportional to the acoustic intensity, so all phase information about the object is lost. When multiple orders are used, the interference between orders will preserve phase information about the object.

Consider for simplicity the case of a phase object on a uniform background, i.e.,

$$S_o = \begin{cases} A, & |y| > y_o \\ Ae^{iX_o}, & |y| < y_o \end{cases} \tag{82}$$

Suppose that the ± 1 orders are used on reconstruction, and $h_i \propto S_o$; then

$$|\Psi|^2 \propto \begin{cases} k_o|h_i|[1 + \cos 2\hat{\eta}_y], & |y| > y_o \\ k_o|h_i|[1 + \cos(2\hat{\eta}_r y - 2X_o)], & |y| < y_o. \end{cases} \tag{83}$$

where it is assumed that the levitation is much less than k_o^{-1} for simplicity. The optical image consists of a fringe pattern with spatial frequency $2\hat{\eta}_r$ superposed on a uniform background. At $|y| = y_o$, the fringe pattern has a abrupt phase shift of $2X_o$, twice the phase of the acoustic object. The fringe pattern will consequently be shifted by an amount $X_o/\hat{\eta}_r$. Assuming that this shift can be detected if it is greater than one-tenth the wavelength of the fringe pattern, then X_o must be $\geqslant 0.2\pi$.

Suppose that the 0 and +1 orders are used to reconstruct an image of the same acoustic object above. The optical intensity is

$$|\Psi|^2 \propto \begin{cases} 1+k_o^2|h_i|^2 + 2k_o|h_i| \sin \hat{\eta}_r y, & |y| > y_o \\ 1+k_o^2|h_i|^2 + 2k_o|h_i| \sin (\hat{\eta}_r y - X_o), & |y| < y_o \end{cases} \tag{84}$$

The image consists of a bright background plus a fringe pattern of spatial frequency $\hat{\eta}_r$. At $y = \pm\, y_o$, the fringe pattern has a phase shift of X_o.

If the shift in the fringe pattern is detectable when is greater than a tenth of the wavelength of the pattern, then X_o must again be $\geqslant 0.2\pi$. Hence, the sensitivity is the same when using either ± 1 orders or the 0 and +1 orders. However, the fringe pattern is more easily seen when using the ± 1 orders, assuming $k_o h_i << 1$. If $k_o h_i$ is $0(1)$, then it can be shown that either set of orders can be used.

If the +1 and +2 orders are used, then for $k_o h_i << 1$,

$$|\Psi|^2 \propto \begin{cases} k_o^2|h_i|^2 \left\{1 + k_o|h_i| \sin \eta_r y + \frac{1}{4} k_o^2|h_i|^2\right\}, & |y| > y_o \\ k_o^2|h_i|^2\left\{1 + k_o|h_i| \sin (\hat{\eta}_r y - X_o) + \frac{1}{4} k_o^2|h_i|^2\right\}, & |y| < y_o \end{cases} \tag{5)}$$

For the simple phase object on a uniform background, the image consists of a uniform background laced with fringes. There is a phase shift of X_o at $y = \pm y_o$, and the phase object will again be detectable if $X_o \geqslant 0.2\ \pi$. The background is given by the term $k_o|h_i|^2$; the fringe pattern by $(k_o|h_i|)^3 \sin \hat{\eta}_r y$. The ratio of the amplitude of the fringe pattern to that of the background is a measure of the contrast. For the +1 and +2 orders, this ratio is $k_o|h_i|$, which is << 1.

When the +1 and -1 orders are used the ratio is unity, so that the contrast is better with these orders. Since the fringe pattern is produced by the interference between orders, its contrast is maximized when the intensities in both orders are equal. This always occurs when using the ± 1 orders.

However, since the intensity in these orders is $k_o^2|h_i|^2$, assumed << 1, it may be desirable to use the 0 and +1 orders to form the fringe pattern in the back focal plane of the second optical lens. Normally, the viewing screen is located there; however, if the screen is removed and a second lens system is inserted to filter out the "d.c." term in (84), then the final optical intensity will be

$$|\Psi_r|^{21} \propto \begin{cases} 2k_o|h_i| \sin \hat{\eta}_r y + k_o^2|h_i|^2, & |y| > y_o \\ 2k_o|h_i| \sin(\hat{\eta}_r y - X_o) + k_o^2|h_i|^2, & |y| < y_o \end{cases} \quad (86)$$

Now the fringe pattern is much more visible than the background.

CONCLUSION

A theoretical study of liquid surface levitation holography has been made, based on and extending the work of Pille and Hildebrand, [1], [2]. This has lead to a set of constraints to be satisfied in order to obtain satisfactory performance for typical levitation systems.

It is found that the liquid surface has a low spatial frequency response (bulge) to the radiation pressure in the reference beam and in the focused acoustic image. The bulge can reach a steady-state value for value typical system parameters, because the rate at which the surface is subjected to pulses of radiation pressure is much less than the frequency of oscillation of the surface, at low spatial frequencies. The non-zero spatial frequency components of the bulge can cause degradation of the reconstructed optical image. To avoid this, it is necessary to constrain the amplitude of the "bulge" (measured from its average value) to be much less than an optical wavelength.

In like fashion, it is found that low spatial frequency distortion of the membrane at the bottom of the minitank can cause degradation of the acoustic image, assuming the minitank fluid to be different from the fluid in which the acoustic object is immersed. The acoustic image results from a superposition of plane waves at the liquid surface; each plane wave must pass through the membrane. If the membrane is deformed into a sinusoidal pattern, every plane wave incident on the membrane will give rise to a set of waves traveling through the minitank. This will be a source of "noise" in recording the acoustic image at the liquid surface. To avoid this, the membrane displacement is made much less than $(k_u - k_l)^{-1}$, with k_u the wavenumber in the minitank, and k_l the wavenumber in the object tank.

For ease in interpretation of the optical image, it is desirable to avoid contrast reversal. For example, areas of high acoustic transmission in the acoustic object should appear as bright areas on optical reconstruction and not vice versa. In the usual method of optical reconstruction, this requires the levitation to be less than $1/k_o$, for k_o the optical wavenumber.

Finally, it is necessary to require that the liquid surface respond linearly to radiation pressure. Nonlinearity of response can lead to loss of fidelity in recording the acoustic image. The criterion for linear response is that the levitation be much less than the wavelength of a gravity-capillary wave propagating on the surface.

The constraints which are applicable depend upon the nature of the acoustic object being imaged. For high-contrast objects, the effect of low-spatial frequency distortion of both the liquid surface bulge and of the membrane can be ignored. High-contrast acoustic objects are those whose transmission function varies by more than an order of magnitude over the field of view of the imaging system.

Low-contrast objects are those having transmission functions varying by less than an order of magnitude over the field of view. The image quality of these objects is affected by both bulge and membrane distortion. The image quality of low-contrast objects can be degraded if the surface becomes nonlinear. It is also desirable to avoid contrast reversal.

These requirements can be put in the form of inequalities to be satisfied by the acoustic intensity in the reference beam and in the focused acoustic image. For typical system parameters, some constraints will be more restrictive than others, and consequently will govern the operation of the levitation system.

A method of extending the imaging capabilities of levitations holography has been demostrated. This consists of using more than one order of light reflected from the liquid surface on optical reconstruction. For amplitude objects, combining various orders can sometimes lead to improved image quality. Diffraction and interference effects occuring in and around the acoustic object can be "washed out" in the optical reconstruction. This occurs when using specularly reflected light as well as higher orders. The specularly reflected light comes from dark areas of the acoustic image, which cause little surface levitation. The higher orders come from bright areas, which correspond to regions of greater levitation. The resulting optical images are "cleaner", since some of the background noise is removed. Acceptable images can be obtained in cases where the usual method of optical reconstruction (using only one reflected order) fails.

The multiple-order reconstruction method can also render the outline of phase objects visible. There will be a fringe pattern in the optical image due to the interference of the various orders. At the egde of the phase object, a shift in the fringe pattern will occur: if this shift is large enough to be detected by eye, it will define the outline of the phase object.

APPENDIX A

It is necessary to calculate the value of the membrane displacement, since the amplitudes of all non-zero orders diffracted by the membrane depend upon it. To keep the problem tractable, a number of assumptions are made. The interaction between the membrane displacement and motion of the liquid surface is ignored. This interaction is treated in some detail in Ref. [3]. The minitank is assumed to extend to $\pm\infty$ in the y-direction and a two-dimensional problem is treated.

The force balance at the membrane gives the differential equation

$$2\rho_M b \frac{\partial^2 w}{\partial t^2} - T \frac{\partial^2 w}{\partial y^2} = \Delta P_s + \Delta P_r \tag{A-1}$$

Here

ρ_M = membrane density

$2b$ = membrane thickness

T = tension/length in membrane

w = displacement of membrane

P_s = pressure induced by membrane motion

P_r = radiation pressure

The quantity ΔP_s is the jump in P_s at the membrane, due to different fluid properties in the minitank and water bath. Likewise, ΔP_r is the jump in radiation pressure at the membrane.

For inviscid, irrotational flow, P_s can be found from the the linearized Bernoulli equation

$$P_s = -\rho_o \dot{\Phi}_s - \rho_o g z_s. \tag{A-2}$$

ρ_o is the equilibrium fluid density and z_s the elevation of fluid particles from their undisturbed state. $\mathbf{v}_s$ is the velocity in the fluid caused by motion of the membrane, and Φ_s is the corresponding velocity potential; i.e., $\mathbf{v}_s = \nabla\Phi_s$. For an incompressible fluid, $\mathbf{v}_s$ satisfies

$$\nabla \cdot \mathbf{v}_s = 0 \tag{A-3}$$

so that

$$\nabla^2 \Phi_s = 0. \tag{A-4}$$

At the membrane,

$$\Delta P_s = \rho_u \dot{\Phi}_u - \rho_l \dot{\Phi}_l + (\rho_u - \rho_l) w \tag{A-5}$$

since $z_s = w$ at the membrane. The subscript u refers to quantities in the minitank; the subscript l refers to those in the water bath.

The radiation pressure in the vertical ($z-$) direction is given by [6]

$$P_r = <\rho_o(\mathbf{v}_z \cdot \mathbf{e}_z)^2> + <P_a> \tag{A-6}$$

where $< >$ denotes a time-average and

$\mathbf{v}_a$ = particle velocity in acoustic field

$\mathbf{e}_z$ = unit vector in z-direction

P_a = pressure in fluid due to acoustic field

It can be shown that [6]

$$<P_a> = -\frac{1}{2}\rho_o<\mathbf{v}_a \cdot \mathbf{v}_a> + \frac{1}{2}\frac{<P_a^2>}{\rho_o c^2} \tag{A-7}$$

where second order quantities in the acoustic variables have been retained. Combining these results,

$$\Delta P_r = \frac{1}{2}(\rho_u - \rho_l) <(\mathbf{v}_a \cdot \mathbf{e}_z)^2>$$

$$-\frac{1}{2}\left[\rho_u<(\mathbf{v}_{au} \cdot \mathbf{e}_y)^2> - \rho_l<(\mathbf{v}_{al} \cdot \mathbf{e}_y)^2>\right]$$

$$-\frac{1}{2}\left[\frac{<P_{au}^2>}{\rho_u c_u^2} - \frac{<P_{al}^2>}{\rho_l c_l^2}\right] \tag{A-8}$$

Rather than deal with this complicated expression for ΔP_r, it will merely be noted that ΔP_r is the same order of magnitude as the radiation pressure at the liquid surface, for fluids of practical interest [3].

ΔP_r acts as a driving pressure which causes membrane motion. If the radiation pressure at the liquid surface is of the form $P_r \propto Rect(\hat{t}/\Delta t)$, then $\Delta P_r \propto Rect(\hat{t}/\Delta t)$ also. If ΔP_r is represented by

$$\Delta P_r = \Delta p_r \cos Ky \; Rect(\hat{t}/\Delta t) \tag{A-9}$$

then w, Φ_u and Φ_l will vary as cos Ky as well; e.g.

$$w = W(t) \cos Ky$$

and similarly for Φ_u, Φ_l.

Combining the kinematic boundary condition

$$\frac{\partial \Phi_u}{\partial z} = \frac{\partial \Phi_l}{\partial z} = \dot{w} \tag{A-10}$$

at the membrane with the expression for ΔP_s and the membrane equation of motion gives

$$(\rho_u + \rho_l + 2\rho_M bK_n) \; \ddot{W}(t) + \left[TK^2 + (\rho_l - \rho_u)g\right] KW(t)$$

$$= -K\Delta p_r \; Rect(\hat{t}/\Delta t). \tag{A-11}$$

This has solution

$$W(t) = \frac{-K\Delta p_r \sin \omega_M t \overset{t}{*} Rect(\hat{t}/\Delta t)}{\omega_M(\rho_u + \rho_l + 2\rho_M b\ K)} \tag{A-12}$$

where the frequency of free vibration, ω_M, of the submerged membrane is given by

$$\omega_M^2 = \frac{(TK^2 + (\rho_l - \rho_u)g)\ K}{\rho_u + \rho_l + 2\rho_M bK} \tag{A-13}$$

When $K \sim 0$ (1 rad/cm), the convolution in (A-12) can be approximated by $2\Delta t \sin \omega_M t$ for $t > 2\Delta t$ and $\omega_M \Delta t << 1$; then $W(t) = |W| \sin \omega_M t$ with

$$|W| = \frac{2(\Delta t)\ (\Delta p_r)\ K}{\omega_M(\rho_u + \rho_l + 2\rho_M b\ K)} \tag{A-14}$$

This can be simplified by noting that $2\rho_M bK$ is $<<$ $(\rho_u + \rho_l)$ for typical values $2b = .05$ mm, $\rho_M \cong 2$ gm/cm^3, and $K \sim 0(1$ rad/cm). Furthermore, since $\Delta P_r \sim 0(P_r)$ then

$$|W| \sim 0\left(\frac{\Delta t P_r K}{\omega_M(\rho_u + \rho_l)}\right) \tag{A-15}$$

References

1. Pille, P., "Real Time Liquid Surface Acoustical Holography," Master's Thesis, University of British Columbia (1972).

2. Pille, P., and Hildebrand, B.P., "Rigorous Analysis of the Liquid-Surface Acoustical Holography System," *Acoustical Holography*, Vol. 5, Plenum Press, New York, N.Y., pp. 335-371 (1973).

3. Clark, A.V., "Liquid Surface Levitation Holography, Part One: Theoretical Analysis," Naval Research Lab. Rep. No. 8205, Naval Research Lab., Washington, D.C. (in publication).

4. Landau, L.D., and Lifshitz, E.M., *Fluid Mechanics Course of Theoretical Physics*, Vol. 6, Pergamon Press, London, (1959).

5. Hildebrand, B.P., and Brenden, B.B., *An Introduction to Acoustical Holography*, Plenum Press, New York, N.Y., (1972).

6. Gol'dberg, Z.A., "Acoustic Radation Pressure," in *High Intensity Ultrasonic Fields*, Plenum Press, New York, N.Y., (1971).

REFLECTED BRAGG IMAGING STUDIES AND PRELIMINARY RESULTS FOR IMPROVED OPTICAL DESIGN

R. La Canna, P. Spiegler, F. Kearly, and J. Rahimian

University of California, Department of Radiological Sciences, The Center for the Health Sciences, Los Angeles, California 90024

ABSTRACT

Because of the right angle geometry between the sound field and the interrogating light wedge, Bragg imaging is ideally suited for the study of reflected images. In order to improve the intensity and the clarity of these reflected images, a pulsed dye laser has been developed and tested. The laser has a variable pulse repetition frequency to a maximum of about 15 pps and operates at a wavelength of 590 nanometers with a maximum power of 5mW. The dye laser is fired through an SCR circuit using a delayed trigger from the transducer generator. Image photographs of various test objects will be presented and system resolution limitations will be discussed.

Efforts have also been directed toward improving the optical design for Bragg imaging systems. The systems that have been described in the literature incorporate several optical components with low f-numbers in order to produce high quality images. Furthermore, these components must have dimensions comparable to the desired image size and for optimal imaging should be diffraction limited. As such, production costs of the individual elements can be excessive. Alternate lens systems have been studied and a design realized which utilizes one large cylindrical triplet lens, one small cylindrical triplet lens, and prisms. Since only the large cylindrical component should offer any production difficulty this optical system would result in substantial cost reduction. Design considerations will be discussed and compared to those of conventional systems.

INTRODUCTION

In a series of earlier papers (1,2,3,4), the feasibility of using Bragg diffraction as a medical diagnostic tool has been amply demonstrated and design criteria have been proposed toward implementing such a system. Efforts have continued in order to show the practicability of such a system in a clinical situation. We will present in this paper the latest results of incorporating a pulsed laser into a Bragg imager and some preliminary results for new optical designs.

As a brief review, the Bragg diffraction imaging system commonly described in the literature is depicted in Figure 1. A laser beam is expanded 50 to 100 times in diameter by lenses L1 and L2 and shaped by a cylindrical lens, L3, into a light wedge within an acoustic cell (usually water filled). The interaction between the light wedge and the sound field (which acts as a volume diffraction grating) provides the mechanism for imaging the sound. Lenses L4 and L5 are crossed cylindrical lenses which correct the image for astigmatism. Finally, lens L6 is a spherical lens that magnifies the corrected image making it observable to the unaided eye.

Most importantly, the right angle geometry of the Bragg system makes it an attractive possibility for imaging, in real time, a reflected sound field from interfaces of biological tissue.

LASER

In order to avoid intolerable background levels in Bragg imaging of reflected sound fields, it is necessary to pulse the transducer and, synchronously, the laser producing the light wedge.

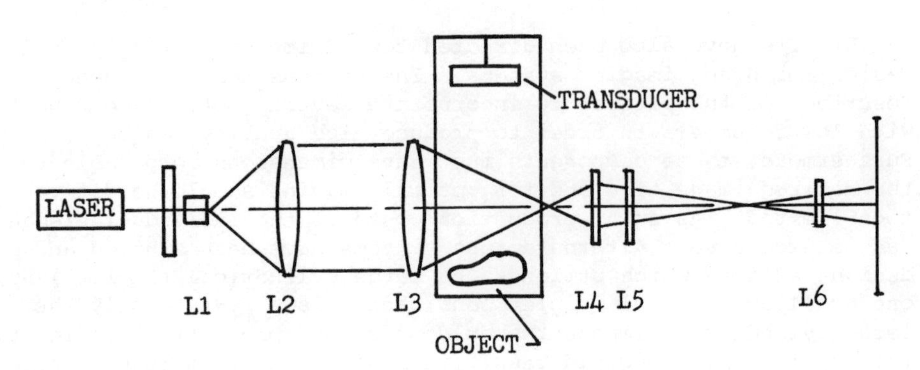

Fig. 1. Schematic diagram of a Bragg imaging system for reflected sound.

Initial efforts at reflection imaging were achieved with a cw He-Cd laser with a wavelength centered at 440 nm and an output of 50 mW. Synchronization between the reflected sound field and the light wedge was done with a beam chopper consisting of a slotted disc. The rotation speed of the disc was variable up to 400 cps, but because of the inherent design of the chopper more than 90% of the effective laser output was lost necessitating the use of a large laser to produce an image of reasonable intensity.

To circumvent this difficulty, a pulsed dye laser was developed and incorporated into the existing system. The design of the laser followed those described in the literature (5,6,7). The laser liquid consisted of a 5 x 10^{-5} molar solution of Rhodamine 6G (mol. wt. 543.02) in ethanol. This dye produced wavelengths between 560 nm and 625 nm with a peak wavelength at about 590 nm. Firing of the dye laser was accomplished with an Xenon flash lamp which had a minimum rated output of 10 joules/pulse. The flash lamp and the (quartz) laser tube were mounted along opposite focal lines of a cylindrical elliptical housing constructed of 3/16" Al. The housing was 3.5" long with major and minor axes of 4.2" and 3.9" respectively. The beam spot diameter of the laser could be varied by changing the distances of the mirrors from the ends of the laser tube. These dimensions were usually adjusted to keep the beam diameter as small as possible which served to reduce loss and distortion through the beam expander assembly.

A block diagram of the system electronics is shown in Figure 2a. An SCR circuit was used to fire the flash lamp through a triggered spark gap. The power supply for the triggered spark gap had a maximum output of 30 kV at 15 mA. The firing repetition rate was controlled with a variable oscillator circuit to a maximum of 15 pps. The oscillator pulses were used as an external trigger for a variable pulse generator which controlled the operation of the transducer. A delayed trigger pulse from the pulse generator to the SCR of the laser circuit completed the firing sequence. Dimensions involved required a delay of roughly 100 μsec between the transducer firing and the laser firing. Transducer pulses of 25-35 μsec were used which more than amply covered the dimension of the light wedge. The maximum light output of the system was approximately 5 mW which is enough to produce a clearly visible image in a darkened room.

Several difficulties were encountered with the use of a pulsed dye laser which would seem to overshadow the relative simplicity of construction. Most important, and unexpected, was that the beam itself was not "clean" in the optical sense, that is, having an intensity distribution which is reasonably constant over the area of the beam spot. Because of this it was found that the use of a filter in front of the beam expanding microscope objective

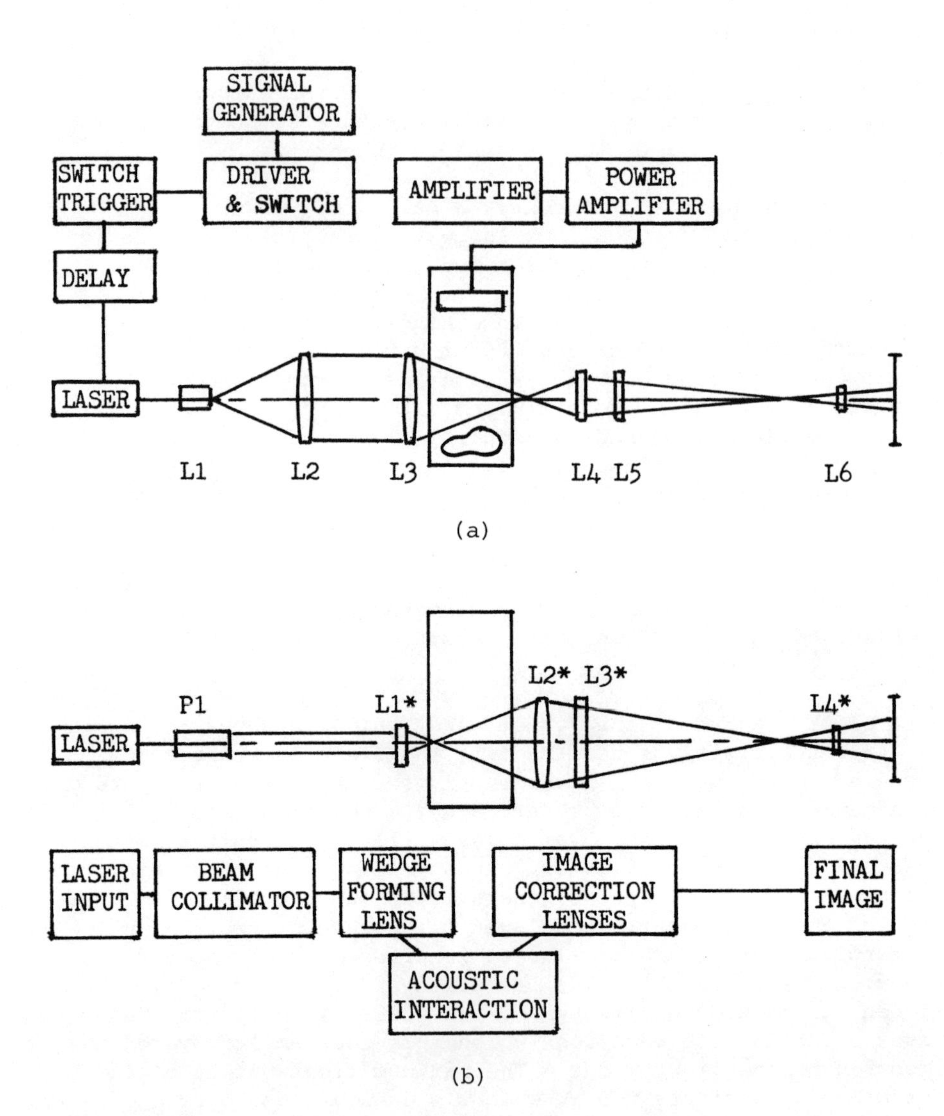

Fig. 2. (a) Converging lens system showing block diagram of accompanying electronics.
(b) Diverging lens system.

was not all that beneficial and could be eliminated without affecting the quality of the images. In addition, there was degradation of the beam intensity with time which was dependent on the firing rate used. Finally, dye lasers require a rather high operating voltage (in the region of 20-30 kV) and, as such, power supply construction is more sensitive to voltage breakdown.

OPTICS

We will present in this section some considerations on designs for improved optical components to be used in conjunction with a Bragg imaging system. In order to produce high quality images, the design of a Bragg unit must include the following criteria:

a) All lenses must be free of spherical aberration along the optic axis.

b) Lenses L1, L2, L3, and L4 (Figure 2a) must have the same relative aperture (f-number) over the area of the beam used to form the light wedge.

c) The relative aperture of these lenses should be close to unity since the resolution in the corrected image for a vertical line source of sound is given by

$$R_v = \Lambda f_n$$

where Λ is the wavelength of sound and f_n is the relative aperture of the lenses (4).

d) The object being imaged should be placed close to the light wedge since the resolution for a horizontal line source of sound is given by

$$R_h = \Lambda d/h$$

where, again, Λ is the wavelength of sound, d is the distance from the line source to the center of the light wedge, and h is the height of the light wedge (Ibid.)

Spherical aberration inherent in lens design can be eliminated either by using lenses with aspheric or acylindrical surfaces or by using a series of thin elements as components for a single compound lens. Since the manufacture of aspheric and acylindrical surfaces is extremely difficult, lens systems consisting of several thin elements provides a practical and available solution.

To achieve an f-number of about 2 and be free of spherical aberration, lenses L1, L2, L3, and L4 must each be at least a triplet configuration, that is, must each consist of three thin elements. Furthermore, to produce an image with a cross sectional area a x a, the cylindrical lenses L3, L4, and L5 must have dimensions of $(a \times a)/f_n$, and the spherical lens L2 must have a diameter greater than $\sqrt{2}$ a. Each lens should also be coated with a high anti-reflection coating since for glass with an index of refraction of 1.75, say, reflection losses are more than 7% per surface. In summary, then, the manufacturing cost of a high resolution optical system capable of producing a large image can be exorbitant.

LENS DESIGN

A method for designing uncemented telescope objectives which are free of spherical aberration for monochromatic light was first outlined by G. S. Fulcher (8). This method, based on the third order thin lens equation and often used in the design of infrared imaging systems (9), uses the following assumptions:

1. All lenses have zero thickness.
2. The spacing between lenses is zero.
3. All lenses are made from the same material and have, therefore, the same index of refraction, N, and the same relative dispersion.
4. Each individual lens element is shaped for minimum spherical aberration.
5. The power (reciprocal focal length) of each element is identical, i.e., the total power of the set of lenses is divided equally among the elements making up the set.

It can be shown (8) that for a set of i thin lenses, the angular blur spot size due to spherical aberration is given by:

$$b = \frac{N\left[i(4N-1) - 4(N-1)^2 \sum_{j=1}^{j=i} j(j-1)\right]}{128\, i^3\, (N-1)^2\, (N+2)\, f_n^{\,3}} \tag{1}$$

where N = refractive index
$f_n = f/2y = 1/2y\phi$ = relative aperture
f = focal length of objective = $1/\phi$ = 1/power
y = semi-aperture (half free lens diameter)

Equation (1) above demonstrates that the blur due to aberration is zero if i=3 and N=1.75 or if i=4 and N=1.5, etc. The curvatures of lens j in a multiplet which would yield the minimum spherical aberration for this lens in the set are then given by:

$$c_{j1} = \frac{N(2N+1) + 4(N^2-1)(j-1)}{2i(N+2)(N-1)} \phi \qquad (2)$$

$$c_{j2} = \frac{N(2N-1) - 4 + 4(N^2-1)(j-1)}{2i(N+2)(N-1)} \phi \qquad (3)$$

With the above equations it was possible to design an optical system for a Bragg imaging unit having a relative aperture of about 2. The strategy employed was to use equations (2) and (3) for the initial design of lenses L3 and L4 for SF4 glass which has an index of refraction of N=1.75. A ray tracing program was then used to remove the spherical aberrations introduced when coupling the optics to an acoustical cell.

The results of the ray tracing program can be summarized as follows:

1. A triplet lens as designed using equations (2) and (3) is essentially free of spherical aberration based on Rayleigh's criterion, viz., that the optical path difference of the emerging ray from a reference spherical wavefront is less than $\lambda/4$, where λ is the wavelength of light.
2. The spherical aberrations introduced by the acoustic interaction cell can easily be removed by altering the curvatures of one of the three lenses.
3. The values computed for the curvatures are not very critical and changes of several percent can be tolerated.
4. The system is easily adaptable to other wavelengths merely by changing the spacing between lenses.

Because of these conditions a lens design was accomplished which could be constructed with relative ease, free of spherical aberration, and follow a conventional approach to the optics of a Bragg imaging system. An alternative, and unusual, design was also achieved with the above format as illustrated in Figure 2b which is drawn approximately to scale in the lens dimensions. In this design lenses L1 and L2 are replaced by **a Brewster telescope** That is, the laser beam is expanded in a single direction by a set of properly constructed prisms. Lens L3 can then be a narrow cylindrical triplet that produces a diverging light wedge which passes through the acoustic interaction cell. Lens L4 is a large cylindrical triplet, roughly 6 times the areal size of L3, and is the only large compound lens in the combined optical system.

Fig. 3. The acoustical interaction cell showing the end viewing window. Lenses L2, L3, L4, L5 are seen from right to left.

RESULTS

Figure 3 is a picture of the acoustic cell with lenses L2, L3, L4, and L5 clearly visible viewed from right to left. The entrance and exit windows to the acoustic cell are 1/4" plate glass with an index of refraction, N=1.5. The end imaging window is constructed of dental dam.

In Figure 4 we present earlier, unpublished, results showing the horizontal and vertical resolution obtained with both transmission and reflection modes using a cw He-Cd laser and beam chopper assembly. In both series of pictures the object viewed was immersed in the water bath. Figures 5 and 6 show images made with the pulsed dye laser. The image of the arrow was also made with the arrow immersed in the water bath. The two views of the hand shown in Figure 6, however, were made through the end membrane with the hand in the positions indicated in the upper photos. The difference in the acoustical image sizes is merely due to photographic enlargement. It should be noted that the photographs of the acoustical images shown in Figures 5 and 6 were made with one laser pulse at approximately 70% maximum power output.

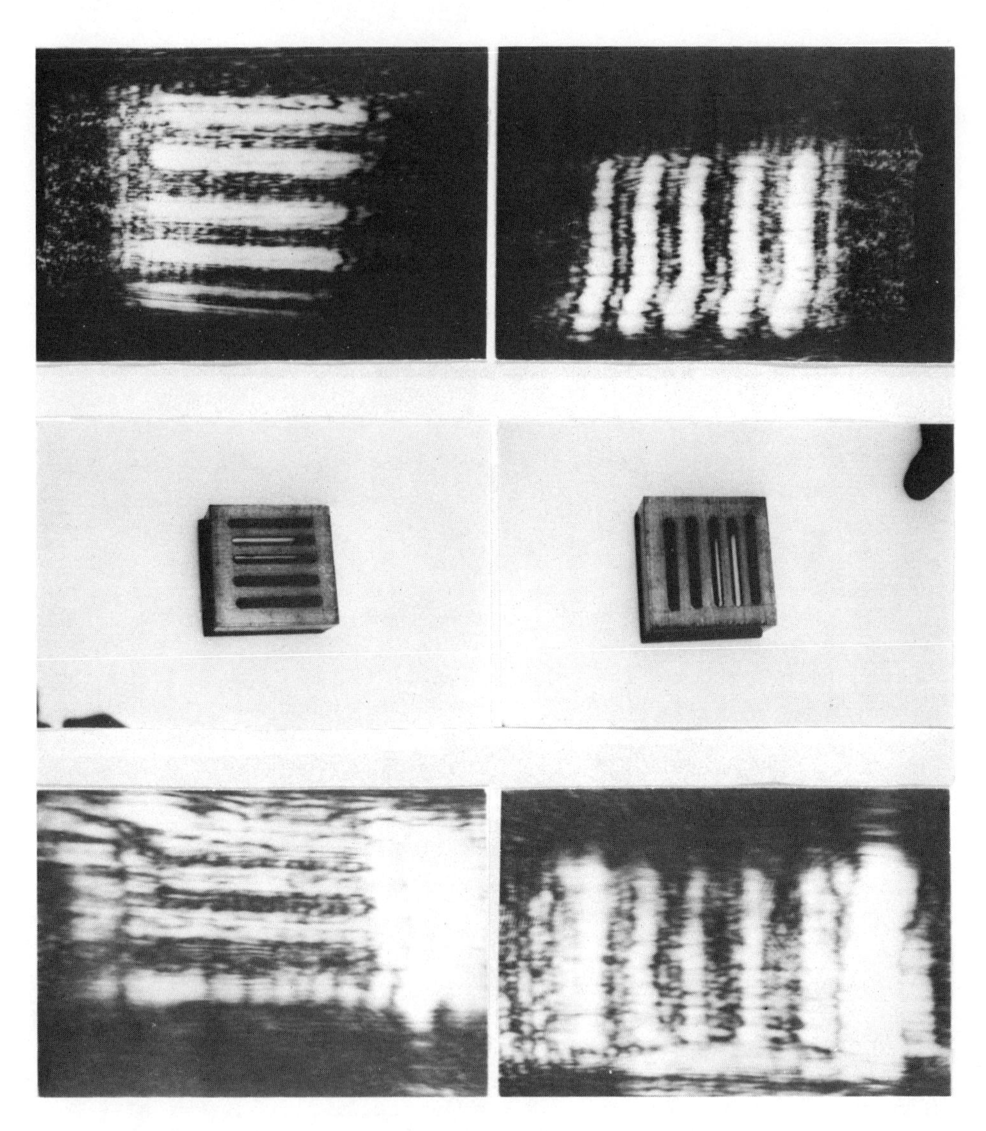

Fig. 4. Acoustical images taken with a cw He-Cd laser. Upper photos are transmission mode and lower photos are reflection mode. (3.2 MHz ultrasound)

Fig. 5. Acoustical image of a wooden arrow taken with the pulsed dye laser. (3.2 MHz ultrasound)

The triplet design specifications for lenses L3 and L4 for the two cases discussed above are tabulated in Tables 1 and 2. Both designs correspond to a wavelength of 587.5 nm. Table 1 are the values for a conventional system in which L3 and L4 are approximately the same size (Figure 2a), and Table 2 are the values for the lens system shown in Figure 2b. Both designs were made for zero spacing between the elements of a given triplet. Included in the design, of course, is the lens triplet formed by the entrance and exit windows and water of the acoustic cell. In both cases, the designs were done for windows of SF14 glass 1/4" thick (N=1.76) and an acoustic cell interior width of 5 cm. Also both designs satisfy the Rayleigh criterion to about $\lambda/10$.

CONCLUSIONS

While it is true that the images taken with the dye laser do not have nearly the resolution hoped for, it is felt that this is primarily due to the nature of the dye laser itself. We have demonstrated, however, that a reasonably bright image can be obtained from a reflected sound field using a pulsed laser with an output $\leq$ 5 mW. Work is presently underway to incorporate a 8 mW pulsed Argon laser which has a very clean beam and an external trigger firing mode with a repetition rate up to 60 pps.

Finally, it should be pointed out that in addition to the laser and the optics of a Bragg imaging system, the ultrasound component,

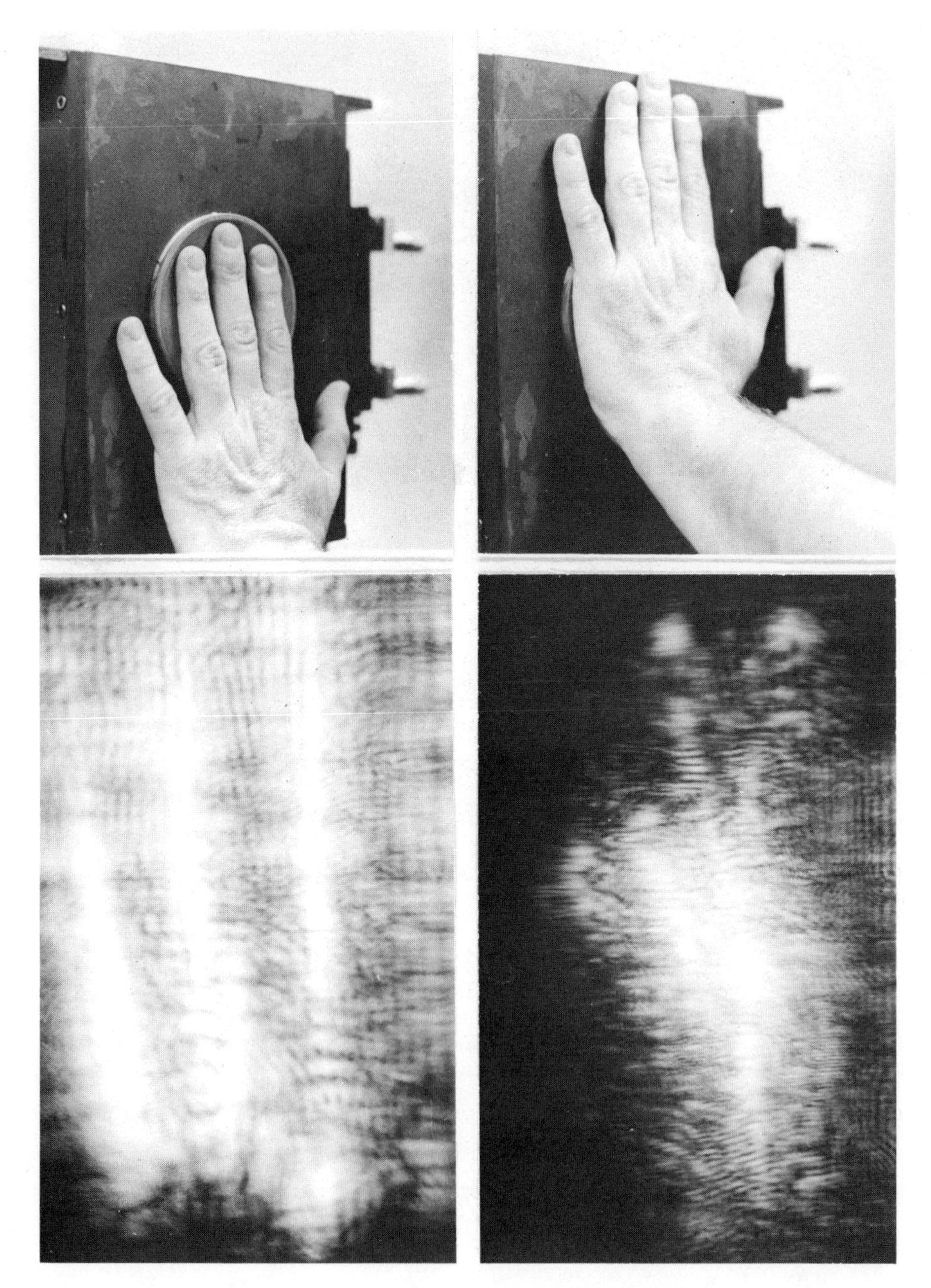

Fig. 6. Acoustical images of the fingers and hand taken with the pulsed dye laser. (3.2 MHz ultrasound)

Table 1

Triplet Lens Design Specifications
Dimensions: L3,L4 = 3.175cm x 3.175cm
Glass: SF4, N=1.75

	Element	Curvature(cm^{-1}) (front/rear)	Thickness(cm)
	1	0.073/0.0038	0.254
L3	2	0.140/0.070	0.254
	3	0.145/0.075	0.254
	1	-0.127/-0.196	0.254
L4	2	-0.040/-0.110	0.254
	3	0.030/-0.040	0.254

Table 2

Triplet Lens Design Specifications
Dimensions: L3 = 0.5cm x 3.175cm,
L4 = 3.175cm x 3.175cm
Glass: SF4, N=1.75

	Element	Curvature(cm^{-1}) (front/rear)	Thickness(cm)
	1	1.09/0.65	0.1
L3	2	0.88/0.44	0.1
	3	1.79/1.35	0.1
	1	-0.08/-0.15	0.254
L4	2	-0.08/-0.15	0.254
	3	0.00/-0.07	0.254

including the design of the acoustical cell, needs to be investigated thoroughly. Careful design of both the cell and the transducer for the reflected mode is essential to improve the resolution and reduce image artifacts.

ACKNOWLEDGEMENT

We would like to express our appreciation to Dr. D. F. Frieda for permission to reproduce the images shown in Figure 4.

REFERENCES

1. A. Korpel, "Optical Image of Ultrasonic Fields by Acoustic Bragg Diffraction", Ph.D. dissertation, University of Delft, Netherlands (1969).

2. J. Landry, H. Keyani, and G. Wade, "Bragg-Diffraction Imaging: A Potential Technique for Medical Diagnosis and Material Inspection", Acoustical Holography, Vol. 4, G. Wade, Ed. (Plenum Press, New York, 1972).

3. N. Tobochnik, P. Spiegler, R. Stern, and M. A. Greenfield, "Assessment of Bragg Imaging and the Importance of Its Various Components by Comparison with Radiographic Imaging", Acoustical Holography, Vol. 6, N. Booth, Ed. (Plenum Press, New York, 1975).

4. D. F. Frieda, P. Spiegler, F. Kearly, M. A. Greenfield, and R. Stern, "Design Criteria for Medical Bragg Imaging", Acoustical Holography, Vol. 7, L. W. Kessler, Ed. (Plenum Press, New York, 1977).

5. W. Schmidt, "Farbstofflaser", Laser, Nr.4, 47 (1970).

6. P. K. Takahashi, "Tunable Organic Dye Laser Irradiation of E-Coli", Ph.D. dissertation, Louisiana State University, Baton Rouge, LA (1979).

7. P. F. Shafer, Ed., Dye Lasers, (Springer-Verlag, New York, 1973).

8. G. S. Fulcher, "Telescope Objective Without Spherical Aberration for Large Apertures, Consisting of Four Crown Glass Lenses", J. Opt. Soc. Amer., 37, 47 (1947).

9. M. J. Riede, "Multiple Thin Lens Objectives for IR Imaging Systems", Electro-Optical Systems Design, Vol. 11, 38 (November, 1975).

ANGULAR SPECTRUM FILTERING OF PARASITIC IMAGES

INTRODUCED BY THE SPATIAL SAMPLING OF ACOUSTICAL IMAGES

C. Bruneel, B. Delannoy, R. Torguet, E. Bridoux
and JM. Rouvaen
Centre Universitaire de Valenciennes

59326 Valenciennes Cédex - France

INTRODUCTION

The power level of the acoustic echoes used in acoustical imaging systems is very low. The signal processing is therefore eased by sampling the reflected acoustic field with a receiving piezoelectric transducer array before subsequent amplification. This spatial sampling introduces parasitic images. By filtering the most inclined part of the angular spectrum, the power level of these ghosts may be severely reduced. A parallel is drawn between the present case and the situation arising in grating optics and the results of this analysis are well supported by the experimental findings.

RECONSTRUCTION OF ULTRASONIC IMAGES

For biomedical applications of ultrasound, the most popular imaging scheme is the so called echographic B-scan mode. Two different techniques are used in order to reconstruct the image.

First, one may work one line by line basis. This, earlier and commonly resorted to, technique is an equivalent of the sonar working principle and has led to manual scanning and linear electronic switching devices (1-3) or sector-scan apparatuses (4-5). The main limiting factor of this technique is the rather long time required for image obtention, equal to the maximum ultrasonic impulse round trip transit delay times the available number of image lines.

Second, the whole image may be formed during a single ultrasonic impulse round trip. A phase sensitive reconstruction scheme of the acoustic informations is then needed in order to take the ultrasonic wave diffraction into account. Several schemes have been used like the transfer on an optical beam (6-7), an acousto-electronic lens (8-9) or a purely electronic processing (10). All these rely on the spatial sampling of the acoustic field reflected by the biological sample using a transducer array. The sampling process introduces perturbations in the reconstructed images, like some parasitic images.

The amplitude of these parasitic images in the image plane of the system has been previously studied (11). The distribution of these images is now studied here inside the Fourier plane, and some criterions are established for filtering this angular spectrum. The calculations are performed in a monodimensionnal situation like that of the echographic B-scan mode, but the results may be readily applied to the more general bidimensionnal case of a transducer mosaïc array. Moreover, owing to the linearity of Fresnel and Rayleigh-Sommerfeld diffraction formulas (12), one may consider, without loss of generality, only the more tractable reconstruction problem for an acoustic point source.

The comparison with the classical results for optical diffraction gratings shows the occurence of specific problems encountered in the acoustic imaging field, particularly bound to the uniform phase over each elementary source of the grating.

DIFFRACTION BY A PLANE OPTICAL GRATING

Let us consider a plane optical grating, with a period a' and a slit width 2b' (see fig. 1)

The spectrum may be defined using the Fourier integral :

$$A\ (\sin i) = \int_{-\infty}^{+\infty} A\ (x)\ \exp\ (j2\pi x \sin i/\lambda)\ dx \qquad (1)$$

where λ is the optical wavelength, x the abcisse of the considered point, A (x) the wave amplitude density and i the angle of the considered direction .

Let us consider an incident, unit amplitude, plane optical wave, impinging on the grating at an angle i mesured from its normal. The diffracted wave amplitude, at an angle Θ, is given then by the classical formula (13) :

$$C\ (\sin\Theta - \sin i) = A_1(\sin\Theta - \sin i).\ A_2\ (\sin\Theta - \sin i) \qquad (2)$$

where :

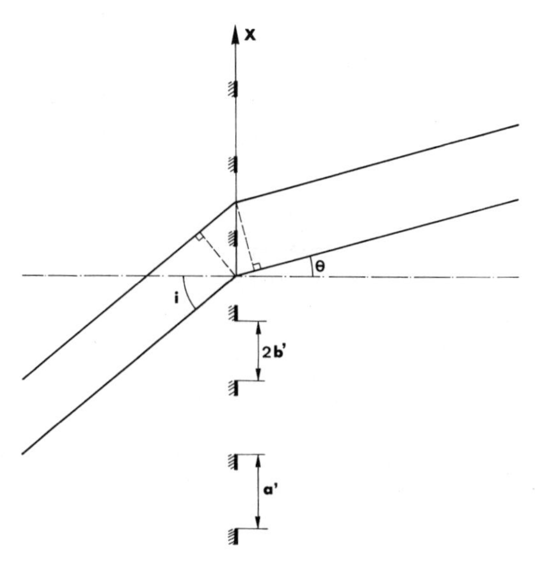

Fig. 1 : Geometry of the grating.

$$A_1\ (X) = \sin\ (2\pi b'X/\lambda)/(2\pi b'X/\lambda) \tag{3}$$

$$A_2\ (X) = \frac{1}{\pi}\ \frac{\sin\left[(2N + 1)\ \pi a'X/\lambda\right]}{\pi a'\ X/\lambda} \tag{4}$$

and 2N + 1 is the number of slits.

When the angular spectrum of the incident beam is very wide and given by the spectral density G (sin i), the diffracted wave spectrum is given by :

$$B\ (\sin\Theta) = \quad C\ (\sin\Theta - \sin i)\ G\ (\sin i)\ d\ (\sin i) \tag{5}$$

so that :

$$B\ (\sin\Theta) = \quad C\ (\sin\Theta) * G\ (\sin\Theta) \tag{6}$$

where the symbol * represents a convolution operator.

When the number of slits becomes very large, it follows that :

$$A_2\ (X) \simeq \sum_p \delta\ (X - p\ \lambda/a') \tag{7}$$

where δ is a Dirac impulse, and the diffracted spectrum takes the form sketched in the fig. 2, where only the main and first order parasitic images have been considered.

When N gets low, the situation is very different since the incident spectrum is limited by the grating. For example, if a

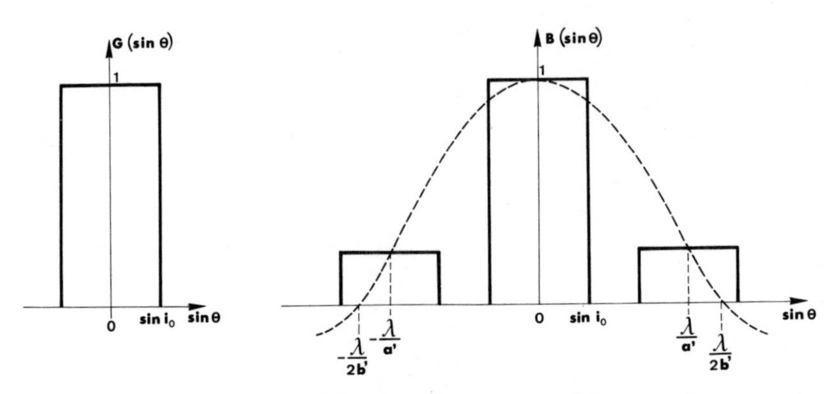

Fig. 2 : Response of a slit grating to a wide angular spectrum.

point source lying at the distance D of the grating plane is considered, all angular spectrum components greater than the limiting angle i_r defined by :

$$tg\ i_r = Na/D$$

are filtered out.

SOURCE GRATINGS

In ultrasonic imaging problems, the situation differs somewhat from the previous one. For an optical grating, phase variations along the slide width are allowed, but, when considering an acoustical sampling transducer array, the phase must remain a constant along each sampling transducer. Let us consider a signal constant in amplitude whose phase varies linearly along the array. By analogy with the preceding optical case, the dephasing between two adjacent sources may be written as :

$$\phi = 2\ \pi a' \sin i/\lambda \tag{8}$$

and the amplitude of the wave launched at an angle Θ will be given by :

$$A(\sin\Theta, \sin i) = A_1\ (\sin\Theta).\ A_2\ (\sin\Theta - \sin i) \tag{9}$$

If the emitted acoustic signals are a linear combination of the preceding signals, with a spectral density G (sin i), the resultant amplitude in the direction Θ becomes now :

$$B\ (\sin\Theta) = A_1\ (\sin\Theta)\ \left[A_2\ (\sin\Theta) * G\ (\sin\Theta)\right] \tag{10}$$

When N is very large, one gets the spectral response shown in fig. 3.

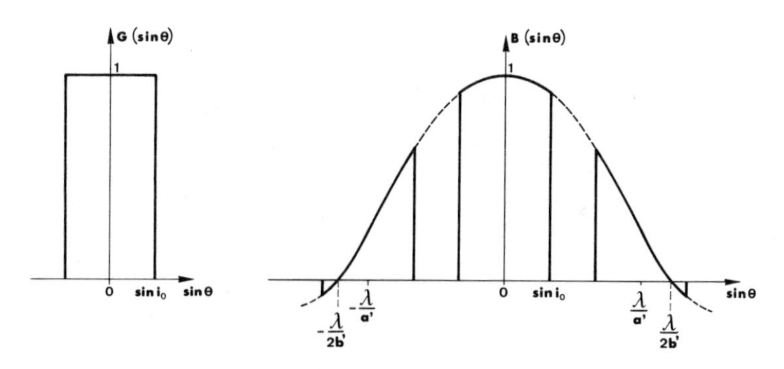

Fig. 3 : Response of an acoustic grating to a wide angular spectrum

The comparison with fig. 2 shows an increase of the energy associated with the parasitic images. For a very wide incident spectrum, the parasitic image level may even become equal to that of the main image.

OPTICAL RECONSTRUCTION OF ACOUSTICAL IMAGES

a) Analogical Calculations

A reconstruction system, which uses the transfert of the amplitude and phase informations about the reflected acoustic field upon an optical beam, has been tested. However, the reflection coefficient for acoustic energy of biological media being generally very low, a direct acoustooptic interaction is not feasible. So, the reflected acoustic field is sampled using a transducer array, then the electrical signals are amplified and finally reconverted to acoustic signals, using a second transducer array cemented to a highly photoélastic material (see fig. 4).

Let a, 2b, r, and Λ be, respectively, the spatial period of the sampling array, the transducer width, the incidence angle and the acoustic wavelength. The dephasing between two adjacent transducers onto which a plane wave impinges at an angle r mesured from the normal, is :

$$\phi = 2\ \pi a \sin r/\Lambda \tag{11}$$

Owing to the equ. (8), this corresponds to an angle i at the reemission stage given by :

$$\sin i = \frac{\lambda a}{\Lambda a'} \sin r \tag{12}$$

For an acoustic point source, which may be viewed as an isotropic radiation, the radiation pattern of the receiving trans-

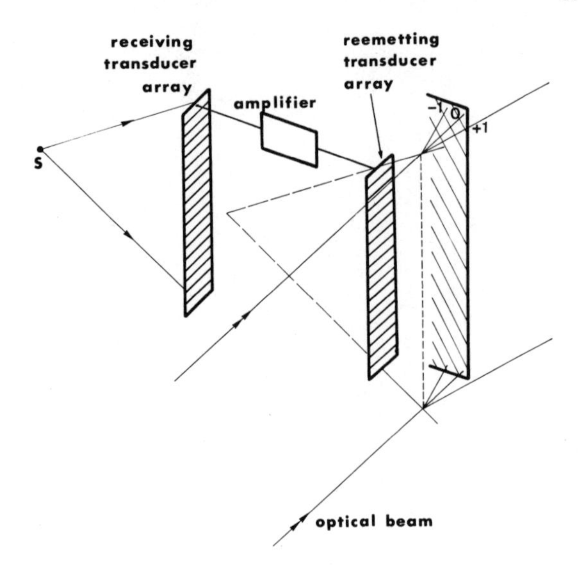

Fig. 4 : Optical system for the acoustical image reconstruction.

ducers of width equal to 2b limits the angular spectrum and acts like a low pass filter. The received signals follow the law :

$$F\ (\sin r) = \frac{\sin\ (2\pi b\ \sin r/\Lambda)}{2\pi b\ \sin r/\Lambda} \tag{13}$$

The reemitted acoustic field may be written as :

$$G\ (\sin i) = \frac{\sin\ (2\pi b\ a'\ \sin i/a\lambda)}{2\pi ba'\ \sin i/a\lambda} \tag{14}$$

Using this relation together with equ. (10) and assuming a very large number (2N + 1) of transducers, the spectrum is represented by :

$$B\ (\sin \Theta) = \sum_p \frac{\sin \beta Y}{\beta Y}\ \frac{\sin\ (Y - pA)}{Y - pA} \tag{15}$$

with $\beta = \frac{b'a}{b\ a'}$; $y = \frac{2\pi b}{\lambda}\ \frac{a'}{a}\ \sin \Theta$, $A = 2\pi\ \frac{b}{a}$

b) Numerical Calculations

The spectrum calculated for a practical example is shown in Fig. 5, where the parameters are taken as :

$A = 0.9\ \pi$, $\beta = 0.8$

and only one parasitic image is shown, the other being easily deduced by a symmetry operation.

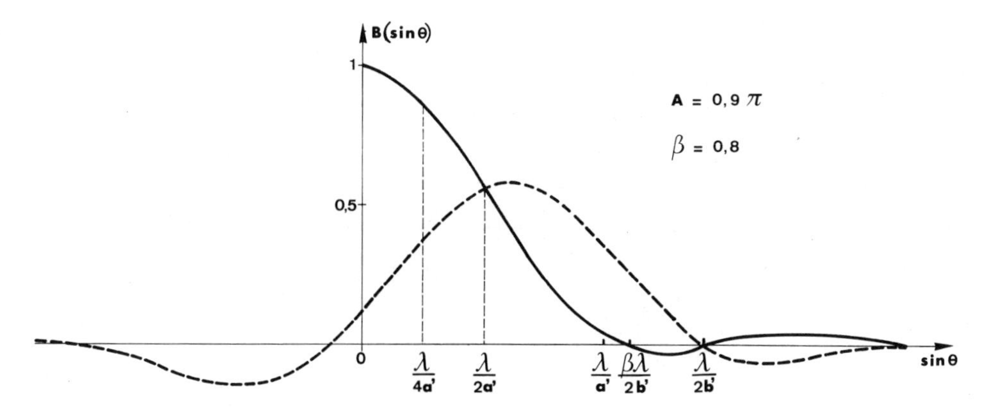

Fig. 5 : Optical spectrum of an acoustic point source.

The relation (15) enables the determination of the parasitic image angular spectrum positions and also that of the energy associated with each parasitic image, since this energy is proportionnal to the integral of the squared spectrum by virtue of Parseval's theorem.

In the following table, the values of the energy of the first parasitic images, normalized to that of the main one, are given for different situations :

- without filtering
- with a $\lambda/2a'$ angular filtering
- with a $\lambda/4a'$ angular filtering

From these figures some conclusions may be drawn :

- without filtering, the parasitic image level is quite high as soon as one of the two arrays involved has a transducer width much smaller than the sampling period.
- with filtering, the parasitic image level raises down very fast and depends only, within the first approximation, of the sampling done by the array with the highest filling ratio.

c) Experimental Results

The spectrum of an optical beam is easily obtained in the focal image plane of a lens.

The filtering operation may therefore be done by setting a slit in the focal plane of a lens lying immediatly behind the source array.

		Without filtering		With a $\lambda/2a'$ filtering		With a $\lambda/4a'$ filtering	
β	A	I_0	I_1/I_0 (dB)	I_0	I_1/I_0 (dB)	I_0	I_1/I_0 (dB)
1	π	1	-8,2	0,946	-11	0,66	-17
1	0,95 π	1	-7,6	0,935	-10,3	0,63	-15,7
1	0,9 π	1	-6,9	0,92	-9,5	0,6	-14
1	0,8 π	1	-5,5	0,87	-7,9	0,55	-10,7
0,9	0,9 π	1	-6,2	0,906	-9,2	0,58	-14
0,8	0,9 π	1	-5,36	0,897	-9,07	0,547	-14
0,01	0,9 π	1	0	0,77	-8,94	0,447	-14

The experiment has been performed using a 20 transducers array at a frequency of 3 MHz inside a water bath, that is for a 0.5 mm wavelength. The transverse cut of an assembly of 8 rods, 2.5 mm in diameter, regularly spaced on a 4 cm diameter circle, has been observed. The spatial sampling period is equal to 2.5 mm and the filling ratio to 0.9 ($A = 0.9\ \pi$) at the reception and reemission stages ($\beta = 1$). The parasitic images are clearly seen on the photograph (see fig. 6) and their level lies nearly 9 dB down the main image. This small departure from the previous theory may be explained by the fact that the array aperture is limited and that the sources are not rigorously ponctual, as assumed.

Some filtering arises then from these features and the parasitic image level is somewhat decreased with respect to the theoretically expected figure.

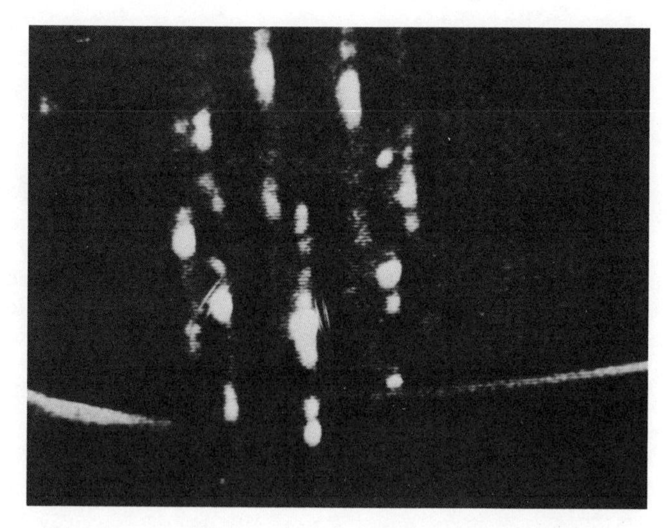

Fig. 6 : Sample image without filtering.

The angular spectrum has been then filtering by using a slit lying in the focal plane of a 15 cm focal length lens. Owing to the a' = 1 mm period of the reemitting array and the λ = 633 nm optical wavelength, a slit width e = 45 µm is needed for a $\lambda/4a'$ filtering. One sees (fig. 7) a very pronounced decrease of the parasitic image level. The price to be paid is a loss of spatial resolution for the system.

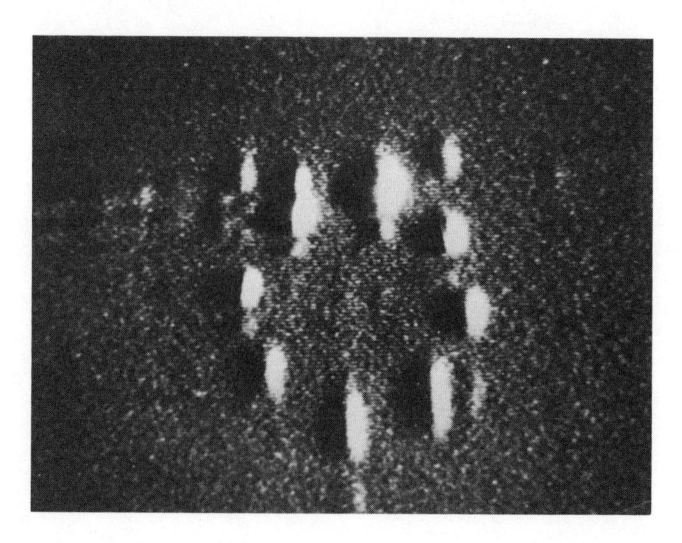

Fig. 7 : Sample image with filtering.

According to the Shannon's criterion, a spatial resolution nearly equal to twice the sampling period is observed for this $\lambda/4a'$ filtering.

CONCLUSION

The present study shows some particular characteristics of an acoustic source array with respect to a classical optical grating and, more precisely, the effect of the phase uniformity over each acoustic source. The theoretical study has led to a spectral filtering method for the parasitic images which has been well confirmed by the experiment

BIBLIOGRAPHY

1. N.Bom, CT Lancee and J. Honkoop "Ultrasonic viewer for cross sectionnal analysis of moving cardiac structures", Bio-med eng 6, 500, 1971.

2. P. Alais, M. Fink "Fresnel zone focussing of linear arrays applied to B and C echography" Acoustical Holography and Imaging Vol. 7, Chicago 1976.

3. M. Berson, A. Roncin, L. Pourcelot "Appareil multitransducteur à focalisation dynamique", S.F.A.U.M.B. Paris 1977.

4. J.C. Somer "Phased array systems", Symposium on echocardiography Rotterdam June 1977.

5. E.B. Miller, F.L. Thurstone "Linear ultrasonic array design for echosonography, J. Acoust. Soc. Am, Vol. 61, N° 6, June 1977.

6. R. Torguet, C. Bruneel, E. Bridoux, JM. Rouvaen and B. Nongaillard, Acoustical Holography and Imaging Vol. 7, Chicago 1976.

7. C. Bruneel, R. Torguet, E. Bridoux, JM. Rouvaen, B. Delannoy and M. Moriamez, I.E.E.E. Ultrasonic Symposium, Phoenix, Oct. 1977.

8. C. Bruneel, B. Nongaillard, R. Torguet, E. Bridoux, JM Rouvaen Ultrasonics Nov. 1977, 263-264.

9. C. Bruneel - thèse de doctorat (1978).

10. B. Delannoy, R. Torguet, C. Bruneel, E. Bridoux, JM. Rouvaen and H. Lasota "Acoustical image reconstruction using parallel processing analogical electronic systems" to be published. J.A.P.

11. C. Bruneel, E. Bridoux, B. Delannoy, B. Nongaillard, J.M. Rouvaen and R. Torguet, J.A.P., Vol. 49, N° 2, 569-573, 1978.

12. J.W. Goodman, Introduction to Fourier Optics. Mc Graw Hill, New York 1968.

13. M. Born, E. Wolf, Principles of Optics,(third Edition) Pergamon Press 1965.

SUPER-RESOLUTION FOR SEPARATING CLUSTERED IMAGES

T. Sawatari and P. N. Keating

Bendix Research Laboratories

Southfield, Michigan 48076

ABSTRACT

A technique for accurate estimation of the positions of objects separated by less than the classical resolution distance is described, where the main emphasis is on rapid processing and noise-tolerance. The technique consists of three steps: signal pre-cleaning (or filtering), deconvolution, and the use of spatial null processing. A computer simulation test showed that this method could determine the location of each of three objects within the resolution limit when a 41-element array was used and the original S/N amplitude ratio at each array receiver was greater than 20 dB. Detailed results of the simulation are discussed.

1. INTRODUCTION

When more than two discrete objects are within the classical limit (one beamwidth) of the resolution of the system, conventional imaging techniques are not able to separate the objects and accurate location-estimates of the object cannot be made.

One approach to identifying objects within the resolution limit of the system is with the use of super-resolution techniques. Here, super-resolution is referred to as a technique to resolve (or restore) information within the resolution limit of a system. Many investigators[1-8] have proposed various super-resolution approaches, yet it remains a difficult problem in practice. The difficulty of implementing such a technique is in its high noise sensitivity.

This paper describes a method which will allow rapid and accurate estimates of the positions of an unknown number of localized objects which are located within the classical resolution distance (i.e., within one beamwidth) of an array system. There is no requirement that the radiation signals from the different objects be statistically independent, so that the approach is applicable to returns from nearby objects at about the same range in active imaging systems or sonar.

The technique developed, a super-resolution technique for clustered objects (SRCO), involves the use of three steps. The first step is a signal precleaning process, where object returns outside a reduced field of view are suppressed by using a spatial bandpass filter. In the second step, a deconvolution algorithm is used to extend the aperture so that approximate locations and numbers of localized objects within the reduced field of view can be identified. In the third step, a spatial null processing algorithm[9-12] is used to minimize the effect of the other nearby objects so that estimates of the position of each object of interest can be made more accurately.

The SRCO technique has been evaluated with computer simulations using the specific object scenario described in Section 3. The results showed the SRCO functions as expected, e.g., three object locations were simultaneously estimated with an accuracy of better than 5% of the resolution limit, for a 41-element uniform array and a signal-to-noise amplitude ratio $\geq$20 dB at each array receiver.

The technique to estimate the positions of clustered objects is described in Section 2 and the results of the computer simulation are described in Section 3. Conclusions are given in Section 4.

2. IMPROVED ESTIMATES OF THE POSITIONS OF CLUSTERED OBJECTS

The super-resolution technique originally proposed by Harris,[2] the analytical continuation method, might appear to be the most suitable approach to our problem, i.e., to estimate accurate object locations when several localized objects are clustered within resolution limits and the radiations from them are not statistically independent. The reason for this is that the techniques described in the literature[5-8] are either time-consuming or require conditions which our object scenario cannot satisfy.

However, the problem with the Harris technique is its high noise sensitivity as discussed in the following paragraphs.

The analytical-continuation method, which is also known as the deconvolution method, is used to restore (calculate) large aperture array information ($x_{s\ell}$; ℓ = 1, 2, . . . L) from the original array information (x_ℓ; $\ell = 1, 2, \ldots L$). The values $x_{s\ell}$ and x_ℓ are the complex amplitudes of the field at each point of the array. In other words, when the original array has a resolution of ΔB, and the field of view is B, we calculate the field of the restored array which has the resolution Δb and a field of view b. For increased resolution, $\Delta b < \Delta B$ and consequently $b < B$ since we have the relationships, $B = (L-1)\Delta B$ and $b = (L-1)\Delta b$. The region where the objects are clustered is defined here as the object region b.

If the desired resolution (Δb), the object region (b), and the desired S/N figure in the restored information are specified, the minimum S/N ratio in the original signal, which we must have to achieve the desired restoration, can be estimated. For example, if we want to increase the resolution 3 times while maintaining the object region identical to the classical limit of resolution of the original array, we require at least a S/N amplitude ratio of 30 in the original signal. In this case, we can only expect the S/N amplitude ratio in the restored information to be around 0.7. This example shows the high noise sensitivity of this technique.

The deconvolution technique requires a high signal-to-noise ratio in the input array data. However, since, in general, the number (M) of the original array elements is much larger than the number of the resolution elements desired by the restored system, we can improve the S/N ratio of the necessary input data and, consequently, can extend the applicability of the deconvolution method. The process of improving the S/N ratio is similar to that of obtaining the array gain in the conventional imaging technique. A specific implementation of this process is described in the following section and in Section 2.2 a deconvolution algorithm developed under this study is described.

The data restored by the deconvolution method are expected to have a large amount of noise as is explained in the preceding section. Furthermore, the sidelobes of the objects continue to interfere with the other objects in the restored array data, even though there is no longer any mainlobe interference. The noise and sidelobes make accurate estimation of each object location difficult. In Section 2.3, we describe an additional processing step which is applied to the restored data, which results in more accurate estimation of object positions.

2.1 Preprocessing of Input Data

In our object scenario, the objects are concentrated within the object region (b). (See Figure 1, where the conventional image distribution of array data is shown.) The field outside the region (b) consists of the white noise included in the array data and the sidelobe component of the objects. By cutting out these fields,

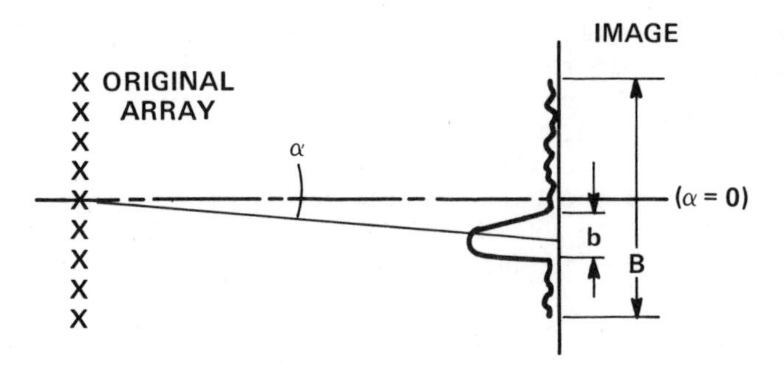

Figure 1. Schematic of array plane and object plane where b is the object region and B is the field of view of the array system.

using a spatial filter, a significant amount of noise will be removed from the data.

Before we describe the implementation of such a spatial filter, we have to perform the following two processing steps on the input array data: (a) the central location (α in Figure 1) of the object region of interest has to be shifted to a particular direction (say, $\alpha = 0$) by multiplying a linear phase term to the original array data, and (b) if there are other interference sources (high peaks) in the field of view (B, in Figure 1) outside the object region b, they have to be removed by the spatial null processing method.[9-12] The process (a) is a simple one. This is, however, very important because by doing this, the filter described below can be determined independently from the input signal. The process (b) is to reduce the sidelobe components of the interference in the filtered signal.

Now we are ready to describe mathematically the filtering process. Let us denote a vector $\underset{\sim}{x}$ of dimension M as input data of an array system of M elements, i.e., the jth element of the vector corresponds to the input signal of the jth array element. The input signal is the amplitude and phase of a particular frequency component of the incident field. The width of the object region is denoted as b, the central location of which has been shifted to the direction ($\alpha = 0$) by the above preprocessing (a).

The filtered signal $\underset{\sim}{x}_F$ is expressed by

$$\underset{\sim}{x}_F = F \cdot \underset{\sim}{x} \qquad (1)$$

where F is the filter matrix, which is defined below. When the filter represents a uniform passband only over the object region extending from -b/2 to b/2 around the center of the object region and the array is uniformly linear, the elements of the matrix F can be uniquely determined as

$$F = (F_{\ell m}) = \left(\frac{\sin(\pi(D\ell - am)b/\lambda}{\pi(D\ell - am)b/\lambda} \right) \quad (2)$$

where λ is the wavelength of the radiation of interest, a is the array element spacing, and D is the spacing between the two adjacent elements of the thinned array (see Figure 2) corresponding to the filtered signal $\underset{\sim}{x}_F$.

The dimension (L) of $\underset{\sim}{x}_F$ is determined from the resolution elements ($b/\Delta b = L-1$) obtained by the deconvolution method described in the following section. As becomes more apparent in the next section, this is due to the fact that the deconvolution algorithm requires data from only L array elements to calculate the field corresponding to a larger aperture of L elements which has the restoration element (L-1) (see Figure 2). Practically speaking, since L

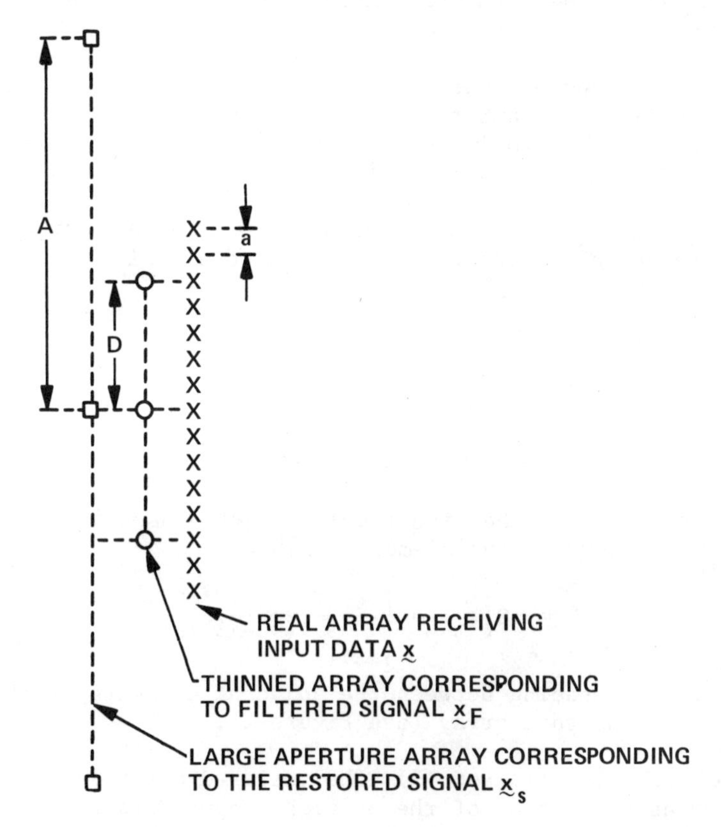

Figure 2. Geometrical relation between real aperture, thinned (filtered) aperture, and large (restored) aperture.

can only be at most 4, due to the noise restriction, the array elements (M) (say, 41) are much greater than L.

The maximum improvement of the S/N amplitude ratio in the input data that can be expected from this filtering operation is $\sqrt{M/L}$. This is obtained by the ratio $\sqrt{D/a}$ which is, referring to Figure 2, $\sqrt{M/L}$.

2.2 Super-Resolution Matrix

In this subsection, we describe a method to calculate a field outside the aperture of the given array to obtain a super-resolution effect. The method adopted is similar to that proposed by Harris,[2] i.e., the so-called analytical continuation approach. Since a detailed explanation of this technique is given in the literature, we describe here only the resultant expression useful for the present application. We use here a matrix notation and refer to the technique as the deconvolution method, since we are dealing with an array system.

For our purposes, the deconvolution technique is used primarily to identify the number of objects lying within the object region and their approximate locations. This information is used in the next step to estimate more accurate object locations.

The restored signal corresponding to a larger array (see Figure 2) can be calculated from the filtered signal as

$$\underset{\sim}{x}_S = S^{-1} \cdot \underset{\sim}{x}_F \qquad (3)$$

where S is a square matrix of dimension LxL; L is the dimension of $\underset{\sim}{x}_F$, the inverse of which we call here the super-resolution matrix, i.e., this process is the so-called deconvolution process.

For the case of the linear uniform array used to obtain Eq. (2), we can determine the elements of S as

$$S = (S_{\ell'\ell}) = \left(\frac{\sin(\pi(D\ell - A\ell')b/\lambda}{\pi(D\ell - A\ell')b/\lambda} \right) \qquad (4)$$

where A is the spacing between the adjacent elements of the synthesized larger aperture array (see Figure 2).

Combining Eq. (1) and (3), the restored array signal $\underset{\sim}{x}_S$ can be expressed as a function of the original input data $\underset{\sim}{x}$;

$$\underset{\sim}{x}_S = S^{-1} F \underset{\sim}{x} \qquad (5)$$

The rectangular matrix $S^{-1}F$ (LxM) is independent of the input signal $\tilde{x}$. This is due to the preprocessing in which the object region of interest is shifted to the predetermined direction ($\alpha = 0$).

The restored signal $\tilde{x}_s$ obtained through Eq. (5) is now beamformed so that the discrete objects within the object region are identified by the super-resolution effect. The imaging process is simply expressed by

$$\tilde{y} = B\tilde{x}_s \tag{6}$$

where B is a rectangular matrix with a dimension of (NxL) and is given by

$$B = (B_{n\ell}) = \left(e^{i\frac{2\pi}{\lambda} A\ell\Delta\beta n} \right) \tag{7}$$

where $\Delta\beta = b/N$.

The vector ($\tilde{y}$) corresponds to the object field at discrete positions. Observing the distribution of the field ($\tilde{y}$), we can identify the number of objects and their approximate locations with an accuracy of one resolution of the large array obtained by the deconvolution technique.

It is to be noted that, since the dimension of the vector $\tilde{x}_s$ is L, only L components out of N elements of the vector $\tilde{y}$ are unique when N>L. This implies that (N-L) information bits contained in the vector ($\tilde{y}$) are interpolated values of the object field distribution. However, this interpolation is very important when we apply the third step which is used, as is described in the next subsection, to increase the estimation accuracy of the object locations.

Furthermore, it is to be noted that the imaging process of Eq. (6) gives us a gain in S/N amplitude ratio, i.e., the so-called "array gain", which is, in a statistical sense, $\sqrt{L}$. Combining this gain with that obtained by the filtering $\sqrt{M/L}$ (see Section 2.1), we have the net gain $\sqrt{M}$. This is identical to that obtained by the conventional imaging process. These results are valid, however, only when the width (b) of the object region is one resolution of the original system.

2.3 Final Noise Reduction

The restored object field given in Eq. (5) is still disturbed by the noise inside the object region, which was not removed by the filtering process explained in Section 2.1. Also the accurate location estimate is difficult to achieve because of the mutual sidelobe effects of all objects within the object region.

We have shown recently[13] that the spatial null processing technique is a powerful algorithm to estimate accurate locations of the objects in the presence of nearby interferences (localized objects) and white noise, if we know the number of objects within the region of interest. Since the deconvolution method described in the preceding section provides information on the number and approximate locations of the discrete objects, we can now apply this process. The process was modified, however, to achieve faster processing times than those which could be achieved using previous iterative approaches.[13] This modification is mathematically described below.

We first describe the complete processing involved in the third step of our SRCO. The approach used here is first to find the approximate location of each object using Eq. (6), i.e., the number of the peaks and the peak locations appearing in the object field distribution ($\tilde{y}$) are assumed to be the number of the objects and their locations. Second, we subtract the field coming from each object direction from the input data, which is the restored input signal ($\tilde{x}_s$). In other words, nulls are located in the direction of all objects. The total power of the remaining field is a function of null locations, i.e., the total power remaining will vary if the null locations are slightly altered around those initially estimated. The best estimates of object locations, as defined in the method,[13] are those locations which minimize the total power of the remaining field.

In the following discussion, we mathematically describe the modified null processing technique.

Let us denote $\tilde{\Theta} = (\Theta_1, \Theta_2 \cdot\cdot\cdot \Theta_j)$ (J<L) as the approximate angular position of the discrete objects (J), which were found by observing the reconstructed object field (y) in Eq. (6). Knowing $(\tilde{\Theta})$, we can construct a complement of the projection operator expressed by[10]

$$N(\tilde{\Theta}) = I - P(\tilde{\Theta}) \tag{8}$$

where I is a unit matrix of dimension (LxL) and P is a projection operator of rank J and is given by[14]

$$P(\tilde{\Theta}) = \sum_{j}^{J} \tilde{k}_j(\tilde{\Theta})\, \tilde{k}_j^{\dagger}(\tilde{\Theta}) \tag{9}$$

where $(\underset{\sim}{k}_j(\Theta))$ is an orthogonal unit vector created from a set of direction vectors $(\underset{\sim}{d}_j(\Theta_j))$ using the Schmidt orthonormalization method. The direction vector is given by

$$\underset{\sim}{d}_j(\Theta_j) = (d_j(\Theta_j)_\ell) = \left(e^{i\frac{2\pi}{\lambda}\Theta_j A\ell} \right) \tag{10}$$

i.e., this corresponds to one row of vectors of the imaging matrix B where $\Theta_j = \Delta\beta_j$.

Using the matrix (N) defined in Eq. (8), the field remaining after j^{th} nulls are located is expressed by

$$\underset{\sim}{x}_R = N\underset{\sim}{x}_s \tag{11}$$

The total power remaining in the image plane is then given by (using Perseval's theorem)

$$Z(\underset{\sim}{\Theta}) = |\underset{\sim}{x}_R|^2 \tag{12}$$

If the best estimates $(\Theta^o = (\Theta_1^{\,o}, \Theta_2^{\,o}, \ldots \Theta_j^{\,o}))$ of the object locations are near the approximate locations $(\underset{\sim}{\Theta})$, we may find the set of best estimates from a straightforward quadratic approximation as follows.

The matrix (N) which minimizes the remaining power (z) may be approximated by taking the first order of Taylor expansion of Eq. (8), i.e.,

$$N(\underset{\sim}{\Theta}^o) = N^o(\Delta\underset{\sim}{\Theta}) = N(\underset{\sim}{\Theta}) + \sum_j^J \Delta\Theta_j \frac{\partial N(\underset{\sim}{\Theta})}{\partial\Theta_j} \tag{13}$$

where

$$\underset{\sim}{\Theta}^o = \underset{\sim}{\Theta} + \Delta\underset{\sim}{\Theta} \quad (\Delta\underset{\sim}{\Theta} = (\Delta\Theta_1, \Delta\Theta_2, \ldots \Delta\Theta_j)) \tag{14}$$

In terms of I and P of Eq. (14), we have

$$N^o(\Delta\underset{\sim}{\Theta}) = I - \left(P + \sum_j^J \Delta\Theta_j \frac{\partial P}{\partial\Theta_j} \right) \tag{15}$$

where

$$\frac{\partial P}{\partial \Theta_j} = \underset{\sim}{k}_j{}' \underset{\sim}{k}_j{}^{\dagger} + \underset{\sim}{k}_j \underset{\sim}{k}_j{}'^{\dagger} \tag{16}$$

and where

$$\underset{\sim}{k}_j{}' = \frac{\partial \underset{\sim}{k}_j}{\partial \Theta_j} \tag{17}$$

Using this matrix (N) in Eq. (13), the remaining power is given by a function of $(\Delta\underset{\sim}{\Theta})$:

$$Z^o(\Delta\underset{\sim}{\Theta}) = |N^o(\Delta\underset{\sim}{\Theta})\underset{\sim}{x}_s|^2 \tag{18}$$

This is a quadratic form in $(\Delta\underset{\sim}{\Theta})$. Since $Z^o(\Delta\tilde{\Theta})$ is minimum for the best estimate of $(\Delta\underset{\sim}{\Theta})$ (i.e., $(\underset{\sim}{\Theta}^o)$ through Eq. (14)), we find the best estimate as

$$\Delta\underset{\sim}{\Theta} = -A^{-1}\underset{\sim}{B} \tag{19}$$

where, from Eq. (15),

$$A = (A_{ji}) = (\underset{\sim}{A}_j{}^{\dagger}\underset{\sim}{A}_i + \underset{\sim}{A}_i{}^{\dagger}\underset{\sim}{A}_j) \tag{20}$$

$$\underset{\sim}{B} = (B_j) = (\underset{\sim}{A}_j{}^{\dagger}\underset{\sim}{c} + \underset{\sim}{c}^{\dagger}\underset{\sim}{A}_j) \tag{21}$$

and where

$$\underset{\sim}{A}_j = \frac{\partial P}{\partial \Theta_j}\underset{\sim}{x}_s \tag{22}$$

$$\underset{\sim}{c} = (I - P(\underset{\sim}{\Theta}))\ \underset{\sim}{x}_s \tag{23}$$

This method of obtaining the best estimate of the object location $(\underset{\sim}{\Theta}^o)$ through Eq. (14) and (19) is rapid since it is not an iterative method, and provides a reliable estimate if $|\Delta\Theta| < (1/2) \times$ (one resolution width).

3. RESULTS OF COMPUTER SIMULATION

The signal processing technique described in Section 2 was evaluated using a computer simulation of the following object scenario. Forty-one equally spaced elements (M = 41) were used to form the array for a coherent imaging system.[4] The objects were discrete (delta function type), located in the far field, and localized within one resolution limit. Interfering objects outside the object region of interest were not considered in the simulation. If there are any such interfering objects, these can be eliminated by the method proposed in References 9-12. The performance of the above algorithm is examined as a function of the amount of random noise added to the simulated signal, and the number and separation of the discrete objects.

Figure 3 shows a conventional point spread function obtained for a single object located in the direction normal to the array ($\alpha = 0$ in Figure 1) where no noise is included. The full scale of the lateral axis (from -20 to +20) corresponds to the width of the resolution limit ($\lambda/a(M-1)$); λ is the wavelength, a is the spacing, and M is the total number of array elements. For simplicity, we have chosen $\lambda = a = 1$. Figure 4 is the result of the super-resolution applied to input data identical to that of Figure 3 (i.e., the result of step 2 only). The large aperture information restored using the super-resolution (Eq. (3)) is a 4-element linear array having a spacing of (40xa). In the figures, A1, α1, and SD are the complex amplitude of the signal, the location of the object (measured by the scale of 40 units per one resolution limit), and the rms amplitude (standard deviation) of noise added to each receiver element.

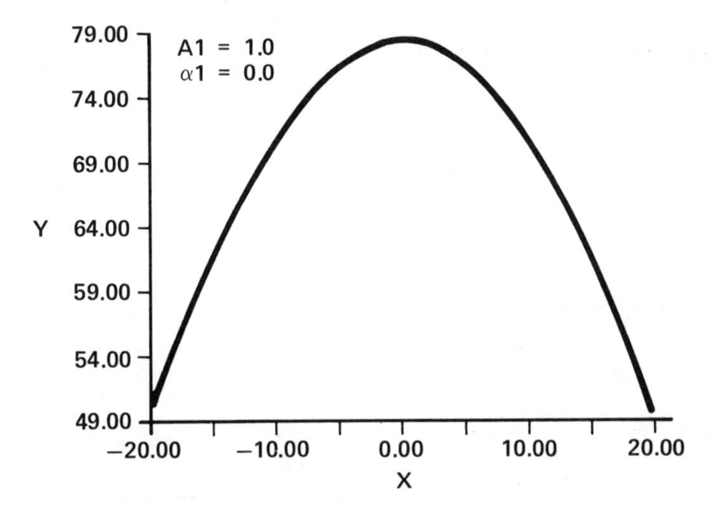

Figure 3. Conventional image of a point object (point spread function).

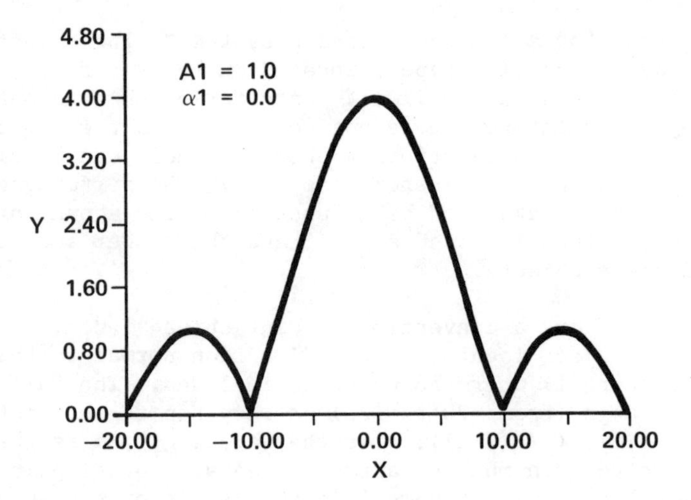

Figure 4. Point spread function after super-resolution.

Figures 5 and 6 demonstrate the white noise reduction effect of the present method (steps 1 and 2) for the same single object on the axis. The dotted lines in the figures are the results of the image pattern obtained from the restored aperture information without using the filtering procedure (step 2 only). Solid lines are similar results obtained using the spatial filter (steps 1 and 2). The white noise added to each input data (specified by the rms amplitude (SD) relative to the signal amplitude on each receiver) has a gaussian probability function and was altered, 0.05 and 0.1

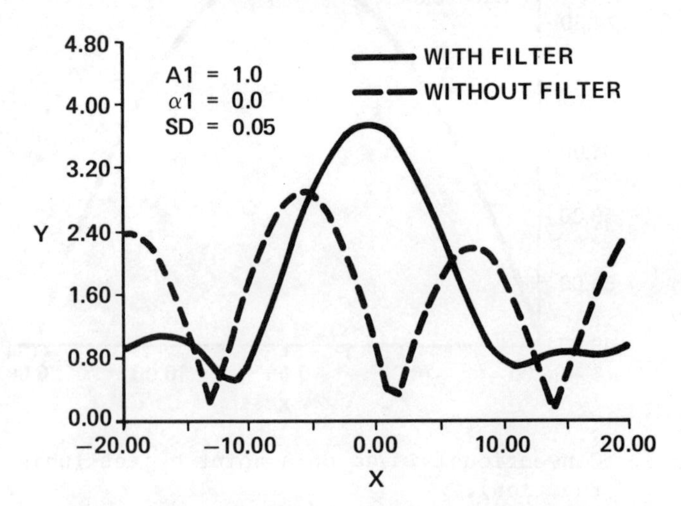

Figure 5. White noise effect for super-resolution (SD = 0.05); dotted line, without noise rejection filter; solid line, with noise rejection filter.

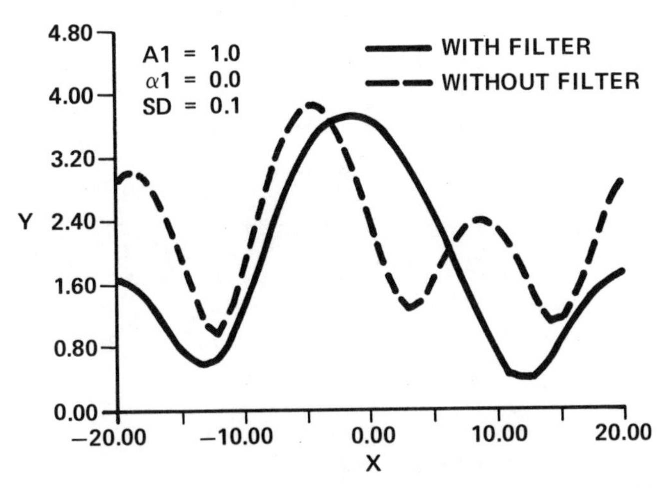

Figure 6. White noise effect for super-resolution (SD = 0.1); dotted line, without noise rejection filter; solid line, with noise rejection filter.

of its rms amplitude, while the amplitude of an object signal was maintained as unity. Two independent random numbers were chosen to create a complex noise field (real part and imaginary part) for each array element.

These results indicate that the noise reduction method of step 1 can reduce the noise effect significantly; the input signal with noise (SD = 0.1) was used for the deconvolution method and reasonable results were obtained. Note that the original deconvolution method was only applicable to the input data with (SD) less than 0.02. The super-resolution operation increases the resolution 3 times that of the conventional imaging technique. This is, of course, due to the fact that the restored larger aperture is 3 times greater than that of the original array.

Tables I and II show typical examples of the final performance of the method, including the third step (fine estimate of object location using a process identical to Eq. (19)), by which we reduce the noise effect and the mutual sidelobe influence inside the object region. In Table I, the results of double objects are summarized; amplitude (A_i) of the two objects are 1 and $e^{i\pi/2}$ respectively, and their locations (α_i) are −8.0 and +8.0 in the horizontal scale (i.e., 40 units are the total scale of one resolution limit). In order to evaluate the performance of the present method, the statistical averages were obtained; the average ($\bar{\alpha}_i$) and the rms (σ_i) of the estimated locations of the object are shown as a function of the amount (SD) of white noise injected. Thirty different white noise patterns belonging to the same standard deviation (SD) have been used to obtain the average value and the rms value of the estimated location (i.e., the signal is unchanged).

Table I. Result of Two Target Simulation.

$A_1 = 1.0$ $\alpha_1 = 8.0$

$A_2 = e^{i\,\pi/2}$ $\alpha_2 = -8.0$*

SD	$\bar{\alpha}_1$	σ_1	$\bar{\alpha}_2$	σ_2
0.0	6.0	0	−6.0	0
0.01	6.4	2.4	−5.7	0.6
0.05	9.3	6.2	−6.8	2.5
0.1	10.0	7.5	−7.3	3.3
0.15	7.3	3.8**	−6.0	4.5

A_i = COMPLEX AMPLITUDE OF OBJECT (i = 1, 2)
α_i = LOCATION OF OBJECT
$\bar{\alpha}_i$ = ESTIMATED LOCATION OF OBJECT (AVERAGE)
σ_i = RMS ERROR OF THE ESTIMATION

*A_i, amplitude of object, α_i, location of object which is measured by the horizontal scale; 40 units (−20 to 20) is one resolution limit; SD, degree of white noise, $\bar{\alpha}_i$, estimated location of the targets (average over 30 different noise patterns); σ_i, rms deviation of the estimated location).

**The decrease of σ_1 in spite of the increase of SD was probably due to the criterion we set for the calculation, which is that if the estimated values (α_1 and α_2) were utterly incorrect, the values were not included in the average. This rejection occurred when SD > 0.1.

Table II shows similar results for three discrete objects of different magnitudes. The amplitude and the location of each object are indicated in the table. The separation between the adjacent objects was 15 in the horizontal scale, which is a little more than 1/3 of one resolution limit.

It should be noted here that, in both cases of two and three discrete objects, if the separation between the objects is less than 1/3 of the resolution limit, the algorithm does not function. This is because the resolution of the super-resolution system is only 3 times that of the original system.

Table II. Result of Three Target Simulation.

$A_1 = e^{i\pi/2}$ $\alpha_1 = 0.0$
$A_2 = 0.8$ $\alpha_2 = 15.0$
$A_3 = -0.7$ $\alpha_3 = -15.0$ *

SD	$\bar{\alpha}_1$	σ_1	$\bar{\alpha}_2$	σ_2	$\bar{\alpha}_3$	σ_3
0	−0.4	0	13.0	0	−13.0	0
0.01	0.28	1.3	13.1	0.7	−13.8	1.3
0.05	1.0	2.6	13.6	2.7	−14.7	2.1
0.1	−0.3	4.3	12.7	5.7	−13.1	4.6
0.15	1.5	5.5	13.5	5.3	−12.0	5.1

*A_i, amplitude of object, α_i, location of object which is measured by the horizontal scale; 40 units (-20 to 20) is one resolution limit; SD, degree of white noise; $\bar{\alpha}_i$, estimated location of the objects (average over 30 different noise patterns); σ_i, rms deviation of the estimated location).

4. SUMMARY AND CONCLUSIONS

A signal processing technique we call Super-Resolution for Clustered Objects (SRCO) was developed and a computer simulation test was performed. This technique was designed to achieve accurate object location estimates when the number of localized objects within one resolution limit is not known a priori and the radiation from each discrete object is not statistically independent. The technique developed aimed at being applicable for noisy input data and capable of rapid processing.

SRCO consists of three basic steps. The first step is to reduce the white noise and/or interference outside the object region of our interest. The second process is to extend the effective aperture using a deconvolution method. Each step requires only a simple matrix multiplication to the original input data to shorten processing time. The third step uses a modified version of the spatial null processing technique to obtain more accurate location estimates from the data obtained by steps 1 and 2. The third step is also designed to be fast.

The computer simulation showed that the SRCO has reduced the noise sensitivity significantly in comparison with the previously known, non-iterative, super-resolution method; e.g., the maximum rms amplitude noise allowed was increased by a factor 10 in comparison with the conventional analytical continuation method. Furthermore, the location estimation of multi-discrete objects in the final step was as good as that of a single object. This indicates that the mutual sidelobe effect, which significantly reduces the location-estimation accuracy of the discrete objects, has been effectively removed.

In spite of the improvement in noise sensitivity, the technique appears to still be limited for wide use in acoustical imaging because a signal-to-noise ratio of about 10 is needed. However, in an active imaging system or active sonar where higher S/N signals are obtainable, this technique can be useful.

ACKNOWLEDGEMENTS

The authors would like to express their appreciation to R. F. Steinberg for his assistance in preparing the manuscript and his encouragement. The authors also thank Paul Singh for his assistance in computer programming.

REFERENCES

1. G. Toraldo Di Francia, J. Opt. Soc. Am., 45, 497, (1955).

2. J. L. Harris, J. Opt. Soc. Am., 54, 297 (1964).

3. C. W. Barnes, J. Opt. Soc. Am., 56, 474 (1966).

4. B. R. Frieden, "Progress in Optics", Vol. IX, 311, ed. by E. Wolf, North Holland (1971).

5. R. W. Gerchberg, Optical Acta., 21, 9, 709 (1974).

6. P. De Santis and F. Gori, Optical Acta., 22, 8, 691 (1975).

7. B. R. Frieden, J. Opt. Soc. Am., 62, 4, 511 (1972).

8. Y. Biraud, Aston. Astrophys., 1, 124 (1969).

9. R. F. Steinberg, P. N. Keating, and R. F. Koppelmann, "Acoustical Holography", Vol. 6, ed. by N. Booth, 539, Plenum Press, New York (1976).

10. P. N. Keating, R. F. Koppelmann, R. K. Mueller, R. F. Steinberg, and G. J. Zilinskas, J. Acoust. Soc. Am., 59, 106 (1976).

11. P. N. Keating and T. Sawatari, "Acoustical Holography", Vol. 7, ed. by L. W. Kessler, 537, Plenum Press, New York (1977).

12. P. N. Keating, "A Rapid Approximation to Optimal Array Processing for the Case of Strong Localized Interferences", to be published.

13. T. Sawatari, P. N. Keating, and R. E. Willey, "Holographic Adaptive Interference Rejection Technique for Low Angle Radar", to be submitted to IEEE Trans. on Aerospace and Electronic Systems.

14. R. K. Mueller, private communication.

THREE DIMENSIONAL IMAGING BY WAVE-VECTOR DIVERSITY

N. H. FARHAT and C.K. CHAN

University of Pennsylvania

200 S. 33rd St., Philadelphia, PA., 19174

ABSTRACT

A method for the three dimensional imaging of objects by wave-vector diversity (frequency and viewing and/or illumination angle diversity) is analyzed using a Fourier Optics approach. The analysis is applicable to two classes of objects of practical interest namely weakly scattering objects and perfectly reflecting objects. It is shown that under frequency swept plane wave illumination the data collected by an array of receivers deployed in the far field of the object represents a 3-D data manifold that is within a quadratic phase factor equal to the 3-D Fourier transform of the object function. Methods for removal of the quadratic phase factor which otherwise can lead to image distortion are discussed. By invoking Fourier domain projection theorems we show that the distortion corrected 3-D data manifold can yield a reconstruction of the object slice by slice or all at once using integral holography. Similarities between wave-vector diversity imaging and possible "imaging" features in certain cetacean and in the bat are pointed out.

1. INTRODUCTION

The development of longwave (acoustic or microwave) holographic imaging systems possessing resolution and image quality approaching those of optical systems is hampered by three factors: (a) prohibitive cost and size of longwave imaging apertures, (b) inability to view a 3-D image as in optical holography and (c) degradation of image quality by speckle noise because of the low numerical apertures attainable with present techniques. For example, a longwave imaging aperture operating at a wavelength of 3 cm should be about 3 km in size in order to achieve image resolution comparable to an ordinary photographic camera. In addition to inconvenient size, the cost of filling such a large aperture with suitable coherent sensors is clearly prohibitive. Furthermore, recall that in conventional longwave holography when optical image retrival is utilized, it is necessary to store the longwave hologram data (fringe pattern) in an optical transparency suitable for processing on the optical bench using laser light. In order to avoid longitudinal distortion* of the reconstructed image, the size of the optical hologram replica must be m (= λlong/λlaser) times smaller than the longwave recording aperture. For the example cited earlier, this means an optical hologram replica of less than a millimeter in size. It is certainly not possible to view a virtual 3-D image through such a minute hologram even with optical aids since these tend to introduce their own longitudinal distortion. As a result, longwave holographers have long learned to forgo 3-D imagery and settled instead for 2-D imagery obtained by projecting the reconstructed real image on a screen. This permits lowering of the reduction factor m and consequently lowering of the resolution requirements of the photographic film. This permits in turn the use of highly convenient Polaroid transparency film for preparation of the optical hologram replica. Because of the small size (measured in wavelength) of longwave apertures attainable in practice and the above method of viewing the real image, speckle noise is always present leading to degradation in image quality.

In this paper we describe and analyze an imaging method that utilizes wave-vector diversity (frequency and viewing and/or illumination angle diversity) that circumvents the limitations discussed above. It is worthwhile to point out that our studies of wave-vector diversity imaging (or frequency swept imaging) are motivated to a large extent by evidence of super-resolved "imaging" capabilities by frequency sweeping in certain cetacean and in the bat which are known to use frequency swept signals in their "sonar". We first present the theoretical principles of the method. Fourier-domain projection theorems are discussed next and utilized to illustrate the ability of the method to provide 3-D image information. This is followed by a presentation of the results of a computer simulation. Finally the features of wave-vector diversity imaging are summarized and the similarities to "imaging" procedures in the dolphin and in the bat are pointed out.

---

*Longitudinal distortion causes for example the image of a sphere to appear elongated in the range direction like a very long ellipsoid.

2. ANALYTICAL CONSIDERATION

In this section we present a Fourier-Optics analysis of wave-vector diversity imaging first of a two dimensional object in which there is no restriction on the object orientation relative to the transmitter and the receiver. This is followed by extension to three dimensional objects.

Consider an isolated planar object of finite extent with reflectivity $D(\bar{\rho}_o)$, where $\bar{\rho}_o$ is a two dimensional position vector in the object plane (x_o, y_o). The object is illuminated as shown in Figure 1(a) by a coherent plane wave of unit-amplitude and of wave vector $\bar{k}_i = k\,\bar{1}_{ki}$ produced by a distant transmitter located at $\bar{R}_T$. The wavefield scattered by the object is monitored at a distant receiver at $\bar{R}_R$ lying in the far field region of the object. The position vectors $\bar{\rho}_o$, $\bar{R}_T$ and $\bar{R}_R$ are measured from the origin of a cartesean coordinate system (x_o, y_o, z_o) centered on the object. The object is assumed to be nondispersive i.e., D is independent of k. However, when the object is dispersive such that $D(\bar{\rho}_o, k) = D_1(\rho_1)D_2(k)$ and $D_2(k)$ is known, the analysis presented here can easily be modified to account for such object dispersion by correcting the data collected as k is changed for $D_2(k)$.

Referring to Figure 1(a) and ignoring polarization effects, the field amplitude at $\bar{R}_T$ caused by the object scattered wavefield may be expressed as,

$$\psi(k, \bar{R}_R) = \frac{jk}{2\pi} \int D(\bar{\rho}_o) e^{-j\bar{k}_i \cdot \bar{r}_T} \frac{e^{-jk\, r_R}}{r_R}\, d\bar{\rho}_o \tag{1}$$

where $d\bar{\rho}_o$ is an abbreviation for $dx_o dy_o$ and the integration is carried out over the extent of the object. Noting that $\bar{r}_T = \bar{\rho}_o - \bar{R}_T$, $\bar{R}_T = -R_T \bar{1}_{k_i}$ and using the usual approximations valid here: $r_R \simeq R_R + \rho_o^2/2R_R - \bar{1}_R \cdot \bar{\rho}_o$ for the exponential in (1) and $r_R \simeq R_R$ for the denominator in (1) where $\bar{1}_R = \bar{R}_R/R_R$ and $\bar{1}_{k_i} = \bar{k}_i/k$ are unit vectors in the $\bar{R}_R$ and $\bar{k}_i$ directions respectively, one can write eq. (1) as,

$$\psi(k, \bar{R}_R) = \frac{jk}{2\pi R_R} e^{-jk(R_T + R_R)} \int D(\bar{\rho}_o)\, e^{-j\,\bar{p} \cdot \bar{\rho}_o}\, d\bar{\rho}_o, \tag{2}$$

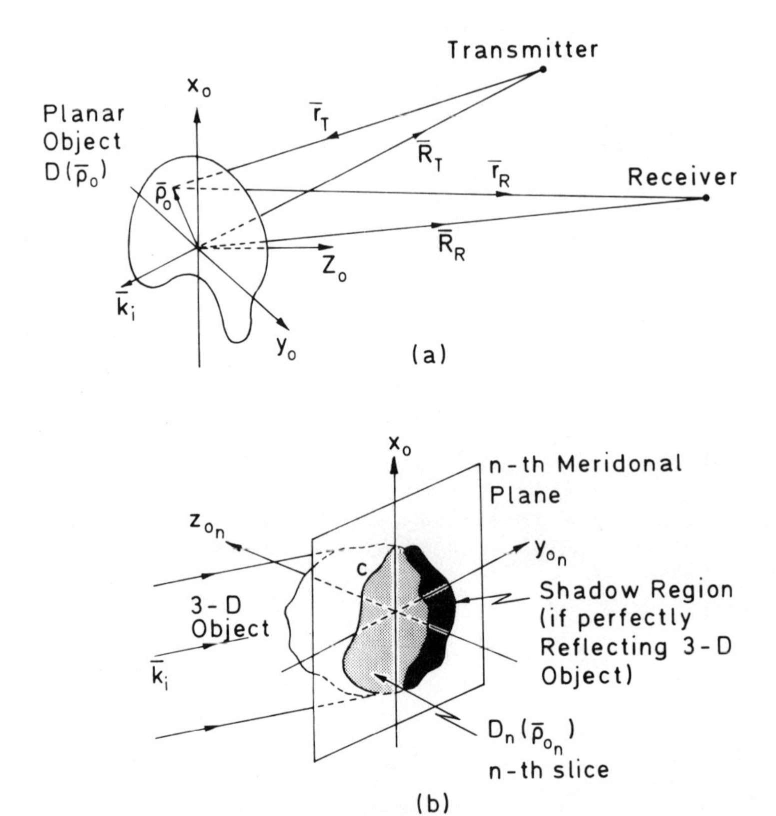

Fig. 1. Geometry for Wave-Vector Diversity Imaging of (a) a two dimensional object and (b) three dimensional object showing a meridonal slice.

where we have used the fact that the observation point is in the far field of the object so that exp $(-j\, k_o^2/2R_R)$ under the integral sign can be replaced by unity. In eq. (2), $\bar{p} = k\,(\bar{1}_{k_i} - \bar{1}_R) \triangleq p_x\,\bar{1}_x + p_y\,\bar{1}_y + p_z\,\bar{1}_z$ is a three dimensional vector whose length and orientation depend on the wavenumber k and the angular positions of the transmitter and the receiver. For each receiver and/or transmitter present, $\bar{p}$ indicates the position vector for data storage. An array of receivers for example would yield therefore as k is changed (frequency diversity)or as $\bar{k}$ $(=k\bar{1}_{k_i})$ is charged (wave-vector diversity) a 3-D data manifold. The projection of this 3-D data manifold on the object plane yields $\psi(k,\bar{R}_T)$ because $\bar{p}\cdot\bar{\rho}_o = \bar{p}_t\cdot\bar{\rho}_o = p_x x_o + p_y y_o$ where $p_x = k(\bar{1}_{k_i} - \bar{1}_R)_x$ and $p_y = k(\bar{1}_{k_i} - \bar{1}_R)_y$ are the cartesian components of the projection $\bar{p}_t$ of $\bar{p}$ on the object plane. Accordingly eq. (2) can be expressed as,

$$\psi(k,R_R) = \frac{jk}{2\pi R_R}\, e^{-jk(R_T + R_R)} \int D(x_o,y_o)\, e^{-j(p_x x_o + p_y y_o)}\, dx_o dy_o \tag{3}$$

Because of the finite extent of the object, infinite limits can be assumed and the integral in (3) is recognized as the two dimensional Fourier transform $\tilde{D}(p_x, p_y)$ of $D(x_o, y_o)$. It is seen to be dependent on the object reflectivity function, the angular positions of the transmitter and the receiver, and on the values assumed by the wavenumber k but is entirely independent of range. Data containing $\tilde{D}$ can thus be collected by varying these parameters. The range information is contained solely in the factor $F = k \exp[-jk(R_T + R_R)]/2\pi R_R$ preceeding the integral. The field observed at $\bar{R}_T$ has thus been separated into two terms one of which, $\tilde{D}$, contains the lateral object information and the other, F, contains the range information. The presence of F in eq (3) hinders the imaging process since it complicates data acquisition and if not removed, gives rise to image distortion because R_R is generally not the same for all receivers. To retrieve an image of the object via a 2-D Fourier transform of eq. (3) (what can be carried out optically), the factor F must first be eliminated. This operation yields $\tilde{D}$ over a two dimensional region in the p_x, p_y plane. The size of this region, which determines the resolution of the retrieved image[1,2] depends on the angular positions of the transmitter and the receiver and on the values assumed by k, i.e. the width and position of the spectral window employed. The later dependence on k implies super-resolution imaging capability because of the frequency synthesized dimension of the 3-D data manifold generated. Because of the dependence of resolution on the relative positions of the object, the transmitter, and receiver (or receiver array), the impulse response is clearly spatially variant. In fact a receiver situated at $\bar{R}_T$ for which $\bar{p}$ is normal to the object plane can not collect any lateral object information because for this condition ($\bar{p} \cdot \bar{\rho}_o = 0$) the integrals in (2) and (3) yield a constant.

Such a receiver is located in the direction of specular reflection from the object where the diffraction pattern is stationary i.e. does not change with k. In this case the observed field is solely proportional to F containing thus range information only. Obviously this case can easily be avoided through the use of more than one receiver which is required anyway when 2-D or 3-D objects resolution is sought.

Several methods for the elimination of F from the collected data appear possible. These include: (a) By furnishing a target derived reference (TDR) to the receivers by means of Porter's method (3) (b) By optical filtering (4), (c) By high speed analog-to-digital conversion and storage of the signals detected by receivers that are furnished with a common reference signal derived from or phase-locked to the transmitter. In this later case, digital correction for F can then be performed using apriori knowledge (obtained by other independent means) of R_R for each receiver. The most attractive of these, the TDR method, is currently under experimental study.

The analysis presented here can be extended to three dimensional objects by viewing a 3-D object as a collection of thin meridonal slices (see Fig. 1 (b)) each of which represents a two dimensional object of the type analyzed here. With the n-th slice we associate a cartesean coordinate system x_o, y_{o_n}, z_{o_n} that differ from other slices by rotation about the common x_o axis. Since the vectors $\bar{p}$, $\bar{R}_T$ and $\bar{R}_R$ are the same in all n-coordinate systems, eq (3) holds. $\psi_n(k,\bar{R}_T)$ is then obtained from projection of the three dimensional data manifold collected for the 3-D object on the x_o, y_{on} plane associated with the n-th slice. An image for each slice can then be obtained as described before. An inherent assumption in this argument is that all slices are illuminated by the same plane wave. This is a reasonable approximation when the 3-D object is weakly scattering and the Born approximation is applicable or when the 3-D object is perfectly reflecting and does not give rise to multiple reflections between its parts. In the later case the two dimensional meriodnal slices D_n ($\bar{\rho}_{o_n}$) deteriorate into contours, such as C in Fig. 1(b) defined by the intersection of the meridonal planes with the illuminated portion of the surface of the object.

Accordingly we can write for the n-th meridonal slice or contour,

$$\psi_n(k) = F \int D_n(\bar{\rho}_{o_n})\ e^{-j\bar{p}\cdot\bar{\rho}_{o_n}}\ d\bar{\rho}_{o_n} \tag{4}$$

We can regard $D_n(\bar{\rho}_{o_n})$ as the n-th meridonal slice or contour of a three dimensional object of reflectivity $U(\bar{r})$ where $\bar{r}$ is a three dimensional position vector in object space. This means that $D_n(\bar{\rho}_{o_n}) = U(\bar{r})\ \delta(z_{o_n})$ where δ is the Dirac delta "function". Consequently eq. (4) becomes,

$$\begin{aligned}\psi_n(k) &= F \int U(\bar{r})\ \delta\ (z_{o_n})\ e^{-j\bar{p}\cdot\bar{\rho}_{o_n}}\ d\bar{\rho}_{o_n} \\ &= F \int U(\bar{r})\ \delta\ (z_{o_n})\ e^{-j\bar{p}\cdot\bar{r}}\ d\bar{r}\end{aligned} \tag{5}$$

where $d\bar{r}$ designates an element of volume in object space and where the last equation is obtained by virtue of the sifting property of the delta function.

Summing up the data from all slices or contours of the object we obtain,

$$\sum_n \psi_n = F \int U(\bar{r})\ e^{-j\bar{p}\cdot\bar{r}}\ dr = \psi(\bar{p}) \tag{6}$$

because

$$\sum_n U(\bar{r})\ \delta\ (z_{o_n}) = U(\bar{r}).$$

Assuming that the Factor F in eq. (6) is eliminated as before by TDR or by other means, equation (6) reduces to

$$\psi(\bar{p}) = \int U(\bar{r})\ e^{-j\bar{p}\cdot\bar{r}}\ d\bar{r} \tag{7}$$

The phase or range corrected data $\psi(\bar{p})$ obtained for example by the TDR method is therefore the 3-D Fourier transform of the object reflectivity $U(\bar{r})$. An alternate formulation to that given above of super-resolved wave-vector diversity e.m. imaging of 3-D perfectly conducting objects is possible by extending Bojarski [3] and Lewis' [4] formulation of the inverse scattering problem to the bistatic case, along lines that are somewhat different than those given by Raz [5].

3. THE FOURIER DOMAIN PROJECTION THEOREMS

The preceeding discussion establishes the existence of a direct 3-D Fourier transform relationship between the object reflectivity $U(\bar{r})$ and the phase (or range) corrected $\bar{p}$ space data $\psi(\bar{p})$ collected by a coherent array of receivers forming an imaging aperture. Image retrieval from the 3-D data $\psi(\bar{p})$ can be carried out digitally using the fast Fourier transform. This requires digital data storage and processing. Due to the inherent two dimensionality of computer displays, the digitally reconstructed image can be presented to the viewer in cross-sections or in perspective. Direct coherent optical processing of the 3-D data $\psi(\bar{p})$ is not feasible because the coherent Fourier transforming property of the convergent lens is confined to 2-D input formats. A hybrid (opto-digital) approach based on the Fourier domain projection theorem can circumvent this difficulty. The motivations for utilization of optical methods are lower cost and high capacity for data storage together with a prospect for real-time operation and true three dimensional display.

The underlying principle for hybrid data processing is the Fourier domain projection theorem. Spatial domain and dual Fourier domain projection theorems have been utilized respectively in astronomy [8] and in x-ray crystallography and electron microscopy [9]. In this section we will review the Fourier domain projection theorem then explain its utilization in deriving 3-D image information from $\psi(\bar{p})$.

Let (p) and U(r) be three dimensional Fourier transform pairs i.e., rewriting eq. (7)

$$\psi(\bar{p}) = \int_V U(\bar{r}) e^{-j\bar{p}\cdot\bar{r}} d\bar{r} \tag{8}$$

where $\bar{r} = x_o\bar{1}_x + y_o\bar{1}_y + z_o\bar{1}_z$ and $\bar{p} = p_x\bar{1}_x + p_y\bar{1}_y + p_z\bar{1}_z$ are position vectors in object space and Fourier space respectively. In the volume integral (8) $d\bar{r}$ designates the element of volume $dx_o dy_o dz_o$ in object space.

The projection of $\psi(\bar{p})$ on the p_x-p_y plane is

$$\psi_{proj.}(p_x,p_y) = \int \psi(p_x,p_y,p_z,)dp_z \tag{9}$$

Substituting (8) in (9),

$$\psi_{proj.}(p_x,p_y) = \int_V U(x_o,y_o,z_o) \left(\int e^{-j(p_x x_o + p_y y_o + p_z z_o)} dp_z \right) dx_o dy_o dz_o \tag{10}$$

If the $\bar{p}$ space basis is sufficiently large the integration of the exponential $e^{-jp_z z_o}$ with respect to p_z yields approximately $\delta(z_o)$ and equation (10) becomes,

$$\begin{aligned}\psi_{proj.}(p_x,p_y) &= \int_V U(x_o,y_o,z_o)\delta(z_o)e^{-j(p_x x_o + p_y y_o)} dx_o dy_o dz_o \\ &= \int\int_{x_o,y_o} D(x_o,y_o)e^{-j(p_x x_o + p_y y_o)} dx_o d_y. \\ &= \int D(\bar{\rho}_o)e^{-j\bar{p}\cdot\bar{\rho}_o} d\bar{p}_o \end{aligned} \tag{11}$$

where $D(x_o,y_o) = D(\bar{\rho}_o)$ is a the slice through the object $U(\bar{r})$ through the x_o,y_o plane.

Equation (11) indicates that the projection of the 3-D Fourier domain data on a given plane is the 2-D Fourier transform of the object central cross-section parallel to that plane. It follows from eq. (11) by the inverse transform,

$$D(x_o,y_o) = \int \psi_{proj.}(p_x,p_y)\ e^{\ j(p_x x_o + p_y y_o)}\ dp_x dp_y \tag{12}$$

which is the required theorem. According to this theorem, the projection of the 3-D transform data on any plane yields a 2-D data manifold from which a central cross-sectional outline of the illuminated portion of the object parallel to that plane can be obtained via a two dimensional Fourier transform which can be carried out optically. Photographic transparency records of the Fourier domain projections are suitable for use as input to an optical bench utilizing laser illumination to execute an instantaneous optical Fourier transform with the aid of a simple convergent lens. This leads to optical reconstruction of the corresponding cross-section of the object.

True 3-D image reconstruction is possible if one can view all projection holograms simultaneously. The most logical way to proceed in this regard is to combine all projection holograms into a single composite hologram which can reconstruct when viewed by an observor all cross-sectional outlines of the illuminated portion of the reflecting object simultaneously with correct angular ordering so that an impression of viewing a true 3-D image is created. One potential method of carrying this out is multiplex or integral holography [10].

Through the use of what we call *Weighted Fourier Projection Theorem* to be described next, it should be possible to reconstruct parallel slices of a 3-D object from the 3-D data manifold $\psi(\bar{p})$. To the best of our knowledge no mention of this theorem is found in the literature.

Let $G_\alpha(p_x,p_y)$ be the weighted projection of $\psi(\bar{p})$ defined by,

$$G_\alpha(p_x,p_y) = \int_{p_z} \psi(p_x,p_y,p_z)\, e^{j\alpha p_z}\, dp_z \qquad (13)$$

Substituting for ψ from eq. (8) yields,

$$G_\alpha(p_x,p_y) = \int_{p_z} e^{j\alpha p_z} \int_{x_o,}\int_{y_o,}\int_{z_o} U(x_o,y_o,z_o)$$

$$e^{-j(p_x x_o+p_y y_o+p_z z_o)}\, dx_o dy_o dz_o dp_z \qquad (14)$$

Carrying out the integration with respect to p_z first assuming the extent of the $\bar{p}$ space data to be sufficiently large, the integral with respect to P_z may be approximated by $\delta(z_o-\alpha)$ Equation (14) can be reduced then to,

$$G_\alpha(p_x,p_y) = \int_{x_o}\int_{y_o} U(x_o,y_o,z_o=\alpha)e^{-j(p_x x_o+p_y y_o)}\, dx_o dy_o \qquad (15)$$

which says that $G_\alpha(p_x,p_y)$ and $U(x_o,y_o,z_o=\alpha)$ are two dimensional Fourier transform pairs. The weighted Fourier domain projection theorem follows from the inverse transform

$$U(x_o,y_o,z_o=\alpha) = \int_{p_x}\int_{p_y} G\,(p_x,p_y)e^{j(p_x x_o+p_y y_o)}dp_x dp_y \qquad (16)$$

Accordingly parallel slices of the object can be reconstructed slice by slice by changing the parameter α .

In practice one expects that digital storage of the corrected 3-D data manifold $\psi(\bar{p})$ produced by a thinned receiver array is more convenient than optical storage. The projections $\psi_{proj}(p_x,p_y)$ and the weighted projections $G_\alpha(p_x,p_y)$ can therefore be carried out digitally. Two dimensional Fourier transform of these projections carried out either digitally or optically (and hence the hybrid nature) permits recovery of the 3-D image information slice by slice.

4. COMPUTER SIMULATION

Computer simulation of wave-vector diversity imaging of two 1m diameter perfectly reflecting spheres was undertaken to verify the method. The simulated geometry is shown in Fig. 2(a) where a circular array of 50 coherent receivers is assumed to be in the far field region of the object. The choice of the object was influenced by the ready availability in the literature of series formula for the field scattered from a perfectly reflecting sphere under plane wave illumination (12). The direction of illumination was assumed to be fixed. Wave-vector diversity was realized by sweeping the illumination wavelength between 7.5 cm and 15 cm (i.e. case of frequency swept imaging). The corrected (TDR) data $\psi(\bar{p})$ collected by the elements of the receiver array was stored in the computer in the proper format specified by the values of $\bar{p} = k\,(\bar{1}_{k_i} - \bar{1}_R)$ assumed for each receiver. The projection of the stored $\psi(\bar{p})$ on the plane of the array was displayed on the computer CRT display and photographed to yield the projection hologram shown in Fig. 2(b). The corresponding optically retrieved image is shown in Fig. 2(c). A central cross-sectional outline image of the two spheres that is parallel to the projection plane is clearly delineated as two adjacent circles of equal diameter as predicted by theory.

To evaluate the feasibility of reconstruction by parallel slices, the weighted Fourier domain projection theorem was applied to the stored data $\psi(\bar{p})$ for the preceeding example.

Preliminary photographs for the images retrieved from weighted projections for values of α equal to zero cm (central slice as in Fig. 2(c)), 30 cm and 40 cm are shown in Fig. 3. Although the quality of these preliminary images is not good they clearly illustrate the ability to reconstruct in parallel slices by means of the weighted Fourier domain projection theorem.

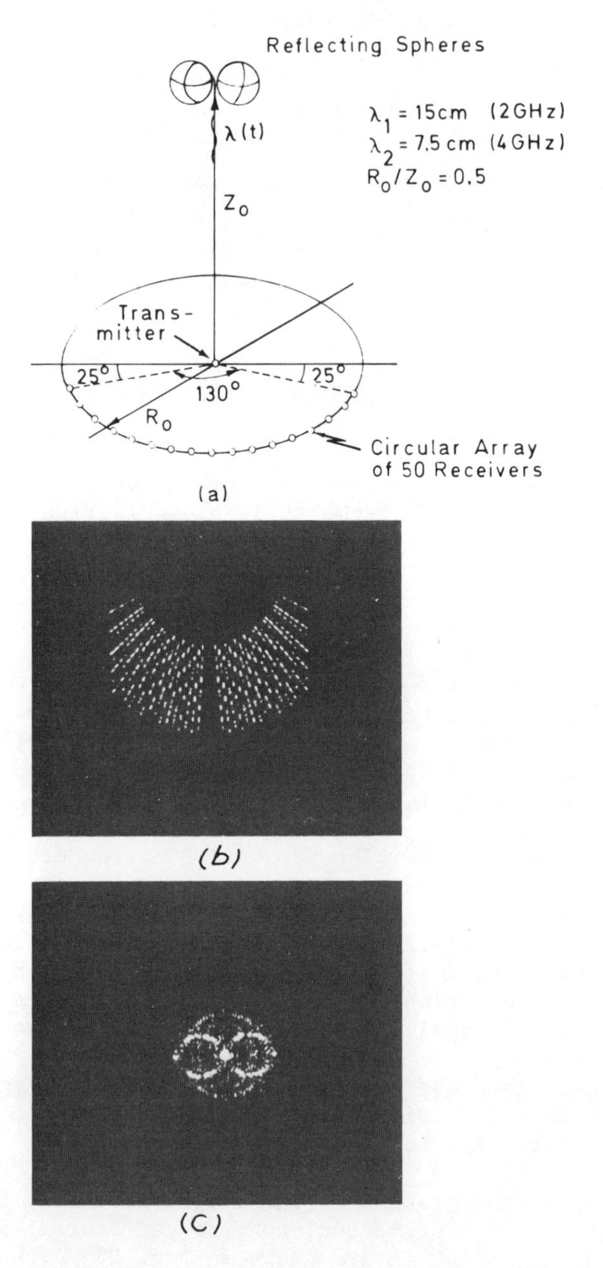

Fig. 2. Computer Simulation of Wave-vector Diversity Imaging, (a) Geometry, (b) Projection Hologram, (c) Retrieved Central Cross-sectional Image.

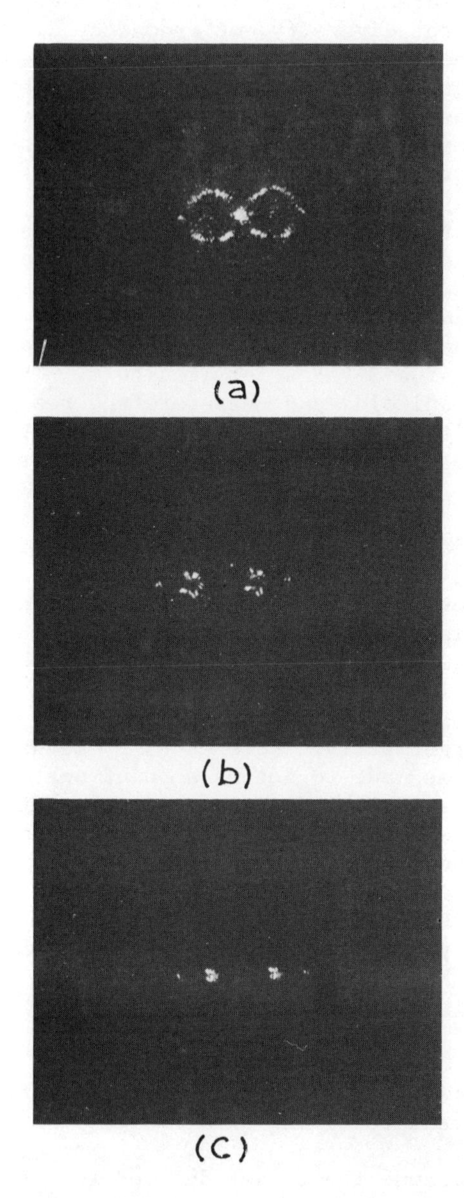

Fig. 3. 3-D Image Retrieval by Parallel Slices for Values of α = 0 cm (a), 30 cm (b) and 40 cm (c).

5. CONCLUSIONS

We have presented the principles of a long-wave imaging method in which wave-vector diversity, or frequency diversity (when the direction of object illumination is fixed), is employed to enhance the amount of object information captured by a broad-band coherent receiving array deployed in the far field of the object. The main features of the method whose basic principles have also been verified by computer simulation are:

(a) the data collected by a thinned coherent array of receivers intercepting the wavefield scattered from a distant 3-D reflecting object, as the frequency of its illumination and/or its direction of incidence is changed, can be stored as a 3-D data manifold in $\bar{p}$ space from which an image of the object can be retrieved by means of a 3-D Fourier Transform. The size and shape of the 3-D data manifold, and therefore the resolution, depend on the relative position of the object the transmitter (illuminator) and the receiving array and on the spectral width of the illumination utilized.

(b) the data collected must be corrected for a quadratic phase term before it is stored in a 3-D manifold and an undistorted image of the 3-D reflecting object reconstructed through the 3-D Fourier transform operation. A bothersome range-azimuth ambiguity is also avoided through elimination of this quadratic phase term.

(c) one method of achieving this correction is through the use of a TDR (Target Derived Reference) by means of which the object or target is made to act as a point scatterer by illuminating it with a sufficiently low frequency signal that is a subharmonic of the imaging frequencies utilized. Harmonic mixing of the amplified and limited signals produced by the reference spherical wavefront generated by the object at the receiving array elements with those produced by the imaging wavefronts scattered from the object should in principle yield the required phase corrected data. The use of a TDR has many additional advantages.

These include:

(i) Elimination of the need for a central local oscillator for the coherent receiving array.

(ii) Because the target derived reference source moves with the target there is greater tolerance to target motion during data acquisition.

(iii) Because TDR results in a recording configuration similar to that of a lensless Fourier Transform hologram, the resolution requirements from the recording device are greatly relaxed. In longwave holography this fact is translated into a significant reduction of the number of receiving elements in the recording aperture. In addition the use of TDR allows us to place all the resolving power of the recording aperture on the target. This means that high resolution images of distant isolated targets should be feasible with array apertures consisting of tens of elements. The ability to synthesize a 2-D receiving aperture with a Wells array (11) consisting of two orthogonal linear arrays one of transmitters and the other of receivers provides further means of reducing the number of stations needed for data acquisition without sacrifice in resolution.

(iv) Greater immunity to phase fluctuations arising from turbulance and inhomogenieties in the propagation medium because both the reference and imaging signals arriving at each receiving element of the aperture travel roughly over the same path.

(v) TDR eliminates the range azimuth ambiguity and excessive bandwidth problems that arise in frequency swept imaging when the reference signal for the array aperture is derived instead from the illumination source or a centrally located local oscillator phase locked to it.

(d) Because the dimensions of the 3-D data manifold $\psi(\bar{p})$ in $\bar{p}$ space are dependent, in addition to geometry, on the spectral range of the illumination, super-resolution (i.e. resolution beyond the classical limit of the available physical aperture) is achieved. This aperture synthesis by wave-vector or frequency diversity helps cut down array cost (since a thinned array can be used) and size.

(e) Fourier domain projection theorems enable the generation of two dimensional holograms from projections (or weighted projections) of the corrected 3-D data manifold $\psi(\bar{p})$ permitting thereby optical image retrieval of the 3-D object in meridonal slices parallel to the projection planes one at a time (or in parallel slices one at a time).

(f) True 3-D image presentation could be possible by combining a set of distinct projection holograms, using multiplex or integral holography techniques, into a single composite hologram which can reconstruct all corresponding cross-sectional outlines of the illuminated portion of the object simultaneously yielding thus a viewable true-3D image.

(g) Unlike conventional Longwave holography no wavelength scaling is required here to avoid longitudinal distortion in optical reconstruction.

(h) Because of the broad-band nature of the illumination and the aperture synthesis by frequency diversity, reduction of speckle noise in the image is expected.

(i) In the Frequency Swept mode of wave-vector diversity imaging (i.e. when $\bar{1}_{k_i}$ = const.), the collected data at each receiver represents a frequency response of the object. Assuming the scattering process is linear, this frequency response is related to the impulse response of the object by a Fourier transform (2), (13). This suggests that when impulse illumination is utilized, instead of frequency swept illumination, a 3-D data manifold $\psi(\bar{p})$ may be generated by Fourier transforming the impulse response at each receiver, correcting the data for the Factor F, and storing the result in accordance to the corresponding $\bar{p}$ for each receiver. The resulting corrected $\psi(\bar{p})$ can then be employed as described in this paper to yield 3-D image information. Impulse illumination is desirable in certain instances of rapid target motion but may be more difficult to implement than frequency swept illumination.

It is difficult to conclude without noting some extremely interesting similarities between wave-vector diversity (or frequency diversity) imaging discussed here and certain features of the sonar system in mammals such as bottle-nosed Dolphins, whales and bats. These features have been deduced or hypothesized by several workers (14)-(17) from observations of the remarkable acoustical behaviour, activity during echo-location, and anatomical studies of these mammals especially the Dolphin. Some of the more pertinent features which we only list here are:

(a) All signals emitted are of broad-band nature. They are either in the form of relatively long chirps (whistles), impulse like pings or clicks of less than 1msec duration and frequency content extending up to 100 kHz, or low frequency barks rich in higher harmonics.

(b) Ability to detect signals buried in noise indicating possible coherent or corellation processing.

(c) Evidence of the presence of a sensor array in the melon surface with sensitivity in the 15-100 kHz which might be utilized for reception of acoustic echos in addition to sensing velocity and temperature. The lens-shaped fatty body of the mellon, which is essentially acoustically transparent, might also be utilized as a variable focus acoustic lens for focusing of emitted sound on selected targets.

(d) Evidence of super-resolution capabilities i.e., resolution exceeding the classical limit of any possible available physical recording aperture, such as the melon even when assumed to be operating at the higher frequency range of sound emissions.

6. ACKNOWLEDGMENTS

This work was sponsored in part by the Army Research Office - Durham under Grant No. DAAG29-76-G-0230 and by Air Force Office of Scientific Research, Air Force Systems Command, USAF under Grant No. AFSOR-77-3256A.

REFERENCES

1. N.H. Farhat, "Frequency Synthesized Imaging Apertures," Proc. 1976 Optical Computing Conference, IEEE Cat. No. 76CH1100-7C, pp. 19-24.
2. N.H. Farhat, "Principles of Broad-band Coherent Imaging", J. Opt. Soc. Am., Vol. 67, Aug. 1977, pp. 1015-1021.
3. R.P. Porter, "A Radar Imaging System Using the Object to Provide the Reference Signal", Proc. IEEE (letters), Vol. 59, February, 1971, pp. 307-308.
4. N. Farhat, "Reply to Comments on Computer Simulation of Frequency Swept Imaging", Proc. IEEE (letters), Vol. 65, pp. 1223-1224, Aug. 1977.
5. N.N. Bojarski, "Inverse Scattering", Final Report to Contract B00019-73-C-0316 Naval Air Syst. Command, February 1974.
6. R.M. Lewis, "Physical Optics Inverse Diffraction", IEEE Trans. on Ant. and Prop., Vol. AP-17, May 1969, pp. 308-314.
7. S. Rosenbaum-Raz, "On Scatterer Reconstruction from Far-Field Data", IEEE Trans, on Ant. and Prop., Vol. AP-24, January 1976, pp. 66-70.
8. R.N. Bracewell and S.J. Wernecke, "Image Reconstruction Over a Finite Field of View", J. Opt. Soc. Am., Vol. 65, November 1975, pp. 1342-1346. See Also: R.N. Bracewell, "Strip Integration in Radio Astronomy", Australian J. Phys. Vol. 9, 1956, pp. 198-217.
9. G.W. Stroke and Maurice Halioua, "Three-Dimensional Reconstruction in X-Ray Crytallography and Electron Microscopy By Reduction to Two-Dimensional Holographic Implementation", Trans. Amer. Crytallographic Assoc., Vol. 12, 1976, pp. 27-41.
10. D.L. Vickers, "The Integral Hologram as a Scientific Tool", Lawrence Livermore Laboratory Report UCID-17035, February, 1976, See Also: "White-Light Holographic Displays", Laser Focus, Vol. 13, July 1977.

11. W.H. Wells, "Acoustical Imaging with Linear Transducer Arrays", in Acoustical Holography, Vol. 2, A.F. Metherell and L. Larmore (Eds.), Plenum Press, New York, 1970. Also See: C.N. Nilsen and D.N. Swingler, "Quasi-Real-Time Inertialess Microwave Holography", Proc. IEEE (letters), Vol. 65, March 1977, pp. 491-492.

12. R.F. Harrington, Time Harmonic Electromagnetic Fields, McGraw Hill, New York, 1961, pp. 292-298.

13. E.M. Kennaugh and L. Moffatt, "Transient and Impulse Response Approximations", Proc. IEEE, Vol. 53, August 1965, pp. 893-901.

14. J. Cunningham Lilly, The Mind of the Dolphin, Avon Books, New York, 1967.

15. J.J. Dreher, "Acoustical Holographic Model of Cetacean Echo-Location", in Acoustical Holography, Vol. 1., A.F. Metherell, H.M.A. El-Sum and L. Larmore (Eds.), Plenum Press, New York, 1969, pp. 127-137.

16. W.N. Kellogg, "Auditory Perception of Submerged Objects by Porpoises", J. Acoust. Soc. Am., Vol. 31, January 1959, pp. 1-6.

17. W.E. Evans and J.H. Prescott, "Observations of the Sound Production Capabilities of the Bottlenose Porpoise: A Study of Whistles and Clicks", Zoologica, Vol. 47, 1962, pp. 121-128.

THE IMPEDIOGRAPHY EQUATIONS

S. Leeman

Department of Medical Physics,
Royal Postgraduate Medical School,
Hammersmith Hospital, London, W12 OHS, U.K.

Of all the ultrasonic imaging techniques applied to medicine, pulse-echo methods which provide a display of the signals backscattered from tissues have proved the most useful. With the advent of grey-scale devices and tissue-characterization techniques, attention has focused more narrowly on the precise form of the scattering interaction between ultrasound pulses and human tissue. For an inhomogeneous medium containing density and elasticity fluctuations, the primary characteristic which generates the backscattered echoes is the variation of the characteristic acoustic impedance along the path of the incoming pulse (J. G. Miller, private communication), and it is probably largely true that not only do grey-scale images approximate to pulse-smoothed acoustic impedance maps of tissue, but also that most attempts to characterize tissue on the basis of backscattered echo information are essentially endeavours to reconstruct acoustic impedance variations. Static tissue characterization techniques yield data of only marginal interest unless some attempt is made to process out instrumental influences, such as limited amplifier band-width and finite pulse length, and a number of attempts (Jones, 1977; Beretsky, 1977; Kak and Fry, 1976) have been made to establish directly the impedance profile from the computed impulse response of the tissue. The implementation of this method - dubbed "impediography" by its originators (Jones and Wright, 1972) - is far from straightforward, since, quite apart from numerical difficulties, the assumed relationship between the impedance and the impulse response is usually based on a simple tissue model, which, for mathematical simplicity, is invariably chosen to be the continuum limit of an idealised layered structure, arranged orthogonally to the beam direction. It is the purpose of this paper to investigate to what extent some of these more restrictive model assumptions may be relaxed.

The Effective Impedance

Tissue is manifestly three-dimensional, and may be considered, quite generally, to be a distribution of reflecting/scattering elements, $S(x,y,z)$. For many applications, the interrogating pulse may be considered to be a modulated plane wave of the form

$$P(x,y,z,\tau) = A(z - c\tau)\, B(x,y)\, \exp\left[ik(z - c\tau)\right]$$

where $A(z)$ is the axial pulse shape, $B(x,y)$ the transverse beam profile, and k the magnitude of the carrier wave vector. The pulse is assumed incident along the z-axis, and to be propagated, without change of shape, with velocity c. The time co-ordinate is denoted τ.

As has been indicated above, the contribution to the back-scattering from each scattering element depends primarily on its acoustic impedance, and it follows that, for the conditions on the interrogating pulse assumed here, the (backscattering) reflectivity at distance z is given by the beam-profile-averaged variations of the reflecting elements. It thus becomes possible to define an "effective impedance"

$$Z(z) \equiv \int_{-\infty}^{\infty}\int_{-\infty}^{\infty} S(x,y,z)\, B(x,y)\, dx\, dy$$

$Z(z)$ is a layered impedance distribution which produces the same backscattering as the tissue (model) as seen from that particular interrogating direction, and obviously conforms with the usual multi-layering approximation made to the medium in previous derivations of the impediography relations.

There is little difficulty in extending the notion of effective impedance to situations in which the beam profile changes with distance, as for focused beams, or when beam-spreading by attenuation becomes significant, but the problem becomes rather more subtle when there are marked deviations from the planar character of the carrier wave fronts. An effective impedance may still be defined in that case, but more advanced concepts have to be invoked when there are strong departures from simple wave front geometries, and these will not be discussed here.

It is convenient to express the effective impedance as a function of the travel time, t, where, quite generally

$$t(x) = \int_0^x x'/c(x')\, dx'$$

where c(x) is the ultrasound velocity at distance x.

Implicit in this definition is the idea that the velocity does not change significantly over distances of the order of the beam width - a necessary condition, in any case, if the wave fronts are to retain a simple geometry throughout the medium.

A fixed direction back-scattering experiment can do no better than to uncover the form of Z(t); and the echo-train at the receiver may be deconvoluted with respect to the incident axial pulse-shape, to reveal the impulse response of the tissue. While this procedure is not trivial, the nub of the impediography technique is to relate the impulse response to the impedance profile, and only this aspect is considered here. The concept of the effective impedance relaxes the restriction of previous treatments to a planar multi-layered medium, and the following will show that the involved limiting procedures inherent in other mathematical derivations may also be obviated.

The First-order Equations

Consider the (effective) impedance profile Z(t), constant outside $0 < t < T$, and equal to Z_O (Z_T) for $t \leqslant 0$ ($\geqslant T$). From the well-known properties of the Dirac delta function, $\delta(x)$, it follows that, for $0 < t < T$

$$Z(t) = \int_0^T Z(t')\, \delta(t' - t)\, dt'$$

Integrating by parts, and dropping the restriction on t,

$$Z(t) = Z_T - \int_0^T \frac{dZ(t')}{dt'}\, U(t' - t)\, dt'$$

where the unit step function,

$$U(x) = 1 \text{ for } x \geqslant 0$$

$$= 0 \text{ for } x < 0$$

The last expression may be rewritten

$$Z(t) = Z_T - \int_0^T \frac{d\ln Z(t')}{dt'} Z(t')\, U(t' - t)\, dt'$$

$$\equiv Z_T - \int_0^T \frac{d\ln Z(t')}{dt'} \mathrm{I}(t;t')\, dt'$$

where $\mathrm{I}(t;t_o)$ is (in the t-domain) an impedance step, located at t_o, with the impedance equal to zero on the one, and equal to $Z(t_o)$ on the other side of the discontinuity. Such an impedance step is a perfect reflector, and would, in the presence of simple attenuation, clearly have the reflection impulse response

$$- \delta(t - 2t_o) \exp(- 2\alpha t_o)$$

where α is the attenuation coefficient (per unit travel time!), assumed constant and independent of frequency.

If multiple reflections and the effect of the medium on pulse propagation are provisionally ignored, the "image" of the impedance profile obtained by interrogation with an impulse, i.e. its impulse response, is immediately seen to be

$$h(t) = \int_0^T \frac{d\ln Z(t')}{dt'} \exp(- 2\alpha t')\, \delta(t - 2t')\, dt' \qquad (1)$$

$$= \frac{1}{2} \frac{d\ln Z(t')}{dt'}\bigg|_{t' = t/2} \exp(- \alpha t)$$

Aliter,

$$h(2t) = \frac{1}{2} \frac{d\ln Z(t)}{dt} \exp(- 2\alpha t)$$

For loss-less media ($\alpha = 0$) this reduces to the well-known first-order impediography equation

$$\int_0^t h(2\tau)\, d\tau = \frac{1}{2} \ln(Z(t)/Z_o)$$

i.e.

$$Z(t) = Z_o \exp\left[2 \int_o^t h(2\tau)\, d\tau\right]$$

In the presence of simple attenuation ($\alpha \neq 0$) the impedance profile cannot be reconstructed exactly from the impulse response function without additional assumptions; however, if the value of the attenuation coefficient is known, the simplest procedure would be to form the integral

$$\int_o^t h(2\tau) \exp(2\alpha\tau)\, d\tau = \frac{1}{2} \ln(Z(t)/Z_o)$$

and to solve for $Z(t)$.

For more complex media, the first-order impulse response is easily seen, by the above arguments, to be

$$h(t) = \int_o^T \frac{d\ln Z(t')}{dt'} f(t - 2t')\, dt' \qquad (2)$$

where $f(t - 2t')$ is the impulse reflection calculated for the ideal plane impedance step described above, but now situated within a homogeneous medium of the required acoustic propagation properties. For even relatively simply dispersive media, $f(t)$ takes on quite a formidable form (Leeman, 1978), and the extraction of the impedance profile from the impulse response would demand complicated inversion procedures. The effect of instrumental distortion may also be included in f, and the expression for $h(t)$ may then be developed into a consideration of the assessment of system performance, particularly the concept of resolution (Gore and Leeman, 1978). The link between A-scans and impediography, contained in eq. 2, has been indicated also by Jones (1977).

Equations of Higher Order

Equation (1) may be re-interpreted to provide a direct and physically appealing method for writing down the impedance profile impulse response when multiple reflections are taken into account. Although the third order equation only will be derived, the method indicated here may be extended as a formal theory to include even multiple scattering (into any direction) events, as well as to incorporate modifying effects of the medium, such as dispersive absorption, on the propagation of the interrogating pulse. In general, these formal manipulations will not lead to expressions that may be calculated without further grossly simplifying assumptions, but the simple medium adopted in the example below allows exact solutions.

Consider a first order reflection at $t = \tau$. The incident impulse, δ_o, is generated at $t = 0$, and travels out to $t = \tau$. On arrival at τ, it may be symbolically written as

$$\delta_\tau = P(\tau,0)\,\delta_o$$

where $P(\tau',\tau)$ is an operator describing the propagation of the impulse from $t = \tau$ to $t = \tau'$. If a reflection is experienced at τ, its effect on the pulse may again be symbolically written as an operator, R_τ, so that the reflected pulse generated at τ is:

$$R_\tau\,\delta_\tau = R_\tau\,P(\tau,0)\,\delta_o$$

Finally, the pulse arrives back at the origin, with the form

$$P(0,\tau)\,R_\tau\,P(\tau,0)\,\delta_o$$

The example of a loss-less, ideal medium is considered here, and then $P(t',t)$ is a simple time-translation operator, leaving the pulse shape unchanged, and with $P(t',t) = P(t,t')$.

Remembering that, since the interrogating impulse is generated at $t = 0$,

$$\delta_o = \delta(t)$$

it is clear that

$$P(\tau,0)\,\delta_o = \delta(t - \tau)$$

An examination of eq. (1) will suggest the choice

$$R_\tau = \frac{d\ln Z(\tau)}{d\tau} \qquad (3)$$

i.e. the reflection operator is a scalar quantity, for the case considered here.

The impulse response of the entire impedance profile, for first order reflections, is thus given by

$$h_1(t) = \int_0^T P(0,\tau)\, R_\tau\, P(\tau,0)\, \delta(t)\, d\tau$$

a result entirely compatible with eq. (1) if the choice (3) is made.

The third-order equation may now be written down immediately

$$h_3(\tau) = -\int_0^T ds \int_0^T dt \int_0^T du\, P(0,s) R_s P(s,t) R_t P(t,u) R_u P(u,0) \delta_0 U(u-t) U(s-t)$$

The step-functions must be inserted to ensure the (physical) condition that the second reflection always occurs at a location of smaller travel time than either the first or third reflection, and the minus sign is a consequence of the condition that the second reflection location is necessarily approached from the direction of decreasing travel time (see Fig. 1).

Substituting the appropriate values for R_τ and $P(\tau,\tau')$,

$$h_3(\tau) = -\int_t^T ds \int_0^T dt \int_t^T du\, \frac{d\ln Z(s)}{ds} \frac{d\ln Z(t)}{dt} \frac{d\ln Z(u)}{du}\, \delta(2[\tau - s + t - u])$$

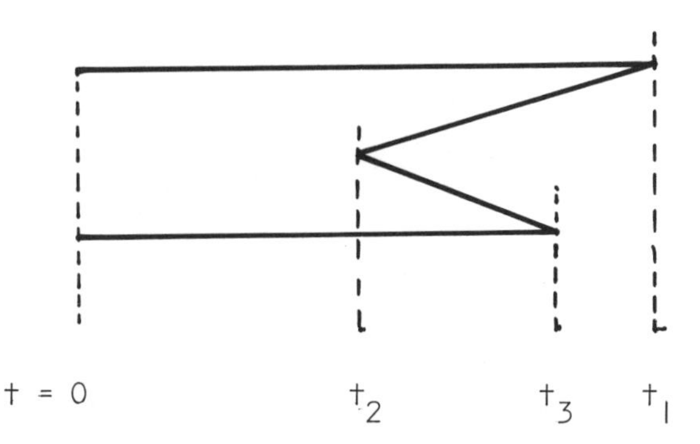

Fig. 1: Schematic diagram of typical contribution to the third order impulse response.

The δ-function may be removed by integrating over τ, and this will introduce a further limitation on the upper limits of the integrals,

$$\int_0^{2\tau} h_3(\tau')d\tau' = -\frac{1}{2}\int_t^{\tau} ds \int_0^{\tau} dt \int_t^{\tau} du \, \frac{d\ln Z(s)}{ds} \, \frac{d\ln Z(t)}{dt} \, \frac{d\ln Z(u)}{du}$$

This is easily calculated to give

$$\int_0^{2t} h_3(\tau)\, d\tau = -\frac{1}{6} \ln(Z(t)/Z_o)$$

This result has been obtained via pulse-bounce diagram techniques by both Jones (1977) and Kak and Fry (1976).

Conclusions

The theoretical description of the impediography method has, in the past, been based on the concept of a relatively uncomplicated, layered medium (impedance a function of one variable only). A derivation of the impediography equations has been presented here, which extends the method to inhomogeneous media, and which highlights the dependence of the reconstructed impedance profile on the parameters of the probing acoustic field. The particular merit of the calculation presented is that it obviates the involved limiting procedures inherent in other derivations, and the extension to more complex media as well as the inclusion of higher order reflections becomes more straightforward.

Acknowledgments

Professor J. S. Orr is thanked for his encouragement, and the Wellcome Trust for making a travel grant available.

References

Beretsky, I. (1977) "Raylography, A frequency domain processing technique for pulse echo ultrasonography". In: Ultrasound in Medicine, Vol. 3B, Ed. D. N. White and R. E. Brown, pp. 1581-1596, Plenum Press, New York.

Gore, J. C. and Leeman, S. (1978) "A method of describing transducer performance". In: The Evaluation and Calibration of Ultrasonic Transducers, Ed. M. Silk, IPC Press, Guildford. In Press.

Jones, J. P. and Wright, H. A. (1972) "A new broadband ultrasonic technique with biomedical implications". Presented at the 83rd Meeting of the Acoustical Society of America, Buffalo, New York.

Jones, J. P. (1977) "Ultrasonic impediography and its applications to tissue characterization". In: Recent Advances in Ultrasound in Biomedicine, Vol. 1, Ed. D. N. White, pp. 131-156, Research Studies Press, Forest Grove.

Kak, A. C. and Fry, F. J. (1976) "Acoustic impedance profiling: an analytical and physical model study". In: Ultrasonic Tissue Characterization, Ed. M. Linzer, pp. 231-251, National Bureau of Standards Special Publication 453.

Leeman, S. (1978) "Pulse propagation and scattering in dispersive media". Presented at meeting on: The Transmission and Scattering of Ultrasound in Human Tissues, Brunel University, London. (Available on request).

TECHNIQUES FOR *IN VIVO* TISSUE CHARACTERIZATION

L. Joynt, R. Martin, M.D., and A. Macovski, Ph.D.

Dept. of Elec. Engrg. and Div. of Cardiology
Stanford University and Stanford Medical Center
Stanford, CA, U.S.A. 94305

ABSTRACT

In vitro tissue characterization studies have shown differences between normal and ischemic or infarcted myocardium. However, the deterministic approach of these methods requires precise knowledge and control of factors precluded by the *in vivo* situation. A technique which avoids these difficulties takes into account the stochastic nature of the diffuse signals from a tissue volume by recording echo ensembles from a clinical instrument in real-time and regarding them as sample functions from a random process. Two experiments which exploit the random nature of the returned echo waveforms are described. An *in vivo* differentiation between normal and damaged myocardium is obtained from amplitude probability distributions of echoes from the hearts of live dogs with induced ischemia which were recorded using a real-time phased array sector scanner. Similar results were obtained from histograms computed from normal human and myocardial infarction patient data. The second experiment applied frequency domain methods to recognizing tissue structure characteristics. Autocorrelations of the frequency spectra of echoes from a sponge, as tissue model, showed peaks corresponding to fiber size. Similar processing of *in vivo* heart and liver data resulted in similar peaks. Thus it appears promising that basic tissue structure information can be derived from *in vivo* data.

INTRODUCTION

Tissue characterization attempts to provide diagnostic information about the state of health or disease of tissue that is not available from the usual ultrasonic imaging display. A very useful differentiation that one would like to make is between normal and ischemic or infarcted heart muscle. The extent of infarction is an important factor in the prognosis and treatment of heart disease. In vitro tissue characterization studies have shown differences between normal and ischemic or infarcted myocardium [1,2,3]. However, the deterministic approach of these methods requires knowledge and control of factors such as orientation of the interrogated tissue volume to the ultrasound beam, the filtering of the acoustic pulse by overlying tissue, and changes in the location of the desired tissue volume due to cardiac motion. In the in vivo situation one cannot control these factors or even know them precisely. Thus, it seems appropriate to look for alternate approaches.

AMPLITUDE DISTRIBUTION ANALYSIS

The stochastic nature of echoes originating from within tissue becomes apparent upon examination of the wave equation for ultrasound propagation in a medium [4]

$$\nabla^2 p - \frac{1}{c_0^2}\frac{\partial^2 p}{\partial t^2} = \frac{1}{c_0^2}\left(\frac{k(r)-k_0}{k_0}\right) + \nabla \cdot \left(\frac{\rho(r)-\rho_0}{\rho(r)}\nabla p\right) \qquad (1)$$

where $p(r,t)$ is the acoustic pressure at the position r at time t, $k(r)$ is the compressibility, k_0 is average compressibility in the region, $\rho(r)$ is density, ρ_0 is average density in the region and c_0 is acoustic velocity, $k = (\rho_0 c_0^2)^{-1}$. Sound waves in a homogeneous medium are thus described by a homogeneous wave equation that predicts that plane waves will propagate without scattering if the acoustic properties are uniform. Biological tissue is not perfectly uniform, rather the density and compressibility of small tissue samples fluctuate about their mean values so that in any region the local acoustic properties differ from the average. Sound waves will be scattered in such an inhomogeneous medium. The echoes arising from within tissue can be attributed to these density/compressibility fluctuations. Echoes returned from the myocardial tissue of the in vivo heart will, in addition, be affected by heart motion, intervening tissue, the transducer orientation, and finite resolution volume to make the signals returning at a specified time interval fluctuate randomly. Thus, one may consider successive echo trains, $x_1(t)$, $x_2(t), \ldots,$ to be sample functions from a random process $X(t)$. The amplitude distribution $P(x_1, x_2, \ldots)$ of the echoes will then reflect the degree and type of inhomogeneity existing throughout the sample volume interrogated. A broad distribution of amplitudes should indicate more inhomogeneities in the insonified medium.

A sequence of digitized sample functions, $X(t)$, from a normal human heart is shown in Fig. 1. Data was recorded from a real-

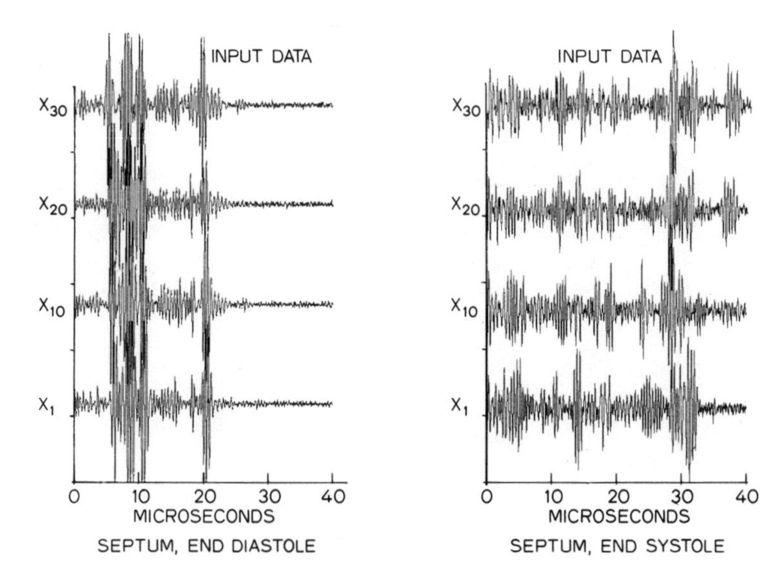

Fig. 1 Two Sequences of Digitized RF Sample Functions Returned From a Normal Human Heart.

time ultrasonic scanner using a microprocessor-controlled data acquisition system described previously [5]. Let L be the effective length of the impulse response of the ultrasound scanner system and T be the sample interval for digitizing the RF signal. Let a sequence of amplitude values $\hat{x}_j$ be defined by

$$\hat{x}_j = \max_{0 \leq nT \leq L} \left| x_j(nT) \right| \tag{2}$$

where the origin is chosen such that the interval [0.L] is included in the tissue volume. A histogram of these amplitude values $\hat{x}_j$ provides an estimate of the probability distribution of the amplitudes of echoes returned from the chosen tissue site.

Amplitude histograms were taken on data from a study of dogs with ischemia induced by coronary artery occlusion and on data from the hearts of myocardial infarction patients and normal subjects. The ultrasound scanner used as a signal source was the Varian V-3000 Real-Time Sector Scanner. A section of "diffuse" echoes from the interior of the septum or posterior wall, located by specular echoes which indicated the interfaces, was chosen to represent the desired tissue site. In the dog study, either the left anterior descending (LAD) coronary artery or the circumflex branch of the left coronary artery was occluded for a short time (10, 20, or 30 minutes), then unoccluded and later occluded. A

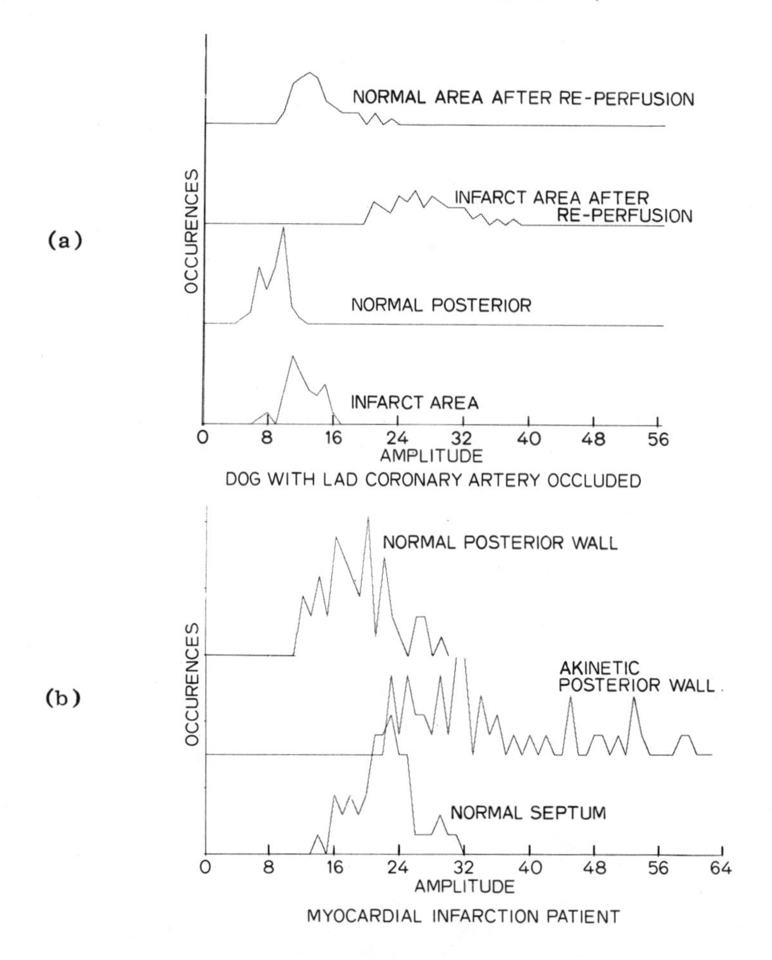

Fig. 2 Amplitude Histograms of Echoes Returned from the Myocardium of (a) a dog with LAD coronary artery occluded and (b) a myocardial infarction patient.

representative histogram from this experiment is shown in Fig. 2(a). The amplitude histograms of echoes from the affected area showed marked changes after reperfusion. A broader distribution results from the echoes returned from the reperfused portion of the myocardium than from normal areas or the affected area before reperfusion. Analysis of data from dogs with occlusion of the LAD coronary artery for up to four hours without subsequent reperfusion did not show this broadening of the amplitude distribution of echoes from the affected site compared to the normal area. Histograms of returned echoes from both sites were very similar.

In a pertinent study, Sommers and Jennings [6] observed histologic and histochemical changes in myocardium subjected to 40 minutes of ischemia followed by up to four hours of reperfusion of the previously ischemic area. These changes were compared to those resulting from occlusion of the same vessel without subsequent reperfusion. In both cases cells were irreversibly injured and cell death occurred. However, the cellular changes following the return of blood supply to areas of transient ischemia, as revealed by electron microscopy, were much more severe than those seen following permanent arterial occlusion. After 20 minutes of recirculation there was extensive disruption of the intracellular structure. Disorganization of myofibrils accompanied by the appearance of dense contraction bands, intracellular edema, inflammatory cell infiltration, calcium deposition, and total loss of stainable glycogen were seen.

These findings are consistent with the observed changes in the dog heart amplitude histogram. Ultrastructural changes in the myocardium could be expected to be reflected in changes in the probability distribution of the amplitude of the returned echoes. Figure 2(b) shows histograms of returned echoes from the heart of a myocardial infarction patient. Histograms of data from the infarcted myocardium showed the same kind of increased spread of amplitude values compared to normal myocardium in the same individual. The infarcted region was found by observing diskinesis on the real-time display of the ultrasonic scanner. The increased width of the histogram appears to be due to the occurrence of higher amplitude echoes from the damaged myocardium. This is consistent with the findings of Gramiak, et. al. [7] and Lele, et. al. [2] of the occurrence of different amplitude reflections from damaged myocardium as compared to normal myocardium.

Since the particular amplitude value of a returned echo is subject to extreme variations due to ultrasonic scanner settings and orientation to the ultrasound beam, the shape of the amplitude distribution is a more robust and consistent indicator of tissue parameters than any simple shift of amplitude values. In order to check that the shape of the amplitude distribution is due to real effects rather than scanner artifact, experiments were done using a variety of ultrasound scanners and transducers. Histograms of echoes returned from the interior of the septum of a normal subject are shown in Fig. 3. Data was obtained using the Varian V-3000 Real-Time Sector Scanner with a 2.25 mHz transducer and a prototype B-scan system used for opthalmic applications, which was designed and built by the Stanford Center for Integrated Electronics in Medicine, with 2.25 and 5 mHz transducers as signal sources. The histograms of the Varian 2.25 mHz data and the CIEM B-scanner 2.25 mHz data greatly resemble each other. The histograms of data recorded using the CIEM B-scanner with the 5 mHz transducer are also similar in shape to those of other normal data. Thus, it seems reasonable to conclude that the shape of the amplitude distribution of the returned echoes can provide a reliable indication of tissue characteristics.

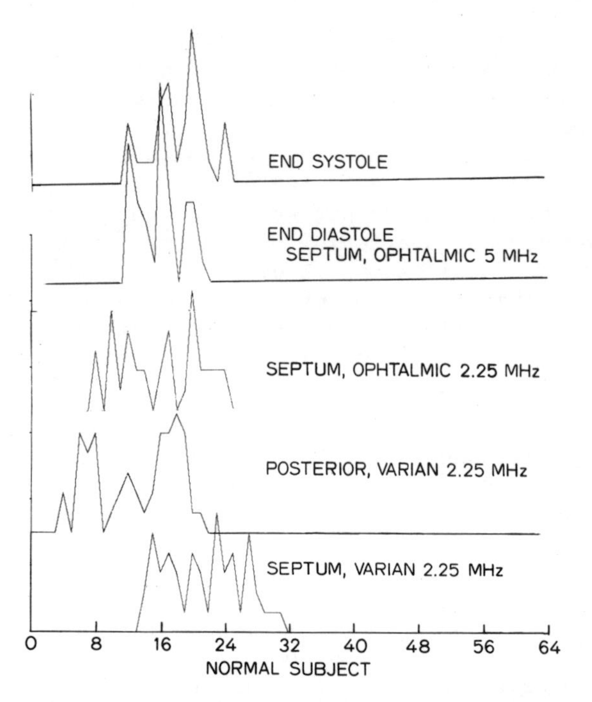

Fig. 3 Comparison of Amplitude Histograms of Echoes From Normal Septum Obtained Using Different Ultrasound Scanners and Transducers.

FREQUENCY DOMAIN ANALYSIS

A second experiment applied frequency domain methods to recognizing tissue structure characteristics. Microscopic structures of many normal and pathological tissues are organized in characteristic patterns. Acoustically, these patterns may be regarded as collections of diffuse scatterers separated by characteristic spacings. For wavefronts incident on a regular array of small reflecting surfaces, interference patterns which depend on array spacing are produced. Constructive interference occurs at frequencies such that the path length difference is an integral number of wavelengths. The frequency spectrum of echoes returned from such a structure will have peaks at these frequencies. The dimension of the organization exhibited will depend on the frequency (frequencies) of the ultrasound beam as well as on the scattering tissue.

In biological tissue, changes in acoustic properties are expected to occur randomly, but spacings may fall within certain limits that can be statistically specified. Hopefully, if the structure is altered by a disease process, the average spacing would also be found to alter. A correlation of these changes in

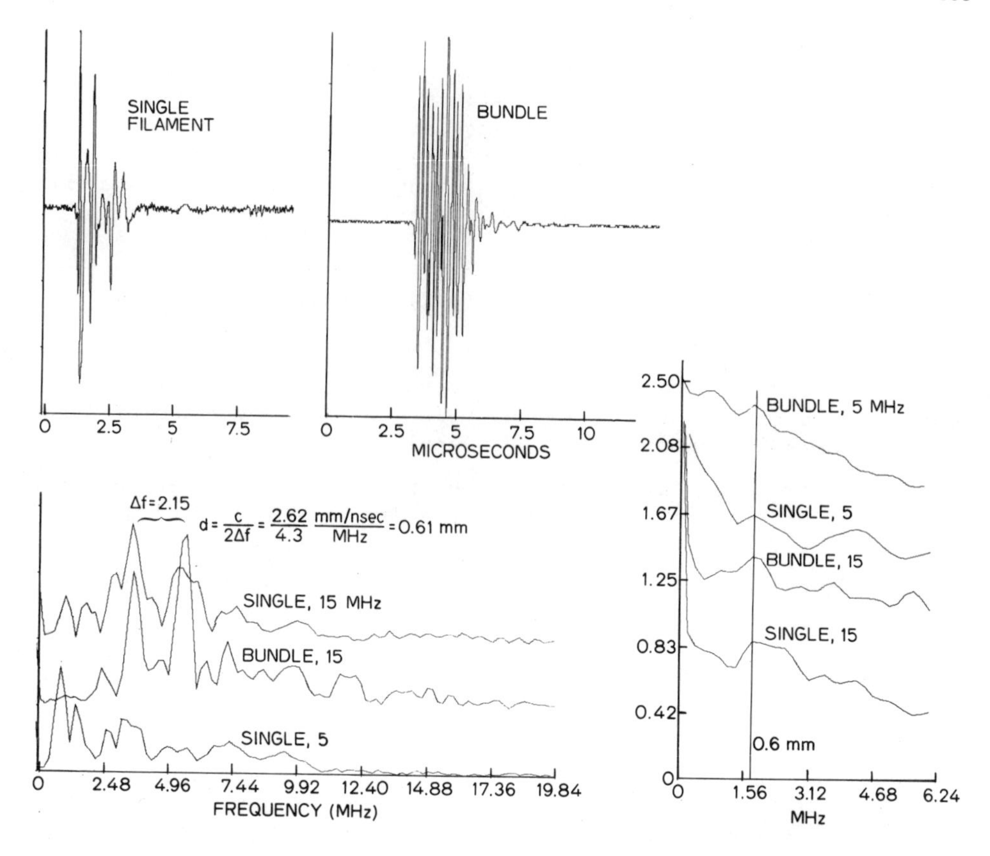

Fig. 4 Recorded RF, FFT Spectra and R(F) Curves for Nylon Monofilament in a Water Tank.

scattering characteristics with specific tissue structures and pathologies could prove to be a valuable diagnostic aid.

A fibrous, sponge-like structure, as a tissue model, was scanned using 5, 10, and 15 mHz wideband transducers in the CIEM B-scan system. The sponge is a random arrangement of random sized holes but with average periodicities in the reflected signal due to fiber size. Discrete Fourier transforms of diffuse echoes returned from the sponge were computed using the Fast Fourier Transform (FFT) algorithm. The resulting frequency spectra were very complex, actually quite similar to those from human tissue. An experiment with nylon monofilament in a water tank reinforced the idea that frequency spacings due to fiber size, at least, should appear. In the signal filament data (Fig. 4), peaks separated by a distance corresponding to the filament diameter of 0.6 mm were seen. Data from a bundle of fibers was even more striking. Several filaments were included in the FFT window and the peaks corresponding to the filament diameter were accentuated by about 2:1. A spatial averaging occurs when more than one filament is included in the FFT window. If one considers the filament diameter information as the signal and all other information as noise in the frequency spectrum, spatial averaging has increased the signal-to-noise ratio. Define the autocorrelation of the frequency spectrum $R(F_i)$ by

$$R(F_i) = \frac{1}{N-k} \sum_{p=1}^{N-k} F(p)F(p+k); \quad k=1,...,N-1 \tag{3}$$

Autocorrelation of the frequency spectrum indicates directly the frequency spacing of the peaks and thus the filament diameter since it represents the spectrum of the signal power.

Consider again the sponge (or human tissue). Movement of the transducer while recording the data results in a series of sample functions $x_1(t), x_2(t),...$ where each of the $x_i(t)$ represents echoes returned from a different location in the sponge. The discrete Fourier transform is defined by

$$F_i(m) = \sum_{k=0}^{N-1} x(kT)e^{-j2\pi km/N}; \quad m=0,...,N-1 \tag{4}$$

where N is the number of points included in the FFT window and T is the sample interval for digitizing the RF signal. Since the sponge fibers are randomly distributed, averaging the $F_i(m)$ results in reinforcement of the frequency components corresponding to fiber size while averaging out fiber spacing information and other noise. Autocorrelation of the FFT for the compressed sponge insonified by a 10 mHz wideband transducer in Fig. 5 shows that the autocorrelation peaks corresponding to fiber size are enhanced by averaging over an increasing number of locations which approaches a limit for large n. The resolution of the resulting $R(F_i)$ depends on the length of the FFT window and the bandwidth of the transducer used since the transducer frequency bandwidth determines the autocorrelation interval. A caveat is that only the first half of the autocorrelation coefficients should be assigned significance. They become increasingly unreliable as fewer points are included in the computation.

Application of this technique to normal human liver in vivo was performed using a 5 mHz transducer with the CIEM B-scan system. The transducer was translated slightly during data acquisition so that the recorded traces $x_i(t)$ were returned from different sample volumes in the liver. The resulting $R(F_i)$ and $R(\overline{F})$ along with representative RF and FFT spectra are shown in Fig. 6. Peaks at frequency spacings indicating a spatial dimension of .8mm were noted.

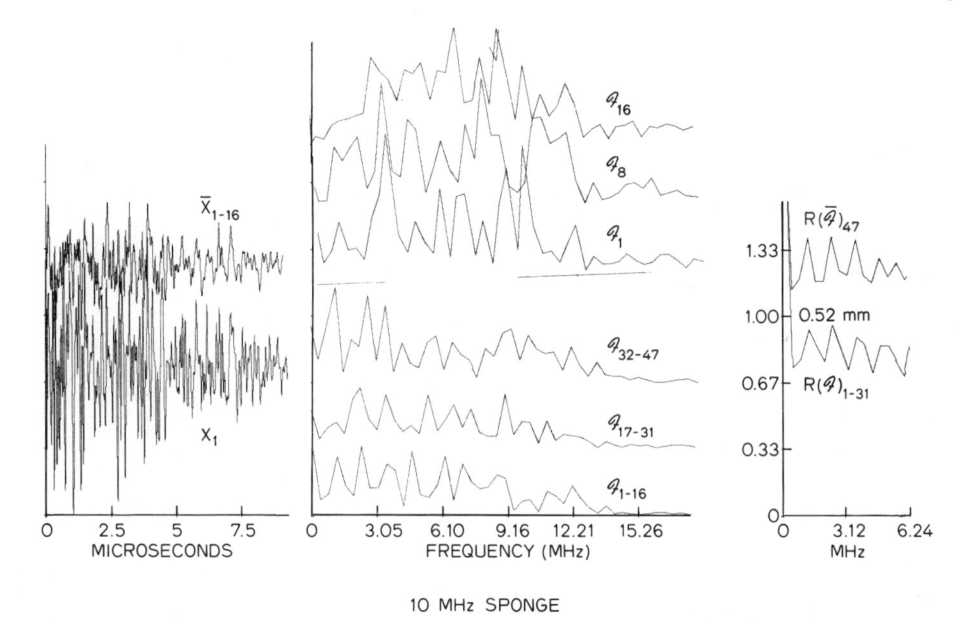

Fig. 5 Recorded RF, FFT Spectra, and R(F) Curves for Compressed Sponge Data Obtained Using a 10 mHz Wideband Transducer.

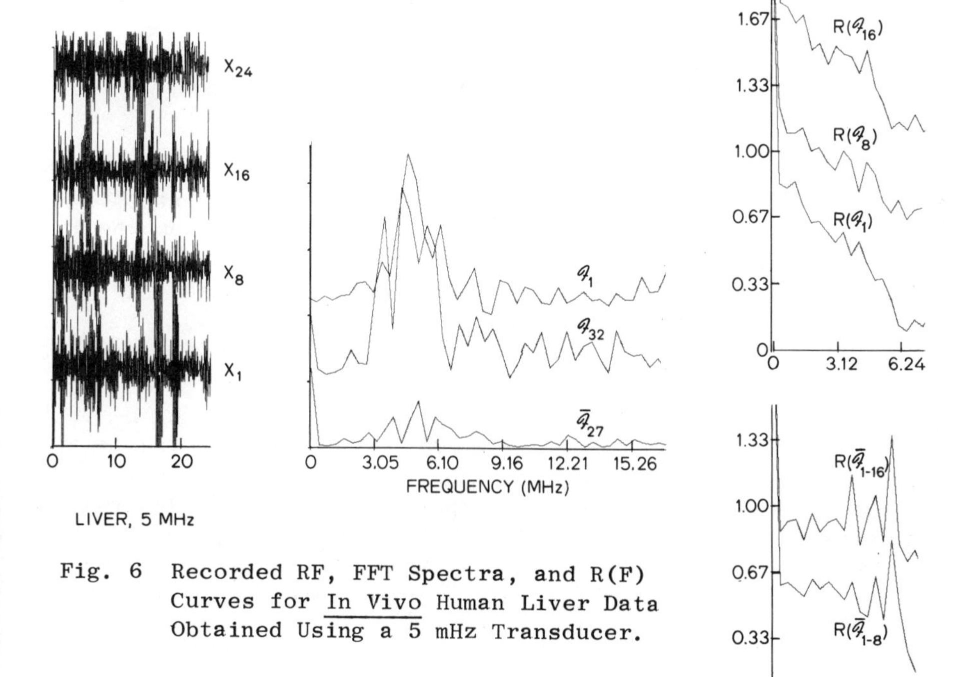

Fig. 6 Recorded RF, FFT Spectra, and R(F) Curves for <u>In Vivo</u> Human Liver Data Obtained Using a 5 mHz Transducer.

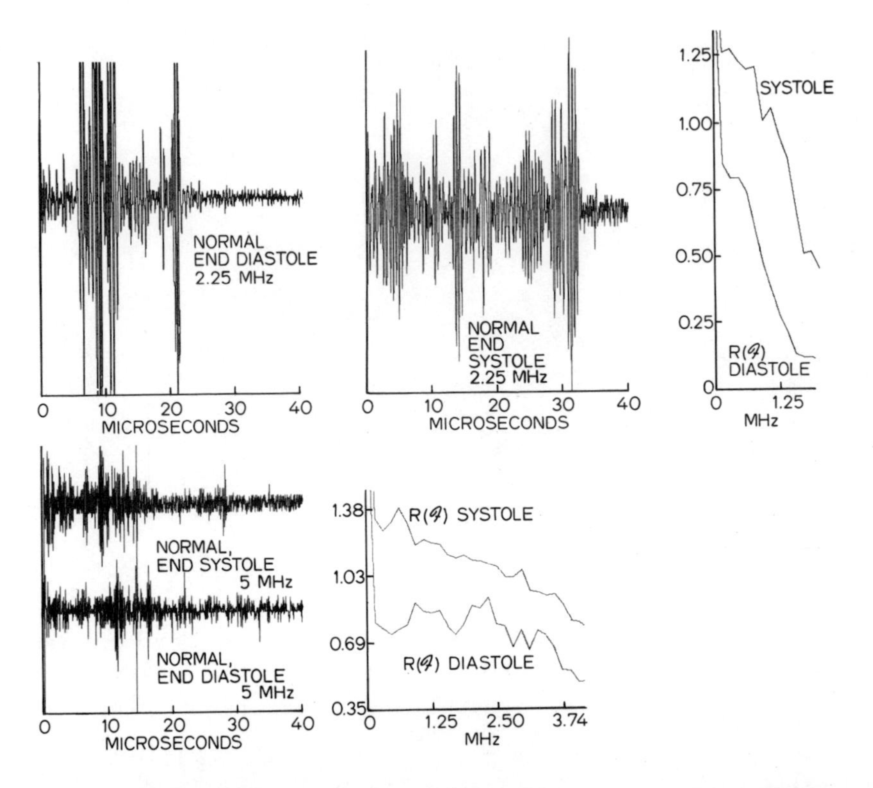

Fig. 7 Recorded RF and Corresponding R(F) Curves for In Vivo Normal Human Heart Data Obtained at End Diastole and End Systole Using 2.75 mHz and 5 mHz transducers.

If a disease process were to alter the tissue structure at this level of organization, it would be discernible through changes in the $R(\overline{F})$ function. We propose to study a variety of normal and cirrhotic livers and examine the corresponding $R(\overline{F})$ functions to see whether a differentiation can be made.

That overall changes in the organization of tissue structure may be discerned from the R(F) function is suggested by an experiment on in vivo cardiac data obtained using the opthalmic B-scanner and the Varian scanner (Fig. 7). The R(F) function was computed from echoes returned from the interior of the septum. The approximately 40% shift in the spatial dimension indicated by R(F) between end diastole and end systole reflects the amount of heart wall thickening between end diastole and end systole. This is directly visible in the 2.25 mHz RF data. However, it is not directly recognizable from the 5 mHz RF although the associated R(F) functions indicate the change.

Data recorded from a patient with severe idiopathic hyper-

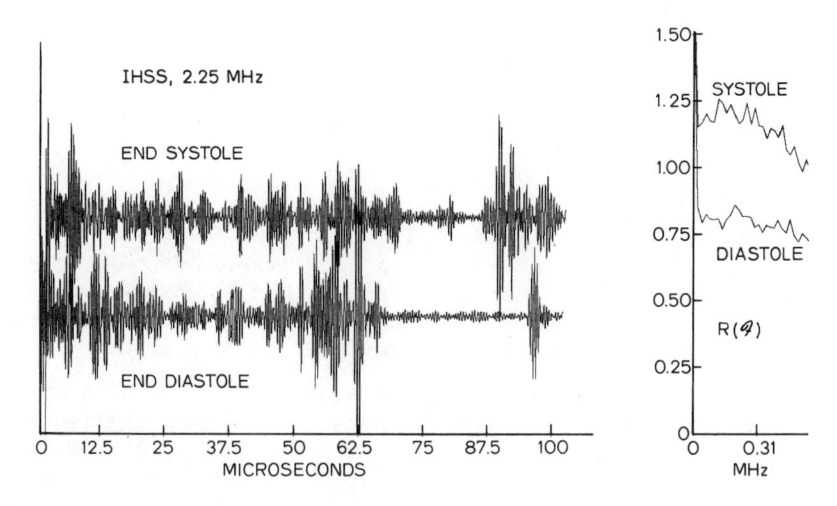

Fig. 8 Recorded RF and Corresponding R(F) Curves for In Vivo IHSS Heart Data Obtained at End Diastole and End Systole Using a 2.25 mHz Transducer.

trophic subaortic stenosis (IHSS) using the Varian real-time scanner is shown in Fig. 8. The abnormally thickened septum shows little change between end diastole and end systole. The corresponding R(F) curves for end diastole and end systole are similar with peaks occurring in approximately the same locations. Thus, changes in the organization of the tissue structure due to the state of contraction of the heart have been reflected in the R(F) function.

CONCLUSION

Recorded echo ensembles can be regarded as sample functions from a random process. The stochastic nature of the diffuse signals from a tissue volume arise from fluctuations of the local density and compressibility about their mean values. In biological tissue, these fluctuations are expected to occur randomly but may fall within certain statistically specified limits. In disease processes, when the structure is altered, the statistical values of the scatterer characteristics could be expected to alter correspondingly. Amplitude distribution analysis provided a differentiation between normal damaged myocardium in vivo in a small preliminary study. Autocorrelation of spatial averages of the frequency spectra of returned echoes yields information about tissue structural organization that is not discernible on the usual ultrasonic scanner visual display. Further analysis of how the tissue interaction with the ultrasound beam is affected by disease processes is required. Diagnostically significant information about tissue structure may well be derivable from signals from clinical B-mode instruments.

The authors wish to acknowledge the support of NIH Grant GM-17940.

REFERENCES

[1] Lele, P.P., A.B. Mansfield, A.I. Murphy, J. Namery and N. Senapti, "Tissue Characterization by Ultrasonic Frequency-Dependent Attenuation and Scattering," Nat'l. Bureau of Stan. Special Pub. 453, Proc. of Seminar on Ultrasonic Tissue Char. held at NBS, Gaithersburg, MD, May 28-30, 1975.

[2] Lele, P.P. and N. Senapti, "The Frequency Spectra of Energy Backscattered and Attenuated by Normal and Abnormal Tissue," in Recent Advances in Ultrasound in Biomedicine, D.N. White, ed., Research Studies Press, Forest Grove, OR.

[3] Yuhas, D.E., J.N. Mimb, J.G. Miller, A.N. Wiess, B.E. Sobel, "Changes in Ultrasonic Attenuation Indicative of Regional Myocardial Infarction," in Ultrasound in Medicine, D.N. White, ed., Vol. 3, Plenum Press, NY, 1977.

[4] Morse, P.M. and K.N. Ingard, Theoretical Acoustics, McGraw-Hill, NY, 1968.

[5] Joynt, L., D. Boyle, H. Rakowski, W. Beaver, "Identification of Tissue Parameters by Digital Processing of Real-Time Clinical Cardiac Data," presented at 2nd Symp. on Ultrasonic Tissue Charac., NBS, Gaithersburg, MD, June 1977.

[6] Herdson, P.B., H.M. Somers, and R.B. Jennings, "A Comparative Study of the Fine Structure of Normal and Ischemic Dog Myocardium with Special Reference to Early Changes Following Temporary Occlusion of a Coronary Artery," Amer. J. of Path., Vol. 46, p. 367, 1965.

[7] Gramiak, R., R.C. Waag, E. Schenk, P.P.K. Lee, K. Thomson, P. Macintosh, "Ultrasonic Detection of Myocardial Infarction by Amplitude Analysis," in Ultrasound in Medicine, Vol. 4, D.N. White, ed., Putnam Press, NY 1977.

[8] Rhyne, T.L., "An Ultrasonic Tissue Characterization for the Lung," presented at the 29th ACEMB, Sheraton-Boston, Boston, MA, Nov. 6-10, 1976.

[9] Sobel, B.E., "Biochemical and Morphologic Changes in Infarcting Myocardium," Chap. 22 in The Myocardium:Failure and Infarction, Eugene Braumwald, ed., Hospital Practice Co., Inc., NY 1974.

IN VIVO CHARACTERIZATION OF SEVERAL LESIONS IN THE EYE USING ULTRASONIC IMPEDIOGRAPHY

Joie Pierce Jones and Catherine Cole-Beuglet

Department of Radiological Sciences, University of California Irvine, Irvine California

All ultrasound systems now in routine clinical service utilize only a portion of the information available in an echo waveform. For example, a conventional A-scan displays the intensity of the echo waveform as a function of acoustic travel time and a conventional B-scan produces an ultrasonic mapping in which the spot intensity is proportional to the echo intensity. Phase information (such as the sign or polarity of the echo waveform) is recorded by the transducer, which is a pressure sensitive device, but is not utilized in present display or measurement schemes.

Several years ago this investigator[1,2,3] proposed a technique for utilizing both phase and amplitude information in a pulse-echo system. The technique, termed impediography, is based on a time-domain characterization of system dynamics in terms of an impulse response function. Impediography refers to a rather general class of signal processing operations which, when applied to input/output signals, yield quantitative information concerning the physical properties of the system under study. Ultrasonic impediography involves two fundamental operations. The first is a processing operation upon both the incident and reflected acoustical signals to obtain the system impulse-response function. The second is a processing operation upon the impulse response to obtain physical properties as a function of acoustic travel time. In present implementations[4,5,6] the impulse response is obtained by deconvolution of incident and reflected waveforms. Simple analytical expressions then relate the integral of the impulse response to the specific acoustical impedance and generate profiles of impedance as a continuous function of acoustic travel time (or position). The term impediography is suggested by the ability of the technique to generate profiles of impedance. The acoustical mapping made with impediography, whether a simple A-scan plot of impedance versus distance or an impedance profile of an arbitrary cross-section, is termed an impedogram.

A number of preliminary experiments have been conducted by this author(6) and several other workers(7) to evaluate the potential impediography might offer as a diagnostic tool and to assess the problems involved in the development of a practical clinical system. The results of these experiments have been quite encouraging: accurate impedograms have been obtained for a number of highly idealized physical models as well as for several samples of biological tissue. The present paper describes what we believe to be the first in vivo results obtained in the clinic with an impediographic measurement system. The study to be described was limited to measurements on the human eye, which is, acoustically, the simplest organ in the body and perhaps the only organ whose geometry of planer or near planer interfaces strictly satisfies the theoretical formalism of impediography presently available. For a more general development of impediography please refer to the paper by Dr. Leeman in the present volume.

A block diagram of the experimental system is shown in Figure 1. A conventional ophthalmic immersion B-scanner was modified so that any single A-scan line of interest could be digitized by a fast A/D converter. The resulting signals were recorded on magnetic media and subjected to impediographic processing using an off line PDP 11/40 computer. A custom designed microprocessor-based system (built around a PDP 11/03) was used to control the data acquisition process. The details of this system will be described in a future publication. It should be noted that the system shown in Figure 1 is designed for the real-time acquisition of ultrasound data but with off-line (or non-real-time) data analysis. We are presently implementing a computerized ultrasound data analysis system in our laboratory at UCI for the real time acquisition <u>and</u> analysis of ultrasound data. One feature of this system, built around a conventional B-scan ultrasound unit, is its ability to record both the phase and amplitude of A-mode waveforms selected through regions of interest on a conventional B-mode ultrasonogram.

Any serious study of the echo waveforms produced by the interaction of ultrasound with tissue requires the accurate characterization and calibration of the ultrasound system. Thus, the receiver's gain transfer function was measured and recorded in computer memory. That is, a table of receiver amplitude outputs as a function of receiver amplitude inputs was generated and employed as a compensation factor for the non-linearity of the receiver. A similar table of receiver frequency response was used to compensate the band limiting characteristics of the receiver. Finally, a table of receiver gain as a function of TGC voltage was recorded in computer memory and used to compensate for gain changes in-

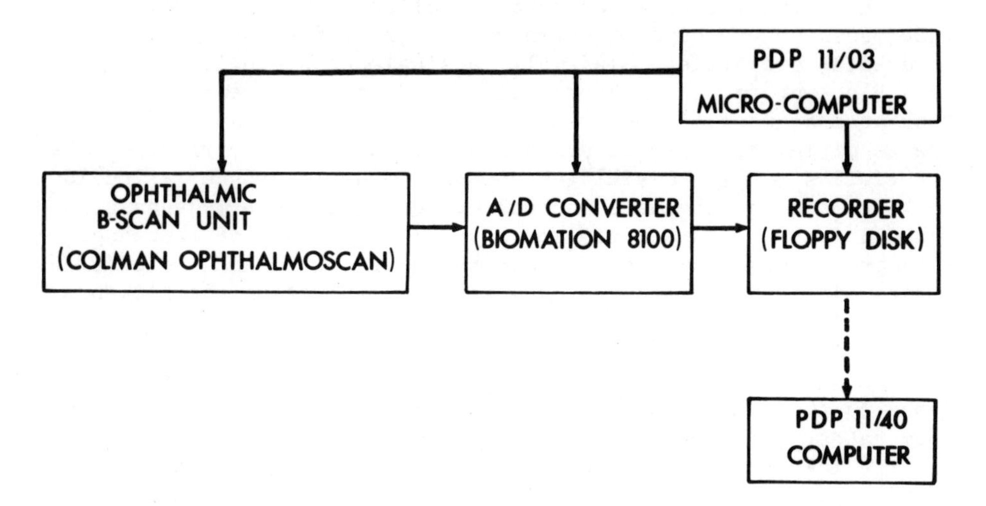

Figure 1. Block Diagram
of the Experimental System

troduced by the instrumentation. Using this calibration scheme it was therefore possible to transform the apparent or measured echo amplitude into a true or real amplitude suitable for quantitative analysis.

To be totally rigorous, the various system parameters should be evaluated as a function of temperature and other environmental factors. Moreover, the transducer beam pattern should be measured under various conditions and used to correct the recorded echo waveform. None of these latter measurements have yet been made although they are planned for the future. However, any corrections implied by these measurements would not be expected to significantly alter the experimental results presented here.

In the next step of our system calibration procedure, we made measurements on a wide range of plane reflectors having known acoustical properties. Initially a single plane-reflector (such as a block of silicon rubber) was placed in our experimental water tank at a fixed distance from the transducer. The amplitude of the resulting echo was measured and corrected by the receiver characteristics tables to yield an absolute measure of the reflection coefficient. In the same manner, measurements were made on a wide variety of plane reflectors over a wide range of transducer/reflector distances. Corrections by the receiving characteristics tables yielded consistent values for the reflection coefficients at all distances.

Finally, to complete our system calibration procedure, we repeated these measurements on plane reflectors using the rf waveform rather than the pulse amplitude. This required implementation of impediographic processing in order to measure the reflection coefficients and to obtain an impedance profile. Thus, the initial pulse (recorded by reflection from a perfect reflector) and a single A-scan echo wave train were sampled by the Biomation 8100 A/D converter at an 80 MHz sample rate and the resulting digital data, together with various system parameters, transferred to a floppy disc. This disc was then transported to a PDP 11/40 computer for off-line analysis.

The first step of the signal analysis process was to convert the measured ultrasound data into a "true" waveform using the receiver characteristics tables in the computer. Next, this "true" data, which was originally sampled at an 80 MHz rate, was converted into an array whose sample interval was equivalent to that selected for deconvolution. Using an initial pulse with a 10 MHz center frequency, a sample interval of 25 nsec (i.e., 40 MHz) was selected. This corrected data base was prepared such that it contained all maxima, minima and inflection points which were in the original data. In addition, curve fitting techniques were applied so that the corrected waveform and its derivative were continuous. If $x(t)$

represents the corrected incident waveform and y(t) represents the corrected echo waveform, then, if reflection is a linear process, y(t) is equal to the convolution of x(t) with the impulse response function, h(t). Thus,

$$y(t) = \int_0^t x(t-\tau)h(\tau)d\tau \equiv x(t) * h(t)$$

The problem then is to calculate h(t) given y(t) and x(t)-this requires deconvolution. The specific deconvolution algorithm utilized was a four-stage iterative procedure. If X(ω) and Y(ω) represent the Fourier transforms of the incident and reflected waveforms respectively, then the initial estimate of the impulse response in the frequency domain is given by

$$H_0(\omega) = \frac{Y(\omega)}{X(\omega)} \cdot \frac{X^*(\omega)}{X^*(\omega)} \cdot W(\omega)$$

where * represents the complex conjugate and W is the system bandwidth. Taking the inverse transform yields the initial estimate of the impulse response. Thus,

$$h_0(t) \equiv \mathcal{F}^{-1}[H_0(\omega)] .$$

We next calculate

$$P_0(\omega) \equiv \sum_i a_{0i} e^{-i\omega t_{0i}}$$

where $a_{0i}(t_i)$ are the relative maxima of $h_0(t)$. The first estimate of the impulse response is then given by

$$H_1(\omega) = H_0(\omega) + [1 + W(\omega)] P_0(\omega) .$$

Continuing this iterative process we find rapid convergence (in two to four estimates) to the impulse response function. Once the impulse response has been obtained, the specific acoustical impedance is computed from the expression

$$\int_0^t h(\tau)d\tau = \frac{1}{2} \ln [Z(t)/Z_0] .$$

The signal processing procedure just described was applied to several plane reflectors. Impedance values obtained by such a process were consistent with the known acoustical properties of these materials and with the earlier echo amplitude measurements.

Confident that our ultrasound system could accurately record profiles of impedance, at least in highly idealized situations, the instrumentation was applied in the clinic. Subjects for the study were selected from the normal caseload at several affiliated hospitals. Since the system was integrated with a conventional B-scan unit, no changes in patient management were required. Further-

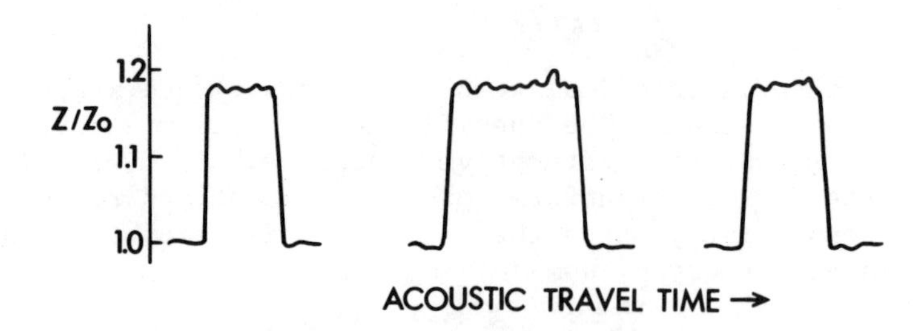

Fig. 2. Normalized Impedograms for Three Retinoblastomas

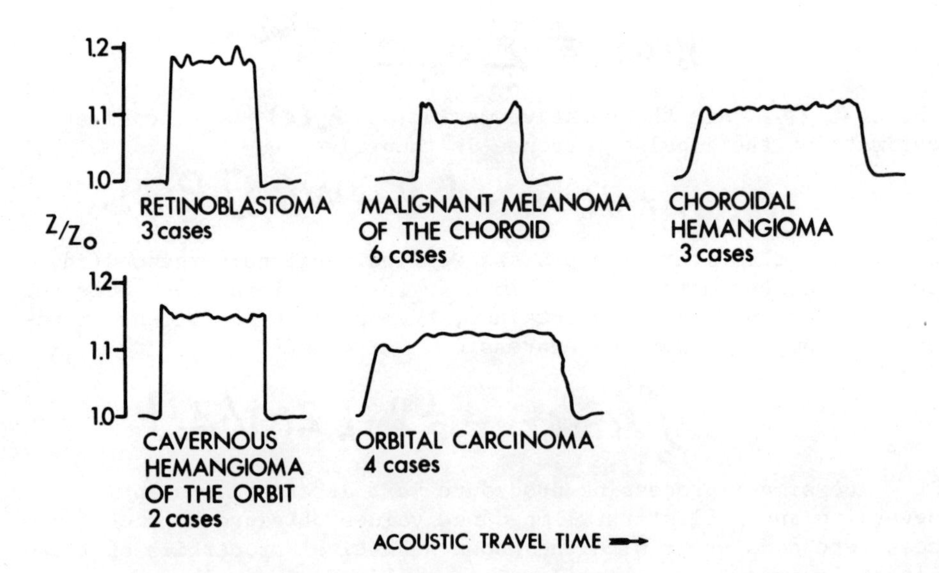

Fig. 3. Typical Impedograms for Several Lesions

more, there were no added requirements or constraints on the sonographer in the form of complex electronic adjustments or difficult scanning techniques. During the course of a standard ultrasound examination, the sonographer could make one or more scans in which a single A-line passing through tissue of interest as monitored on the B-scan could be brightened and photographed. When the single A-scan line passing through the tissue of interest was brightened, the special hardware automatically recorded the ultrasound waveform along this line in a manner previously described. When the scanning session was completed, the digital information recorded on the floppy disc was taken to an off-line computer for data reduction and processing. The processed data in the form of impedograms were then compared with the B-scans and with other available patient data.

Figure 2 shows impedograms of three lesions which were confirmed to be retinoblastomas. The retinoblastoma is a space-occupying mass lesion and is perhaps the most important intraocular tumor occurring in infants and children. The impedograms have been normalized so that they appear to be surrounded by distilled water. Such a normalization enhances the display of the impedance profile signature. Although the three lesions clearly have different dimensions, their impedogram signatures are quite similar.

Figure 3 summarizes the results of our preliminary study by presenting typical impedograms for several lesions in the eye. The typical impedogram for a retinoblastoma is based on the 3 cases shown in Figure 2. The typical impedogram for a malignant melanoma of the choroid is based on six cases. This lesion is also a space-occupying mass and is perhaps the most important intraocular tumor occurring in adults. The typical impedogram for a choroidal hemangioma is based on three cases. This is also an intraocular tumor.

The typical impedogram for a cavernous hemangioma of the orbit is based on two cases. This orbital lesion is a relatively frequent benign type of tumor. The typical impedogram for an orbital carcinoma is based on four cases.

Clearly, the results shown in Figure 3 are based on far too few cases to be conclusive. However, they suggest that a differential diagnosis of several lesions in the eye might be made using impediography in which the ultrasonic signature is based on the structure of the impedogram as well as the impedance level. We would encourage further work in this area to verify our present findings and to develop, if possible, definitive diagnostic schemes for tissue characterization.

REFERENCES

1. J. P. Jones, "Toward an Ultrasonic Technique to Indirectly Measure Tissue Abnormalities Within the Human Body," Technical Memorandum, Bolt, Baranek and Newman, Incorporated (February, 1971).

2. J. P. Jones and H. A. Wright, "A New Broad Band Ultrasonic Technique with Biomedical Implications, I: Background and Theoretical Discussion," presented at the 83rd Meeting of the Acoustical Society of America, Buffalo, New York (April, 1972).

3. J. P. Jones, "Impediography: A New Approach to Non-invasive Ultrasonic Diagnoses," Proceedings of the 25th Annual Conference on Engineering in Medicine and Biology, Bal Harbour, Florida (October, 1972).

4. J. P. Jones, "Impediography: A New Ultrasonic Technique for Diagnostic Medicine," in Ultrasound in Medicine, Volume I, (D. N. White, Editor), Plenum (1975).

5. J. P. Jones, "A Preliminary Experimental Evaluation of Ultrasonic Impediography," Ultrasound in Medicine, Volume 1, (D.N. White, Editor), Plenum (1975).

6. J. P. Jones, "Ultrasonic Impediography and its Application to Tissue Characterization," Recent Advances in Biomedicine, Volume 1, (D. N. White, Editor), Research Studies Press (1977).

7. I. Beretsky, G. Farrell and B. Lichtenstein, "Raylography-A Pulse Echo Technique with Future Biomedical Implications," Ultrasound in Medicine, Volume 2, (D. N. White & R. Barnes, Editors), Plenum (1975).

HIGH RESOLUTION IMAGING OF FILTERED ULTRASONIC ECHO RESPONSES

EDWARD HOLASEK, WAYNE D. JENNINGS, EDWARD W. PURNELL

Lorand V. Johnson Laboratory for Research in Ophthalmology
Department of Surgery, Division of Ophthalmology
Case Western Reserve University, School of Medicine
Cleveland, Ohio 44106

Several years ago we developed a high resolution ultrasonic B-scanner for general purpose diagnosis, but specifically oriented for use in ophthalmology.[1] In the special case of ophthalmology, high resolution is necessary because of the small size of the eye and the amount of detail required for the diagnosis. High axial resolution (.3mm) was achieved through the use of wideband transducers operating at a nominal frequency of 10MHz. Adequate lateral resolution (approximately 1mm) resulted through the use of focused ultrasonic beams.

Although clinical utility has been achieved through high resolution broadband (gray-scale) echo responses of soft tissue, we realized that more information was available than that being utilized in gray-scale imaging alone.

A second clinically oriented system has been developed for real-time B-mode imaging of filtered signals of the received echo complex.[2] The broadband transducer response was divided into three bands through the use of band pass filters to provide spectral information. When used as an adjunct to high resolution gray-scale imaging of clinical tissue, this signal processing technique provided both low spectral and low temporal resolution, color-coded, echo patterns which appeared to be specific for certain diseases.[3]

This study was supported in part by Public Health Grant EY 00224 from the National Eye Institute, the Ohio Lions Eye Research Foundation, and Research to Prevent Blindness, Inc.

Signal processing techniques for spectral characterization are being pursued by others in the field using either standard spectrum analyzer techniques or the averaging of spectra from a number of adjacent echo beam positions.[4] Often computerized digital signal processing is used for spectral characterization of stored video signals.[5,6]

Our current research in B-mode imaging utilizes the use of analog filtering techniques applied directly to received echo trains. These techniques include the concept of an instantaneous power spectrum applied to ultrasonic imaging. The theoretical analysis of instantaneous power spectra as applied to tissue characterization is covered elsewhere.[7,8,9]

It is the intention of this paper to describe the instrumentation and method used to display instantaneous power spectra. This paper also indicates the applicability of instantaneous power spectra to high resolution tissue characterization by some simple physical model studies conducted in the laboratory. These studies illustrate the instrumentation principle leading to tissue characterization not only in ophthalmology, but in other areas of the anatomy as well.

The instrumentation for displaying the instantaneous power spectrum of ultrasonic echo waveforms has two display modes. The first mode is a color-coded B-scan display of the instantaneous power spectrum, with 2MHz resolution in the frequency domain. The second mode displays the instantaneous power spectrum with higher resolution in the frequency domain for a single echo pattern in graphical form.

The basic principle used to display the instantaneous power spectrum is illustrated in Figure 1. It has been shown that the instantaneous power in a limited frequency band can be produced by multiplying a symmetrically filtered (band limited) time domain waveform by the unfiltered (broadband) time domain waveform.[7] The resultant signal is the band limited instantaneous power of the original waveform. This process is analogous to producing the broadband instantaneous power of a signal by squaring.

This signal multiplication technique is the basis for both display modes for instantaneous power spectra. The technique produces a signal that contains both spectral information and high resolution temporal information about the echo waveform.

The color-coded B-scan display of the instantaneous power spectra provides low resolution spectral information about an ultrasonic sector scan while retaining high axial and lateral resolution necessary for a good ophthalmic diagnostic imaging system.

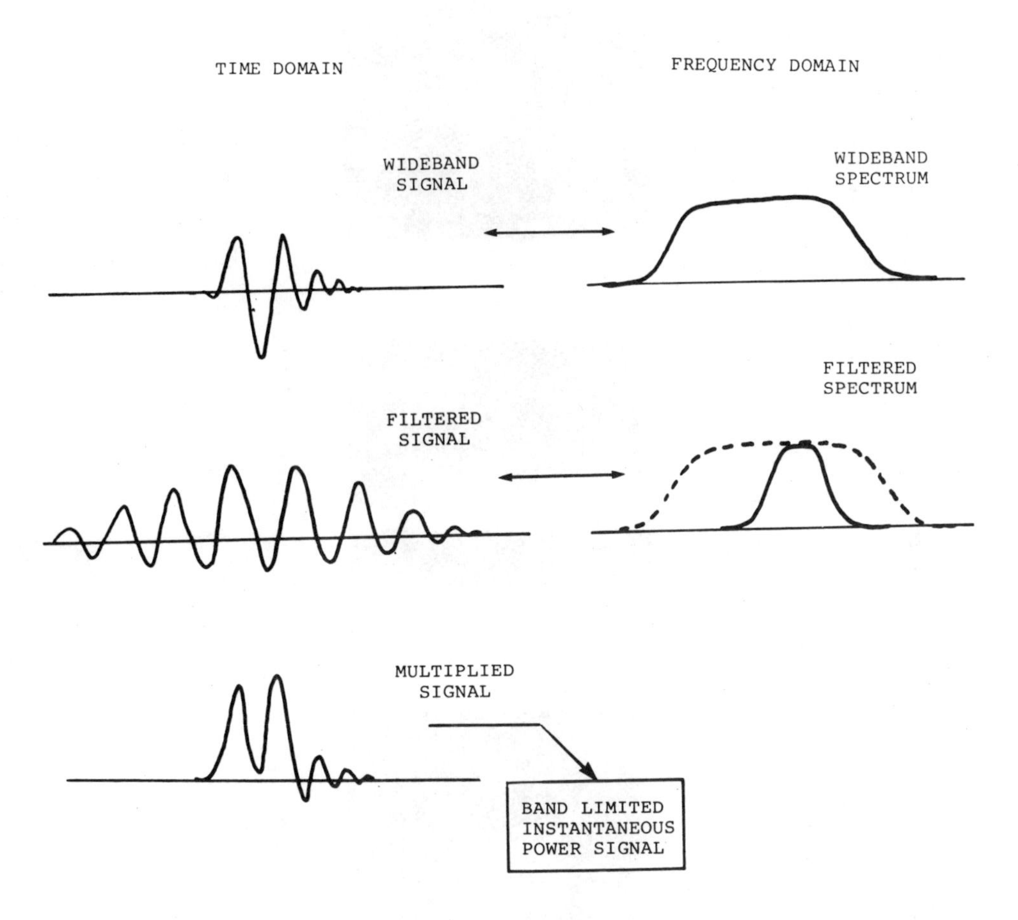

Figure 1. Basic principle of bandlimited instantaneous power. Broadband and filtered signals are shown in both time and frequency domains. The filtered waveform is noncausal with respect to the broadband signal. This is a result of symmetrical filtering which is required by the theory of band limited instantaneous power. The filtered waveform is also delocalized in time, due to the reduction in its bandwidth in the frequency domain. Multiplication of the filtered and broadband signals results in a signal that represents the band limited instantaneous power of the broadband signal.

The spectral resolution of the color-coded display is 2MHz. The three filters used in the system are 6-pole Butterworth design, with relatively square shape in the frequency domain.

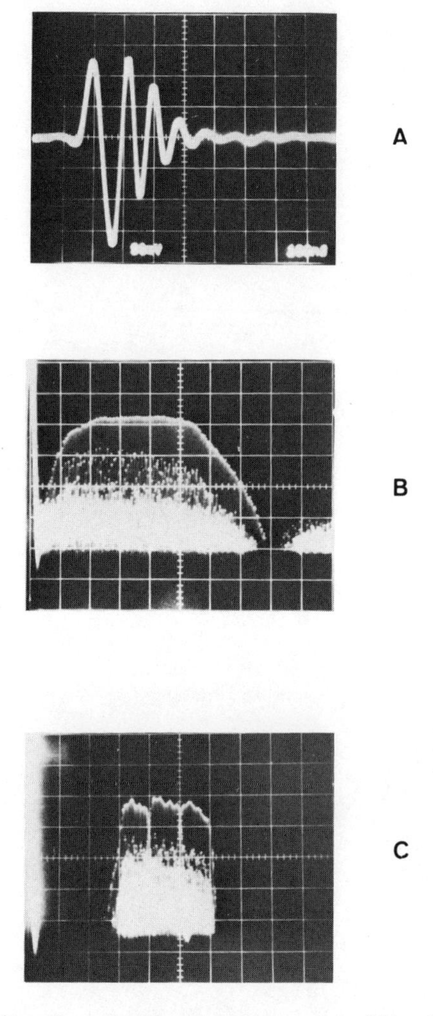

Figure 2. The broadband transducer spectrum (B) shows the range of frequencies, 2MHz/Div., contained in the ultrasonic pulse (A). Three superimposed filtered spectra (C) used for color coding the display are shown.

The block diagram of the color-coded display system (Figure 3) illustrates the filtering delay line, multiplication and switching circuits necessary to produce the band-limited instantaneous power for the three filter bands. The filtering system sequentially filters the broadband echo signal to produce three filtered waveforms. Simultaneously with filtering, the broadband signal is delayed by coaxial delay lines corresponding to the delay the signal experiences as it passes through the filter. This delay line compensation is necessary to produce the symmetrical filtering effect required to produce the band-limited instantaneous power of the signal. The delay lines assure that the filtered signals and the unfiltered signals arrive in phase at a double balanced mixer for signal multiplication. Another set of delay lines must then be added after signal multiplication to compensate for delay variations among the filters. These final delay lines produce three band limited instantaneous power signals that are in complete registry for overlay of the resultant color-coded B-scan images.

Normalization of the three instantaneous power signals is performed by a programmable gain amplifier. The amplifier compensates for variations in total power among the three bandpass regions in the pulse transmitted by the transducer.

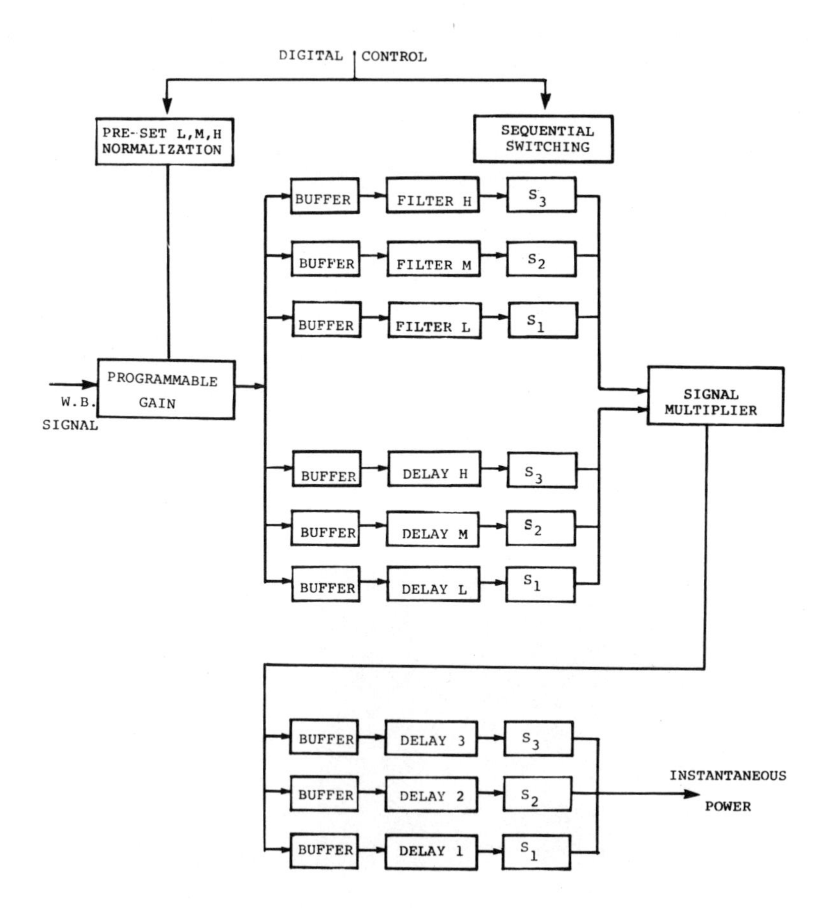

Figure 3. Block diagram of digitally controlled instrumentation method for generation of the instantaneous power signal for the color-coded display mode.

The three different gain settings are presettable and can be adjusted to approximately normalize any transducer spectrum.

The signal processing steps described above produce three B-scan images, one for each of the three filters. The images are displayed sequentially on a high resolution cathode ray tube, and photographed through three color filters on a single 35mm frame of color film. The composite color B-scan ultrasonogram which results is a low resolution [2MHz] color-coded representation of the instantaneous power spectrum of all the echoes in the sector scan. Because of the delay line compensation for variations among the filter delays, the three colored images overlay, and appear white when the normalized instantaneous power spectrum is flat over the 6 to 12MHz range. When the instantaneous power spectrum has a particular feature of an accentuated or missing frequency band, the corresponding spot on the composite color ultrasonogram appears colored. Resolution, however, is limited to resolving instantaneous spectral bands separated by 2MHz or more, and will not display the finer structure of the instantaneous power spectrum. The limitation is partly due to the color coding process, which becomes complicated when more than three colors are used. Higher frequency resolution can be obtained, however, by using a larger number of narrower bandpass filters and by using a second, graphical, display mode.

The signal processing steps required for the graphical display mode parallel those required for the color-coded display mode. A voltage controlled filter (V.C.F.) is used to sequentially filter the broadband signal into many frequency bands, the number of which is determined by the voltage staircase applied to the filter. (Figure 4.)

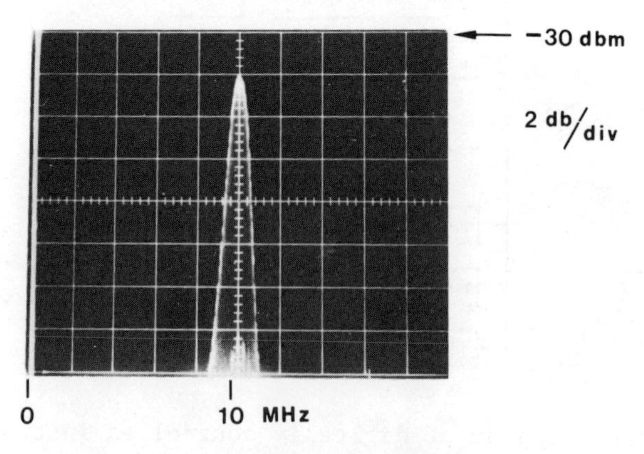

Figure 4. Voltage controlled filter response shown at a typical setting (10MHz) within the 6MHz to 12MHz filter range.

Due to the specific construction of the voltage controlled filter, the filter center frequency is a nonlinear function of the applied voltage. Conversly, a nonlinear voltage staircase is required to produce a linear frequency sweep for the graphical display. The nonlinear sweep is generated by a digital to analog converter (D/A) from stored calibration data in a read only memory (ROM), (Figure 5.)

Favorable alternatives to the V.C.F. include a large number of fixed filters or a programmable transversal filter using a multiply tapped delay line. The disadvantages of the particular V.C.F. used in this work include lack of symmetry in both the frequency and time domains. These disadvantages could be eliminated with either the set of fixed filters, or the programmable transversal filter. Practical experience with the V.C.F., however, indicates its adequacy as an initial system filtering device.

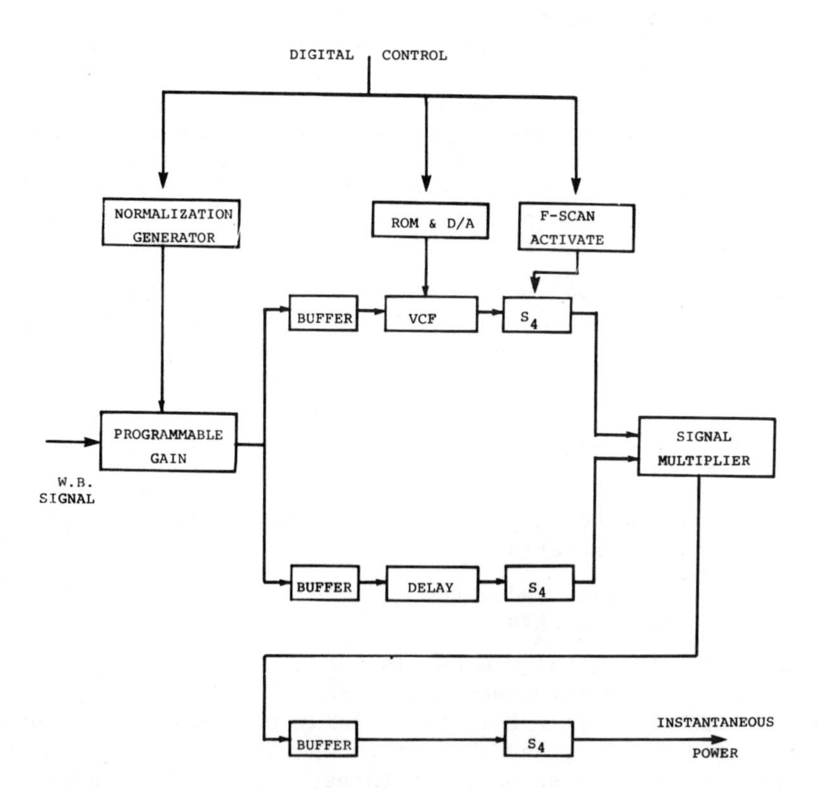

Figure 5. Block diagram of digitally controlled instrumentation method for generation of the instantaneous power signal for graphic display mode, frequency scan.

Compensation for the filter delay is accomplished by a single broadband delay line that approximates the average delay of the filter. Thus, the filtered and unfiltered signals are approximately in phase at the double balanced mixer for signal multiplication.

Normalization of the graphical display mode is accomplished through the use of a programmable gain amplifier driven by a function generator. The function generator produces a gain control signal that specifically matches the spectrum of the transmitted ultrasonic pulse, using operational amplifiers and biased diodes.

In the graphical display mode, the transducer remains in a fixed position, pointed at a particular region of interest,and the instantaneous power spectrum is displayed as a function of both frequency and time by intensity modulation. The display is generated by a series of vertical sweeps on the C.R.T.. The display contains 1024 vertical lines, clocked at a 1.4 KHz rate. The horizontal positon of each sweep corresponds to the frequency of the voltage controlled filter, and the vertical scale is time, or range of ultrasonic echoes. The band limited instantaneous power of the echo waveform, produced at the double balanced mixer, is used to intensity modulate the appropriate vertical sweep on the CRT, for each setting of the voltage controlled filter. The resultant display resembles a time-motion display, with the horizontal axis as a frequency axis rather than a slow time axis of the time-motion display. A complete sweep from 6 to 12 MHz requires about .7 seconds.

The two display modes for instantaneous power spectra of ultrasonic echoes have been incorporated into a clinically oriented scanning system based on a high resolution broadband ultrasonograph that has been described earlier.[8] Three types of display are possible with this system:

1) conventional broadband high resolution sector scan ultrasonograph
2) color-coded sector scan display of instantaneous power spectra
3) intensity modulated graphical display of instantaneous power spectra

Switching among the display modes is facilitated by a central digital control. In the color scanning mode, the sector sweep is synchronized with the filter selection, delay line switching, optical color coding apparatus, and the camera shutter. In three successive sweeps across the screen, the three filtered B-scans are superimposed on a single photographic film, with each scan appropriately colored by optical filters automatically interposed between the camera and the CRT.

The high resolution graphical display mode is prepared for activation by first placing the system in the conventional B-scan mode. As the broadband sector is displayed, a bright line appears within the sector at an angle determined by the setting of a potentiometer. (Figure 6). Two other controls determine the length of the line, and its range delay from the origin of the sector (transducer face). Thus any segment of any individual echo line within the sector can be selected and displayed as a brightened line on the B-scan display. When the graphical display mode is activated, the transducer is automatically stepped to the position indicated by the brightened line. After the transducer has reached the proper position, the filter and delay lines are electronically switched into the circuit, and the pre-determined echo line segment is processed to display its instantaneous power spectrum. The time required to switch between the sector scan mode and the graphical display mode is approximately equal to the time required

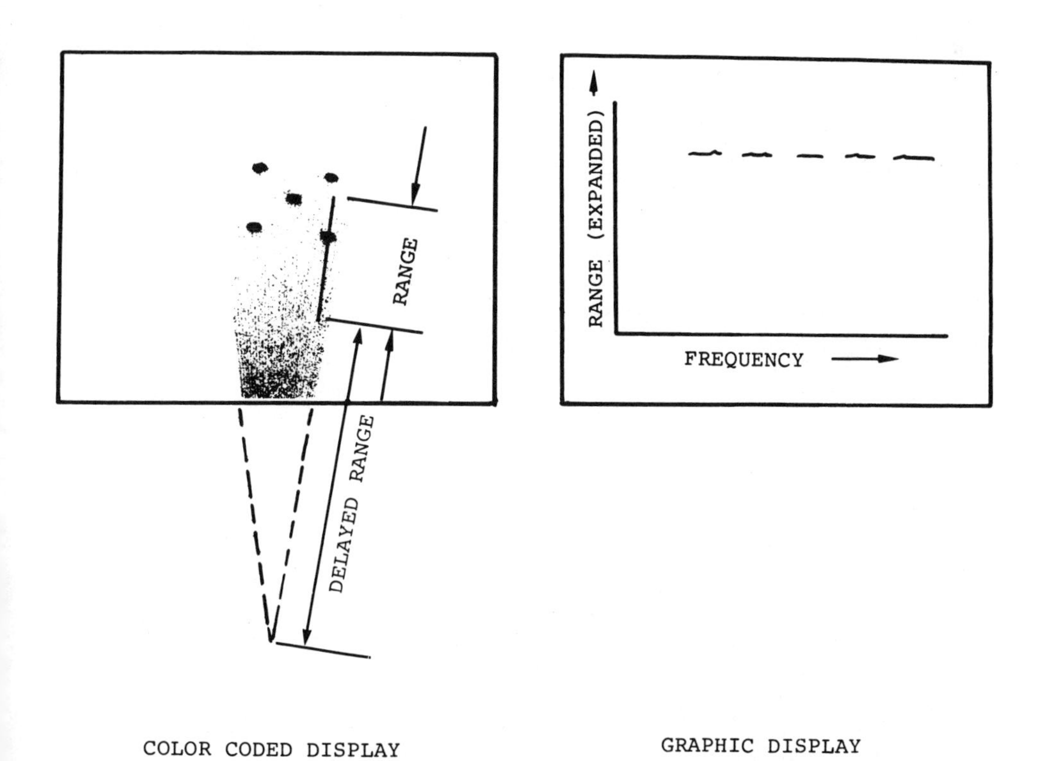

Figure 6. Illustration of two display modes for scan of ophthalmic test structure (left) with intensified line in preparation for graphic display mode (right).

for the transducer to find its proper position from any random position it held at the time the mode switch occurred. Typical mode switching time is about one second.

SYSTEM TESTS AND RESULTS

The system described above has been used to display the instantaneous power spectrum of some simple structures immersed in water. The structures studied were:

1. A thick glass block
2. A 0.38mm thick plastic strip (echo time separation - 340nsec.)
3. A 0.66 thick plastic strip (echo time separation - 520nsec.)
4. A 0.2mm diameter nylon wire.

In each case, the structure to be scanned was placed at the center of the focal zone (60 to 80mm) of the ultrasonic beam in a tank of distilled water. The scanning mechanism was coupled to the tank by a water filled rubber sleeve. This is similar to the way the mechanism is coupled to goggles worn by a patient in clinical applications.

The first target represents the simplest possible structure for analysis, the single specular reflecting interface. The target is used for normalizing the instantaneous power spectra with respect to the spectral content of the transmitted ultrasonic pulse. With proper normalization, the color-coded scan of the glass block produces a single white echo. The normalized video signals for each of the three bands of the instantaneous power spectrum are shown in Figure 7. (Only video signals are shown since color pictures cannot be reproduced in this publication.) For the glass block target, the graphical display of the instantaneous power spectrum produces a smooth, continous, frequency independent trace representing the relatively featureless instantaneous power spectrum of the transmitted pulse. (Figure 7C)

The parallel plate interfaces produce a more complex instantaneous power spectrum. The Fourier spectrum of a double echc contains peaks and valleys associated with constructive and destructive interference of the individual echoes with each other. These interference effects are also present in the instantaneous power spectrum of the double echo. The spacing between the interferences fringes in the spectrum are inversely proportional to the time spacing between the echoes. The three video signals for the color-coded display mode represent the instantaneous power spectrum averaged over each of the three filter bandwidths. Variations in these average values become variations in color when the sector

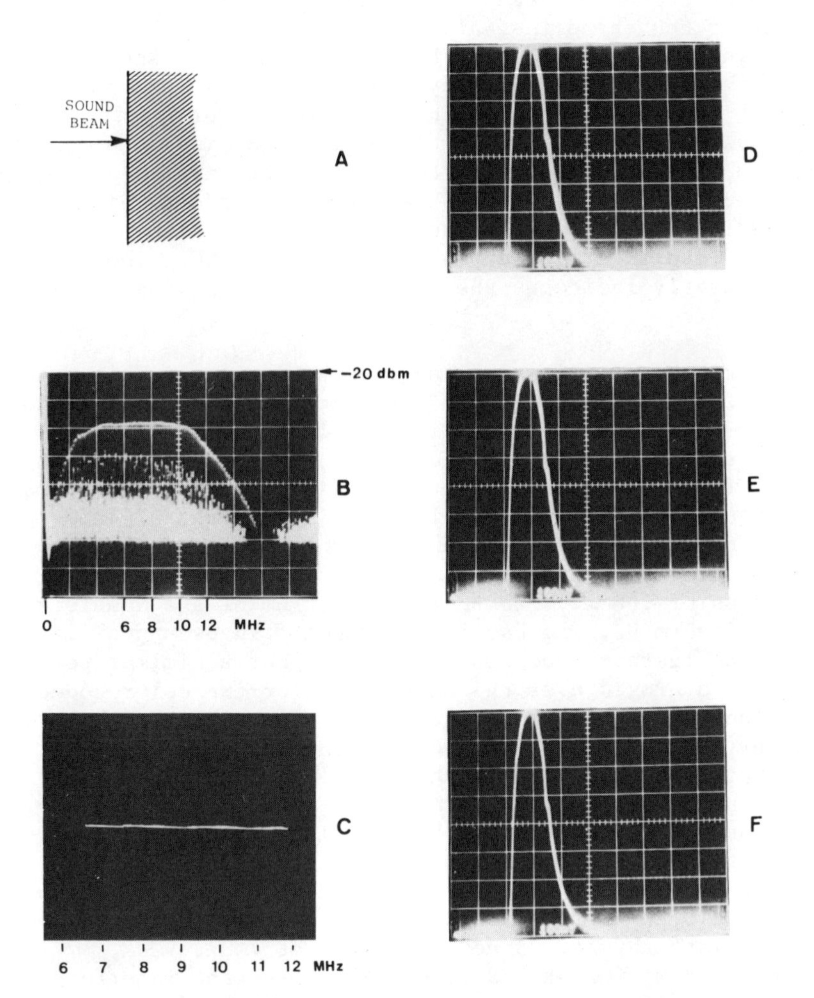

Figure 7. Glass block Test Target used for calibration.
7A. Glass Target placed perpendicular to the sound beam for normalizing the system.
7B. Fourier Spectrum of echo response.
7C. Graphic display of instantaneous power spectra.
7D. Video signal for blue color-coding (6-8MHz band).
7E. Video signal for green color-coding (8-10MHz band).
7F. Video signal for red color-coding (10-12MHz band).

scan is displayed. The first parallel plate interface studied was a thin [.38mm] plastic strip. Figure 8, shows the Fourier power spectrum of the double echo from this interface, and the video signals for the three instantaneous power spectra bands. Due to the thickness of the sample, the Fourier spectrum shows a distinct minimum in the middle frequency range [8-10MHz], and the corresponding video signal for the instantaneous power spectrum in that band is also reduced. Thus, the color-coded display of the instantaneous power spectrum of this target appears as a combination of red and blue, with a noticable absence of green. In the graphical display mode, the instantaneous power spectrum of the .38mm parallel plate interface appears as in Figure 8C. The graphical display clearly indicates the periodicity of the instantaneous spectrum.

The graphical display mode gives a continuous, high resolution display of the instantaneous power spectrum of a single echo line. The color-coded display mode displays only the average instantaneous power in three relatively broadbands. Due to this averaging, it is possible for a target to appear white in the color-coded display mode, yet have significant variations in the instantaneous power spectrum of its echo. An example of this situation is shown in Figure 9. The structure is another parallel plate interface, with a thickness of .66mm. The Fourier spectrum of the double echo is shown in Figure 9B, and has interference fringes approximately 2MHz apart. The instantaneous power spectrum has a similar periodicity. Thus, when averaged over the 2MHz bands for the color-coded display, the instantaneous power spectrum appears equal for each band. The resultant video signals (Figures 9D, 9E, 9F) produce a white display after color coding. In the graphical display mode, however, the instantaneous spectrum retains the interference pattern. The high frequency resolution of the graphical display system permits visualization of instantaneous spectral features that are lost in the lower resolution color coding process. Figure 9C shows the graphical display of the instantaneous power spectrum of the .66mm parallel plate target. Spectral maxima at 7MHz, 9MHz, and 11MHz and minima at 8MHz and 10MHz match the interference fringes observed in the Fourier spectrum of the double echo.

The fourth object used to test the system was a 0.2mm diameter nylon wire. The reflection, scattering, and interference effects are more complicated for the cylindrical wire target than for the parallel plate structures. The Fourier spectrum of the echoes from a single nylon wire are shown in Figure 10B. The interference pattern is considerably more complicated than the simple scallops observed with the parallel plate interface. Video signals for the instantaneous power spectrum displayed in the color-coded mode are shown in Figure 10D, 10E, 10F. The corresponding graphical display is shown in Figure 10C. Both display modes reveal the complex spectral character of the echoes. Both modes also resolve

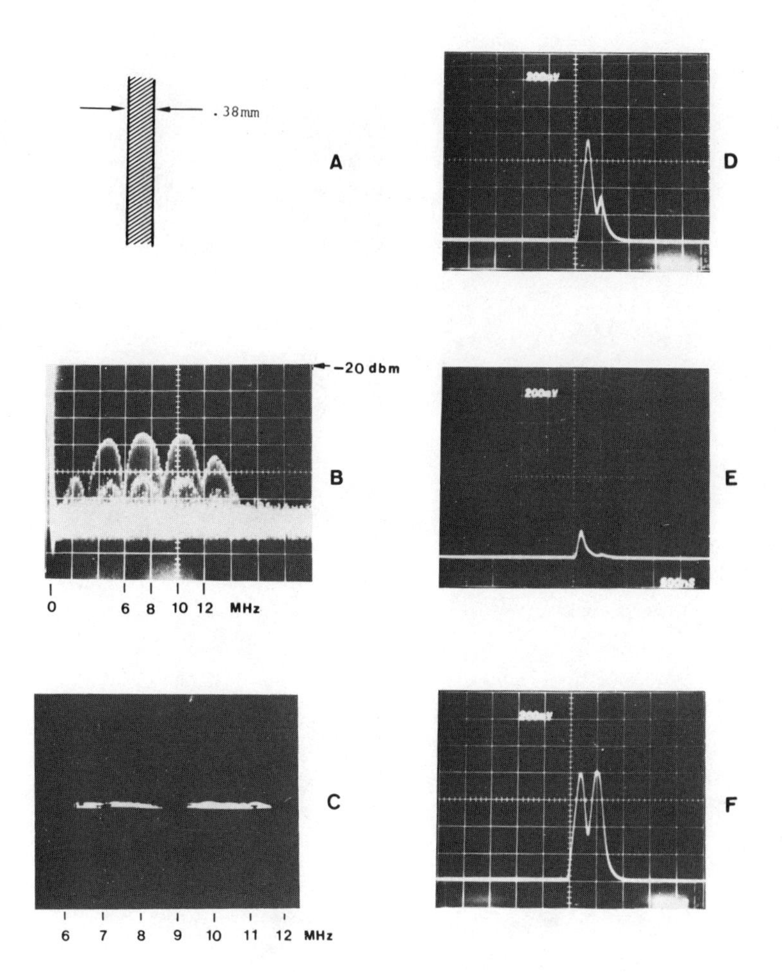

Figure 8. Thin parallel plate test target for instantaneous power spectral display.
8A. Thickness chosen to demonstrate destructive interference phenomenon.
8B. Fourier spectrum of ehco response.
8C. Graphic display of instantaneous power spectrum.
8D. Video signal for blue coding.
8E. Video signal for green coding.
8F. Video signal for red coding.

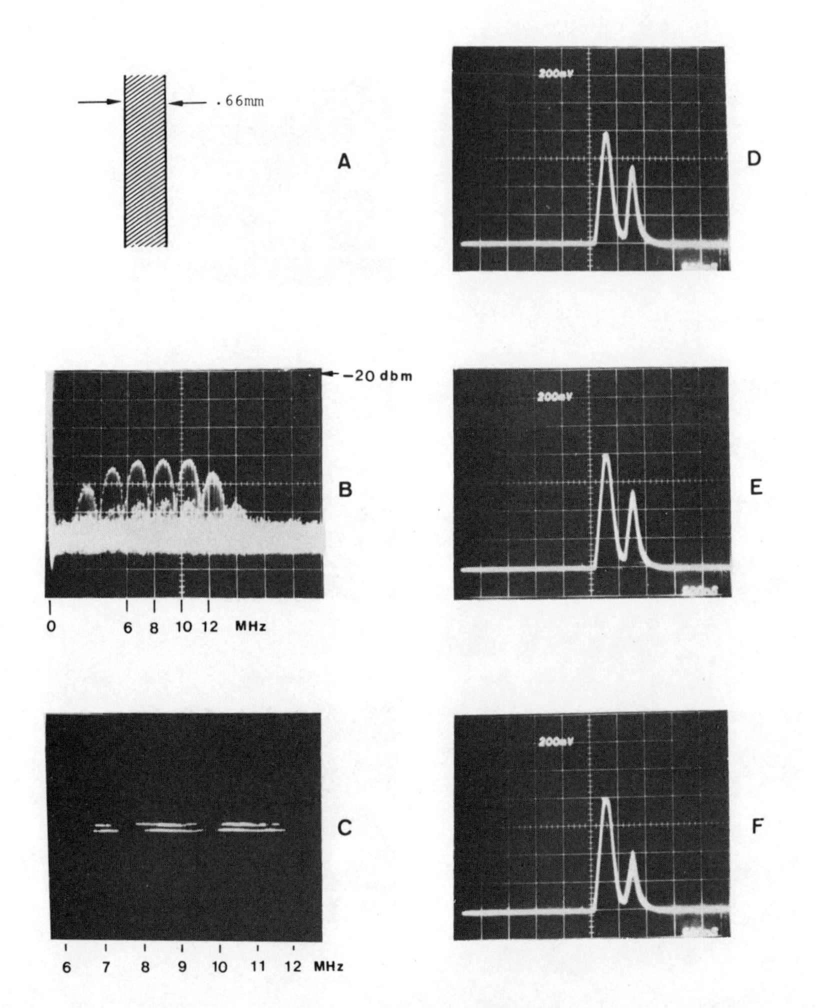

Figure 9. Thicker parallel plate target for instantaneous power spectrum display.

9A. Thickness chosen to demonstrate the frequency resolution of both display modes.

9B. Fourier spectrum of echo response.

9C. Graphic display of instantaneous power spectrum.

9D. Video signal for blue coding.

9E. Video signal for green coding.

9F. Video signal for red coding.

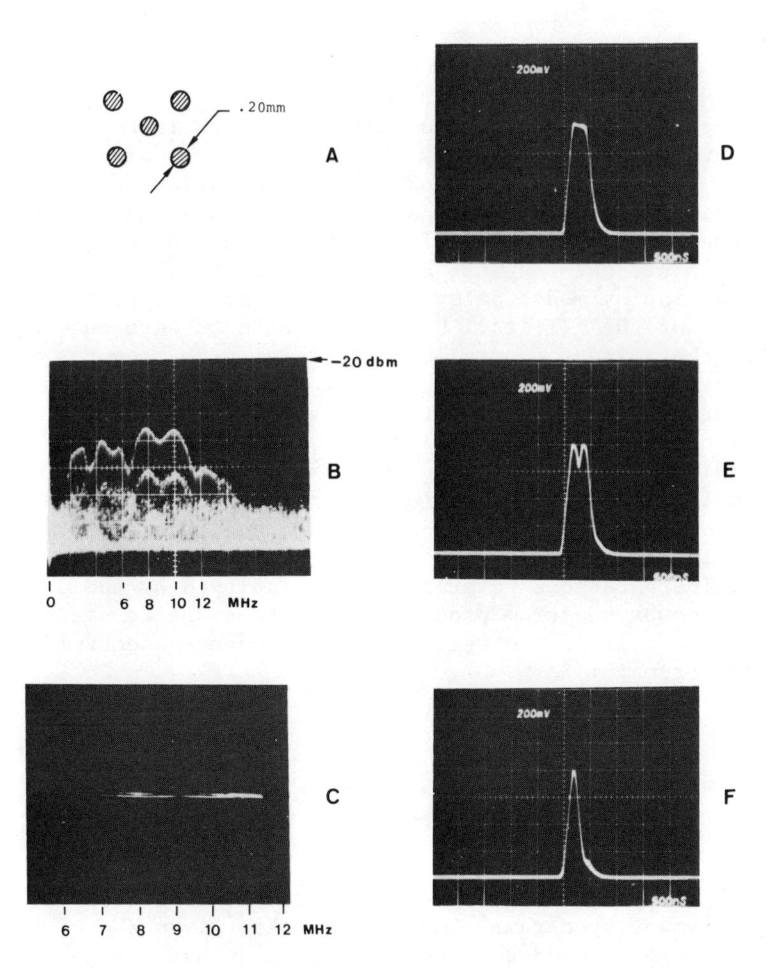

Figure 10. Nylon wire test target.
10A. Wire size chosen to demonstrate time resolution of both display modes.
10B. Fourier spectrum of echo response.
10C. Graphic display of instantaneous power spectrum.
10D. Video signal for blue video.
10E. Video signal for green video.
10F. Video signal for red video.

the front and back sides of the .2mm diameter wire, illustrating the high temporal resolution of the system based on the instantaneous power spectrum.

DISCUSSION - CONCLUSIONS

The accuracy of the spectral data contained in the two display modes is determined primarily by the delay line compensation for the group and phase delays introduced by the various filters in the system. Inaccurate delay compensation produces signals that are out of phase when multiplied. Inaccurate band limited instantaneous power signals are the result. For the three filters used in the color-coded display mode, delay lines were selected to match the group delays of the filters. Dispersive effects were not considered. For the voltage controlled filter, a further approximation was made by using a single delay line for the group delay of the filter, averaged over its operating range of 6 to 12MHz. Thus, the color-coded display mode is probably more accurate than the graphical display mode, assuming the fixed filters and the voltage controlled filter have comparable dispersions. The results of system tests described above indicate good qualitative accuracy of the system in both display modes. The assessment is based on comparing the instantaneous power spectrum displayed by the system to the Fourier power spectrum produced by another analog spectrum analyzer. The relation between the instantaneous spectrum and the Fourier spectrum is:

$$|F(\omega)|^2 = \int_{-\infty}^{\infty} \rho_s(t,\omega)\, dt$$

where $|F(\omega)|^2$ is the Fourier power spectrum, and $\rho_s\,(t,\omega)$ is the instantaneous power spectrum. For the examples given above, the system appears to give good qualitative agreement between the instantaneous power spectra and the Fourier power spectra, based on Eq. (1).

Another important consideration in instrumentation design is the relation between temporal resolution and dynamic range of the system. In general, signal processing techniques have opposite effects on these two parameters, except in infinite bandwidth systems. When the dynamic range of a system is extended by nonlinear amplification of signals, such as logarithmic amplification, the time resolution of the system is reduced. Conversly, the use of techniques that increase the time resolution result in decreased dynamic range of the system. Thus, the signal multiplication technique used in the instantaneous power spectrum display system, which increases the temporal resolution of the echo signals, limits the dynamic range of the system. The resultant dynamic signal is about 10db.

System modifications are in progress to improve the accuracy of the instantaneous power spectrum display and to increase its dynamic range. The modifications include the use of a symmetrically tapped transversal filter for the graphical display mode, and a logarithmic amplifier for both display modes.

REFERENCES

[1] Holasek, E. and Sokollu, A.: Direct Contact, Hand-held Diagnostic B-Scanner, Proceedings of the 1972 Symposium of IEEE (72CHO-708-8SO) pp. 38-43, 1972.

[2] Holasek, E., Gans, L.A., Purnell, E.W. and Sokollu, A.: A Method for Spectra-Color B-Scan Ultrasonography, Journal of Clinical Ultrasound, Vol. 3, pp. 175-178, 1975.

[3] Purnell, E.W., Sokollu, A., Holasek, E. and Cappaert, W.E.: Clinical Spectral Ultrasonography, Journal of Clinical Ultrasound, Vol. 3, pp. 187-189, 1975.

[4] Coleman, D.J. and Lizzi, F.L.: Acoustic Spectral Analysis of Ocular Tissues, Ultrasound in Medicine, Vol. 3B, p. 2075, 1976.

[5] Robinson, D.E. and Williams, B.G.: Computer Analysis of Ultrasonic Pulse Echo Signals, Ultrasound in Medicine, Vol. 3B, pp. 1443-1453, 1976.

[6] Gramiak, R., Waag, R.C., Naada, N.C. and Astheimer, J.: Computer Processing of Cardiac Ultrasound Images, Ultrasound in Medicine, Vol. 3B, pp. 1815-16, 1976.

[7] Jennings, W.D., Holasek, E. and Purnell, E.W.: Theoretical Analysis of Instantaneous Power Spectrum as Applied to Spectra-Color Ultrasonography, 2nd International Symposium on Ultrasonic Tissue Characterization, N.B.S., Pub. 453, 1977.

[8] Jennings, W.D., Holasek, E. and Purnell, E.W.: High Resolution Instantaneous Ultrasonic Spectrum Analysis I: Theoretical Considerations, Ultrasound in Medicine, Vol. 4B, 1977.

[9] Jennings, W.D., Holasek, E. and Purnell, E.W.: Frequency Scanning, in press.

ACOUSTIC ECHO COMPUTER TOMOGRAPHY

Glen Wade, Scott Elliott, Ibrahim Khogeer, Gail Flesher,
Joseph Eisler, Dean Mensa, N.S. Ramesh, and Glen Heidbreder

Dept. of Electrical Engineering and Computer Science,
University of California, Santa Barbara, CA 93106

ABSTRACT

Acoustic tomography systems described in the past were designed to provide a cross-sectional image of an object by employing transmission of a beam of ultrasonic energy. The data thus obtained were processed by a computer using a suitable algorithm. This resulted in an image based upon either the acoustic velocity or the attenuation characterisic of the object. This paper describes two novel imaging techniques which are based on the reflection of acoustic waves from the distribution of scattering centers within the object. One of the primary advantages envisioned for the proposed techniques is the use of a single collimated beam of ultrasound in the plane of the tomogram in lieu of slow mechanical scanning or multiple transducer arrays. Results are presented of a laboratory experiment which was performed to test the validity of one of the proposed concepts.

INTRODUCTION

Internal imaging of objects by acoustic means is carried out using either transmission or reflection of acoustic waves. Previous investigators have examined the use of the principles of X-ray computer tomography (CT) for processing images made from acoustical transmission data [1,2]. In this paper we will explore the applicability of the CT data processing technique to two novel methods which utilize reflection of ultrasonic waves from acoustic discontinuities within the object. One of these methods is based on pulse-echo data and the second is based on Doppler-shift data. The purpose of this study is to explore the possiblity of improving

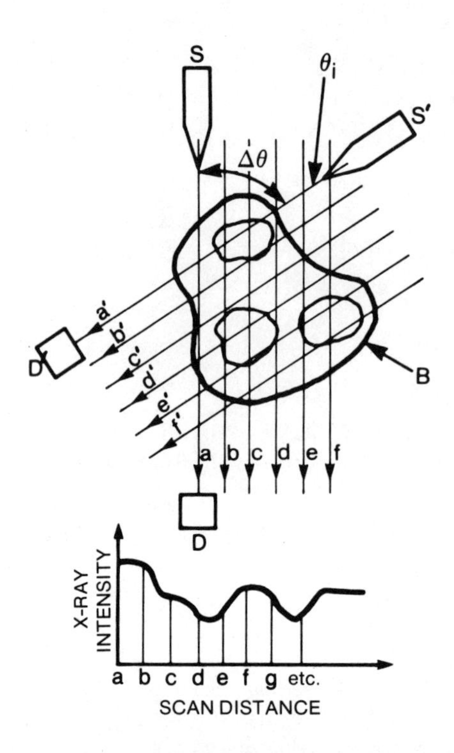

Fig. 1. Data acquisition in x-ray tomography.

the quality of acoustic images in the manner achieved in the X-ray application of CT.

In order to place acoustic reflection tomography in proper perspective, a brief review of the principles of X-ray computer tomography will be presented. X-ray tomography hinges on the fact that the decrease in intensity of a finely collimated X-ray beam depends on the integral of the X-ray density of the material traversed by the beam [3,4], as follows;

$$\ln I/I_0 = - \int \alpha(x,y)\,d\ell \tag{1}$$

where I_0 is the beam intensity at the source, I is the beam intensity at the receiver, and $\alpha(x,y)$ is the X-ray density along path ℓ. The method of data collection is illustrated in Fig. 1. A finely collimated X-ray beam is produced by X-ray source S, which can be translated so that the beam moves in a plane parallel to the paper. The X-ray intensity from the source is attenuated along path a and is detected at D. Successive measurements are taken for paths b, c, etc. in a single plane, until a complete scan has been made. The sequence of these measurements, called a projection, is shown in the lower portion of Fig. 1. These data are stored in a computer. Source S and detector D are rotated by $\Delta\theta$ degrees to a new position S'D' in the plane of the scan and a new projection is obtained. This is repeated, using small increments $\Delta\theta$, until 180° is spanned. Using a reconstruction algorithm, the computer calculates values of $\alpha(x,y)$ from the projection data and a gray scale image of $\alpha(x,y)$ is constructed. The resulting image represents values of $\alpha(x,y)$ in a thin slice of the object. (It has been implicitly assumed that $\alpha(x,y)$ is a scalar property of the object.) The quality of the image depends on the number of samples in a projection, the number of projections used, and the deblurring function used.

The following sections describe proposed methods of applying CT to ultrasonic reflection data. These concepts are presented here in the hope of stimulating their consideration by the reader.

PULSE-ECHO TOMOGRAPHY

The principles of X-ray computer tomography are applicable to acoustic imaging when acoustic data represent a line integral of a specific acoustic property of the object, similar to the integral given in Eq. (1). Consider the pulse-echo arrangement in Fig. 2a. A planar acoustic transducer emits a sheet acoustic beam aligned with the plane of the paper, thereby insonifying a thin slice of the object. The beam is made sufficiently wide to encompass the object's largest lateral dimension. A short acoustic pulse is emitted by the transducer which also acts as the receiver. Scattering centers in the object produce echo pulses which are detected by the receiver. Singly scattered echoes from scatterers at the same range arrive at the same time at the receiver and their sum appears in the output voltage. Such a sum can be viewed as the integral of the echoes from that range. Because the echoes from successively greater ranges arrive at the receiver at successively later times, the received echo train can be considered as a projection of these integrals where range has been replaced by time. One transmitted pulse generates data for one complete projection. The transducer is rotated around the object by an angle $\Delta\theta$ and another projection is recorded. This is repeated for the desired angular span. Using the projections so obtained, the computer reconstructs an image of the distribution of reflection centers in the insonified slice of the object. It is to be noted that the line integrals which we identify as projection data are in reality the sum of acoustic echoes as they return from the scatterers in the object. The echo from a given scatterer depends not only on the scattering strength but also on the round trip attenuation. The way each echo enters into the summing integral depends on the

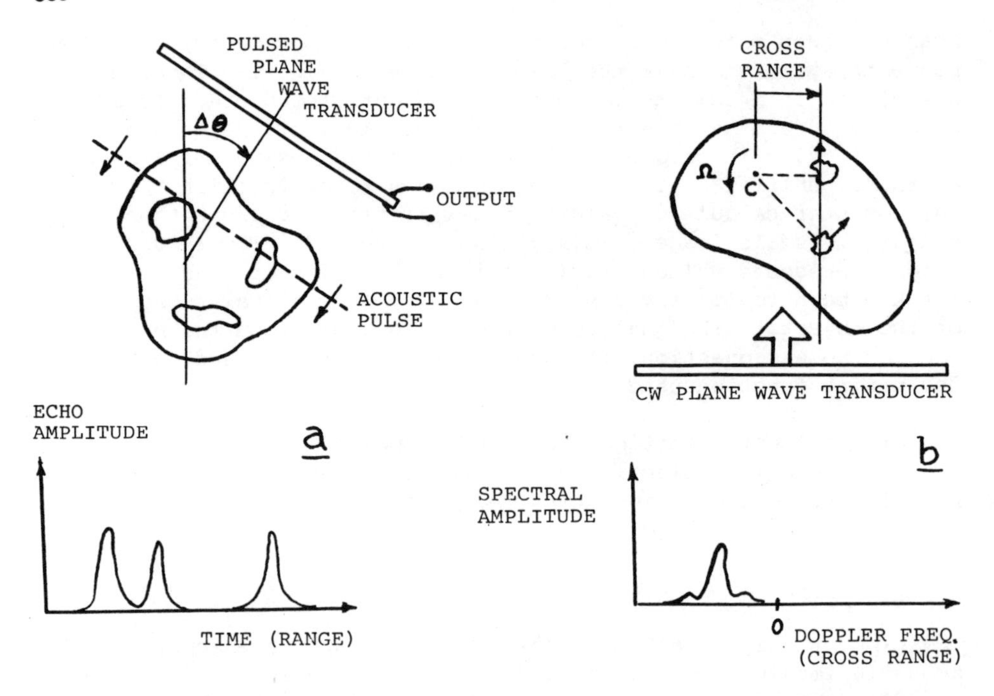

Fig. 2 Data acquisition in a) pulse echo tomography; b) Doppler reflection tomography.

shape of the pulse and the round trip travel time. In turn, these factors and others depend strongly on the material structure of the object being imaged. All such factors must be evaluated carefully and the experimental conditions chosen to minimize the deviation of the projection data integrals from the form of integral required in Eq. (1). Another approach which merits investigation considers using the randomness in the data caused by the factors mentioned above and including it purposely in the imaging scheme. Because the reconstruction process is linear, the summing of projections during reconstruction may produce the necessary averaging to negate the effects of the randomness. The latter approach is apparently used in a particular technique of moon imaging [7].

Because the line integrals for successively larger ranges are produced as a time sequence in the transducer output voltage, electronic sampling and storing of the data can be carried out. It is estimated that 100 or more line integral samples per projection are needed for a good quality image. In the method described, the electronic data collection can be at least two orders of magnitude faster than that achievable using a point-by-point mechanical scan. Furthermore, only a single transmit-receive transducer is required.

DOPPLER REFLECTION TOMOGRAPHY

When an object is in motion with respect to a stationary acoustic transmitter, scattering centers within the object return echos that are Doppler shifted in frequency by an amount that depends on the velocity of the individual scatterer. We propose to use this frequency shift to locate the scattering centers [6]. The data collecting principle is illustrated in Fig. 2b. The transmitting transducer is identical in output pattern to that in the pulse-echo tomographic system described above except that it emits a steady (CW) sinusoidal wave at frequency f_0. The object which is insonified by this sheet beam rotates at a uniform angular velocity of Ω radians/sec about an axis perpendicular to the plane of the beam. It is easy to show that scattering centers lying on a line of constant cross-range all have the same component of velocity normal to the transducer. Thus, at a given instant, each of these scattering centers reradiates the same Doppler-shifted frequency toward the receiving transducer which we place at the same position as the transmitting transducer. Measuring the receiver output at that frequency yields the sum or the line integral of the scattered radiation at that cross range. By measuring the amplitude of other frequencies which are in the receiver output, the line integrals are obtained for the scatterers at the corresponding cross ranges. The ensemble of integral data is interpreted as a projection. Allowing the object to rotate to a new angle, another projection is measured. As before, many such projections are taken until a body has rotated through 360°. The CT algorithm uses these projection data to produce an image of the distribution of scattering centers in the insonified slice of the object. (The earlier comments concerning the nature of the line integrals in the pulse echo case apply here as well).

To determine the constraints on these measurements, consider an object that can be enclosed in a circle of radius R and which is divided into N cross-ranges in each of which the Doppler-shift is to be measured. As the scatterers move through a cross-range strip of width 2R/N, they will radiate, on average, the Doppler-shifted frequency for that cross-range. In order for the scatterers to remain in the same cross-range strip during the measurement interval Δt, the shortest allowable crossing time must be greater than the measuring time. Thus,

$$\frac{2}{N\Omega} > \Delta t \tag{2}$$

Because one side of the object gives positive Doppler-shifts while the other side gives negative Doppler shifts, the overall span of Doppler-shift frequencies is $4R\Omega/\lambda$ Hz. Dividing this span into N values gives the resolution in frequency:

$$\Delta f = \frac{4R\Omega}{\lambda N} \tag{3}$$

To measure to this resolution in the interval Δt, we require

$$\Delta t > \frac{1}{2\Delta f} = \frac{\lambda N}{8R\Omega} \tag{4}$$

The number of projections to be taken will usually be greater than 100. Each Δt gives one projection, hence in one complete rotation of the object,

$$\Delta t < \frac{2\pi}{100\Omega} \tag{5}$$

This condition has the same variation with Ω as in Eq. (2). Sample calculations suggest that Eq. (2) will usually provide the dominant upper limit.

Finally, to insure that all scatterers in a cross-range band add together in the receiver over the greatest amount of the allotted measurement interval Δt, the differential travel time of all reflections in that band must be much less than this interval. Thus,

$$\Delta t > 10\,\frac{2R}{c} \tag{6}$$

where c is the acoustic wave velocity in the object.

The upper limit of Eq. (2) can be combined with the lower limits of Eqs. (4) and (6) to obtain the following upper bounds:

$$N < 4(R/\lambda)^{1/2} \tag{7}$$

$$\Omega < \frac{1}{10}\,\frac{c}{RN} \tag{8}$$

As an example, if R = 20 cm, λ = 0.05 cm (f_0 = 3 MHz), and c = 1500 m/s, then N must be less than 80. The bound in Eq. (8) requires that the rotational rate be less than 1.5 revolutions per second. This rotational rate represents a reasonable upper bound.

DEMONSTRATION EXPERIMENT

An experiment was performed to demonstrate the validity of the principles underlying the pulse-echo technique described above. The experiment consisted of measurements of reflections obtained from simple geometrical metal objects submerged in water

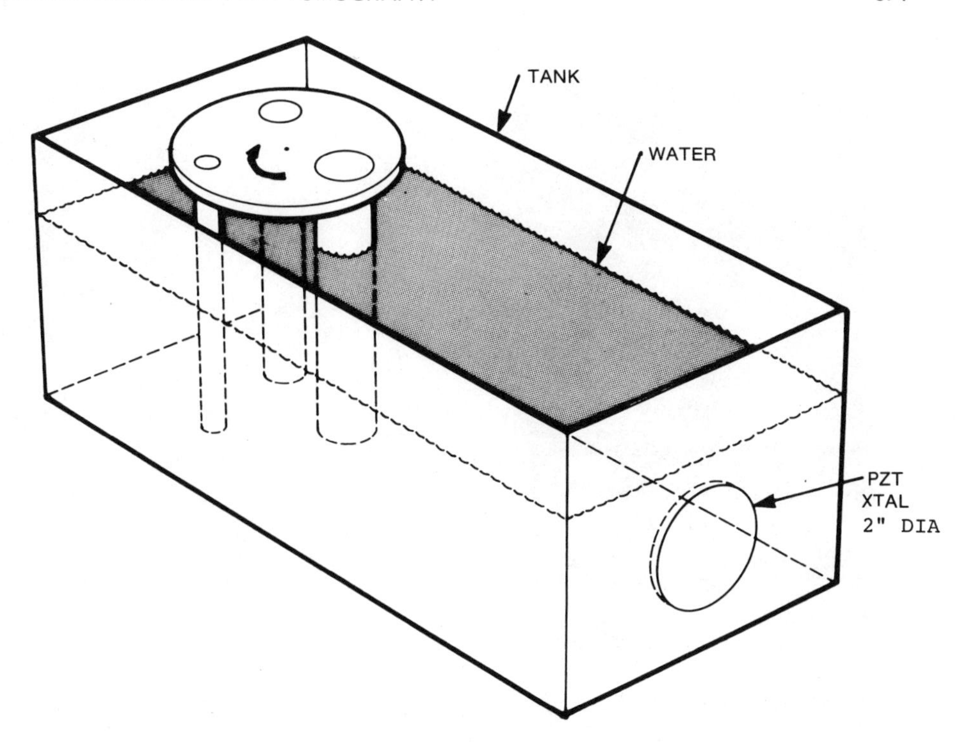

Fig. 3. Experimental arrangement to demonstrate pulse-echo tomography.

and insonified by short ultrasonic pulses. Both analog and digital data were taken and examined or used for computer processing. Image reconstructions were plotted by the computer in the form of gray scale tomograms and three-dimensional isometric plots.

The schematic view of the experimental set-up is shown in Fig. 3. The metal objects consisted of three right-circular brass cylinders having outside diameters of 1.27, 0.635, and 0.318 cm. The cylinders were 15 cm long and were mounted on a radius of 1.58 cm on a horizontal turntable with a scale calibrated in degrees. The cylinder assembly was placed in a tank, filled with water. At one end of the tank was a PZT-5 transducer which acted both as a source and a receiver of ultrasonic waves. The transducer had a diameter of 5 cm and had a natural resonant frequency of about 3 MHz.

The transducer was driven with a narrow pulse, and with electrical loading, a damped oscillatory pulse about 1 microsecond long was produced. The transmitted plane acoustic wave created

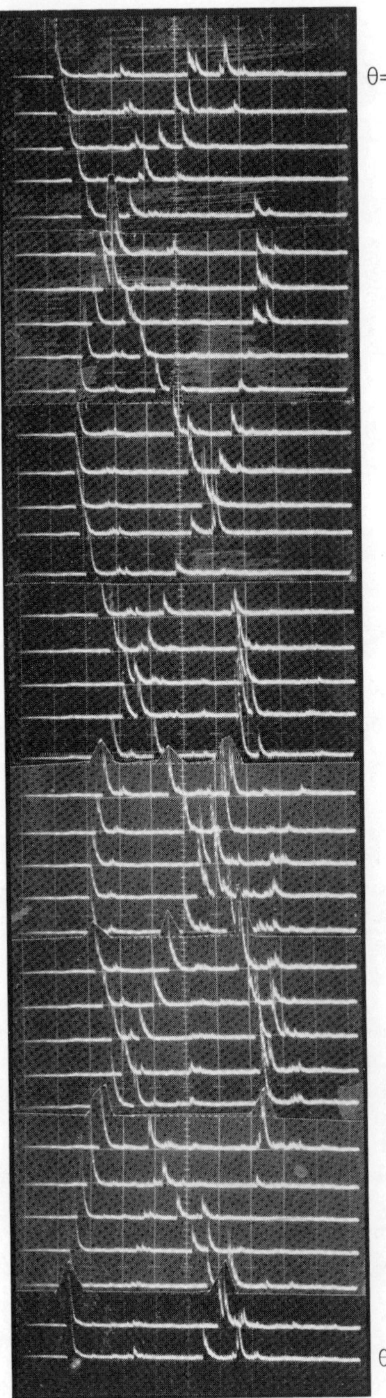

Fig. 4. An ensemble of 36 experimental reflection projections for the experimental objects shown in Fig. 3. A complete revolution of 360° is covered in 10° steps. The time scale of these oscilloscope traces is 7.7 microseconds per division which represents a range equivalent of 0.58 cm per division. The sinusoidal variation of range with turntable rotation is easily observed.

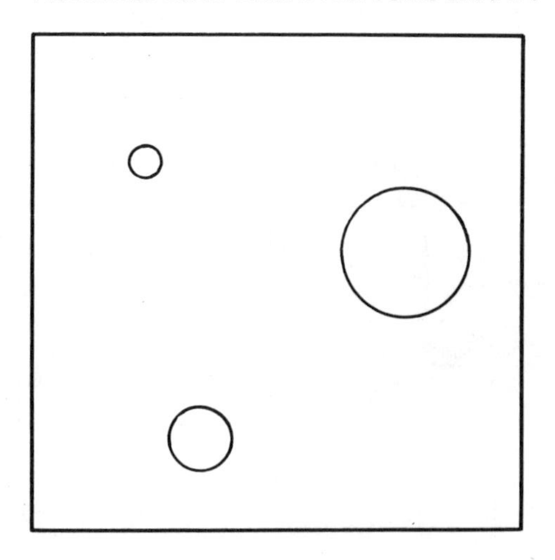

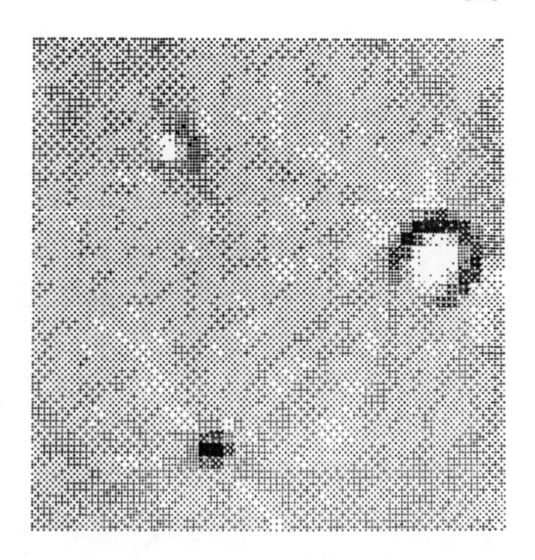

Fig. 5. Left: Ideal reflectivity tomogram for the experimental object. Right: Computer calculated tomogram using the experimental reflection data.

strong specular reflections from the cylindrical surfaces facing the transducer. (Due to the nearly perfect acoustic reflectivity of brass in water, the rear half of each cylinder was nearly completely shadowed by the front half during each acoustic pulse. Consequently, reflection data from the rear halves are missing in each projection, whereas, for proper image reconstruction, these data should be present.) The received echoes were detected with a peak-following detector. Fig. 4 shows an ensemble of analog records representing the reflected echoes obtained for a complete 360° rotation of the object assembly at incremental angles ($\Delta\theta$) of 10°. Each trace is interpreted as a projection. Inspection of these 36 traces shows the change in the echo time (range) with the position of the cylinders, the eclipsing of one cylinder by another, and echo overlapping. Additionally, the echos from a given cylinder vary due to changing range, non-uniform transducer sensitivity, and a slight tilt of the cylinders. The data were not corrected for any of these effects.

Computer data was taken using $\Delta\theta = 3°$, for a total of 120 projections. The projections were electronically sampled at 128 points along the time axis and quantized to 64 discrete levels in amplitude. The projections were used to calculate a 40×40 pixel image, following the tomographic algorithm of Shepp and Logan [5]. Ideally, the tomogram should appear as shown in the left of Fig. 5. The actual tomogram is shown in the right of Fig. 5. A 32-level gray scale is simulated by printing up to 64 black dots in each pixel. The locations of the three cylinders can be seen, but the loss of data has

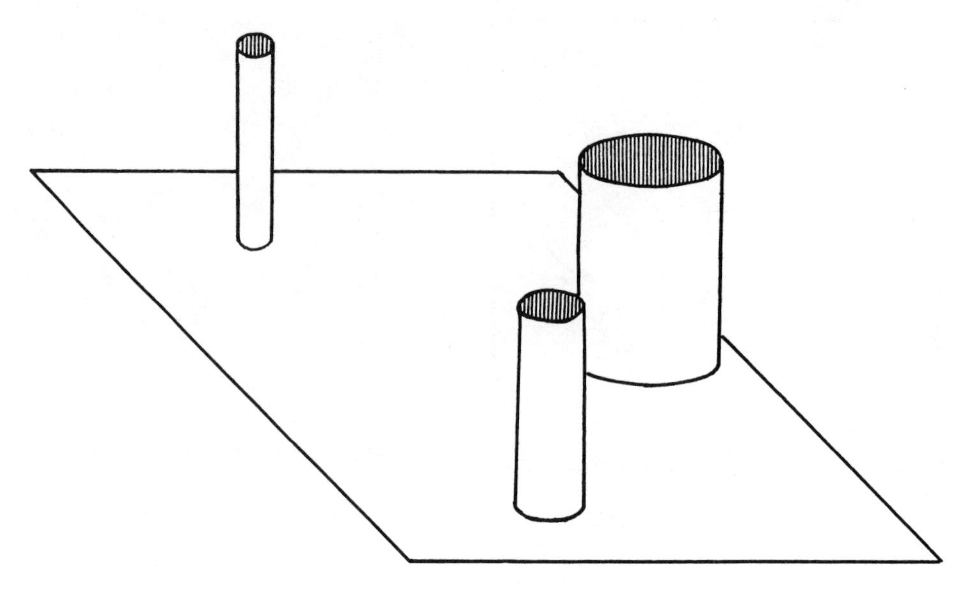

Fig. 6. Ideal isometric presentation of the reflection tomogram. The vertical coordinate represents the reflectivity distribution.

blurred their outlines, changed their size, and artifact noise is evident in the intervening water volume. It is informative to present the same data as an isometric plot, where the vertical coordinate represents the reflectivity distribution as a function of position. The ideal isometric reconstruction should be as shown as in Fig. 6. Using the experimental data, the computer calculates the isometric image as shown in Fig. 7. Using a non-linear compression of the vertical coordinate, the image is computer-processed by clipping the highest vertical amplitudes and compressing the lowest amplitudes to provide the image presented in Fig. 8. In Figs. 7 and 8, the 40 pixel data have been interpolated using a quadratic function and 960 points plotted to define each of 160 contours. These figures show a large amount of background artifact noise, the spreading of the walls of the cylinders, and a general departure from the ideal cylindrical shape. The large cylinder has a gap in one side where data were lost due to eclipsing by the medium sized cylinder. The smallest cylinder is spread out due to excessive eclipsing by the other two cylinders.

CONCLUSIONS

Principles have been presented for obtaining projection-like data using reflections of acoustic waves from scattering centers from within insonified objects. The particular algorithm used in

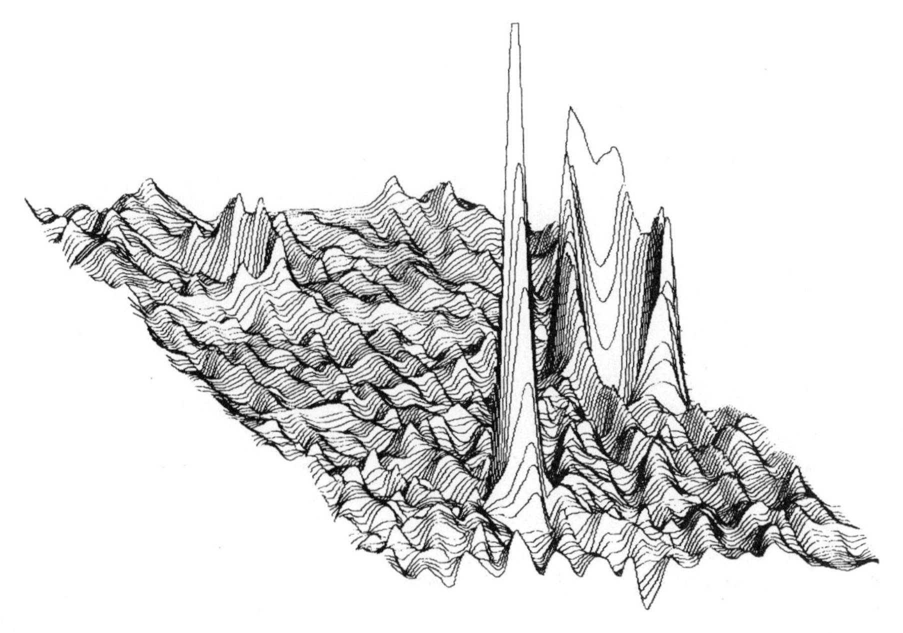

Fig. 7. Computer calculated isometric presentation of the reflection tomogram using experimental data.

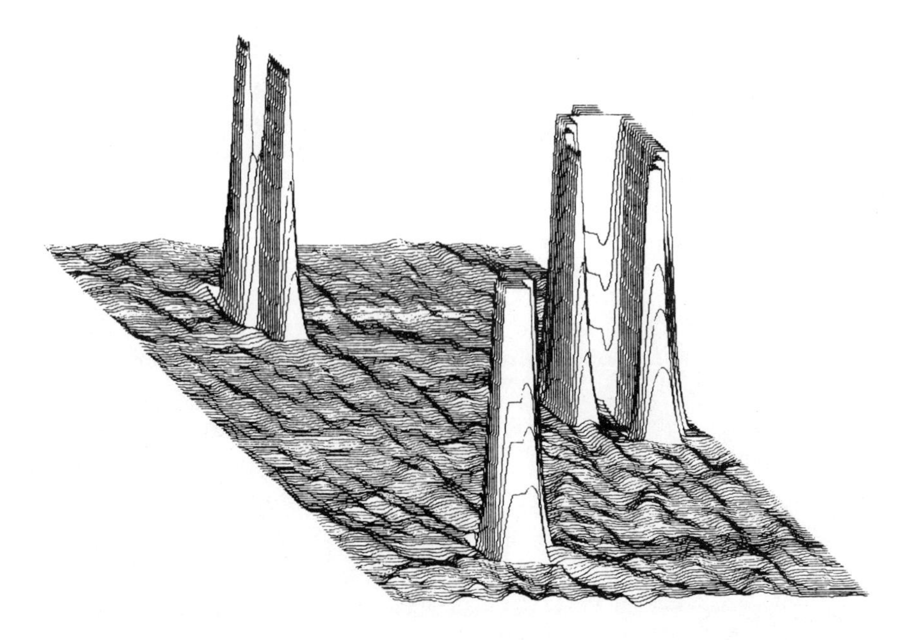

Fig. 8. Computer calculated isometric presentation of the experimental tomogram after simple image processing.

the X-ray CT technique does not appear to be directly applicable to all cases of reflection acoustic CT. Thus, the results of the preliminary pulse-echo experiment show considerable departure from ideal imaging and suggest that further development of specific algorithms for processing acoustic data is needed.

REFERENCES

[1] J.F. Greenleaf, S.A. Johnson, S.L. Lee, G.T. Herman and E.H. Wood, "Algebraic Reconstruction of Spatial Distributions of Acoustic Absorption Within Tissue from their Two-Dimensional Acoustic Projections," Acoustical Holography, Vol. 5, P.S. Green, Editor, Plenum Press, New York, 1974, pp. 591-603.

[2] J.F. Greenleaf, S.A. Johnson, W.F. Samayoa and F.A. Duck, "Algebraic Reconstruction of Spatial Distribution of Acoustic Velocities in Tissue from their Time of Flight Profiles," Acoustical Holography, Vol. 6, N. Booth, Editor, Plenum Press, New York, 1975, pp. 71-90.

[3] H.L. Baker, J.K. Campbell, D.W. Hauser, D.F. Reese, P.F. Sheedy and C.B. Holman, "Computer Assisted Tomography of the Head - An Early Evaluation," Mayo Clinic Proceedings, Vol. 49, January 1964, pp. 17-27.

[4] B. Gordon, G.T. Herman and S.A. Johnson, "Image Reconstruction from Projections," Scientific American, October 1975, pp. 56-68.

[5] L.A. Shepp and B.F. Logan, "The Fourier Reconstruction of a Head Section," IEEE Transactions of Nuclear Science, Vol. NS-21, June 1974.

[6] T. Hagfors and D.B. Campbell, "Mapping of Planetary Surfaces by Radar," Proc. IEEE, Vol. 61, No. 9, 1219-1225, Sept. 1973.

[7] A.K. Fund and R.K. Moore, "Effects of Structure Size on Moon and Earth Radar Returns of Various Angles," J. Geophys. Res., Vol. 69, pp.-1075-1081, March 1964.

ALGEBRAIC AND ANALYTIC INVERSION OF ACOUSTIC DATA FROM PARTIALLY OR FULLY ENCLOSING APERTURES

Steven A. Johnson, James F. Greenleaf, Balasubramanian Rajagopalan, and Mitsuo Tanaka

Department of Physiology and Biophysics
Biodynamics Research Unit, Mayo Foundation
Rochester, Minnesota

ABSTRACT

Several methods are presented for inversion of both reflection and transmission data (i.e., produce an image from the data) collected from transducers distributed on a sphere or portions of a surface surrounding a distribution of scatterers. These methods are based on finding approximate forms of the true generalized inversion of the linear operator which transforms the particular distributions of scatterers, transducer arrangement, and transmitted waveforms into the corresponding collected data. Thus, reflection and transmission data collection may be written as linear operations AMx=b where A is an attenuation operation and M is a time varying spatial integrating operator, b is data and x is the desired image. One such iterative method is presented which uses a Kacmarz-like inversion technique which is analogous to the ART reconstruction algorithm used in x-ray reconstructive tomography. This iterative method is well suited for partially enclosing apertures. Progress in formulation of non-iterative or analytic methods is presented. These methods are analogous to the Radon inversion formulae for x-ray density reconstruction. A comparison is made of the images produced by these new methods and the "synthetic focus" method previously developed in our laboratory.

INTRODUCTION

A new method for obtaining images of distribution of acoustic parameters has been formulated (1). This method is applicable for use with arbitrary apertures. The new method which we propose, however, does not model any image forming process but rather models the processes by which the scattered acoustic energy, which is detected or received, is produced by the interaction between incident waves and the object to be imaged. In this paper, we seek an inverse transformation which maps the incident energy fields and the experimental data into an image of the object. The accuracy and fidelity of the resulting image will depend on several factors: 1) the reliability of the data collection apparatus; 2) the accuracy and completeness of the interaction model; 3) the nature of the inverse transform or its approximation; and 4) the accuracy of the inversion algorithm. The process of data formation may be represented by the following equation:

$$E(\Psi)x = d \tag{1}$$

where $E(\Psi)$ is the experimental set of incident acoustic energy fields used to probe the object x, and d is the corresponding set of experimental data. The inversion transform $[E(\Psi)]^{-1}$ may be used to find x in terms of d and this operation is written:

$$x = [E(\Psi)]^{-1}d \tag{2}$$

Methods for obtaining x from $E(\Psi)$ and d, belong to one of two classes: 1) closed form solutions such as an explicit form for $[E(\Psi)]^{-1}$ and 2) iterative methods of solving the equation $E(\Psi)x = d$ for x.

Imaging methods based on data modeling would possess the following advantages over focus modeling techniques: 1) imaging of all ranges is possible and no distinction need be made of near or far field regions; 2) deconvolution of the effect of change in the frequency content of the data due to frequency dependent attenuation and scattering is possible in the image forming process; and 3) the method permits a priori information and side contraints to be introduced by additional equations. For example, the equation $E(\Psi)x = d$ may be supplemented by boundry condition or constraint equation $Cx = b$. By this means, continuity of surface and areas, limits on values of parameters, the modeling of noise, and other features may be incorporated into imaging process. For example, the continuity equation for non-compressible flow $\nabla \cdot \vec{V} = 0$ could be used to provide improved accuracy in Doppler flow measurements.

The task of producing images by the data modeling method can be divided into two important steps; 1) accurate modeling of the acoustic interaction and data collection process and 2) the use of accurate, and hopefully effecient, inversion algorithms. The first of these two steps will be explored next.

THEORY FOR MODELING OF ACOUSTIC INTERACTION AND DATA ACQUISITION

All data acquired is dependent on the acoustic interaction between the incident field and the geometrical configuration and other properties of the data collection system. These considerations should be incorporated into the mathematical model. We begin with the assumption that the D'Alembertian wave equation is a reasonable approximation for most imaging problems (extensions to more complete versions of the wave equation will be treated briefly later in this discussion).

$$\nabla^2 \Psi(\vec{r}) - \frac{1}{C(r)^2} \frac{\partial^2 \Psi(r)}{\partial t^2} = -4\pi S(r,t) \quad (3)$$

where $\Psi(\vec{r})$ is the total acoustic field $C(\vec{r})$ is the speed of sound and $S(\vec{r},t)$ is the distribution of point sources. We shall make use of a mathematical device to find an approximate solution to the problem of scattering from a distribution of point scatterers. This device is simply to replace each point scatterer with a point source from which a spherical wave is retransmitted whose phase is identical to the incidence wave as measured at the point scatterer. By this means, Equation 3 above may be made to model scattering under the assumption that the correct value of the incident wave at each scattering point is given by the unperturbed incident wave. This assumption is the essence of the Born approximation (suggestions for accounting for the removal of energy from the incident wave by absorption and attenuation due to scattering will be treated later in our discussion). Under the above assumption, the solution to Equation 3 is well known and is

$$\Psi_o(\vec{r},t') = \Psi \text{ incident} + \int G_o(\vec{r}',t',\vec{r}_o',t_o') \cdot S_o(\vec{r}_o',t_o') d^3\vec{r}_o dt_o' \quad (4)$$

where G_o is the Green's function given by

$$G_o(\vec{r}',t',\vec{r}_o,t_o') = \delta\left(t_o' - \left(t' - \frac{|r_o - r'|}{C}\right)\right) \Big/ |\vec{r}_o' - \vec{r}'| \quad (5)$$

and where G_o is the solution to Equation 1 when

$$S(\vec{r}',t') = \delta(\vec{r}' - \vec{r}_o')\,\delta(t' - t_o') \quad (6)$$

The function $t = (t' - |\vec{r}_o' - \vec{r}'|/C)$ is the retarded time and represents the time t when a wave which is observed at $\vec{r}'$ at time t' actually starts from source point at $\vec{r}_o$ for a homogeneous wave propagation velocity $C(\vec{r})$.

Referring now to Figure 1, we derive the expression for the signal received at an arbitrary receiver when an arbitrary transmitting transducer transmits into a distribution of scatters.

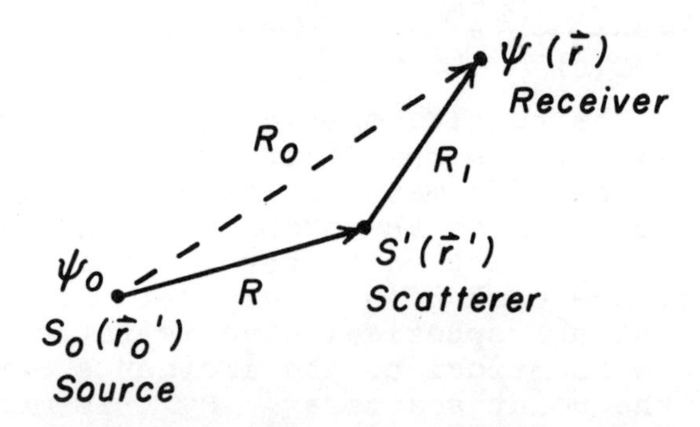

Figure 1 - Illustration of the geometrical relationship between a source point S_o at $\vec{r}_o'$ which generates input wave Ψ_o, a scattering point S' at $\vec{r}'$ and a receiver at point r which records wave $\Psi(\vec{r})$ produced by S' as well as original wave Ψ_o produced by S_o. Note $(\vec{R}+\vec{R}_1)/C = [(R^2+R_o^2-2RR_o \cos(R,R_o))^{1/2}]/C$ is retarded time interval.

We first sum over all sources as seen at point $\vec{r}'$. Equation 4 is written in the proper notation to describe this operation and next sum over all scatters as seen at receiver point r. The results is

$$\Psi(\vec{r},t) = \Psi_o(t,t) + \iint S'(\vec{r}',t')G_1(\vec{r},t;\vec{r}',t') \cdot S_o(\vec{r}_o',t_o')G_o(\vec{r}',t';\vec{r}_o',t_o')d^3\vec{r}'dt'd^3\vec{r}_o'dt_o' \quad (7)$$

We next include the effect of finite length (non-impulsive) transmitted waveforms. This is done by convolving the solution $\Psi(\vec{r},t)$ with actual transmitted waveform $b(t)$. The resulting received pressure $P(r,t)$ at r is given by

$$P(\vec{r},t) = \int \Psi(\vec{r},t_2)b(t-t_2)dt_2 \quad (8)$$

If a finite area receiving transducer D is used then the time signal from this detector is given by

$$D(t) = \int P(\vec{r},t)d^3\vec{r} \quad (9)$$

These integrals may be combined in a straight forward manner to give a complete expression for $D(t)$ in terms of $S'(r)$ and the transmitted waveform. For the case of a point transmitter and a point receiver, the expression is

$$P(\vec{r},t) = \int \Psi_o(\vec{r},t_2)b(t-t_2)dt_2 + \int S'(\vec{r}')g_o(\vec{r}_o-\vec{r}') \cdot g_1(\vec{r}'-\vec{r})\delta\left(t' - \left[t_2 - \frac{R+R_1}{C}\right]\right)b(t-t_2) \cdot dt_2 d^3\vec{r}' \quad (10)$$

where $g_o = \delta(|\vec{r}_o-\vec{r}'|)/|\vec{r}_o-\vec{r}'|$ and $g_1 = \delta(\vec{r}'-\vec{r})/|\vec{r}-\vec{r}'|$

We direct our attention to deriving a discrete or matrix version of the above equations which may be more easily implemented on a digital computer.

DISCRETE VERSION OF ACOUSTIC INTERACTION MODEL FOR LOSS LESS, HOMOGENEOUS REFRACTIVE INDEX MEDIA

The above integral representations will now be placed into a discrete form which may be represented in matrix formation. Equation 10 may be written:

$$\Delta P_{j,m,t} = P_{j,m,t} - \sum_{t_2} b_{t-t_2} \quad \Psi j,m,t_2$$

$$= \sum_{t_2} \sum_{t'} \sum_{k} b_{t-t_2} \quad S_k \, g_{o,m,k} \, g_{1,m,k} \, \delta_{t' - t_2 - \frac{R+R_o}{C}} \tag{11}$$

Here indices j,m,t,k refer respectively to receiver position, transmitter position, receiver time, and scattering point position. The right hand side of Equation 11 may be rewritten as

$$\Delta P_{m,j,t} = \sum_{k} \sum_{t_2} \sum_{t'} g_{o,m,k}, \, g_{1,k,j} \, b_{t-t_2} \cdot \left(\delta_{t'} - \left(t_2 - \frac{R + R_o}{C}\right)\right) S_k \tag{12}$$

The above equation may be written in matrix form by combining the amplitude terms g_o and g_1 into one matrix A_1 and combining the time operators b and δ into one matrix T_1. Let the measured pressure data P be given by d_1. And the unknown scattering distribution S_k by x. Then we may write

$$A_1 T_1 x = d_1 \tag{13}$$

Clearly, d_1 may be written as a column vector whose indices vary most rapidly in time, then with receiver position, and lastly with transmitter position. Also, x is a column vector whose indices describe all points in a three-dimensional grid superposed on the object to be imaged. A focusing model solution to Equation 13 has been given previously by Johnson et al. and was termed synthetic focus imaging (1-5).

While Equation 13 may be solved by iterative methods such as the Kacmarz (6-8), the Gauss-Jordan (9) or other methods, it may be more economical of computation time to modify the waveforms in vector d_1 to correspond to an impulse transmitted waveform in order to increase the sparseness of matrix T_1. Increased sparseness of T_1 allows a Kacmarz method such as a curved line ART (method) to operate at greater time effeciency. Let the data with impulse waveform be represented by d_o and the corresponding time selection matrix by T_o, then we may write for this special case

$$A_1 T_o x = d_o \tag{14}$$

It is clear that $d_1 = d_o * b$ and $T_1 = T_o * b$ where b is any finite duration transmitted waveform and the operation * represents convolution. It is also clear that d_o may be generated from d_1 by an appropriate deconvolution technique. One suggested deconvolution technique which has been tested in our laboratory with several test waveforms in the presence of noise is a modified ART technique which solves the equation $b * d_o = d_1$ for d_o, given b and d_1, by successive iterations as the relaxation parameter is allowed to converge to zero like terms of a series such as (1, 1/2, 1/3, 1/4, ...) or (1, 1/2, 1/4, 1/8, ...). The Kacmarz method tends to converge to a least squares, noise tolerant solution by gradually reducing the iterative corrections to the solution (10). Images obtained from solving Equations 13 and 14 for simulated data are given in a later section.

EXTENSION OF MODEL TO INCLUDE ATTENUATION AND INHOMOGENEOUS REFRACTIVE INDEX

Such an extension should start from a wave equation which includes dissipation, however, we proceed by heuristic means and modify the solution given by Equation 10. We note that the distance $|\vec{r}'-\vec{r}|$ may be generalized to $\int ds$ integrated along a ray path from $\vec{r}'$ to $\vec{r}$ and note that the factor $(1/|\vec{r}-\vec{r}'|)$ which describes loss of amplitude at $\vec{r}$ due to spreading of a spherical wave originating at $\vec{r}'$ may be multiplied by a term $\exp(-\int \alpha ds)$, with the integral taken from r to r'. Dissipative losses due to attenuation is included by linear attenuation coefficient α. Here we have omitted the frequency dependence of attenuation. This may not cause great difficulties in modeling narrow or medium bandwidth systems. The modified Green's functions corresponding respectively to G_o and g_o:

$$G_2(\vec{r}',t';\vec{r}_o',t_o') = g_2(\vec{r}_o'-\vec{r}')\,\delta\left(t_o - \left[t' - \int_{\vec{r}_o'}^{\vec{r}'} \frac{ds}{c}\right]\right). \tag{15}$$

$$g_2(\vec{r}_o-\vec{r}') = \exp\left[-\int_{\vec{r}_o'}^{\vec{r}'} \alpha(S)\,dS\right] \Big/ \left(\int_{\vec{r}_o'}^{\vec{r}'} ds\right)^{n/2} \tag{16}$$

Here n = 0 for plane waves, n = 1 for cylindrical waves and n = 2 for spherical waves. G_3 and g_3, the new G_1 and g_1, respectively, are given by similar expressions except that only spherical waves are scattered by a point (i.e., n = 2 for G_1 and g_1). Discrete equations similar to Equation 10 may be written using the generalized Green's functions of Equations 14 and 15. Matrix equations similar to Equations 13 and 14 may also be written with $A_2(g_2, g_3)$ replacing A_1 and $T_2(b, \delta(\int ds/C))$ replacing T_1. Thus,

$$A_2 T_2 x = d_1 \tag{17}$$

A solution to Equation 17 based on a focusing model which has been tested with exemplary images from simulated data has been given previously by Johnson et al. (1-5) and is termed "synthetic focusing". This solution technique is analogous to the convolution backprojection technique and this analogy has been described by Johnson et al. (4,5). More recently other writers (11-13) have also described this analogy and presented similar algorithms for the case of non-dissipative constant refractive media. Extension to the case of frequency dependent attenuation has also been treated and a suggested solution based on a focusing model has been given by Johnson et al. (1,5).

EXPERIMENTAL EVALUATION OF IMAGING METHODS WITH SIMULATED DATA

The above theory has been partially tested by producing images from simulated data of distributions of point scatters. The simulated data was assumed to be noise free. The method of matrix inversion is presented in the appendix. The two-dimensional amplitude impulse response function of a single scatterer is shown in Figure 2 and the corresponding amplitude trace through the center of the impulse response along a line parallel to the x axis is shown in Figure 3. The data were obtained by simulating a circular array of transducers arranged around 360° in a single plane and assuming the scattering point lies in this plane. Transmission and reception from a common transducer was assumed. The transmission of a unit impluse was assumed. Thus, the equation $A_1 T_o x = d_o$ was solved for the image x. Only one iteration over all the data was made in forming the solution images in Figures 2 and 3. A Kacmarz algorithm was used which updated the image by adding corrections along wave fronts for all discrete time samples and for all transducer positions. The sampling rate was assumed to be 10 samples per microsecond.

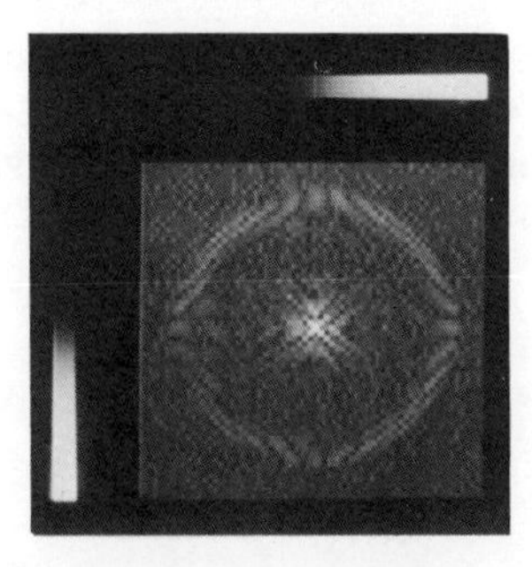

Figure 2 - Computer generated gray scale image of the modulus (absolute value) of the solution vector x in the equation $A_1 T_o x = d_o$ as defined in the text. The image is 64 x 64 pixels in area and the length of the side of the image is 1.92 cm. One iteration of ART-like algorithm is shown. The white line corresponds to the sample amplitude shown in the next figure. Data collection geometry and features are given in the text.

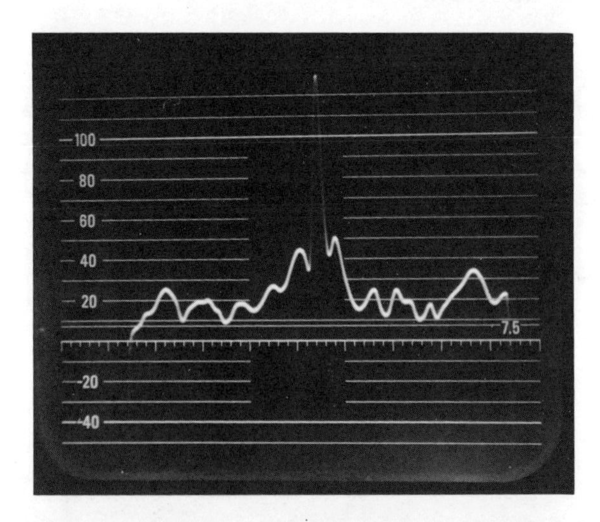

Figure 3 - Impulse response function of the image of a single scattering point as shown in Figure 2.

The effect of imaging multiple scatters and a study of limiting resolution was also tested using the ART-like algorithm. The simulated data was noise free. Simulated data was generated to correspond with six scattering points separated by 0.6, 0.9, 1.5, 3.0, and 6.0 millimeters, respectively, and arranged on a single line is shown in Figure 4. A sampling rate of 10 samples per microsecond was also assumed. A wide band 1.0 MHz transducer was simulated. Sampled image magnitude along this line of points is shown in Figure 5. All scattering points are resolved in the figures.

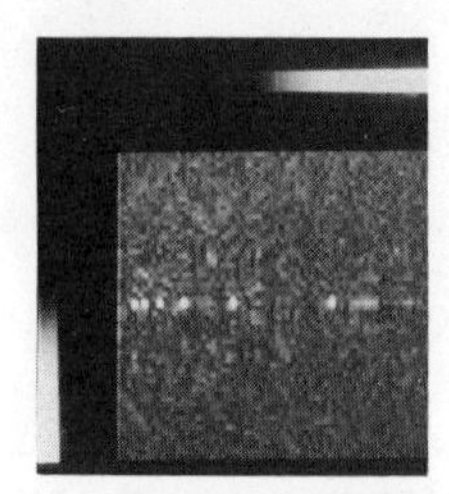

Figure 4 - Computer generated gray scale image of the modulus (absolute value) of the solution vector x in the equation $A_1T_ox = d_o$ as defined in the text, for six scattering points on a common line. The same data collection geometry and features as that used for Figure 2 are assumed.

The effect of inconsistancy between the waveform in the data d and the waveform b as contained in matrix T was explored. Data for a damped sine wave $\exp(-t/t_d) \cdot \sin(\omega t)$, with $t_d = 4$ µs and $2\pi/\omega = 1$ µs was used for data d, while an impulse wave was assumed for matrix T.

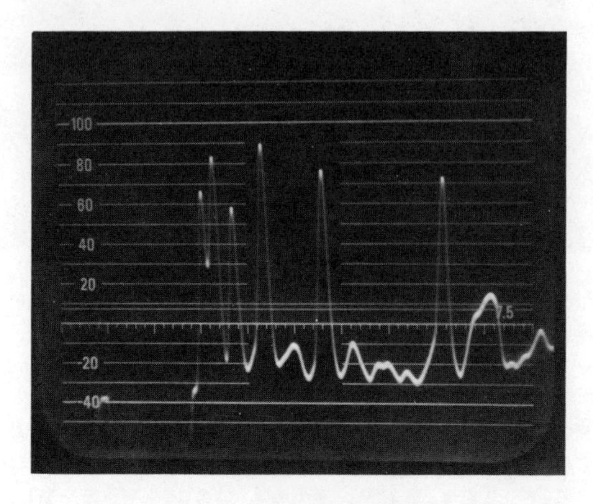

Figure 5 - Sampled magnitude of the image in Figure 5 as sampled along the line which is common to all six points.

Following the previous notation this inconsistency is expressed by the equation $A_1T_ox = d_1$. When a solution vector x for this equation is found by the ART-like (i.e., Kacmarz) method, the corresponding image may be displayed. Data was simulated for five scattering points arranged on a common line. This arrangement is identical to the arrangement in Figures 4 and 5 except the left most point (the end point with the nearest neighbor) was not included. The amplitude image of the solution vector x for $A_1T_ox = d_1$ is shown in Figure 6. The five scattering points are clearly visible but the inconsistancy between T_o and d_1 has introduced a textured background of amplitude about 20% of the peak values at the scattering centers. This experiment suggest that slight to moderate inconsistancies between T_o and d_1 may be tolerated in some cases when using the iterative inversion algorithms.

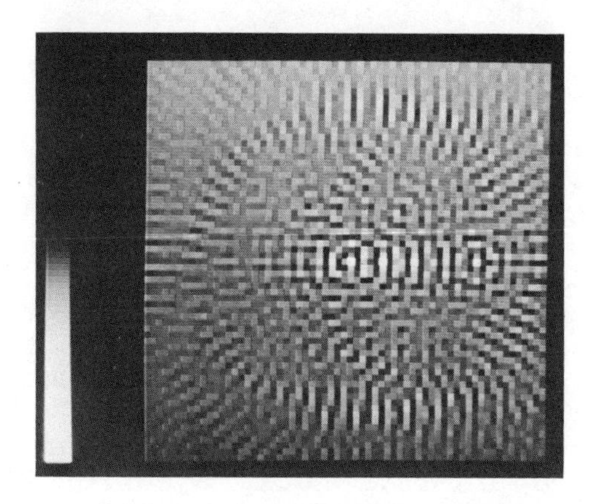

Figure 6 - Computer generated gray scale image of a linear arrangement of scatters made with an inconsistant model of the transmitted waveforms. A 360° transducer arc and the same data collection geometry and features as used in Figures 2 and 4 are assumed.

The images and point reponse functions shown above which are generated by inverting an interaction model should be compared with those produced by applying a focusing model imaging algorithm to the same data. The image of the impulse response function of our synthetic focus algorithm when applied to the same data used to generate Figure 2 has been reported earlier but is shown here in Figure 7 for completeness (1,5). Our synthetic focus algorithm was adapted using two different kernels, the Ram. Lak. kernel (4) and the Tanaka-Iinuma kernel (1,5). The data d_o was convolved with these kernels and backprojected by use of the synthetic focus algorithm. Images from this experiment as shown in Figure 7 also show the difference between displaying magnitude or amplitude image of the same picture vector.

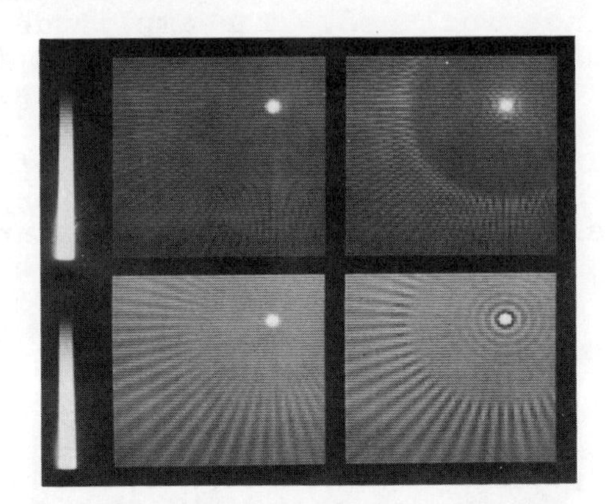

Figure 7 - Images of the two-dimensional point response functions produced by a reflection tomographic technique termed "synthetic focus imaging" by our laboratory. The algorithm is based on a model for producing focused images which backprojects the reflection data along circular wave fronts. A common transducer for both transmission and receptions is assumed for 60 transducer positions arranged uniformly on the circumference of a circle inscribing the scattering point. The scattering point lies in the plane of the transducer circle. The data is modified prior to backprojection to correspond to that which would be produced by convolution of the data produced by use of a unit impulse with a deblurring kernel. Left column: images from Tanaka-Iinuma kernel. Left column: images from Ram. Lak. kernel. Top row: modulus images. Bottom row: amplitude images.

PARTIALLY ENCLOSING APERTURES

The effect which selection of kernel type has on impulse response function using partially enclosing apertures is shown in Figure 8. In this study, the effect of transmitting the Tanaka-Iinuma kernel, a damped sine wave kernel, and the Ram. Lak. kernel is shown by the resulting images of amplitude and modules

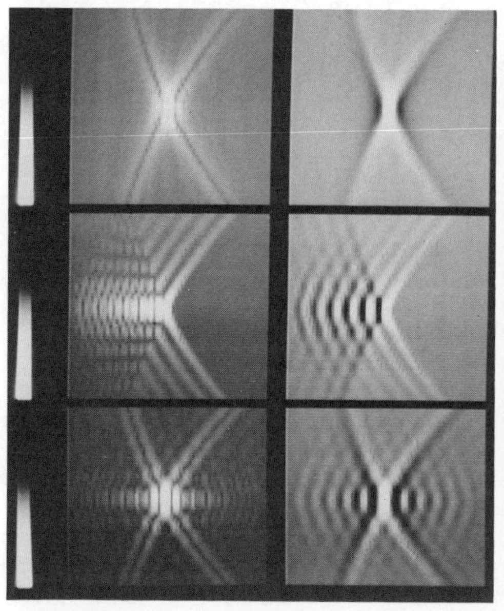

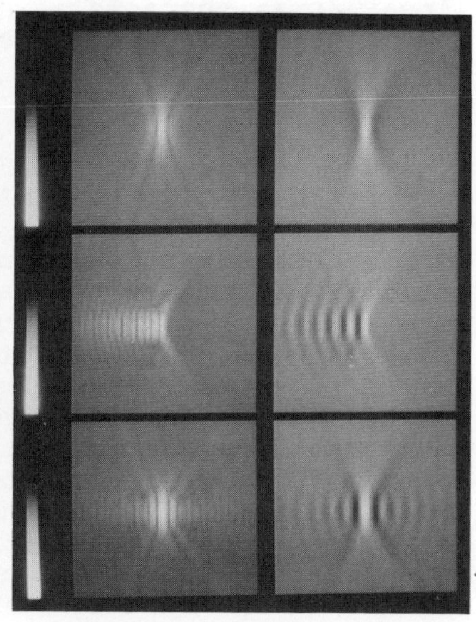

Figure 8 - Computer generated images of the 2-D impulse response function for a scatter located on the perpendicular bisector, one array length from the center of a linear array. Top row: Tanaka-Iinuma kernel. Middle row: damped sine wave. Bottom row: Ram. Lak. kernel. Left column: modulus with no apodizing. Middle left column: amplitude with no apodizing. Middle right column: modulus with apodizing. Right column: amplitude with apodizing. Each image 64 x 64 pixels and 0.96 cm per side. The apodizing function is a cosine function which is unity at center of array and zero at one element beyond ends of array.

of a single scatterer. The scatterer is located 2.4 cm from a linear array of 32 elements and of length 4.8 cm. A maximum frequency spectral content of 1 MHz is assumed for the Ram. Lak. kernel. A frequency of 1 MHz for the damped sine wave and an average frequency of 1 MHz (2 MHz upper limit) for the Tanaka-Iinuma kernel is assumed to provide nearly equal full width at half max, for the respective impulse response functions. The array is assumed to be at the right of each image. The algorithm used is of the focus modeling type (synthetic focus) and is identical to that used to form the images of Figure 8. The effect on the image point response function of apodizing the linear array response is also shown in Figure 8. Apodizing suppresses the edges of the bow tie point response at the cost of a slight loss of resolution.

NON-ITERATIVE TECHNIQES FOR INVERSION OF ACOUSTIC INTERACTION AND DATA ACQUISITION MODELS

Models of acoustic interaction based upon perturbation solutions of the wave equation which have closed form inversion formulas for plane wave illumination have been given by the authors (5). These formulas are based on similar techniques developed for use with optical data (14). Meuller and Kaueh have also proposed these and other perturbation solutions to the wave equation as a means for obtaining refraction and diffraction free transmission tomographic images of complex refractive index distributions (15,16). Ball and Johnson have obtained a preliminary perturbation solution to the wave equation and a corresponding refractive index inversion formula in three dimensions for incident spherical waves (17).

SUMMARY

Ultrasonic imaging schemes may be classified into two groups, namely those based on applying a focusing model and those based on inversion of an acoustic interaction data collection model. In this paper we have directed our attention to imaging techniques which rely upon sampling the time history of waves which are incident upon an arbitrary aperture consisting of a dense network of sampling locations. Several years ago the authors developed an imaging technique for sampled apertures which was based upon a focusing model and this technique was termed "synthetic focusing" (2). We later showed that the synthetic focus technique can be thought of as an analog to the x-ray reconstruction tomography (i.e., x-ray CT) algorithm known as convolution filtered backprojection (4). Other workers have also recognized this analogy and have given their own versions of this method a second name (i.e., reflection tomography) based more directly on the analogy (11,12). Images from real data have also been obtained in our laboratory by use of this algorithm (1-5). The application of this algorithm has been extended through trials with simulated data in our laboratory to permit corrections for refraction and attenuation (1,5).

The previously developed focus modeling algorithms are now supplemented by acoustic interaction and data collection modeling algorithms which are described in this paper. This technique is analogous to the ART method used in x-ray reconstruction tomography (which chronologically preceeded the convolution filtered backprojection technique. Thus, the analogy between x-ray and ultrasound imaging is strengthened by the addition of the second analogous method. The ultrasound ART-like iterative method may find application in those cases where additional constraint equations are required, where the waveform must be deconvolved from the image (i.e., obtaining optimumly sharp resolution), or where limited numerical aperture (i.e., non-enclosing apertures) must be used.

APPENDIX

The Kacmarz Method

The Kacmarz method is a procedure for solving a finite system of linear equations $E_j f = g_j$ in a Hilbert space by starting with an initial guess f_o and successively improving the guess by the use of the equations individually. If the present guess is f_n and the next equation to be used is $E_j f = g_j$, the prescription for improvement is as follows: if N_j is the null space of E_j, then f_{n+1} is the orthogonal projection of f_n on $f+N_j$. The procedure converges to the orthogonal projection of the initial guess f_o on the plane f+N, where N is the intersection of all the null spaces. Thus, in the usual case where $f_o = 0$ the procedure converges to the minimum norm solution of the system of equations. The carrying out of the procedure depends, or course, on finding formulas for the individual planes $f+N_j$. Let $N_j^{\perp}$ represent the orthogonal complement of N_j. See Figure A1 for the vector space relationships between these entities for specific source j = a.

It follows from the definitions that $(f_{n+1}-f_n) \in N_j^{\perp}$, while $(f-f_{n+1}) \in N_j$, so that

$$f-f_n = (f-f_{n+1}) + (f_{n+1}-f_n) \tag{A1}$$

gives the orthogonal decomposition of $f-f_n$ into a part in N_j and a part in $N_j^{\perp}$. Consequently, a second characterization of f_{n+1} is that

$f_{n+1}-f_n$ is the orthogonal projection of $f-f_n$ on $N_j^{\perp}$. (A2)

Thus, the computability of f_{n+1} from f_n involves two things: I) An explicit formula for the orthogonal projection of an arbitrary vector h on $N_j^{\perp}$. II) The fact that when the formula is applied to $h = f-f_n$, f itself is involved only through the known data $E_j f$.

In the present ultrasound application the situation is as follows. There is given a fixed bounded open set B inside which all objects under consideration lie, and the Hilbert space is the space $L_0^2(B)$ of square integrable functions vanishing outside B. It is assumed that each object is characterized by a point scattering function f in this Hilbert space and that the ultrasound scan data arising from the source a (using a, both as the source and receiver) is the function

$$[E_a f](r) = \int_{S^{n-1}} f(a+r\theta)A_a(a+r\theta)d\theta \quad \text{(A3)}$$

where S^{n-1} is the unit sphere, θ is the generic point of this sphere, and $d\theta$ is the element of surface area. For example, if the dimension n is 2, S^{n-1} is a circle in the plane, and $d\theta$ is arc length on the circle. Thus, r is a scalar while a and θ are vectors. The vector a defines the source and the vector $r\theta+a$ is a point on the wave front. The function $A_a(r,\theta)$, which depends upon the ultrasound attenuation, is assumed, as a first approximation, to be known and to be independent of f. We consider the specific step in the Kacmarz method where data from source a is to be used to update the image of unknown f. Set j = a then $E_a f$ is the data from source a. See Figure A2 for the geometrical relationships between these variables in a typical experimental apparatus.

It is easily shown by the methods of (6-8) that if N_a is the null space of E_a, then $N_a^{\perp}$ consists of all function h' in $L_0^2(B)$ of the form

$$h'(a+r\theta) = c(r)A_a(a+r\theta) \text{ when } a+r\theta \in B, \quad \text{(A4)}$$

and, moreover, that if h is any function in $L_0^2(B)$, then the orthogonal projection of h on $N_a^{\perp}$ is given by (A4) with

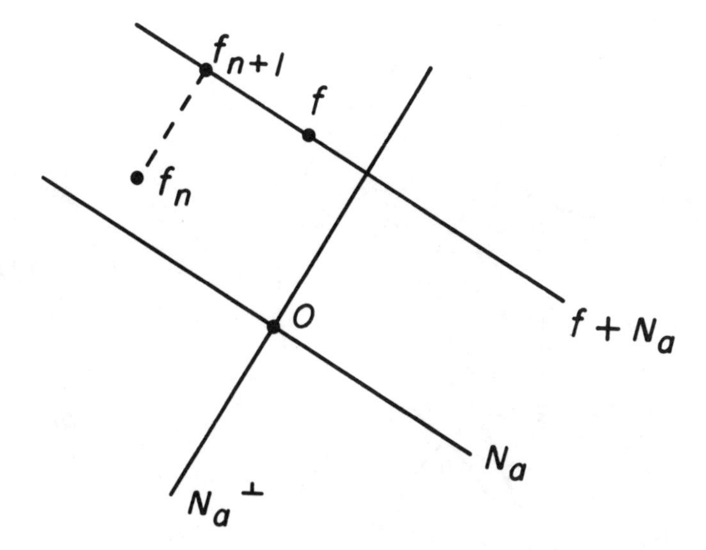

Figure A1 - Representation of null space N_j, the orthogonal complement $N_j^{\perp}$, the origin O, the actual solution f, the orthogonal projection from guess f_n to update guess f_{n+1} for one iterative complete step in the Kacmarz method.

$$c(r) = \frac{[E_a h](r)}{\int_{S_{a,r}} A_a(a+r\theta)^2 d\theta}, \text{ where} \tag{A5}$$

$$S_{a,r} = \left\{ \theta \varepsilon S^{n-1}: \quad a+r\theta \ \varepsilon B \right\}. \tag{A6}$$

It follows from this that the Kacmarz correction called for by the use of the source a is given by

$$f_{n+1} = f_n + h' \tag{A7}$$

where h' is defined by formulas (A4) and (A5) with $h = f - f_n$. The requirement II) above is clearly satisfied.

Note that $E_a h = E_a(f-f_n) = g_a - E_a f_n$. Thus, $E_a f$ is the data. $E_a f_n$, called the pseudo projection of the trial image f_n, is a computed rather than an experimental projection. Thus, h' corresponds to the backprojected correction to f_n. The backprojection is weighted by $A(a+r\theta)$ along the backprojection path.

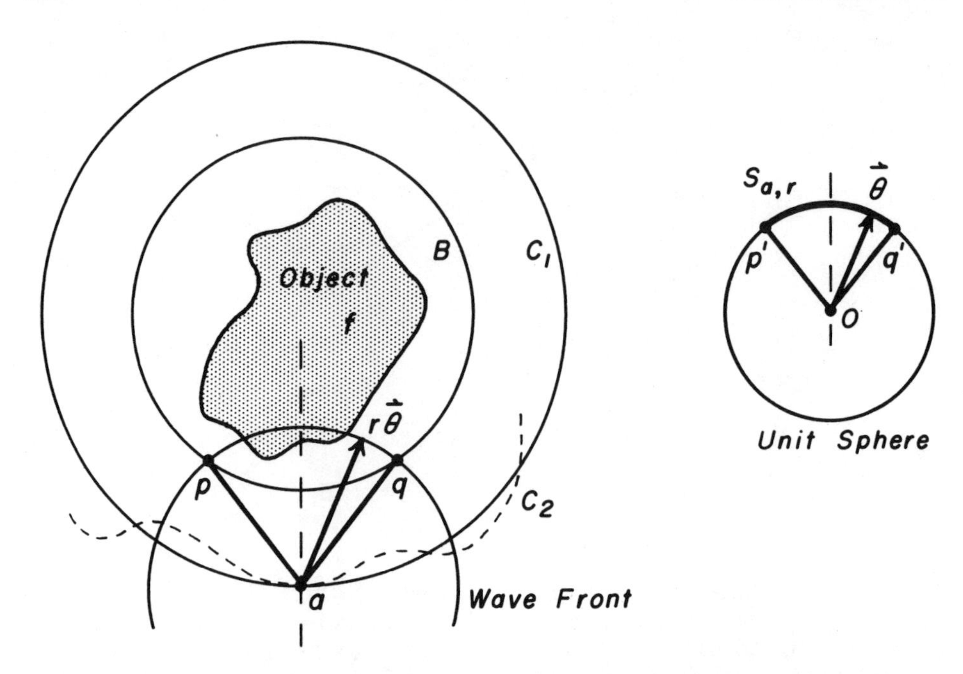

Figure A2 - Illustration of data collection geometry for application of Kacmarz iterative matrix inversion method. Object consists of point scattering function f. A typical source and receiver is located at point a. Speed of sound is C. At time t = 2r/C the received signal at a will correspond to the integral of function f along arc pq. Function f vanishes outside region B. Vector $r\vec{\theta}$ on pq is parallel to unit vector $\vec{\theta}$ at region $S_{a,r}$ on unit sphere. Lines Op' and Oq' are parallel to lines ap and aq, respectively. $S_{a,r}$ defines angular extent of backprojection since backprojection of correction information occurs only on arc pq. Sources on circle C_1, but could be located on any boundary such as C_2. Vector θ in text is equivalent to $\vec{\theta}$ in above figure.

This path of backprojection is the wave front given by the set $\{r\theta\}$, where $\theta \varepsilon S_{a,r}$.

In the case of parallel x-ray data, the Kacmarz method for some implementations is often identical to the method known as additive ART (18). For divergent 2D x-rays, the function $A(a+r\theta)$ in Equation A4 is replaced by 1/r, however, for the reflection ultrasound tomographic backprojection, the 1/r (2-D) or $1/r^2$ (3-D) wave spreading factor is cancelled by the Jacobian of the generalized surface element and only an integral over the surface vector θ is required as given by Equations A3 and A5. Thus, only the attenuation factor $A_a(a+r\theta)$ is required as a weight. For each point $r\theta$ on wave front of radius r the weight may be modeled by the line integral of twice the attenuation coefficient along the ray connecting a and point $(a+r\theta)$. Thus,

$$A_a(x) = \exp[-2\int_0^{x-a} \alpha(a+\frac{s(x-a)}{|x-a|})ds] \tag{A8}$$

where ds is arc length along ray connecting a and $a+r\theta$.

In general, α includes the effect of both absorption and scattering from the function f. If α is known, this will probably create no problem. However, if α is not known, then a more general Kacmarz problem with two unknowns, α and f, is suggested with corresponding more generalized data.

ACKNOWLEDGEMENTS

The derivation of the Kacmarz formula in the Appendix is due in large part to Professor K. T. Smith who is presently a visiting scientist in the Physiology and Biophysics department at the Mayo Clinic. The support from Dr. Erik L. Ritman and Dr. Earl H. Wood and the staff of the Biodynamics Research Unit, Mayo Clinic, is appreciated. The secretarial and graphic help of Mrs. Pat Gustafson and co-workers is also appreciated. The experimental help from Mr. Ken Liang and Chris Hansen is appreciated. The synthetic focus data collection program used for circular geometries was written by Mr. William Samayoa and his help is appreciated. Photographic services were provided by Mr. Leo Johnson. The support of Dr. Curtis C. Johnson, former Chairman, and Dr. J. D. Andrade, present Chairman of the Department of Bioengineering, University of Utah, is greatly appreciated.

This research was supported in part by NIH grants HL-00170, HL-00060, HL-04664, HL-01111, HV-7-2928, and NCI grant CB-64041.

REFERENCES

1. Johnson, S. A., J. F. Greenleaf, M. Tanaka, B. Rajagopalan, and R. C. Bahn: Reflection and transmission techniques for high resolution quantitative synthesis of ultrasound parameter images. 1977 IEEE Ultrasonics Symposium Proceedings, IEEE Cat. #77CH1264-1SU.

2. Johnson, S. A., J. F. Greenleaf, F. A. Duck, A. Chu, W. F. Samayoa, and B. K. Gilbert: Digital computer simulation study of a real-time collection, post-processing synthetic focusing ultrasound cardiac camera. In: Acoustical Holography, Volume 6 (Newell Booth, editor), Plenum Press, New York, 1975, pp 193-211.

3. Duck, F. A., S. A. Johnson, J. F. Greenleaf, and W. F. Samayoa: Digital image focusing in the near field of a sampled acoustic aperture. Ultrasonics 15(2):83-88, 1977.

4. Johnson, S. A., J. F. Greenleaf, B. Rajagopalan, R. C. Bahn, B. Baxter, and D. Christensen: High spatial resolution ultrasonic measurement techniques for characterization of static and moving tissues. Proceedings of the 1977 Ultrasound Tissue Characterization Symposium (In Press).

5. Johnson, S. A., J. F. Greenleaf, M. Tanaka, B. Rajagopalan, and R. C. Bahn: Quantitative synthetic aperture reflection imaging with correction for refraction and attenuation: Application of seismic techniques in medicine. Proceedings of the 1978 San Diego Biomedical Symposium 17: 337-349, 1978.

6. Smith, K. T., D. C. Solmon, and S. L. Wagner: Practical and mathematical aspects of reconstructing objects from radiographs. Bulletin of American Mathematical Society 1:1227-1270, 1977.

7. Smith, K. T., D. C. Solmon, S. L. Wagner, and C. Hamaker: Mathematical aspects of divergent beam radiography. Proceedings of the National Academy Science (June) 1978.

8. Hamaker, C., K. T. Smith, D. C. Solmon, and S. L. Wagner: The divergent beam x-ray transform. Rocky Mountain Journal of Mathematics (1978), Special issue dedicated to N. Aronszajn.

9. Conrad, V. and Y. Wallach: Iterative solution of linear equations on a parallel processor system. IEEE Transactions on Computers C-26(9):838-847, 1978.

10. Private communications with Dr. Arnold Lent, Department of Computer Science, SUNY, Buffalo, New York. It has been learned that similar results have been reported earlier by: Whitney, T. M. and R. K. Meany, SIAM Journal of Numerical Analysis 4:109, 1967.

11. A preliminary experiment in this area was performed as early as mid-1976 by G. Wade, S. Elliott, I. Khugeer, G. Flesher, J. Eisler, and D. Mensa at the University of California at Santa Barbara; private communication. (See also these Proceedings).

12. Norton, S. J. and M. Linzer: Tomographic reconstruction of reflecting images. Third International Symposium on Ultrasonic Imaging and Tissue Characterization, June 5-7, 1978, National Bureau of Standards, Gaithersburg, Maryland (In Press).

13. Corl, P. D., G. S. Kino, C. S. DeSilets, and P. M. Grant: A digital synthetic focus acoustic imaging system. Acoustical Imaging, Volume 8 (Alex Metherell, editor), 1978 (In Press).

14. Iwata, K. and R. Nagata: Calculation of refractive index distribution from interferograms using the Born and Rytov's approximations. Japanese Journal of Applied Physics, Volume 14 (1975) Supplement 14-1.

15. Mueller, R. K.: A new approach to acoustic tomography using diffraction techniques. Acoustical Imaging, Volume 8 (Alex Metherell, editor), 1978 (In Press).

16. Mueller, R. K. and M. Kaueh: Ultrasonic tomography via perturbation techniques. Third International Symposium on Ultrasonic Imaging and

Tissue Characterization, June 5-7, 1978, National Bureau of Standards, Gaithersburg, Maryland (Abstract).

17. Ball, J. and S. A. Johnson: Departments of Physics and Bioengineering, respectively; University of Utah, Salt Lake City, Utah, May 1978, unpublished work.

18. Brooks, R. A. and G. DiChiro: Principles of computer assisted tomography (CAT) in radiographic and radioisotopic imaging. Physics in Medicine and Biology 21(5):689-732, 1976.

BREAST IMAGING BY ULTRASONIC COMPUTER-ASSISTED TOMOGRAPHY

James F. Greenleaf, Surender K. Kenue,
Balasubramanian Rajagopalan, Robert C. Bahn*,
and Steven A. Johnson

Department of Physiology and Biophysics,
Biodynamics Research Unit, Mayo Foundation,
Rochester, Minnesota and *Department of
Pathology and Anatomy, Mayo Foundation,
Rochester, Minnesota

INTRODUCTION

The purpose of our applying acoustic tomographic methods in the breast is directed toward the early detection of carcinoma of the breast. Our rationale is based on the principle that early detection of cancer is not only dependent upon the spatial resolution of the physical method of detection but also depends upon sensitivity for detecting changes in basic properties of tissue which can be logically related to the normal and abnormal histologic elements and spatial organization of the organ under investigation. For the breast, detection of such changes demands methods which are capable of delineating the basic elements of the normal breast such as fat, inter- and intralobar connective tissue, ducts and acinar tissue as well as the localization of common lesions of the breast such as fibrocystic disease, cysts, fibroadenomas, medullary carcinoma, and scirrhous carcinoma. Development of such capabilities of histologic and spatial tissue differentiation should allow one not only to find an established advanced lesion, but also to detect early changes in the morphology of the normal breast by detecting focal changes in the acoustic properties of normal tissue constituents which may imply the presence or the potential of development of a neoplasm.

Methods of acoustic computerized transmission and echo tomographic imaging (as studied by our laboratory) have been described previously (1-9). It has been noted (10) that reconstruction methods should allow the acquisition of multiple separate images formed from the relationship to the received signal of such acoustic properties as attenuation, speed, and reflection. This type of image acquisition is powerful and efficient since the data for multiple independent images can be collected with a single instrument. Quantitative values representing specific properties of tissue may be derived from the multiple images. The conjoint use of multiple properties for the characterization of biological images has been previously reported by the authors (10).

The feasibility of ultrasonic differential imaging of the histologic structures of the breast has been demonstrated in studies of the attenuation of ultrasound by specific tissue components (11-14). Essentially, these studies have shown that the least attenuation was associated with fatty tissue, intermediate attenuation by parenchymatous elements and the greatest attenuation by collagenous and ultimately calcified tissues. Ultrasound echographs of the breast (13-15) have demonstrated qualitatively the geometry of the general mammary contour, the rib cage, pectoral musculature, the nipple, fatty and parenchymatous regions and various pathologic lesions including cysts, fibrocystic disease, hematoma, fibrosis, fibroadenoma, scirrhous carcinoma, anaplastic carcinoma, and mucoid carcinoma. The wide range of variations and the multiple modalities of interactivity of ultrasound energy with tissues provides the basis for the approach to breast diagnosis described in this report.

Several potential roles exist for ultrasound imaging in the clinical management of breast disease. These include: 1) a method of breast examination for young women in whom x-ray mammography is contraindicated, 2) a method of premammography screening for all women, 3) a supplemental method of imaging of the breast in patients in whom the results of x-ray mammograms are ambiguous, and 4) a method of serial temporal surveillance of breast lesions in patients in whom multiple, short-interval surveys by x-ray is contraindicated.

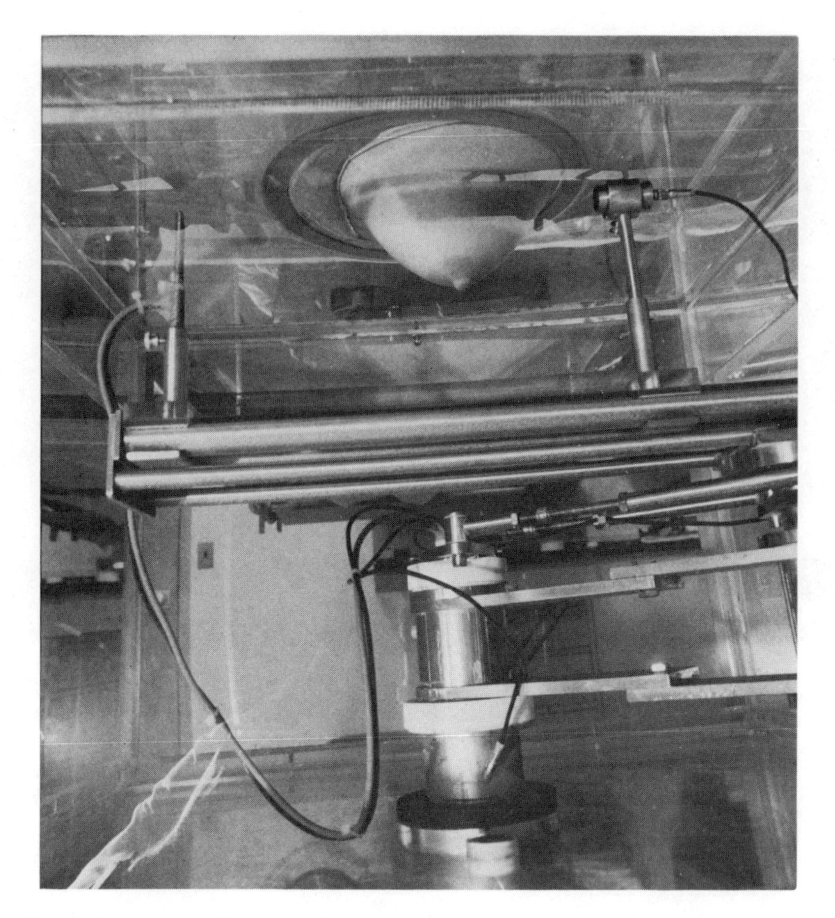

Fig 1 - Breast descends through circular aperture into water tank. Raster scan of transducers at one angle of view, as the scanner moves down the breast, results in acoustic mammogram (Figure 2). Transducers on fan beam scanner are then scanned around the breast to obtain data required for ultrasonic computer-assisted tomogram (Figure 3). (Reproduced with permission from J. F. Greenleaf, et al.: 1977 Ultrasonics Symposium Proceedings, IEEE Cat. No. 77 CH 1264-1SU:989-994, 1977.)

PATIENT SCANNING AND DATA ACQUISITION

The patient is placed prone on the scanner with the breast in the water tank (Figure 1). The scanner then obtains an "acoustic mammogram" by scanning the breast repeatedly from one angle of view as the scanner arm

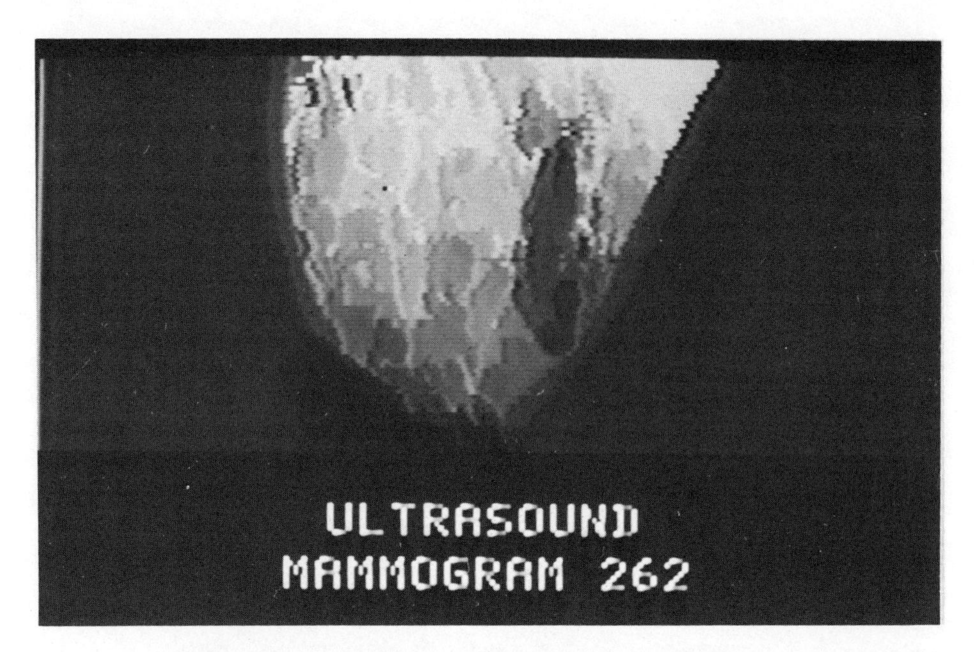

Fig 2 - Acoustic mammogram of *in vivo* breast. Image intensity represents time of arrival of acoustic energy propagating through the breast. Bright areas are late arrival and dark areas are early arrival times. Regions of early arrival correspond to dense fibrous tissue or carcinoma - late arrival corresponds to regions which are mostly fat.

descends from the base to the tip of the breast. An ultrasound mammogram obtained in this way is shown in Figure 2. This image represents time of arrival of the acoustic pulse through the breast, brighter regions being later and darker regions being earlier arrival times. The ultrasound mammograms are used to determine the optimal levels for scanning the breast for tomography to minimize the number of cross section images required for complete coverage of the breast. The efficacy of the acoustic mammograms in diagnosis is currently being evaluated.

Detailed methods of the scanning and signal analysis methods have been described previously (10). Currently, arrival time is measured by setting bipolar thresholds above and below the noise levels on the received signal (after a preamplifier having 46 dB of gain). The time at which the signal crosses either the positive or negative threshold is measured to within ± 20 nanoseconds over an input amplitude range of 70 dB using a 16-bit 100 MHz pulse arrival timer. Four hundred samples of arrival time are obtained for each sweep of 60° and 60 to 120 sweeps are used for each scan.

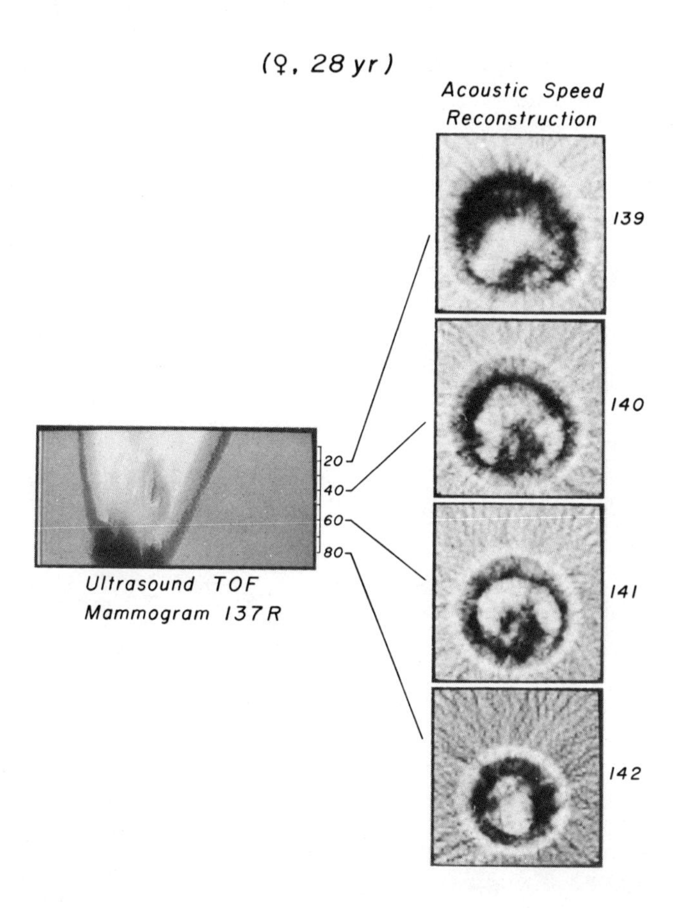

Fig 3 - Acoustic mammogram and associated computer-assisted tomograms of acoustic speed through indicated levels of the *in vivo* breast of a 28-year-old woman with carcinoma. Regions of higher acoustic speed (bright areas in tomograms) were found by histology to be associated with ductile carcinoma and were identified by pattern recognition program (Figure 7).

Examples of computer-assisted ultrasonic tomograms and associated mammograms from a 28-year-old woman with two regions of carcinoma in the right breast are shown in Figure 3. The dark region in the acoustic mammogram extending from scan lines 20 to 60 is the region of cancer. The tomograms obtained at that level indicate regions of high acoustic speed with a characteristic "grainy appearance. The associated x-ray tomogram obtained at about the same level of the breast on the General Electric Tomographic Breast Scanner (currently undergoing evaluation at Mayo) (20) is shown in Figure 4. One can see that the acoustic images give approximately the same appearance and geometry as the x-ray tomogram after contrast injection. The acoustic speed image is probably related to collagen content of scar and connective tissue (16) while the x-ray image apparently measures tissue mass density. Whether or not a contrast agent similar to that used in roentgenography could or should be utilized in acoustic tomography is a question to be answered after further study.

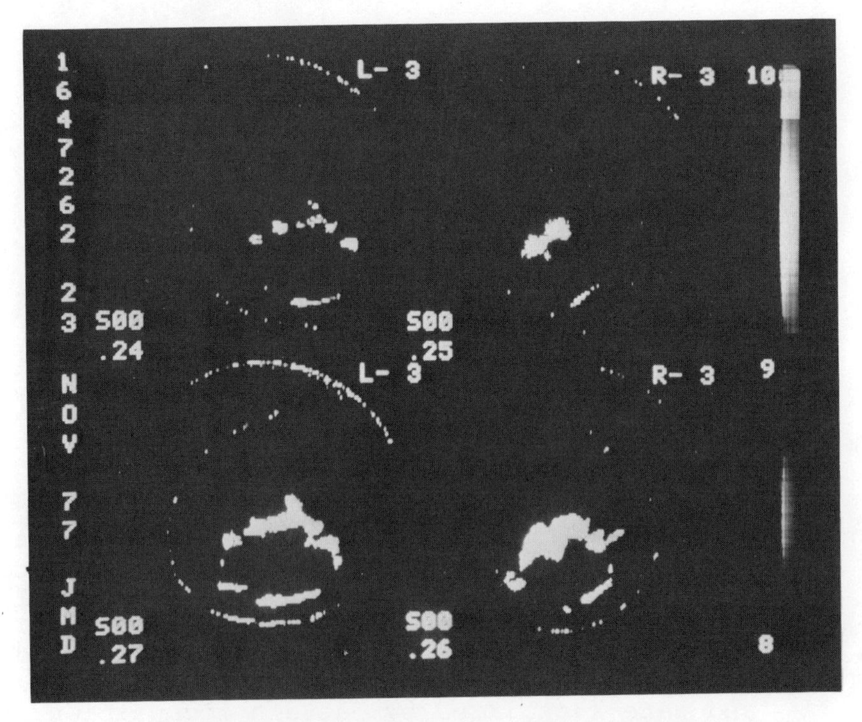

Fig 4 - X-ray computer-assisted tomogram of breast. Computer-assisted tomograms through the left breast (left side) and right breast (right side) before (top) and after (bottom) intravenous injection of x-ray contrast material. Also indicated on the right is the range of CT numbers from black to white in the image. The slice thickness of these images is approximately 1 cm and thus direct correlation between these images and the 3 mm thick ultrasound images is extremely difficult. However, the lower right image appears very much like the computerized ultrasonic tomogram of level 20 in Figure 3.

METHODS OF HISTOLOGIC EVALUATION OF THE BREAST

It must be emphasized that because of the extreme interactivity of acoustic energy with tissues, and because of the potential of our scanner for obtaining several images of independent acoustic properties of tissue, careful and accurate histologic evaluation and identification of the tissue after acoustic scanning is absolutely necessary.

The breast is an extremely difficult organ to examine histologically in a complete and exhaustive manner. It is principally composed of fatty, fibrous, and epithelial tissue. The distributions of these elements are not random. For example, a simplified description of the morphology of the breast is that it consists of a fatty mass within which is embedded a fibrous mass within which is embedded an epithelial ductal system. Our current impression concerning the ultrasound images is that the characteristics of the bright regions (high acoustic speed) appear to be related to the degree of "concentration" of the connective tissue elements of the breast, to the qualitative changes within the connective tissue elements, and to the state of proliferation of the epithelial elements.

We have extended the traditional methods of histologic examination of the breast to include submacroscopic examination of stained and cleared 2 mm thick coronal planes of the entire breast allowing evaluation of the fatty, fibrous and epithelial elements of the breast while at the same time preserving the natural spatial relationships of the histologic elements which prevailed in the intact living breast.

The procedure is greatly superior to so-called "routine" methods of breast examination. We have already encountered significant variability among the descriptions of breast specimens obtained from surgical, mammographic, pathologic and ultrasonic evaluations. These differences precipitate major clinical problems in surgical management when the lesion is small and not definitely palpable. It is our goal to provide complete and consistent descriptions of the three-dimensional morphology of the examined breast using acoustic imaging methods.

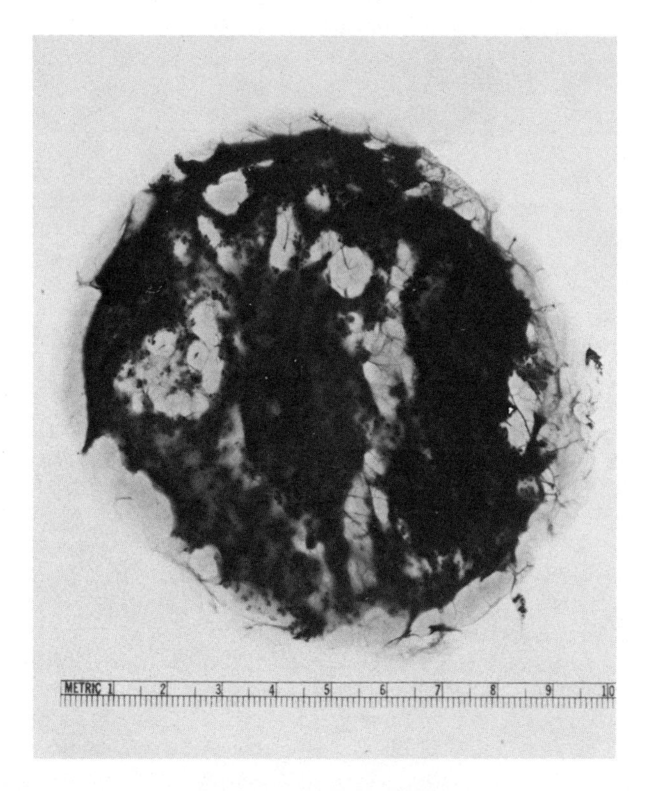

Fig 5 - Histologic preparation of coronal plane of normal breast of 18-year-old patient. Note confluent fibrous macrolobules embedded in fatty tissue, and darker microlobules embedded in fibrous tissue. Preparations of this type can be compared directly to computerized ultrasonic tomograms.

An example of the present state of our histologic work is seen in Figure 5. Each breast is fixed in formalin, mechanically sliced into 2 mm thick coronal sections, defatted, stained, for instance with azure-eosin (pH 4.5), dehydrated, cleared and mounted between sheets of plastic. The preparation can be photographed, examined under the dissecting microscope and under lower powers of the transmitted light microscope. Regions of tissue may subsequently be rehydrated, embedded in paraffin, cut, stained and examined histologically in the traditional manner.

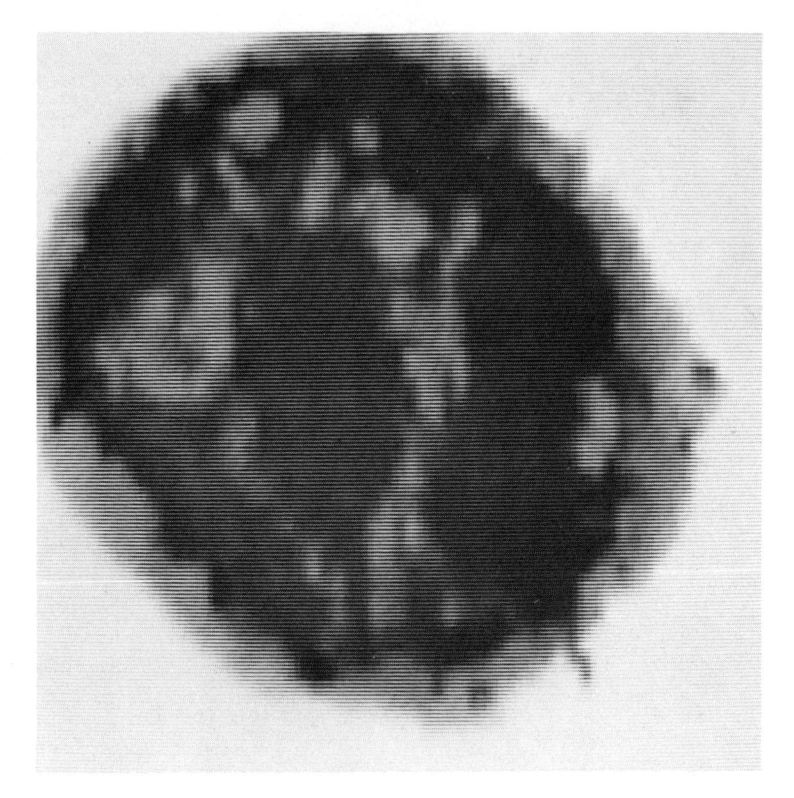

Fig 6 - Digitized image of gross specimen. Black and white optical photograph of specimen shown in Figure 5 was digitized with resolution of 64 x 64 pixels to demonstrate best resolution possible with acoustic tomography reconstructions having 64 x 64 pixels.

An estimate of the resolution to be expected in the 64 x 64 pixel reconstructions of acoustic speed can be seen in Figure 6 which was obtained by digitizing the black and white photograph of a 2 mm thick slice of a fixed and cleared excised breast of Figure 5. This degree of resolution is the maximum resolution to be expected under our current reconstruction methods although we expect ultimately to achieve resolution equivalent to 128 x 128 or 256 x 256 pixel images using previously described methods (4) for correcting refraction index aberrations in the reconstructed images.

AUTOMATED ANALYSIS OF IMAGES

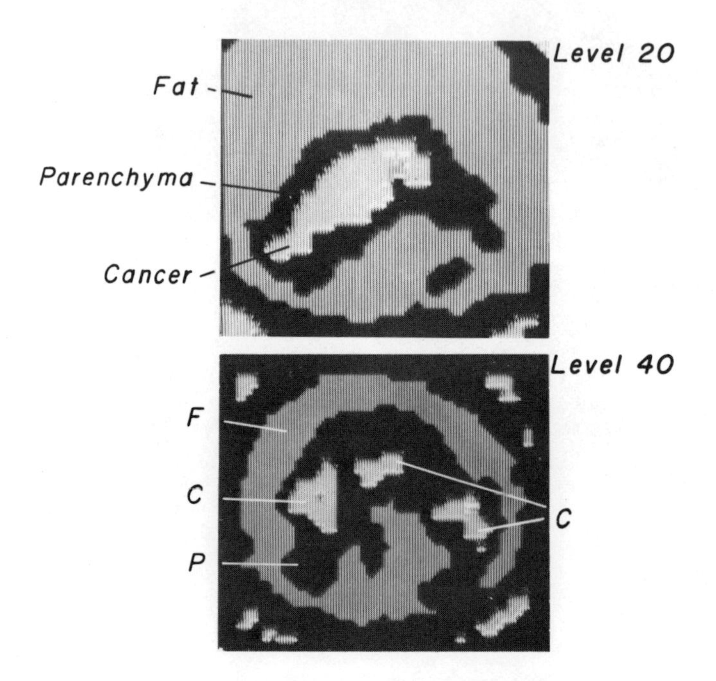

Fig 7 - Pattern recognition of computerized ultrasound tomographs of *in vivo* breast. Pattern recognition programs were applied only to the breast portion of the images of level 20 and level 40 in Figure 3. The resulting images, somewhat enlarged relative to Figure 3, contain tissue identifications of regions of fat, parenchyma, and cancer. The regio of cancer in the image of level 20 is much the same as the region in the x-ray tomograms of Figure 4 which increased in x-ray density after injection of contrast material. The additional region of cancer at 3 o'clock seen at level 40 was not seen on the x-ray mammogram or x-ray tomogram, but was found to be present in the excised breast.

Analysis of the acoustic images by pattern recognition methods, developed by Drs. R. C. Bahn and S. K. Kenue, resulted in the images shown in Figure 7. The regions of cancer are correctly identified and compare very favorably with those regions in Figure 4 (obtained by the x-ray tomographic scanner) in which increases in x-ray density were exhibited after injection of x-ray contrast material. The region of cancer shown at 3 o'clock in the pattern recognition image of Figure 7 was not identified by palpation or on either x-ray tomograms or x-ray mammograms. However, this region was later identified as cancer by histologic examination of the excised breast. This result suggests that ultrasound tomographic mammography may be capable of detecting some cancers which are occult by currently used diagnostic methods.

Our current pattern recognition programs were applied to a training set of 92 acoustic tomographic images from 15 breasts containing 88 separate regions of histologically identified tissue regions separated into 181 different "biopsies" resulting in the identification matrix shown in Table 1. The results of this first pattern recognition study, conducted entirely on computerized acoustic tomography images of acoustic speed from living patients, resulted in two false-positives (cancer called fibrocystic disease) seen in column 4 and resulted in four false-negatives seen in the row representing histologically identified cancer. The addition of computer-assisted tomographic images of acoustic attenuation and the addition of compound B-scan images would greatly increase sensitivity and specificity of these encouraging preliminary results.

DISCUSSION

The major aberrations in the computerized ultrasonic tomographic images are due to 1) bending of acoustic rays caused by refraction; 2) diffraction effects; and 3) phase cancellation effects at the transducer caused by arrival of wave fronts from multiple directions.

The refraction of acoustic rays causes ray bending and since the reconstruction is done assuming straight rays, aberrations occur in the image. Under assumptions that significant alterations in refraction index occur only over a scale much larger than the wavelength of sound, Iwata and Nagata (17) have derived an inversion formula for calculating refraction index from measure-

RESULTS OF CLASSIFICATION OF 181 CENTRAL

PIXELS OF TYPICAL REGIONS (3 x 3)

TISSUE	N	CLASSIFICATION 1	2	3	4	5	TOT	SENSITIVITY
Fat	1	40	0	0	0	0	40	1.0
Parenchyma	2	0	29	3	0	0	32	.91
Fibrocystic Disease	3	0	0	34	2	0	36	.94
Cancer	4	0	2	2	24	0	28	.86
Dense Parenchyma	5	0	6	3	0	36	45	.80
TOTAL		40	37	42	26	36	181	--
SPECIFICITY		1.0	.81	.81	.92	1.0	---	--

Table 1 - Table of initial results from application of pattern recognition methods to acoustic speed images obtained from intact breasts. Five tissues were classified from 181 regions of 92 computer-assisted acoustic speed tomograms and compared to the histologic evaluations of sections of the excised breasts. The columns represent the computerized classification of the regions obtained from the ultrasound images and the rows represent the classification by histology from the gross specimens. Of 28 regions of cancer, the pattern recognition method found false-negatives in four regions and false-positives in two regions.

ments of scattered waves resulting from insonification with parallel waves but to our knowledge it has not yet been applied to acoustic reconstruction. Perhaps another approach is to use the initial reconstruction of values of acoustic speed obtained from straight rays to determine the curved ray paths to be used in subsequent reconstructions, thus iterating on the final corrected image, as previously suggested (4).

Diffraction occurring within the wave as it travels around the edge of the breast and from the breast surface to the receiver, causes an alteration of the wave phase and amplitude not accounted for in the reconstruction algorithms. This effect may be greatly decreased by using the amplitude and phase of the wave, as measured along the receiver locus, and the Fresnel equations (19) to "back propagate" the wave and determine mathematically the correct phase and amplitude of the wave as it exited from the insonified object.

Phase cancellation of the signal over the surface of the receiver transducer is very difficult to eliminate since the use of a small receiving aperture decreases sensitivity as does the use of phase insensitive receivers (18). A trade-off between small receiver apertures and phase cancellation effects seems to be required.

The clinical use of a high resolution scanner will require special display methods to facilitate visual analysis by the radiologist of the resulting large number of images. One method of facilitation of the evaluation procedure would be to use automatic pattern recognition methods to "screen" the images and "flag" those most likely to provide the radiologist with information directly related to diagnosis. The pattern recognition methods mentioned in this report are a first step toward automated localization and characterization of breast lesions. In addition, image pattern analysis methods, when combined with very careful and extensive tissue characterizations by histology, provide an automatic method for relating patterns and other characteristics of the acoustic image to the tissue types under study. Digitized images of tissues stained to emphasize selected histochemical characteristics, such as Figure 6, could be automatically compared to the associated acoustic image, thus determining the weighting function for relating various histochemical characteristics to the various acoustic properties. For instance, histologic images of tissues stained for emphasizing distributions of connective tissue should correlate very well with images of acoustic speed since the high collagen content of connective tissue should produce correspondingly high acoustic speed (16).

Complete characterization of the tissues of the breast, and probably tissues of the remainder of the body, will require measurement of the multimodal interactions of acoustic energy with tissues. Methods of measuring such data in the breast should include computerized ultrasonic tomography of acoustic speed and attenuation and perhaps in addition, compound B-scans.

ACKNOWLEDGMENTS

The authors gratefully acknowledge the assistance of Mrs. Pat Gustafson and her assistants for clerical help and Mrs. Donna Balow and co-workers for art assistance. Paul Thomas wrote reconstruction programs and Chris Hansen constructed high speed electronics. Dr. David F. Reese provided the x-ray tomogram. This research was supported in part by grants HL-04664, RR-00007, HV-7-2928, HL-00060, HL-00170, and HL-07111 from the National Institutes of Health and NCI-CB-64041 from the National Cancer Institute.

REFERENCES

1. Greenleaf, J. F. and S. A. Johnson: Algebraic reconstruction of spatial distributions of refractive index and attenuation in tissues from time-of-flight and amplitude profiles. Proceedings of Seminar on Ultrasonic Tissue Characterization. National Bureau of Standards Special Publication 453:109-119, 1976.

2. Greenleaf, J. F., S. A. Johnson, S. L. Lee, G. T. Herman, and E. H. Wood: Algebraic reconstruction of spatial distributions of acoustic absorption within tissue from their two-dimensional acoustic projections. In: Green, Philip S.: Acoustical Holography. New York, Plenum Press, 1974, Volume 5, pp 591-603.

3. Greenleaf, J. F., S. A. Johnson, W. F. Samayoa, and C. R. Hansen: Refractive index by reconstruction: use to improve compound B-scan resolution. In: Kessler, L.: Acoustical Holography. New York, Plenum Press, 1977, Volume 7, pp 263-273.

4. Johnson, S. A., J. F. Greenleaf, A. Chu, J. D. Sjostrand, B. K. Gilbert, and E. H. Wood: Reconstruction of material characteristics from highly refraction distorted projections by ray tracing. Proceedings of Image Processing for 2-D and 3-D Reconstructions from Projections: Theory and Practice in Medical and Physical Sciences. Stanford, California, August 4-7, 1975, pp TUB2-1 - TUB2-4.

5. Johnson, S. A., J. F. Greenleaf, W. F. Samayoa, F. A. Duck, and J. D. Sjostrand: Reconstruction of three-dimensional velocity fields and other parameters by acoustic ray tracing. 1975 Ultrasonic Symposium Proceedings IEEE Catalog No. 75 CHO994-4SU:46-51, 1975.

6. Johnson, S. A., J. F. Greenleaf, C. R. Hansen, W. F. Samayoa, M. Tanaka, A. Lent, D. A. Christensen, and R. L. Woolley: Reconstructing three-dimensional fluid velocity vector fields from acoustic transmission measurements. In: Kessler, L.: Acoustical Holography. New York, Plenum Press, 1977, Volume 7, pp 307-326.

7. Johnson, S. A., J. F. Greenleaf, F. A. Duck, A. Chu, W. F. Samayoa, and B. K. Gilbert: Digital computer simulation study of a real-time collection, postprocessing synthetic focusing ultrasound cardiac camera. In: Booth, Newell: Acoustical Holography. New York, Plenum Press, 1975, Volume 6, pp 193-211.

8. Greenleaf, J. F., F. A. Duck, W. F. Samayoa, and S. A. Johnson: Ultrasonic data acquisition and processing system for atherosclerotic tissue characterization. 1974 Ultrasonic Symposium Proceedings IEEE Catalog No. 74 CH896-1SU:738-743, 1974.

9. Greenleaf, J. F., S. A. Johnson, W. F. Samayoa, and F. A. Duck: Algebraic reconstruction of spatial distributions of acoustic velocities in tissue from their time-of-flight profiles. In: Booth, Newell: Acoustical Holography. New York, Plenum Press, 1975, Volume 6, pp 71-90.

10. Greenleaf, J. F., S. A. Johnson, and A. Lent: Measurement of spatial distribution of refractive index in tissues by ultrasonic computer assisted tomography. Ultrasound in Medicine and Biology 3:327-339, 1978.

11. Chivers, R. C. and C. R. Hill: Ultrasonic attenuation in human tissue. Ultrasound in Medicine and Biology 2:25-29, 1975.

12. Goldman, D. E. and T. F. Hueter: Tabular data of the velocity and absorption of high-frequency sound in mammalian tissues. Journal of the Acoustical Society of America 28:35-37, 1956.

13. Kossoff, G., E. K. Fry, and J. Jellins: Average velocity of ultrasound in the human female breast. Journal of the Acoustical Society of America 53:1730-1736, 1973.

14. Wells, P. N. T.: Absorption and dispersion of ultrasound in biological tissue. Ultrasound in Medicine and Biology 1:369-376, 1975.

15. Jellins, J., G. Kossoff, F. W. Buddee, and T. S. Reeve: Ultrasonic visualization of the breast. Medical Journal of Australia 1:305-307, 1971.

16. O'Brien, W. D., Jr.: The role of collagen in determining ultrasonic propagation properties in tissue. In: Kessler, L. W.: Acoustical Holography. New York, Plenum Press, 7, 1976, pp 37-50.

17. Iwata, Koichi and Ryo Nagata: Calculation of refractive index distribution from interferograms using the Born and Rytov's approximation. Japanese Journal of Physics 14-1:379-383, 1975.

18. Klepper, J., G. H. Brandenburger, L. T. Busse, and J. G. Miller: Phase cancellation, reflection, and refraction effects in quantitative ultrasonic attenuation tomography. 1977 IEEE Ultrasonic Symmposium Proceedings IEEE Catalog No. 77 CH 1264-1SU:182-188, 1977.

19. Morse, Philip M. and K. Uno Ingard: Theoretical Acoustics. New York, McGraw-Hill, 1968, 927 pp.

20. Gisvold, J. J., P. R. Karsell, D. F. Reese, and E. C. McCullough: Clinical evaluation of computerized tomographic mammography. Mayo Clinic Proceedings 52:181-185 (March) 1977.

A NEW APPROACH TO ACOUSTIC TOMOGRAPHY USING DIFFRACTION TECHNIQUES

R. K. Mueller, M. Kaveh, and R. D. Iverson

Department of Electrical Engineering

University of Minnesota, Minneapolis, MN 55455

ABSTRACT

A method is presented for the three-dimensional reconstruction of objects from their two-dimensional profiles obtained by ultrasonic imaging techniques. This method uses a perturbation approximation to the propagating field to solve for the ultrasonic velocity distribution based on the wave equation. In this technique, no assumptions are made about the ultrasonic ray geometries. Furthermore, the reconstruction is carried out in the frequency domain, making the method also computationally efficient. Some numerical simulation results are presented.

I. INTRODUCTION

This paper suggests a method for the reconstruction of three-dimensional velocity distributions from two-dimensional projections based on the wave equation for the propagating sound field under a perturbation approximation. This technique is useful in cases such as ultrasonic imaging, where reconstruction is based on the velocity distribution within objects and the measured data are phase and amplitude of the transmitted beam rather than intensity signals as, for example, in x-ray photography. In this method, no assumptions on the ray geometries are made. A similar idea was put forward by Iwata and Nagata [1] and Wolfe [2] for reconstructing three-dimensional optical refractive-index distributions from interferograms. The method discussed makes extensive use of Fourier domain calculations making the computations also relatively efficient.

Reconstruction of three-dimensional objects from their projections has been the subject of investigations in many fields such as radio-astronomy [3], electron microscopy [4] and x-ray photography [5]. More recently reconstruction techniques have been used in conjunction with ultrasonic imaging of internal organs of the body [6], [7]. The basic problem is the visualization of the internal structure of an object from a multitude of its projections obtained, for example, by a series of distinct x-ray photographs.

There have been basically two approaches to the reconstruction problem. One involves reconstruction in the Fourier domain, based on the fact that the Fourier transform of a projection of an object corresponds to a slice of the Fourier transform of the object. Thus, with enough projections the Fourier transform of the object can be reconstructed. This method has been reviewed in detail by Mersereau and Oppenheim [8].

The second method is in the spatial domain and includes the algebraic reconstruction technique, ART [9]. **In this method,** cross sections of the object are approximated by N x N grids, with the desired parameter to be reconstructed, assumed an unknown constant in each block of the grid. Sufficient number of projections are then used to solve simultaneous linear equations in the N^2 block values. This method is computationally less efficient than the Fourier transform technique. A priori information can, however, be used with ART to yield good reconstruction with iterative solutions of the equations. Furthermore, this approach has been useful in the application involving reconstruction of ultrasonic images, where the velocity of sound in tissue has been found to be the most accessible and reasonable parameter to be reconstructed [4].

II. THE NEW METHOD

II.1 Preliminaries

The basis for the reconstruction method under study is the wave equation governing the propagation of sound field through the inhomogeneous medium of interest. In the following, two-dimensional reconstructions from one-dimensional projections are considered. The wave equation for single frequency operation is given by

$$C^2(\vec{r})\ \nabla^2\psi(\vec{r}) + \omega^2\ \psi(\vec{r}) = 0 \qquad (1)$$

where $C(\vec{r})$ is the ultrasonic velocity at position $\vec{r}$, $\psi(\vec{r})$ is the wave function, $\vec{r} = \{x,y\}$ and ω is the angular frequency. It is assumed that the sound velocity in the medium can be approximated as a constant with a small perturbation. More precisely, the square of the velocity is modeled as

$$c^2(\vec{r}) = [1 + F(\vec{r})]\, c_o^2 \tag{2}$$

where $F(\vec{r}) << 1$ and C_o^2 is the square of the velocity of propagation in the medium surrounding the object under study. If plane wave ($\psi_o = e^{i\vec{k}\cdot\vec{r}}$) is incident on the object, one can write the solution $\psi(\vec{r})$ of equation (1) in the form

$$\psi(\vec{r}) = \psi_o(\vec{r})(1 + \psi_1(\vec{r})) \tag{3}$$

where $\psi_1(\vec{r})$ is a complex modulating field, which is of the order of the velocity disturbance $F(r)$.

If one introduces equations (2) and (3) into the wave equation (1) and retains only first order terms in $\psi_1(\vec{r})$ and $F(\vec{r})$, one obtains a first order approximation to ψ_1 as:

$$(\nabla^2+k^2)\ \psi_o(\vec{r})\ \psi_1(\vec{r}) = +k^2\psi_o(\vec{r})\ F(\vec{r});\ k = \omega/c_o \tag{4}$$

$F(\vec{r})$ and $\psi_1(\vec{r})$ can be described in a given neighborhood of the object by a Fourier integral.

$$F(\vec{r}) = \int_{-\infty}^{\infty} a(\mu,\nu)e^{i\vec{\Lambda}\cdot\vec{r}}d\mu d\nu;\ \psi_1(\vec{r}) = \int_{-\infty}^{\infty} b(\mu,\nu)e^{i\vec{\Lambda}\cdot\vec{r}}d\mu d\nu \tag{5}$$

with

$$\vec{\Lambda} = \{\mu,\nu\}$$

Combining equations 4 and 5 gives:

$$b(\mu,\nu) = -\frac{k^2}{(\vec{\Lambda}-\vec{k})^2-k^2}\ a(\mu,\nu) \tag{6}$$

or if one assumes plane waves $\psi_o^{(n)}(\vec{r})$ impinging from different directions $\vec{k}^{(n)}$ one obtains

$$\psi_1^{(n)}(\vec{r}) = \int_{-\infty}^{\infty} b^{(n)}(\mu,\nu)e^{i\vec{\Lambda}\cdot\vec{r}}d\mu d\nu \tag{7}$$

with

$$b^{(n)}(\mu,\nu) = -\frac{k^2}{(\vec{\Lambda}-\vec{k}^{(n)})^2-k^2}\ a(\mu,\nu)$$

where $a(\vec{\Lambda})$ is the Fourier transform of the disturbance and by definition independent of the wave vector $\vec{k}^{(n)}$ of the impinging plane wave $\psi_o^{(n)}(\vec{r})$. Equation (7) is a formal solution of the wave equations expressed in terms of the spatial frequency spectrum $a(\mu,\nu)$ of the disturbance.

Our objective is to determine the spectrum $a(\mu,\nu)$ of the disturbance from the observation of the scattered field $\psi_1^{(n)}(\vec{r})$. The measurement geometry is shown in Fig. (1).

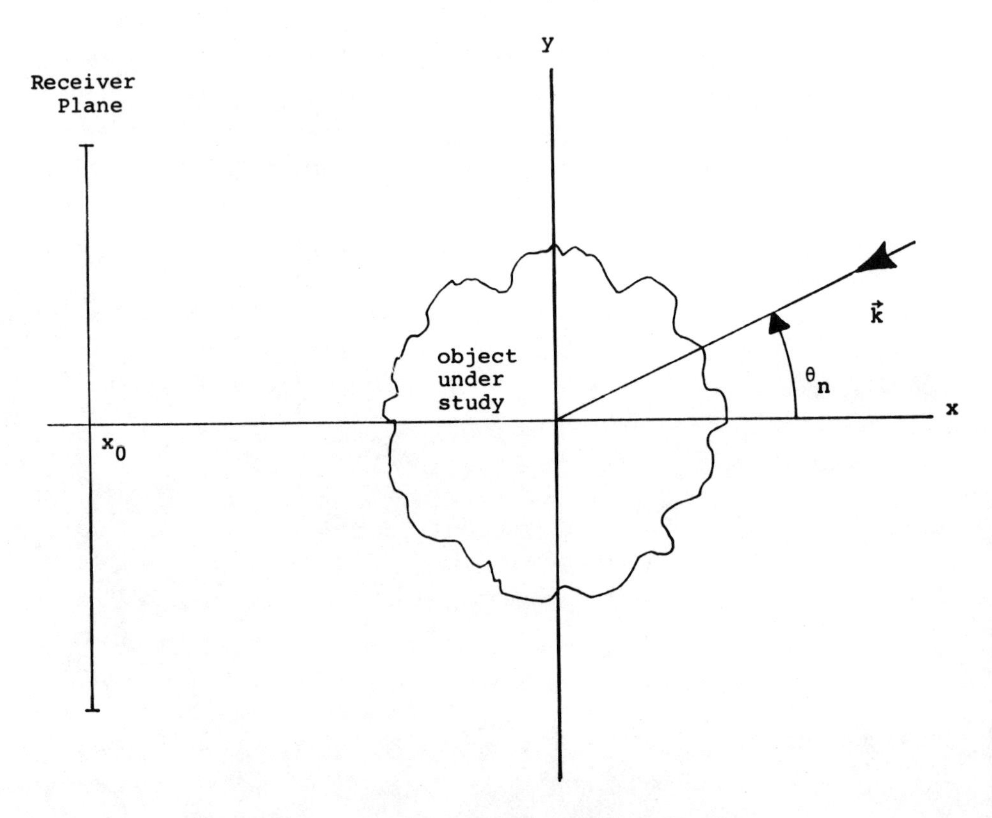

Fig. 1. Measurement geometry.

We observe the total field $\psi^{(n)}(r)$ along a line $x = x_o$ and derive from it the corresponding modulating field $\psi_1^{(n)}(x_o,y)$

$$\psi_1^{(n)}(x_o,y) = [\psi^{(n)}(x_o,y) - \psi_o^{(n)}(x_o,y)]/\psi_o^{(n)}(x_o,y) \quad . \tag{8}$$

One can observe this field, however, only over a finite aperture A. This can be accounted for by introducing a window function g(y). The observable quantity $d^{(n)}(x_o,\nu)$ is then given by

$$d^{(n)}(x_o,\nu) = \int_{-\infty}^{+\infty} g(y)\ \psi_1^{(n)}(x_o,y)e^{-i\nu y}\ dy \tag{9}$$

Integrating over y and introducing equation 5 we obtain

$$d^{(n)}(x_o,\nu') = \int_{\nu=-\infty}^{+\infty} G(\nu'-\nu) \int_{\mu=-\infty}^{+\infty} e^{i\mu x_o}\ b^{(n)}(\mu,\nu)\ d\mu d\nu \tag{10}$$

where g(y) is a shading function and $G(\nu-\nu')$ is given by

$$G(\nu-\nu') = \int_{-a}^{a} g(y)\cdot e^{-i(\nu-\nu')y}dy \tag{11}$$

The convolution with $G(\nu)$ in equation 10 takes into account the finite window $-a<y<a$ and any shading function g(y) used to observe $\psi_1^{(n)}(x_o,y)$. The integral over μ which now relates the $d^{(n)}(x_o,\nu)$ with the unknown spatial frequency spectrum $a(\mu,\nu)$ is

$$I(\nu,n) = \int_{-\infty}^{\infty} e^{i\mu x_o}\ b^{(n)}(\mu,\nu)\ d\mu = -k^2 \int_{-\infty}^{+\infty} \frac{\exp(i\mu x_o)\ a(\mu,\nu)}{(\vec{\Lambda}+\vec{k}^{(n)})^2 - k^2}\ d\mu \tag{12}$$

The denominator in the integrand is a quadratic expression in μ and can be written as

$$(\vec{\Lambda}+\vec{k}^{(n)})^2-k^2 = (\mu-\mu_1)\cdot(\mu-\mu_2) \tag{13}$$

The two roots μ_1 and μ_2 are given functions of the parameters ν and the two components, $k_x^{(n)}$ and $k_y^{(n)}$, of the wave vector $\vec{k}^{(n)}$ of the impinging wave $\psi_o^{(n)}(\vec{r})$. The roots are

$$\mu_{1,(2)} = -k_x^{(n)}\ (\mp)\ \sqrt{(k_x^{(n)})^2 - \nu^2 - 2k_y^{(n)}\nu} \tag{14}$$

where $\mu_{1(2)}$ is real for $|\nu + k_y^{(n)}| < k$.

We assume that the spectrum, $a(\mu,\nu)$, of the velocity disturbance is well behaved, that is, $a(\mu,\nu)$ has no singularities for finite μ and ν and is band limited in the sense that there exists a constant $\bar{\mu}$ such that

$$\frac{a(\mu,\nu)}{(a(\mu,\nu))_{max}} < e^{-\left(\frac{\mu}{\bar{\mu}}\right)^2} \quad \text{for } \mu > \bar{\mu} \tag{15}$$

It can be shown that for any spatially limited disturbance the integral along the arc in Figure (2) in the limit $a \to \infty$ is arbitrarily small.

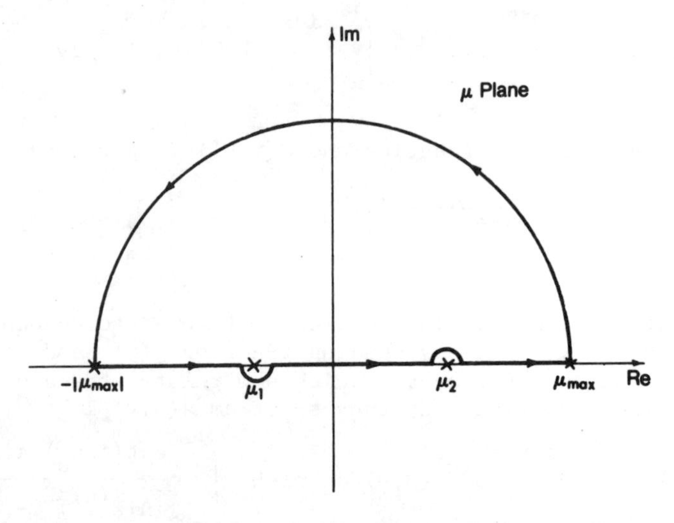

Fig. 2. Integration contour in complex μ plane.

We can, therefore, replace the integral along the real axis by the contour integral along the contour which joins $-\infty$ on the real axis by an arc in the upper half of the complex μ plane to the point $+\infty$ on the real axis. The integration path goes through the two singularities $\mu_{1,2}$ which lie on the real axis. One can show that inclusion of μ_1 and exclusion of μ_2 from the interior of the contour is equivalent to considering only outgoing scattered waves. The integral in equation (12), therefore, reduces to:

$$I(\nu,n) = -\frac{k^2 e^{i\mu_1 x_o}}{2\pi i} \cdot \frac{a(\mu_1,\nu)}{\mu_1 - \mu_2} \tag{16}$$

which finally gives the relation

$$d^{(n)}(x_o,\nu) = \frac{-k^2}{2\pi i} G(\nu) * I(n,\nu) \tag{17}$$

Equation (17) represents a convolution with a narrow pulse like function $G(\nu)$. We obtain, therefore, a good approximation to this equation if we take the slow varying function

$\frac{a(\mu_1,\nu)}{\mu_1 - \mu_2}$ out of the integral:

$$d^{(n)}(x_o,\nu) = \frac{-k^2}{2\pi i}\frac{a(\mu_1,\nu)}{\mu_1 - \mu_2} G(\nu)*e^{i\mu_1 x_o} \tag{18}$$

Equation (18) is now the basis for reconstructing the velocity disturbance $F(\vec{r})$. In the simulated reconstructions discussed in the following section, the window function $G(\nu)$ was replaced by a Dirac δ function, thus:

$$d^{(n)}(x_o,\nu) \simeq \frac{-k^2}{2\pi i}\frac{a(\mu_1,\nu)}{\mu_1 - \mu_2}\, e^{i\mu_1 x_o} \tag{19}$$

III. SIMULATION RESULTS

The initial task of this research involved the generation of test data and preliminary verification of the proposed reconstruction schemes. Thus, as a first step, the scattered wave from a cylindrical inhomogeneity was calculated and numerically implemented. A brief explanation of the generation method is given in the Appendix.

The scattered wave generated in this way was used to calculate $\psi_1(\vec{r})$ and through discrete Fourier transformation, $b^{(n)}(\vec{\Lambda})$. It is obvious from equation (14) that, for a given $\vec{k}^{(n)}$, the permissible (the real valued) $\vec{\Lambda}$'s lie on a circle shown in Figure (3).

It is paramount for computational efficiency, however, to reconstruct $a(\vec{\Lambda})$ on a grid of equi-spaced frequencies in order to take advantage of fast Fourier transform algorithms. Figure (4) shows the loci of available $\vec{\Lambda}$'s (indicated by the circles) superimposed on a rectangular grid.

As a first step in our simulation, for each ν belonging to a grid point, the value of $a(\mu_1,\nu)$, calculated from equation (19), is assigned to a grid point (μ,ν) for which $|\mu_1-\mu|$ is minimum. This has proved to be successful in our preliminary simulations. The assignment of the calculated $a(\mu,\nu)$ to the nearest grid point introduces a zero order interpolation error. The nature of this error is of interest and its effect on the reconstructed

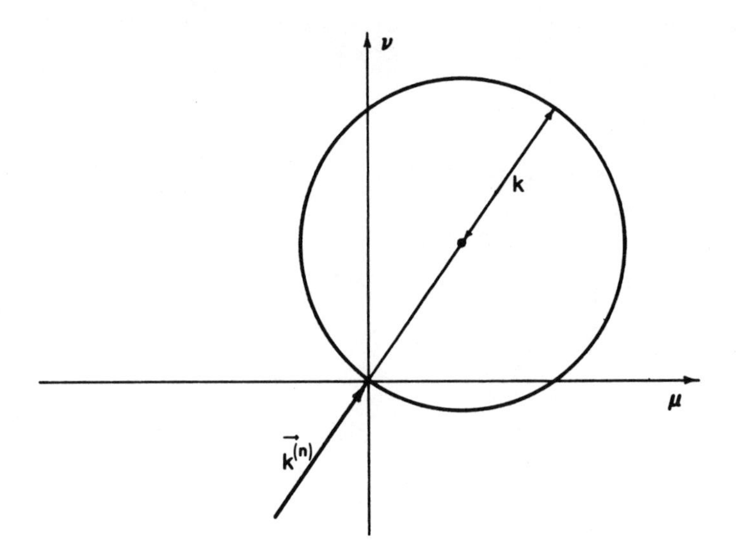

Fig. 3. Locus of permissible Λ's for a given $\vec{k}^{(n)}$.

distribution needs further investigation. It is worth noting that this method of assigning $a(\vec{\Lambda})$ values to grid points can be improved on by higher order interpolation. However, due to computational efficiency of the zero-order reconstruction described above, this method is considered more attractive. Figures (5a) - (5d) show density plots of reconstructed cross sections of uniform cylindrical inhomogeneities, with various diameters and relative velocity differences from the surrounding medium. The reconstruction scheme is zero-order as explained earlier, using a 64 x 64 grid and spatial sampling of $\Delta x = \Delta y = 0.2\lambda$. The noise apparent on the reconstruction is believed to be the interpolation noise plus errors due to aliasing and other phenomena. Since the frequency assignments in the low frequency portion are more accurate (note non-uniform density of the loci of available frequencies in Figure (4)), the reconstructed $a(\vec{\Lambda})$ was windowed with a raised cosine to reduce the high frequency noise as well as aliasing errors. This was done, of course, at the expense of the resolution of edges.

Figures (6a) and (6b) show plots of the same reconstructed velocity distributions as in Figures (5a) and (5b) with windowing in frequency domain prior to inverse transformation. It is noted that, as expected, better reconstruction is obtained of disturbances with smaller effective perturbation, i.e., where the perturbation approximations are more valid.

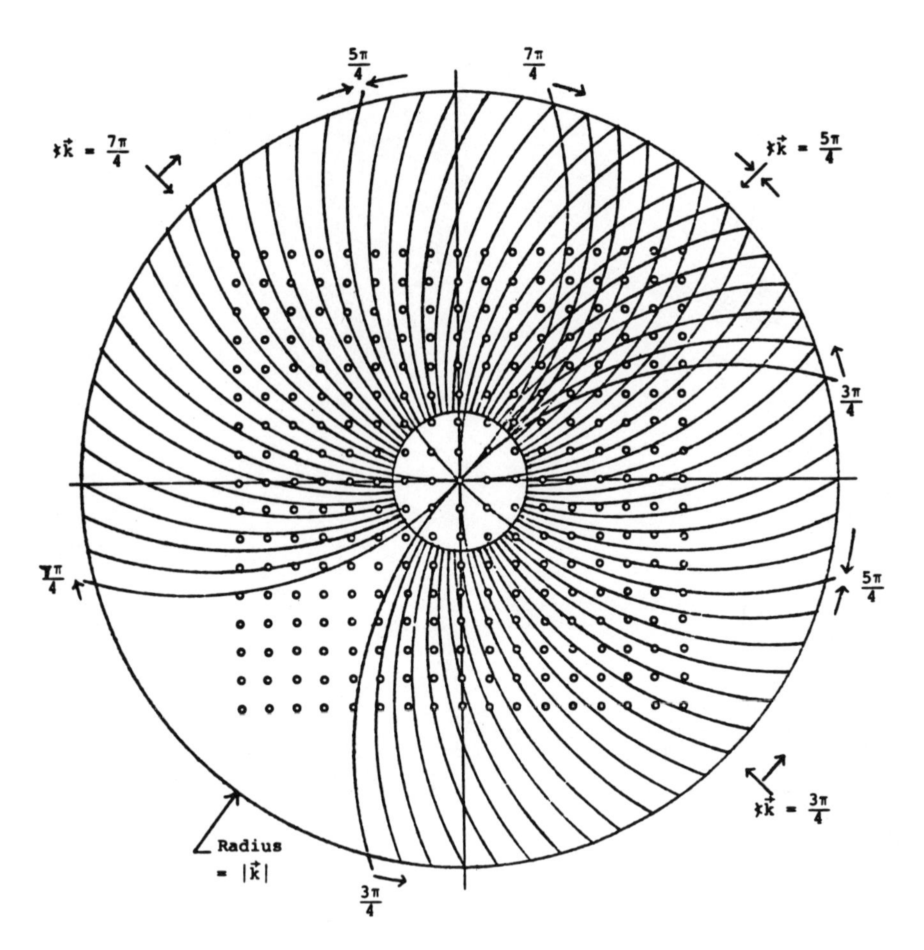

Fig. 4. Loci of available Λ's superimposed on a uniform grid of spatial frequencies.

Further work along these lines will involve computer simulation of scattered waves from more complex scatterers including off-center and multiple cylindrical disturbances and attempts at their reconstruction. Also the nature of the interpolation noise and smoothing strategies will be investigated. Further developments into the theory of reconstruction and its region of validity will continue.

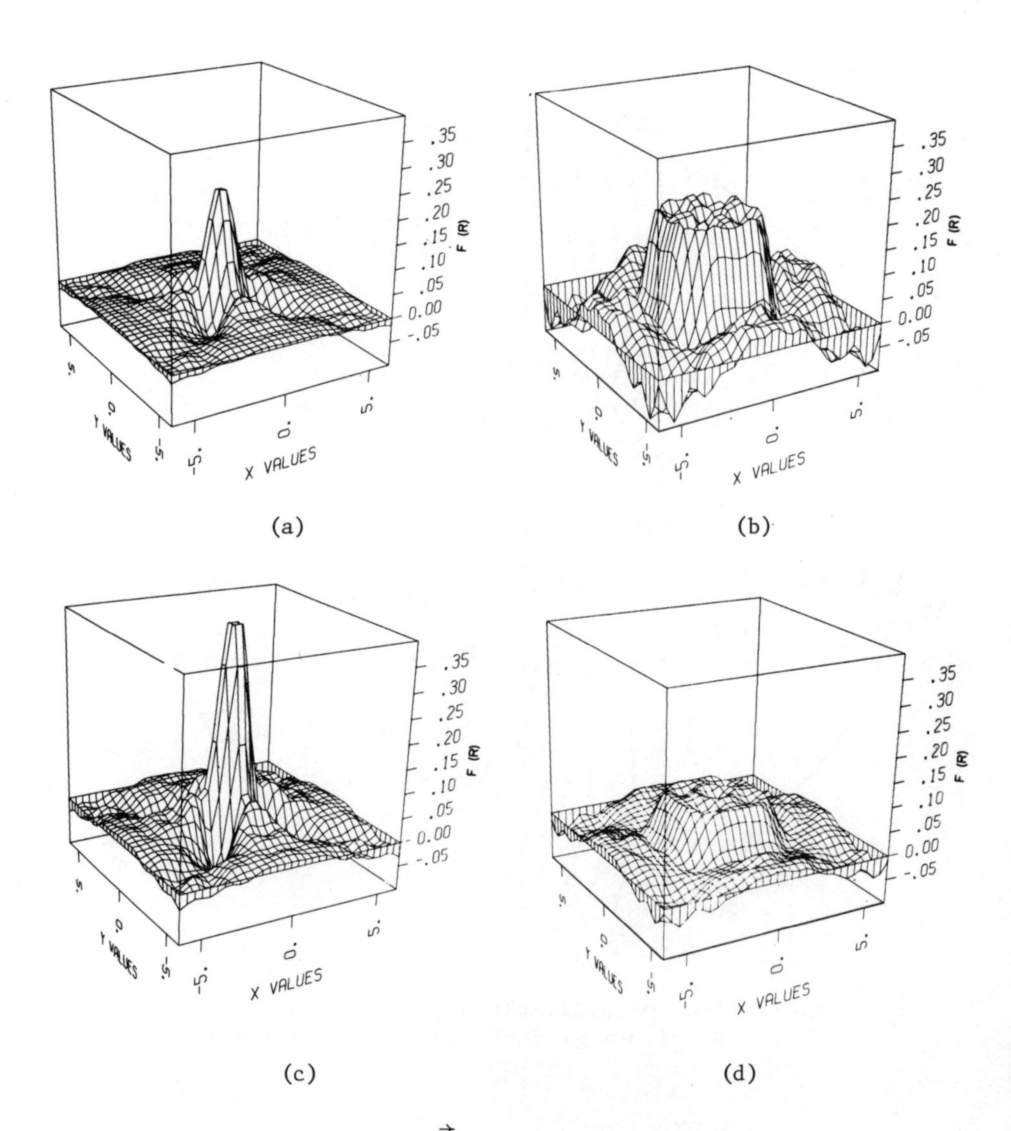

Fig. 5. Reconstructed $F(\vec{r})$ without smoothing for velocity differences $\Delta c/c_o$ and radii a. The wavelength is 1 cm. (a) a=1 cm., $\Delta c/c_o=0.025$; (b) a=3 cm., $\Delta c/c_o=0.025$; (c) a=1 cm., $\Delta c/c_o=0.05$; (d) a=3 cm., $\Delta c/c_o=0.01$.

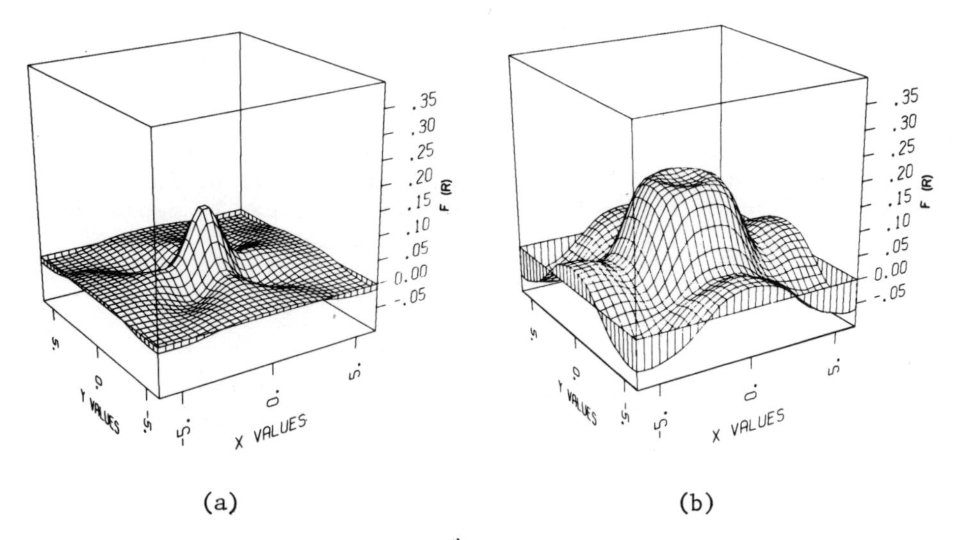

Fig. 6. Reconstructed $F(\vec{r})$ with smoothing for velocity differences $\Delta c/c_o$ and radii a. The wavelength is 1 cm. (a) a=1 cm, $\Delta c/c_o$=0.025; (b) a=3 cm, $\Delta c/c_o$=0.025.

IV. CONCLUSION

We have demonstrated by computer simulation a new method to reconstruct two (or three) dimensional ultrasonic velocity distributions from observed one (or two) dimensional phase and amplitude distributions of a transmitted or scattered sound field. This new method of ultrasonic tomography is based on direct solutions of the wave equation which governs the sound-propagation through the test object and not on geometric ray approximations as other tomographic methods discussed in the literature. The significance of our new approach for acoustic tomography lies in the fact that one does not have to make the assumption, implicit in the geometric ray approach, that structural elements of interest are large compared to the wavelength of the radiation used to obtain the image.

ACKNOWLEDGMENT

This work was supported by the National Science Foundation under Grant ENG 76-84521.

APPENDIX

SUMMARY OF DATA GENERATION SCHEME

The scattering of ultrasonic waves by a cylindrical object of the same density but different sound velocity as the surrounding medium can be described by the velocity potentials of incident and scattered waves. The incident plane wave (ψ_o) can be represented as

$$\psi_o(r,\theta) = e^{i\vec{k}\cdot\vec{r}} \tag{A-1}$$

The propagation vector can be written in terms of the propagation constant of the medium, ($k = \omega/c$) and the angle of propagation (θ) as

$$\vec{k} = k(\cos\theta\ \hat{x} + \sin\theta\ \hat{y}) \tag{A-2}$$

and the field vector ($\vec{r}$) can be written as

$$\vec{r} = r(\cos\phi_1\ \hat{x} + \sin\phi_1\ \hat{y}) \tag{A-3}$$

as indicated in Figure (A-1).

Allowing the cylinder to be centered at the origin and denoting the J-Bessel and Hankel function of the first kind by $J_m(\alpha)$ and $H_m^{(1)}(\alpha)$ respectively we can represent the incident wave (ψ_o) and the scattered wave (ψ_s) as [10]

$$\psi_o(r,\phi) = J_o(kr) + 2\sum_{m=1}^{\infty} i^m \cos(m\phi)\ J_m(kr) \tag{A-4}$$

and

$$\psi_s(r,\phi) = \sum_{m=o}^{\infty} A_m \cos m\phi\ H_m^{(1)}(kr) \tag{A-5}$$

The interior wave (ψ_s') can also be expressed in terms of the propagation constant (k') within the cylinder as

$$\psi_s'(r,\phi) = B_o J_o(k'r) + 2\sum_{m=1}^{\infty} B_m\ i^m \cos(m\phi)\ J_m(k'r) \tag{A-6}$$

Applying the boundary conditions at r = a

$$\psi_o(a,\phi) + \psi_s(a,\phi) = \psi_s'(a,\phi) \tag{A-7}$$

and

$$\frac{\partial}{\partial r}\,[\psi_o(r,\phi)+\psi_s(r,\phi)]_{r=a}\;\frac{\partial}{\partial r}\,\psi_s{}'(r,\phi)\Big|_{r=a} \tag{A-8}$$

the coefficient scattered and interior wave are determined as

$$A_m = \varepsilon_m \frac{k'J_m(ka)J_{m+1}(k'a)-kJ_m(k'a)J_{m+1}(ka)}{kJ_m(k'a)H^{(1)}_{m+1}(ka)-k'H^{(1)}_m(ka)J_{m+1}(k'a)} \tag{A-9}$$

where

$$\varepsilon_o = 1 \text{ and } \varepsilon_m = 2i^m \tag{A-10}$$

and

$$B_m = ika\,\frac{J_{m+1}(ka)y_m(ka)-J_m(ka)y_{m+1}(ka)}{k'aJ_{m+1}(k'a)H^{(1)}_m(ha)-kaJ_m(k'a)H^{(1)}_{m+1}(ka)} \tag{A-11}$$

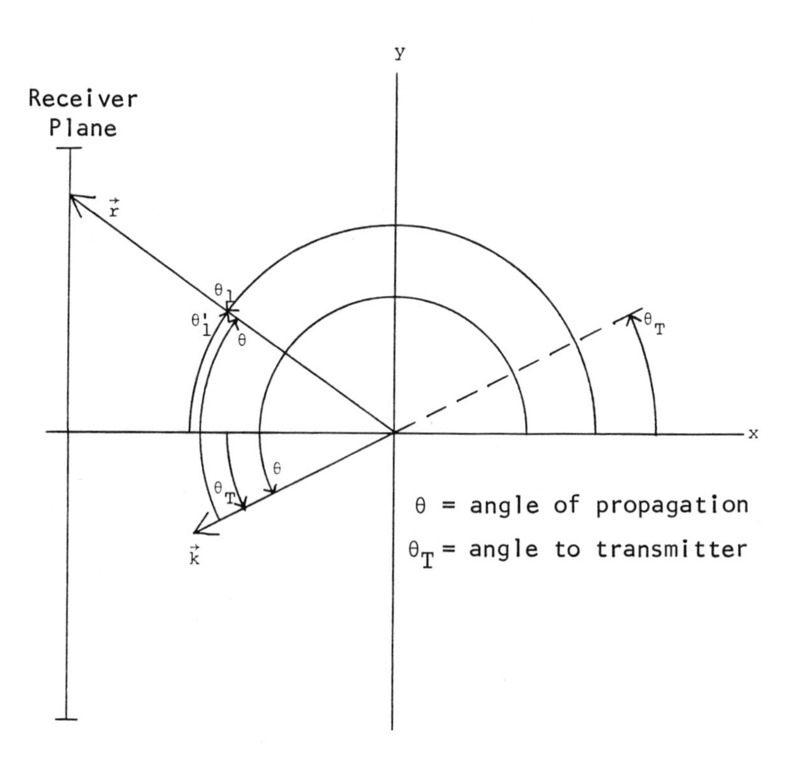

Figure A-1. Simulated signal geometry

REFERENCES

1. K. Iwata and R. Nagata, "Calculation of Three-Dimensional Refractive Index Distribution from Interferograms", J. Opt. Soc. of America, Vol. 60, pp. 133 - 135, 1970.

2. E. Wolf, "Three-Dimensional Structure Determination of Semi-Transparent Objects from Holographic Data", Optics Communications, Vol. 1, pp. 153 - 156, 1969.

3. R. Bracewell and A. Riddle, "Inversion of Fanbeam Scans in Radio Astronomy", The Astrophys. J., Vol. 150, pp. 427 - 434, 1967.

4. D. DeRosier and A. Klug, "Reconstruction of Three-Dimensional Structures from Electron Micrographs", Nature, Vol. 217, pp. 130 - 134, 1968.

5. R. Gordon, R. Bender and G. Herman, "Algebraic Reconstruction Techniques (ART) for Three-Dimensional Electron Microscopy and X-Ray Photography", J. Theor. Biol., Vol. 29, pp. 471 - 481, 1971.

6. J. Greenleaf et al., "Algebraic Reconstruction of Spatial Distributions of Acoustic Absorption within Tissue from their Two-Dimensional Acoustic Projections", Acoustical Holography, Vol. 5, pp. 591 - 603, 1974.

7. J. Greenleaf et al., "Algebraic Reconstruction of Spatial Distributions of Acoustic Velocities in Tissue from their Time-of-Flight Profiles", Acoustical Holography, Vol. 6, pp. 71 - 90, 1975.

8. R. Mersereau and A. Oppenheim, "Digital Reconstruction of Multidimensional Signals from their Projections", IEEE Proceedings, Vol. 62, pp. 1319 - 1338, 1974.

9. G. Herman and S. Rowland, "ART: Mathematics and Applications. A Report on the Mathematical Foundations and on Applicability to Real Data of the Algebraic Reconstruction Techniques", J. Theor. Biol., Vol. 43, pp. 1 - 32, 1973.

10. S. Bezuszka, "Scattering of underwater plane ultrasonic waves by liquid cylindrical obstacles", Journal of the Acoustical Society of America, Vol. 25, November 1953.

SCANNING ACOUSTIC MICROSCOPE OPERATING IN THE REFLECTION MODE

B. Nongaillard, J. M. Rouvaen, E. Bridoux, R. Torguet
and C. Bruneel

Laboratoire d'Optoacoustoelectrionique ERA n° 593 CNRS
Université de Valenciennes, 59326 Valenciennes-Cedex
France

The microhardness, acoustical absorption, and other mechanical properties of thin samples may be imaged using the techniques of acoustical microscopy (1). A widely studied scheme is the scanning acoustic microscope, inititially proposed by professor Quate (2). With this confocal microscope one can obtain a number of images in the transmission mode, with a diffraction limited resolution (3). Some results have also been given on the reflection mode of operation (4).

Our aim is to visualize structural details under the apparent surface of thick specimens, using a reflection acoustic microscope. The applications cover the areas of metallurgy, petrography and non-destructive testing. The focussing of the acoustic beam under the apparent surface of a thick sample has been studied. For this purpose, the propagation equation has been solved in the various media encountered: the delay line, the fluid coupling medium, and the sample (see Fig. 1).

A computer calculation, using the Rayleigh-Sommerfield Diffraction integral, has been performed to evaluate the acoustic field in the vicinity of the focal plane. In the first stage the velocity potential is computed inside some planes lying at different distances from the acoustic lens center of curvature in the liquid coupling medium (5).

$$\Phi(M) = \frac{k}{j\pi} \int_{\theta=0}^{\theta_m} \int_{\psi=0}^{\pi} A(\theta) \exp\left[jkR(1-\cos\theta)/n\right] \frac{\exp(-\alpha+jk)r}{r} \cos(\vec{n},\vec{r}) \, R^2 \sin\theta \, d\theta \, d\psi$$

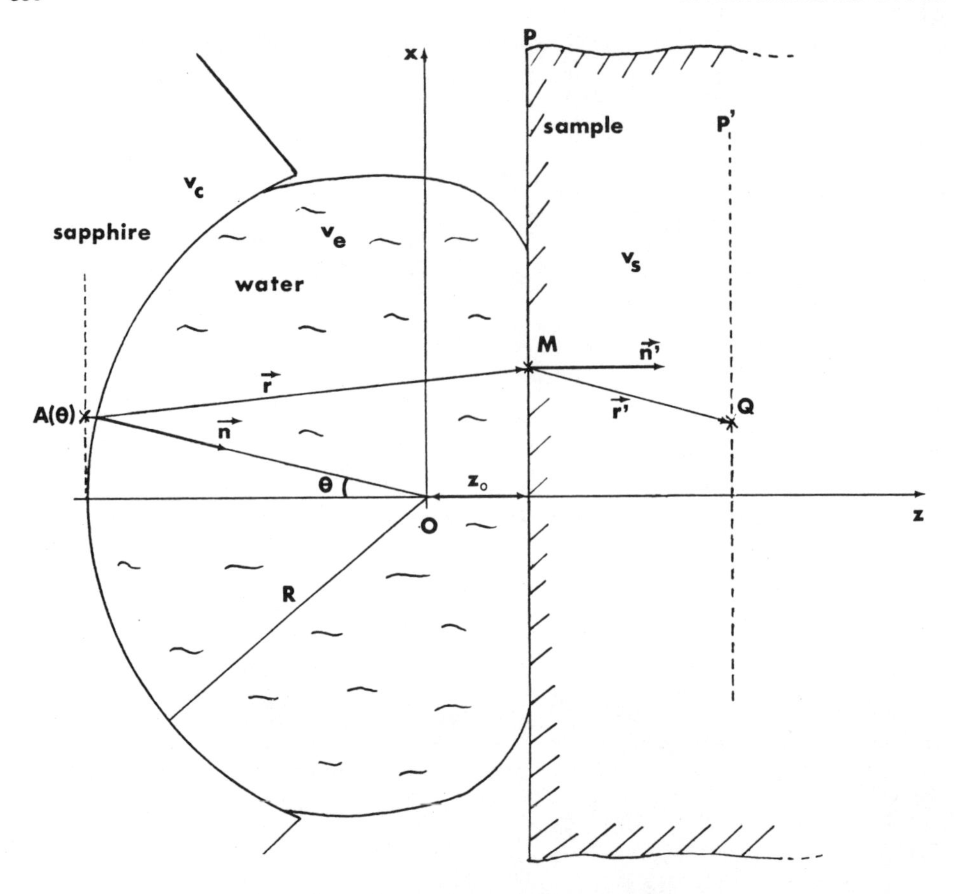

Fig. 1. Geometry of the system.

where A (θ) is the incident field on the lens, α is the attenuation coefficient of the coupling medium (in Neper per meter unit), $n = v_c/v_e$, the refractive index of sapphire with respect to the water, R and θ_m are the radius of curvature and the angular aperture, respectively of the lens, and k is the wave number in water. The velocity potential is then computed in the bulk of a solid sample, with any absorption neglected to a first approximation:

$$\phi(Q) = \frac{k}{j\pi n'} \int_{\theta=0}^{\pi} \int_{x=0}^{x_m} \phi(M) \frac{\exp(jkr'/n')}{r'} \cos(\vec{n'}, \vec{r'})\, x\, dx\, d\theta$$

where $n' = v_s/v_e$ is the refractive index of the sample with respect to the water.

The validity of the results is demonstrated by taking the refractive index of the sample with respect to water equal to unity, satisfying the Huyghens principle quite well. The axial and transverse potential variations against the sample refractive index have been studied. It is seen here (Fig. 2) that the diffraction limited acoustic spot width is nearly proportional to the refractive index. The decrease of the numerical aperture resulting from the total reflection phenomenon on the sample surface is negligible and the spatial resolution is given by the classical formula

$$r = \frac{\lambda}{2NA}$$

where λ is the wavelength in the sample.

The longitudinal variations (see Fig. 3) give the depth of focus and the distance from the sample surface to the focal point. The former obeys the law

$$d' = d/n'$$

which is seen by using simple geometrical acoustics.

The acoustic wave velocity is 4000 meters per second in typical rocks as compared to 1500 meters per second in the water coupling medium. Using a 130 megahertz operating frequency, the attainable spatial resolution is nearly 6 micrometers at the surface and 20 micrometers in the bulk of the sample. The acoustic lens is ground in an 8 millimeter long sapphire rod with a 1 millimeter radius of curvature and an angular aperture of 60 degrees. The occurence of parasitic reflected signals which may mask the useful information carrying signal has led us to an impulse mode of operation. The parasitic signal F (see Fig. 4) arises from the reflection of electric energy at the imperfectly matched piezoelectric transducer. Moreover, the difference in the acoustic impedances of the delay line and coupling fluid produces a strong reflected acoustic signal R at the acoustic lens spherical surface. The useful signal E which suffers attenuation inside the coupling liquid during transmission through the boundaries is typically 30 decibels lower than the parasitic signals. A delayed gate system (see Fig. 5) enables the separation of the useful pulsed signal from the parasitic ones. A sample and hold is used to get a bright continuous image.

In order to check the sensitivity and spatial resolution of our system, test grids have been imaged (see Fig. 6).

The ability to obtain a transmission image by using a reflector plane under the studied specimen has been demonstrated. The sample was a 6 micrometer thick mylar sheet. When water is present behind the back surface, only some defects and dust are

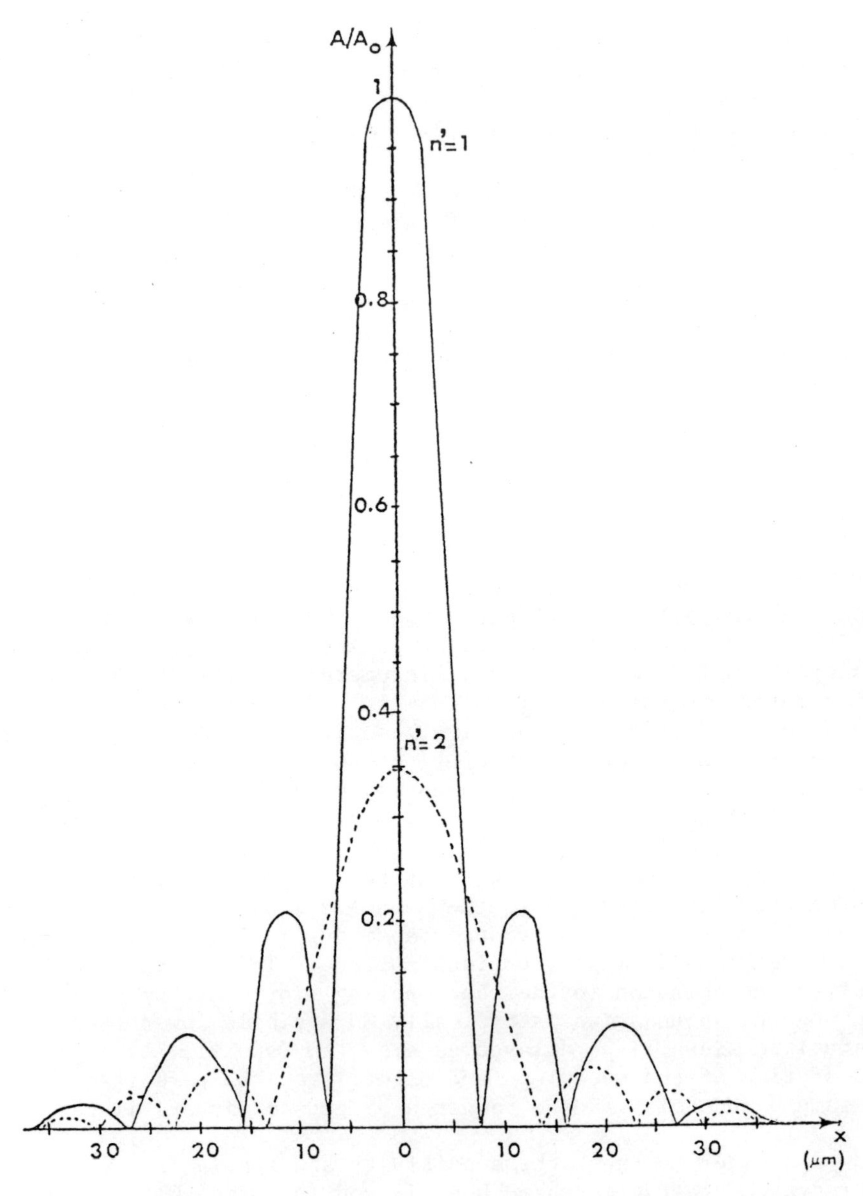

Fig. 2. Transverse potential variations in the focal plane.

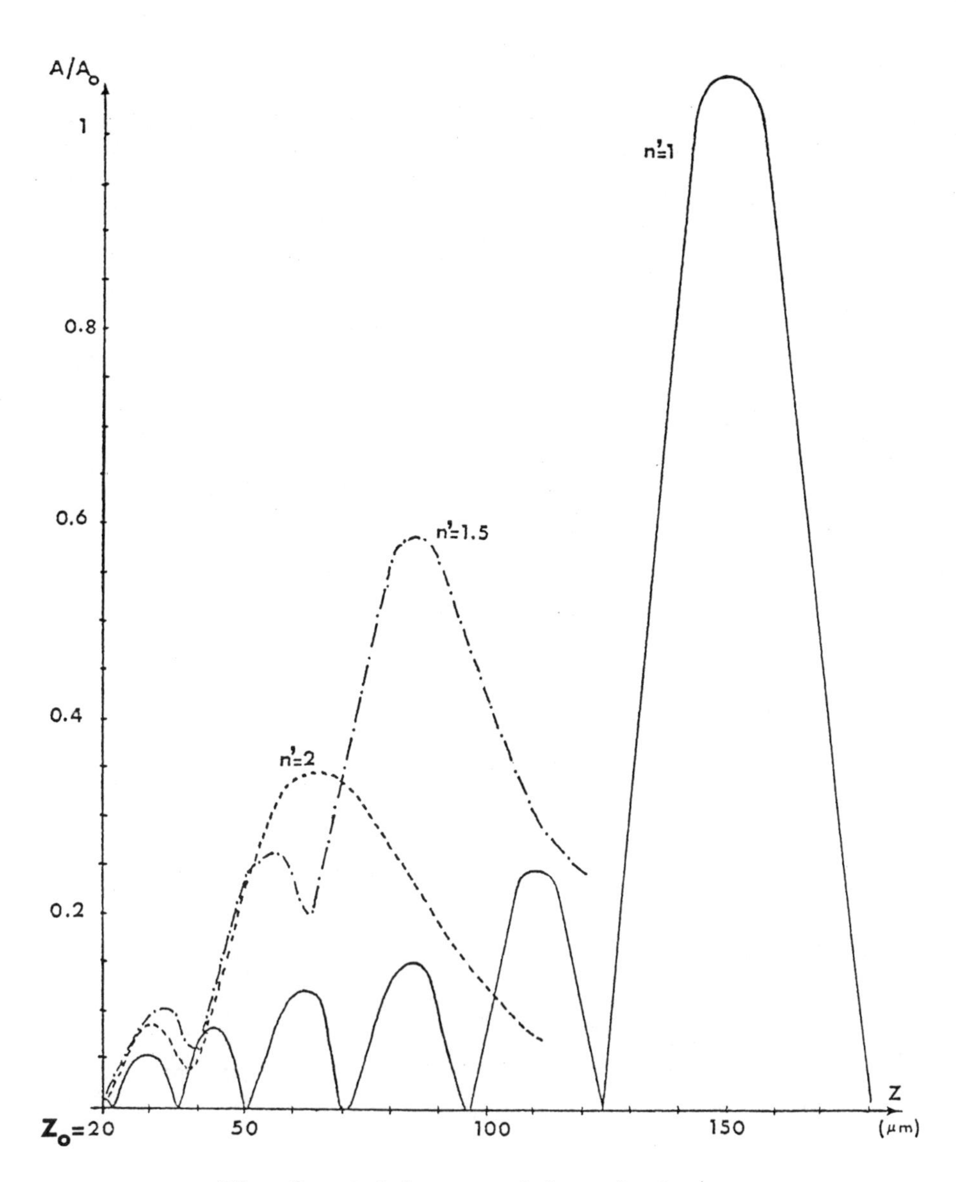

Fig. 3. Axial potential variations.

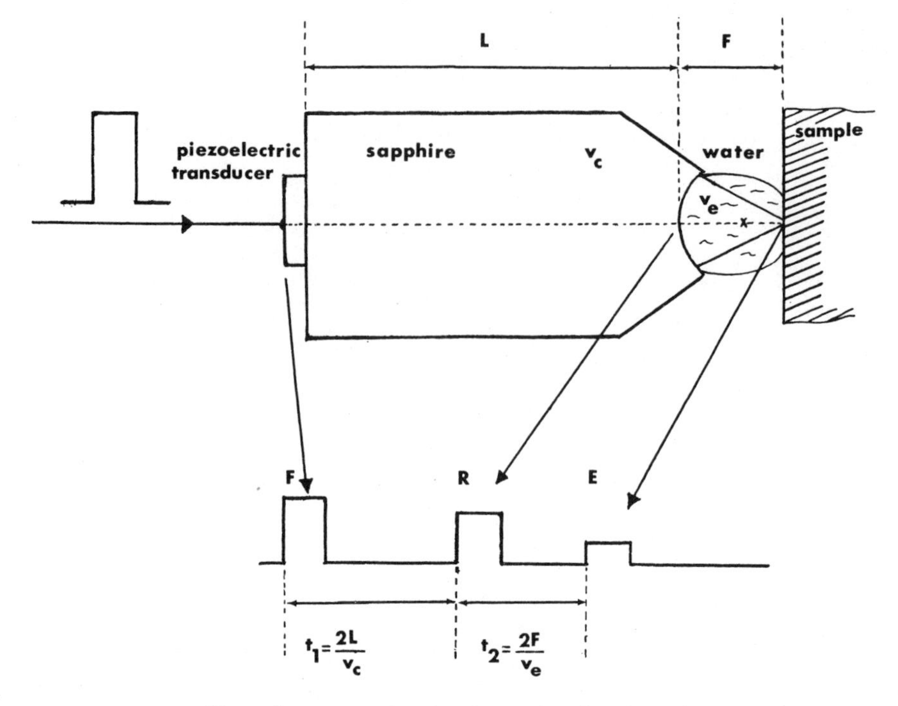

Fig. 4. Received electrical signals

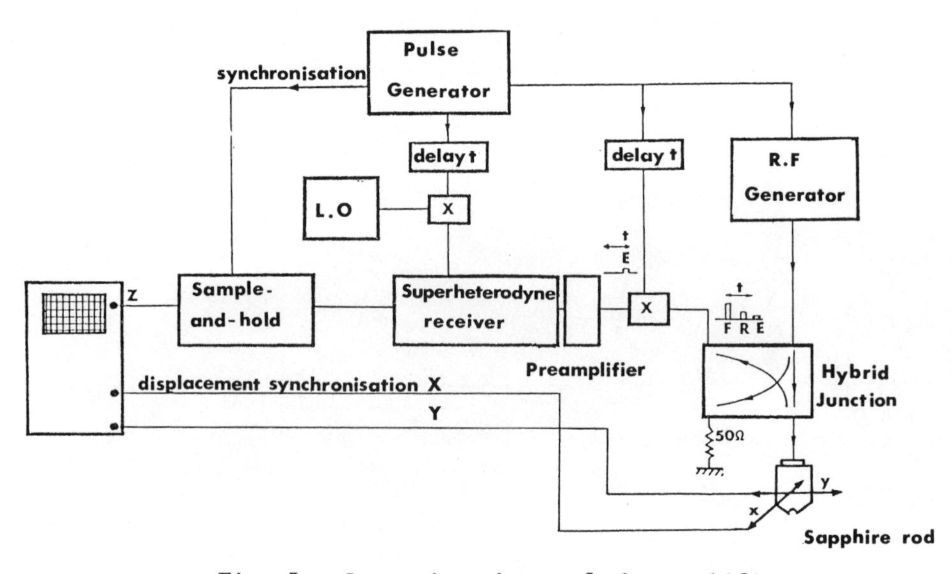

Fig. 5. Synoptic scheme of the amplifier.

Fig. 6. 12 micrometer bars.

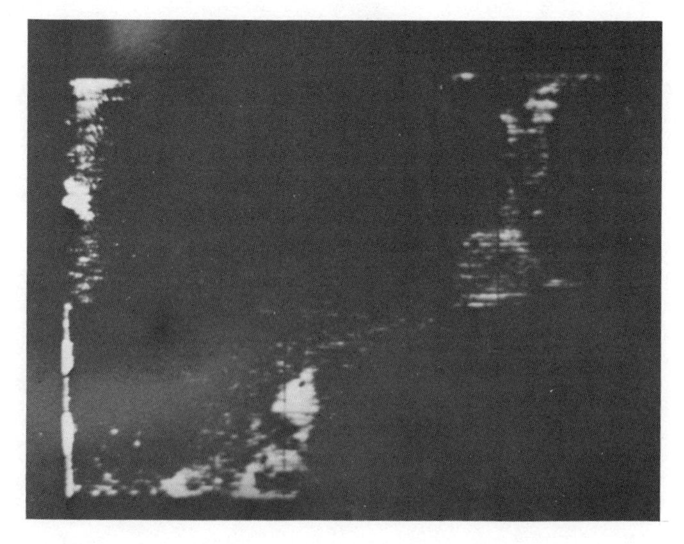

Fig. 7. 6 μm thick mylar sheet with water behind the back surface.

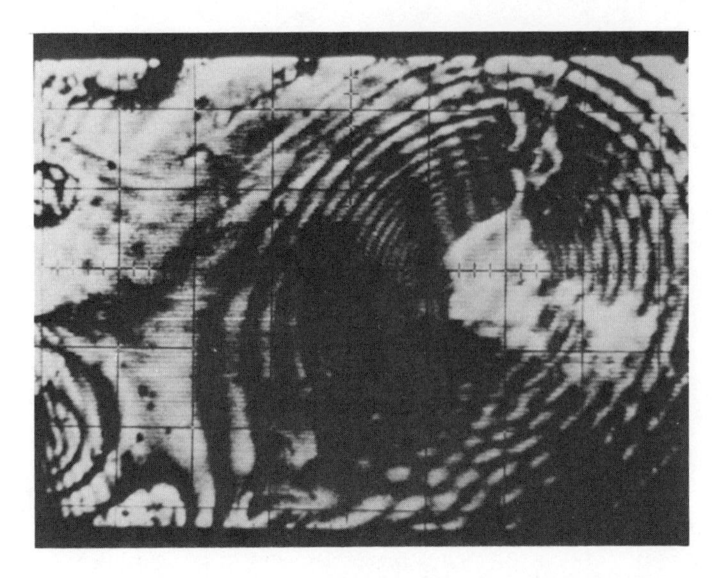

Fig. 8. The mylar sheet with air behind the back surface.

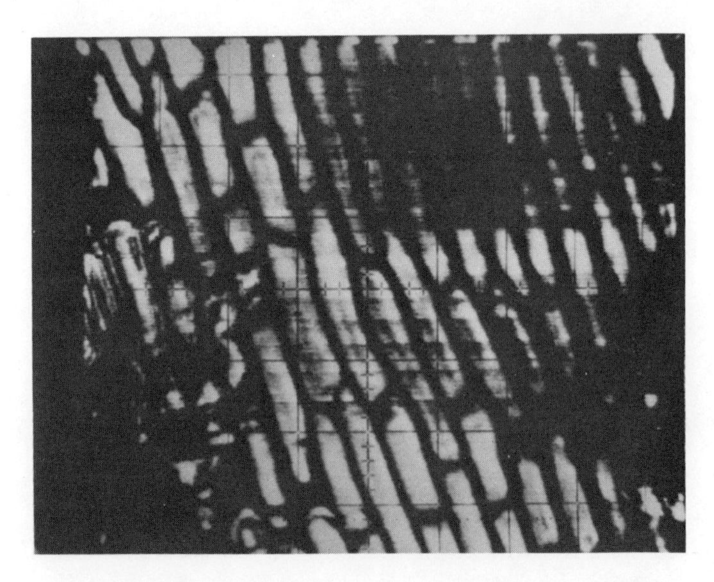

Fig. 9. Onion skin with a reflector plane behind the back surface.

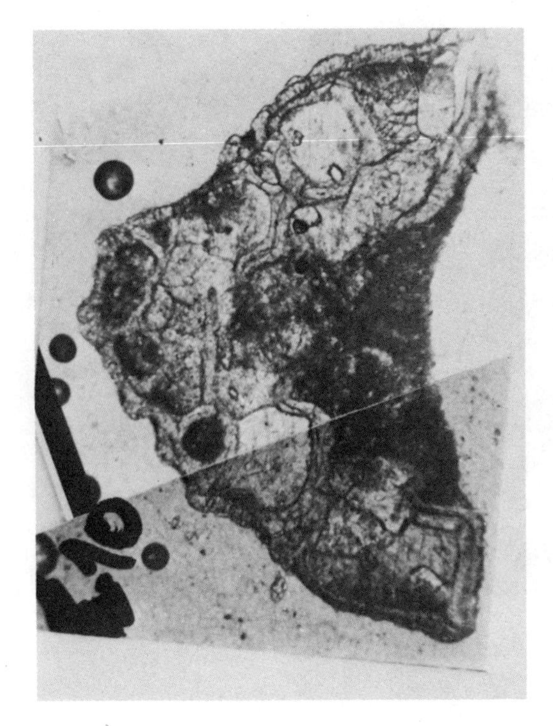

Fig. 10. Opitcal image of a fossil, 0.5 mm long.

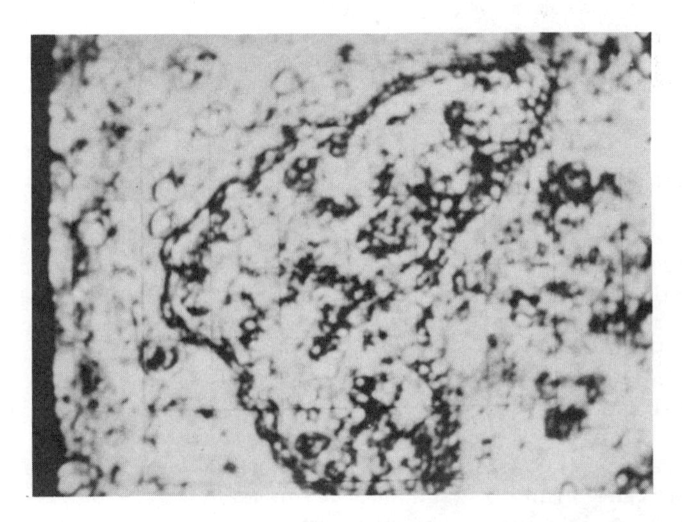

Fig. 11. Acoustical image of the sample surface.

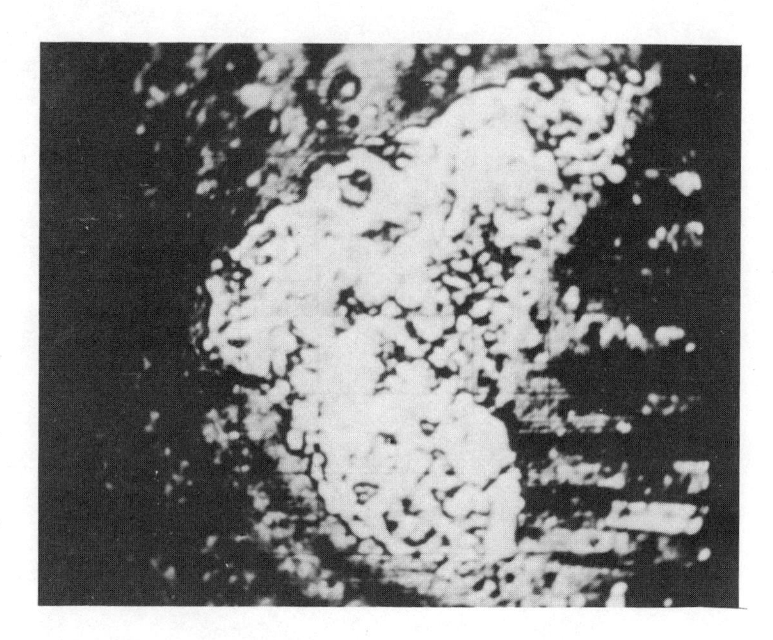

Fig. 12. Acoustical image of a plane lying at 100 μm depth.

visible (see Fig. 7). If this surface is loaded only by air, the contrast is enhanced and the dust is seen more clearly (Fig. 8). A reflection acoustic microscope may therefore be easily converted to the transmission mode, a desirable feature for the observation of biological poorly-reflecting samples. As an example, the highly contrasted image of an onion skin is given here. (Fig. 9).

Thin rock slices have also been observed. A typical optical image of a fossil is given here (Fig. 10) and the acoustical image of the sample surface is shown (Fig. 11) together with that of a plane lying at nearly 100 micrometers depth.

In order to visualize structures lying near the surface of a sample, pulses of very short duration must be used. The acoustic echoes from the surface and the studied plane may then be separated in time. When studying more deeply inside the specimen, the acoustic lens radius of curvature must be correctly chosen according to the nominal observation depth. Applications to the localization of very small diameter defects may therefore be devised in the non-destructive testing area. The very fast conversion from the reflection to the transmission operation mode may prove fruitful in the study of thin biological specimens.

REFERENCES

1. Sokolov, S., "The Ultrasonic Microscope", Akademia Nauk SSSR, Doklady, 64, 333 (1949).
2. R. A. Lemons and C. F. Quate, Appl. Phys. Lett., 24, 163 (1974).
3. R. A. Lemons and C. F. Quate, "A Scanning acoustic microscope," 1974 Ultrasonics Symposium Proceedings, IEEE Cat. 41, 44 (1974).
4. R. A. Lemons and C. F. Quate, Appl. Phys. Lett., 25, 251 (1974).

A PRACTICAL LINEAR ARRAY IMAGING SYSTEM FOR NONDESTRUCTIVE TESTING APPLICATIONS

James M. Smith

Army Materials and Mechanics Research Center

Watertown, Massachusetts 02172

ABSTRACT

A prototype ultrasonic inspection system, using a sequentially fired transducer array, has been built that is much faster than conventional single-element tranducer systems. This system was assembled with off-the-shelf components and would appear to be an attractive solution to nondestructive testing problems that require fast inspection rates and have resolution and sensitivity requirements that are consistent with the use of unfocused transducer elements. The fast inspection of a number of Army components, such as artillery shells, projectile rotating bands, and tank track pads, has been demonstrated.

INTRODUCTION

In the last few years there has been a growing interest in the development of ultrasonic inspection systems that are capable of high inspection rates. Within the Army there is a strong need for fast ultrasonic inspection systems for large-caliber projectiles, for gun tubes, for critical aircraft and tank components, and for a number of other products that must be tested in large numbers. In many cases the required inspection rate cannot be achieved in a practical fashion with single-element transducers, and the use of transducer arrays becomes attractive. This paper will discuss a linear array acoustical imaging system that was built at AMMRC with off-the-shelf electronic components in order to keep the costs down.

This system is fast in comparison to conventional nondestructive testing equipment, and is an effective inspection tool for a number of Army components.

PRINCIPLES OF OPERATION

The heart of the AMMRC acoustical imaging system is a pulser-receiver, built by the Advanced Diagnostic Research Corporation [1], that sequentially excites a 64-element piezoelectric transducer array. This instrument became available commercially about three years ago and was originally designed to give physicians a real-time B-scan [2] display primarily of the abdominal region. It has been a valuable tool for the study of fetal activity in pregnant women. It became apparent to AMMRC that this unit could be successfully applied to nondestructive testing applications in a much more useful C-scan [2] format if the ultrasonic data were processed and displayed differently. The block diagram, shown in Figure 1, illustrates the approach that was taken to adapt the ADR instrument to nondestructive testing applications.

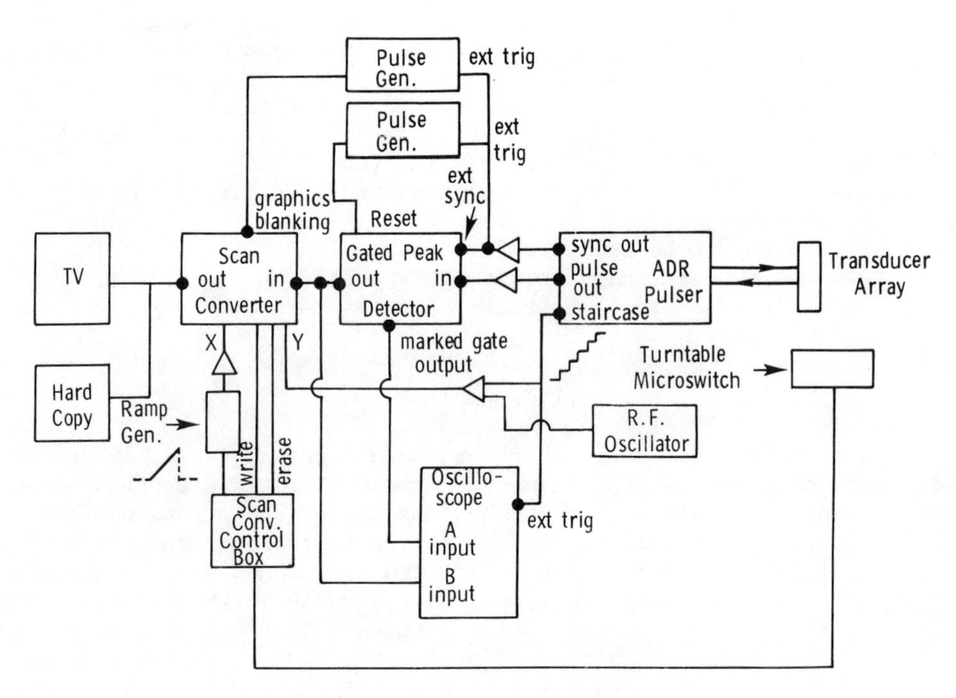

Figure 1. Block diagram.

The operation of this system begins with the pulser-receiver and the transducer array. The ADR pulser-receiver, which operates in a pulse-echo mode, excites four adjacent transducer elements at one time and then listens for the returned echoes before the next set of four elements is fired. For example, in Figure 2, the first four elements (1, 2, 3, 4) are excited; then elements 2, 3, 4, 5; then 3, 4, 5, 6, and so on until all of the 64 elements in the array have been fired in this fashion. Each set of four elements is treated as one larger, single-element transducer. It takes 1/40 of a second to sweep across the entire array in the above manner, which corresponds then to the time required to generate one line of the acoustic image. A complete image is formed by moving the array or the object along a line perpendicular to the axis of the array or, for the case of a cylindrical symmetric object, by rotating the object through 360 degrees.

The ultrasonic echoes received by each set of four elements are time gated, peak detected, and then stored in an analog scan converter. The timing requirements for the gated peak detector (Panametrics Model 5052GPD-1) and the scan converter (PEP-500) are accomplished by driving these instruments synchronous with the ADR pulser-receiver. The gated peak detector holds the detected acoustic

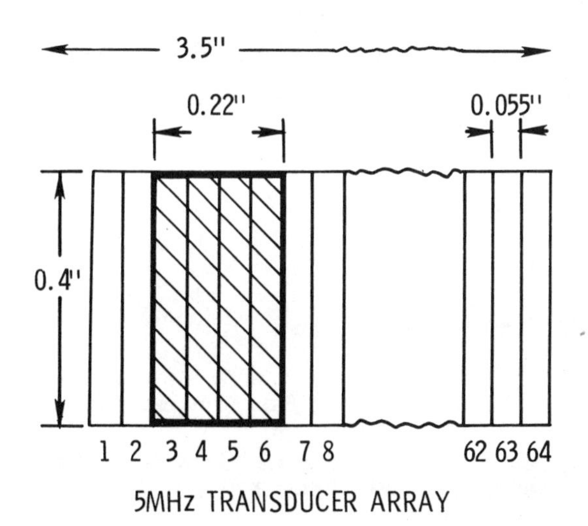

Figure 2. Diagram of transducer array.

echo for approximately 20 μsec and is then reset to zero in preparation for the next transducer excitation. In order to reduce the background noise level of the acoustic image, the scan converter is switched into the writing mode (through the graphic blanking input) only during this same period of time when a signal from the peak detector is expected.

The X and Y input signals to the scan converter provide the proper positioning of the electron beam within the scan converter in order to produce an acoustical image. The X input is a linear ramp voltage synchronous with the "X" position of the component under examination. The Y axis is the staircase output from the ADR instrument which indexes the scan converter in the vertical axis for each transducer excitation. However, added to each step is a small RF signal (1 MHz), which accomplishes the task of smearing together the index lines, thereby producing a more continuous image. At the completion of the scan, the stored image in the scan converter is read into a TV monitor or into a hard copy unit. The hard copy produces a permanent gray scale record on an 8-1/2×11" sheet of paper.

EXPERIMENTAL RESULTS

The fast inspection of a number of Army components has been demonstrated at AMMRC with the linear array system. Some typical examples are shown in Figure 3 and are listed below (from right to left in the figure) with the corresponding inspection problem:

	Inspection Component	Inspection Problem
1.	An 8-inch artillery shell motor body	Bonding of the copper rotating band to the steel
2.	A 155-mm artillery shell warhead	Crack detection
3.	A graphite epoxy compact tension specimen	Delamination between plies
4.	A tank track pad	Rubber to metal bonding

A. Rotating Band Inspection

The array imaging technique has been successfully applied to the inspection of brazed and welded overlay copper rotating bands bonded to 8-inch and 155-mm artillery shell motor bodies. The problem is one of determining whether the band is adequately bonded to the motor body. The motor body is placed on a small turntable and the transducer array is positioned inside the shell with the axis of the array pointed in the vertical direction, as shown in Figure 4.

Figure 3. Army components inspected with prototype linear array system.

Figure 4. Transducer array and rocket motor body.

Only one full rotation of the shell is required to generate an acoustic image of the entire rotating band region. This procedure takes less than 10 seconds, which is considerably faster than the time required (3 to 5 minutes) to inspect a band with a conventional single-element transducer.

The acoustic image of an 8-inch brazed copper rotating band is shown in Figure 5. The dark patches on the TV monitor represent unbonded areas between the 8-inch motor body and the rotating band. This particular motor body, which contained intentional unbonded regions produced by placing Grafoil disks between the shell body and the band before the brazing operation, had previously been used as a standard for conventional ultrasonic tests of rotating bands. The C-scan of Figure 5 compares very favorably with the conventional scans as documented in Reference 3.

An illustration of the close agreement between conventional C-scans and the array imaging approach, as applied to the rotating band problem, is shown in Figure 6. The top image is a conventional scan (with an unfocused 10 MHz transducer) of a welded overlay band on an 8-inch rocket motor body, and the lower C-scan was generated by the fast imaging technique using a 5 MHz transducer array. The five flat-bottom holes that were drilled through the band, ranging from 1/8" to 1/4" diameter, are clearly visible in both scans. The long dark region on the left indicates a poorly bonded section of the band that was previously documented in Reference 3.

The results with the linear array technique have been so encouraging that funding for a production-line station to inspect rotating bands for a particular 155-mm round has been approved. This equipment will be built and evaluated during Fiscal Year 1978.

B. Crack Detection in Warheads

To demonstrate the crack detection capability of the linear array technique, a number of shadow axial saw cuts were made in the exterior sidewall (ogive, bourrelet, and base region) of a 155-mm warhead. The depth of these cuts ranged from 10 to 60 mils and their lengths were approximately one-half inch. All notches were easily detected with the linear array system using 45° circumferential shear waves which were generated by laterally offsetting the transducer array away from the normal direction. The top image of Figure 7 demonstrates the detection of three notches (20, 40, 60 mils deep) near the bourrelet. The sensitivity of the imaging system was turned down in this example. As the sensitivity is increased (the lower two images), a number of minor surface blemishes in the shell sidewall are also detected. This technique is also sensitive

Figure 5. C-scan of brazed copper rotating band.

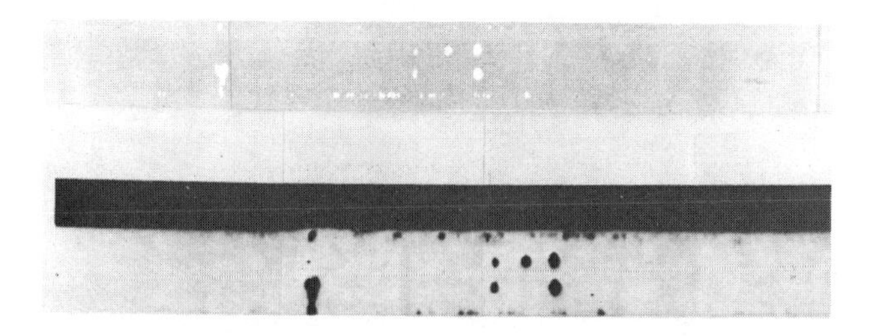

Figure 6. C-scan of welded overlay rotating band. The top image is a conventional scan, the bottom image was produced with the array imaging technique.

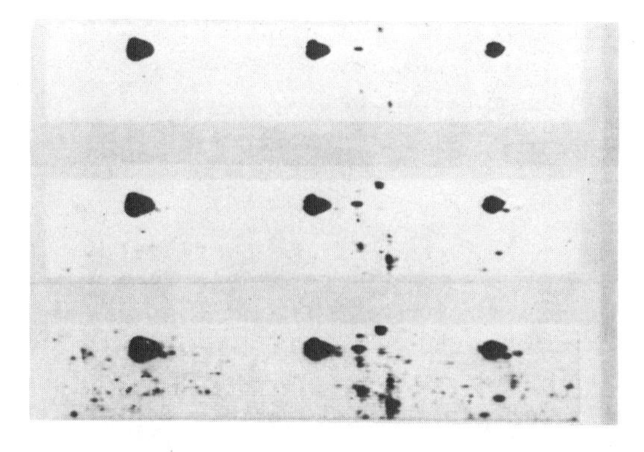

Figure 7. Three acoustical images of notches in a 155 mm projectile. The sensitivity of the imaging system was increased in going from the top image to the lower one.

to large inclusions and is currently being used in conjunction with metallographic examinations to study the inclusion content in a recently developed 155-mm round.

C. Inspection of Tank Track Pad and Graphite Epoxy Specimens

Figures 8 and 9 show a comparison between conventional C-scans (top) and the images produced with the linear array technique (bottom) for a tank track pad and graphite epoxy compact tension specimens. Both the track pad and the graphite epoxy specimens were imaged in less than 5 seconds with the linear array. The track pad was inspected from the steel side and the ultrasonic echoes were gated to detect unbonds between the steel and the rubber. In the graphite epoxy specimens, the problem was one of detecting delaminations between plies. These two examples are interesting because the conventional acoustical images, produced with a focused 15 MHz transducer, are representative of the best that is possible with conventional NDT equipment. As expected, because the array elements are unfocused, the images produced with the linear array do not match the resolution and sensitivity of these conventional C-scans. However, for many Army nondestructive problems, the resolution and sensitivity of the linear array technique is more than adequate.

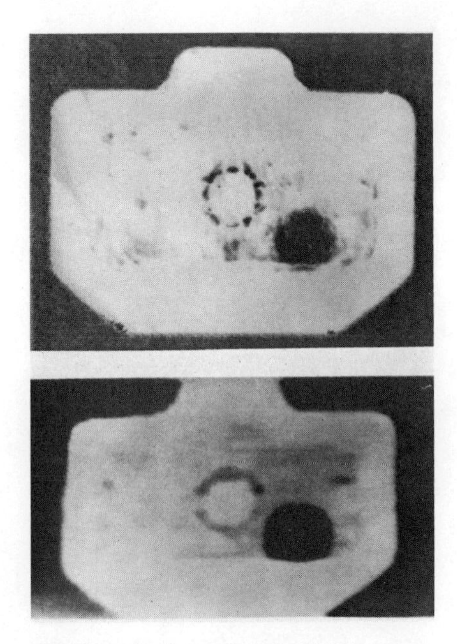

Figure 8. Comparison between a conventional C-scan (top) and that produced with the linear array (bottom) for a tank track pad.

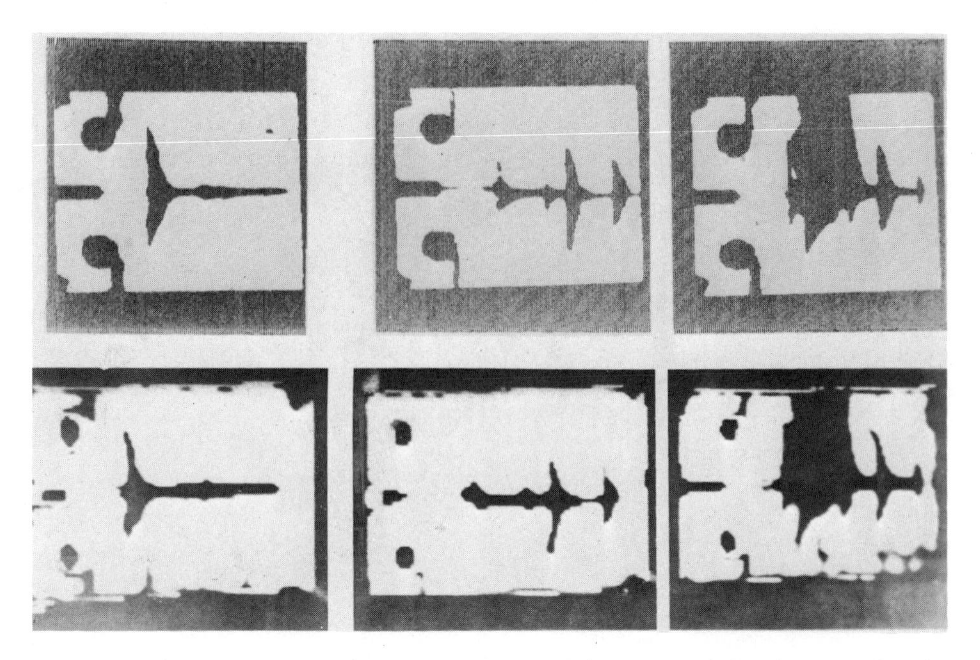

Figure 9. Comparison between conventional acoustic images (top) and those produced with the linear array (bottom) for graphite epoxy specimens.

CONCLUSIONS

A prototype ultrasonic inspection system, using a sequentially fired transducer array, has been built that is much faster than conventional single-element transducer systems. This system was assembled with off-the-shelf components and would appear to be an attractive solution to nondestructive testing problems that require fast inspection rates and have resolution and sensitivity requirements that are consistent with the use of unfocused transducer elements. The essential electronic components used in this system are easily available, and the cost of applying this technique to production line testing is not much more than conventional UT.

ACKNOWLEDGMENT

The author would like to thank R. H. Brockelman and D. J. Roderick for many useful discussions concerning the inspection of rotating bands in large caliber projectiles.

REFERENCES

1. Advanced Diagnostic Research Corporation, 2202 South Priest Drive, Tempe, Arizona 85282.

2. A description of B-scan and C-scan formats is given in the AMMRC Report (MS 77-7) entitled "A Survey of Advanced Techniques for Acoustic Imaging" by J. Smith.

3. AMMRC TR 76-20, "Ultrasonic Inspection of Brazed and Welded Overlay Rotating Band Attachment on Artillery Shells" by F. S. Hannon, R. H. Brockelman, J. M. Quigley, and D. J. Roderick.

EXPERIENCES IN USING ULTRASONIC HOLOGRAPHY IN THE LABORATORY AND IN THE FIELD WITH OPTICAL AND NUMERICAL RECONSTRUCTION

V. Schmitz and M. Wosnitza

Institut für zerstörungsfreie Prüfverfahren
Universität, 6600 Saarbrucken 11, FR Germany

1. INTRODUCTION

The holographic method was described and demonstrated by Gabor, Greguss, Leith-Upatnieks, Thurstone and Metherell[2] among others. For its application there is a need for coherent radiation. In conventional imaging the intensity distribution is recorded and in holography the complex amplitude distribution is recorded. There are two ways of realizing holographic reconstruction: optical and computer processing. In this paper experiences with both methods in the laboratory and in-situ gathered in the last two years are reported, in using ultrasonic holography as a modern tool for imaging flaws in thick-walled sections and in inspection of reactor pressure vessels.

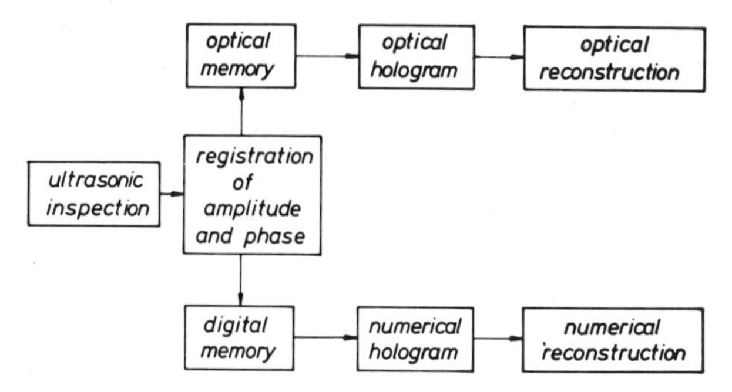

Fig. 1. Basic arrangement for optical and numerical reconstruction.

2 PRINCIPLES OF OPERATION

We applied a system with simultaneous scanning of source and receiver. That means, a single probe radiates an ultrasonic longitudinal or shear wave into the material and receives the flaw signal. The probe scanning over the surface measures the phase and amplitude in each position. After mixing the object signal with a reference signal the complex amplitude and phase are stored. Fig. 2 shows the system we are working with[1].

Fig. 2. Equipment for ultrasonic holography with optical and numerical reconstruction.

2.1 OPTICAL RECONSTRUCTION

A thorough description of the system was given in earlier volumes of the series "Acoustical Holography"[2]. Thus we will discuss no details in this paper. The principle is described in the following scheme. For registration of phase and amplitude an inclined plane reference wave is simulated by a phase shift of the reference pulse depending on the locus of the probe. Having mixed the object and reference signals electronically, the resulting signal is used to modulate the intensity of a storage scope or a light emitting diode mechanically coupled to the motions of the probe. The hologram is formed synchronously with the scanning of the probe. Copied on a transparency and with a monochromatic light passing through it the picture of the flaw is presented on a monitor.

A basic question in NDT systems is for detectability and resolution of small flaws. In imaging systems, flaw lateral magnification, flaw longitudinal magnification, dependency of frequency, flaw depth, flaw inclination and system parameters are of great importance. Details of the concept are published by B. P. Hildebrand

and B. B. Brenden[3]. In this section we will only discuss those formula which are important for practical NDT.

The lateral (Δx) and axial (Δr) resolution is compared with the resolution of the probe,

$$\Delta x = 1{,}22 \cdot \frac{c}{f} \cdot \frac{FL}{D} \qquad \Delta r = 2 \cdot \frac{c}{f} \cdot \left(\frac{FL}{D}\right)^2 \tag{1}$$

where

c = velocity of ultrasonic wave
f = ultrasonic frequency
FL = focal length of the probe
D = diameter of the probe

With a short focal length and a large probe, it is possible to get a lateral resolution of 1 wavelength. Thus two flaws which are separated by a distance of length can be distinguished. In metal with a frequency of 5 MHz this distance is about 1 mm.

The axial resolution is not satisfactory. In the above mentioned example, this distance is greater than 10 mm. Therefore, we determine the depth by time-of-flight measurements. With the pulse-echo overlap method it is possible to determine the depth to better than 0.1 mm. Figure 4 shows the principle. After performing the hologram we position the probe above the middle of significant flaws. This has been painted out by the white line. In the A-scan, first there is the transmitted pulse and after the time of flight through the water coupling the surface echo and then the flaw echo appear. With the aid of a dual beam oscilloscope Tektronix 7844

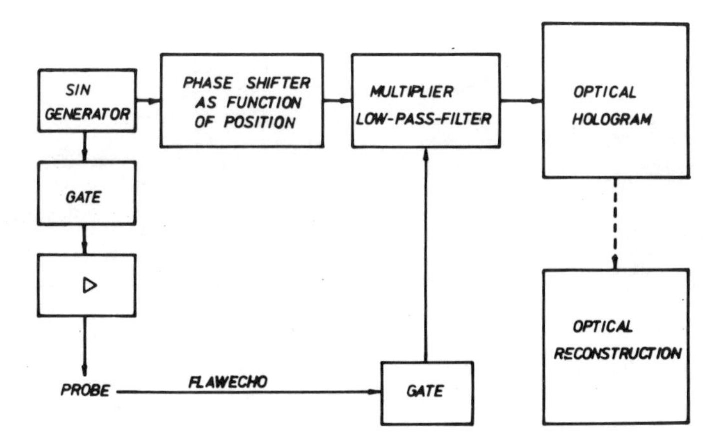

Fig. 3. Ultrasonic holography with optical reconstruction

and two times bases 7B85 and 7B80 we duplicated the HF-signal and shifted the flaw echo beneath the surface echo. An expanded view is seen in figure 4c where we used not the envelope of the signal but the HF signal. The shift can be done to an accuracy of one tenth of the wavelength and is read out as Δ. This can be performed for different parts of the hologram in a few minutes. By line scanning there is the possibility to get the depth profile of an extended flaw like a crack. The size of the flaw is determined by

$$F = I / M \tag{2}$$

where F = flaw size, I = image size, M = magnification.

a) Position for time measurement

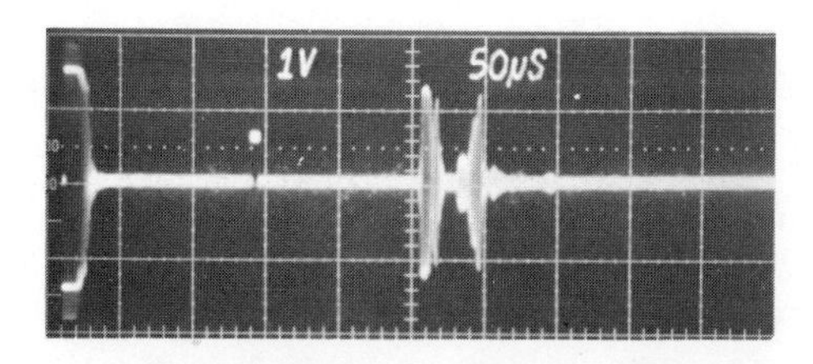

b) Transmit pulse surface echo flaw echo

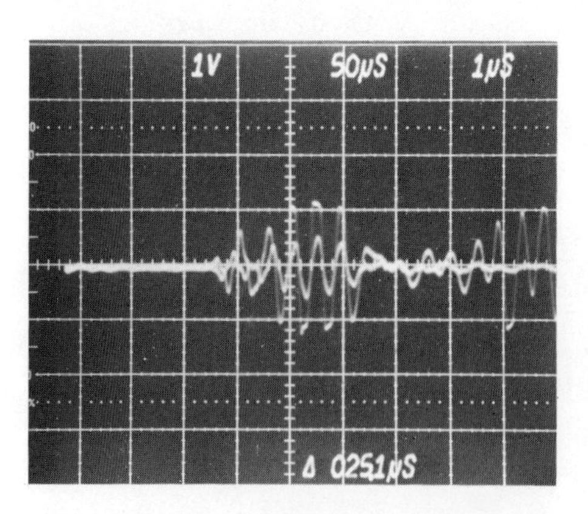

c) Pulse-echo overlap method for flaw depth measurement

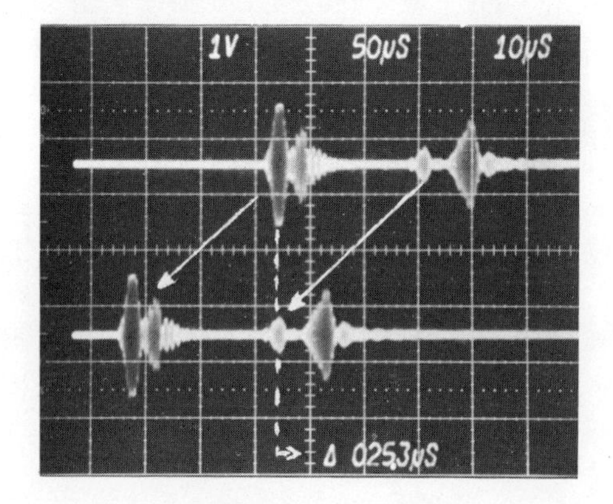

d) Δ = time between surface pulse and flaw-echo

Fig. 4. Methods for determining the flaw depth.

This magnification depends on system parameters, wavelength λ and flaw depth r.

$$M = \frac{const}{\lambda \cdot r} \qquad (3)$$

It follows that the image size is inversely proportional to the wavelength and the flaw size. Flaws at different depths are imaged with different sizes and it is necessary to use formula (3). The distance of the flaw image behind the hologram depends upon the ultrasonic and light wavelength and the reduction factor m of the scanned aperture to the size of the transparency and the flaw depth r:

$$r_{HB} = \frac{\lambda_S}{\lambda_L} \cdot \frac{1}{2m^2} \cdot r \qquad (4)$$

In practice, the right hand side of equation (4) is known _a priori_ or by measurement. At the correct distance HB behind the hologram, the well focused image of the flaw is formed. By measuring the size of the image I and utilizing formula (3), the size of the flaw may be determined.

2.2. NUMERICAL RECONSTRUCTION

For numerical reconstruction the system is modified according to Fig. 5. By switching off the phase shifter the reference signal becomes now a sine-oscillation. In a parallel channel the sine-oscillation is shifted in phase by constant 90° so as to get a co-sine-oscillation. After multiplying the flaw signal on the one side by sine and on the other side by cosine, we obtained the real and imaginary parts of the holographic signal. This is performed in the lower part of the electronic processor in the middle of Fig. 1 with the aid of a charge coupled device (CCD) controlled by a frequency synthesizer. Via an ADC the data are fed into a PDP 11/34 computer. The ADC we used is a 100 MHz Biomation model 8100. By this means we could change the sample interval over a wide range. The reconstruction of the flaw image is presented on a Calcomp plotter. The process of recording amplitude and phase by the outlined method has been outlined by Deschamps[5], K. Magurak[6], R. Diehl[7], R. Mueller[8]. J. Kutzner, H. Wustenberg described equipment for linear acoustical holography in NDT with numerical reconstruction based upon this method[9].

The mathematical formulation of hologram reconstruction is based upon the scalar diffraction theory. U(x,y) is the complex field in a given aperture Σ. The complex field at a distance r behind this aperture is given by the integral transform [10]

$$U(x_B,y_B) = \iint_{\Sigma} h(x_B,y_B,x_H,y_H) \cdot U(x_H,y_H)\, dx_H\, dy_H \tag{5}$$

with the weighting function

$$h(x_B,y_B,x_H,y_H) = \frac{1}{i\cdot\lambda}\frac{1}{r_{HB}} \cdot e^{ikr_{HB}} \cos(\vec{n}, \vec{r}_{HB}) \tag{6}$$

and with $U(x_H,y_H)$ identical to zero outside the aperture.

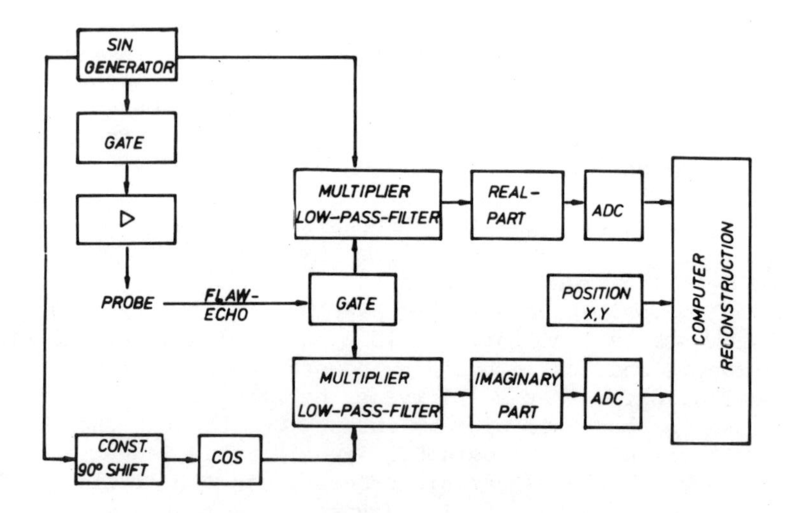

Fig. 5. Ultrasonic holography with numerical reconstruction.

Assuming the distance between aperture and observation plane to be much greater than the maximum linear dimension of the aperture, and the image in the x_B,y_B plane to be extended only over a small finite region about the z axis, then

$$\cos(\vec{n}, \vec{r}_{HB}) = 1 \tag{7}$$

By introducing the Fresnel approximation we can simplify the weighting function

$$h = \frac{1}{i \cdot \lambda} \cdot \frac{1}{r_{HB}} \cdot e^{ikr_{HB}} \tag{8}$$

$$\sim \frac{1}{i \cdot \lambda} \cdot \frac{1}{z_{HB}} \cdot e^{ikz_{HB}} \cdot e^{i \frac{k}{2z}[(x_B - x_H)^2 + (y_B - y_H)^2]}$$

Rewriting the last expression as a convolution of $U(x_H, y_H)$ with h, we get

$$U(x_B, y_B) = \frac{1}{i\lambda z} \cdot e^{ikz} \iint_{-\infty}^{+\infty} U(x_H, y_H) \cdot e^{i \frac{K}{2z}[(x_B - x_H)^2 + (y_B - y_H)^2]} dx_H dy_H \tag{9}$$

$$= \frac{1}{i\lambda z} \cdot e^{ikz} \cdot U(x_B, y_B) * e^{i \frac{k}{2z}(x_B^2 + y_B^2)}$$

The calculation can be performed by application of the convolution theorem. Using

$$F^{\pm} \left\{h(x)\right\} = H(\varsigma) = \int_{-\infty}^{+\infty} h(x) \cdot \epsilon^{\mp 2\pi i x \varsigma} dx \tag{10}$$

equation (9) can be written as

$$U(x_B, y_B) = \frac{1}{i\lambda z} \cdot e^{ikz} [F^- \left\{F^+[U(x_H, y_H)] \cdot F^+[e^{i \frac{k}{2z}(x_H^2 + y_H^2)}]\right\}] \tag{11}$$

Concerning the holographic process according to equation (11), the convolution theorem requires in sequence the Fourier transform of the two-dimensional field of the experimental data $U(x_H, y_H)$ and of the calculated exponential function. After multiplication of both data in the Fourier domain the next step in calculating equation (11) is the back-transform to the spatial domain. Alternatively, the quadratic terms in the exponent of equation (8) can be expanded to

$$U(x_B,y_B) = \frac{1}{i\lambda z} \cdot e^{ikz} \cdot e^{i\frac{k}{2z}(x_B^2 + y_B^2)} \iint_{-\infty}^{+\infty} g(x_H,y_H) \cdot e^{-2\pi i \frac{1}{\lambda z}(x_B x_H + y_B y_H)} dx_H dy_H \tag{12}$$

with

$$g(x_H,y_H) = U(x_H,y_H) \cdot e^{i\frac{k}{2z}(x_H^2 + y_H^2)} \tag{13}$$

that is

$$U(\frac{1}{\lambda z}x_B, \frac{1}{\lambda z}y_B) = \frac{1}{i\lambda z} e^{ikz} \cdot e^{i\frac{k}{2z}(x_B^2+y_B^2)} \cdot F^+ \left\{U(x_H,y_H)\cdot e^{i\frac{k}{2z}(x_H^2+y_H^2)}\right\} \tag{14}$$

The complex exponential data are multiplied with the experimental data $U(x_H,y_H)$ and then Fourier-transformed. The term $1/\lambda z$ assures correct scaling in the image plane. In the image plane we do not observe the complex amplitude but the inensity. Therefore, we have to multiply with the conjugate complex. After evaluating the convolution there remains

$$\left|U(x_B,y_B)\right| = \left| F^- \left\{ F^+[U(x_H,y_H)] \cdot F^+[e^{i\frac{k}{2z}(x_H^2+y_H^2)}]\right\} \right|^2 \tag{15}$$

and for the Fourier transform

$$\left|U(\frac{x_B}{\lambda z}, \frac{y_B}{\lambda z})\right| = \left| F^+ \left\{ U(x_H,y_H)\, e^{i\frac{k}{2z}(x_H^2 + y_H^2)} \right\} \right|^2 \tag{16}$$

The fast Fourier transform can only handle positive values of the aperture. Hence we have to perform a coordinate transformation

$$x = x + L/2 \qquad\qquad y = y + L/2 \tag{17}$$

Instead of (15) we have

$$\left| U(x_B,y_B) \right| = \left| F^- \left\{ F^+[U(x_H,y_H)] \cdot F^+[e^{i\frac{k}{2z}((x_H-\frac{L_x}{2})^2 + (y_H-\frac{L_y}{2})^2)}] \right\} \right|^2 \tag{18}$$

and

$$U(\frac{x_B}{\lambda z}, \frac{y_B}{\lambda z}) = \left| F^+ \left\{ U(x_H,y_H) \cdot e^{i\frac{k}{2z}((x_H-\frac{L_x}{2})^2 + (y_H-\frac{L_y}{2})^2)} \right\} \right|^2 \tag{19}$$

Now let us introduce the discrete fast fourier transform. In doing this, we first divide the aperture length L_x by M and L_y by N

$$\Delta x = \frac{L_x}{M} \qquad \Delta y = \frac{L_y}{N} \tag{20}$$

$$x = m \cdot \frac{L_x}{M} \qquad y = n \cdot \frac{L_y}{N} \qquad \begin{array}{l} m = 0, 1, \ldots, N-1 \\ n = 0, 1, \ldots, M-1 \end{array}$$

Equation (19) can be rewritten

$$\left| U(x_B,y_B) \right| = \frac{1}{M} \cdot \frac{1}{N} \left| \sum_{m=0}^{M-1} \sum_{n=0}^{N-1} U(x_H,y_H) \cdot e^{i\frac{\pi}{\lambda z}[(x_H-\frac{L_x}{2})^2 + (y_H-\frac{L_y}{2})^2]} \cdot e^{-2\pi i(\frac{m\cdot\mu}{M} + \frac{n\cdot\nu}{N})} \right|^2 \tag{21}$$

$$\mu = 0, \ldots, M-1$$
$$\nu = 0, \ldots, N-1$$

with

$$x_B = \pm \frac{\mu \cdot \lambda \cdot z_{HB}}{L_x} \qquad \mu = 0, \ldots, M-1$$

$$y_B = \pm \frac{\nu \cdot \lambda \cdot z_{HB}}{L_y} \qquad \nu = 0, \ldots, N-1 \tag{22}$$

with (20)

$$x_B^{max} = \frac{\lambda \cdot z_{HB}}{\Delta x} \quad ; \quad y_B^{max} = \frac{\lambda \cdot z_{HB}}{\Delta y} \tag{23}$$

where

z_{HB} = distance image to hologram
Δx = sample interval in x-direction
Δy = sample interval in y-direction
λ = reconstructing wavelength

Arriving at this point we compare our result with the formula used for optical reconstruction. Instead of the optical wavelength we use the ultrasonic one in the numerical reconstruction. The magnification m can be set to unity. Then the image-to-hologram distance r_{HB} is equal to flaw-distance for separated emitter-receiver scanning and equal to 0.5 flaw-distance for simultaneous emitter-receiver scanning of the flaw-to-hologram distance

$$r_{HB} = r/2.$$

The image size depends linearly on the wavelength and the flaw depth and is inversely proportional to the sampling interval (Eq. 23). The sampling interval Δx is determined by the scanning velocity v (mm/s) and the sample Δt (m/s):

$$\Delta x = v \cdot \Delta t.$$

3. EXPERIENCES WITH OPTICAL RECONSTRUCTION

In the last two years we investigated systematically the capability of ultrasonic holography in nondestructive testing. We examined

(a) ferritic and austenitic materials with drilled flaws of different depths, sizes and orientations,

(b) vacuum diffusion welded specimens with artificial plane and globular flaws and natural flaws.

The results of defect sizing are compared with *a priori* knowledge of the flaws, with macrographs and with magnetic particle testing results. Other authors have reported about their experiments on this subject, too[11], [12]; similar work has previously been reported.

3.1 RESOLUTION CAPABILITY

In our first experiment we examined the resolution capability with a special test specimen. 3 mm flat bottom holes were drilled to a depth of 100 mm below the scan surface. Their distances varied between 1 mm in steps of 1 mm to 6 mm. We used a focused probe with 25 mm diameter and 100 mm focal length in water. The US-frequency was 5 MHz. For this arrangement the lateral resolution Δx, and the axial resolution, Δr, were 1.4 mm and 9.2 mm respectively.

hologram

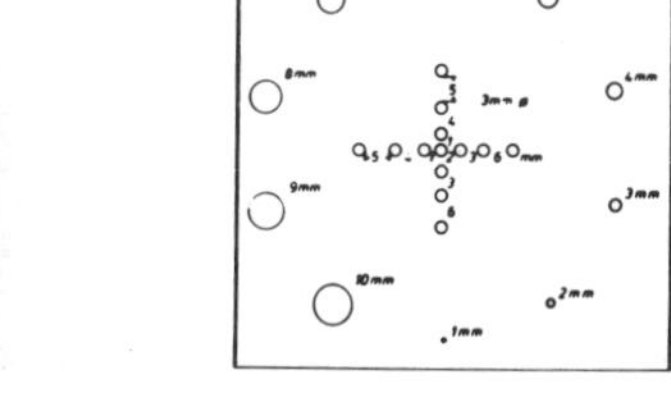

flawpattern

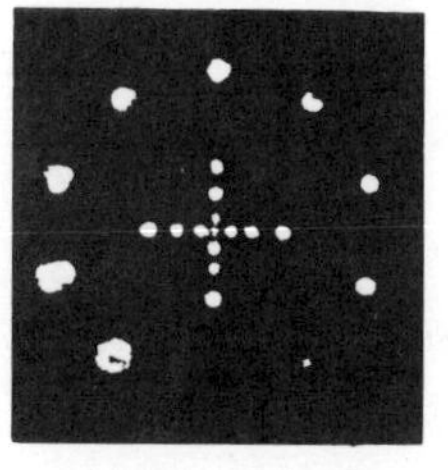

reconstruction

Fig. 6. Test specimen for demonstrating the resolution capability of ultrasonic holography flaw depth: 100 mm US-frequency: 5 MHz.

All flaws are clearly distinguished as was expected from theoretical considerations. In addition the circle in Fig. 6 demonstrates that the sensitivity has been sufficient to image all flaws from 10 mm to 2 mm.

3.2. DETERMINATION OF FLAW SHAPES

As mentioned before, the axial resolution of ultrasonic holography is poor. In most cases it is not possible to determine the shape of the flaws with one insonification angle but only the effective cross section. The next experiment will demonstrate this aspect.

A specimen was prepared with three 10 mm holes of different base shapes; one was flat, the second a cone of 30° angle and the third a cone of 45° angle, as illustrated in Fig. 7. The reconstruction in Fig. 7 points out clearly that the diameter of all three flaws has been estimated correctly.

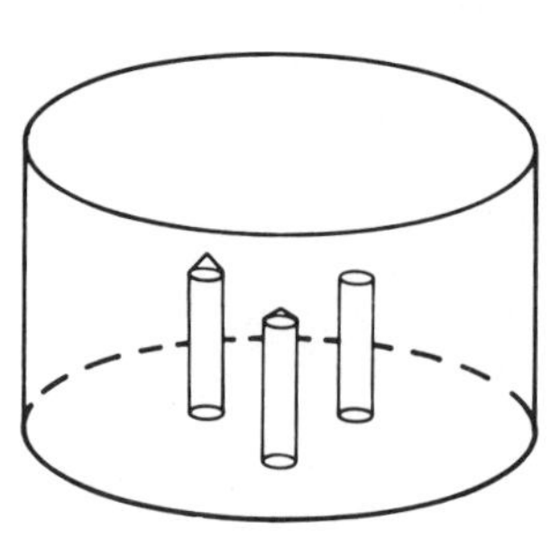

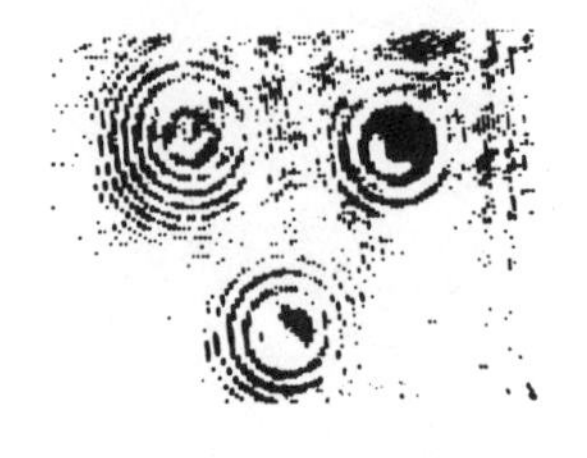

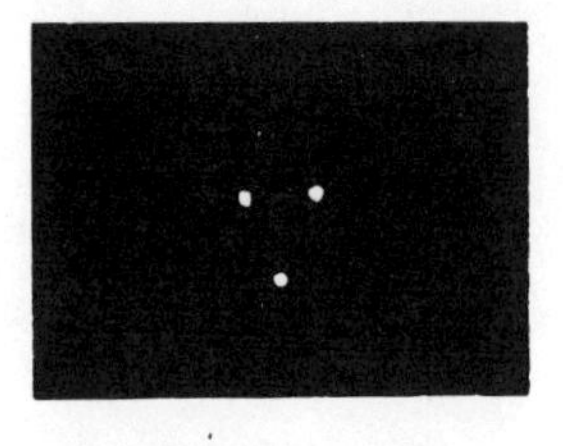

hologram

reconstruction
flaw size: 10 mm
flaw depth: 50 mm

parameter: angle of
slope: 0°, 30°, 45°

Fig. 7. Test specimen for control of shape determination.

Yet there is no information concerning the shape. There are two ways to get this:

- time of flight measurements with the pulse-echo overlap method,
- insonification by another angle (45°), i.e. in different directions.

3.3. DEPENDENCE OF INCLINATION

In this section the dependence of flaw inclination has been investigated. The flaws diverge from the acoustic axis (45° to the normal) by up to $\pm 15^{o}$. The ultrasonic holography uses wide angle beams. We expect therefore the danger of total deflection of the ultrasonic beam to be low. The reflected energy depends upon the quotient of flaw size to wavelength and upon the structure of the flaw. In the test specimen of Fig. 8 we expect a better detectability of the 5 mm than that of the 10 mm flaws. This has been confirmed. The reconstruction of the 30° inclined 10 mm flaw revealed only the contour of the flaw caused by diffraction, whilst the surface has deflected the total energy. This problem will be investigated in more detail in section 4.

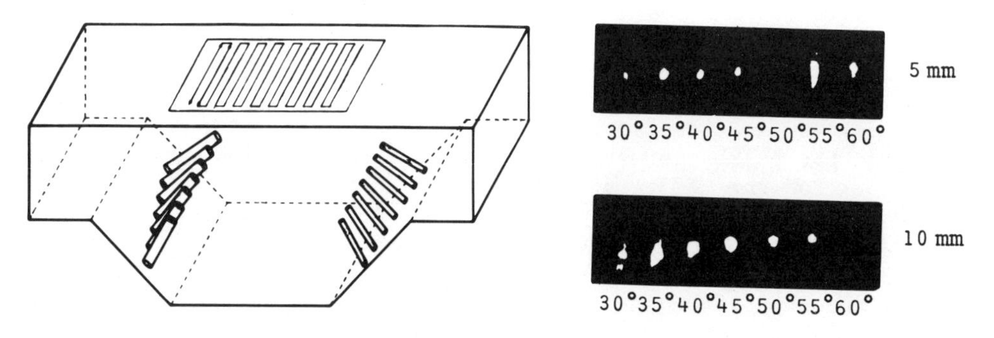

Fig. 8. Test specimen with inclined flaws 30° - 60°
flaw diameter: 10 mm left, 5 mm right side
flaw depth : 100 mm
material: 23 NiMoChr 747
dimensions: 500 x 250 x 190 mm^3

3.4. INFLUENCE OF PROBES AND PROBE ARRANGEMENT

The choice of the probe and the homogeneous beam pattern of such a probe is one of the keys to good imaging.

We used focused probes with lenses, focused probes with curved piezoelectric plates in the immersion technique and small probes in the contact technique. In the near future we will investigate focused contact probes and arrays.

For flaws of small depth we prefer small probes with planar membranes because of the open directiveity pattern and the homogeneity beyond the near field length. At greater depths the signal-to-noise ratio demands a better ratio of sound field-width to flaw-width at the corresponding depth. This can be easily handled by exchanging probes with different focal lengths. To gain a better insight into this subject we give an example.

In Fig. 6 we obtained a better image with a 4 MHz probe (6 mm diameter) than with a 4 MHz focused probe with 100 mm focal length in water.

For maximal flaw response to an impinging wave, the angle between incidence and reflection should be small. Based upon prior knowledge of the flaw orientations, we selected angles between 0° and 45° for the pulse echo technique. The tandem technique was also used. These arrangements are well known in NDT. The next examples demonstrate holography with 45° shear wave, holography in tandem technique with contact technique probes and holography in pulse echo at 0° with water column coupling.

In the test specimen (Fig. 9) we machined three flat bottom holes of 7 mm diameter with 0°, + 10°, -10° inclination to the left front surface. In a first experiment the smaller probe was both emitter and receiver. In the second experiment the bigger one emitted the pulse and with the correct distance to it the smaller one detected the reflected pulse via the back wall and flaw reflection. Both reconstructions show the three defects but do not reveal any influence of inclination. Because of the better arrangement of the tandem test, the amplification of transmit power could be diminished by 20 dB.

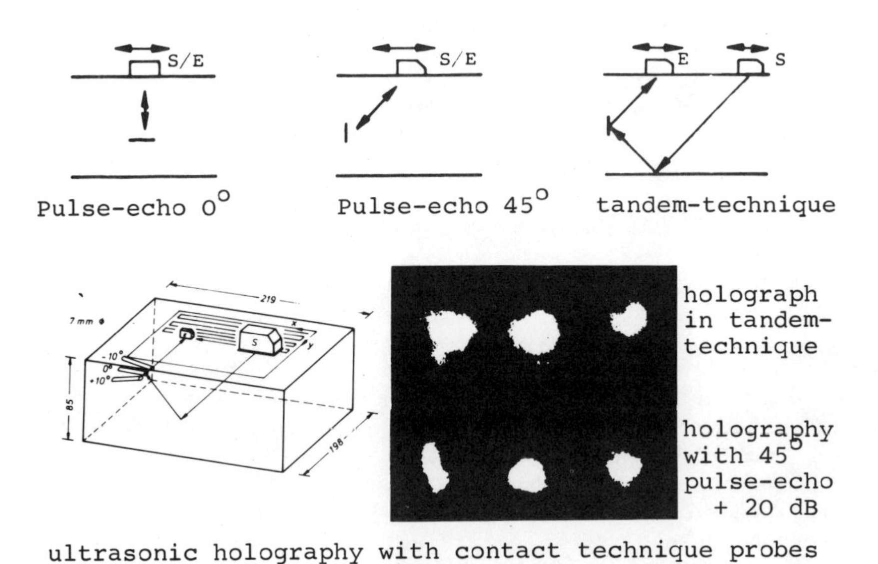

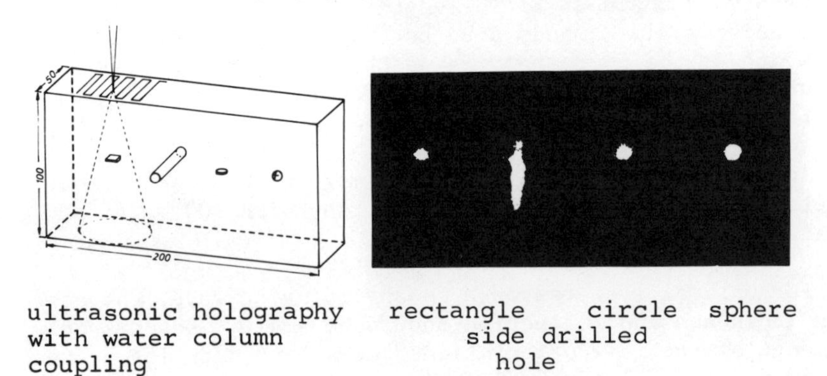

Fig. 9. Ultrasonic holography with contact technique probes in different arrangements and with water column coupling.

In the lower test specimen there are 3 mm planar and globular flaws introduced with the aid of vacuum diffusion welding. From left to right there is at a depth of 50 mm, a rectangle, a cylindrical side drilled hole, a disc and a sphere. At a frequency of 2.5 MHz (λ = 2.4 mm) the measured flaw sizes are correct to within 20%.

3.5. T-WELDED SPECIMEN

In a T-welded specimen with the dimensions 500 x 450 x 370 mm^3 we cut out a piece 220 mm x 220 mm x 200 mm. We investigated this specimen from all sides with a focused probe of 200 mm focal length in water and a frequency of 2.25 MHz. Fig. 10 shows the positions of the specimen and the six reconstructions.

We found two significant instances of lack of side wall fusion and a great number of flaws in the root. Combining the measured sizes of flaw number F_{15} (the flaw beside F_{11}), the following values resulted: 10.5 x 9.4 x 6.7 mm^3. After destructive testing a macrograph showed the maximum dimensions to be 4 mm x 3.4 mm x 1.3 mm. This means that the holographic estimated value is too high. The reason for this lies in the fact that the wavelength of 2.7 mm is not small compared with the dimensions of the flaw. To diminish the flaw size estimation error one has to examine with a higher frequency. In addition, a magnetic particle testing showed that the root defects exist as an array of cracks with lengths between 4 mm and 15 mm.

3.6. HOLOGRAPHIC EXAMINATION OF HEAVY SECTION STEEL TEST BLOCKS

Here we will present some of our examinations using both compressional and shear wave holography. The first one is the so called Electroslag = PISC 2. We investigated a band 75 mm wide on both sides of the weld over a length of 750 mm. We show in Fig. 11 the flaw reconsturctions at a depth between 53 mm and 88 mm.

The ultrasonic frequency was 2.25 MHz, the sensitivity corresponds to the detection of a 1 mm flat bottom hole at a depth of 100 mm, the resolution capability was 6 mm. We used 4 angles to examine this volume.

The flaw image of each examination and the schematic diagrams are seen in Fig. 11. The length of the flaws parallel to the surface is less than 11 mm and perpendicular to the surface less than 10 mm. Combined with the image reconstructed by pulse echo at 0° we can decide that there are planar and not globular flaws.

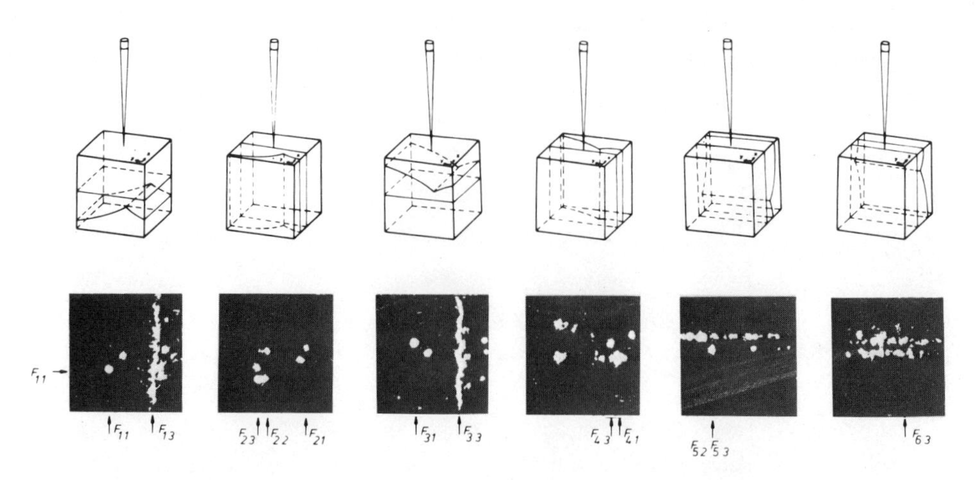

flaw reconstruction in 6 different views

macrograph

magnetic particle testing

Fig. 10. T-welded specimen with root defects and lack of side wall fusion.

We examined the test block "submerged arc" at various depths. In Fig. 12 the weld is running from left to right. There are flaws both in the basic material and in the side walls. The maximum size was 10 mm.

In the "nozzle weld" test block we found many flaws in the middle of the basic material. Three of them at depths of 100 mm, 103 mm and 10.8 mm respectively, have the dimensions 10.6 mm, 16 mm and 23 mm. The flaws are lying parallel to the surface. We compared these results with those obtained by automatic testing, by hand testing and by focal probe testing. The results agree very well, but a final judgement can only be given after destructive testing of the samples.

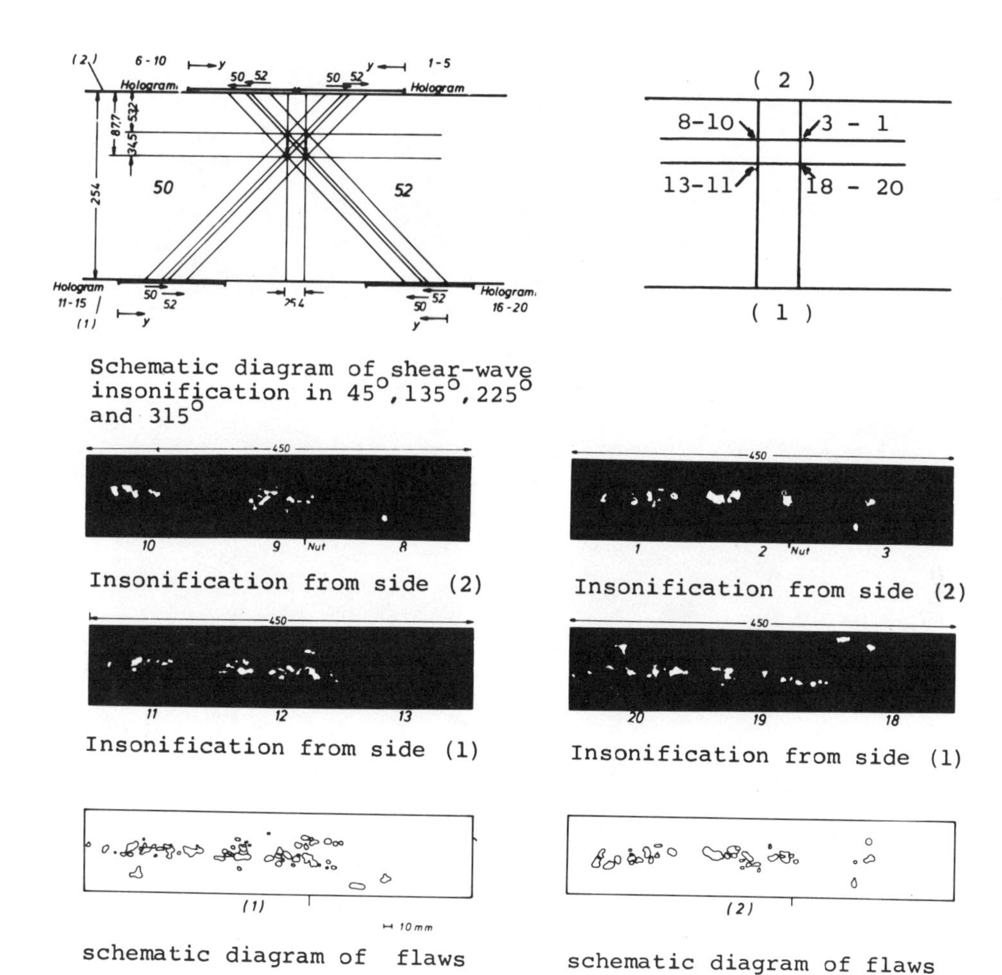

Fig. 11. Electro Slag - welded specimen 50-52.

3.7. INFLUENCE OF CLADDING

The test specimen for this examination was doubly cladded. The material was 22 NiMoCr 3.7 and the cladding 23 Cr/11 Ni. The claddings were 60 mm wide and offset 30 mm against each other.

The testing frequencies were 1.6 MHz, 3 MHz and 4 MHz. The flaw pattern at a 100 mm depth is the same as in the first specimen described above. The roughness of the surface before smoothing was 65 μm. On the left side of Fig. 14 the flaw images are distorted

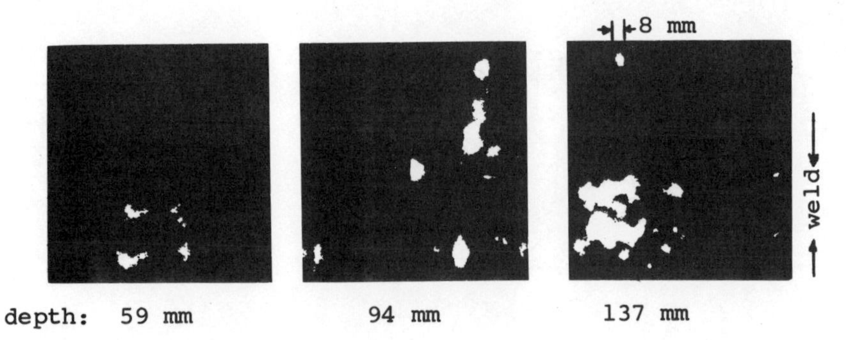

Fig. 12. Submerged Arc - welded specimen 51 - 53
Examination of the weld by longitudinal
wave holography with 2,25 MHz.

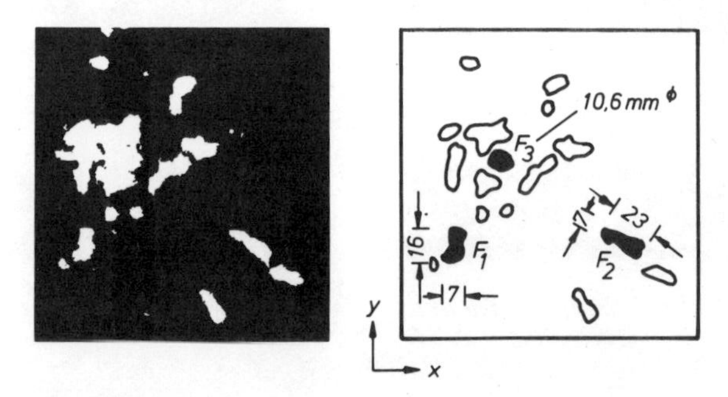

Fig. 13. Examination of nozzle weld
Flaws are lying in the plate base material
US-frequency: 2.25 MHz.

by the phase distortion caused by the cladding. A recognition of the flaw pattern is possible only with the lowest ultrasonic frequency of 1.6 MHz. Smoothing the surface of 2.5 μm allows the formation of a clear image the resolution of which increases as the frequency increases. That means the phase distortion is only caused by the surface roughness and not by the second cladding or by the interface.

Summarizing all our experiments we can state that the holographically determined dimensions were close to the true dimensions. The response of conventional ultrasonic systems alters the measured dimensions on a large scale. The presentation of images is better suited to the human eye than data or tables. One important factor is the redundancy of holography. That means, if some information is lost the remaining part of the hologram still allows the reconstruction of the image. The disadvantage of the ultrasonic

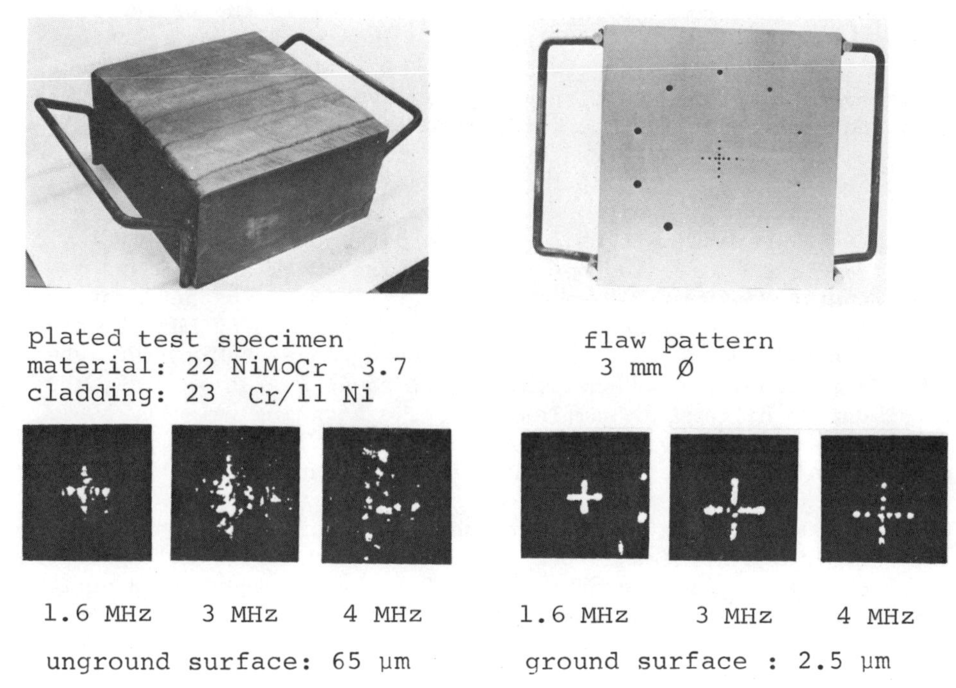

Fig. 14. Influence of cladding.

holography in the present stage is the long time required to generate a hologram. This results in long inspection times. In most of the institutes there are efforts to improve the system by electronically scanning and focusing.

4. EXPERIENCES WITH NUMERICAL RECONSTRUCTION

The performance of numerical reconstruction has some advantages compared with optical reconstruction:

1) The choice of wavelength is not limited to the optical wavelength.
2) Furthermore, there arises only one image of the flaw.
3) All the possibilities of signal processing can be used and the reconstruction can be done automatically.

Some of these methods are explained below.

4.1. DEFLECTING AND REFRACTING BEHAVIOUR OF LARGE EXTENDED FLAWS

One of the problems in ultrasonic testing is the deflecting behaviour of large extended flaws. The numerical reconstruction is a well suited tool for quantitative investigation of this phenomena. It is sufficient to detect the complex wave field along one single line and to reconstruct the intensity distribution in the flaw plane. The equipment of Kutzner and Wustenberg or the equipment in Fig. 2 are well suited.

We will demonstrate this on large horizontal and vertically situated flaws. One flaw has the dimension 20 mm x 30 mm and is located at a depth of 140 mm respectively 56 mm. In the first case (see Fig. 15) a small contact technique probe of 3.62 MHz is scanning in simultaneous emitting and receiving mode at 0°. The repetition frequency was 430 Hz, the signal-to-noise ratio 20 dB, the total number of complex samples 1024, the scanning velocity 50 mm, and the sample interval 2 ms for the real and for the imaginary part. The reconstruction is shown in the right part of the Fig. 15. With regard to correct scaling (equation 23) we measured a flaw width of 30 mm with a 20 dB decrease in maximum intensity. There are some irregularities in the intensity distribution which will be seen more clearly in the next case. For the large extended vertical flaw we used first shear waves at 45° in pulse-echo. The reconstruction indicates a trough in the intensity distribution. The contour of the flaws are emanating spherical or cylindrical waves which are detected well. The surface of the flaw is deflecting the total energy. Without this knowledge we would say that there are two separated flaws. This is the reason for the German philosophy of prefering the tandem method in investigating vertically expected flaws. The reconstruction with this method is sharp and precise. Instead of a deflection of the energy on the front surface of the flaw the sound wave is directly reflected to the receiver. Both reflections and diffractions contributed to the reconstruction of the flaw. In all three experiments we obtain the true width of the known flaw by measuring the width of the intensity distribution with a 20 dB decrease in maximum intensity.

For comparison with the optical reconstruction, at the bottom of Fig. 15 there are the presentations of puse-echo at 0°, 45° and with the tandem technique. The puse-echo at 45° reveals the depression also. By using the tandem technique it is possible to image the left front side of the specimen.

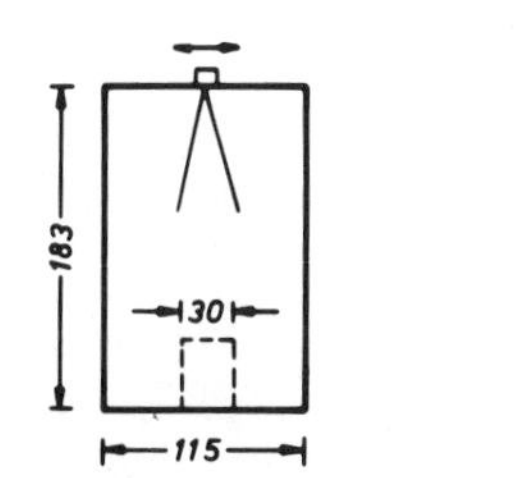

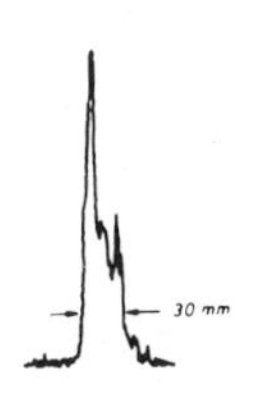

Reconstruction of large flaws lying horizontally with IE 0°

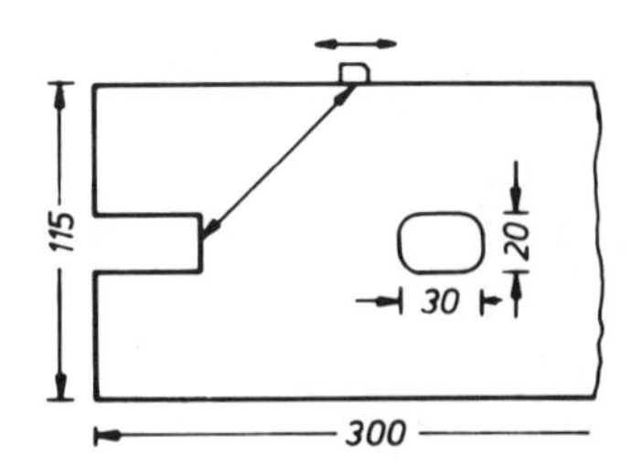

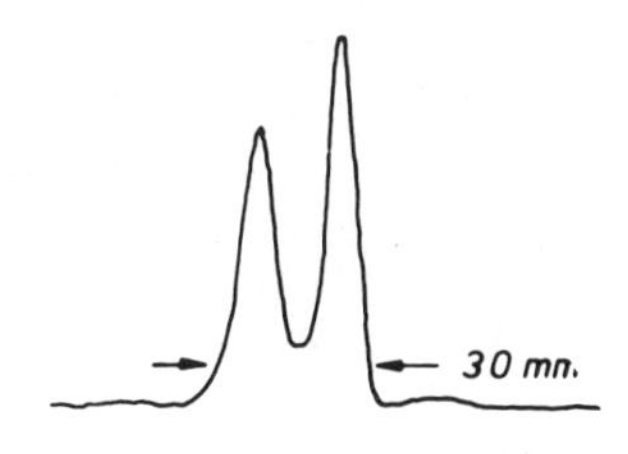

Reconstruction of large flaws lying vertically with IE 45°

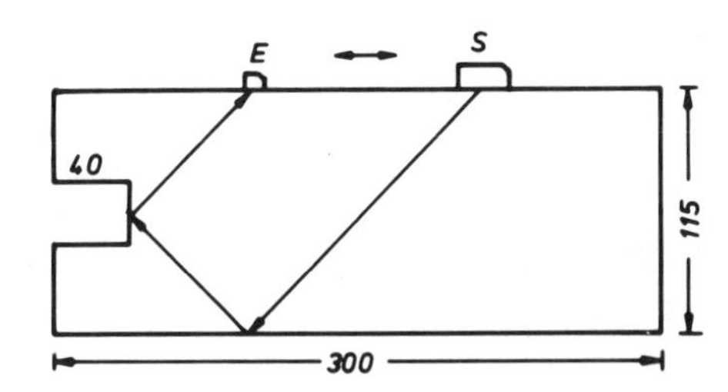

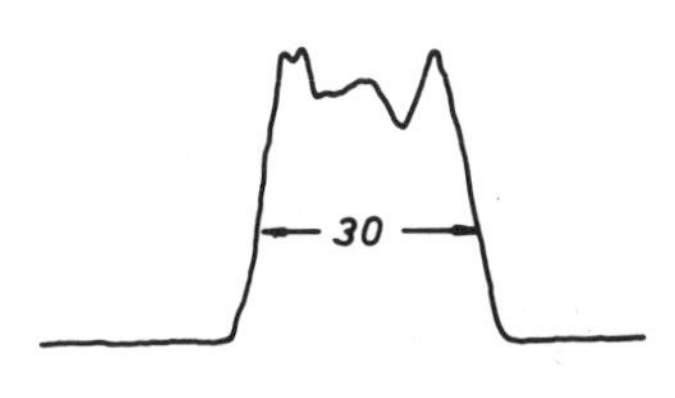

Reconstruction of large flaws lying vertically with tandem

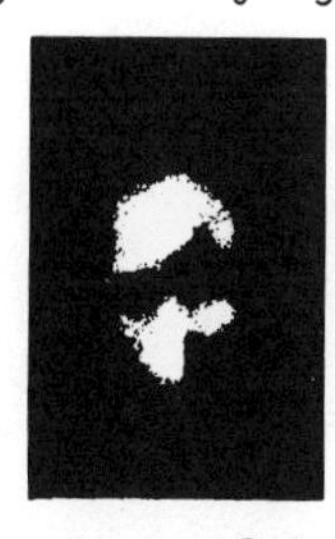

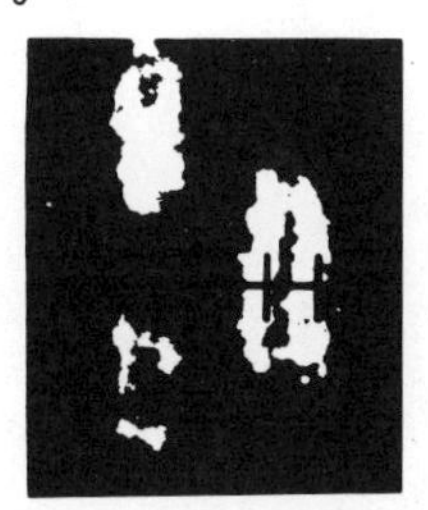

IE 0° IE 45° tandem technique

Optical reconstrcution

Fig. 15. Deflecting and refracting behaviour of large extended flaws lying horizontally and vertically.

4.2. SIGNAL AVERAGING IN ULTRASONIC HOLOGRAPHY

It is useful to introduce the method of signal averaging to compensate peaks in the reconstruction caused by uncertainties in the experimental arrangement. In the first case, Fig. 16, we used five different frequencies (2.8 MHz, 2.9 MHz, ..., 3.2 MHz). The sample interval was 0.1 mm, the reconstruction time needed was 36 seconds. In the first three reconstructions, the left slope is overemphasized, the last two show a more equalized behaviour of both slopes but with a trough in the middle. The averaging of the fiver reconstructions is shown on the right side. The intensity distribution is easier to interpret but the number of signals is not sufficient. The number can be increased drastically by developing a system with faster scanning and an almost realtime reconstruction. One of the possible ways to perform this is by making use of piezoelectric arrays with electronic scanning and focusing and by applying fast array processors to shorten the time needed for the numerical reconstruction.

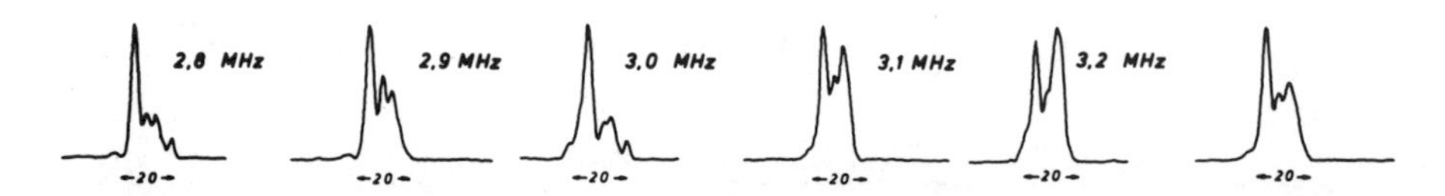

Fig. 16. Linear acoustical holography with signal averaging field of application: flaws in austenitic structure.

4.3. NUMERICAL RECONSTRUCTION AND ITS APPLICATION TO RESOLUTION MEASUREMENTS

In Fig. 6 we showed the resolution capability of ultrasonic holography with optical reconstruction. The image presented there can give a more or less qualitative decision about the resolution. In the next experiment we show by shear wave excitation at 45° a comparison between optical and numerical reconstruction. The flaws have a diameter of 7 mm and are situated at a depth of 45 mm. The frequency was 2.5 MHz. On top of Fig. 17 the optical hologram and the reconstruction is shown. The signals below represent the real and imaginary part of the numerical hologram and the reconstruction. The flaws are separated by 1 to 4 wavelengths. In both cases the flaws are separately imaged. Between the three flaws on the right side there are interferences, so that the second one is reconstructed with a lower amplitude.

4.4. SAMPLE INTERVAL AND QUANTIZATION LEVEL

In digitizing analog signals we have to take care of the well known sampling theorem. In the scope of this theorem we investigated experimentally the effect of the variation of the sample interval.

The test specimen used was a vacuum diffusion bonded probe with an elliptical flaw of 7 x 12 mm^2. The full information content is preserved when each transmitted and received pulse is sampled. The repetition rate for the transmit pulse depends on the shape and the scattering of the material. After 4 ms we could emit a new transmit pulse. With the scanning velocity we had a sample distance of 0.1 mm. We started with a sample interval of 0.1 mm and repeated with 0.25 mm, 0.5 mm and 1 mm.

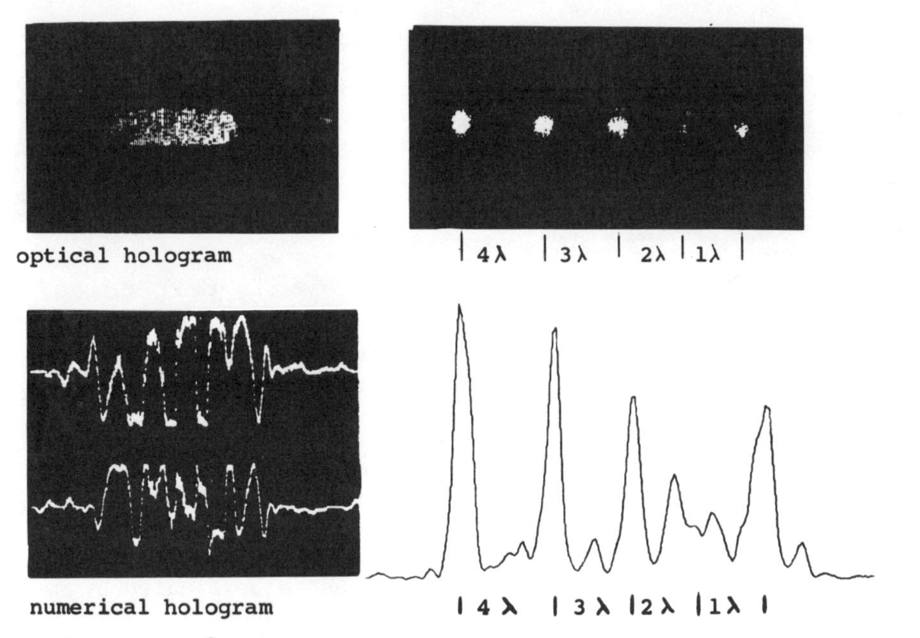

Fig. 17. Shear wave holography and resolution measurement.

In all cases shown in figure 18 the sample-to-sample distance has been sufficient and causes no overlapping of the signals. An exact analysis shows that working with shear waves at 45°, the sample intervals in both directions (x-axis and y-axis) have to be chosen differently. The distance between scanning probe and flaw varies rapidly if the scanning direction is perpendicular to the flaw plane, and so the phase fluctuations are more rapid. As the two pictures on the bottom of Fig. 18 show, the shortest period is

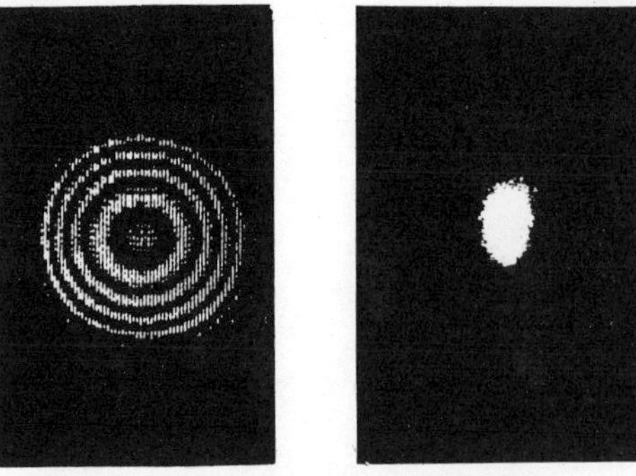

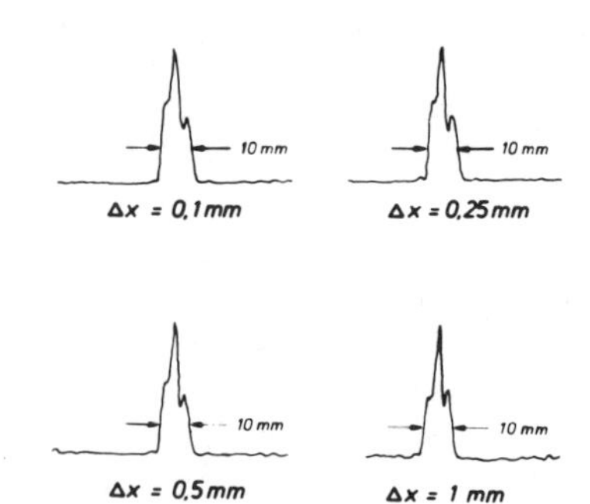

elliptical flaw 7x12 mm
diffusion bonded specimen

transmit-pulse rate 0.1mm
variation of sample -
interval

shear wave holography at 45°

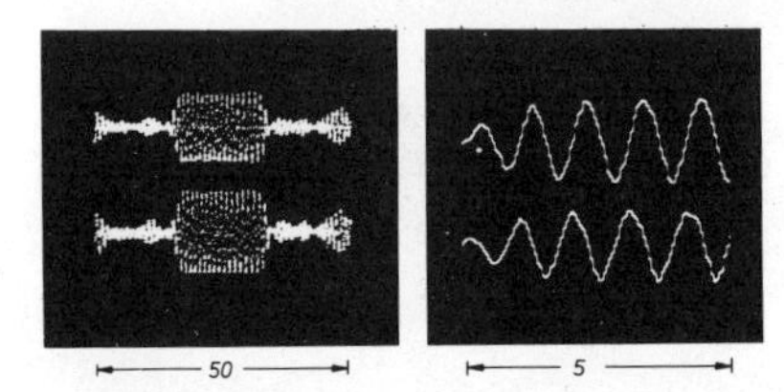

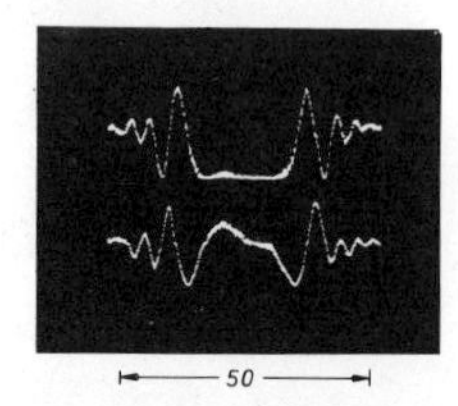

scan axis parallel to flaw plan axis perpendicular

Fig. 18. Dependency of numerical reconstruction on the sample interval in two perpendicular directions.

5 mm on the left side and 1 mm on the right side. With 4 samples per period we have a minimal sample distance of 1 mm on the side and 0.2 mm on the other side. Instead of the 1024 samples used only 64 and 248 were necessary. So in two-dimensional numerical reconstruction a field of n x m with n = m has to be handled by computer.

Concerning the dependency on the number of quantization steps we refer to Jones and Kesner[13]. They stated that in the presence of strong and weak signals with a variation of signal amplitude of 20 dB, the phase must be stored in more than 15 quantization steps and the amplitude in more than 7 quantization steps. In the case of one single flaw 2 or 3 steps would be sufficient. In all our experiments we worked with a biomation model 8100, that means with 256 steps (8 bits).

4.5. TEST SPECIMEN WITH SLAG INCLUSIONS

Furthermore, we examined a test piece with natural flaws, which contained two rows of slag inclusions consisting of many individual particles. We investigated this specimen with water coupled probes and with 5.5. MHz longitudinal waves. The arrangement is shown in Fig. 19 together with a macrograph, the numerical reconstruction along one line and the numerical two-dimensional reconstruction.

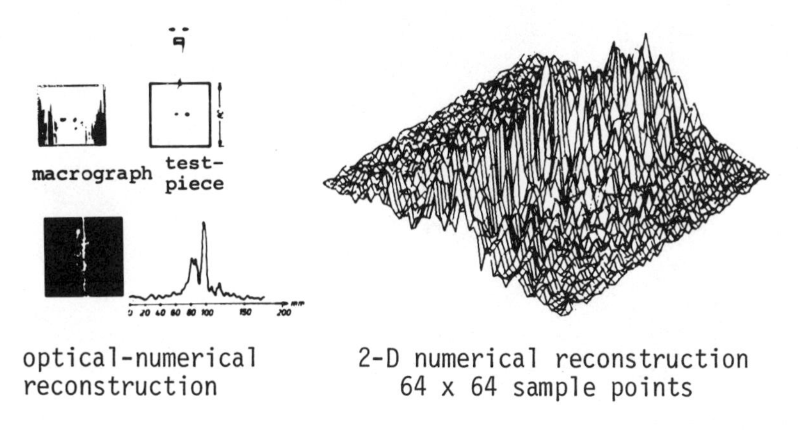

Fig. 19. Test piece with slag inclusions in 40 mm depth.

The slag inclusions have dimensions of about 1.5 mm and are situated at a depth of 40 mm. The sample interval was 0.5 mm in the x-direction and 0.6 mm in the y-direction. We investigated an area of 32 mm x 39 mm with 64 x 64 samples. The time needed for reconstruction was 720 s. The calculated intensity distribution in the flaw plane was rotated and tilted to present the result in a 3D-projection. The time needed can be diminished by improving the software and the hardware. Our current investigations will compare the advantages and disadvantages of the three diffraction techniques: Fresnel integral (14), Fresnel integral (11) and frequency domain approach for small hologram - flaw distances.

4.6. 3D-PRESENTATION OF FLAW-IMAGE PLANE

In displaying different rotational 3D views we present at the top of Fig. 20 the simulation of the reconstruction of two separate flaws, rotated about 90°. Fig. 20 (center) shows the image of a 10 mm hole at a depth of 100 mm in side view and in 3D view. Fig. 20 (bottom) is the image of a 2 mm hole at a depth of 100 mm. The number of samples are 64 x 64 and the sample interval is 0.5 mm. The possibilities of 2D numerical reconstruction in connection with 3D projection are clearly shown.

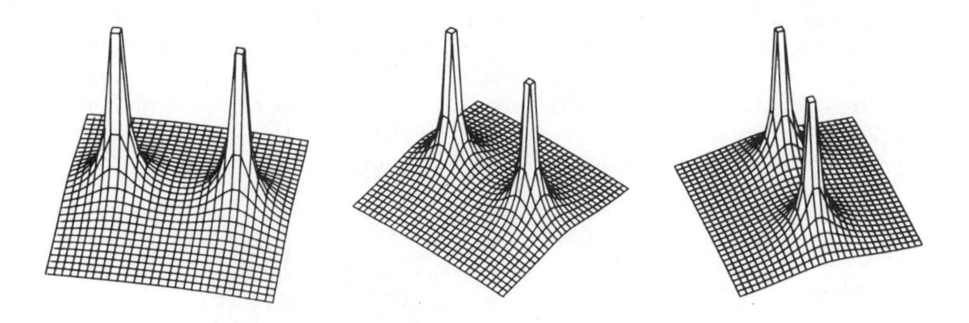

3D-projection of computer simulated flaws

3D-projection - 10 mm flaw in 100 mm depth - side view

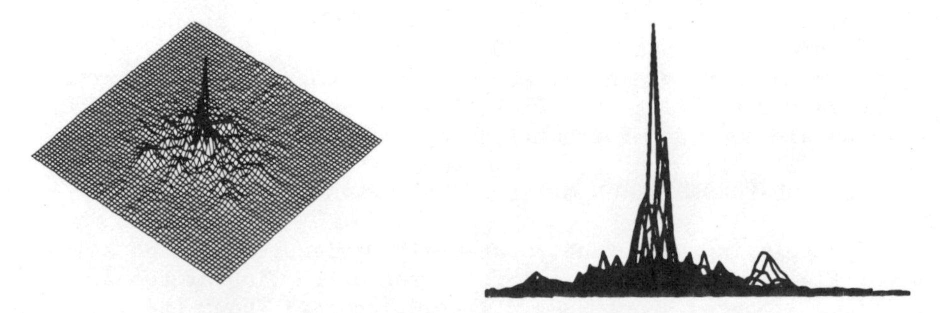

3D-projection - 2 mm flaw in 100 mm depth - side view

Fig. 20. 3D - Presentation of reconstructed flaws.

4.7. AN AUSTENITIC WELD CONTAINING A FATIGUE CRACK

In this case we investigated a 20 mm thick specimen with an austenitic weldment and with two cracks separated by 25 mm in both side walls of the weld. We used shear wave holography with 2.34 MHz, a scanning velocity of 25 mm/s and a sample interval of 20 ms. With 64 x 64 samples we got the following reconstruction in which only one crack is seen along a distance of 30 mm. In the picture beside, both cracks and the right corner of the test specimen are represented. The crack proceeds from the back surface. In interpreting the detected signal we recognize the crack by direct reflection and by double reflection of the ultrasonic wave via the back-wall. So the true width, h, is calculated from the measured width, b, by $h = 0.7 \cdot b$.

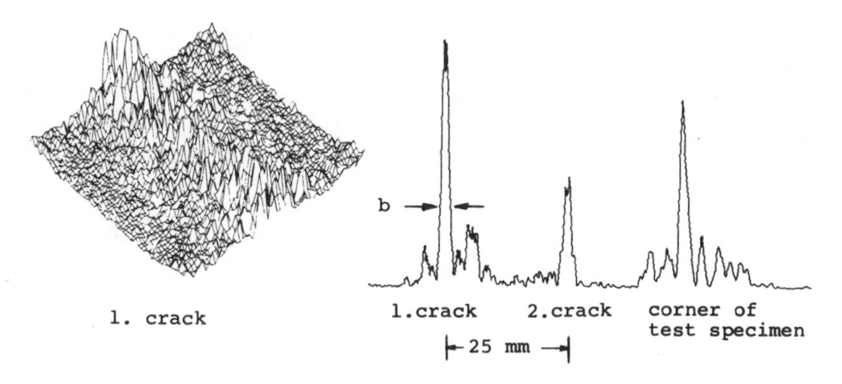

Fig. 21. Austenitic weldment with two cracks caused by vibrations.

5. OTHER APPLICATIONS AND FURTHER DEVELOPMENTS

5.1. INVESTIGATION OF VIBRATING SURFACE OF AN ARRAY

Another possible application for ultrasonic holography is the investigation of the vibrating surfaces of probes especially those of piezoelectric arrays. Similar investigations have been conducted by M. S. Lang[14] and M. Ueda[15]. In our present projects we are developing electronic steering and focusing systems with the aid of arrays, so we are most interested in the physical behaviour of such arrays.

The 2 MHz array we used is made up of 20 elements, each consisting of only three electronically connected bars of width 1.5 mm and length 10 mm. We excited different elements of the array and picked up the field with a small probe of a distance z. The arrangement is seen in Fig. 22.

From the complex field at distance z we calculated back to the surface of the probe. In the first picture on the right side we excited only 1 element. It can be clearly distinguished that each bar consists of three parts. The next picture shows the result of exciting the first element and elements No. 10 and 11. In addition to these two the twentieth element is vibrating because a surface wave is running along the front of the array. In the last example we excited the elements 1, 3, 5, 7, ..., 19. Here the resolution capability of 3 mm is reached, but we can distinguish the 10 different elements. There are some of them which are preferably excited.

5.2. EXPERIENCES WITH IN-SITU TESTS

We used the ultrasonic holography equipment with water column coupling to investigate the shell-weld of a reactor pressure vessel with vertical insonification and a weldment in a pressurizer with shear wave insonification. The mechanical scanner was attached to the wall with magnetic feet. The time needed to create a hologram was 17 minutes, but we did not have problems with the mechanical or electronical stability. Some problems were caused by air bubbles in the water column which created strong echoes, and the alignment of the probe holding device parallel to the surface. We feel a need for smaller manipulators because of the space restrictions around the pressure vessels. Concerning the probe holding device, we constructed two new ones. It is possible to manipulate contact technique probes which are moved with two cardan shafts along the surface and pressed against it by spring loading or by compressed air. The other one can handle probes with column water coupling with excitation of compression or shear waves. Summarizing we can state that ultrasonic holography is an important tool in analyzing the detected flaws.

5.3. INVESTIGATION OF CURVED SPECIMEN

All investigations of the previous sections were done on plane or nearly plane surfaces. In this case the distance between the surface and the scanning plane is constant and the insonification angle is constant, too. In the case of a curved specimen the probe must be moved along the curved surface. Therefore, the hologram plane is curved, too. In reconstructing such a hologram the curvature which differs from object must be taken into consideration. In each position of the curved hologram we suggest that the phase of the object signal be shifted so that the curved hologram is transferred to a plane hologram. The amount of phase shifting depends upon the wavelength in the material, the curvature of the material and the length of the aperture. In a cylindrical specimen we get for $\Delta\Phi$

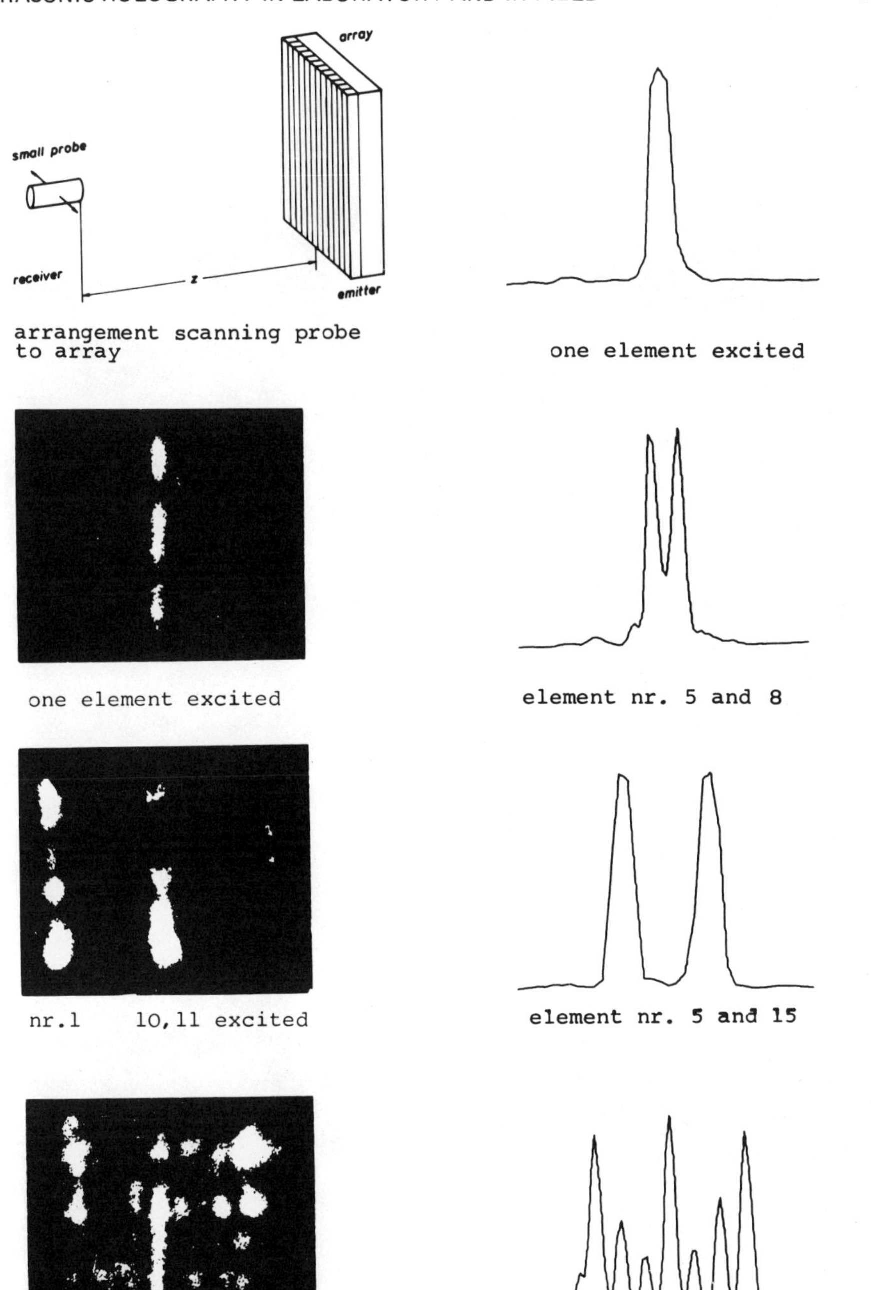

Fig. 22. Reconstruction showing radiation pattern from a 20 element array of 30 x 30 mm^2 and 2 MHz.

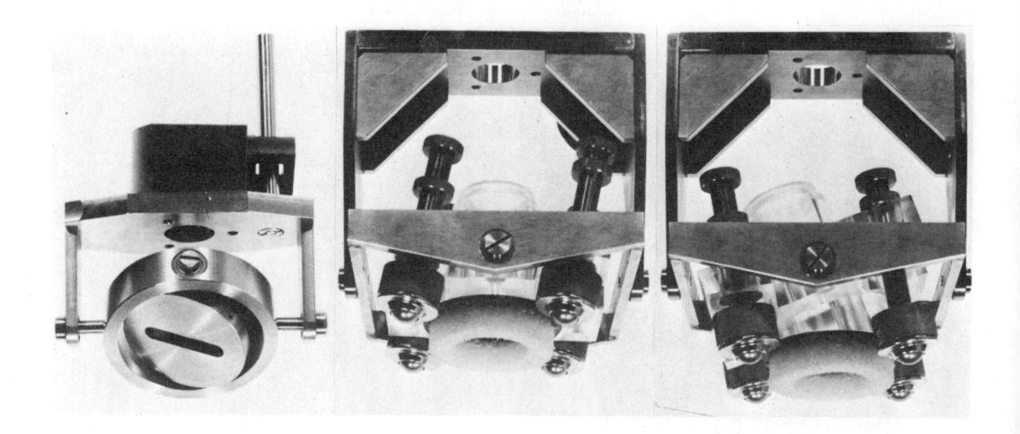

contact technique probe holder for shear wave 45°

water column coupling for longitudinal wave insonification 0°

water column coupling for shear wave insonification 45°

Fig. 23. Probe holding devices for contact technique probes with water column coupling. ⊢—⊣ 5 cm

Fig. 24. In-situ test on reactor pressure vessel.

This phase can be performed by hardware. The amount of $\Delta\Phi$ varies proportional to x according to the formula above. After phase shifting of the flaw signal the two reference signals are mixed in the described manner. For better understanding imagine a pipe with radius of 250 mm. The probe should be moved along 60 mm. With a wavelength of 2.6 mm we have a maximum phase shift of 244°. This phase shift can easily be done with charged coupled devices.

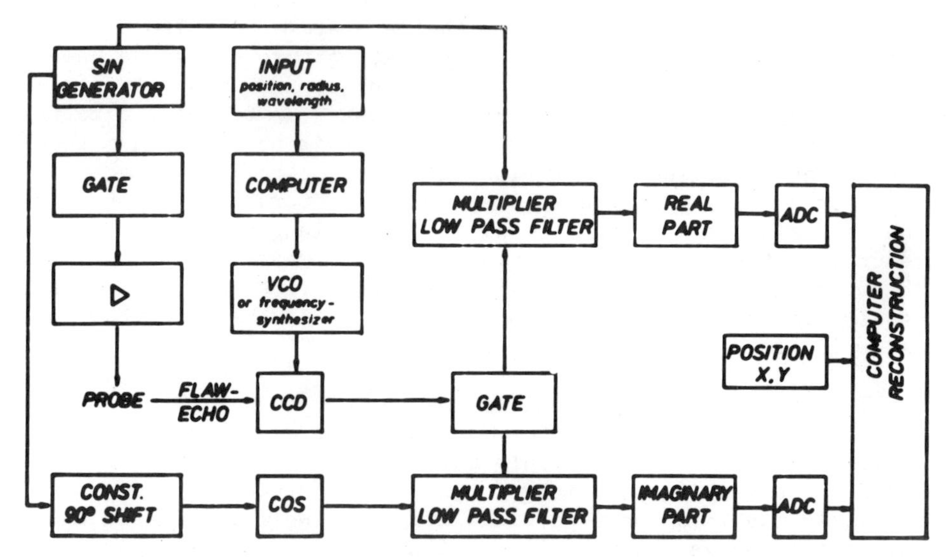

Fig. 25. Schematic diagram for ultrasonic holography on curved specimen.

The phase shift of such devices is a function of the clock frequency. The clock frequency is calculated by the computer and can be performed by VCO or frequency synthesizer.

6. SUMMARY

We have discussed the basic arrangement for ultrasonic holography with optical and numerical reconstruction on plane and curved surfaces. The most important parameters such as lateral and axial resolution, image magnification and dependency on wavelength and flaw depth have been briefly reviewed. The flaw depth is determined by pulse-echo-overlap method, whereby the flaw position is controlled in the hologram itself.

The formula for numerical reconstruction in the Fresnel distance is shown as a Fourier transform where the quadratic terms in the exponent have been expanded, and alternatively as a convolution of the holographic data with this exponent. With the aid of the discrete fast Fourier transform, the calculations are performed. The correct scaling factor, i.e. the image magnification factor, i.e. the image magnification factor, depends upon the wavelength, the flaw depth and the sample interval.

The results obtained with optical reconstruction on various test specimens with artificial and natural flaws, are very encouraging. Experiments were done not only in the laboratory but also in the field. We have shown that a large variety of probes can be used including probes using the contact technique and with water column coupling. The arrangement of the probes can be adapted to the testing problem by pulse-echo at 0° and 45° or by the tandem technique. The determination of flaw sizes is controlled on macrographs or on flaws with known shapes. The accuracy of sizing depends upon the relation between flaw size and wavelength. For a precise measurement the wavelength should be small compared with the flaw, otherwise the flaw is over estimated. The influence of cladding was also investigated. We found that the main reason for the phase distortion lies in the surface roughness.

The numerical reconstruction was performed on a PDP 11/34 computer. The advantage of working with sine and cosine references results in the formation of only one image. The numerical method offers the possibility of introducing signal processing and quantitatively examining many physical effects. We investigated the behaviour of large extended flaws situated perpendicular to or parallel to the surface. The local image intensity of such flaws depends upon the deflecting behaviour of the flaw surface and the diffracting behaviour of the contour of the flaw. The possibility of signal averaging is shown to be a promising method for investigating welds in austenitic structures. In shear wave holography we demonstrated that the minimum sample distance is not the same in both directions. In most of our experiments we worked with 1024 sample points, but 256 points would have been sufficient. On flaws like slag inclusions, 2 mm and 10 mm flaws at a depth of 100 mm, and cracks in austenitic weldments we performed two-dimensional numerical reconstructions. The presentation of these images was done by a 3D-projection which was able to tilt and rotate the image.

All the experimental work was performed with a slow scanning mechanical system. The use of faster recording devices such as transducer arrays is needed for a more widespread in-service application of ultrasonic holography. This can perhaps be done with some tradeoff between multiplexing and parallel processing. For a faster reconstructing device, array processors with shorter cycle times and more parallel working memories might be used.

7. ACKNOWLEDGEMENTS

This work is supported by the Federal Ministry of Research and Technology within the scope of reactor safety program RS 102-20 and RS 315.

8. REFERENCES

1. Holscan 200, Holosonics Inc., Richland, Washington 99352
2. "Acoustical Holography", Volumes 1-7, 1967-1976 New York, Plenum Press
3. B. P. Hildebrand, B. B. Brenden, "An Introduction to Acoustical Holography", Plenum Press, New York, 1972
4. B. P. Hildebrand, H. D. Collins, Evaluation of Acoustical Holography for the Inspection of Pressure Vessel Section, Mat. Res. Stand. 12, No. 12
5. G. A. Deshamps, Some remarks on radio-frequency holography, Proceedings of the IEEE, 55(4):570-571 (1967)
6. K. Magurak, "Probleme de holographischen Abbildung im Mikrowellenbereich", Forschungsbericht Nr. 6-72, lfd. Nr. 11, Werthoven, Dez. 1972, Forschungsinstitut fur Hochfrequenzphysik
7. R. Diehl, "Ein Beitrag zur Auswirkung stochastischer Storungen auf die akustisch-holographische Abbildung von Objekten", Diss. TU Hannover, 1973
8. N. H. Farhat, "Advances in Holography", Volume 1, R. Mueller, p. 1-96, Marcel Dekker, New York and Basel, 1975
9. J. Kutzner, H. Wustenberg, Akustische Linienholographie, ein Hilfsmittel zur Fehleranzeigeninterpretation in der Ultraschallprufung, Materialprufung 18 (6): 189 (1976)
10. J. W. Goodman, "Introduction to Fourier Optics", McGraw-Hill, 1968
11. A. E. Holt, W. E. Lawrie, "Acoustical Holography", Volume 7, 1976, p. 599-609
12. E. E. Aldridge, A. B. Clare, M. I. J. Beale, D. A. Shepherd, "Ultrasonic Holography for the Inspection of Thick Steel Specimens", Research Agreement No. 6210.GA/8/801 ECC-AERE-G 1002, 1976
13. C. H. Jones, J. W. Kesner, Comparison of signal-processing methods, J. Acoust. Soc. Am., 62 (5): 1226-1238 (1977)
14. M. S. Lang, "Computer Generated Wavefront Reconstructions", Penn. State, Box 30, State College, Pa. 16801
15. M. Ueda, T Katayama, T. Sato, Measurement of Vibration Amplitude Distribution of Ultrasonic Transducer by Holographic Technique, Bulletin Research Laboratory, Precision Machinery Electronics, 1976, No. 38, p. 9-13

DIGITAL SIGNAL PROCESSING IN ACOUSTICAL FOCUSED IMAGE HOLOGRAPHY

Fuminobu Takahashi, Katsumichi Suzuki and
Takahiro Kanamori

Energy Research Laboratory, Hitachi Ltd.
1168 Moriyama, Hitachi, Ibaraki, Japan

ABSTRACT

This paper presents a new concept of signal processing in acoustical focused image holography, named "digital accustical holography". Presented here is a method based on time coincidence measurement between digitized reflected pulses and digital clock pulses for phase detection instead of heterodyne phase detection used in conventional acoustical holography. This digital phase detection permits the use of a wide band (spike shaped) transmitted pulse which increases the depth resolution remarkably. Furthermore, fringe representation corresponding to the deviation in the propagation path length of less than half a wave length can be easily achieved by changing the clock pulse frequency independently of the carrier frequency of the transmitted waves.

A prototype signal processor based on the above mentioned method was produced and its function was tested by imaging artificial defects fabricated in a 10 mm thick stainless steel plate and located at a distance of 5 mm from the weld. A focus type transducer was used and its resonance frequency was 3 MHz. Images of artificial defects can be distinguished clearly from images of the weld by the use of a wide band transmitted pulse. The height of each defect is evaluated by counting the number of fringes appearing in the focused image hologram. The deviation of the estimated height from the true value is about 30% for the defect of 1 mm height with 6 MHz clock pulses, while the corresponding deviation is about 50% by the conventional focused image acoustical holography.

INTRODUCTION

Acoustical focused image holography is an attractive method for size evaluation and imaging of flaws within a metal structure in the field of non-destructive testing. Acoustical holography has at least two advantages over a conventional ultrasonic pulse-echo method. One is quantitative size evaluation of a flaw and the other is an imaging capability of flaws.

An image of a flaw obtained by the focused image holography consists of interference fringes or contour lines across the flaw surface. Fringe separation represents half wave length deviation in ultrasonic propagation path length between a transducer and the flaw surface, and the length of the fringes corresponds to the width of the flaw. Thus, it is possible to evaluate flaw size by measuring numbers and lengths of the fringes appearing in the focused image hologram[1].

A heterodyne phase detection method has been used to obtain a focused image hologram in conventional acoustical holography. This detection method, however, has two disadvantages; 1) relatively poor depth resolution caused by the use of sinusoidal mode transmitted pulse of more than 3 μs in width which is necessary for the heterodyne phase detection, 2) impossibility of fringe representation corresponding to the deviation in the propagation path length of less than half a wave length since sinusoidal mode reference and transmitted waves with identical frequency are used.

This paper describes a new method of signal processing in acoustical focused image holography to overcome the above mentioned disadvantages of the conventional acoustical holography. This new method is named as "digital acoustical holography", since phase difference between reference and reflected waves is detected by a digital manner instead of analogue phase detection. Experimental verification of the new method is also presented in this paper.

SYSTEM DESCRIPTION

Principle

In this section, a principle of the new method is presented by comparing with the principle of the conventional acoustical holography with a simultaneous source-receiver scanning.

We suppose to make a focused image hologram of an oblique flat plate as shown in Fig. 1a. Direction of a reference wave is supposed to be normal to the scanning plane of a transducer.

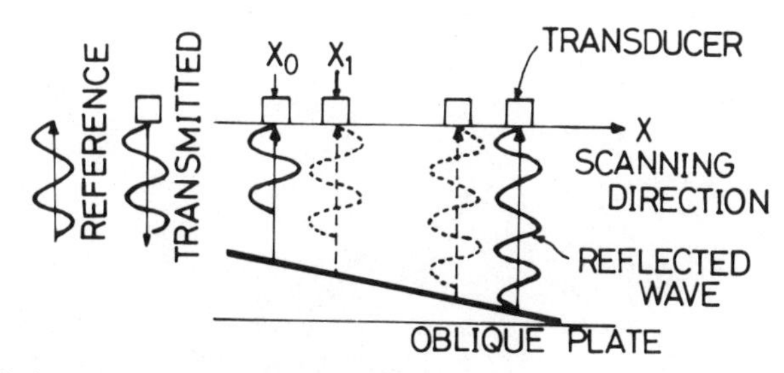

(a) Geometrical condition

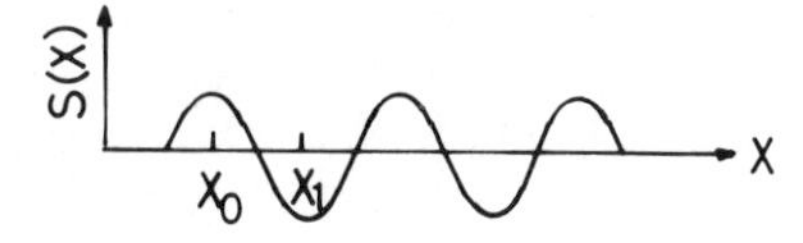

(b) Interference signal

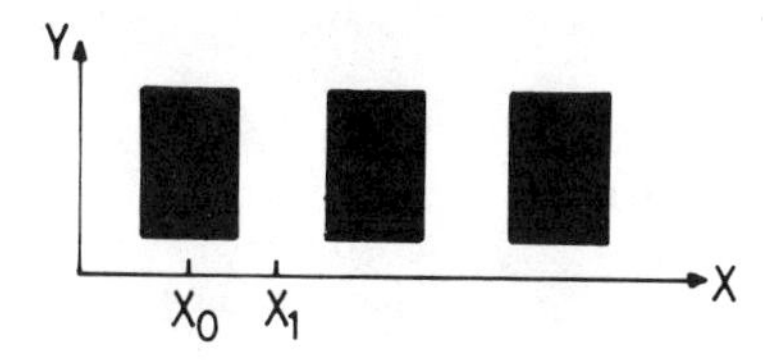

(c) Focused image hologram

Fig. 1. : Principle of the conventional acoustical focused image holography.

In the conventional acoustical holography, a sinusoidal mode pulse with relatively long duration in time is transmitted and the reflected wave from the plate is interfered with the reference wave to detect the phase difference between the two waves. Let the X-axis be in the scanning direction of the transducer. The reflected wave S_d from the plate at the transducer and the reference wave are represented by:

$$S_d = A_d(x)\exp(iwt + \phi_d(x))$$
$$S_r = A_r(x)\exp(iwt + \phi_r(x)) \quad (1)$$

respectively, where $A_d(x)$ and $A_r(x)$ are amplitudes of the two waves, $\phi_d(x)$ and $\phi_r(x)$ are the phase of the reflected and reference waves. In the present case, $\phi_r(x) = 0$ since the direction of the reference wave is supposed to be normal to the scanning plane. An interference term $S(x)$ of the two waves is then:

$$S(x) = 2A_d(x)A_r(x)\cos(\phi_d(x)), \tag{2}$$

which describes the periodic spatical variation of intensity in the interference term. Since the phase ϕ_d is expressed as

$$\phi_d(x) = \frac{2\pi}{\lambda} r,$$

where λ is wave length and r is propagation path length of the acoustic wave at the position X, the periodicity of S(x) is governed by the propagation path length of the acoustic wave. Thus, the intensity of S(x) is maximum at the position X_0, where the reflected wave is in phase with the reference wave, namely the propagation path length of the acoustical wave equals to $n\lambda$, where n is integer. On the other hand, S(x) takes the minimum value at the position X_1, where the reflected wave is out of phase with the reference wave, as shown in Fig. 1b. A corresponding image hologram would be obtained as shown in Fig. 1c.

In this method, the ingerference term S(x) given by Eq. (2) is obtained by a heterodyne detection method which consists of electronic multiplication of $S_d(x)$ and $S_r(x)$ and low-pass filtering of the multiplied signal[2]. Because of using the heterodyne detection method, the sinusoidal wave with relatively long duration in time is used to drive the transducer. Furthermore, frequency of the reference and transmitted waves should be identical. These two facts cause the following two disadvantages; (1) depth resolution is limited by the transmitted pulse duration, (2) impossibility of fringe representation corresponding to the deviation in the propagation path length of less than half a wave length.

In the new method, digital clock pulses with 50% duty ratio are employed as a reference signal instead of the sinusoidal waves used in the conventional method as shown in Fig. 2a. A transmitted pulse is generated synchronizing with the clock pulse, and a reflected pulse from an object is digitized. A leading edge of the digitized reflected pulse would appear at the high state of the clock pulse as shown in Fig. 2b, since the propagation path length is equal to n , where n is integer, at the position X_0 in Fig. 1a. On the contrary, the digitized reflected pulse rises up at the low state of the clock pulse at the position X_1 in Fig. 1a. Then, a time coincidence measurement between the clock pulse and the leading edge of the digitized reflected pulse is performed and a coincidence signal is generated when the leading edge of the digitized reflected pulse coincides with the high state of the clock pulse, as shown in Fig. 2c. The high state of the coincidence signal is kept until the coincidence between the two pulses does not happen to occur.

The high state of the coincidence signal means that the phase difference between the reflected wave and the reference wave exists in a range of 0 and π radians, and the low state of the signal indicates that the phase difference exists in a range of π and 2π radians. This coincidence signal can be used to represent fringes indicating the deviation in the propagation path length in the focused image hologram

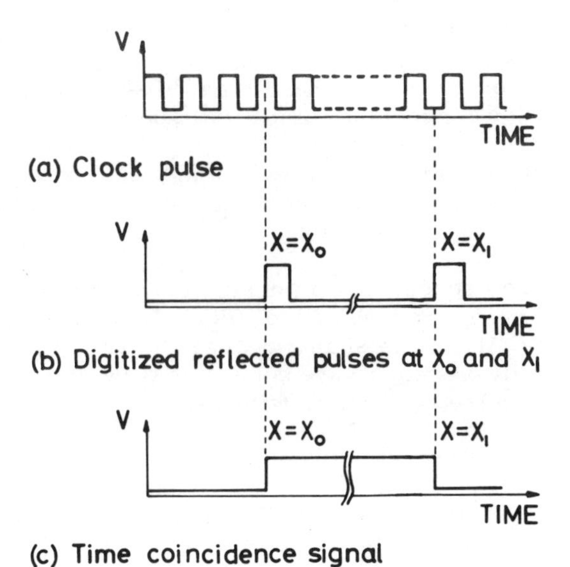

Fig. 2. : Principle of the new signal processing method

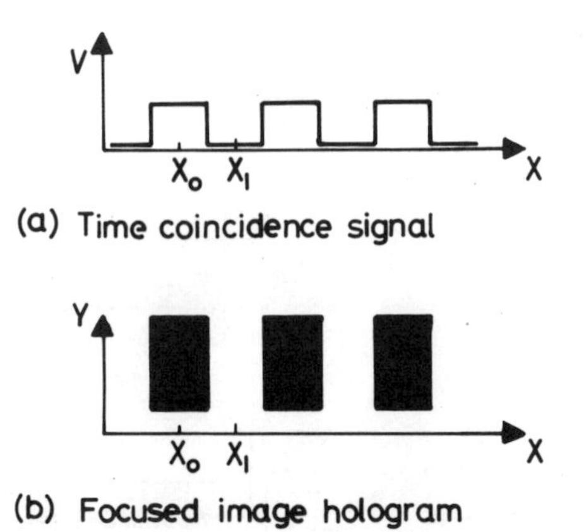

Fig. 3. : A time coincidence signal and corresponding image hologram

as shown in Fig. 3.

This detection method has two advantages over the heterodyne phase detection method. The first one is that high depth resolution is available by the use of a wide band transmitted pulse of which width is narrower than the sinusoidal pulse width, and the second is that fringe representation corresponding to the deviation in the propagation path length of less than half a wave length can be easily achieved by increasing the clock pulse frequency independently of the carrier frequency of the transmitted waves.

An inclined reference wave in acoustical scanned holography has been simulated by shifting phase of the electronic reference signal with respect to the velocity of the scanning transducer[3]. In the present method, the inclined reference wave is simulated by delaying the timing of the transmitted pulse generation with respect to the position of the transducer which is available from a scanner of the transducer.

Digital Signal Processor

A prototype signal processor based on the present method is produced and its block diagram is shown in Fig. 4.

A clock pulse generator provides 12/N MHz clock pulses, where N = 1, 2, 3, ... , 16, which are obtained by dividing an output of a 24 MHz quartz oscillator. A spike pulse generator feeds a wide band pulse to a transducer to generate an acoustic pulse of about 1 μs in width.

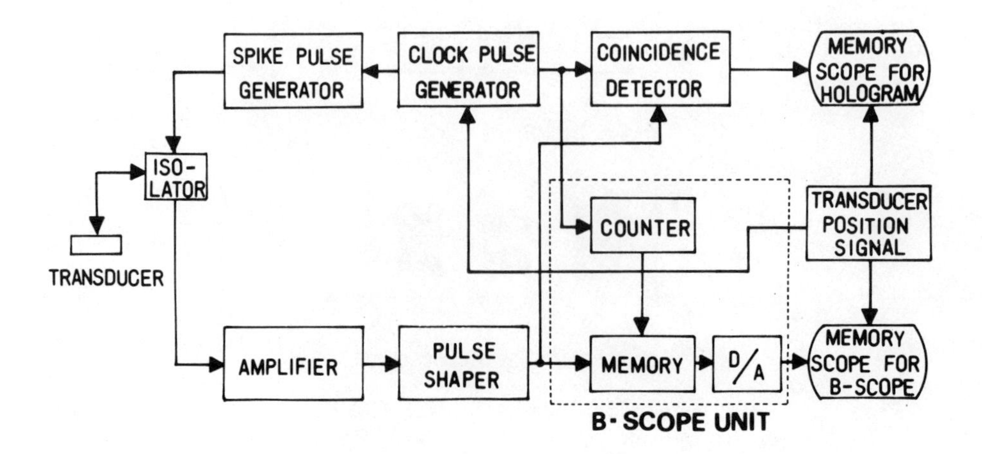

Fig. 4. : Block diagram of the prototype signal processor

Generation of the acoustic pulse is synchronized with the clock pulse in case of simulating the on-axis reference wave, and is delayed with respect to the position of the transducer to simulate the off-axis reference wave.

A reflected pulse from an object is amplified and transformed to a digital pulse in a pulse shaper. The digital reflected pulse is fed into a coincidence detector to detect the phase difference of the reflected pulse to the clock pulse. Finally, a coincidence signal generated in the coincidence detector is used to represent the focused image hologram on a storage monitor scope with transducer position signals which are supplied from the scanner.

The signal processor is also able to generate signals for B-scope representation that indicates depth information on the object by measuring propagation time of the reflected wave from the object. The propagation time measurement is easily achieved by counting number of clock pulses which falls between the transmitted and the reflected pulses. The B-scope unit is mainly composed of a counter, a memory and a digital-analogue converter, as shown in Fig. 4.

The prototype signal processor with the B-scope unit is shown in Fig. 5. Weight of the processor is about 15 kg and is reduced to be one third of a conventional acoustical holography signal processing unit by adoptation of digital signal processing.

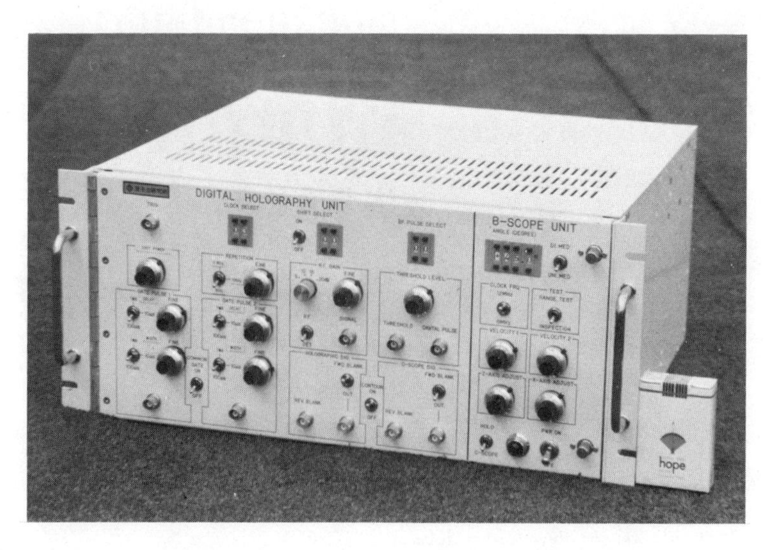

Fig. 5. : Picture of the prototype signal processor which includes the B-scope unit

EXPERIMENTAL RESULTS

Experiments are performed in order to confirm the concept of the digital signal processing described in the previous section.

Two kinds of transducers used in the present experiments are both focused type, 25.4 mm in effective diameter and have 101 mm focal length in water. Their resonance frequencies are 1.0 and 3.0 MHz, respectively. All of the experiments are performed with the on-axis reference wave. Details of each experiment is explained in the corresponding section.

Imaging of An Oblique Flat Plate

A focused image hologram of an oblique plate in water is constructed by 1 MHz acoustic waves with 1-6 MHz clock pulses. The obtained hologram is compared to a hologram of the same object made by the conventional acoustical holography system with 1 MHz acoustic waves.

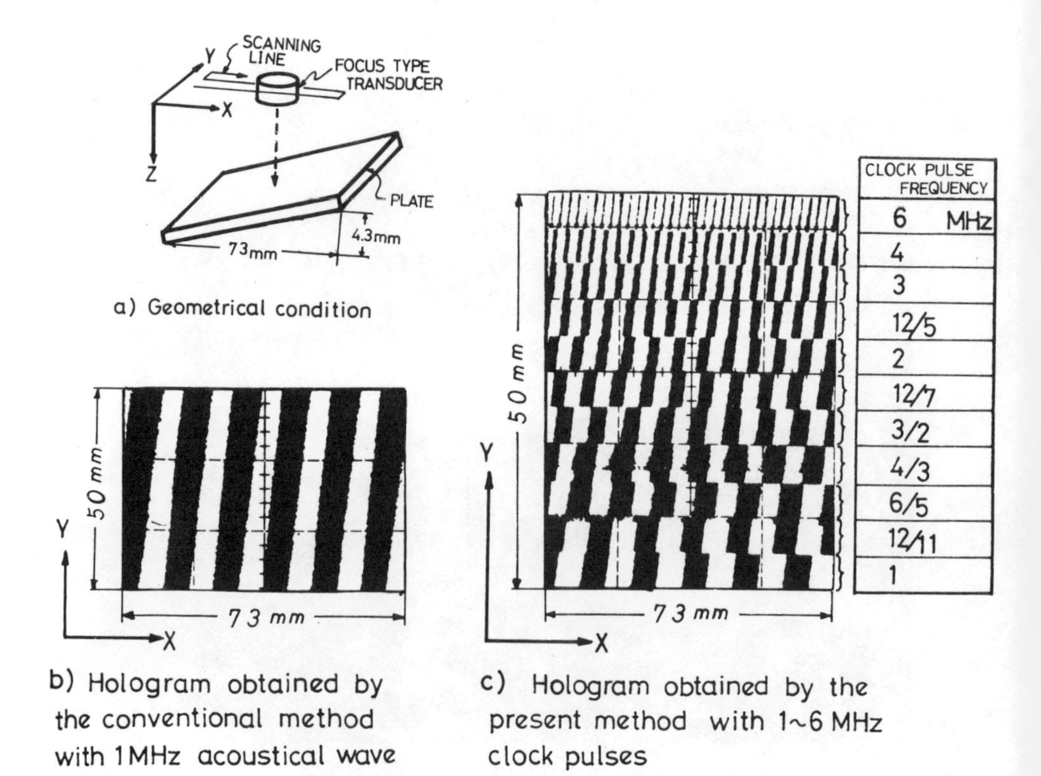

Fig. 6. : Holograms of the inclined plate in water obtained by the conventional and the new methods

Hologram construction geometry is shown in Fig. 6a. Inclination of the plate to the scanning plane of the transducer is 58.8 mrad (3.4°). The hologram obtained by the conventional method is shown in Fig. 6b, while Fig. 6c shows the hologram obtained by changing the clock pulse frequency. Fringe spacing (13.1 ± 0.8 mm) of the image hologram obtained with 1 MHz clock pulse agrees well with the result obtained by the conventional holography (13.3 ± 0.7 mm).

Relations between period of the digital clock pulse and the depth deviation Δh corresponding to half a wave length is shown in Fig. 7. Circles in the figure indicate experimental data obtained from the hologram shown in Fig. 6c. A dotted line indicates a value corresponding to half a wave length of 1 MHz acoustic wave in water. Figures 6 and 7 show that the digital signal processor works as expected and fringe representation corresponding to the deviation in the propagation path length of less than half a wave length is achieved by changing the frequency of the digital clock pulse regardless of the frequency of the transmitted wave.

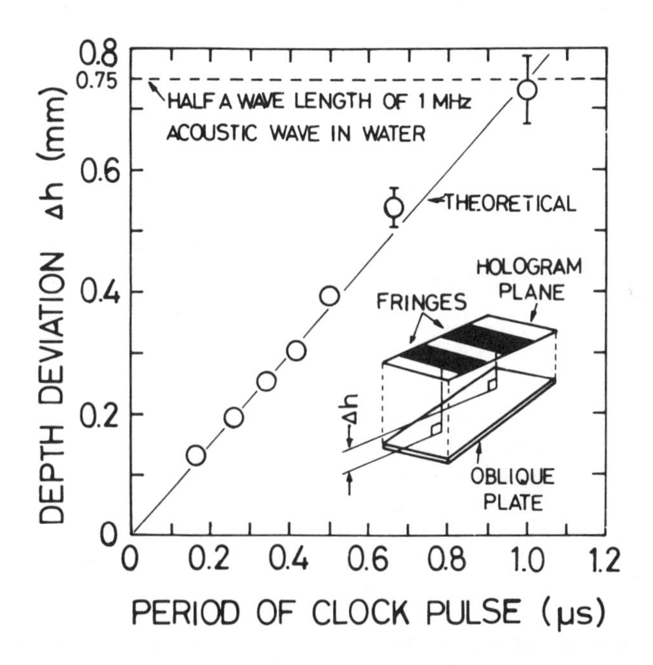

Fig. 7. : Relations between period of the digital clock pulse and the depth deviation corresponding to half a wave length

Imaging of Back Surface of A Plate

The present method is applied to image oblique back surface of a stainless steel plate which is fabricated to simulate inner surface of welding region of a pipe. Imaging geometry and shape of the object are shown in Fig. 8a. A transducer with resonance frequency of 3 MHz was used in this experiment.

Images obtained by the focused image holography and B-scope method are shown in Fig. 8b and 8c. The image obtained by the B-scope method indicates cross section of the object. Inclined region of the back surface is represented by fringes of which spacing becomes narrower in proportion to the clock pulse frequency.

Imaging of Artificial Vertical Flaws

Five artificial vertical flaws fabricated in a 10 mm thick stainless steel plate are imaged by the present method as shown in Fig. 9. Artificial flaws of which height is ranged from 1 to 4 mm are located at a distance of 5 mm from the weld. A 3 MHz focused type transducer was used.

Holograms of the flaws obtained by the present method with 1, 3, and 6 MHz clock pulses are shown in Fig. 10a to 10c. A hologram made by the conventional method is also shown in Fig. 10d. Comparison of these figures verifies that depth resolution is improved in the new method, since all images of the flaws in Fig. 10a to 10c are separated from the image of the weld, while the images in Fig. 10d are not. Fringe spacing in the holograms shown in Fig. 10a to 10c become narrower in proportion to the increase of the clock pulse frequency.

Height of each flaw is evaluated by counting a number of fringes appearing in the hologram. Relations between the evaluated size and the fabricated size of the flaws are shown in Fig. 11. Deviation of the estimated height obtained by the new method from the fabricated value is about 30% for the flaw of 1 mm in height with 6 MHz clock pulses. On the other hand, the corresponding deviation in the conventional method is about 50%.

CONCLUSION

It has been shown that the new signal processing method for acoustical holography is verified experimentally with imaging several objects. Experimental results show that the new method has two features over the conventional method in the focused image holography mode. The first feature is that high depth resolution is available by the use of a wide band transmitted pulse. The second is that fringe representation corresponding to the depth deviation in the propagation

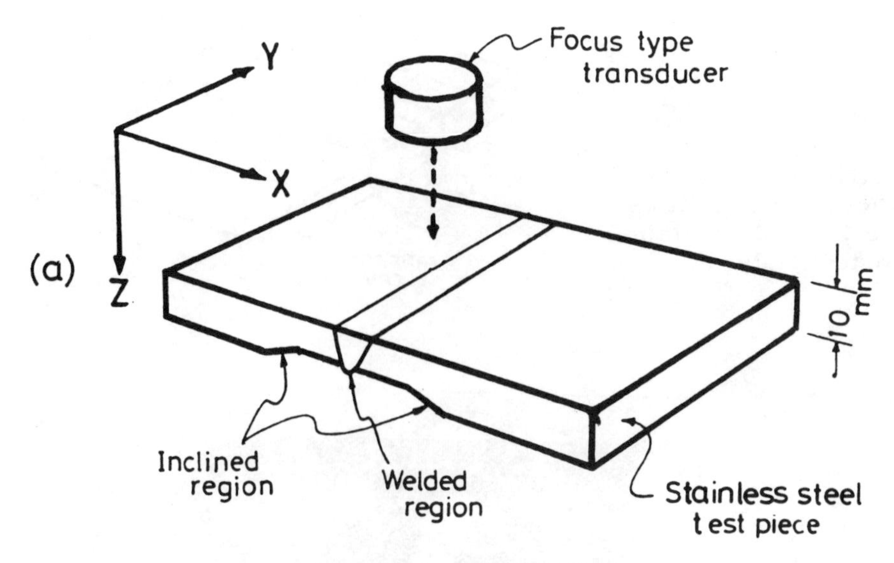

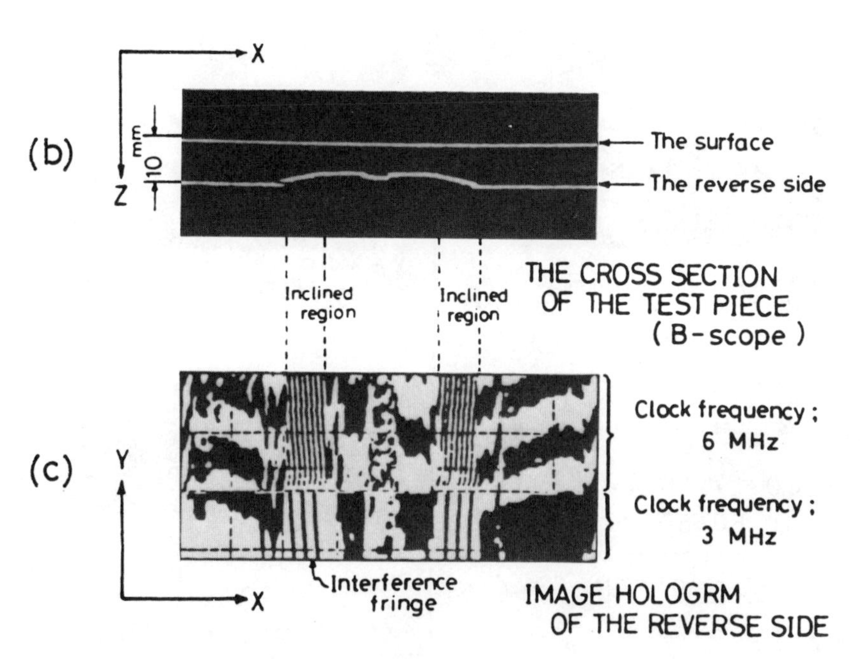

Fig. 8. : Images of back surface of 10 mm thick stainless steel plate obtained by B-scope method and by the present method

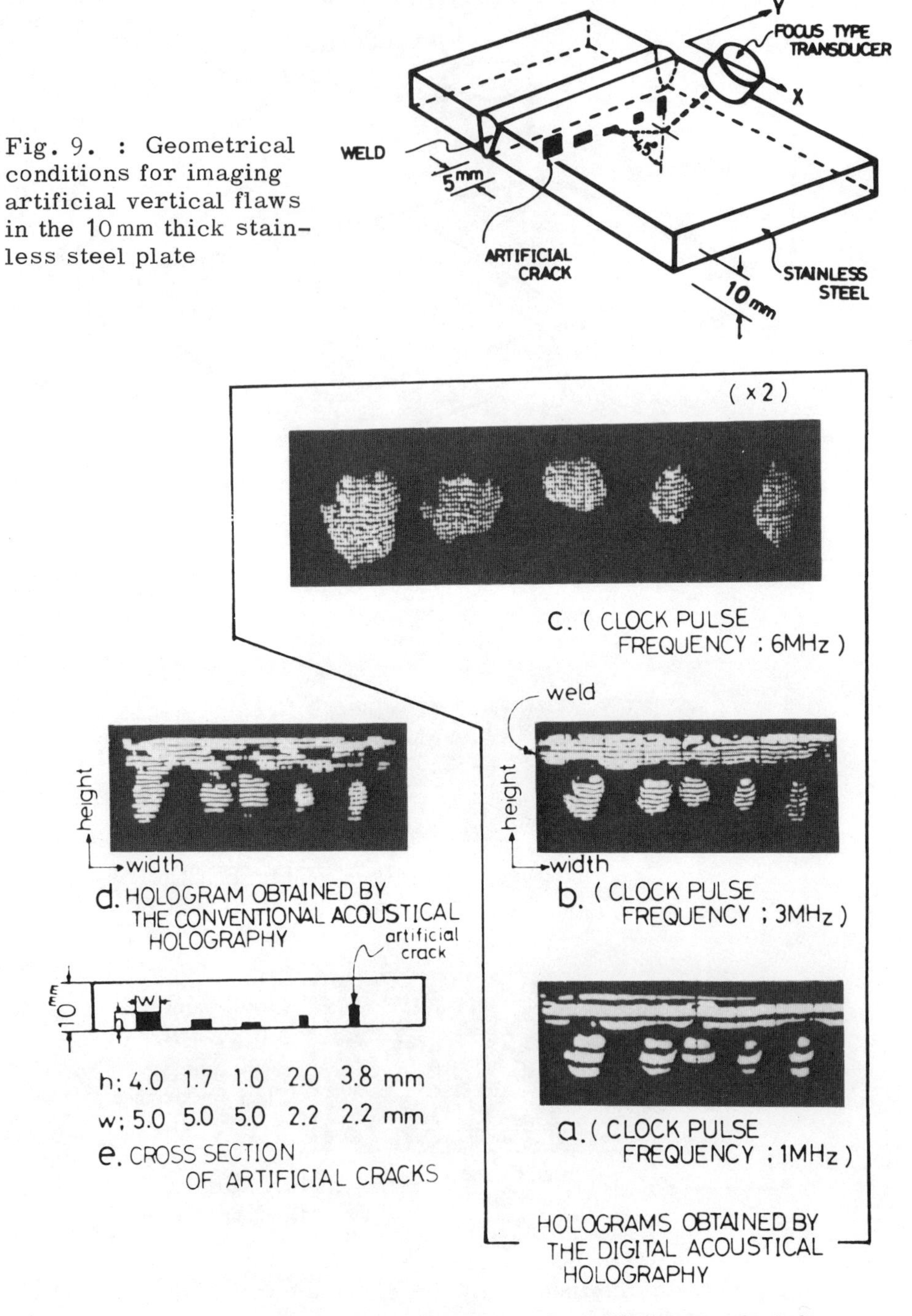

Fig. 9. : Geometrical conditions for imaging artificial vertical flaws in the 10 mm thick stainless steel plate

Fig. 10. : Focused image holograms of artificial vertical flaws obtained by the conventional and the new method

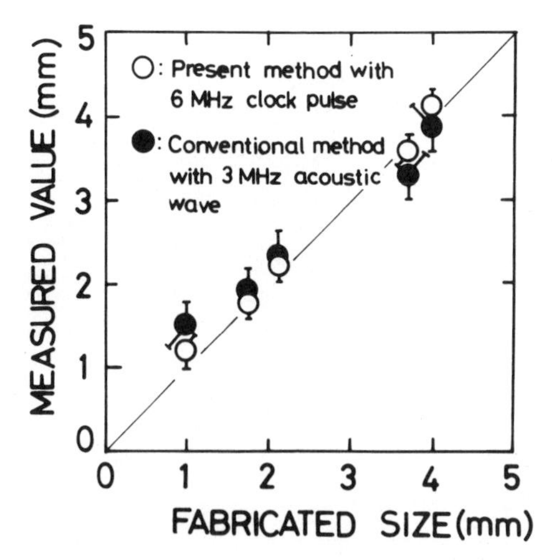

Fig. 11. : Comparison of the measured height of vertical artificial flaws with the fabricated size.

REFERENCES

1. K. Suzuki, F. Takahashi and Y. Michiguchi, "Application of Shear Wave Focused Image Holography to Nondestructive Testing", Acoustical Holography, Vol. 7 edited by L.W. Kessler, Plenum Publishing Company, 1977.

2. B.P. Hildebrand and B.B. Brenden, An Introduction to Acoustical Holography, Plenum Press, New York, 1972.

DYNAMIC IMAGING OF THE AORTA IN-VIVO WITH 10 MHz ULTRASOUND

D.J. Hughes, L.A. Geddes, J.D. Bourland, and C.F. Babbs

Biomedical Engineering Center, Purdue University

West Lafayette, Indiana 47907

INTRODUCTION

This paper describes a new technique for measuring the inner and outer radii of deep blood vessels *in vivo* using three 10 MHz ultrasound transducers mounted at the tip of a 7 French intravascular catheter. The method permits dynamic study of the vascular radii as functions of time and pressure throughout the cardiac cycle. Only minor surgery is required, and the vessel under study need not be excised or dissected from its connective tissue bed. The measurements obtainable with this system promise to be useful in the calculation of Young's modulus of elasticity of blood vessels, the evaluation of animal models of atherosclerosis and related arterial pathologies, and the study of fundamental mechanical properties of arteries *in situ*.

METHODS AND MATERIALS

The catheter-borne ultrasonic transducer used to measure vascular dimensions in living animals, contains three piezoelectric elements, each with a resonant frequency of 10 MHz. The transducers are spaced equiangularly (120 degrees) around the catheter tip in a plane perpindicular to the long axis. Each piezoelectric element measured 2 mm in diameter and is constructed from lead metaniobate covered with an epoxy lens. The iso-echo amplitude field pattern of the transducer is shown in Figure 1 along with a sketch of the catheter.

The three piezoelectric transducers are activated sequentially by a Panametrics 5050PR pulser which also provides preamplification

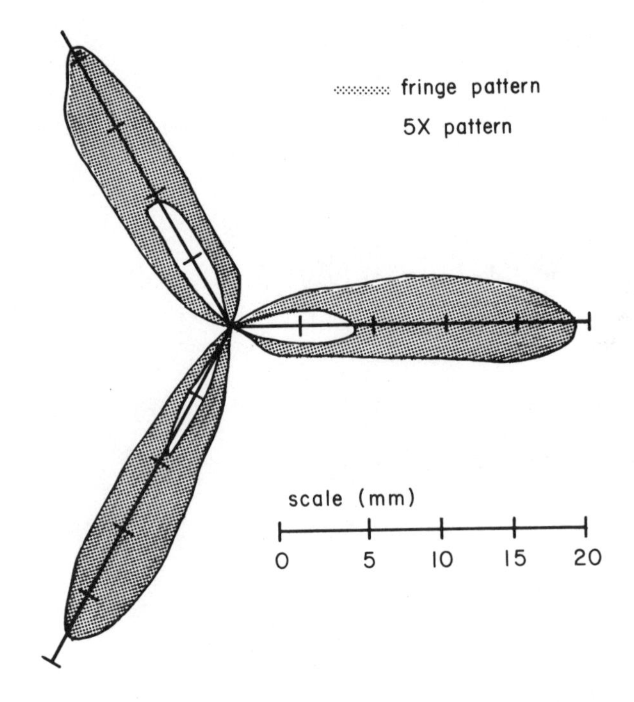

Figure 1 TRIAXIAL CATHETER ISO-ECHO AMPLITUDE PATTERN

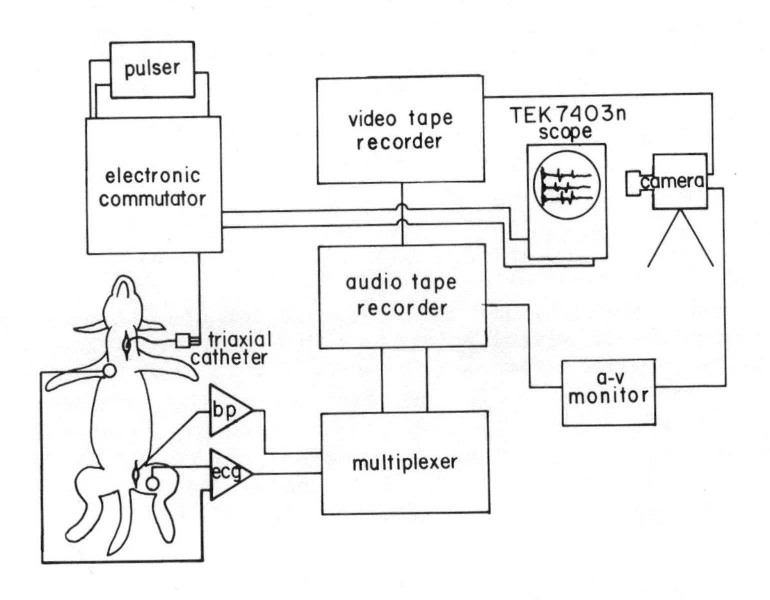

Figure 2 EXPERIMENTAL APPARATUS

of the echo signals. This unit is used in conjunction with an electronic commutator designed to scan each piezoelement at a rate which is adjustable from 0 - 333 Hz. The system operates at a pulse repetition frequency (PRF) of 1 kHz. An RLC band-pass filter within the commutator attenuates high and low frequency noise. The echoes from each element are displayed simultaneously in real time on a Tektronix 7403 high frequency oscilloscope.

With a transducer resonant frequency of 10 MHz, the wavelength of the transmitted pulse in most biological media is about 0.15 mm. In principle, this wavelength allows resolution of objects down to about 0.075 mm in size.

Figure 2 diagrams the experimental apparatus. Dogs anesthetized with sodium pentobarbitol (30 mg/kg, i.v.) served as subjects. The triaxial ultrasonic catheter was advanced down the left carotid artery into the desired position in the aorta. Simultaneously, a Millar catheter-tip blood pressure transducer was advanced up the femoral artery into the aorta and located adjacent to the tri-axial catheter tip. Echoes from the pressure catheter-tip identified the proper location which was verified upon post-mortem examination.

As shown in Figure 2 the ultrasonic data were obtained via the triaxial catheter and displayed on the high frequency oscilloscope. Using a video camera, this display was recorded on video cassette tape. Blood pressure and ECG data were multiplexed for recording on a Revox audio tape recorder. An audio dub was recorded on both audio and video tape as a timing reference for synchronizing the data during replay and analysis.

By replaying the video recorded echo signals frame-by-frame on a stopframe video cassette recorder (Sony VO-2850) and obtaining the time-of-flight for transmission of the pulses to and from the vessel wall, the distances from each piezoelectric transducer to the vessel wall were calculated, using previously determined values (Hughes et al, 1978) of the speed of propagation (C_b) for 10 MHz ultrasound in blood. Therefore, the acquired data, $C_b t_1/2$, $C_b t_2/2$, and $C_b t_3/2$, are the distances from the transducers to the inner wall of the vessel, where t_1, t_2, and t_3 are the round-trip times-of-flight. To each of these values the 1.125 mm radius of the catheter was added to obtain the true intra-arterial distances d_1, d_2, and d_3 from the long axis of the catheter to the vessel walls as shown in Figure 3. The corresponding distances to the outer wall of the vessel given by $d_1 + C_a \Delta t_1/2$, $d_2 + C_a \Delta t_2/2$, and $d_3 + C_a \Delta t_3/2$ where Δt denotes the time between inner and outer wall echoes and C_a is the velocity of sound in the aorta (Hughes, 1978).

The quantities actually sought are the effective inner and outer radii (R & R') of the aorta. Assuming circular symmetry, Figure 3 shows the points of intersection of the ultrasonic axes with the vessel walls (1, 2, 3) which are joined to form a triangle with sides S_{12}, S_{23}, S_{31}. Radial lines are then drawn from the points 1, 2, 3 to the center of the vessel. Applying the inscribed polygon law to the perimeter of the triangle gives

$$P = S_{12} + S_{23} + S_{31} = 3R\sqrt{3}.$$

Then, from the law of cosines applied to each side of the inscribed triangle,

$$S_{12}^2 = d_1^2 + d_2^2 - 2(d_1 d_2)\cos 120^\circ$$

$$S_{23}^2 = d_2^2 + d_3^2 - 2(d_2 d_3)\cos 120^\circ$$

$$S_{31}^2 = d_3^2 + d_1^2 - 2(d_1 d_3)\cos 120^\circ$$

Combining these expression yields the equation for calculating the vessel radius (inner or outer) when the times-of-flight are known:

$$R = \frac{1}{3\sqrt{3}} \quad \sqrt{d_1^2 + d_2^2 + d_1 d_2} + \sqrt{d_2^2 + d_3^2 + d_2 d_3} + \sqrt{d_1^2 + d_3^2 + d_1 d_2}$$

Figure 4 shows various catheter locations in the aortic lumen and the resulting instantaneous echo displays. The ideal case shows a concentric location of the catheter within the lumen. Note the equality of time-of-flight in each channel (e.g. $t_1 = t_2 = t_3$). Cases A, B, and C depict various off-center catheter locations.

The equation for the vessel radii is applicable for any arbitrary location within the vessel provided echo signal intensities are sufficient to acquire three pairs of wall echoes. However, it is important to emphasize that these calculations apply only to vessels which are circular in cross section. In practice this assumption tends to limit the technique to examination of arteries and may exclude examination of some veins. It is also assumed that the long axis of the catheter parallels the long axis of the vessel, as could be confirmed in the present studies at autopsy. Swaying of the catheter within the vessel was minimized by orienting the catheter downstream with the flow of blood, and has not proved to be a hindrance in obtaining echo information or in calculating vessel radii.

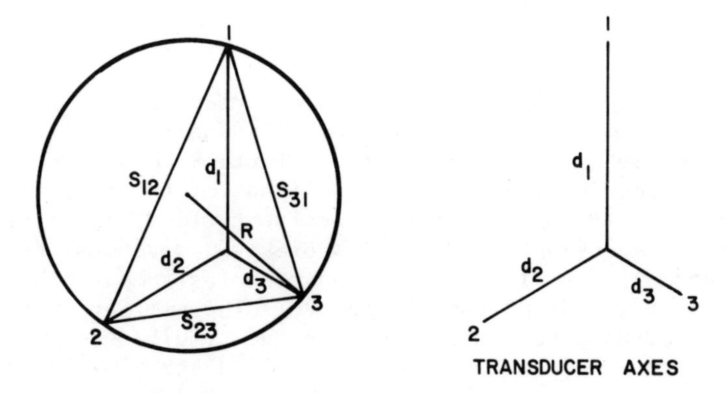

FOR THE INSCRIBED TRIANGLE, THE PERIMETER IS:

$$S_{12} + S_{23} + S_{31} = 6R\sin 60° = 3R\sqrt{3}$$

COMBINING YEILDS,

$$R = \frac{1}{3\sqrt{3}}\left[\sqrt{d_1^2 + d_2^2 + d_1 d_2} + \sqrt{d_2^2 + d_3^2 + d_2 d_3} + \sqrt{d_3^2 + d_1^2 + d_3 d_1}\right]$$

Figure 3 CALCULATION OF VESSEL DIMENSIONS IN-VIVO

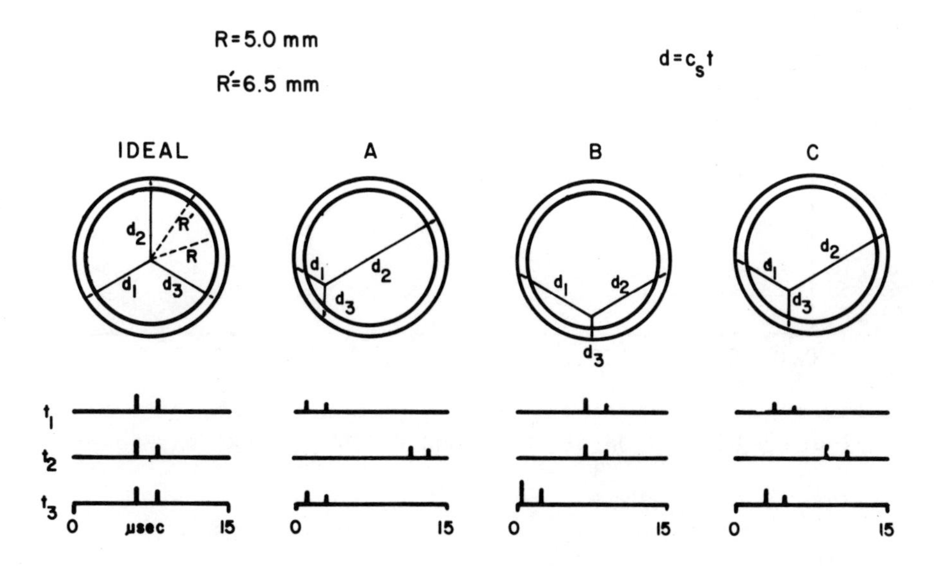

Figure 4 ECHO PATTERNS FOR VARIOUS CATHETER LOCATIONS IN AORTIC LUMEN

RESULTS

Frame-by-frame analysis of video recorded time-of-flight information allowed reconstruction of the dynamic wall motion of the aorta in living animal subjects. Figure 5 shows a typical reconstruction showing the radial dimensions of the inner wall of the aorta of an 18 kg dog for one cardiac cycle. The bottom record is the simultaneously recorded blood pressure waveform, reproduced on a galvanometer oscillograph (Midwestern Instruments Model 805) and transposed to the ultrasonic record. The similarity between the vessel radius waveform and that of the simultaneously recorded blood pressure waveform that causes vessel distension is striking. The systolic radius was 5.53 mm and the diastolic radius was 5.10 mm, indicating 8 percent change in radius during the cardiac cycle. The sampling interval for the radial dimensions was 33 msec.

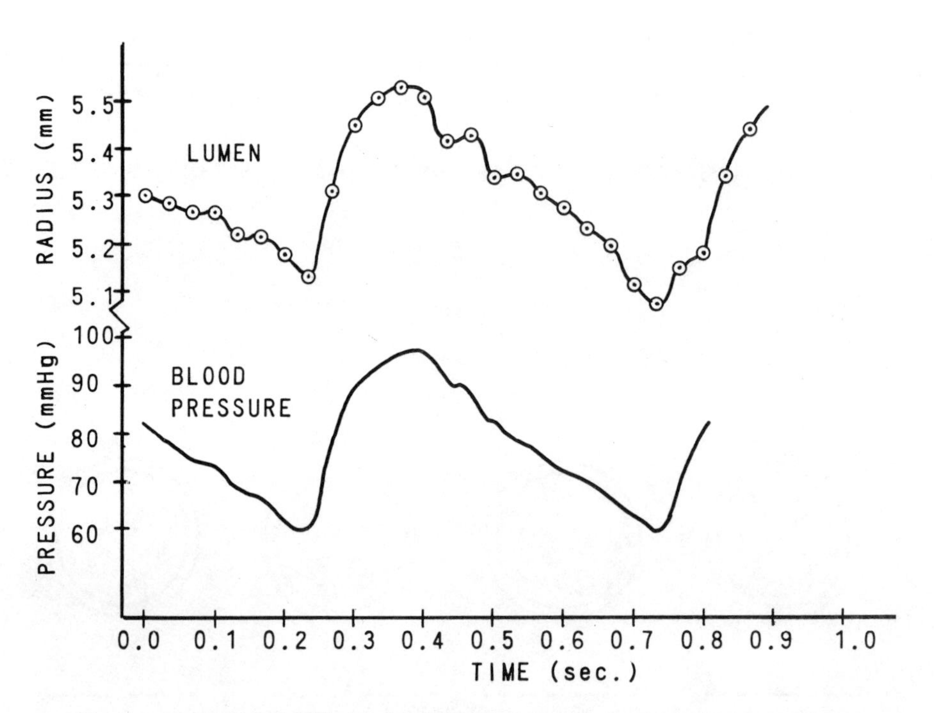

Figure 5 COMPARISON OF LUMEN AND BLOOD PRESSURE WAVEFORMS

In some animals it was possible to obtain inner and outer wall echoes, thereby allowing computation of both the inner and the outer wall radii. Figure 6 depicts this relationship in the descending thoracic aorta of a 16 kg dog with a mean blood pressure of 110 mmHg. The characteristics evident in the inner wall waveform were also present in the record of the outer wall waveform. The difference between outer wall and inner wall radii is the vessel wall thickness. The wall thickness was approximately constant throughout the cardiac cycle. Systolic outer wall radius was 6.12 mm and inner wall radius 5.58 mm. Diastolic radii were 5.72 mm and 5.20 mm respectively. This represents a 7 percent change in outer wall radius and an 8 percent change in inner wall radius over the cardiac cycle.

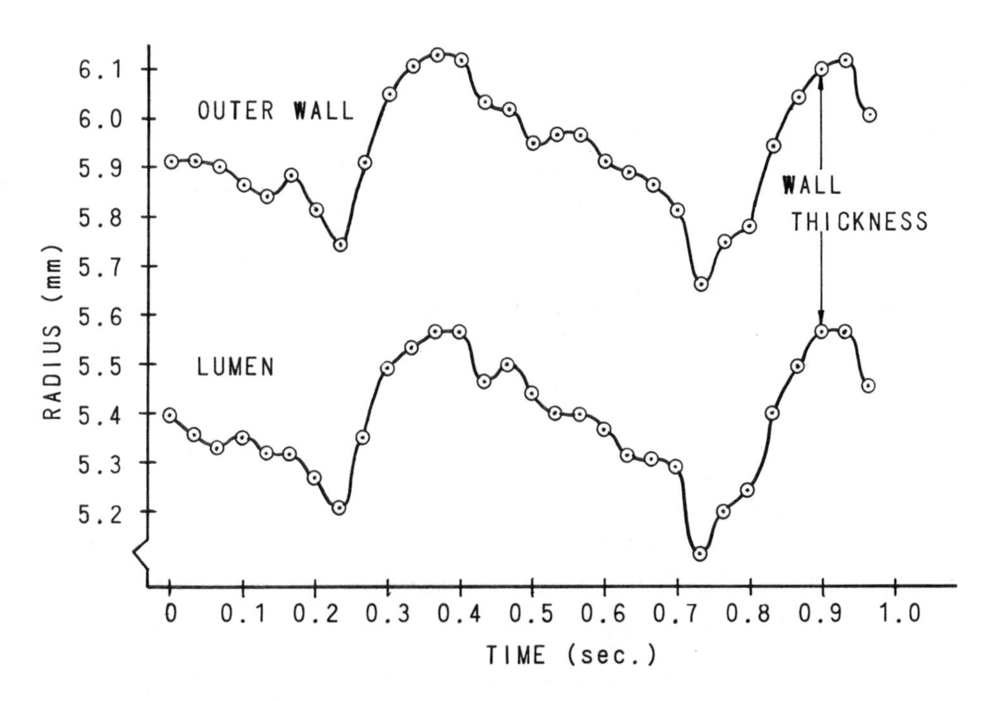

Figure 6 DYNAMIC AORTIC DIMENSIONS IN-VIVO

DISCUSSION

The similarity between the vessel radius waveform and intraluminal pressure waveform has been reported previously. Patel et al, (1960) recorded pressure and diameter in the main pulmonary arteries in dogs utilizing an electrical strain gauge caliper transducer. These workers reported a 10-20 percent change in pulmonary arterial diameter during a cardiac cycle. In this study, the animals were subjected to thoracotomy and the arteries were dissected from the surrounding tissue to enable implantation of the transducer. Patel et al, (1961) were also able to employ the caliper transducer to monitor aortic diameter changes in thoracotomized dogs. The descending thoracic aorta exhibited a 4 percent change in diameter while only a 1 percent change in diameter was recorded in the abdominal aorta. Others (Barnett et al, 1961; Greenfield, 1962; Peterson et al, 1960) have also obtained pressure and diameter records in the aortae of living dogs using a caliper transducer. Peterson found a 2 percent change in carotid artery diameter during a single cardiac cycle.

Bertram (1977) has recorded arterial diameter changes in vivo in living dogs using an ultrasonic transit time system. By transmitting an ultrasonic pulse across the vessel to the receiving transducer, the transit time was multiplied by the propagation speed in blood to obtain the arterial diameter. This technique yielded a 4 percent change in femoral artery outer diameter.

All of these previously reported techniques have involved thoracotomy and/or dissection of the connective tissue surrounding the vessel. The triaxial catheter system has the unique advantage of being able to record aortic radius changes without disrupting the vascular geometry. The importance of intact vascular geometry on the dynamic behavior of the aorta has not been determined. However, we believe that the differences which exist between pressure and radius records obtained through extensive dissection and those records obtained via the triaxial catheter in the intact animals merit further investigation.

SUMMARY

The dynamic wall motion of the descending aorta of anesthetized dogs has been recorded by an ultrasonic intra-aortic triaxial catheter-tip transducer advanced into the aorta via a carotid artery. Aortic pressure and inner and outer radius changes in situ were recorded over long periods of time in intact animals. An 8 percent change in aortic inner wall radius and a 7 percent change in aortic outer wall radius were observed during the cardiac cycle. Comparison of dynamic inner and outer wall waveforms with blood pressure waveforms reveals a striking similarity. Slight differences between inner and outer wall motion may suggest viscoelastic behavior of the aortic wall.

BIBLIOGRAPHY

Hughes, D.J., L.A. Geddes, J.D. Bourland, C.F. Babbs, and V.L. Newhouse "Attenuation and Velocity of Ultrasound in Blood at 37°C", Proc. of 3rd Symposium on Ultrasonic Tissue Characterization, NBS, Gaithersburg, MD, M. Linzer ed. 1978.

Hughes, D.J. and B. Snyder unpublished results 1978.

Bertram, C.D., "Ultrasonic transit-time system for arterial diameter measurement", Med. Biol. Engr. Comp., vol. 15 no. 5, 489- 499, 1977.

Patel, D.J., D.P. Schilder, and A.J. Mallos, "Mechanical properties and dimensions of the major pulmonary arteries", J. Appl. Physiol., 15: 92, 1960.

Patel, D.J., A.J. Mallos, and D.L. Fry, "Aortic mechanisms in the living dog", J. Appl. Physiol., 16: 293, 1961.

Barnett, G.O., A.J. Mallos, and A. Shapiro, "Relationship of aortic pressure and diameter in the dog", J. Appl. Physiol, 16: 545, 1961.

Peterson, L.H., R.E. Jensen, and J. Parnell, "Mechanical properties of arteries in-vivo", Circulation Res., 8: 622, 1960.

THE FOCUSSING OF ULTRASOUND BEAMS THROUGH HUMAN TISSUE

F.S. Foster and J.W. Hunt

The Ontario Cancer Institute and
Department of Medical Biophysics, University of Toronto,
Toronto, Canada M4X 1K9.

I. INTRODUCTION

As a focussed ultrasound beam converges in tissue each point of the wavefront experiences velocity and attenuation fluctuations due to the structure and composition of the tissue at that point. These fluctuations lead to a spreading of the beam at the focal zone and steering effects (1,2) where the whole focal zone is moved off axis. Since beam distortion is directly related to loss of resolution in pulse-echo systems it is important to characterize this phenomenon and establish, if possible, the transducer parameters which maximize system resolution. Using a unique variable aperture, variable focal length ultrasound system, the beam distortion in the focal zone has been studied as a function of f-number (focal length/diameter), intervening tissue type, and intervening tissue thickness. The transducer diameter varied from 11 mm to 50 mm and thus the results apply to both wide aperture and conventional transducers.

II. EQUIPMENT

The most important component of the experimental system is the transmitting transducer (AR1). This device (Figure 1) consists of a six-ring annular array with a maximum diameter of 50 mm. Various annular sections are excited simultaneously such that different diameters of the transducer are obtained. The transmitted ultrasound beam is concentrated by a novel variable-focus liquid lens. The lens consists of liquid Dupont Freon 114B2 (supplied by Columbia Organic Chemicals, Columbia, South Carolina), trapped between the surface of the transducer and a 100 μm Mylar membrane. The focal length of the beam is easily altered by forcing

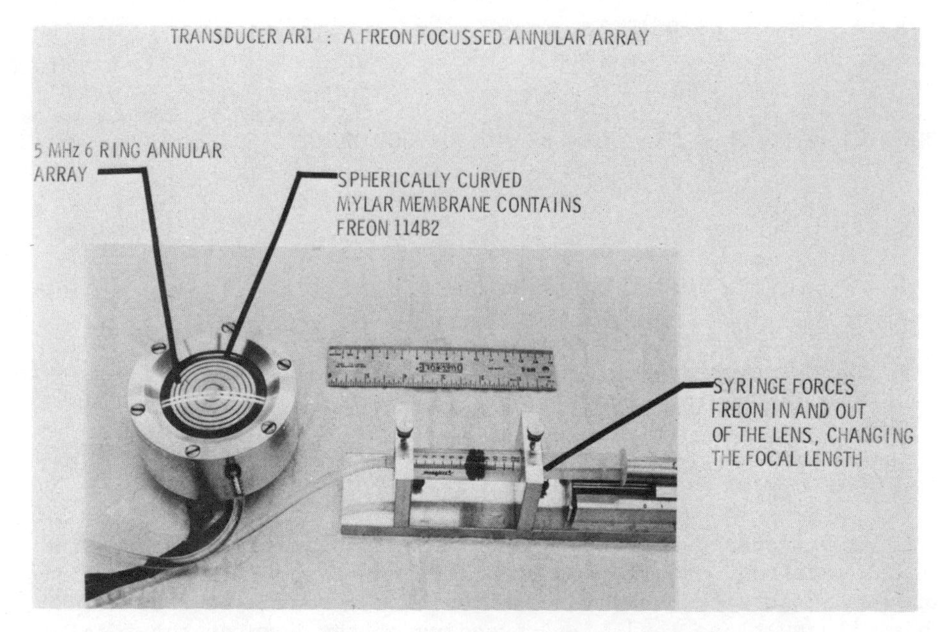

Figure 1. Transducer AR1: a Freon focussed annular array

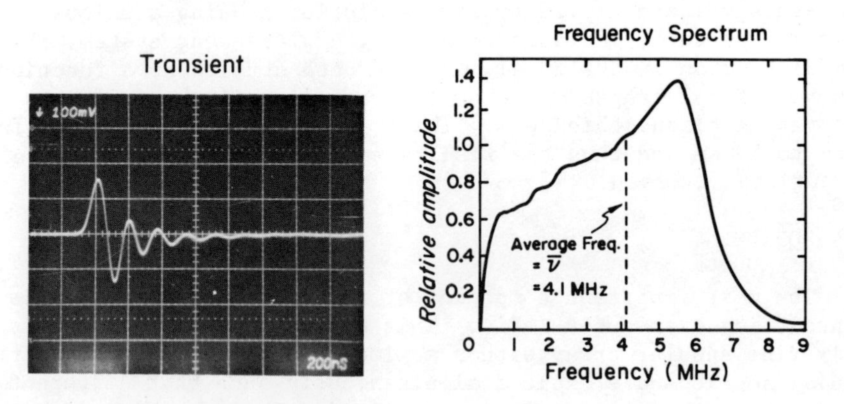

Figure 2. The characteristics of the pulse transmitted by AR1

the Freon in or out of the lens using a syringe. This liquid Freon lens is extremely effective for two reasons: 1) The speed of ultrasound in the Freon is extremely low (590 M/s), therefore a simple thin lens can be used to focus very large apertures. For example, a Freon lens with a thickness of only 3 mm will bring the ultrasound from a 5 cm diameter source to a focus at 10 cm (i.e., f/2). 2) Because the density of the Freon is high (2.06×10^3 Kg M^{-3}), the acoustic mismatch between the lens and the water is small (reflected amplitude of the signal is <10%).

As shown in Figure 2, the transducer is highly damped; therefore the Fourier frequency spectrum is broad. The measurement of the pulse shape was performed by a thick receiver (3 μs transit time) in the far field of the center transducer element, as discussed in detail by reference (3). Good fidelity of the transmitted pulse was obtained in this manner. The frequency spectrum has a bandwidth of 5.1 MHz (-6 dB points) with an average frequency of 4.1 MHz.

During the experiments tissue samples were confined between two Mylar membranes, as shown in Figure 3. The profiles of the ultrasound beam were obtained by an automatic scanner. The receiving microphone consisted of a 0.8 mm diameter, 5 MHz transducer mounted in the tip of a hypodermic needle. The microphone was scanned across the focal zone by means of a variable-speed motor. The position of the microphone was adjustable in three dimensions within 0.05 mm. The profile of the ultrasound beam in the focal zone was obtained using a X-Y oscilloscope, for which the 'Y' input represented the amplitude of the signal detected by the microphone, and the 'X' input was a voltage proportional to the microphone position. An example of one profile is shown in Figure 4. The full-width at half-maximum (FWHM) of the amplitude distribution was chosen as a readily measurable index of resolution at the focal zone.

III. EXPERIMENTAL STUDIES

a) Focussing Experiments in Water

The system was calibrated by measuring the amplitude distribution for various focal lengths and diameters in distilled water. Figure 5 shows a plot of FWHM vs f-number for this experiment. Since the f-number is directly proportional to the resolution, it is expected that equivalent f-numbers at different focal lengths will have the same FWHM's. As Figure 5 shows, this is true to a reasonable extent. Note that the calibration curve lies somewhat above the theoretical line predicted using simple Fraunhofer diffraction. The equation used in this calculation is:

$$\text{FWHM} = 1.41\ (c/\bar{\nu})\cdot(z/d) = 1.41\ c\ (\text{f-number})/\bar{\nu} \qquad (1)$$

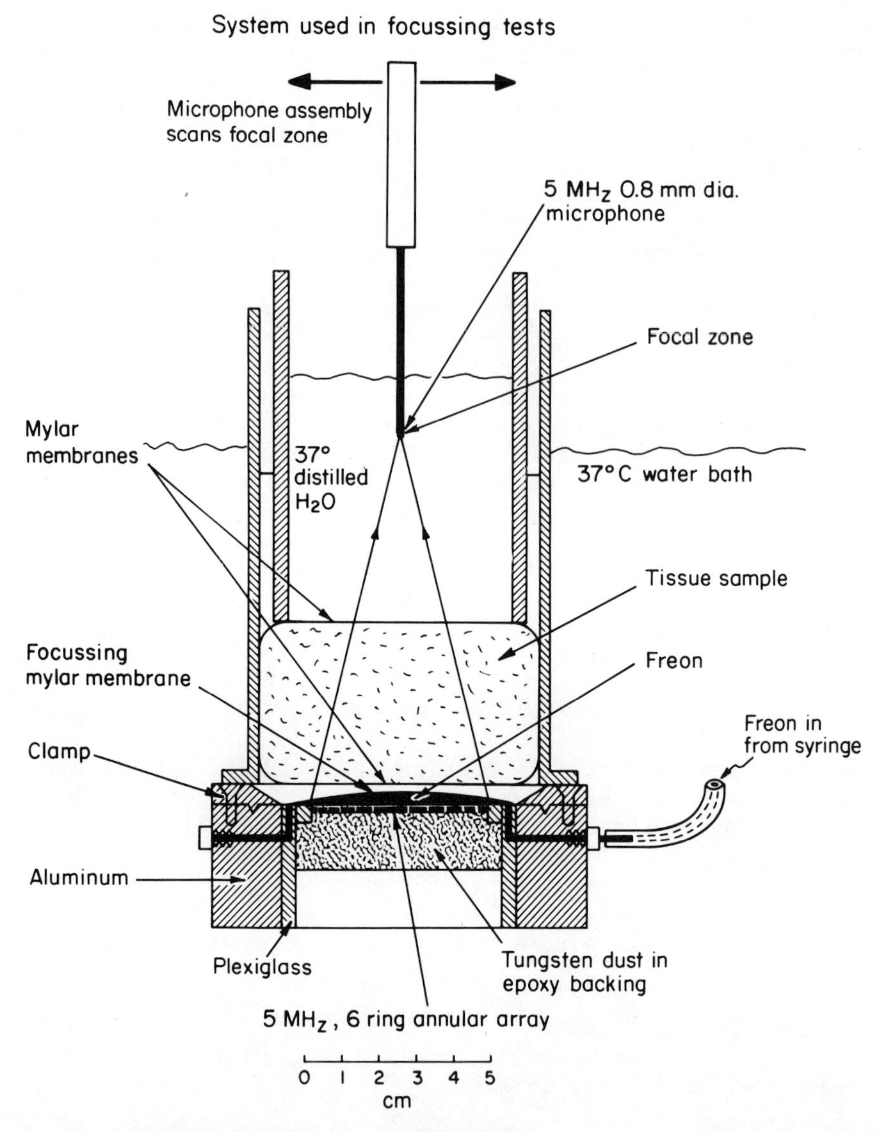

Figure 3. Cross section of transducers and tissue sample holding apparatus.

where c is the velocity of ultrasound in water, $\overline{\nu}$ is the average frequency of the pulse spectrum, z = focal length, and d the diameter of the transducer. The discrepancy between the experimental and theoretical values is probably due to lens aberrations and the wide bandwidth of the transmitted pulse. The spacings between rings (.5-.8 mm) and the width of the microphone are involved to a lesser extent.

Example of a Beam Profile

Focal length = 10.0 cm
Focal number = 2.0
Intervening tissue = 1.2 cm human liver
Temp. = 37°C

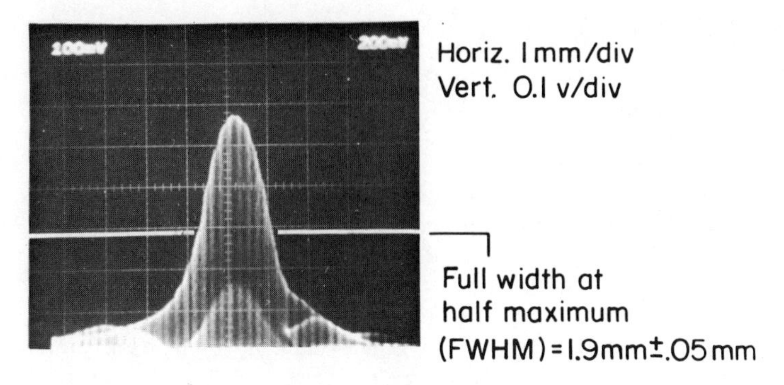

Figure 4.

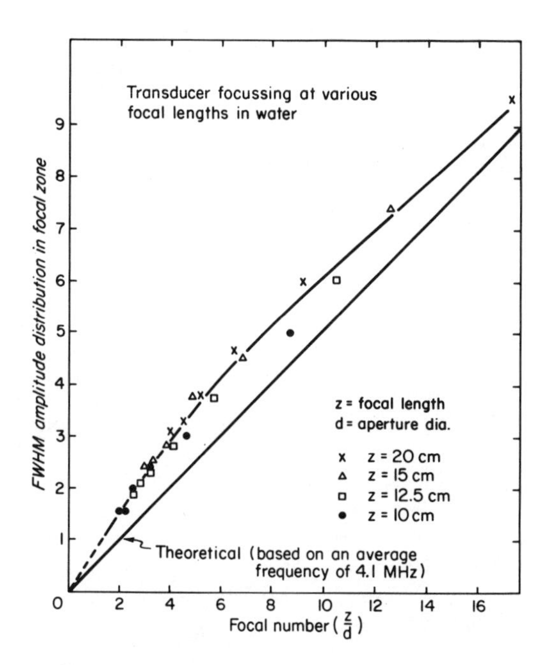

Figure 5. Test of transducer focussing in water. The focal length was adjusted between 10 cm and 20 cm by changing the amount of Freon in the lens. At each focal length FWHM's were obtained for each of the six available apertures. The theoretical curve was calculated using equation (1).

b) Tissue Experiments

Fresh tissue samples were obtained at the time of autopsy and submerged in PBS (phosphate buffered saline). Samples were stored at 4°C for no more than 36 hours and then heated to 37°C for the experiments. Figure 6 shows lateral resolution as a function of depth in normal liver tissue as measured with the system shown in Figure 3. In this experiment liver slices were added in approximately 1.5 cm increments to a total depth of 8.5 cm. After each slice, polaroid photographs were made of the ultrasound amplitude distribution for each of the six apertures available. The special advantage of this technique is that a wide range of apertures can be tested without changing transducers or altering tissue geometry. The results (Figure 6) indicate that focussing ability decreases linearly with increasing thickness of intervening tissue. Also, since the slopes of the lines are approximately equal, it appears that the rate of defocussing is independent of aperture. The fluctuations in the FWHM's for the small apertures were greater than those for the large apertures. The large fluctuations in focussing ability for small apertures were also reported by Banjavic _et al_ (1) and Halliwell (2). Presumably the improvement noted at larger apertures is due to averaging effects imposed by the tissue on the propagating pulse.

Focussing as a function of tissue depth was compared for several tissues with the system set at f/2. The results are shown in Figure 7. The three liver samples and the tumor [a] sample all displayed similar defocussing characteristics. The shaded area depicts this trend. Based on the slope of the shaded area 20 cm of this type of tissue would reduce the resolution by a factor of 2. The results for the abnormal sample of liver did not appear to deviate significantly from those of the normal.

Figure 8 shows a plot of resolution _vs_ f-number for 5 cm samples of various tissues. The focussing response of liver is shown as the lower shaded area and again the results for abnormal and normal liver are similar. The tumor sample demonstrates a more rapid loss of resolution at high f-numbers when compared with liver. Both the liver and tumor plots indicate that better resolution can be obtained by lowering the f-number below f/2. The results for three samples of breast tissue are shown as the upper shaded area. Here, again, focussing improves with decreasing f-number. Note that the beam tends to spread much more in breast than in liver. This is due to the increased inhomogeneity and attenuation of breast tissue.

---

[a] Retro-peritoneal mass associated with Hodgkin's disease (type undefined), firm in consistency, white in color; microscopic analysis not yet complete.

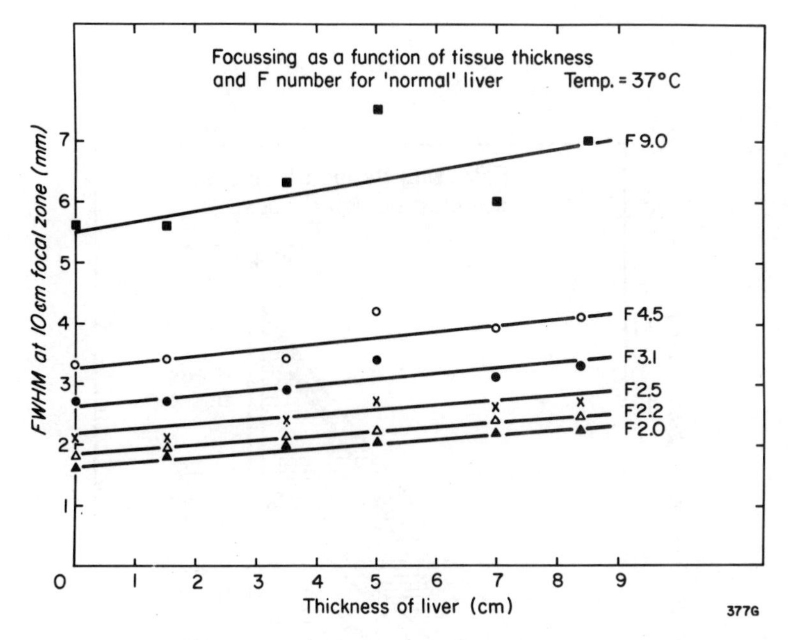

Figure 6. Focussing for normal human liver. The focal length was set at 10 cm. As slices of tissue were added the FWHM's were measured for each aperture of the annular array.

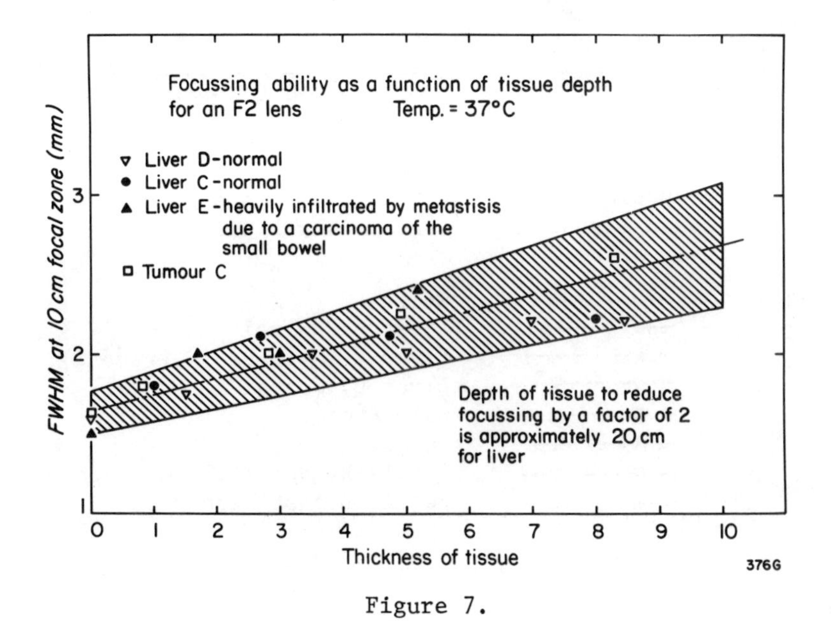

Figure 7.

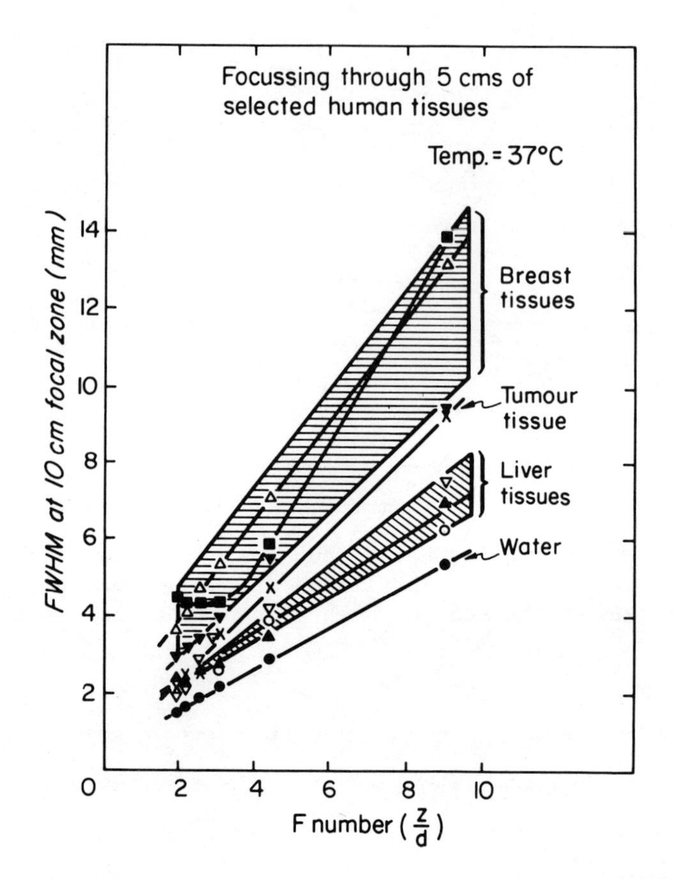

Figure 8. Focussing of the wideband pulse through 5 cm of selected human tissues. The focal length was 10 cm and the aperture was varied.

Legend:

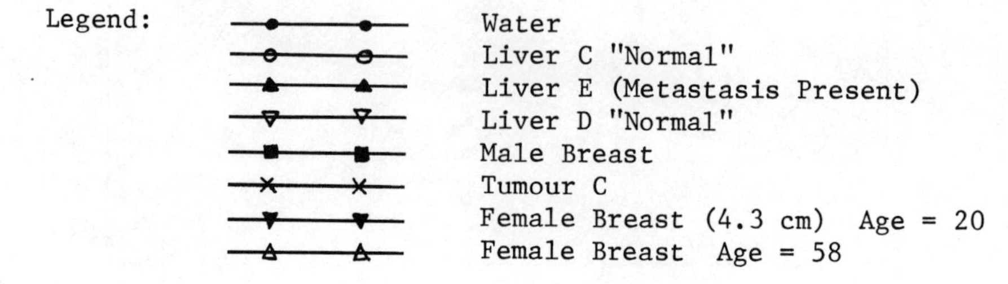

Attenuation plays an important role in the defocussing of wideband ultrasound pulses. Since attenuation increases rapidly with frequency, the spectrum of the pulse is shifted towards lower frequencies as it propagates through the tissue. The greater the attenuation of the tissue sample, the more pronounced is the degradation of the focussed beam. Note that the results presented here are only valid for the pulse shown in Figure 2. Work on the effect of bandwidth on beam focussing in attenuating media is being carried out at present. Also, the effects of tissue geometry and velocity mismatches between the coupling medium and tissues are being tested.

IV. CONCLUSIONS

A variable aperture, variable focal length, wideband transducer has been employed to study the defocussing effects of human tissue on ultrasound. The results of this work are:

1) The optimum f-number for focussing in liver and other relatively homogeneous tissues is less than f/2. For breast, an inhomogenous and highly attenuating tissue, focussing ability is reduced compared to liver but seems to improve linearly with decreasing f-number. Some breast samples have shown nonlinear focussing responses such as leveling off of the FWHM between f/3 and f/2 and anomalous focussing or defocussing at higher f-numbers. The optimum f-number for focussing in breast tissue may be between f/3 and f/2 or perhaps lower.

2) Increasing thickness of tissue resulted in a linear loss of focussing in liver. The rate of defocussing appeared to be independent of aperture. It was estimated that 20 cm of liver would be required to reduce the resolution by a factor of 2 for an f/2 system.

3) Lower f-numbers (larger apertures) produce a more stable focus in human tissue.

4) There is a general relationship between tissue attenuation and beam defocussing. This is mainly due to a shifting of the frequency spectrum towards longer wavelengths as the pulse propagates into the tissue.

ACKNOWLEDGEMENTS

We would like to acknowledge the financial support of the National Cancer Institute of Canada and the Medical Research Council of Canada for different aspects of this work. We would also like to thank Drs. J.S. Carruthers and T.C. Brown of the Dept. of Pathology, Princess Margaret Hospital, for the tissue samples.

REFERENCES

1) R.A. Banjavic, J.A. Zagzebski, E.L. Madsen, R.E. Jutila, (no title), A.I.U.M. Abstracts, p. 89, 1977.

2) M. Halliwell, Ultrasonic beam distortion by the normal female breast _in vivo_, A.I.U.M. Abstracts, p. 94, 1977.

3) F.S. Foster and J.W. Hunt, The design and characterization of short pulse ultrasound transducers, Ultrasonics, in press.

MYOCARDIAL BLOOD FLOW: VISUALIZATION WITH ULTRASONIC CONTRAST AGENTS

Balasubramanian Rajagopalan, J.F. Greenleaf,
P.A. Chevalier and R.C. Bahn

Mayo Foundation, Rochester, Minnesota 55901

Ischemic heart disease is one of the most common health problems in the contemporary American society. Arteriosclerotic narrowing of the coronary arterial tree and the associated myocardial ischemia can impair myocardial contractility and ventricular performance. This in turn may reduce arterial pressure and hence coronary perfusion pressure leading to further ischemia and potential enlargement of the necrotic region. This vicious cycle may result in death. Ischemic heart disease accounts for more than 675,000 deaths per year in the U.S.A. Techniques that assess regional myocardial perfusion and detect acute myocardial infarction promise to be useful in the detection and evaluation of coronary artery disease. Besides they would be useful in the assessment of therapies aimed at limiting the degree of ischemia and the extent of tissue necrosis. Several diagnostic methods are being developed for myocardial perfusion imaging; positron scanning [1] and radioactive thallium 201 scanning [2] being the two most promising methods. In this paper we present experimental evidence that indicates that ultrasound backscattering from myocardial tissue with suitable contrast agents in the blood may be helpful in the visualization of the regional blood flow in the heart tissue.

Ultrasound backscattering is normally used to visualize heart structures. In echocardiology, the real time two dimensional cross sectional pictures produced by the various scanners give strikingly detailed pictures of the boundaries of cardiac anatomy [3,4,5,6]. Little work has been done in studying the scattering from the myocardial tissue itself to characterize the status of the myocardium. In this report the backscattering from the myocardial tissue *in vivo* has been investigated using a high resolution B-scanner. The backscattering from the myocardial tissue mass is enhanced significantly if some contrast agent such as indocyanine green is injected into the left ventricle or at the aortic root. This contrast enhancement seems to be related to the blood flow since by depriving some myocardial region of blood supply, one observes concomitant reduction in the backscattering from the regions of deficient perfusion.

EXPERIMENTAL DETAILS

Six dogs weighing between 9 and 12 kg were anesthetized with intravenous Nembutal (sodium pentobarbital, 25 mg/kg). Anesthesia was maintained by the intermittent intravenous injection of pentobarbital. Respiration was maintained via a cuffed endotracheal tube with a mechanical ventilator. A left thoracotomy was performed and the heart supported in a pericardial cradle. A snare was placed around the left anterior descending coronary artery (LAD) approximately 2 cm from where this vessel emerges from under the left atrial appendage. This snare was tightened during the later part of the experiment to occlude the LAD and cause regions of deficient perfusion. A catheter (#6, diameter 1.19 mm) for injecting the contrast agent was introduced into the left ventricle through an incision in the carotid artery. Two plastic beads 2 mm in size were sutured on the anterior wall of the left ventricle about 1.5 cm apart. These markers scatter ultrasound significantly and hence were good indicators of the position of the ultrasound scanner head with respect to the heart. The B scan equipment used was a research instrument developed jointly by Mayo Foundation and SRI International [7]. The scanner incorporates a 10-M Hz focused transducer and the real time B scan images are produced by the rapid linear translation of the scanner head. The real time images produced correspond to a 3 cm by 3 cm square region of tissue at a repetition rate of 15 fields per second. The scanner head was placed on the beating heart and the head was moved manually to scan the various longitudinal and transverse cross sections of the heart. The real time images were recorded using a 2" videotape recorder for later analysis. After several experiments with contrast injection ($\simeq$ 4 cc) had been recorded, the snare around the LAD was tightened to produce an ischemic region in the left ventricle. The ultrasound B scan images of the left ventricle with the contrast injection were again recorded on 2" videotape.

Since there could be strong collaterals supplying blood to the region perfused by the LAD, a radioactive microsphere experiment was performed to delineate the ischemic regions. Radioactive technitium-labelled human albumin microspheres (size 30 $\pm$ 20 microns) were injected into the left atrium to assure proper mixing. These microspheres were distributed in the heart tissue through coronary circulation and provided a quantitative measure of the perfusion in the different regions of the heart muscle. The animal was sacrificed with an intravenous injection of Nembutal and the heart was sliced and scanned by an Anger gamma camera. Autoradiograms and x-ray transmission images were also obtained.

RESULTS AND DISCUSSION

Fig. 1 is the picture of a slice of the dog heart showing the sewn on beads. The right and left ventricular chambers are on the left and right sides, respectively. The top of the picture corresponds to the anterior region of the heart. The square region marked with dotted lines approximately corresponds to the region of B-scan image. The enhanced return of acoustic energy from the heart tissue with the injection of indocyanine green into the left ventricle is seen in Fig. 2. The top of the picture is the anterior surface of the left ventricle and the clear region in the center is the left ventricular cavity with the catheter for contrast injection at the center. Even without the contrast material, there are faint echoes from within the myocardium as can be seen in the frame on the left. The picture on the right was taken about 10 seconds after the introduction of green dye into the left ventricular chamber to allow the intensely scattering material in the chamber to clear. As the contrast agent was pumped out of the left ventricle and perfused the myocardium, the bright flush appeared in the myocardial mass. The time variable gain settings were the same for both the pictures so that a qualitative comparison of the brightness can be made between the two images. The bright flush in the right frame is obvious when corresponding regions are compared with those in the left frame. In the real time images of the beating heart, the increased brightness of the myocardial tissue with the injection of the contrast agent can be seen quite dramatically. Similar results were obtained when the contrast agent was introduced at the aortic root instead of into the left ventricular chamber.

Fig. 3 illustrates the effect of the occlusion of the LAD on ultrasound B scan. The B-scan image of the left ventricle with contrast injection after ligation of the LAD is shown on the right. A similar image taken before the ligation is shown in the left frame for comparison. The lack of brightness corresponding to the lack of strong backscattering from the anterior region of the left ventricle is apparent. This effect is also observed more dramatically in the real time video images. From Fig. 3, one would infer that the occlusion of the LAD produced zones of deficient perfusion in the anterior region of the left ventricle.

This inference is supported by the radioactive microsphere distribution in the sliced heart muscle. The heart was sliced parallel to the B-scan cross sections and scanned for γ activity using an Anger gamma camera. The six slices in Fig. 4 show the distribution of microspheres in the occluded dog heart. Though

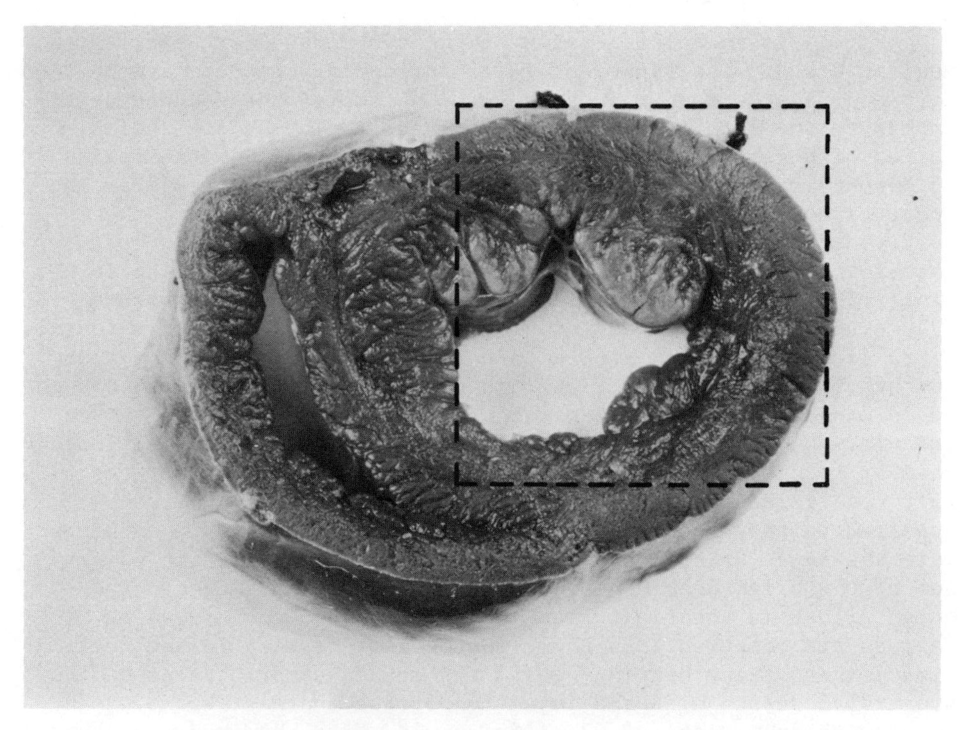

Figure 1 Photograph of the transverse cross section of the excised dog heart. Top and bottom of the picture correspond to the anterior and posterior regions of the heart. Left ventricle with its thick wall is on the right and the right ventricle is on the left. The semicircular tissue on the upper right side of the left ventricular chamber is a papillary muscle. The two marker beads sewn on the surface of the left ventricle can be seen at the top of the picture. The square region marked by dotted lines approximately corresponds to the 3 cm by 3 cm area imaged by the B-scanner.

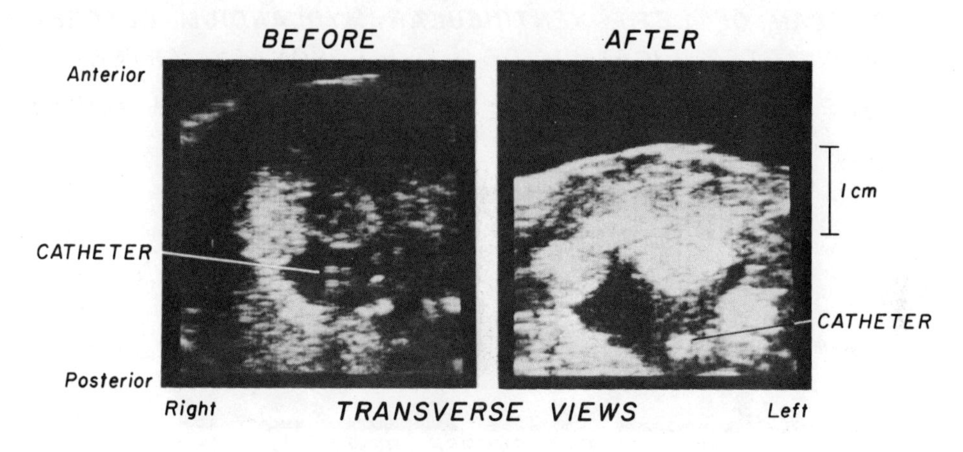

Figure 2 B scan image of the left ventricular myocardium in vivo using a 10 MHz focused B-scan instrument. The scanner head is in contact with the anterior surface of the beating heart. The image approximately corresponds to the dotted square region in Fig. 1. Left frame is the B-scan picture taken before the injection of indocyanine green into the left ventricle. The reflections from the catheter which was introduced through an incision in the carotid artery can be seen in the center of the chamber. The right frame is the picture of the same region after the injection of indocyanine green (≃4 cc). Because of the movement of the heart due to respiration and change in dimensions due to the heart beat, the two pictures are not exactly superposable; but they correspond to the same cross section. The bright flush in the myocardium after contrast injection is seen clearly in the right frame.

B-SCAN OF LEFT VENTRICULAR MYOCARDIUM BEFORE AND AFTER LIGATION OF LAD CORONARY ARTERY

(Dog, 9 kg, Open Chest, 30 sec after Green Dye, LV Injection)

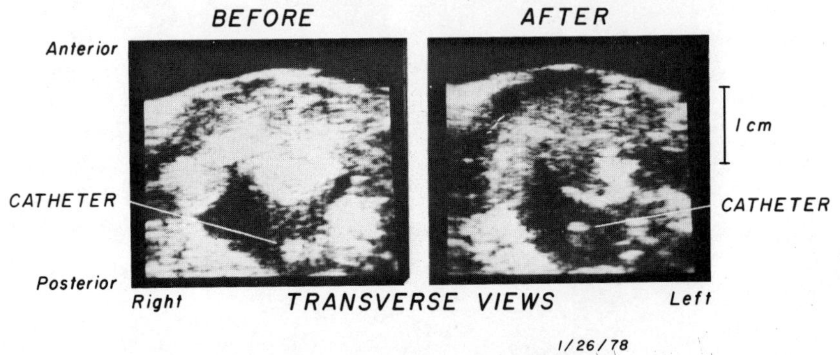

Figure 3 Comparison of B-scan images with contrast injection (≃4 cc) obtained before and after the ligation of the left anterior descending coronary artery (LAD). Though contrast was introduced in both the cases, after ligation enhancement of the backscattering is not uniform over the whole tissue. The absence of enhanced backscattering from the anterior region after the ligation of the LAD is seen by comparing the right frame with the picture taken before the ligation as seen in the left frame.

(Dog, 10 kg, Microspheres Injected after Ligation of LAD Coronary Artery)

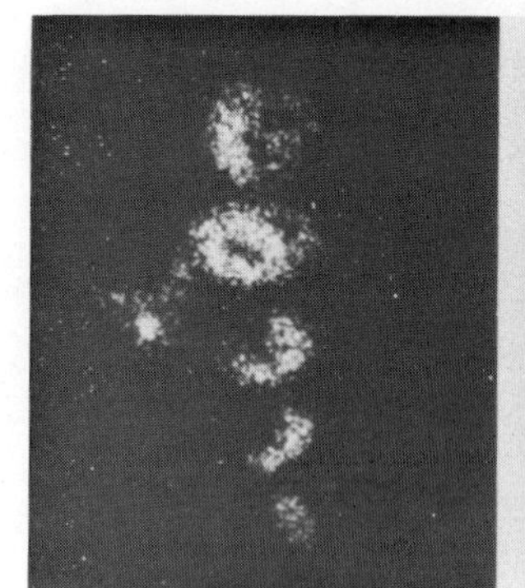
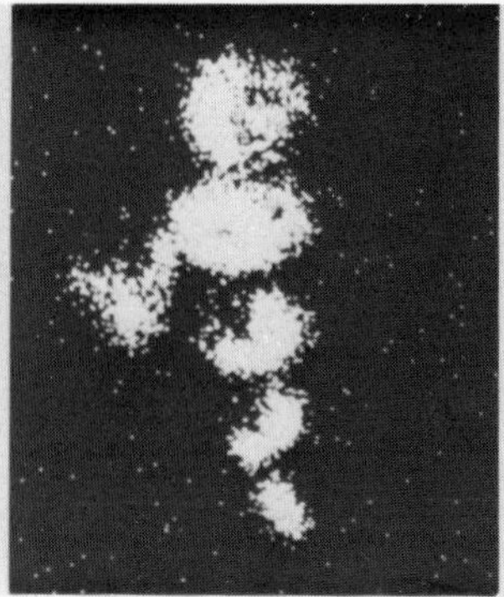
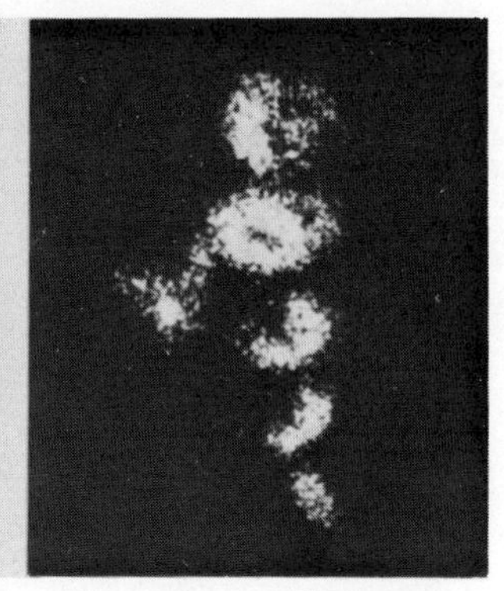

4/13/78
MAYO 1978

Figure 4 Anger gamma camera scan of the sliced dog heart after the ligation of the LAD. Radioactive technitium labelled human albumin microspheres were injected into the left atrium before the animal was sacrificed by intravenous injection of Nembutal. The heart was immediately removed, washed and sliced parallel to the B-scan images. The six slices were scanned using an Anger gamma camera. The brightness of the picture is proportional to the amount of gamma activity and hence related to the blood perfusion. The three frames are pictures with different maximum count levels to facilitate the observation of contrast in saturated areas. The slice on the left in each frame corresponds to the base of the heart. The five slices shown vertically correspond to parallel cross sections arranged progressively toward the apex of the heart. The left ventricle is on the left and the anterior region is on the top. Lack of radioactivity corresponding to the lack of blood perfusion in the anterior region of the left ventricle is seen clearly in the last three slices. The other three slices are above the ligation of the LAD and hence the perfusion is unaffected.

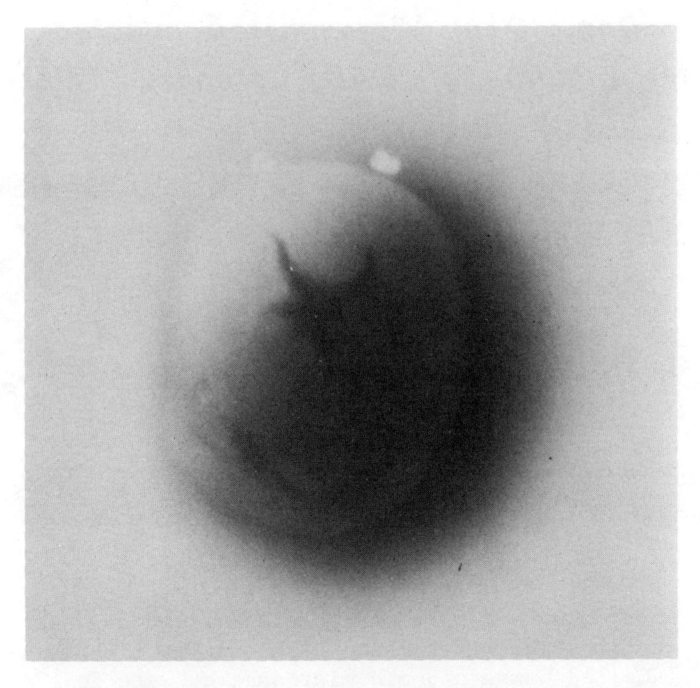

Figure 5 Composite picture of the x-ray transmission image and the autoradiograph of a sliced dog heart. This picture corresponds to the second slice from the bottom in Fig. 4. The dark shadow corresponds to the regions of radioactivity. The two marker beads on the surface of the left ventricle are seen in the x-ray transmission picture because of their opacity. The left and right ventricular chambers are also seen in the x-ray picture. Lack of radioactivity in the anterior region as observed in Fig. 4 is seen as the absence of a dark shadow in the top of the picture.

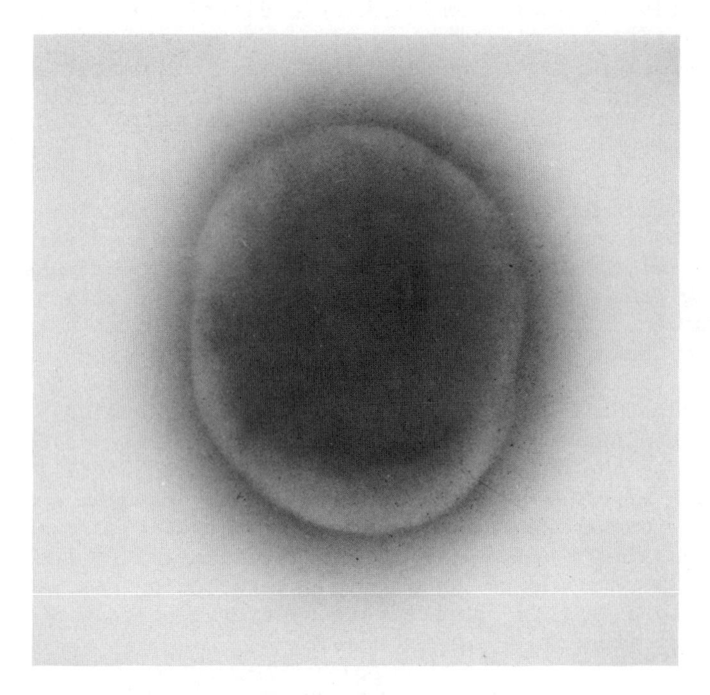

Figure 6 Composite picture of the x-ray transmission image and the autoradiograph of a sliced dog heart. This picture is similar to Fig. 5. The slice corresponds to the third slice in Fig. 4 and is above the ligation of the LAD. The whole image is dark, indicating that the perfusion is unaffected at this level.

the images are small, the lack of radioactivity in the anterior regions can be seen very clearly in the last three slices. These are the regions that tend to become poorly perfused with the occlusion of the LAD. Fig. 5 is a composite picture of x-ray transmission of a slice and its autoradiograph. The dark regions correspond to radioactive microspheres which were distributed locally in the myocardial tissue. The clear anterior region with no radioactivity shows the lack of perfusion in that area. This picture is different from Fig. 6 which is a similar composite image of a slice above the level of the LAD occlusion. The whole slice is dark, indicating that the perfusion above the snare is unaffected.

Intracardiac injections of an acoustic contrast agent has been used for the identification of heart structures by several investigators [8,9,10]. In this report we observe that the enhanced backscattering from the myocardial tissue with the injection of indocyanine green is related to the blood perfusion. Backscattering from microbubbles carried by the coronary circulation may explain these observations. These microbubbles may be caused by cavitation [11,12] and stabilized by the green dye. The enhanced backscattering was not observed if the contrast agent was introduced into the right side of the heart. This observation is supportive of the microbubble hypothesis for the origin of the enhanced scattering since the microbubbles probably would not be able to pass through the lung capillary bed unless they are very small (≃ 1 micron). Additional work needs to be done in understanding the basic mechanisms involved in this enhanced backscattering before a conclusive report about its origin can be made.

ACKNOWLEDGMENTS

The principal author would like to thank Dr. E. H. Wood, Dr. E. L. Ritman and Dr. T. C. Evans for encouragement, Dr. M. Dewanjee for technical help with the gamma camera scan and Dr. M. G. St. John Sutton and Mr. E. A. Hoffman for experimental assistance. This work was supported by the following grants: HL-04664, RR-00007, HV-7-2928, HL-00060, and HL-07111 from the National Institutes of Health and NCI-CB-64041 from the National Cancer Institute.

REFERENCES

(1) Ter-Pogossian, M. M., Phelps, M. E., Hoffman, E. J., Mullani, N. A.: A positron emission transaxial tomograph for nuclear imaging (PETT). Radiol. 114:89, 1975.

(2) Lebowitz, E., Green, M. W., Bradley-Moore, P., Atkins, H., Ansari, A., Richards, P., Belgrave, E.: ^{201}Tl for medical use. J. Nucl. Med. 14:421, 1973.

(3) Kloster, F. E., Roelandt, J., Ten Cate, F. J., Bom, N. and Hugenholtz, P. G.: Multi scan echocardiography. II. Technique and initial clinical results. Circulation 48: 1075, 1973.

(4) Griffith, J. M., and Henry, W. L.: A sector scanner for real time two dimensional echocardiography. Circulation 49:1147, 1974.

(5) Kisslo, J., Von Ramm, O. T., and Thurstone, F. L.: Cardiac imaging using a phased array ultrasound system. II. Clinical technique and application. Circulation 53: 262, 1976.

(6) Tajik, A. J., Seward, J. B., Hagler, D. J., Mair, D. D. and Lie, J. T.: Two dimensional real time ultrasonic imaging of the heart and great vessels: Technique, image orientation, structure identification, and validation. Mayo Clin. Proc. 53:271, 1978.

(7) Green, P. S., Taenzer, J. C., Ramsey, S. D., Jr., Holzemer, J. F., Suarez, J. R., Marich, K. W., Evans, T. C., Sandok, B. A. and Greenleaf, J. F.: A real time ultrasonic imaging system for carotid arteriography. Ultrasound Med. Biol. 3: 129, 1977.

(8) Gramiak, R., Shah, P. M., and Kramer, D. H.: Ultrasound cardiography: Contrast study in anatomy and function. Radiology 92:939, 1969.

(9) Feigenbaum, H., Stone, J. M., Lee, D. A., Nasser, W. K., and Chang, S.: Identification of ultrasound echoes from the left ventricle by use of intracardiac injections of indocyanine green. Circulation 41:615, 1970.

(10) Weyman, A. E., Feigenbaum, H., Dillon, J. C., Johnston, K. W., and Eggleton, R. C.: Noninvasive visualization of the left main coronary artery by cross sectional echocardiography. Circulation 54:169, 1976.

(11) Bove, A. A., Adams, D. F., Hugh, A. E., and Lynch, P. R.: Cavitation at catheter tips. A possible cause of air embolus. Invest. Radiol. 3:159, 1968.

(12) Kremkau, F. W., Gramiak, R., Cartensen, E. L., Shah, P. M., and Kramer, D. H.: Ultrasonic detection of cavitation at catheter tips. Am. J. Roentg. 110:177, 1970.

COMPUTER ANALYSIS OF GREY SCALE TOMOGRAMS

D. NICHOLAS, A. BARRETT*, J.M.G. CHU[a], D.O. COSGROVE,
P. GARBUTT, J. GREEN*, S. PUSSELL[b], and C.R. HILL
Institute of Cancer Research/Royal Marsden Hospital,
Sutton, Surrey, U.K.
*National Institute of Medical Research, Mill Hill East,
London NW7, U.K.
[a]Dept. of Nuclear Medicine, Royal Prince Alfred Hospital
Sidney, Australia.
[b]Supported by the Coppleson Postgraduate Medical
Institute.

INTRODUCTION

Present day interpretation of clinican B-scans is entirely dependent upon the expertise of the clinician involved and the capabilities of the machine he is operating. Although much useful and accurate information is obtained it is still derived from a subjective and non-quantitative evaluation of two-dimensional images. Recent improvements in ultrasound technology have enabled diagnosticians to discern macroscopic changes in tissue morphology, the complex nature of these changes and their ultrasonic appearance together with incomplete specification of tumour type have resulted in some contradictory descriptions of the ultrasonic appearances associated with various cancerous conditions. In particular, the echo patterns associated with primary and metastatic intrahepatic lesions have received varying descriptions from different authors. Taylor et al[1,2] described the typical appearance as areas of low-level echoes within the normal liver. Conversely, McArdle[3] found, from a study of 21 patients, that the most frequent patterns for liver metastases consisted of echogenic nodules and anechoic regions with an echogenic parenchyma. Recently, Chu et al[4] have suggested that the ultrasonic appearance of liver metastases may be dependent upon the site of the primary neoplasm. These findings indicate the need for a quantitative evaluation of ultrasonic images and a rigorous description (where possible) of related pathology.

Although the ultrasonic appearances of focal liver metastases are inconclusive the determination of their existence is usually reliable. The major problems occur when attempting to differentiate between small tumour nodules and cross-section images of small blood vessels or ducts. The diffuse disorders associated with the liver present a much greater problem for ultrasonic diagnosis. With the possible exception of cirrhosis, which is usually recognised by the overall increase in the intensity of the echoes from within the liver parenchyma, the majority of diffuse disorders are only registered (if at all) by a slight spatial change in the general pattern of parenchymal echoes, as for example in the case of fatty infiltration.

Some of the questions that are now being posed by ultrasound, as an aid to diagnosis of liver disease, are as follows :

(1) can secondary focal deposits be classified with respect to their primary type by either the intensity or spatial distribution of the internal echo producing structures ?

(2) Can tumour response to chemotherapy and/or radiotherapy be monitored by an ultrasonic evaluation ?

(3) Can diffuse infiltrations of hepatic tissues be characterised by quantifying the 'parenchymal' echoes ?

Solutions to these questions are already being considered using a subjective visual evaluation of B-scan images and have been shown[4] to be capable of providing a limited classification of disease states. Our aim is to approach such evaluation on a rigorous quantitative basis thereby greatly reducing operator bias.

APPARATUS

Digital analysis of clinical B-scans has only received scant attention in the past[5,6], although notable work in the analysis of transcranial tomographic B-scans has been achieved by King and Wong[7]. Previous tentative approaches required a fast analogue to digital conversion rate and a large data repository in which to store the digitised B-scan. The advent of the scan conversion memory in recent ultrasonic diagnostic machines, however, has eliminated many of the problems previously associated with fast data acquisition by providing a temporary data store from which subsequent transfer is made to a computer or permanent magnetic storage.

Our initial system is illustrated schematically in Fig.1. As our primary interest is in analysing small, preselected, portions of the B-scan we can avoid digitising the whole B-scan and hence

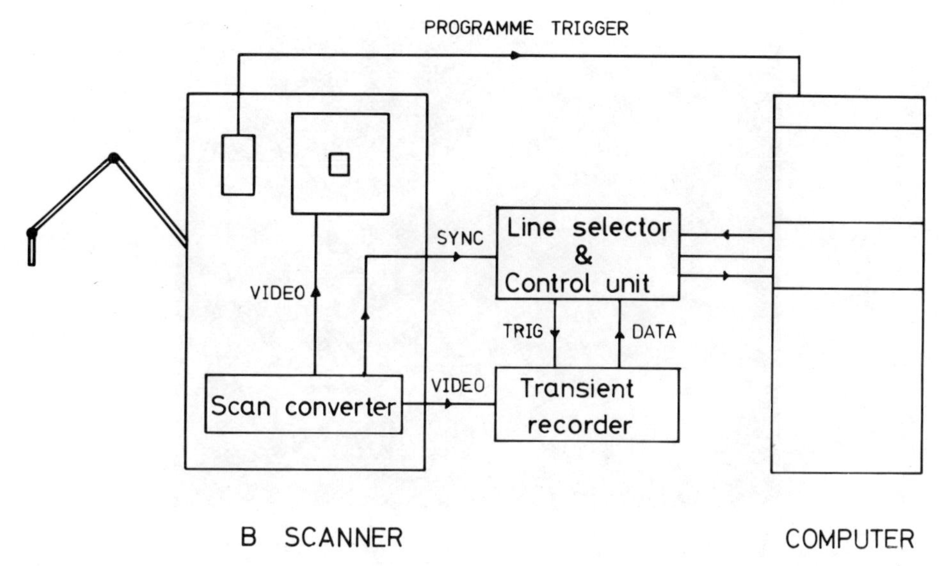

Fig.1 Schematic diagram of data capture procedure

limit our data to a small matrix with high resolution. A line selector enables specific lines (corresponding to the 625 lines of video information fed to the standard monitor) to be accessed for digitisation. This is accomplished by reading the relevant line of information directly from the analogue scan converter into a 20 MHz 8-bit transient recorder with 4k byte memory. This speed allows us to read at the output video rate of 25 MHz and digitise alternate pixels of information. Only 64 specific words of information are retained and buffered onto magnetic disc associated with a PDP 8/e mini-computer. By taking 64 consecutive lines of information we have limited our stored data to a 64 x 64 matrix corresponding to a 1.5 x 2.5 cm rectangle of information (in real space) positioned in the centre of the field of view on the monitor. The data collection area is depicted on the monitor by a semi-transparent box superimposed on the displayed B-scan (as shown in Fig. 2) Since this digitised region is fixed in relation to the video signal to the monitor, different regions of the B-scan can only be analysed by using the 'pan-pot' control on the scan converter to move the picture of the B-scan with respect to the superimposed 'box'.

To date only a fixed 64 x 64 matrix of digitised data is stored although different sizes of image can be investigated either by utilising the 'zoom' feature incorporated in the analogue scan converter or by re-scanning at a different magnification. Larger matrices could easily be stored although our present requirements do not necessitate this.

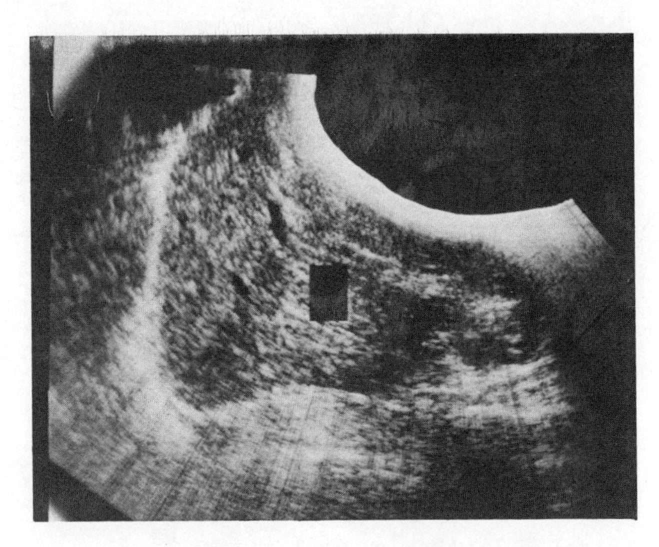

Fig.2 Typical longitudinal B-scan of human liver with region to be digitised indicated by superimposed rectangle

PILOT *IN VIVO* STUDY

Before embarking upon quantifying B-scan data it is necessary to ensure that any differences in echo pattern are due solely to tissue variations and are not associated with the scanning apparatus. Visual inspection of B-scans from normal liver tissue (see Fig.3) suggests that the echo pattern is relatively 'fine' when close to the transducer and coarsens as the distance from the transducer increases. If the region of the B-scan under investigation is within the near-field of the transducer then the echo patterns will be dependent upon the beam characteristics. For this reason we do not attempt to analyse echo patterns which lie within 3 cm of the transducer.

Variations also occur within the far-field in that the echo width is directly related to the beam-width, which increases with distance from the transducer. Although this has the effect of 'blurring' the picture, as illustrated in the digitised images portrayed in Fig. 4a and Fig.4b, the actual number of discernable echoes remains unaffected when analysed quantitatively. As a check of this several normal liver B-scans were digitised at varying distances from the transducer and the number of discernable discrete echo peaks plotted as a function of distance. Fig.5 gives the results of these tests for normal liver tissue and for an infiltrated liver, and demonstrates that beam width effects can be ignored provided that investigations are limited to the far-field of the transducer.

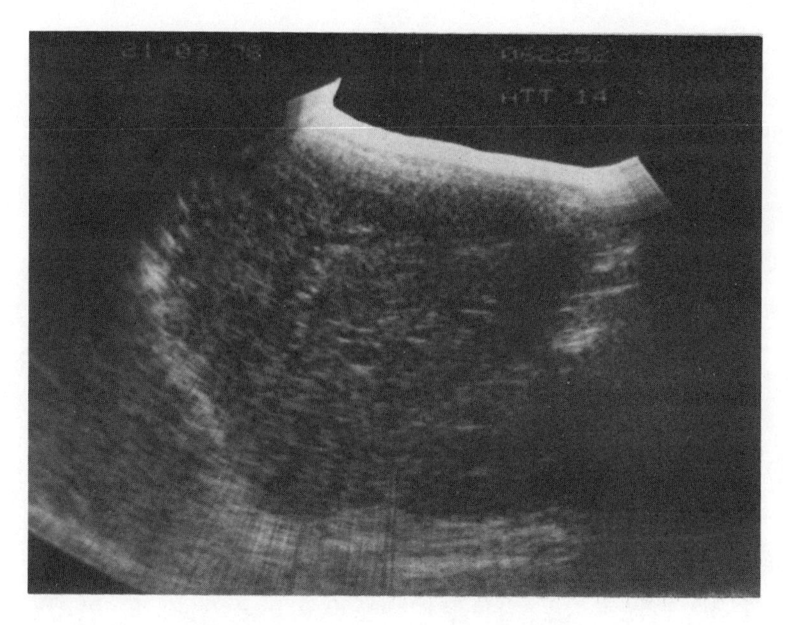

Fig.3 Typical longitudinal B-scan of normal liver tissue showing pattern variation with depth from skin surface

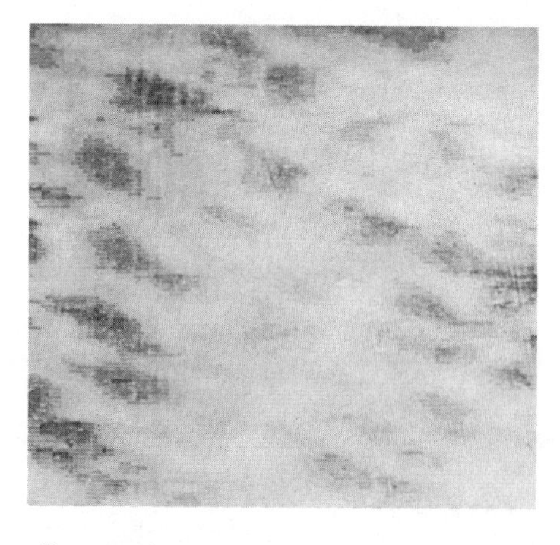

a

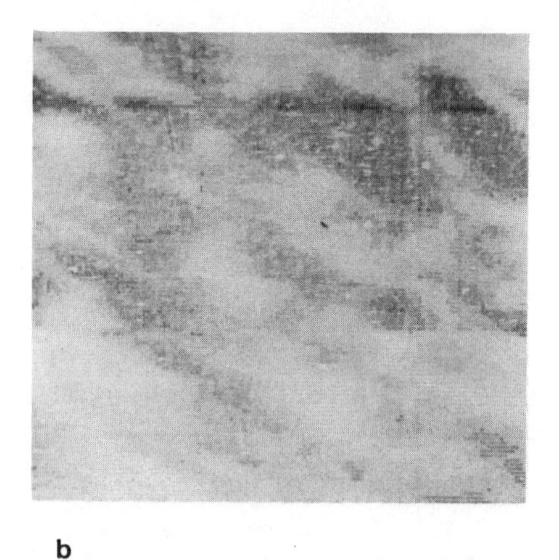

b

Fig. 4 Digitised versions of a portion of Fig.3 at :
(a) 3 cm from the skin surface (b) 9 cm from the skin surface

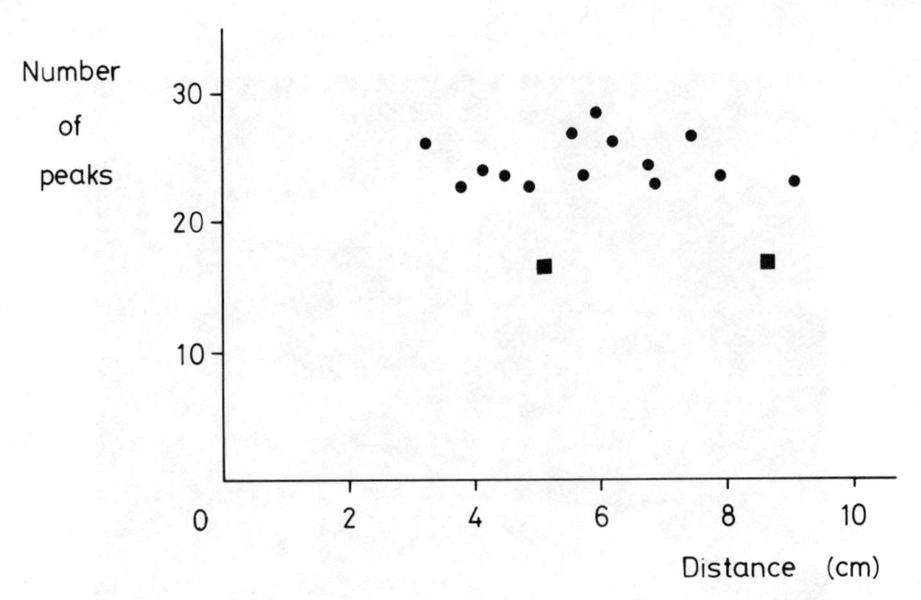

Fig.5 Analysis of the number of discrete echoes within the digitised region as a function of distance from transducer/skin surface. ● - values associated with normal liver parenchyma ; ■ - values associated with a diffusely infiltrated liver

Quantitation of B-scans

To date 62 B-scans associated with various liver conditions have been digitised by the above method. The stored data has subsequently been analysed in an attempt to quantify the scans using simple parameters. Two simple techniques which will be reported here involve estimating the number of echo peaks within the digitised area and replotting the original data in the form of a histogram of echo amplitude against frequency of occurrence. Fig.6a and Fig.6b are examples of the histograms associated with normal liver parenchyma and an adenoma, respectively. From these histograms the mean echo amplitude and coefficient of variation have been extracted as potential measures for characterising differing tissue pathologies.

At present only relative measures are meaningful as the absolute value of the mean depends upon the settings of the diagnostic ultrasound machine and patient variability, and this can only be achieved satisfactorily where focal deposits are being investigated. The method then involves comparing areas of suspected involvement with adjacent presumed normal regions. Table 1 summarises the results for various liver conditions; only rough estimates of the mean value can be suggested where the disease

TABLE I. SUMMARY OF THE QUANTITATIVE ANALYSIS OF LIVER B-SCANS

HEPATIC DISEASE		NO. OF SCANS	NO. OF DISCRETE ECHOES	HISTOGRAM VALUES	
				Mean*	Coefficient of Variation
NORMAL		35	21--28		10--15
FOCAL DEPOSITS	TERATOMA	8	16 ± 1	+ / -	10--12
	ADENOMA	3	33 ± 1		9--10
	CARCINOMA	8	17 ± 2	-20 ± 9	11--13
	FIBROSARCOMA	3	15 ± 1	-15 ± 2	10--11
DIFFUSE INFILTRATIONS	LYMPHOMA	2	23 ± 1	-	14--15
	CIRRHOSIS	3	30 ± 1	+	11--14

*Expressed as a difference from the mean value associated with normal tissue.

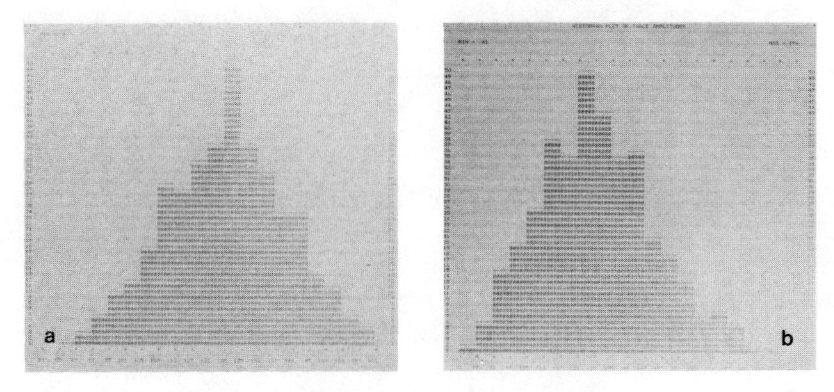

Fig.6 Histogram of echo amplitude against frequency of occurrence for : (a) normal liver parenchyma (b) a cancerous deposit (adenoma)

involes a diffuse infiltration throughout the liver.

It is apparent that the number of discrete echo peaks within a scan is a significant parameter, with most of the secondary focal neoplasms exhibiting significantly fewer echo peaks than is usual for normal liver tissue. The notable exception to date is the adenoma which produced more echoes than normal tissue. Of the diffuse conditions the infiltrated liver (lymphoma) indicated a possible overall decrease in echo amplitude, whilst the cirrhosis showed a distinct increase in the number of echo peaks. Until an absolute measure can be placed upon the mean echo amplitude its use as a variable for tissue differentiation is limited. Similarly the coefficient of variation has yet to show any marked significance as a measure of abnormality. Although only a few preliminary examinations have been conducted the results suggest that quantification has an important role in aiding clinical diagnosis using ultrasound.

Spectral analysis. A more advanced method of analysing the spatial distribution of echoes can be achieved by converting to the frequency domain and extracting periodicities and predominant frequency components. Our method has utilised the Fast Fourier Transform (FFT) which has been implemented on our on-line PDP 8/e computer. An example of an image and its corresponding two-dimensional Fourier transform are illustrated in Fig.7a and Fig.7b. Visually, periodic structures are difficult to pick out, but close inspection of these actual data values shows periodicities corresponding to 4 peaks (per 64 sampling interval) in the x-direction and 5 peaks in the y-direction indicating an overall peak separation of 4 to 5 mm in real space. The different spacings in

(a)

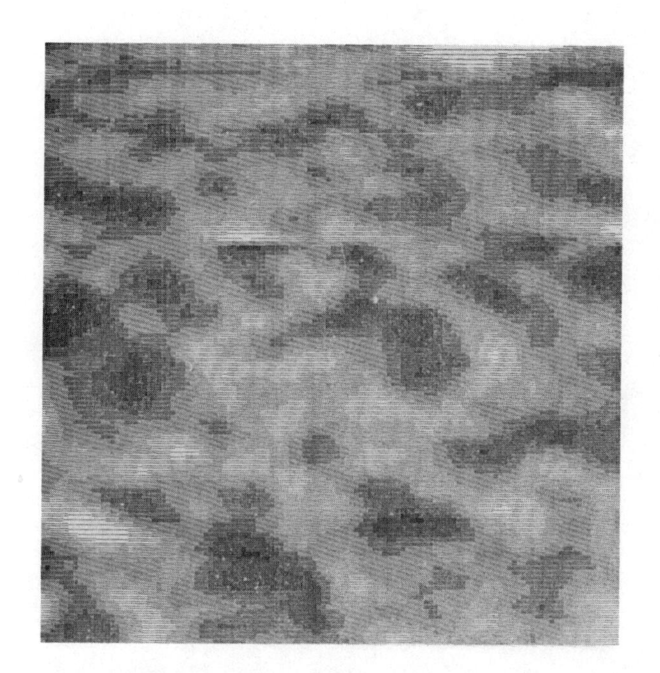

(b)

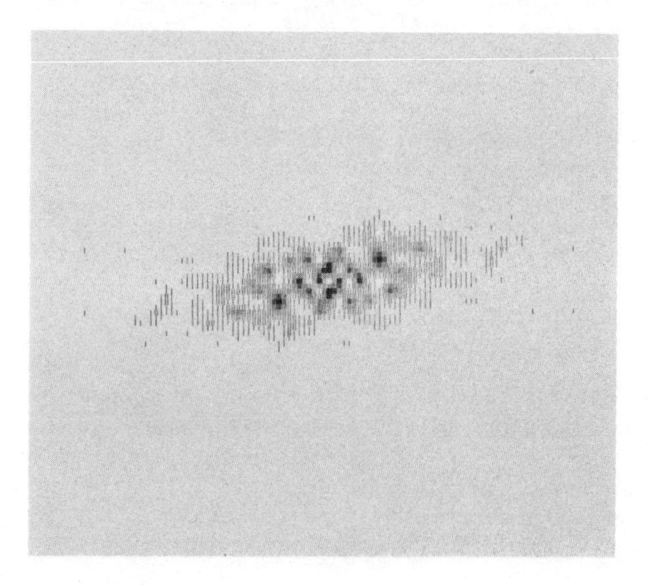

Fig.7 (a) a digitised image and
(b) the magnitude of the Fourier transform

the x and y directions are due to the different sampling intervals when digitising in these directions.

Table II portrays the original peak values from Table I together with the predominant spacial separations (where they exist) obtained by Fourier analysis. By presenting the information in the frequency domain the randomness of the echo peak distribution can be evaluated and used as a further parameter for possible tissue characterisation.

Feature Extraction and Image Enhancement

The shapes and outlines of many neoplasms, when imaged ultrasonically, can vary according to the disease type and state (especially when responding to treatment). We suspect that the visual interpretation of B-scans would benefit by utilising computer applications of image enhancement techniques which emphasise specific tumour features. This aspect of B-scan analysis has various applications depending upon the requirements imposed by disease conditions and the region of anatomy under investigation. Our present requirements are seen as threefold :

(1) to differentiate between the images produced by small deposits, cysts, abscesses and blood vessels or ducts;

(2) to extract relevant tissue features from background;

(3) to provide a semi-automatic aid to B-scan interpretation.

Although all three are inter-related, their application and complexity depends upon the degree of operator/computer interplay.

At the simplest level, display of the digitised data in a grey scale format provides more visual information than that depicted by polaroid recording of the analogue signals displayed on a monitor. Fig.8a and Fig.8b illustrate this noticeable difference which is primarily due to the gamma curves associated with the monitor phosphor and the polaroid film. In this example visual inspection of the digital print out provides useful evidence that the region of interest is a cross sectional image of a blood vessel and not a cancerous deposit.

We are aiming our research towards an interactive operator-machine system to aid diagnosticians in the analysis of B-scans. Towards the realisation of this goal, we are designing a hierarchic tumour recognition system which follows a decision tree to produce tumour directed analysis of B-scans. The decisions are based upon the following features which are used to indicate subtle

TABLE II. Quantitative analysis of liver B-scans in terms of the spatial distribution of echoes

HEPATIC DISEASE		NO. OF SCANS	NO. OF DISCRETE ECHOES	PREDOMINANT PEAK SEPARATION mm
NORMAL		35	21--28	3.6--4.2
FOCAL DEPOSITS	TERATOMA	8	16 ± 1	4.8
	ADENOMA	3	33 ± 1	3.3
	CARCINOMA	8	17 ± 2	4.2
	FIBROSARCOMA	3	15 ± 1	-
DIFFUSE INFILTRATIONS	LYMPHOMA	2	23 ± 1	-
	CIRRHOSIS	3	30 ± 1	3.6

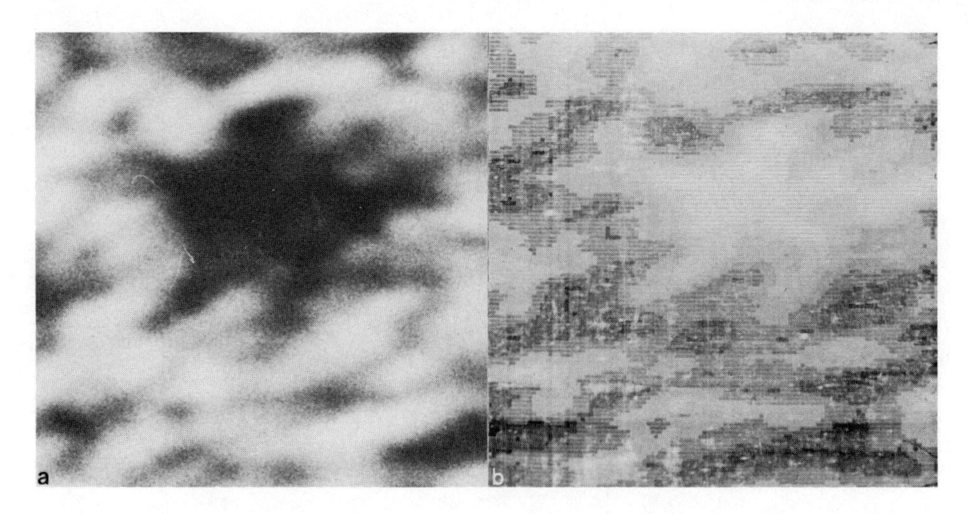

Fig.8 (a) a polaroid picture of a blood vessel in cross-section
(b) a computer display of the digitised image (reversed grey scale)

boundary variations and tumour characteristics :

(1) the maximum of the slope density function associated with the tumour boundary;

(2) the prominence of the boundary;

(3) a measure of the speculation;

(4) a measure of the centre density and average density.

The analysis is tumour directed in the sense that its goal is the identification of likely tumour sites on each B-scan. These sites will then be presented to the diagnosticians along with a suggested diagnosis.

The decision tree method, based on the work of Ballard and Sklansky[8], has the following features :

(1) a number of different edge detection criteria are available with which to classify the tumour/cyst/abscess candidates;

(2) it is possible to choose the resolution best suited for each segment. This allows us to represent large easily found objects with lower resolutions, and smaller more obscure objects with higher resolutions (limited by the original digitisation of the B-scan).

Our approach will result in a coarse-to-fine ordering of these resolutions.

(3) On progression down the decision tree, the procedures are able to restrict their search to small areas in the B-scan such that the area of the B-scan selected by each procedure is a subset of the area input to that procedure. This restriction of data allows us to implement powerful algorithms which would be too time consuming to use on the whole B-scan.

We are currently at the stage of implementing computer programmes to achieve the classification of tumour sites previously selected by the clinician. This involves finding the tumour boundary and classifying it with regard to certain features. Provided this classification of potential tumours is feasible we envisage digitising whole B-scans into 512 x 512 8 bit word matrices and extending our programming to provide semi-automatic interpretation of B-scans.

DEVELOPMENTS IN HAND

Our present system involves the on-line use of a mini-computer for data collection and storage. This is unsatisfactory in that the computer is permanently occupied with this one project - there being no time-sharing facility. We have, therefore, devised a dedicated system for digitising the data from the scan converter and storing it directly onto magnetic tape for subsequent off-line analysis. Our system is based on a microprocessor controlled file orientated storage device utilising 4 track 1/4" magnetic tape cartridges with 450,000 byte storage per track. Thus for a 64 x 64 matrix of 8 bit bytes we can store up to 400 scans on a single cartridge. By operating at 9600 baud the time for data transfer of a single matrix from the scanner to magnetic tape is approximately 5 s. Furthermore by making this system portable it can be linked to any of the ultrasonic scanners in use without inconveniencing patients or clinicians and obviating unwanted on-line connections to the computer.

CONCLUSION

Our involvement in digital B-scan analysis has just commenced and we are still at the development stage regarding apparatus and computer programming. However, our preliminary investigations have shown that even simple quantification can be an extremely useful aid in characterising potential neoplastic liver conditions. The ultimate aim of semi-automatic analysis of B-scans, though still in the future, is at least an exciting possibility.

REFERENCES

1. Taylor, K.J.W., Carpenter, D.A. and McCready, V.R. "Grey scale echography in the diagnosis of intrahepatic diseases," J. Clin. Ultrasound 1 (1973); pp.284-287

2. Taylor, K.J.W. "Ultrasonic pattern of tumours of the liver," J. Clin. Ultrasound 2 (1974); pp.74-77

3. McArdle, C.R. "Ultrasonic diagnosis of liver metastases," J. Clin. Ultrasound 4 (1976); pp. 265-271

4. Chu, J.M.G., Bloomberg, T.J., Cosgrove, D.O., North, L.P. and McCready, V.R. "Liver metastases : the ultrasound B-scan appearances correlated with the site of the primary tumour and with chemotherapy," to be published.

5. Milan, J. "An improved ultrasonic scanning system employing a small digital computer," Br.J. Radiol. 45 (1972); pp. 911-916

6. Kay, M., Shimmins, J., Manson, G. and England, M.E. "A computer interface for digitising ultrasonic information," Ultrasonics 13 (1975) ; pp. 18-20

7. King, J.C. and Wong, A.K.C. "Computer analysis of transcranial sonotomographic B-scans," Comput. Biomed. Res. 5 (1972) ; pp. 190-204

8. Ballard, D.H. and Sklansky, J. "A ladder-structured decision tree for recognising tumours in chest radiographs," IEEE Trans. Computers 25 no.5 (1976); pp.503-513

THE INFORMATION CONTENT OF B-SCAN ANOMALIES

D. NICHOLAS

Institute of Cancer Research/Royal Marsden Hospital

Sutton, Surrey, U.K.

INTRODUCTION

The immense improvement in ultrasonic B-scan picture quality over the last decade has been matched by the ever finer degree of clinical interpretation placed upon these pictures. Present day resolution capabilities have enabled the clinician to visualise anatomical structure of the order of a few wavelengths (typically λ = 0.5 mm for abdominal B-scanners). Since the information is of such a highly detailed character it is imperative that a greater understanding of the interaction of ultrasound and human soft tissues should be attempted. Typical sector B-scan pictures (see figure 1) often depict more than just a simple map of the anatomical placement of tissue structures. These 'extra' information, best described as B-scan anomalies, are often classed as unwanted 'artefacts' and disregarded when evaluating the information content of the picture. Although this is often justified, useful diagnostic information occasionally can be gained by a more rigorous appraisal of the anomalies.

REFLECTIONS AND REVERBERATIONS

Clinical awareness of the complexity of ultrasonic sector scans has been noted by Cosgrove et al[1] where the specific example has been reported of echoes appearing as mirror artefacts following reflection at the diaphragm. Information apparently arising from interfaces above the diaphragm are routinely displayed on B-mode sector scans of the right lobe of the liver. The apparent structure can be confusing and lead to misreporting especially when one considers that lung tissue is strongly attenuating to ultrasound (40 dB/cm/MHz) [2] and no echoes can usefully be obtained from it.

An explanation of this phenomenon depends upon the diaphragm acting as an acoustic mirror. Given the appropriate geometry, an ultrasound pulse can be reflected from the diaphragm back into the liver, and the back-scattered echoes associated with some anatomical structure be re-reflected back to the transducer. This is most easily demonstrated when a large acoustic scatterer, such as an hepatic lesion, is positioned close to the diaphragm. Due to the curved geometry of the diaphragm several complex situations can arise. In Fig.1 two longitudinal sector scans of the human liver are illustrated, both of which exhibit echo complexes above the diaphragm which are similar to the parenchymal echoes associated with normal liver tissue. In Fig.1a an apparent blood vessel is existing across the diaphragm for which there is no visible acoustic structure, within the liver, to relate to. The ray tracing indicates that the 'parabolic' geometry of the diaphragm could result in an apparent image of a blood vessel imaged in cross-section. To further complicate the situation, angulation of the diaphragm is likely to give rise to imaginary images from out of plane acoustic structure - as seems likely in this example. This out of plane imaging is also stressed in Fig.1b where a real blood vessel fails to produce an imaginary image across the diaphragm.

The clinical significance of this phenomenon is apparent when pleural pathology in the right base is imaged, such as pulmonary consolidation or pleural tumours. It is clear that the ultrasonic information obtained under these conditions is clinically meaningful, whereas that normally obtained when the lung is aerated must be ignored.

Although the above discussion illustrates a possible over-interpretation of B-scan information it is more usual for useful information to be missed. A further example of the importance of geometry on acoustic reflection is depicted in Fig.2. Illustrated here is a rectilinear B-scan showing a cross-section of the human neck, performed using a water bath stand off. From the grey scale picture (upper right) we can immediately identify some of the acoustic structure with known anatomy (upper left). A significant amount of acoustic information (lower left) has been labelled 'artefact' in that it does not exhibit an immediate relationship with known anatomy, yet careful consideration of the anomalous structure not only improves our conception of ultrasonic B-scanning but yields extra anatomical information as well as eliminating false conclusions regarding the position of real structure.

First, we notice that doubling the distance between the skin line and the transducer path produces the lower 'artefact'. This is a 'twice around' phenomenon where the acoustic pulse has initially been reflected at the skin surface and further reflected back at the transducer to produce a spurious, delayed transmission

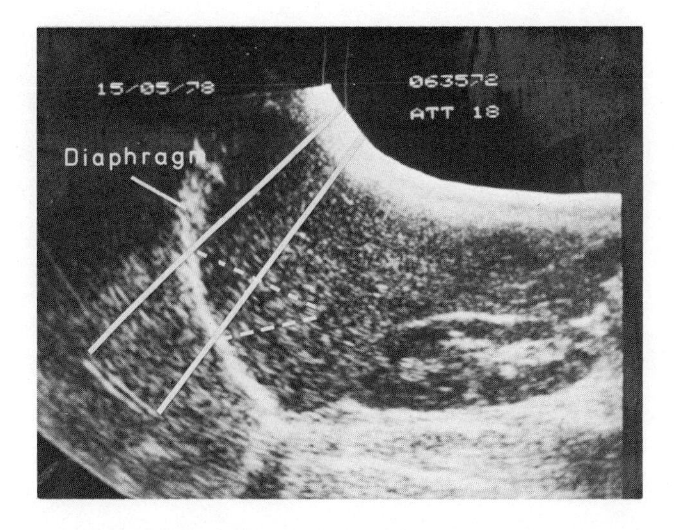

(a)

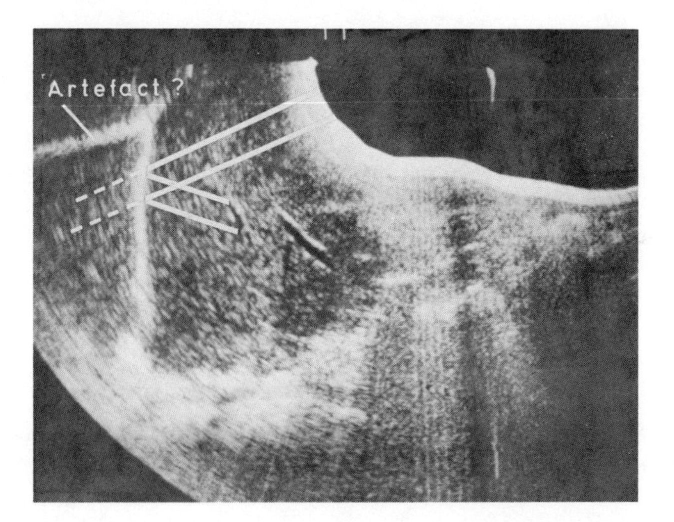

(b)

Fig.1 Longitudinal sector B-scans of the human liver exhibiting apparent structure across the diaphragm.

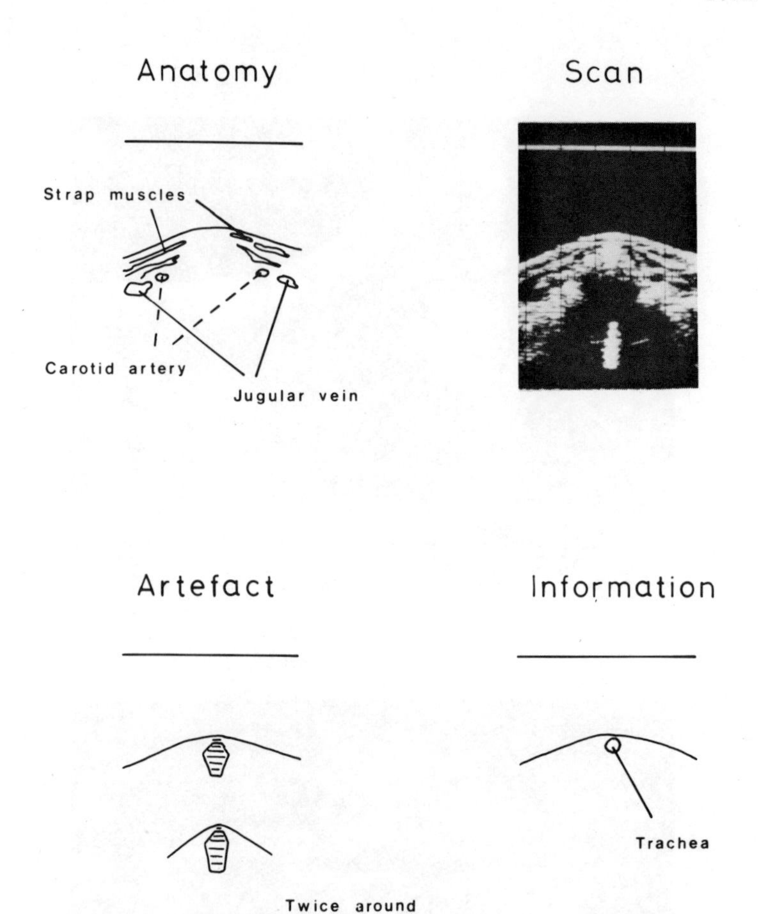

Fig.2 Transverse, rectilinear B-scan of the human neck, illustrating anatomy and artefact.

pulse to image the tissue with. Thus the original image is duplicated, later in time, with a diminished intensity which can be associated with the reflection losses at the acoustic interfaces. 'Twice arounds' are most commonly noticed when a water bath stand off method of scanning is employed and care must be taken not to confuse these secondary images with information obtained from a single pulse-echo reflection.

Secondly, the banded structure positioned close to the skin surface is due to echoes arising from the trachea. At the anterior surface of the trachea the tissue/air interface acts as a perfect reflector to ultrasound. As the skin/water interface is also a strong reflector the impinging acoustic pulse will be reflected back and forth between these two interfaces. This will produce the banding effect which decreases in intensity since the skin/water interface transmits a large proportion of energy at each interaction with the reflected pulse from the trachea. By measuring the 'banding' spacing, the distance of the anterior surface of the trachea from the skin-surface can be determined.

Finally, the lateral dimension of this 'artefact' is indicative of the size and shape of the trachea. Although the size and position of the trachea may not be of importance clinically, this example illustrates the probable cause of many of the artefacts witnessed in B-scanning;

1. the'banding' associated with multiple internal reflections ;
2. the secondary images caused by twice around effects

and

3. the importance of the geometry of reflecting surfaces.

With a familiarisation of these aspects of ultrasound imaging greater confidence can be placed upon its use as a diagnostic tool.

In sector B-scans of the liver a variety of 'flares' or 'bright ups' are occasionally noticed which can sometimes be explained by simple geometric considerations which relate to known or 'suspected' anatomy. Figures 3 and 4 show two familiar forms of artefact which occur when scanning this region of the patient. Bright streaks or 'banding' can occur which appear to originate at the diaphragm (Fig.3) or some portion of the gastro-intestinal tract (Fig.4).

The flaring phenomenon witnessed in Fig.4a is explained, diagramatically, by Fig.5. One often has air pockets present in the transverse colon which act as perfect reflectors to ultrasound, thereby producing shadows on the B-scans corresponding to the regions posterior to the gas where the ultrasound has been unable to penetrate. If, however, the colon is fluid filled, then transmission of the ultrasound waves can occur. One can envisage cases where the colon is fluid filled but has small gas bubbles trapped in it. These bubbles will act as very strong acoustic reflectors and scatter the waves in all directions. When there are two such discrete gas bubbles, multiple scattering between the bubbles with corresponding reception by the transducer will produce a banding effect, where the banding separation is related to the spacing between the gas bubbles. By extrapolating this situation to encompass the existence of 'froth' in the colon we can expect a very complex scattering situation involving very many ray paths (see Fig.6) The almost infinite arrangement of scattering paths will produce multiple echoes resulting in a bright streak in which any banding will be too fine to resolve.

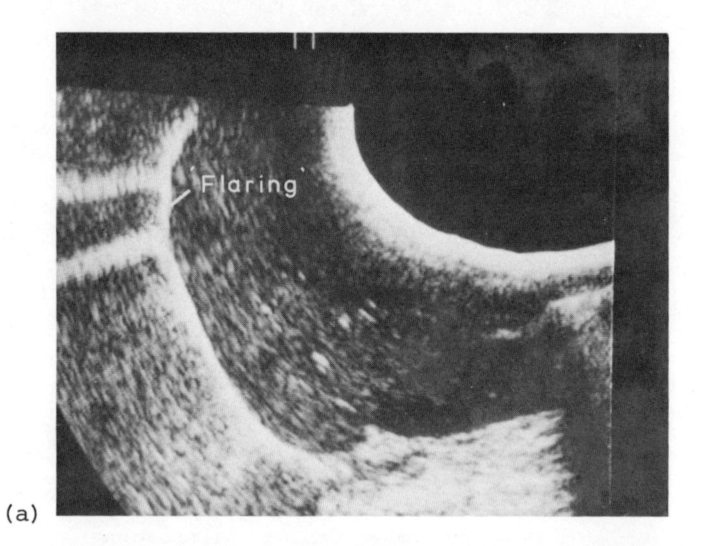

(a)

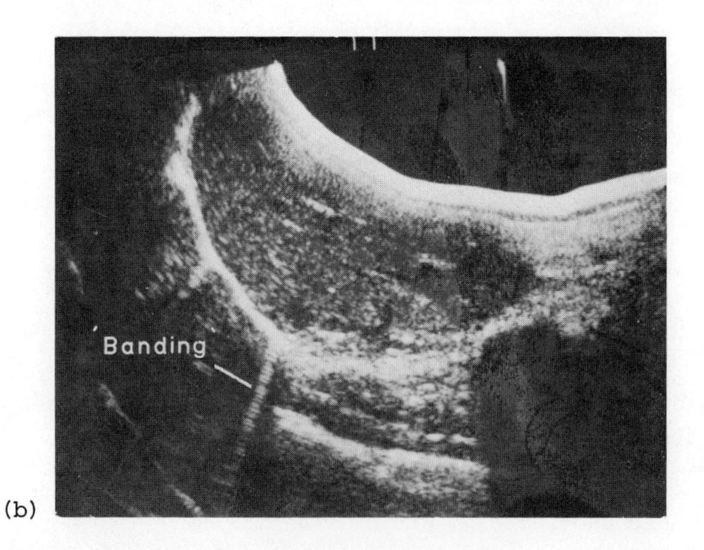

(b)

Fig.3 Longitudinal sector B-scans of the human liver exhibiting bright 'streaks' originating at the diaphragm :

(a) 'flaring'
(b) 'banding'

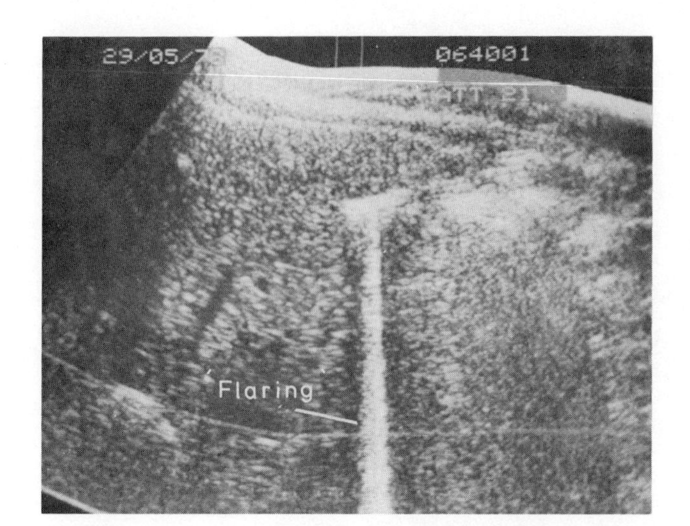

(a)

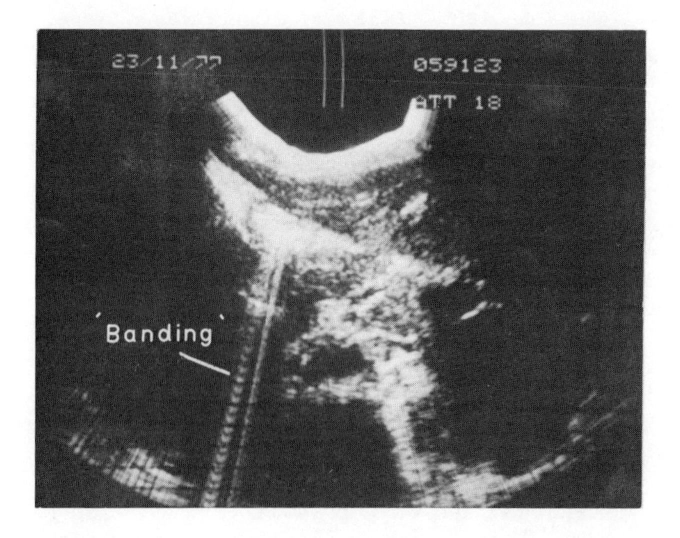

(b)

Fig.4 (a) Longitudinal sector B-scan exhibiting a 'flare' which originates at the transverse colon;
(b) Transverse sector B-scan exhibiting 'banding' which originates at the pyloris.

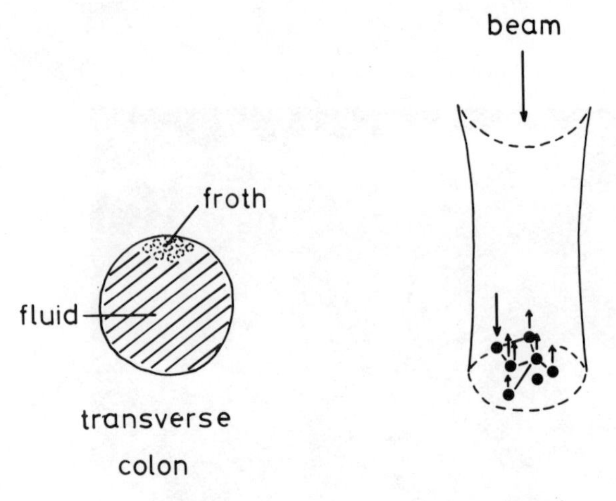

Fig.5 Illustration of how froth in the transverse colon can produce multiple echoes.

These possible explanations are also applicable to the diaphragm where the small aveolae present at the pleural surface, whether few or many, can be considered as potential acoustic scatterers.

The above discussions attempt to explain B-scan features which may not have specific clinical significance. However, without such an attempt at understanding the complex origins of these so-called 'artefacts' useful diagnostic information may be overlooked or misinterpreted.

GALL STONE APPEARANCE

The usual ultrasonic appearance of gall stones shows a bright object (the gall stone) preceding an acoustic shadow. The accepted explanation is that the gall stone, as a solid object, strongly reflects the ultrasound and only permits limited transmission through it. Thus, objects which lie posterior to the stone are only insonified by low intensity sound waves. Fig.7 illustrates that other acoustic pictures can be associated with the presence of stones, examples show 'no acoustic shadowing' behind the stone and an apparent increased intensity ('flaring') behind the object.

These variations in appearance are likely to be due to variations in the transmission of ultrasound through or around the gall stone and suggest that the increase can be related to a

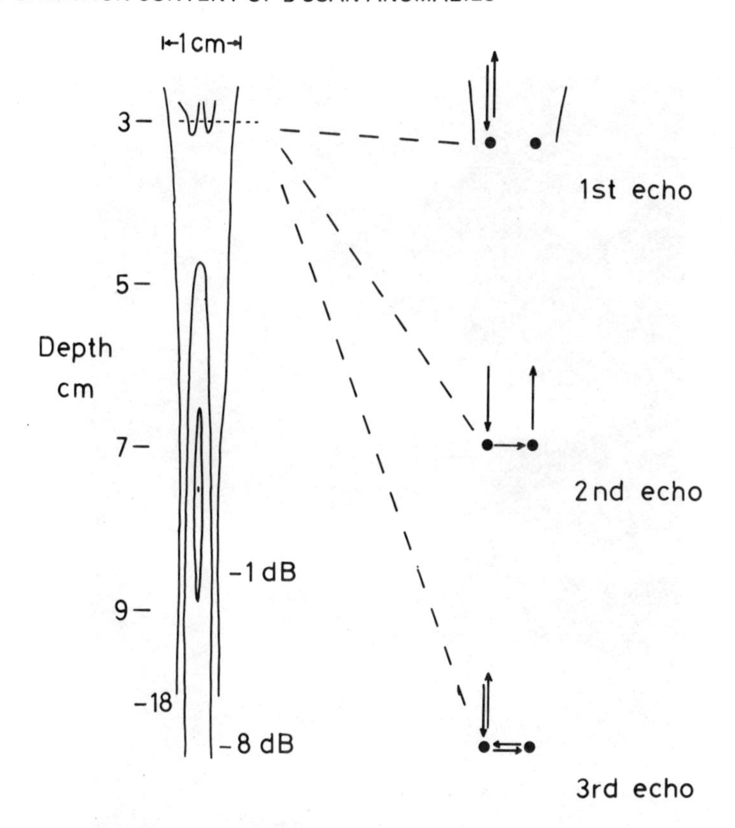

Fig.6 Illustration of multiple echoes originating from two discrete gas bubbles.

decrease in the 'attenuation' properties of the stone. An initial assumption may be to suppose that the absorption of sound within the stone is changing. Although this is possible it is unlikely to be able to account for the large variations in attenuation necessary to produce a shift from 'shadowing' to 'flaring'. The other major component of attenuation is scattering and it is suggested that the phenomenon may correspond to the fact that the acoustic scattering by a small solid object varies with scatterer size[3].

The theoretical evaluation of acoustic scattering from a rigid sphere has previously been described for plane monochromatic waves[4], where the polar variations of scattering are shown to depend upon the relationship of the incident wave number 'K' to the radius 'a' of the spherical scatterer. Fig.8 shows examples of the theoretically derived scattered intensity from a rigid sphere. It is suggested that this formalism may be appropriate for the scattering of ultrasound by gall stones provided certain criteria are satisfied :

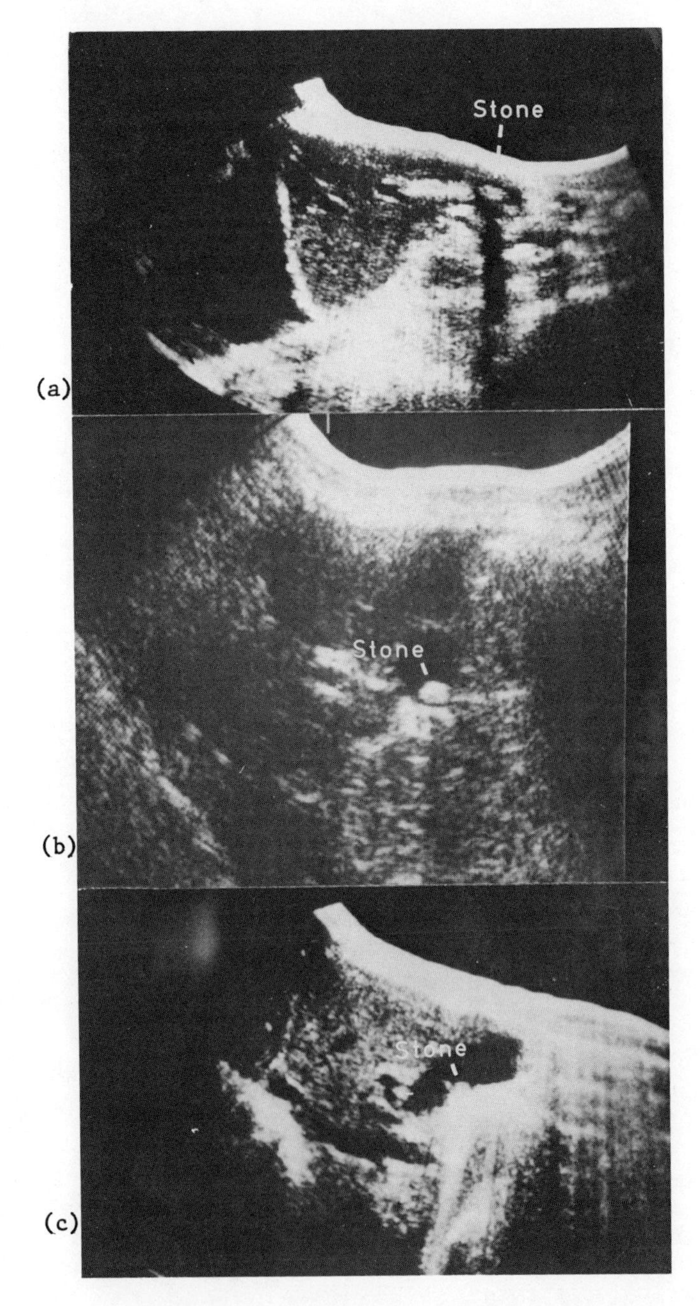

Fig.7 Longitudinal sector B-scans showing different images associated with gall stones.

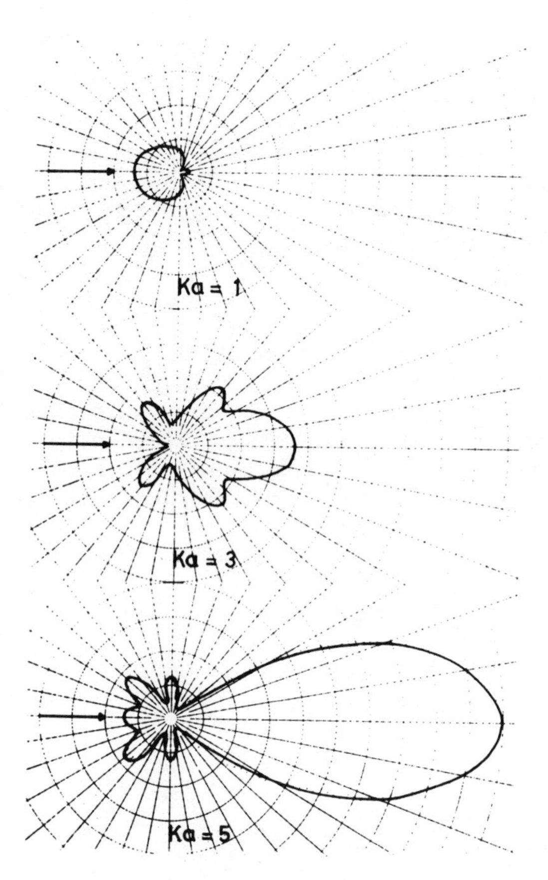

Fig.8 Polar plots of the scattered intensity from a rigid sphere for various values of Ka.

1. the size of the stone must be less than the beam-width of the incident ultrasound (to ensure the applicability of plane wave theory);
2. the value of Ka should be as large as possible, with a lower limit of Ka = 2.

These requirements are satisfied for a gall stone with a diameter of between 0.4 and 3 mm when investigating with an ultrasonic frequency of 3MHz. Stones in this range of size would be expected to exhibit a predominant forward scattering component, which could reinforce the onward travelling wave to produce an apparent decrease in the attenuation of the sound waves by the gall stone. As illustrated in Fig.8, an increase in stone size by a factor of 2 can result in the forward scattering component increasing by a factor of 9. Once the stone becomes larger than the beam-width the scattering contribution alters, as the stone no longer acts as a discrete spherical scatterer.

Should this explanation prove valid then an application of single body scattering theory will permit an estimation of gall stone size, density and compressibility, parameters which may aid in deciding whether to remove the stone.

CONCLUSION

Work is presently in hand to determine the feasibility of these arguments and their clinical significance. It is suggested that information exists in B-scan pictures which has yet to be fully understood and evaluated, though the descriptions emphasised in this paper are only possible explanations of some of the acoustic oddities observed in this highly complex field.

ACKNOWLEDGEMENTS

The author greatly appreciates the help and cooperation of the clinical members of the ultrasound department at the Royal Marsden Hospital in supplying the B-scan pictures. Special thanks are due to Drs. Cosgrove and Pussell for many useful discussions.

REFERENCES

1. Cosgrove, D.O., Garbutt, P. and Hill, C.R. "Echoes across the diaphragm" Ultrasound in Med. and Biol. 3 (1978); pp.385-392

2. Dunn, F. "Attenuation and speed of ultrasound in lung" J. acoust. Soc. Am. 56 (1974); pp. 1638-1639

3. Nicholas, D. "An introduction to the theory of acoustic scattering by biological tissues" Ch.1 in Recent Advances in Ultrasound in Biomedicine (Ed. D.N. White) (1978); Research Studies Press

4. Morse, P.M. and Ingard, K.V. Theoretical Acoustics (1968); McGraw-Hill, New York.

INCENTIVES FOR USING HIGHER FREQUENCY IN ULTRASONIC IMAGING

D.B. Boyle, J.D. Meindl, and A. Macovski.

Center for Integrated Electronics in Medicine

Stanford University, Stanford, CA 94305

ABSTRACT

Improved resolution is only one of the advantages of using higher frequencies in ultrasonic imaging. Other effects may provide a significant increase in the diagnostically useful information provided by an ultrasound scan. Images made with 2.25 MHz systems are dominated by specular echoes. The extreme angular sensitivity of the specularly reflected signal causes "specular noise," which limits the contrast and effective resolution of the image. A tissue model is developed for cardiac imaging which assumes Rayleigh scattering in the myocardium. Model predictions for image improvements with higher frequency as specular noise is reduced by the increased proportion of diffusely reflected signal are supported experimentally. Methods for tolerating the increased attenuation at higher frequencies are discussed. The results of computer processing cardiac and liver data recorded *in vivo* show the feasibility of coherent signal enhancement techniques. A coherent detector which allows temporal averaging and is relatively unaffected by the cardiac motion is defined and demonstrated.

INTRODUCTION

Ultrasonic imaging systems are usually designed to use a frequency which is chosen according to a design rule which considers only the attenuation for a given path length [1,2] and the resolution. After the frequency is chosen, other parts of the system are designed within that constraint. Many of the efforts to improve system performance have been directed toward higher spatial resolution and have been evaluated on that basis. To overcome the limited trade-off between resolution and the depth of the near-field offered by collimated systems, acoustic lenses have been developed [3]. Electronic beam steering [4], focusing [5], and synthetic aperture [6] techniques have also been devised to improve lateral resolution. These improvements have the desirable feature of increasing the signal and therefore the signal-to-noise ratio (S/N) by increasing the aperture. However, attempts to improve axial resolution have usually employed methods which decrease the S/N. These include increased transducer damping, shorter pulses, and signal processing such as deconvolution [7] and differentiation.

Evaluation of the usefulness of these techniques should be based on an understanding of the characteristics of tissue structures. Implicit in many beam pattern plots used to demonstrate system performance is the assumption that the strength of the echo received is proportional to the importance of that structure. Actually, many structures of interest, such as the lumen of a blood vessel, a chamber of the heart, or interior of a tumor, are observed as a lack of signal. Thus a rejection ratio plot of the spatial response of the system might be more appropriate. Such a plot would show the ability of the instrument to reject clutter from outside the main beam which would otherwise obscure the structure of interest. Schemes for improving axial resolution usually assume a one-dimensional, layered model [8]. In clinical applications, this model is frequently not appropriate and at best requires careful alignment to get the beam perpendicular to the layers.

The hypothesis to be developed in this paper is that better clinical system performance could be achieved by using higher frequencies than those selected by the usual design rules. It is obvious from diffraction theory that using higher frequency will allow higher lateral resolution for the same aperture. More interesting is the possibility of increasing the proportion of the signal which is due to diffuse reflection. The rejection of off-axis responses would be improved by reducing the relative strength of specular reflectors, as well by improved beam patterns. By decreasing the lateral extent of the beam, fewer reflectors will be included in a resolution element. Under many conditions, this will be a more effective way to separate echoes than axial deconvolution. Another advantage would be that more of the signal would be received from structures most likely to contain tissue characterization information.

Evaluation of the hypothesis will be divided into three parts. First, some of the problems caused by specular reflectors will be illustrated. Second, the use of higher frequencies will be studied as a method of reducing problems from specular reflectors. This will be a discussion of a tissue model for cardiac imaging and will include other advantages of higher frequencies. Finally, the prospects for tolerating increased attenuation will be evaluated.

METHODS AND EQUIPMENT

Images and data for this paper were obtained from several excised dog hearts and one normal adult male subject. Due to the small sample size and the limitations of the experimental setup, many variables were uncontrolled. Consequently, even when results are presented in a quantitative form, they should be considered as corroboration, but not necessarily verification of effects predicted theoretically.

Two clinical B-scan systems were used for this work. A Varian V-3000 Real Time Sector Scanner was selected as representative of the state-of-the-art in electronically beam steered systems. The center frequency for this scanner is 2.25 MHz. To provide flexibility in the frequency of the transducer, a wide-band system designed for ophthalmology was chosen. The system used is a prototype developed by the Center for Integrated Electronics in Medicine (CIEM). It is a compound B-scan system and was used with 2.25, 5 and 10 MHz transducers.

Data was digitized and recorded for subsequent computer processing by a microcomputer-based data acquisition system [9].

SPECULAR REFLECTORS

As noted by Gore and Leeman [10], the signal received is sensitive to the shape of the specular reflector and its orientation to the ultrasound beam. In cardiac imaging, the motion of the heart changes the orientation of reflecting surfaces. This causes temporal variation in the received signal, as shown in Figure 1. These variations might result from either translation or rotation of the heart. To attribute this change in signal to translation would require lateral movement of the septum of at least half a beamwidth, some 5 mm in 45 milliseconds. Since the velocity of 100 mm/s is much larger than one would expect to encounter in the septum, a change in orientation is probably the cause. Using data from [10], a 5 degree change in the angle of incidence would explain the 18 dB amplitude change in Figure 1.

The change in the amplitude of the signal is not useful information. Since it results from changes in shape which occur on a scale smaller than the scanner resolution, or from a rotation, which is also unobservable, the amplitude change cannot be used to improve the estimate of the position and shape of the object. In fact, the fluctuation in amplitude decreases the detectability of the object because it can exceed the dynamic range of the imaging system. These considerations indicate that the changes in amplitude should be treated as noise and suggest the term "specular noise." We shall define specular noise as any aspect of specular reflection

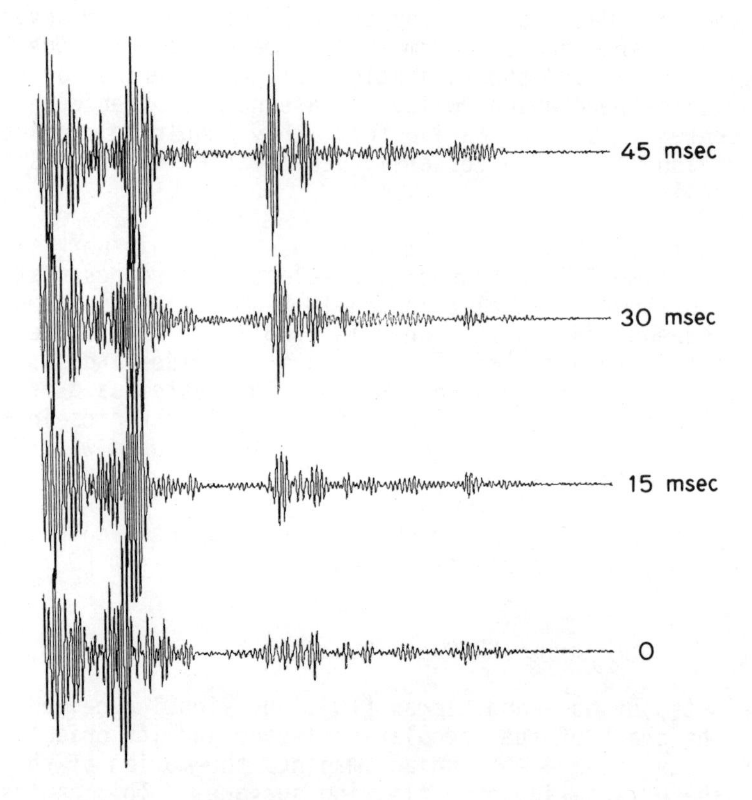

Figure 1. Echo amplitude as a function of time. In this cardiac data recorded _in vivo_, the specular echo from the septum increases by 18 dB in approximately the interval of a real-time scanner frame.

which reduces the detectability of important features of the object being imaged and adds no useful information.

One effect of specular noise on cardiac imaging is to cause gaps in the image of the endocardium. Since the endocardium is an elliptical surface, it is not surprising that the orientation of some parts of the endocardium is unfavorable for the reception of a specular echo. Indeed, visualization of most of the endocardium may depend on the irregularity of surface orientation provided by the trabeculae carnae. The scale of the surface roughness of the trabeculae is large enough to allow considerable variation in amount of signal reflected back to the transducer. When this variation exceeds the dynamic range of the system, gaps occur in the image.

Increasing the system gain allows utilization of lower level echoes, but reveals another consequence of specular noise. The combination of a wide dynamic range of reflectivity of the object to be imaged and off-axis responses in the beam pattern of the imaging system causes "clutter." This clutter reduces the image contrast and limits the use of diffuse echoes for image formation.

These problems are particularly severe for electronically beam steered systems containing side lobe responses. Figure 2 compares the performance of the Varian system with the CIEM system using a 2.25 MHz transducer. The image from the CIEM system is much better "filled in" than the image from the Varian phased-array system. As shown in the top trace, the borders of the septum are, in many places, clearly delineated by strong specular echoes at the boundaries. Close examination of the lower trace shows that the diffuse echoes are detectable above the noise so that the edges of the septum can be found even when specular echoes are not present. In fact, the image could be made using just the diffuse echoes. The trace through the specularly reflecting region from the Varian phased-array system shows considerably less dynamic range than the 35 dB found with the CIEM system. When there are no specular echoes, as shown in the bottom trace, the edges of the septum become difficult or impossible to detect. As much signal is received from the region where the ventricle chambers should be as from the region where the septum should be. Since the clutter signal is about as strong as the diffusely reflected signal, the image from the Varian scanner cannot be improved by increasing the gain. The ventricle chambers would "fill in" as quickly as the gaps in the walls. In this way, the specular noise prevents the use of the diffusely reflected signal. The proposed solution to this problem is to increase the relative strength of the diffusely reflected signal by increasing the frequency and is suggested by the characterics of a tissue model.

TISSUE MODEL

The tissue model used here was developed for cardiac imaging. It divides the sources of backscattered ultrasonic energy into two distinct classes: specular and diffuse reflectors. Specular reflectors are primarily associated with the surfaces of the heart, the epicardium and the endocardium. The myocardium produces diffuse reflections. Since its backscattering coefficient is so much lower than tissue, the blood in the heart may be ignored.

The model assumes that the reflection coefficient of a specular reflector is independent of frequency. The diffuse reflectors are frequency dependent. As shown by Reid [11], the backscattering coefficient for myocardium approaches f^4 dependence and thus is due primarily to Rayleigh scattering. Signal which results from Rayleigh

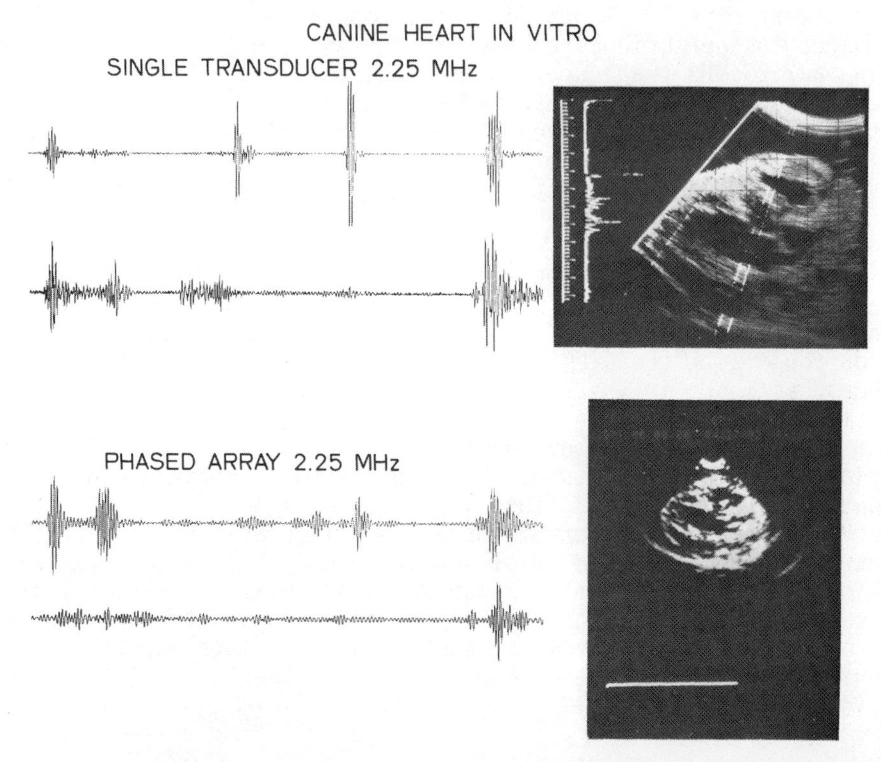

Figure 2. Comparison of single transducer and phased-array systems. Since the phased-array system is more susceptible to off-axis echoes, the rf traces show more clutter than the single transducer system. Setting the threshold to eliminate clutter also eliminates diffuse echoes, causing gaps in the image.

scattering has some desirable properties. Rayleigh scattering is isotropic; this is a significant improvement over the highly orientation dependent specular reflection. Since measurements indicate that the scatterers are less than 30 microns in diameter, it is reasonable to assume that a uniformly large number of them will be present in a resolution element anywhere in the myocardium. Thus the resolution will not be limited by the reflector spacing, as it is with the specular reflectors. The Rayleigh scattered signal will be uniformly present in the regions containing tissue, except for the perturbations caused by speckle noise [12].

Experimental evidence of these effects is provided in Figure 3. These images were made with the CIEM system and show a canine heart in a water tank. Note the unusual orientation. This is a long axis cross-section of the left ventricle, but the posterior wall is on

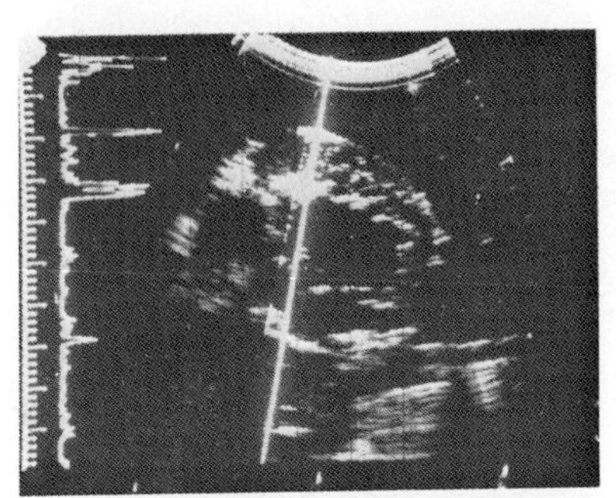

2.25 MHz

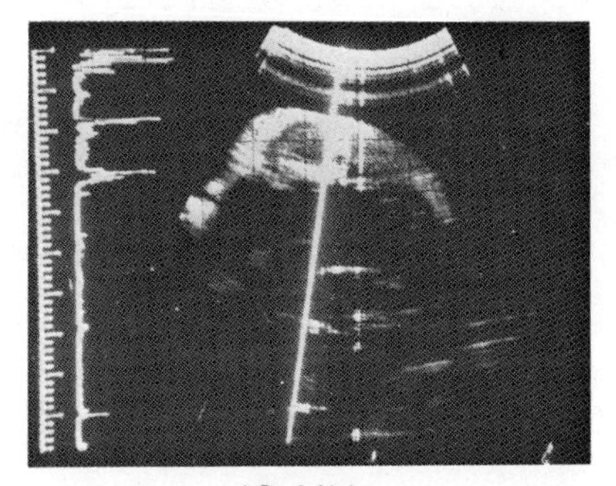

10 MHz

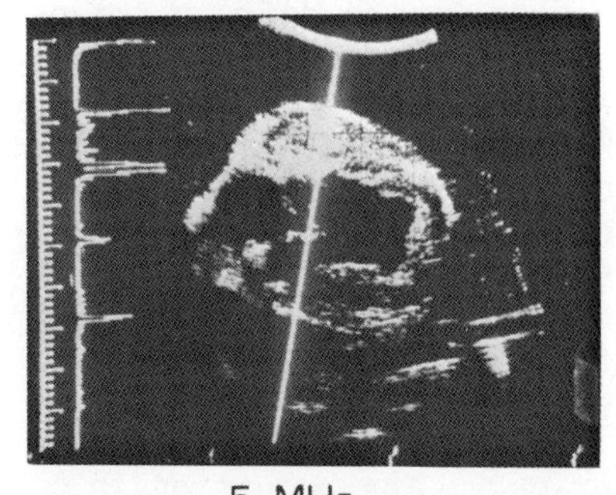

5 MHz

Figure 3. Image quality as a function of frequency. Increased Rayleigh scattering at higher frequency clearly shows the extent of the myocardium.

the top, the septum on the bottom, apex on the right, and the base on the left.

The image made with the 2.25 MHz transducer is typical of those obtained in vivo (which are almost always made at 2.25 MHz). The epicardium is well visualized only in the one place where it is perpendicular to the beam. The endocardium is spotty, particularly at the apex. Some diffuse echoes may be found in the myocardium of the posterior wall. However, the image consists mostly of specular echoes and suffers accordingly.

The 5 MHz image is considerably better; visualization is more complete. This image was made with the same overall system gain. Echoes are now obtained all through the myocardium. The lack of specular reflection from the epicardium is not important because the edge is well defined by the diffuse echoes.

The 10 MHz image of the posterior wall shows more of these desirable characteristics, despite the fact that the transducer sensitivity is so low that the available system gain was about 15 dB less than for the other images. That a useful image could be made attests to the increase in backscattering coefficient with frequency. Since the scanner used does not have time-varied gain, attenuation compensation could not be made. The signal from the septum contains only a 2 MHz component since the 10 MHz signal has been attenuated by the overlying tissue. Note that the image of this area is very similar to that made with the 2.25 MHz transducer.

The results of this experiment agree well with the predictions of the tissue model. Specular reflectors dominate the images at 2.25 MHz, but are much less important at the higher frequencies. While the wide spacing of favorably oriented specular reflectors limited the visualization at the low frequency, the diffuse reflectors observed at the higher frequencies clearly showed the interior as well as the boundaries of the myocardium. After FFT processing to obtain a single component of these wide-band signals, good agreement with Reid's results [11] was found. This near f^4 dependence would not be expected from rough surface type diffuse reflection. Thus the Rayleigh scattering model is supported by this finding. The high frequency images showed that the myocardium is relatively free of specular reflectors, which confirms the association of specular echoes with the endocardium and epicardium.

Higher frequency has various advantages. The reduction in specular noise eases the requirements for the front end, both in terms of the rejection of off-axis echoes, and in the dynamic range of the desired signal. Increasing the proportion of diffusely reflected signal used provides images which clearly show the interior, as well as the edges of the heart walls. Since the resolution-S/N product is a function of the wavelength [12], higher frequency provides the opportunity to reduce the speckle noise by increasing the resolution-S/N product. Increased resolution is a non-trivial advantage since many objects of interest are probably not resolvable with the usual longer wavelength. The significance for tissue characterization is that a higher proportion of the signal will be from cell-size structures within the tissue of interest, rather than from membranes at the interfaces. Also, effects that are difficult to detect at 2.25 MHz become stronger at high frequency [13]. Wide bandwidths, important for studying the scattering properties of periodic structures, are easier to achieve at higher frequency.

ATTENUATION CONSIDERATIONS

The third phase of this investigation considers methods of tolerating the increased attenuation at higher frequencies. In this discussion, the characteristics and performance of the CIEM system will be used as a standard.

Measurements made in vivo with a 2.25 MHz transducer in the CIEM system yield a S/N of greater than 40 dB for the stronger specular echoes and about 15 dB for diffuse echoes (where they are not obscured by specular echoes). Measurements made with a 5 MHz transducer give a S/N = 20 dB for diffuse echoes in the adult septum and indicate that the diffuse echoes from the posterior wall are just below the noise. Thus, only slight improvements will be required to use twice the usual operating frequency. Extending this to 10 MHz, however, is likely to require some novel solutions. An additional 45 dB of attenuation is expected of which only 25 dB is compensated by the higher backscattering coefficient.

The first item to consider is the transmitter. Since the standard pulser has a power spectrum which decreases exponentially with frequency, it is unsuitable for high frequency applications. An increase in power can be achieved by using sinusoidal pulse bursts. Since the structures being imaged generally do not present asymmetric resolution requirements, it is reasonable to sacrifice axial resolution for lateral resolution, at least to the point that an approximately spherical resolution element is obtained. Alternatively, the transmit voltage could be increased. Up to 30 dB improvement could be obtained within current safety guidelines. Sophisticated control of the total dosage and its distribution could minimize possible hazards caused by the increased power.

In the receiver, we consider improvements in detection. Since envelope detection is not optimum for noisy signals, the possibility of improvement seems significant. To study the prospects for signal enhancement and to suggest an improved detector, we introduce the coherence function.

The coherence function is commonly used to study the properties of systems [14]. It is defined as:

$$\gamma^2 = \frac{\overline{|G_{yx}|^2}}{\overline{G_{xx}}\ \overline{G_{yy}}}$$

where G_{yx} is the cross-power spectrum, while G_{xx} and G_{yy} are the auto-power spectra of the input and the output, respectively. This function can be used to compute the S/N as a function of frequency:

$$S/N = \frac{2}{1 - \gamma^2}$$

For the analysis of ultrasound signals, further definitions are useful.

Let COHX be defined such that:

INPUT $G_x = \mathfrak{F}\{\text{Trace } 1\}$

OUTPUT $G_y = \mathfrak{F}\{\text{Trace } 2, 3, \ldots, M\}$

COHX will be used to predict the results of coherent signal averaging. It assumes that the target, and therefore the signal, is not changing with time

Let COHR be defined such that

INPUT $G_x = \mathfrak{F}\{\text{Trace } T_i\}$

$i = 1, M-1$

OUTPUT $G_y = \mathfrak{F}\{\text{Trace } T_{i+1}\}$

The motivation for the definition of COHR is based on an observation noted previously [9] that motion of the ultrasonic target causes variation in the amplitude of the frequency components of the received signal. For short intervals, a straight line approximation to this variation can be made. Since COHR allows a constant ratio of the amplitudes of the frequency components of successive pairs of traces without reduction of the coherence measured, it will be less affected by motion of the ultrasonic target than COHX. However, since the estimate of the input, as well as the output, has the same noise added, COHR is more affected by noise.

A single output, suitable for use as a detector output, can be generated from the coherence spectrum by taking the average coherence in the signal band. By comparing the outputs of COHR and COHX, the effects of motion and the usefulness of coherent signal enhancement techniques can be evaluated. Plotting the coherence as a function of the number of traces used to form the estimate gives some additional information about the relative importance of errors in the estimate and of changes in the signal due to motion while the data was being acquired.

Data was obtained in vivo from the heart and liver using the 5 MHz transducer in the CIEM system. The computed results of COHR and COHX are shown in Figures 4 and 5.

As shown in Figure 4, motion will be a substantial consideration in the processing of cardiac data. While the S/N computed from measurements of the signal strength and the system noise is about 20 dB, much lower values are obtained from the coherence. At diastole, COHX yields a value of only 0 dB while the more rapid motion at systole reduces this to -20 dB. These results suggest that attempting coherent signal averaging will cause signal loss in the diastole data and severe signal attenuation at systole. Confirmation of these predictions may be found in Figure 6. The values

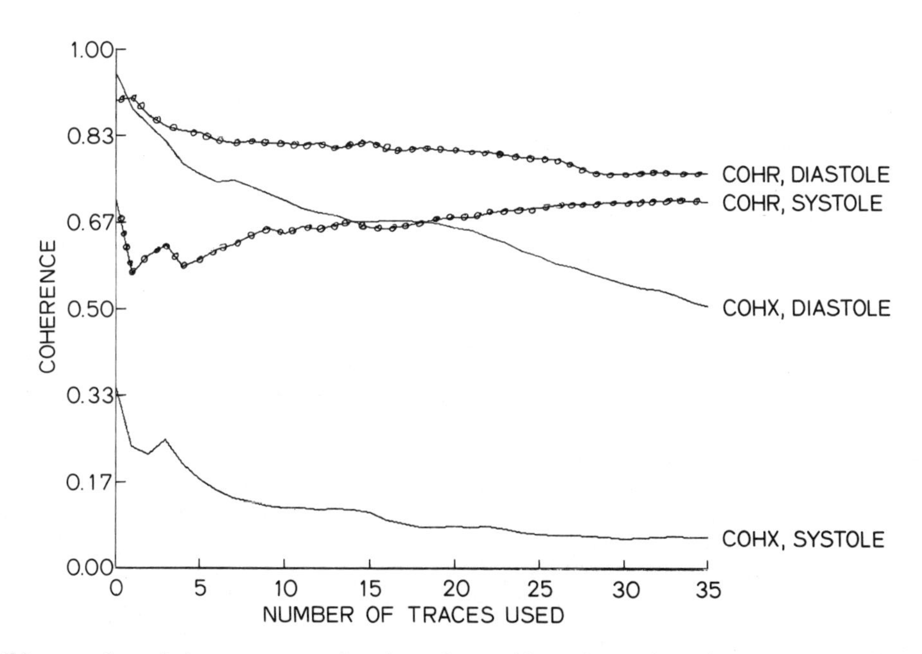

Figure 4. Coherence analysis of cardiac data *in vivo*. The relatively high values for COHR indicate substantial compensation for the motion that is causing the low values of COHX.

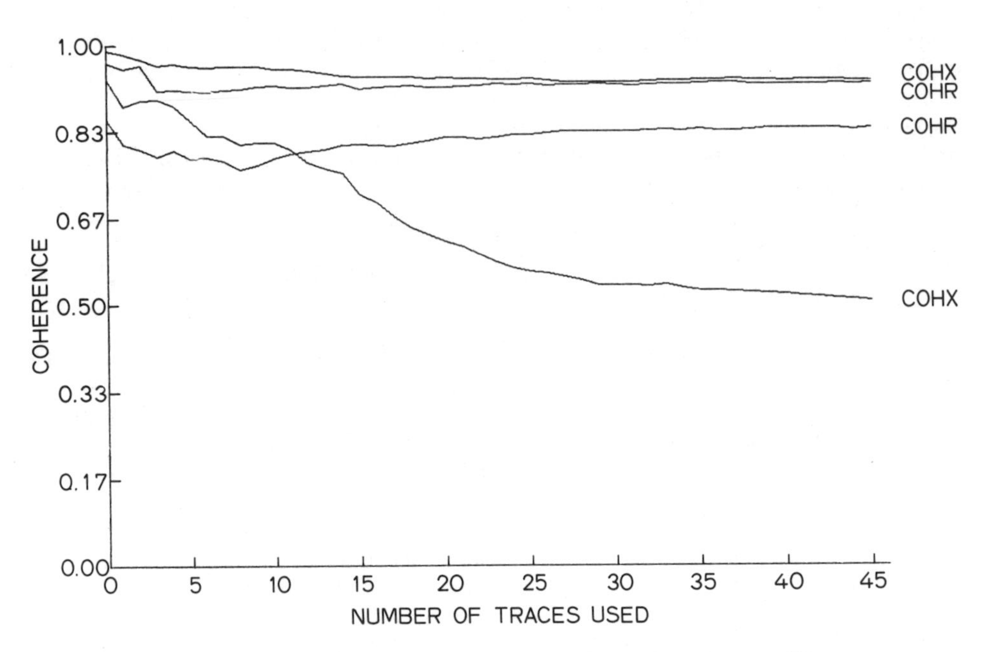

Figure 5. Coherence analysis of liver data *in vivo*. The upper curves are from a data set with little attenuation or motion. In the other data set, increased attenuation (lower S/N) is indicated by the lower values for COHR and significant motion is indicated by the drop in COHX.

obtained for COHR, however, indicate that signal enhancement is a reasonable possibility if a compensation is made for motion. Although the coherence is substantially reduced at systole, indicating inadequate compensation, a S/N of 12 dB is computed from COHR at diastole. There are two factors which cause the COHR estimate of the S/N to be less than the S/N estimated from the system noise. Noise on the estimate of the input accounts for 6 dB, which means that only 2 dB is lost because of motion. Thus at diastole, even a simple scheme like COHR can effectively minimize the effects of tissue motion. The results of preliminary computer simulation studies show that a signal at S/N = -20 dB can be recovered using COHR as a detector. With inclusion of more variables to allow second or third order changes in the amplitude of the frequency components, and more *a priori* information, this performance might be achieved at systole.

Motion is much less of a problem when processing liver data. Figure 5 shows the coherence analysis of two data sets from the

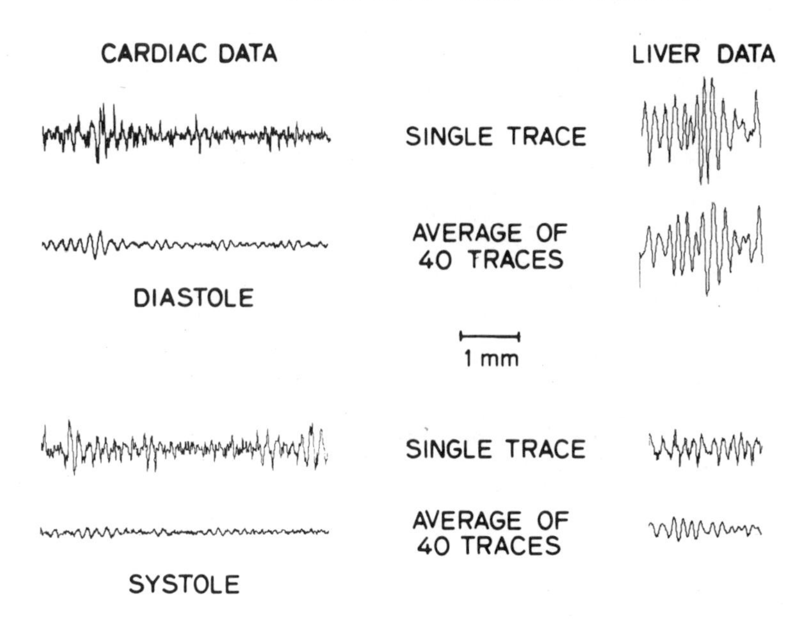

Figure 6. Coherent signal averaging *in vivo*. As predicted by Figure 4, simple averaging gives poor results for cardiac data. Liver data, as shown in Figure 5 and confirmed here, can be sufficiently stationary to give good results.

liver, *in vivo*. The very high coherence values of the first set reflect both the high S/N and the lack of significant motion. The values obtained for COHX will be greater than those for COHR only when motion is insignificant. In the second data set, which was recorded at a depth of 30, rather than 20 mm, motion has significantly reduced the values obtained for COHX. COHR was relatively unaffected by the motion, but does show the effects of increased attenuation from the greater amount of overlying tissue. The results of coherent signal averaging, as shown in Figure 6, are as expected. Since the motion in the second data set was probably caused by the cardiac pulse, ECG timing could be used to minimize this problem. Under these conditions, coherent signal averaging could be used to provide signal enhancement for liver data in vivo. Improvement should follow the usual $\sqrt{N}$ rule and with certain hardware improvements, should be applicable for at least N = 460, for a 26 dB gain. Spatial averaging by a detector such as COHX could boost this to 50 dB.

The results of processing heart and liver data obtained *in vivo* indicate that coherent signal processing is possible. Thus the higher attenuation encountered at higher frequencies can be accommodated by signal enhancement and improved detection. Further investigation will be needed to determine the optimum frequency for a particular depth.

CONCLUSIONS

Ultrasonic scanners designed according to the usual hardware limitation employ a frequency which is lower than optimum as evidenced by the deleterious effects of specular noise. The requirements for the front end are severe for both dynamic range and for suppression of off-axis responses. Because of the scattering properties of tissue, these problems are alleviated by increasing the frequency. Increasing the frequency also improves the prospects for tissue characterization. Possible solutions to the problem of increased attenuation are increased transmit power and improved detectors. Preliminary results show that coherent signal enhancement techniques can easily be applied to liver data and, with compensation for motion, to cardiac data.

Considering the advantages over and above linear increases in resolution, and with some suggestions for the solution to the attenuation problem, rapid progress may be expected in the technology needed to use higher frequencies in ultrasonic imaging.

REFERENCES

1. Devey, G.B., P.N.T. Wells, "Ultrasound in Medical Diagnosis," *Scientific American*, Vol. 238, No. 5, 1978.

2. Erikson, K.R., F.J. Fry, J.P. Jones, "Ultrasound in Medicine - A Review," *IEEE Transactions on Sonics and Ultrasonics*, Vol. SU-21(3), 144-169, 1974.

3. Dameron, D.H., "Diagnostic Improvements in Medical Ultrasonic Imaging Systems," Ph.D. Thesis, Stanford University, 1978, Technical Report 4956-4, Chapters 2, 3.

4. Anderson, W.A., J.T. Arnold, L.D. Clark, W.T. Davids, W.J. Hillard, W.J. Lehr, L.T. Zitelli, "A New Real-Time Phased-Array Sector Scanner for Imaging the Entire Adult Human Heart," *Ultrasound in Medicine*, Vol. 3B, D. White and R.E. Brown, eds., Plenum Press, 1547-1558, 1977.

5. Maginness, M.G., J.D. Plummer, W.L. Beaver, J.D. Meindl, "State-of-the-Art in Two-Dimensional Ultrasonic Transducer Technology," Medical Physics, Vol. 3(5), 312-318, 1976.

6. Burckhardt, C.B., P.A. Grandchamp, H. Hoffmann, "An Experimental 2 MHz Synthetic Aperture Sonar System Intended for Medical Use," IEEE Transactions on Sonics and Ultrasonics, Vol. SU-21(1), 1-6, 1974.

7. McSherry, D.H., "Computer Processing of Diagnostic Ultrasound Data," IEEE Transactions on Sonics and Ultrasonics, Vol. SU-21 (2), 91-97, 1974.

8. Papoulis, A., I. Beretsky, "Improvement of Range Resolution of a Pulse-Echo System," Ultrasound in Medicine, Vol. 3B, D. White and R.E. Brown, eds., Plenum, New York, 1613-1627, 1977.

9. Joynt, L., D.B. Boyle, H. Rakowski, W.L. Beaver, R. Popp, "Identification of Tissue Parameters by Digital Processing of Real-Time Clinical Cardiac Data," presented at the 2nd Intl. Symposium on Tissue Characterization, NBS, Gaithersburg, MD, June 1977.

10. Gore, J.C., S. Leeman, "Echo Structure in Medical Ultrasonic Pulse-Echo Scanning," Phys. Med. Biol., Vol. 22(3), 431-443, 1977.

11. Reid, J.K., K.K. Shung, "Quantitative Measurement of Scattering of Ultrasound by Heart and Liver," 2nd Intl. Tissue Characterization Symposium, NBS, Gaithersburg, MD, June 1977.

12. Burckhardt, C.B., "Speckle in Ultrasound B-Mode Scans," IEEE Transactions on Sonics and Ultrasonics, Vol. SU-25(1), 1-6, 1978.

13. Senapati, N., P.P. Lele, "Ultrasonic Frequency-Domain Analysis and Studies on Acoustical Scattering for Diagnosis of Tissue Pathology," Proc. of the 3rd New England Bioengineering Conf., Tufts University, 347-352, May 1975.

14. Roth, P.R., "Effective Measurements Using Digital Signal Analysis," IEEE Spectrum, Vol. 8(4), 62-70, 1971.

ACKNOWLEDGMENT

This work was supported by the Department of Health, Education, and Welfare under PHS Research Grant P01 GM17940.

LIST OF PARTICIPANTS

Robert C. Addison, Jr.
Rockwell Science Center
1049 Camino Dos Rios
Thousand Oaks, CA 91360

Mahfuz Ahmed, M.D., Ph.D.
Dept. of Radiological Sciences
University of California
Irvine Medical Center
101 City Drive S.
Orange, CA 92668

Pierre Alais
Universite Paris VI
Institut de Mecanique
Theorique et Appliquee
Laboratorire de Mecanique
Physique
2, place de la Gare de Ceinture
78210 Saint-Cyr-L'Ecole
France

Paul Amazeen, Ph.D.
W277 N1080 Woodside Drive
Waukesha, WI 53186
General Electric Company

Charles B. Andes
Advanced Diagnostic Research
Corp.
P.O. Box 28400
Tempe, AZ 85282

Professor Jonny Andersen, Ph.D.
Dept. of Electrical Eng.
University of Washington
Seattle, Washington 98195

Dale Ballinger
4800 East Dry Creek Road
Honeywell, Inc.
Denver, CO 80217

Ralph W. Barnes, Ph.D.
Dept. of Neurology
Bowman Gray School of Medicine
Winston Salem, N.C 27103

Richard A. Banjavic
Medical Physics Division
University of Wisconsin
3321 Sterling Hall
Madison, WI 53706

Kenneth N. Bates
Ginzton Lab
Stanford University
Stanford, CA 94305

Gilbert Baum
Albert Einstein College of Med.
Bronx, N.Y. 10461

Mr. M.I.J. Beale
AERE Harwell
Didcot, Oxon, U.K.
Abingdon 24141 Ext 4776
U.K.A.E.A.

S.D. Bennett
Dept. of Electronic Eng.
University College London
Torrington Place
London NWIE TJE England

Irwin Beretsky, M.D.
Director, Community Hospital
Ultrasound Research
766 N. Main Street
Spring Valley, N.Y. 10977

Richard Bernardi
Picker Corporation
12 Clintonville Road
Northford, CT 06472

Jason C. Birnholtz, M.D.
Harvard Medical School
Boston, MA 02115

David E. Boyce
General Electric Company
Electronics Park Bldg. 7
Syracuse, N.Y.

Douglas B. Boyle
Center for Integrated Elec-
tronics in Medicine
AEL 121, Stanford University
Stanford, CA 94305

Christian Bruneel
Centre Universitaire
59326 Valenciennes
France

Michael J. Buckley
DARPA/MSO
1400 Wilson Blvd.
Arlington, VA 22209

Chi Keung Chan
University of Pennsylvania
1329 Lombard Street, Apt. 608
Philadelphia, PA 19147

Alfred V. Clark, Jr.
Code 8431 Naval Research Lab
Washington, D.C. 20375

Arthur E. Clark
Naval Surface Weapons Center
Silver Spring, MD.

W. Thomas Cathey
Dept. of Electrical and
Computer Engineering
University of Colorado
Denver, CO 80202

Douglas Corl
53-D Escondido Village
Stanford University
Stanford, CA 94305

Dr. H. Dale Collins
Holosonics, Inc.
2400 Stevens Drive
Richland, WA 99352

Neil Collings
Dept. of Electrical Engineering
Burlington, VT 05401

Professor A. J. Cousin
University of Toronto
Dept. of Electrical Engineering
35 St. George Street
Toronto, Ontario, Canada

David Dameron
Stanford University
Durand 339
Stanford, CA 94305

Philippe Defranould
Thomson-CSF
Cagues/mer
France

Professor Jean-Luc Dion
Department d'Ingenierie
Universite' du Quebec a Trois-
Rivieres
Case postale 500
Trois Rivieres, Quebec G9A 5H7

George P. Dixon
221 Kennedy Street
State College, PA 16801
KB Aerotech

Richard E. Doherty
Albert Einstein College of Medicine
1300 Morris Park Avenue
New York, N.Y. 10461

Gregory L. Duckworth
Mass. Institute of Technology
33 Preston Road
Somerville, MA 02143

Michael D. Eaton
Stanford University
AEL 121
Stanford, CA 94305

George J. Eilers
Unirad Corporation
4765 Oakland Street
Denver, CO 80239

Donald C. Erdman
Erdman Instruments, Inc.
1179 Romney Drive
Pasadena, CA 91105

Kenneth R. Erikson
Rohe Scientific Corporation
2722 South Fairview
Santa Ana, CA 92704

Nabil H. Farhat
University of Pennsylvania
200 S. 33rd Street
Philadelphia, PA 19174

John B. Farr
Western Geophysical
1409 Upland Drive
Houston, TX 77043

James T. Fearnside
Hewlett Andover Division
1776 Minuteman Road
Andover, MA 01810

Rainer Fehr
F/zfE
Hoffman LaRoche
Basle, Switzerland

Dave Feinstein
909 Stratford Ave.
Stratford, CT 06497

Howard Fidel
Electronics for Medicine
One Campus Drive
Pleasantville, N.Y. 10570

Mathias Fink
Lab de Mecanique Physique
2 Place de la gare de ceinture
78210 St. Cyr l'Ecole, France

Gail T. Flesher
EECS Dept.
University of Calif. Santa Barbara
5652 Cathedral Oaks
Goleta, CA 93017

Stuart Foster
Ontario Cancer Institute
500 Sherbourne Street
Toronto, Ontario

Hendrik G. Freie
"Oldelft"
Delft, The Netherlands
Van Miereveltlaen 9

Francis J. Fry
Indiana University
7350 N. Pennsylvania St.
Indianapolis, IN 46240

Michael J. Granelli
Searle Ultrasound
One Concord Lane
Yardely, PA 19067

Dr. Wolfgang Gebhardt
IZFP Fraunhofer-Gesellschaft
66 Saarbrucken
Germany

George A. Gilmour
Westinghouse Oceanic Division
Annapolis, MD 21404

J. F. Greenleaf
Mayo Clinic
Rochester, MN 55901

Morris S. Good
Mechanical Engineering Dept.
Drexell University
32nd & Chestnut St.
Philadelphia, PA 19104

Marvin Guter, M.D.
Univ. of Miami Medical School
4195 Braganza
Miami, FL 33133

Wolfgang Haigis
University Eye Hospital Ultrasound Lab
D-8700 Werzburg, Germany

Amin Hanafy
Hewlett Packard
Andover Division
1776 Minuteman Road
Andover, MA 01810

John Hart
Hewlett Packard
Andover Division
1776 Minuteman Road
Andover, MA 01810

B. P. Hildebrand
Battelle Northwest
Richland, WA

Ronald Hileman
1431 Peace Haven Road
Clemmons, NC 27012
Accusonics, Inc.

Wayne Hillard
Varian Ultrasound Division
2341 South 2300 West
Salt Lake City, Utah 84119

M. Luetkemeyer-Hohmann
Krupp Atlas Elektronik
P.O. Box 4485
28 Bremen 44, Germany

Edward Holasek
Case Western Reserve University
4617 Archmere
Cleveland, OH 44109

John F. Holzemer
SRI International
#2 Elizabeth Lane
Menlo Park, CA 94025

David J. Hughes
227 N. 500 W.
Purdue University
Biomedical Engineering
W. Lafayette, IN 47906

Dr. John W. Hunt
Ontario Cancer Institute
Dept. of Medical Biophysics
500 Sherbourne St.
Toronto, Ontario, Canada M4X 1K9

George Jahn
Varian Associates, Inc.
611 Hansen Way
Palo Alto, CA 94303

Steven A. Johnson
Dept. of Physiology
Mayo Clinic
Rochester, MN

Kenneth W. Johnston
Ultrasound Research Labs/ICFAR
URL, 1100 West Michigan, A-32
Indianapolis, IN 46202

Joie P. Jones, Ph.D.
Dept. of Radiological Sciences
University of California, Irvine
Irvine, CA 92717

Linda Joynt
Information Systems Lab,
Stanford University
Stanford, CA 94305

Jerry L. Jackson
Southwest Research Institute
315 Palm Drive
San Antonio, TX 78228

Dr. Ronald J. Jaszczak
Searle Diagnostics, Inc.
2000 Nuclear Drive
Des Plaines, IL 60018

Wayne D. Jennings
Case Western Reserve University
2065 Adelbert Road
Cleveland, OH 44106

Murali P. Kadaba
Wenner-Gren Res. Lab, Rose St.
Lexington, KY 40506

Ed Karrer
Hewlett Packard
1501 Page Mill Road
Palo Alto, CA 94304

Dr. P.N. Keating
Bendix Research Laboratories
20800 Civic Center Drive
Southfield, MI 48076

Lawrence Kessler
Sonoscan Inc.
720 Foster Avenue
Bensenville, IL 60106

Mr. Roy Kopel
Advanced Dx. Research Corp.
P.O. Box 28400
Tempe, AZ 85282

Mr. R. Koppelmann
Bendix Research Labs.
20800 Civic Center Drive
Southfield, MI 48076

Frederick W. Kremkau
Department of Medicine
Bowman Gray School of Medicine
Winston-Salem, NC 27103

Lewis Larmore, Ph.D.
Director of Science
Office of Naval Research
Pasadena Branch Office
1030 East Green St.
Pasadena, CA 91106

John Larson
Hewlett Packard
1501 Page Mill Road
Palo Alto, CA 94304

C.T. Lancee
Erasmus University
Dept. C.V.R.
P.O. Box 1738
Rotterdam 3000 Dr. Netherlands

Ralph LaCanna
University of California
Los Angeles, California
Dept. of Radiology

Dick Lackmond
4300 Alton Road
Miami Beach, FL 33140
Mt. Siani Hospital

Marie-Therese Larmande
Paris University
Labo Mecanique Physique
2 Place de la gare de Ceinture
F78210 St. Cyr l'Ecole France

Colin Lanzl
369 Congress St.
Rensselaer Polytecnic Institute
Troy, N.Y. 12180

Jack M. Lawry
7647 SW 102 Place
Miami, FL 33173

C.Q. Lee
University of Illinois,
Chicago Circle
Information Eng. Dept.
Chicago, IL

John N. Lee
Harry Diamond Labs
2800 Powder Mill Road
Adelphi, MD 20783

S. Leeman
Dept. of Medical Physics
Royal Postgraduate Med. School
Hammersmith Hospital
London, WIZ OHS, U.K.

Dr. George Lewis
Searle Diagnostic, Inc.
2000 Nuclear Drive
Des Plaines, IL 60018

Odd Lovhaugen
Central Institute Ind. Res.
Forskningsv 1, Blindern
Oslo 3, Norway

Rolf K. Mueller
University of Minnesota
Minneapolis, MN

Sam Maslak
1501 Page Mill Road
Palo Alto, CA 94304
Hewlett Packard

Ronald E. McKeighen, Ph.D.
Searle Diagnostic Inc.
2000 Nuclear Drive
Des Plaines, IL 60018

Roger Melen
McCullough 126 S.E.L.
Stanford University
Stanford, CA 94305

A. F. Metherell, Ph.D., M.D.
Radiological Sciences
U. of California, Irvine
101 City Drive S.
Orange, CA 92668

Thomas E. Michaels
Westinghouse Hanford
P.O. Box 1970
Richland, Washington 99352

Thomas J. Moran
Air Force Materials Lab
AFML/LLP
WPAFB, OH 45433

P.E. Moreland
Sperry Products Division
Automation Industries, Inc.
Box 3500 Danbury, CT 06810

Dr. D. Nicholas &
Dr. A. W. Nicholas
76 Tonfield Road
Institute of Cancer Research
Sutton, Surrey, U.K.
(Hammersmith Hospital)

Anant K. Nigam
HRL, Inc.
2880 W. Oakland Park Blvd.
Ft. Lauderdale, FL 33311

Kazuhiko Nitadori
OKI Electric Industry, Co.
4-10-12 Shibaura, Minato-Ku
Tokyo, Japan 108

Howard D. Noble, Jr.
General Electric Company
R & D
P.O. Box 8
Schenectady, N.Y. 12301

Bertrand Nongaillard
Centre Universitaire
de Valenciennes
59326 Valenciennes Cedex
France

Yukio Ogura
Hitachi Const. Machinery
650 Kandatsu, Tsuchiura, Ibaraki
Japan 300

William O'Brien
Bioacoustics Research Lab
University of Illinois
Urbana, IL 61801

Dr. Charles Olinger
Stroke Research Lab Rm. 4304
Medical Science Building
Cincinnati, OH 45267

Jonathan Ophir
Univ. of Kansas Medical Center
Radiology
39th & Rainbow
Kansas City, KS 66103

E. Papadofrangakis
General Electric
R & D
P.O. Box 8
Schenectady, N.Y. 12301

Hernando Pedraza, M.D.
Radiologist
ACR
4 Sycamore Court
Newton, KS 67114

Mr. E. J. Pisa
Rohe Scientific Corporation
2722 South Fairview
Santa Ana, CA 92704

Kiven Plesset
Picker Corporation
12 Clintonville Road
Northford, CT 06472

J. D. Plummer
Stanford University
McCullough 114
Stanford Elect. Labs
Stanford, CA 94305

Dr. Balu Rajagopalan
Mayo Foundation
Rochester, MN 55901

Bruno Richard
CHU Cochin
Paris, France

H.A.F. Rocha
General Electric Company
Research Development Center
Schenectady, N.Y.

George Sackman
Naval Postgraduate School
Monterey, CA

Takuso Sato
Faculty of Science and Eng.
Tokyo Institute of Technology
4259 Nagatsuda Midori-ku
Yokohoma-shi, Japan

Jaroslav Satrapa
Kretztechnik GmbH, Tiefenbach 20
ZIPF, Austria A-4871

Takashi Sawai
2-17 Senkawa
Toshima-ku
Tokyo 171 Japan

Dr. T. Sawatari
20800 Civic Center Drive
Southfield, MI 48076
Bendix Research Lab

Mahesh K. Shah
8848 Kenneth, #2B
Des Plaines, IL 60016

James M. Smith
Army Research Center
Watertown, MA 02172

Lee Smith
4765 Oakland Street
Denver, CO 80239
Unirad Corp.

Jakob Stamnes
Forskningsvn 1, Blindern
Oslo 3, Norway
Central Institute for
Industrial Research

Jerry L. Sutton
Naval Ocean Systems Center
4271 Mt. Henry Avenue
San Diego, CA 92117

Katsumichi Suzuki, Ph.D.
Energy Research Lab
Hitachi Ltd.
Moriyama, Hitachi, Ibaraki
Japan

Robert G. Swartz
Mc 118, SEL
Stanford University
Stanford, CA 94305

Roger H. Tancrell
Raytheon Research
28 Seyon Street
Waltham, MA 02154

Larry Tepper
Unirad Corporation
4765 Oakland Street
Denver, CO 80239

Charles E. Thomas
General Electric Company
Bldg. 37 Rm 559
P.O. Box 43
Schenectady, N.Y. 12301

Kai E. Tomenius
Picker Corporation
12 Clintonville Road
Northford, CT 06422

Fredrick L. Thurstone
Biomedical Engineering Dept.
Duke University
Durham, NC 27706

D. Vilkomerson
1101 Corporate Road
N. Brunswick, N.J. 08902

H.K. Wickramasinghe
610 Gerona
Stanford, CA 94305

Robert C. Waag, Ph.D.
Electrical Engineering
University of Rochester
204 Hopeman Hall
Rochester, N.Y. 14627

James T. Walker
Stanford University
224 Seale Avenue
Palo Alto, CA 94301

Keith Y. Wang
Dept. of Electrical Engineering
University of Houston
Houston, TX 77004

Tom Waugh
Varian Ultrasound Division
2341 South 2300 West
Salt Lake City, Utah 84119

R. D. Weglein
Hughes Research Lab
6317 Drexel Avenue
Los Angeles, CA 90048

Greg A. White
Rensselaer Polytechnic Inst.
Hirai 032 Box 401
R.P.I.
Troy, NY 12181

INDEX